In the country of the blind
the one-eyed man is King.

In the land of the insane
the half-wit is hanged.

The Genetics of Bacteria and their Viruses

A large-scale model of a DNA molecule, showing the position of various atoms. (Reproduced by courtesy of the BBC).

The Genetics of Bacteria and their Viruses

STUDIES IN BASIC GENETICS AND MOLECULAR BIOLOGY

BY

WILLIAM HAYES

F.R.S.

Professor of Molecular Genetics,
University of Edinburgh, and
Honorary Director of the Medical Research Council's
Microbial Genetics Research Unit,
Department of Molecular Biology,
University of Edinburgh

SECOND EDITION

JOHN WILEY & SONS INC
NEW YORK

Library of Congress Catalog Card Number 69–16699

FIRST PUBLISHED 1964
REPRINTED FEBRUARY AND OCTOBER 1965,
MAY 1966, JANUARY 1967
SECOND EDITION SEPTEMBER 1968

RUSSIAN EDITION 1965

Printed in Great Britain

This book is dedicated to
NORA AND MICHAEL,
with affection,
for their tolerance and patience,
and to
MAX DELBRÜCK
for his friendship

Contents

PART VI · THE PHYSICO-CHEMICAL MECHANISMS OF HEREDITY

PART VII · THE PHYSIOLOGY AND GENETICS OF BACTERIOPHAGE

Preface to Second Edition

This second edition is presented in the hope that it will prove as useful to advanced students of molecular genetics as the first edition seems to have done. I stress the word 'advanced' because I am aware that many of the recent ideas in this field, such as the polarity of the operon, or the possible mechanisms of prophage induction or chromosome transfer, although simple to state dogmatically are, in fact, derived from complex lines of reasoning based on data from many different types of experiment and even disciplines. They are sometimes hard enough to grasp thoroughly even when one has grown up with them! The book, then, is not for beginners for whom many excellent, but cheaper and more elementary, volumes are now available.

The four years which have elapsed since this book was first published have witnessed great advances in our knowledge of the molecular basis of life. The most striking has been the elucidation of the genetic code. In the first edition, the summary of the chapter on this subject states that, although we understood the general nature of the code, 'we do not yet know the code which any organism employs to specify even a single amino acid'. Now, just fifteen years after the beginning of what is certainly one of the most exciting periods of discovery and advance in the whole history of biology, the genetic code has been solved in detail and in its entirety, mainly by the development and use of ingenious new chemical techniques. Other fields which have seen striking and important, if not such dramatic, progress are those concerned with the conformation of protein and RNA molecules, the repair of radiation damage, the mechanism of recombination, the mode of replication of viral single-stranded DNA and RNA molecules, the provirus state and its control, the genetic regulation of biosynthesis, and the nature and behaviour of bacterial plasmids. All these recent advances have been treated in considerable detail, and most of the chapters, except some of the introductory ones, have been extensively re-written. Nevertheless I have to admit that the pace of the advance and the sheer volume of emerging information is now so great that a number of new and important findings have been published too late for inclusion. However this book is intended to be a textbook and a work of reference, and not a compendium of recent advances to which access may be had through other sources.

The increased scope of the edition may be judged from the fact that the bibliography has been nearly doubled by the addition of more than 700 new references, there are 18 new figures and ten new plates, while the text has been increased by about 140 pages. It is hoped that the considerably expanded index, together with extensive cross-referencing and, as in the previous edition, the citation in the bibliography of the pages on which each paper is quoted, will enable the reader to collate the maximum information on any particular topic. However I must stress the fact that the list of references is far from complete (it could hardly be otherwise) and that there is inevitably a strong personal (though, I hope, not too biassed) factor in the selection of experiments recorded here. To any who might justifiably feel a little aggrieved that their work has not been mentioned or, even worse, has been misinterpreted, I would like to offer my regrets, and the excuse of my ignorance and the limitations naturally imposed on a book of this scope and pretension when it is written by a single individual.

In fact this edition, like the first, is far from being the product of my own unaided efforts. I owe a debt to many people for the help and advice they have generously given me. Among these is the late Professor J.B.S.Haldane who read the first edition while recuperating from his operation some months before his death, and who wrote me many long letters of invaluable comment and criticism, one of which opened characteristically with the words, "Dear Hayes, In this letter I am going to give you hell"—and did! Most of the alterations in the early chapters stem from his kindly suggestions and I am indeed grateful. I must also acknowledge my indebtedness to many colleagues and friends who have helped me greatly by advice and discussions about those topics on which they are expert, and/or by reading and criticising what I subsequently wrote about them. Foremost among these are Dr O.Darlington, Dr K.W.Fisher, Dr S.W.Glover, Dr J.D.Gross, Professor D.A.Hopwood, Dr D.Karamata, Dr Elinor Meynell, Professor M.H.Richmond, Dr D.A.Ritchie, Dr K.A. Stacey and Professor N.Symonds. I am particularly obliged to Dr.

J.A.Shapiro for the generous and incisive help he gave me in preparing the sections on lysogeny and on the genetic regulation of biosynthesis. To all these, and to others who have assisted in lesser but useful ways, I offer my sincere thanks, for without their support the task of writing this edition would have been much more burdensome.

In addition, I am again most grateful to those who have embellished this book by permitting me to reproduce their photographs, and particularly to those who provided me with originals. Those who have thus given new material for this second edition are Dr D.P. Allison, Dr J.Cairns, Dr L.Caro, Dr R.H.Epstein, Dr D.T.Hughes, Professor F.Jacob, Dr E.Kellenberger, Dr A.Lawn, Dr D.A.Ritchie and Dr A.Ryter. Finally I wish to pay tribute to my publishers, and especially to Mr Per Saugman and Mr J.L.Robson, for the kindly, tolerant and efficient manner with which they have dealt with the many delays in the production of this book which I have inadvertently inflicted on them.

Edinburgh, July 1968 William Hayes

POSTSCRIPT

Since the publication of the first edition of this book, two geneticists of outstanding distinction, who helped to lay the foundations of molecular genetics by their research and whom I am honoured to have known personally, have died—Milislav Demerec in the fullness of his years, and Harriet Ephrussi-Taylor with tragic prematurity. I felt I could not allow this edition to go to press without paying tribute not only to their key contributions to molecular biology, but also to their endearing qualities of personality and their many acts of kindness for which they will long be remembered by their friends. Their published work will ensure that they are not forgotten by posterity, as I hope the pages of this book make manifest.

W. H.

Preface to First Edition

This is intended to be a rather advanced text book on the genetics of bacteria and bacteriophages. It covers the major advances which have been made in the field during the last few years, and their interpretation at the molecular level. The structure of the book is based on what I conceive to be the requirements of an increasing number of postgraduate workers in related fields, who desire to master the concepts of microbial genetics and molecular biology well enough to be able to apply them to their own problems, to comprehend original papers and to assess future developments with understanding.

My experience of teaching postgraduate courses in microbial genetics, in which biochemists, bacteriologists, virologists and physicists predominate, is that the main barrier to perception is lack of knowledge of classical genetics and of the language (often misconstrued as jargon!) through which genetical principles are expressed. I have therefore devoted the early part of the book to a simple account of the facts and ideas upon which the theory of genetic analysis rests. There are two other reasons for adopting this approach instead of using the physico-chemical structure of the genetic material (DNA) as a starting point. Firstly, this book is primarily about what genes are and how they are inherited and expressed, and it is informative to trace the evolution of our concepts about them as an historical sequence. For example, the fact is insufficiently appreciated that all the essential features of the fine structure of the gene, as well as the idea that genes specify the structure of polypeptide chains rather than of proteins, were deduced from the results of genetic analysis alone and in no way depended on a knowledge of DNA architecture. Secondly, scientific truth is never absolute but only relative to what has gone before, so that the present state of our knowledge can only properly be assessed against its ideological background. The apex of a pyramid is significant only when it crowns the completed structure.

Much of this book is concerned with presenting the various novel genetic systems which have been found in microorganisms, and especially in bacteria and their viruses, against the background of more classical behaviour. However, the most exciting and revolutionary aspect of microbial genetics has come from the integration of genetic studies with biochemical and physical analyses of the synthesis, structure and function of nucleic acids and proteins, which are the main macromolecular constituents of cells. This collaboration has produced an explosion of new knowledge about the fundamental nature of genetic and other vital phenomena, in terms of the structure and behaviour of molecules. The molecular basis of the genetic code, its alteration by mutation, and its transcription and translation into specific protein synthesis are now well understood, not only in principle but also in considerable detail. Only the mechanism of genetic recombination has so far eluded analysis. The book describes and discusses these advances in molecular biology, although my limited knowledge and experience of the methods of chemistry and physics has necessarily imposed a presentation which may appear unsophisticated to the general reader and naïve (though not, I hope, misleading) to the expert. I trust that at least the fundamental concepts of molecular biology have been set forth in a clear and understandable way.

This book should have been completed three years ago. In many ways the delay has been fortunate, for otherwise large and important sections of it would now be out of date while I, instead of enjoying my new-won freedom, might well be back in prison rewriting it. As it is, the last three years have witnessed so many fundamental advances that the outlines of a rather clear picture are now beginning to take shape, while the book has grown to three times the size originally projected for it. Although there are still many problems to be solved and many details to be filled in, and although many of our current ideas may have to be modified as exceptions are discovered to the rules we now envisage, it seems probable that most of our present concepts will remain fruitful for some time to come. Apart from the mystery of recombination, I would hazard a guess that the fundamental advances of the future will probably spring from the application of these existing notions to the complex problems of the regulation and coordination of biochemical processes, of differentiation, and of development.

To keep the book within reasonable bounds, a high degree of selection has been necessary with respect to the experiments chosen to illustrate the development of various principles, and this selection is, of course, reflected in the bibliography. I make no claim that my choices are the best that could have been made. They were chosen mainly on the basis of personal familiarity or appeal, and because I considered that they illustrated well the points I wished to make. The degree of engagement involved in writing a book of this scope has precluded the simultaneous, critical reading of the deluge of papers, not to mention pre-prints, which now threaten to engulf us. The easiest way to escape drowning is to float on the surface of the flood and grasp what drifts within one's reach. In this process I have been greatly helped by my friends and colleagues, who have cast their nets more deeply and have presented me with the best of their catch.

William Hayes

London, July 1963

Acknowledgements to First Edition

It gives me great pleasure to acknowledge my indebtedness to all my friends and colleagues who have given me invaluable help, in one way or another, in the preparation of this book. Foremost among these are Dr R.C.Clowes, Dr K.W.Fisher, Professor R.H.Pritchard and Dr N.Symonds, who not only read and criticised large parts of the book in manuscript and proof but who, in many discussions, generously provided me with expert advice on the presentation of their own specialities. I would particularly like to express my gratitude to Professor Pritchard who patiently guided me past many pitfalls in the field of classical genetics into which I (and my readers!) would otherwise have fallen, and to Dr Symonds who, with equal patience and lucidity, initiated me into the mysteries of bacteriophage genetics. I hope that what they taught me will seem as clear at second hand. In addition, I must thank Dr J.Beckwith, Dr S.W.Glover, Dr J.D. Gross, Mr D.A.Ritchie, Dr S.D.Silver and Mr J.G.Scaife, each of whom provided me with frank and useful comments on one or more chapters.

I regard the photographic plates as a valuable adjunct to the book, since they help to link my rather theoretical approach with actuality. I am most grateful to the many authors, and to the editors of the journals concerned, who have permitted me to reproduce previously published photographs, the source of which is acknowledged in the respective legends. However, I would like to express my special gratitude to Dr T.F.Anderson, Dr J.Cairns, Mrs Maureen de Saxe, Dr G.W.Fuhs, Dr R.W.Horne, Dr E.Kellenberger, Dr J.C.Kendrew, Dr A.K.Kleinschmidt, Miss Janet Mitchell, Dr G.E.Palade, Dr H. Slizynska and Dr D.R.Stadler, all of whom kindly provided me with original photographs or negatives.

In writing the introductory chapters on classical genetics I have leaned heavily on *Principles of Genetics* by Sinnot, Dunn and Dobzhansky (McGraw-Hill), while my task has been much alleviated by the many excellent reviews which are referred to in the text and to whose authors I am indebted. Dr Alice Orgel and Dr J.Gordon, whose Ph.D.Theses I had the opportunity to read, helped me to clarify my views about the molecular basis of mutagenesis.

I also wish to thank my secretary, Miss Deirdre Nadal, for relieving me, in the most competent and helpful way, of so much administrative work which would otherwise have seriously interfered with the writing of this book. Finally I must pay tribute to my publishers, and in particular to Mr Per Saugman and Mr J.L.Robson, for the patient, efficient and charming manner in which they have cooperated with me in this enterprise.

Part I
An Introduction to Genetics

CHAPTER 1

The Anatomy of Inheritance

Heredity means the process whereby men, mice, flies, plants, fungi, bacteria, viruses and all forms of living things reproduce themselves, or at least something unmistakably like themselves. If we had to define the phenomenon of life by a single one of its manifestations, this capacity for self-reproduction of an organised system would undoubtedly be the one that we would choose, for although we know of many processes of recreation and adaptation at the molecular level, the word 'life' suggests to us much more than this. It involves the idea not only of growth, organisation and continuity of existence over long periods of time, but also of change, of the capacity for modification of form in response to environmental changes which has expressed itself in evolution. Only heredity encompasses and unites all these ideas within a single category. Thus the science of *genetics*, which may be defined as the study and analysis of heredity in all its aspects, is the fundamental biological science, the focal point upon which all other aspects of biology necessarily converge.

The scientific method consists essentially in the observation and comparison (which is usually quantitative) of differences between things or phenomena which are otherwise equivalent; then follows the formulation of an hypothesis to correlate and explain the differences and, finally, the testing, by experiment, of predictions based on the hypothesis. If all the predictions prove correct, the hypothesis is elevated to the status of a theory which offers a new perspective from which to make further observations. The difficulty about science is to know what to observe and whether the differences we do observe are, in fact, between comparable things. The more complicated is the system we wish to study, the harder it is to make correct decisions on these matters, which is one of the reasons why knowledge of the mechanisms of evolution and heredity made little progress during the twenty-one centuries that elapsed between the time of Aristotle and that of Darwin and Mendel.

To study heredity we must begin with those differences that we can observe between organisms and their progeny. These differences are called *character* differences and they arise through *variation* which we will talk about later. The nature of the characters used in genetic analysis differs widely and may range from such obvious features as the colour of the eyes or skin, or the length and arrangement of limbs, wings, bristles or hair in vertebrates, to the ability or inability of fungi or bacteria to perform a single biochemical step in the synthesis of a particular amino acid or vitamin, or in the fermentation of a particular carbohydrate. We are more likely to understand correctly how a motor car works and is made if we are allowed to watch the individual parts being constructed and assembled, than if our observation is restricted to the completed product as it leaves the factory. For the same reason, one of the major advantages of microbial genetics is, as we shall see, the opportunity that microorganisms offer for the study of the inheritance of characters that we know to be simple and unitary. Confronted with gross morphological features in complex organisms, such as wing shape or bristle distribution in the fruit fly, *Drosophila*, which emerge as the end product of a long series of developmental events, we have no *a priori* way of telling just how complex the character really is and whether or not the same final difference results from alterations at the same or different points in the production line. To extend the analogy of the motor factory, if we have watched every step of manufacture and assembly and then observe that one of the cars leaving the factory will not go, we may know that the defect is a faulty carburettor. On the other hand, if we have only been permitted to look at the final products in a shop window, we cannot tell whether the fault lies in the carburettor, the ignition, the clutch or elsewhere; we only know that the car will not go and are unable to distinguish it from any other new car of the same make that will not go. Having made this point we must go on to state that the fundamental laws of genetics, and the greater part of our present detailed knowledge of inheritance, have come from the observation of overt characters in complex organisms, beginning with Mendel's study of such characters in peas as the surface texture and colour of the seeds, and the colour and position of the flowers. Conversely, many of the character differences employed in bacterial genetics may be imprecise. For example, the inheritance of a genetic alteration resulting in inability to ferment a carbohydrate such as lactose, or in resistance to an antibiotic, may be studied without knowing whether the alteration involves a change in the permeability of the cell wall or in one of several possible metabolic steps.

GENOTYPE AND PHENOTYPE

We have seen that genetics involves the study of character differences between parents and offspring, and the manner in which such differences are inherited. It is important to realise that what is inherited is not the character itself but *the potentiality to express it*, and that this potentiality may be profoundly affected by the environment to which the organism is exposed. The analogy between the mechanical operations of a factory and the biochemical activities of a cell is usually such an apt one that we shall often invoke it, though we will try not to give undue publicity to the motor industry. In the present instance, what we inherit are the blueprints which determine the specifications of the products, and the tools and jigs to make them. Before anything can be made, however, raw materials, a source of energy and suitable conditions for the workers are required and these come from the environment; variations in the environment may greatly alter the output and quality of the products.

There are many examples in nature of the influence of environment on the expression of the hereditary constitution. A striking example is the effect of temperature on hair pigmentation in Himalayan rabbits. The fur of the feet, ears and tails of these animals is black while that of the body is white, and this characteristic is faithfully inherited. If, however, the white fur is plucked from a part of the body and the animal is kept in the cold while new fur is growing, this new fur is black instead of white; conversely, if the black fur is removed and the depilated extremity is kept warm during regeneration, then the new fur is white instead of black. This experiment shows that what is inherited is not the pattern of distribution of black and white fur, but a dependence upon temperature of the ability to synthesise black pigment; the temperature of the feet, ears and tail is normally sufficiently lower than that of the rest of the body to permit pigment synthesis to occur.

Microbial genetics offers numerous demonstrations of this kind of effect, some obvious and others not so obvious. Among the more obvious effects is that of environment on certain morphological features of bacteria. Addition to the culture medium of phenol or ethanol in concentrations too low to affect growth, or even cultivation on a rather dry solid medium, may suppress completely the synthesis of flagella which are the whip-like organs of locomotion, while excess of calcium salts inhibits spore formation, but these features rapidly reappear in all the cells when they are transferred to a normal medium.

A great variety of variant strains of fungi and bacteria have been isolated which, unlike the original or *wild type* strain from which they were derived, are unable to synthesise one or more amino acids, or other factors required for growth, from elementary precursors; such strains are known as *auxotrophs* (Latin: *auxilium*=help; Greek: *trophē*=nourishment) while the nutritionally independent, wild type strains are called *prototrophs* (Greek: *protos*=first). An auxotrophic strain may be identified by its inability to grow and multiply in a synthetic medium lacking its specific growth requirements, in which the wild type, prototrophic strain flourishes. If, however, the synthetic medium is supplemented with the growth factor which the auxotroph cannot synthesise for itself, then auxotroph and prototroph may grow equally well in it and be indistinguishable. Similarly, strains of bacteria which have become resistant to the lethal action of antibiotics or bacterial viruses may appear identical to the sensitive, wild type strains unless the drug or virus is present in the environment to reveal the difference between them. In the fungus *Neurospora crassa* several types of variant have been described which react differently from the wild type with respect to temperature. One of these requires the amino acid tryptophan for growth at 25°C, but is prototrophic within the temperature range 30° to 35° when it produces the enzyme tryptophan synthetase which catalyses the condensation of indole and L-serine to form L-tryptophan. No active enzyme is made when the cells are grown at 25° in the presence of tryptophan, but a protein can be extracted from them which, although enzymically inactive, is antigenically related to the active enzyme; further purification of this protein restores some of its enzymic properties, apparently due to removal of zinc which acts as an inhibitor.

Another more complicated example of environmental influence, is the case of an inducible enzyme system in the bacterium, *Escherichia coli*. Strains of this organism may fail to ferment lactose because of one or the other of two inheritable defects: one defect involves the capacity to synthesise the enzyme β-galactosidase which initiates the fermentation of lactose by splitting it into glucose and galactose; the other prevents synthesis of a second enzyme, called permease, which mediates the penetration of lactose into the cell. If, now, we examine lactose-fermenting strains of *E. coli* for their capacity to produce β-galactosidase, we find that they can be divided into two types; one can produce the enzyme only when lactose or some other galactoside is present in the environment, while the other can produce it whether lactose is present or not. The former type is said to be 'inducible' and the latter 'constitutive'.

This character difference is inheritable. We can thus define two genetic types which appear identical, in their ability to yield β-galactosidase, when lactose is present in the environment but different when lactose is absent (see p. 710 *et seq.*). A comparable state of affairs has been described with respect to production of the enzyme penicillinase, which destroys penicillin, by *Bacillus cereus* and *Staphylococcus pyogenes* (p. 783).

From what has been said so far it is evident that the observed expression of the characters of an organism, known as its *phenotype* (Greek: *phainomai*=I appear), may result from interaction between the environment and its hereditary potential or *genotype*. Thus two individuals which differ genetically, that is in genotype, may display the same phenotype, as when streptomycin-sensitive and streptomycin-resistant bacterial strains are grown in a medium without streptomycin, or when the colonies of prototrophic and auxotrophic strains are compared on a medium containing the growth factor required by the latter. Conversely, genotypically identical, inducible strains of *E. coli* will differ in phenotype with respect to β-galactosidase synthesis depending upon whether lactose is present in the medium or not.

MECHANISMS OF INHERITANCE

The characteristic appearance and behaviour whereby we recognise an individual, which may be a single cell like a bacterium or a highly complicated organism compounded of billions of cells of different kinds like a higher animal or plant, results from the phenotypic expression of its hereditary constitution or genotype, which is made up of the sum total of all the character determinants that the individual inherited from its ancestors. Whereas the phenotype may vary widely with the surroundings in which the individual finds itself, the genotype remains constant. The genotype is therefore the important thing with which genetics is concerned and about which we may now pose three important question: What is the structural basis of genotype? What is the mechanism of its transmission from generation to generation? What kinds of relationship exist between the determinants of the many different characters that together make up the genotype? More or less definitive answers to these questions have come from three very diverse kinds of investigation. The first, chronologically, was Mendel's (1865) study of inheritance in peas from which he inferred the fundamental laws of segregation and of independent assortment of characters which form the basis of the science of genetics; but Mendel's discoveries lay hidden for 35 years before their significance was recognised. About this time, the recognition of the cellular basis of life, and the development of the microscope and of accessory staining techniques, stimulated independent researches on the visible structure and behaviour of the cell which led to discovery of the rod-like chromosomes within the nucleus, and of the intricate sequence of events, known as *mitosis*, which accompany its division. Knowledge of the details of mitosis prompted the conclusion that only the mechanism of inheritance could demand such precision. Subsequent cytogenetical correlation of Mendel's laws of heredity with the operation of an observable physical mechanism was a profound advance which genetical research, during the first half of this century, has been mainly concerned to consolidate and extend in detail.

Very recently the science of genetics has experienced another important and vitalising impetus from the integration of two new developments. The first is the elucidation of the physico-chemical nature of the genetic material itself. The second is the greatly increased resolving power which study of the genetic systems of micro-organisms has given to genetic analysis. We must deal with all these advances in some detail in order not only to establish a clear background of orthodox genetical behaviour against which the peculiarities and complexities of bacterial genetics may be seen in proper perspective, but also to define and learn the meaning of specific words and symbols—the language of genetics—without which it is difficult to talk about the subject at all with any clarity and precision. Our first purpose, therefore, is to attempt to outline the main principles and features of so-called 'classical' genetics. In order to do this briefly and without confusing the reader with details which are irrelevant to the primary theme of this book, it is necessary not only to be dogmatic but also to make many generalisations which may irritate the critical geneticist. Accordingly we must point out that there are exceptions to almost every general statement which we will make, for simplicity of exposition, in this and the following three chapters; moreover it was from the study of just such exceptions that many important discoveries in genetics were made.

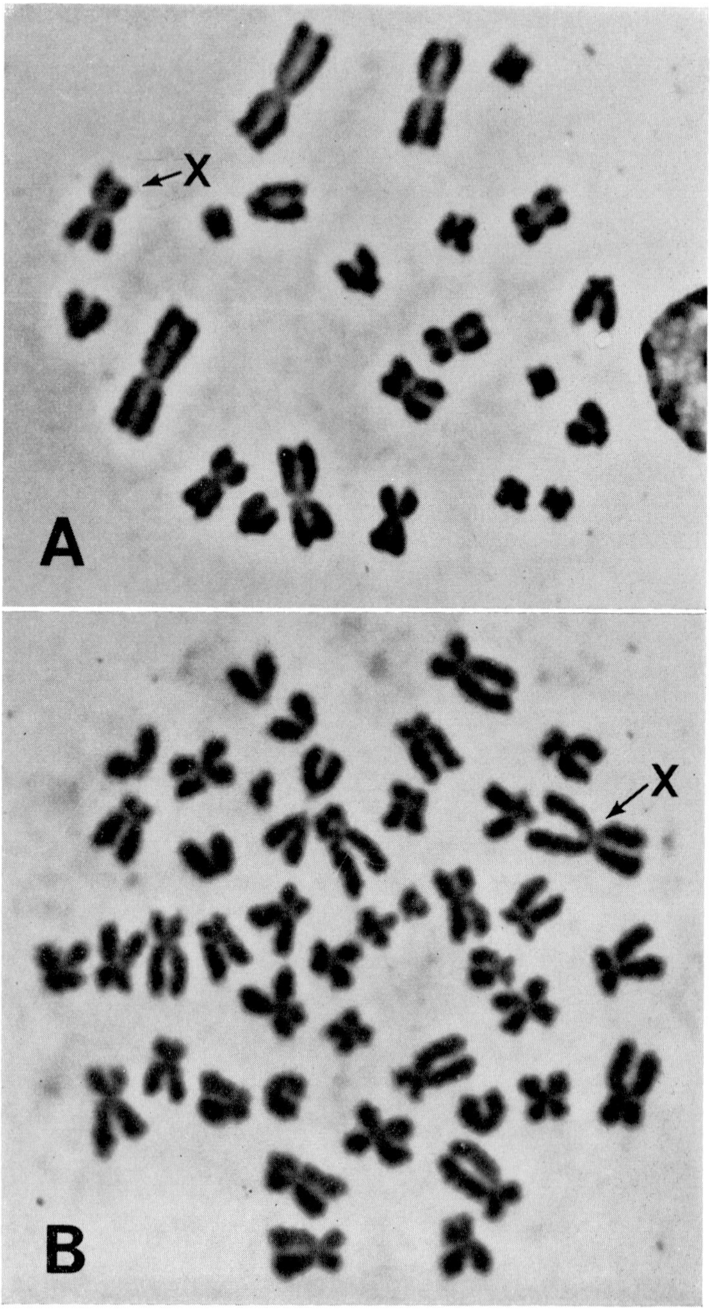

facing p. 9

Since most people seem to find it easier to understand theoretical principles if these can be visualised in terms of some kind of mechanical working model, we will first describe briefly the cytological mechanisms of inheritance before progressing to the more inferential studies which gave birth to Mendel's laws.

ASEXUAL REPRODUCTION

Mitosis

The word 'mitosis' (Greek: *mitos*=thread) connotes the replication and division of the nucleus which precedes and accompanies cellular multiplication. When the undividing, or 'resting', cell is stained with basic dyes, the nucleus appears as a finely reticulated body, enclosed by a membrane and more or less devoid of structure except for one or two bodies, rich in ribonucleic acid (RNA), known as *nucleoli* (Fig. 1A). At the onset of mitosis the nucleoli disappear and the fine nuclear reticulum resolves itself into a number of densely staining threads, the *chromosomes* (Greek: *chroma*=*colour*; *soma*=body). The chromosomes shorten and thicken by forming spirals and are seen to be split longitudinally into two sister *chromatids*, except at one constricted region where the chromosome is usually bent and which is the site of a minute body called the *centromere* (Fig. 1B). This first stage of mitosis, during which the chromosomes make their appearance, is known as *prophase* and is terminated by disappearance of the nuclear membrane and release of the chromosomes into the cytoplasm. During the next stage, *metaphase*, fibres begin to appear which are denser than the surrounding cytoplasm and which radiate from two points at opposite poles of the cell, like the lines of longitude on a globe, to form a structure known as the *spindle*. The chromosomes now come to lie in one

---

PLATE 1. Metaphase chromosomes showing sister chromatids and centromeres. The morphological pattern of the normal chromosome set is characteristic of each species, as plates A and B show.

A. 22 metaphase chromosomes of a male Chinese hamster (Cricetulus griseus).

B. 44 metaphase chromosomes of a male Syrian hamster (Microcricetus auratus), which are not a simple multiple of the Chinese hamster set, since many of the chromosomes differ structurally e.g. the X sex chromosomes (see arrows) are quite different in size in the two species. It is possible that the Syrian hamster was derived from the Chinese hamster by a combination of doubling the chromosome number (polyploidisation) and chromosome structure mutations, during evolution (by courtesy of Dr. D. T. Hughes, The Chester Beatty Research Institute, London, S.W.3).

plane midway between the two poles of the spindle to form the *equatorial plate* (Fig. 1C). The third stage, *anaphase*, is ushered in by division of the centromere of each chromosome which is accompanied by complete separation of the two sister chromatids, one of which remains attached to each daughter centromere. The two daughter centromeres then migrate along the spindle fibres, or are pulled by them, towards opposite poles of the cell, each trailing its chromatid behind it Thus the two identical halves of each chromosome are separated and reformed into two groups at different extremities of the cell (Fig. 1D). During the final stage of mitosis, *telophase*, a new nuclear membrane is formed around each group, the chromatids uncoil and elongate again into fine threads which lose their staining reaction, and new nucleoli appear (Fig. 1E, 1F). Thus two new, identical resting nuclei emerge. At about this time, a new cell membrane or wall begins to appear in the region of the equatorial plate so that the cell is ultimately divided into two halves which may then separate completely. The term 'resting phase' appears to be rather an unfortunate one since it now seems likely that it is, in fact, a period of great metabolic activity during which new chromosome material is synthesised and replication of the chromosomes, which appear double at the next prophase, occurs (Fig. 1).

If we examine chromosomes during early metaphase when they are displayed at their shortest and thickest on the equatorial plate, a number of important and characteristic features become apparent:

1. While the number of chromosomes in the somatic cells of a given species tends to be rather constant, when we compare the cells of different species this number may differ markedly (Plate 1).

2. Chromosomes have individual appearances which enable them to be identified and followed from generation to generation. Some are long, others short. In addition, each type has a characteristic bent shape which appears to be due to the position of attachment of the centromere along its length (Plate 1).

3. In all the somatic cells, of higher animals and plants especially, there are nearly always *two separate chromosomes of each kind*, which split and divide independently of one another at mitosis. That is, each chromosome is represented twice. One of each chromosome pair comes from the father and the other from the mother. Cells of this sort which have two similar sets of chromosomes, one paternal and the other maternal, are called *diploid*. In contradistinction to the diploid somatic cells, the germ cells or *gametes* (Greek: *gamos*=marriage) of all species

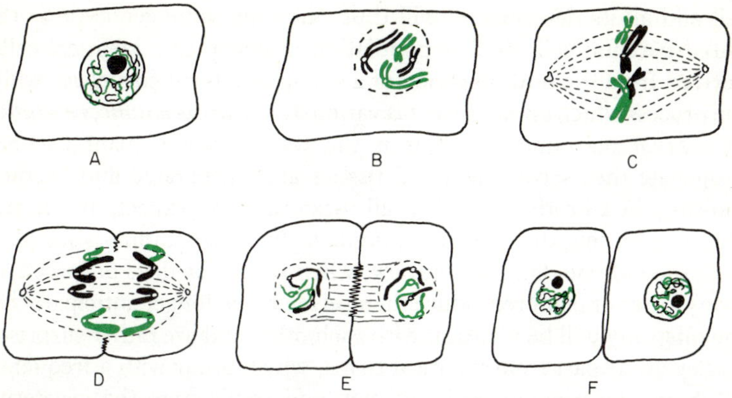

FIG. 1. Diagrammatic representation of mitosis in an imaginary, diploid plant cell.

The cell is shown as having two pairs of chromosomes which are morphologically different; paternal chromosomes are black, maternal green. The minute, open ring joining pairs of chromatids is the centromere. A, resting cell; B, prophase; C, metaphase; D, anaphase; E, telophase; F, daughter cells. The large, black regions in the nuclei of A and F represent nucleoli.

possess only one set of chromosomes, that is, half the diploid number, and are termed *haploid*. The haploid number of chromosomes is 23 in man, 22 in rabbits, 20 in mice, 12 in tomatoes, 10 in maize, 8 in the fungus *Aspergillus nidulans*, 7 in *Neurospora crassa* and 4 in the fruit fly *Drosophila melanogaster*; in other species of *Drosophila* the number varies from 3 to 6. The bacterium *Escherichia coli* is known from genetic analysis to possess only a single chromosome as is also the case in bacteriophage.

It is worth noting, however, that there are frequent exceptions to the rule that somatic cells are diploid. For instance the male sex chromosome is represented only once in man while the somatic cells of most, if not all, male hymenoptera are haploid; on the other hand it is not uncommon for somatic cells, especially of plants, to acquire many more sets of chromosomes than the usual diploid number, for which the term *polyploidy* is used (see p. 40).

Mitosis provides a precise and elegant mechanism for ensuring that two daughter cells each acquire a set, or sets, of chromosomes identical with those of the parent cell. Since, as we shall see, the chromosomes are indeed the bearers of the hereditary determinants, each daughter cell and all its descendants will thus carry the same genotype as the initial parental cell. Such a population of genetically identical cells, derived from the multiplication of a single cell, is called a *clone*, while the process which gives rise to it is variously known as *mitotic*, *vegetative* or *asexual* reproduction. This is the way in which plant cuttings propagate themselves, damaged tissues are regenerated and bacteria multiply in a broth culture; in all these cases we expect, and nearly always find, that the new plant, tissue or bacterial culture is identical with the old one. On the contrary, when people or dogs or insects or many kinds of plant reproduce themselves we no longer anticipate that the offspring will be identical with each other and are rather surprised if they are. Apart from identical twins, which occur with a frequency of about one per 320 births in man and result from the vegetative division of a single fertilised ovum, offspring of the same parents are in fact *never* identical. The reasons for this variability are inherent in the mechanism of *sexual reproduction* which we must now consider.

SEXUAL REPRODUCTION

The essential feature of sex is the pooling of the genetic material of two parental haploid cells in a single cell which thus becomes diploid and is called a *zygote* (Greek: *zugon*=yoke, i.e. joining two things together). This process is known as *fertilisation*. When the two parental cells are genetically different, the zygote and its diploid progeny are said to be *hybrid* for those characters in which the parents differ. In order to be able to repeat the cycle it is necessary that the zygote cell or its descendants should *segregate* again into two haploid cells; as Weismann predicted in 1887, this reductive process is a logical necessity if a doubling of the number of chromosomes in the zygote at each generation, which would rapidly clog the system, is to be prevented. Organisms which reproduce sexually thus have a cycle in which the haploid and the diploid states alternate (Fig. 2). Nature has many variations of this basic theme in her repertoire.

One possibility for variation is the preeminence of one or the other phase of the cycle. In higher animals, for example, the entire body of the organism consists of diploid cells derived from multiplication of the zygote, and only the gametes (spermatazoa and ova) are haploid. At the opposite extreme, most algae, fungi and protozoa, as well as bacteria, are haploid organisms; when zygotes are formed they undergo reductive division at once, or after only a few generations, to yield the predominant haploid cells. In the mosses the diploid phase still forms

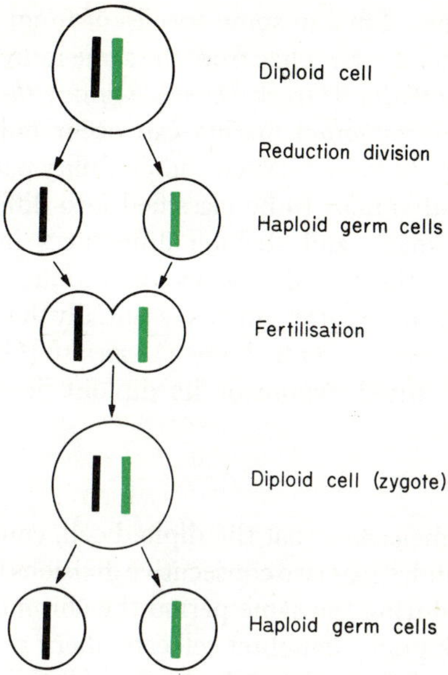

FIG. 2. Schema of the alternation of haploid and diploid states which comprise the sexual cycle.

The chromosome set derived from one parent is shown black, that from the other parent green.

an inferior, secondary part of the structure of the plant and it is not until we reach the ferns that the diploid becomes the dominant phase.

Another kind of variation concerns specialisation or differentiation of the haploid gametes. In the simplest forms of life, sexual union may be between two gametes which are similar in form and structure (*isogamy*—Greek: *isos*=equal; *gamos*=marriage). In higher organisms the gametes may be markedly different in appearance (*heterogamy*—Greek: *heteros*=different), as is seen in the spermatazoa and ova of animals or the pollen and ovules of plants, and there is usually a division of labour between them. Thus spermatazoa are merely stripped down, motile vectors of the male genetic material, while ova are large cells, especially in oviparous mammals, with an abundant cytoplasm rich in nutrients to serve the needs of the developing embryo.

A third kind of variation involves sexual compatibility or incompatibility. Distinctions of this sort are perhaps most subtly exemplified in microorganisms. Thus in some species of fungi, zygotes may arise from a fusion of haploid hyphae from the same individual; such species are termed *homothallic* (Greek: *homos*=same; *thallo*=I bloom). In other species (*heterothallic*) mating can occur only between certain pairs of different individuals; such compatibility patterns are heritable and allow the individuals to be classified into different *mating types*. Despite all the many and striking differences in detail which are apparent between the sexual processes of various animals, plants and microorganisms, one feature remains essentially the same in all of them. This is the sequence of events, known as *meiosis* (Greek: *meion*=less), whereby the fertilised zygote or its diploid descendants segregate haploid gametes.

Meiosis

The essence of meiosis is that the diploid cell, containing two sets of chromosomes, undergoes two consecutive divisions to yield four grand-daughter cells: during the same period the chromosomes divide only once so that each grand-daughter cell can inherit only half the number of chromosomes of the initial diploid cell and is therefore haploid. The stages observed during the first meiotic division bear a superficial correspondence to the stages of mitosis but in fact the two processes differ from the very beginning. It will be remembered that individual chromosomes are distinguishable from one another by their size and shape. In diploid cells, with two sets of chromosomes, each chromosome has a morphologically identical twin; such corresponding pairs are known as *homologues*, one being paternal and the other maternal in origin. In mitosis, as we have seen, these homologous chromosomes divide and are inherited by daughter cells quite independently of each other. The *prophase of the first meiotic division* commences with the appearance of the diploid number of chromosomes as thin, elongated threads which, unlike the chromosomes in the prophase of mitosis (Fig. 1B), appear to be single. Homologous chromosomes are then attracted towards one another and come to lie together in intimate contact when they can be seen to correspond exactly. This is called *pairing* or *synapsis* (Fig. 3B). In this way a haploid number of homologous pairs of chromosomes, called *bivalents*, are formed which proceed to shorten and thicken. The attraction between the pairs then suddenly subsides so that they separate along most of their length but usually remain attached at one or more points; at this stage it can be seen that each chromosome has split longitudinally into two

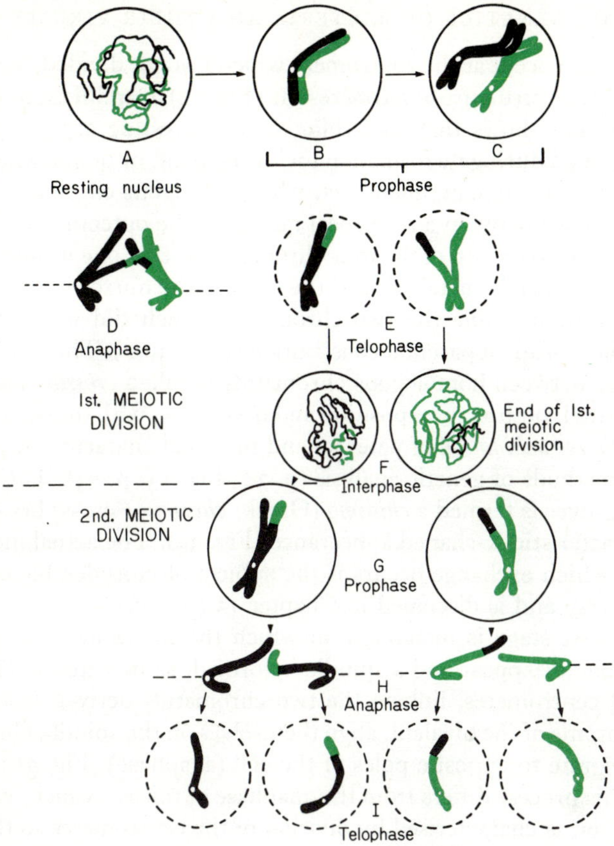

FIG. 3. Diagrammatic representation of Meiosis.

Only a single pair of homologous chromosomes is shown, the paternal chromosome in black and the maternal in green. The minute, open rings in the chromosomes are centromeres.

The circles enclosing the chromosomes represent nuclear membranes.

1st Meiotic Division: A, resting diploid nucleus; B, pairing of homologous chromosomes; C, chiasma formation; B and C together comprise prophase; D, anaphase; E, telophase; F, daughter interphase nuclei.

To simplify the diagram, the arrangement of the chromosomes on the equatorial plate at metaphase is omitted (see Fig. 1, p. 11).

Note that the two daughter cells emerging from the 1st meiotic division are still diploid.

2nd Meiotic Division: This is straightforward mitosis which results in the segregation to a separate cell of each of the four chromatids, two parental and two recombinant, formed during the 1st meiotic division. This 2nd division is, therefore, the reductive one from which the four haploid gamates emerge.

For simplicity, metaphase and the final resting nuclei are not portrayed (see Fig. 1, p. 11, and text).

chromatids, except at the centromere which has not divided, so that a four-stranded structure or *tetrad* results (Fig. 3C). Careful examination of this tetrad shows that something very interesting and important has happened during its formation for, at the point or points where the four chromatids remain in contact, two homologous chromatids have exchanged segments in a reciprocal manner. The outcome of such an exchange between two of the four chromatids is that one is now partly paternal and partly maternal, and the other is the mirror image of this in constitution, while the two chromatids which did not exchange retain their original parental constitutions (Fig. 3C,D). This process of exchange between homologous chromatids is called *crossing-over* and is correlated, both as to its position and frequency, with the occurrence of *genetic recombination* of paternal and maternal characters as judged by the methods of genetic analysis (p. 32; Fig. 13, p. 53). The site of crossing-over is termed a *chiasma* (Greek: *chiasma*=cross) because of its characteristic X-shaped appearance (Fig. 3C). The actual mechanism by which exchange occurs is the subject of considerable current controversy and is discussed in Chapter 15 (p. 376).

The next stage is metaphase in which the nucleolus and nuclear membrane disappear and a spindle is formed, as in mitosis. The undivided centromeres, linking the two chromatids derived from each chromosome of the bivalent, align themselves on the spindle fibres and then migrate to opposite poles of the cell (anaphase) (Fig. 3D). Once again this process differs from the anaphase of mitosis which, you will remember, is characterised by division of the centromeres so that the two chromatids, resulting from the splitting of each chromosome, are separate (Fig. 1D); in the anaphase of meiosis the centromeres do not divide so that the effect is merely to separate homologous paternal and maternal chromatid pairs except that, due to crossing-over, some of the chromatids will be recombinant instead of parental in constitution. Thus, at the end of the first anaphase each half of the cell contains the diploid number of chromatids, some of them recombinant for part of their length, connected in sister pairs by the centromeres. We cannot usually distinguish paternal from maternal chromosomes or chromatids by simply looking at them although a few cases are known where this can be done (p. 38); this kind of distinction can generally only be made by studying the way in which those characters in which the two parents differ are inherited among the progeny, as we will see in Chapter 2 (p. 21). Nevertheless it may be helpful now to anticipate the results of genetic analysis by stating that, at the end of the first meiotic division, there is an entirely random distribution of paternal and maternal chromosomes between the two daughter cells. That is to say, each daughter cell now has only one complete haploid set of chromosomes (though each is double) and, among these, there is an equal probability that any particular one will be paternal or maternal in origin. In genetical terms this process can be described as the segregation and independent assortment of chromosomes according to chance.

After a brief interphase (Fig. 3F), the *second meiotic division* begins. This, unlike the first, is almost indistinguishable from an ordinary mitotic division except that the chromosomes are already separated into two chromatids, joined only at the centromere (Fig. 3G) and do not divide again. At the second anaphase (Fig. 3H) the centromeres divide and migrate to opposite poles pulling their chromatids with them. Thus four cells are formed (Fig. 3I) which have the following features:

1. Each contains half the number of chromosomes of the initial diploid cell, that is, one set of chromosomes only.

2. The distribution, within each set, of chromosomes of paternal and maternal origin is nearly always random.

3. If a single chiasma has arisen during the synapsis of any chromosome pair at the first meiotic division, resulting in a single cross-over (Fig. 3C) then, so far as that chromosome is concerned, one of the four cells will be pure paternal and another pure maternal, while the remaining two will be reciprocal recombinant in type. The genetic constitution, in terms of chromosome inheritance, of each of the four products of meiosis is therefore different.

It is normally the case that all the chromosome pairs (bivalents) show the formation of at least one chiasma; moreover, a chiasma may arise at any one of the numerous regions of the bivalent where the two chromosomes make contact. Since the genetic constitution of the recombinant chromosomes varies with the position of the chiasma, it is clearly most improbable that the recombinants produced by any two identical, diploid mother cells will be alike.

Multiple Crossing-over.

To avoid confusion we have so far considered recombination as a function of the formation of only a single chiasma. We must now discuss briefly what may happen when more than one chiasma arises so that two or more cross-overs occur at different regions within the same bivalent. We have seen that, in a single event, four chromatids or *strands* are involved, two of which undergo reciprocal exchange while the other two remain parental. The first thing to note is that there is no restriction as to which strand of each pair of homologous strands will participate in the exchange. The four strands, marked 1,2 and 3,4, are shown diagrammatically in Fig. 4. You will see that there are four possible types of exchange, between 1 and 3(A), 1 and 4(B), 2 and 3(C) and 2 and 4(D). Since the two strands of each homologous pair (1,2 or 3,4) arise from division of a single chromosome and are identical, it does not matter which of the four types of exchange occurs; all will yield the same pair of reciprocal recombinants, in addition to the two parental types, among the gametes (Fig. 4, lower left). Now let us imagine that a second cross-over occurs at the same time at some other region of the bivalent. The same four possibilities exist for recombination between non-sister strands as in the first cross-over, but what happens here will profoundly affect the outcome. This can best be understood if the examples given in the lower, right-hand diagrams of Fig. 4 are followed by tracing, in turn, along the strands shown in the upper diagram, switching to another strand wherever the vertical dotted line indicates a cross-over. It will be found that when the same two strands (for example, 1,3) are involved in both events (2-strand doubles), two parental and two reciprocal recombinant gametes arise as in a single cross-over, but both recombinants are of double type. When one of the strands is the same but the second strand is different (1,3 and 1,4) in the two events (3-strand double), three of the resulting gametes are recombinant and only one, inheriting the uninvolved strand (2), is parental; of the recombinants, one is of double type while two are of single type and reciprocal. When both of the strands participating in one event are different from those involved in the other (1,3 and 2,4), all four progeny are recombinant and of single type, as if two had arisen from a single cross-over in region 1 and the other two from a single cross-over in region 2 (see pp. 51, 74).

It is possible, of course, that crossing-over may also occur between the two strands of one chromosome, that is, between *sister strands*, but since these strands are genetically identical the outcome of such an event cannot be observed (see p. 392). Nevertheless sister strand exchanges, if they do occur, could modify the effects of double cross-overs of the kind we have described. Take, for example, the 2-strand double involving strands 1 and 3 in Fig. 4, from which two parental and two reciprocal recombinant gametes emerge. If an exchange were to occur between sister strands 1 and 2, in the interval between the

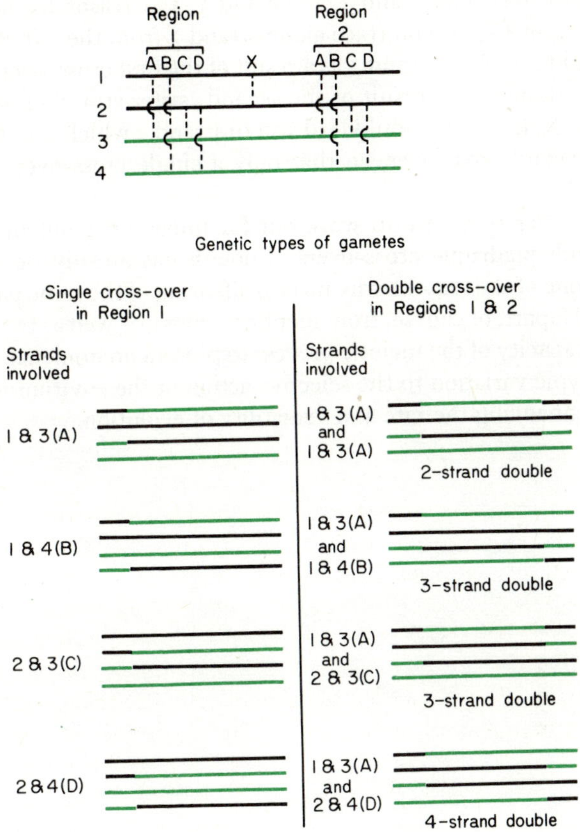

FIG. 4. The effect of single and double cross-overs, involving different pairs of strands, on the genetic constitution of gametes.

The four strands of a bivalent are shown, the strands from one parent being black, and from the other parent green. Two regions of crossing-over, 1 and 2, are indicated; A, B, C, D show the various combinations of strands between which genetic exchange may occur in regions 1 and 2.

The genotypes of the four gametes, resulting from genetic exchange between all the possible combinations of pairs, are represented by the green and black segments in the lines of the lower part of the figure.

The dotted, vertical line at X represents the possible occurrence of a cross-over between sister strands.

(For explanation, see text.)

two cross-overs, its effect would be to mimic a 3-strand double cross-over between strands 1 and 3, and 2 and 3. The reason for this will be evident if, in Fig. 4, you trace along strand 3 from the left. Normally, you would switch to strand 1 as a result of the first cross-over and then back to strand 3 as a result of the second; sister strand crossing-over at region X, however, would lead you to strand 2 which is not involved in the second cross-over, so that only a single cross-over would be observed.

The reader may care to work out for himself the potentialities of triple and quadruple cross-overs! Enough has already been said to explain one of the reasons why no two offspring of the same parents are identical (apart, of course, from identical twins), as well as to reveal the unique capacity of the meiotic process to present an immense diversity of genotypic variation to the selective action of the environment, thus greatly enhancing the rate and versatility of evolution.

CHAPTER 2

Mendel's Laws of Inheritance

Johann Gregor Mendel was born in 1822, the son of a peasant family and, as a boy, entered the Augustinian monastery at what is now the Czechoslovak town of Brno. He became a monk at the age of twenty-five and, shortly afterwards, was sent by his order to study natural science for two years at Vienna University where, it was said, he showed no outstanding ability. He then returned to Brno to teach science and four years later, in 1854, commenced his concise studies of inheritance in the garden pea, the results of which were published in 1865 and subsequently widely distributed throughout the scientific world. The significance of these results, and of the laws he deduced from them, was neither appreciated nor understood by his contemporaries. In 1868 he became abbot of his monastery, the administration of which occupied him increasingly until his death in 1884. It was not until 1900 that his work was rediscovered and acclaimed by other investigators who had obtained similar results.

Mendel's predecessors in the study of heredity had made the mistake of attempting to interpret the manner in which the *whole* phenotype, that is, the sum total of all the characters, of two different parents was inherited, and they were baffled by the immense variety which the hybrid offspring displayed. Mendel's success was due to his introduction of methods of scientific analysis which had not previously been applied to this field but which, today, no one with any pretensions to be a geneticist would dream of disregarding. These were the use of pure-breeding lines as parents, the examination of only a single, cleancut character difference at a time, and the recording of the *numbers* of each type of progeny.

THE PRINCIPLE OF SEGREGATION

Mendel studied seven well defined pairs of characters in the pea, but here we will restrict our attention to three of these—the colour of the flowers (purple or white), and the colour (yellow or green) and shape (round or wrinkled) of the seeds. The pea is a naturally self-fertilising plant, the stamen (male organ) shedding its pollen on the stigma (female organ) of the same flower. The stamen can be removed before pollination, however, so that self-pollination cannot occur, and the plant then crossed with another plant of a different variety by dusting pollen from the second on the stigma of the first. Mendel's first procedure was to cross, in this manner, two plants which differed in the character under study, to sow the resulting seeds and to observe the phenotypic expression of the character in the *hybrid* or *first filial* (F_1) *generation* of plants which grew from them. He found that all these hybrid plants were identical in appearance and resembled one of the two parents. For instance when one of the parents had purple flowers and the other white, the flowers of all the hybrid (F_1 generation) offspring were invariably purple (Fig. 5). Pigmentation was thus shown to be a *dominant* character and whiteness a *recessive* character in hybrids. Similarly when the characteristics of the seeds were investigated it was found that the round shape was dominant to the wrinkled, and the yellow colour dominant to green.

The next step was to cross two hybrids, or allow them to undergo self-fertilisation, to plant the seeds, and then to examine and *count* the distribution of each pair of characters among a large number of the second generation, or F_2, plants. It turned out that, whatever pair of characters was studied, three-quarters of the F_2 generation plants displayed the dominant character and one quarter the recessive character. With regard to flower colour, for example, Mendel found that of a total of 929 F_2 plants, 705 (75·9 per cent) had purple flowers and 224 (24·1 per cent) had white flowers. Again, when the seed shape of 7,324 F_2 plants was analysed, 5,474 (74·74 per cent) yielded round seeds and 1,850 (25·26 per cent) wrinkled seeds. This 3:1 ratio was subsequently confirmed by other workers for a wide variety of characters in various plants and animals.

The reappearance of recessive characters among the F_2 progeny clearly indicated that these characters are neither lost nor blended with the dominant character in the hybrid, but are preserved intact. It follows that the two potentialities which the hybrid inherited from the two parents, such as the ability to yield coloured or white flowers, must be due to physical determinants which can be separated again, or *segregated*, in the F_2 generation. This is *the principle of segregation* and its importance lies in the inherent implication that the expression of a character depends on a particulate factor which is transmitted unchanged from generation to generation through the gametes. This factor is now called a *gene*, originally defined as any particle to which the properties of a mendelian factor may be attributed. Genes which determine characters, such as pigmentation and whiteness in flowers, which are *alternative* to one another in hybrids but which can segregate and express themselves in the F_2 generation, are now called *allelic genes* or, simply, *alleles* (Greek: *allelo*=each other).

Having deduced the existence of allelic pairs of genes as the ultimate determinants of differences of character, Mendel then propounded a general theory as to how the genes were inherited, based on the 3:1 ratio of dominant to recessive phenotypes in the F_2 generation. His interpretation may best be understood by designating two allelic genes determining, say, purple or white flowers, as *A* and *a*, of which *A* is dominant and *a* recessive. Since the plant arises from the union of two gametes, and since a hybrid inheriting the constitution *Aa* yields purple flowers, it may be assumed that the genetic constitution of the white-flowered parent is *aa* and of the purple-flowered parent *AA*. The gametes produced by these two parents may then be represented as *a* and *A* respectively. Fusion of these gametes will give rise to *Aa* hybrid individuals only, which will all show purple flowers since *A* is dominant to *a*. When these hybrids, in turn, come to produce gametes, *A* and *a* will segregate so that roughly half the gametes will have *A* and the other half *a*. Now let us see what we may expect the constitution of the F_2 generation to be if these gametes unite at random during self-fertilisation of the hybrids. There are obviously three possible combinations, *AA*, *Aa* and *aa*, and these will arise in the ratio 1:2:1 as Fig. 5 shows. But since *A* is dominant, *AA* and *Aa* plants will both bear purple flowers, so that the ratio of purple-flowered to white-flowered plants will be 3:1. As we have seen, this is precisely the result found by experiment. It will be realised, of course, that since the heredity of the individuals of the F_2 generation is determined by random mating events, the experimental results will be subject to the same kind of statistical variation as is found in the tossing of coins or the throwing of dice, so that we may only expect a really close approximation to the theoretical value when the numbers analysed are large.

We mentioned at the beginning of this book that the truth of a theory can best be judged by its ability to predict the outcome of further experiments. Mendel subjected his theory of segregation to such a test. We have seen (Fig. 5) that, according to the theory, the 75 per cent of

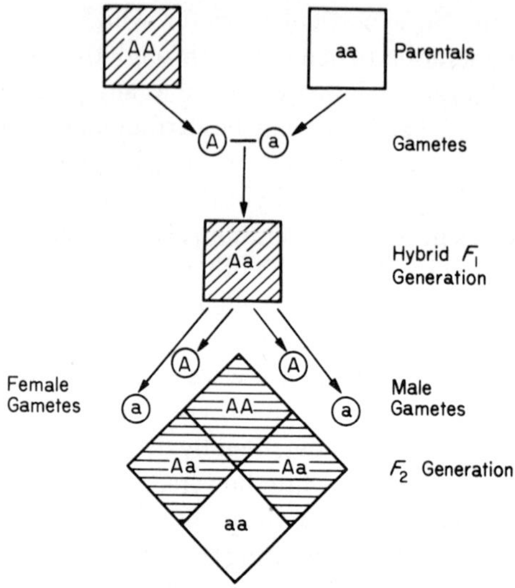

FIG. 5. Schematic representation of how Mendel's principle of segregation determines the 3:1 ratio of dominant to recessive phenotype among the F_2 progeny.

A is the dominant gene and *a* its recessive allele.

The shaded squares represent the dominant phenotype, and the unshaded squares the recessive phenotype.

phenotypically identical purple-flowered plants of the F_2 generation should actually be made up of two genotypic types, *Aa* and *AA*, in the ratio of 2:1. Mendel investigated whether or not this was so by studying the progeny which the purple-flowered, F_2 generation plants produced after self-fertilisation. Plants of *AA* genotype should breed true like the original purple-flowered parent stock, while those of *Aa* genotype should behave like the F_1 hybrids and segregate purple- and white-flowered progeny in the ratio of 3:1. The white-flowered F_2 plants should, of course, breed true like the *AA* plants and yield only white-flowered offspring. These predictions were confirmed by the experiment. Another result which can be predicted by the theory, and which was also confirmed by experiment, is the outcome of a *back-cross* between an F_1 or an F_2 hybrid and the recessive, white-flowered parent. As Fig. 6 shows, the offspring of such a cross should, and do, consist of roughly equal numbers of purple- and white-flowered plants.

In a statistical analyis of Mendel's findings, Fisher (1936) discovered, in the case of some of these latter experiments where only ten seeds

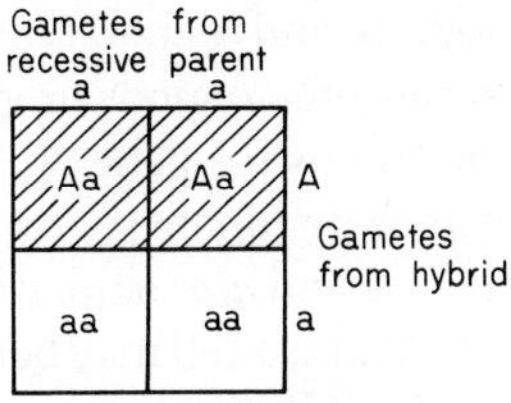

FIG. 6. Diagram showing the phenotype of progeny obtained by back-crossing an F_1 or F_2 generation hybrid to the recessive parent.

A is the dominant gene, and *a* its recessive allele. The genotype of the hybrid is *Aa*; that of the recessive parent *aa*.

The shaded squares represent the dominant phenotype.

were grown from each plant to be tested, that the numerical results were too good to be true since their approximations to the theoretical expectation were far better than chance would allow. The probability of equalling or bettering Mendel's figures as a whole turned out to be about one in 8000. Fisher's conclusion was that Mendel conceived the idea of particulate inheritance first, and worked out its implications theoretically before devising experiments to test it. Thus his assistants, knowing what results to expect, may well have exercised unintentional bias in scoring doubtful phenotypes. This theory is supported by the thoroughness with which Mendel expressed his findings in mathematical and generalised form and, if anything, enhances our appreciation of his genius (see de Beer, 1964, 1966).

We first broached the distinction between genotype and phenotype (p. 5) by pointing to the modifications which the environment may impose on phenotypic expression. We now see that phenotype may also depend on the interaction of two allelic genes in a diploid cell in so far as a recessive gene can only express itself in the absence of its dominant allele as, for example, in *aa* cells. This brings us to two very important words in the language of genetics which will recur frequently throughout this book and which we must now define. These words are *homozygote* and *heterozygote*, and they describe the genetical constitution of diploid cells. A homozygote is a zygote, or other diploid cell, derived from the union of two gametes of identical genotype; a heterozygote is derived from the union of dissimilar gametes. In practice, since the demonstration of heterozygosity requires evidence from genetic crosses that particular pairs of genes are allelic, and since it is obviously impossible to prove that *all* the thousands of pairs of genes in what appears to be a homozygous individual are in fact identical, the words usually have a narrower connotation and denote the relationship between specific, homologous chromosome regions or genes. Thus we may say that a diploid cell is homozygous for gene *A*, meaning that it has the constitution *AA* (homozygous dominant) or *aa* (homozygous recessive) or, as it is usually written, *A*/*A* and *a*/*a*; similarly a cell of genotype *A*/*a* is heterozygous for gene *A*. Again, a cell may be homozygous for gene *A* but heterozygous for gene *B* (*AB*/*Ab*).

FUNCTIONAL ASPECTS OF DOMINANCE

When a pure dominant-recessive relationship exists between two alleles, as was the case with all the pairs of alleles investigated by Mendel in the pea, the distinction between a homozygote (*A*/*A*) and a heterozygote (*A*/*a*) which display the same phenotype can only be made by genetic crosses which permit the recessive character in the heterozygote to express itself in homozygous recessive (*a*/*a*) progeny. However, only a minority of allelic relationships are of this type: in most heterozygotes the expression of the character lies intermediate between that displayed by each allele in homozygotes. In the snap-dragon (*Antirrhinum*), for example, the gene controlling red flower

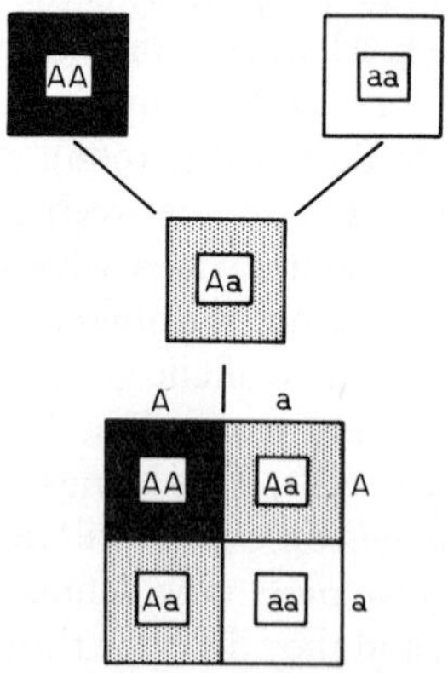

FIG. 7. Illustrating the inheritance of flower colour in *Antirrhinum*, in which the gene determining red colour (*A*) shows no dominance over its allele (*a*) determining white. The hybrid flowers are pink.

colour shows no dominance over that controlling white colour, so that the flowers of the hybrid F_1 generation are intermediate, or pink, in colour. Among the F_2 progeny three types of flower are found—

red, pink and white—and the relative frequencies with which they occur are in the ratios 1:2:1 (Fig. 7). Thus in the snapdragon we can distinguish the heterozygotes from the homozygotes by direct observation and find that they again arise in the proportions predicted by Mendel's theory of segregation.

The presence or absence of dominance has no bearing on Mendel's laws and has nothing to do with the gene itself as an entity, but depends on the nature of the particular biochemical process which the gene controls. Where dominance is found it is a general rule that the wild type gene is the dominant one and the altered gene recessive, but this is simply an expression of the fact that wild type genes have been selected in evolution because they do something useful for the cells that carry them, which usually means that they determine the synthesis of some cytoplasmic substance whose presence or activity is what we observe. Alteration of the gene to an allelic form usually involves loss of the capacity to produce the substance, so that the altered gene is functionally recessive.

As we shall see later, most characters are not determined by single genes but result from the expression of one or more biosynthetic pathways which are mediated by many genes. For example, the production of a plant pigment may involve a series of synthetic steps, one of which yields a colourless precursor. Alteration of a gene which blocks the pathway after this point will prevent synthesis of pigment but not of precursor. Plants homozygous for this allele will bear colourless flowers. However in the case of heterozygotes the colourless precursor may compete with the pigment for incorporation into the flowers which, therefore, appear distinguishably paler than those of the wild type plant.

A good example of the functional interrelationship between alleles, which has a known biochemical basis, concerns the inheritance of sickle-cell anaemia in man. This disease is due to alteration of a gene controlling the synthesis of haemoglobin. An abnormal haemoglobin is synthesised which crystallises when the partial pressure of oxygen is reduced, so producing the characteristic sickle-shaped distortion of the red blood cells. Sickling of the cells may be triggered by anything which lowers the oxygen tension of the tissues, such as chilling or exposure to a rarified atmosphere in an unpressurised aeroplane (p. 152). The fully developed disease, which is usually fatal in childhood, is found in individuals homozygous for the defective allele, in whom all the haemoglobin is abnormal. Heterozygous individuals, however, who have inherited one normal and one abnormal gene, do not manifest the fully developed condition but only what is called the 'sickle-cell trait', and usually survive into adult life to pass on their defective gene to a proportion of their offspring. The red cells of these heterozygous persons can be shown to contain both normal and abnormal haemoglobin in about equal proportions. Sickle-cell anaemia also presents a peculiar example of interaction between environment and genotype, for it turns out that the presence of abnormal haemoglobin in the red blood cells protects them against infection by malaria parasites so that the biological disadvantage imposed on heterozygous individuals in counterbalanced by their increased immunity to malaria. Thus in regions where malaria is endemic, the hereditary transmission of the defective gene is more favoured than in non-malarious regions and sickle-cell anaemia is correspondingly prevalent (Allison, 1954) (cp. p. 768).

THE PRINCIPLE OF INDEPENDENT ASSORTMENT

Having observed and explained how single character differences are inherited, Mendel went on to investigate what happens when the parents differ by more than one character. The experimental procedure was exactly as before. He crossed two pure-breeding plants, one of which yielded yellow, round seeds, and the other green, wrinkled seeds, and examined the colour and shape of the hybrid F_1 seeds produced. It will be remembered that when the inheritance of colour alone is studied, yellow and green behave as alleles, yellow being dominant; similarly, round and wrinkled are allelic characters of which round is the dominant one. As might be expected, all the hybrid seeds were the same—yellow and round. These seeds were then planted to give hybrid F_1 plants, which were then crossed or allowed to fertilise themselves. The resulting F_2 generation of seeds were finally examined and scored for phenotype. The use of such seed characters, provided they are determined by the genotype of the embryo, considerably simplifies analyses of this kind since they are observed directly; the seeds need not be planted. In addition to the original two types of seed (yellow-round and green-wrinkled), two new *recombinant* types, green-round and yellow-wrinkled, were found. It will perhaps make things easier at this stage if we use the bold type symbols **A** and **a** to represent the dominant yellow and recessive green colour, and **B** and **b** for the dominant round and recessive wrinkled shape, respectively (Fig. 8). The four types of F_2 seeds can thus be represented as the parental types **AB** and **ab**, and the two new

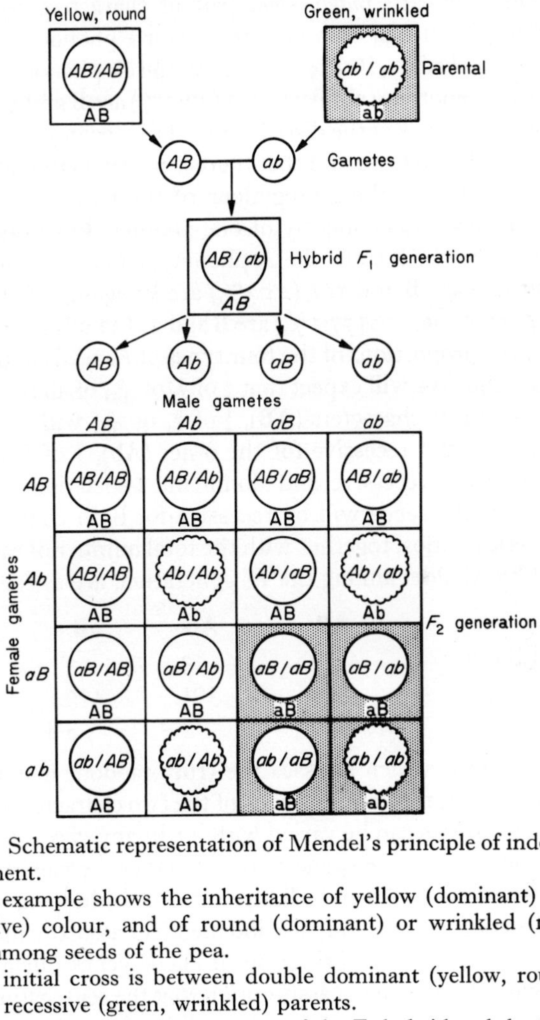

FIG. 8. Schematic representation of Mendel's principle of independent assortment.

The example shows the inheritance of yellow (dominant) or green (recessive) colour, and of round (dominant) or wrinkled (recessive) shape among seeds of the pea.

The initial cross is between double dominant (yellow, round) and double recessive (green, wrinkled) parents.

The genotype and the phenotype of the F_1 hybrid and the F_2 generation progeny are shown.

Phenotype is indicated by bold-type letters (**A**, **a**, **B**, **b**), and genotype by italicised letters (*A*, *a*, *B*, *b*).

A = yellow; a = green; B = round; b = wrinkled.

(For explanation, see text.)

types **Ab** and **aB**. The numbers of each type which were found were not the same. Out of a total of 556 F_2 seeds, 315 were phenotypically **AB**, 101 were **Ab**, 108 were **aB** and 32 were **ab**. When we look at these results to see how each *independent* pair of characters segregate we find the same 3:1 ratio of dominant to recessive phenotype as before; thus of the 556 seeds, 416 (74·8%) show the **A** phenotype and 140 (25·2%) the **a** phenotype; similarly 423 (76·1%) are **B** and 133 (26·9%) are **b**. But when we look at the distribution of the second pair of characters in terms of the first, we find that there is no correlation whatsoever between them; that is, the segregations of the two pairs of characters occur entirely independently of one another. For example if we take all the seeds which show the **A** phenotype (416) we find that 315 (75·7%) of them are **B** and 101 (24·3%) are **b**; again, among the 140 plants of **a** phenotype, 108 (77·1%) are **B** and 32 (22·9%) are **b**. We can now express the proportions of the four types of F_2 seeds in the form of a theoretical ratio. We will expect that $\frac{3}{4}$ of $\frac{3}{4}$, or $\frac{9}{16}$, of all the seeds will show both dominant characters (**AB**), $\frac{1}{4}$ of $\frac{3}{4}$, or $\frac{3}{16}$, will be dominant for one character and recessive for the other (**Ab**), $\frac{3}{4}$ of $\frac{1}{4}$, or $\frac{3}{16}$, will have the reciprocal recessive and dominant association (**aB**), while $\frac{1}{4}$ of $\frac{1}{4}$, or $\frac{1}{16}$, of the seeds will be recessive for both characters (**ab**). These theoretical ratios, together with the total number of each type of seed scored by Mendel among the F_2 generation, are as follows:

Phenotype	AB		Ab		aB		ab
Theoretical Ratio	9	:	3	:	3	:	1
Number	315		101		108		32
Ratio found	9·8		3·2		3·4		1·0

An impressive fact, which stresses the truly random manner of the reassortment, is that the combinations of the two character differences in the original parents can be varied without in any way affecting the results obtained from the F_2 generation. Whether the parental phenotypes are **AB** and **ab**, or **Ab** and **aB**, the outcome is the same.

The explanation of how and why independent assortment occurs is based, once again, on the segregation of allelic genes during gamete formation, with the additional assumption that the segregations of two pairs of allelic genes are entirely random with respect to one another. Fig. 8 should make what happens clear. The essential feature is the genetic constitution of the gametes that arise from self- or cross-fertilisation among the F_1 hybrids which are, of course, genotypically and phenotypically identical. It will be seen that if segregation of all the alleles is random and independent, four types of gamete, representing the four possible combinations of the four alleles, will be produced in approximately equal numbers. Of them roughly one half will be male and one half female. Random fertilisation among them will therefore beget roughly equal numbers of the sixteen genotypic varieties of F_2 generation seeds shown in Fig. 8. Because dominance operates for both characters, however, some of the different genotypes will display the same phenotype. For example, all the plants which have one *A* and one *B* gene will produce yellow (**A**), round (**B**) seeds irrespective of whether their genotype is *AB*/*AB*, *AB*/*Ab*, *AB*/*aB*, *AB*/*ab* or *Ab*/*aB*. In fact there are four possible phenotypes, as shown in Fig. 9, and if the number of each is added up they will be found to occur in the ratio

AB		**Ab**		**aB**		**ab**
9	:	3	:	3	:	1

which is that found by experiment.

The general validity of the theory can be further checked by the same methods as were used to substantiate the principle of segregation as, for example, by back-crossing the F_1 hybrid plants to the double-recessive parental type (Fig. 6) or by studying the F_2 phenotypes which arise when neither, or only one, of the characters studied shows dominance between alleles (Fig. 7). The expectations of the back-crossing experiment are given in Fig. 9; the four possible phenotypes with

Parental genotypes

AB/AB *ab/ab*

Gametes from F_1 hybrid

AB	*Ab*	*aB*	*ab*	
AB/ab	*Ab/ab*	*aB/ab*	*ab/ab*	Gametes from doubly recessive parent all *ab*
AB	Ab	aB	ab	Phenotype of progeny
1:	1:	1:	1	Ratio

FIG. 9. Distribution of phenotypes among the progeny of a back-cross of F_1 hybrids to the double recessive parental type.

Phenotypes are indicated by bold-type letters, genotypes by italicised letters.

Characters A and B assort independently. A is dominant to a, and B is dominant to b.

respect to the two characters, **AB**, **Ab**, **aB** and **ab**, should (and do) appear with equal frequency, the double dominant phenotype (**AB**) amounting to only $\frac{1}{4}$ the total. The phenotypes to be found in the F_2 generation in the absence of dominance will, of course, be more numerous since the homozygotes and the various heterozygous types will all appear different (Fig. 7, p. 26).

LINKAGE

The simplest and most direct test of independent assortment is, as we have seen in Fig. 9, to cross the F_1 hybrids with a double recessive individual. If assortment is independent, the F_1 hybrid should produce all four possible phenotypic types of gamete, **AB**, **Ab**, **aB** and **ab**, in equal proportions, irrespective of whether the two parents are of genotype *AB*/*AB* and *ab*/*ab*, or *Ab*/*Ab* and *aB*/*aB*. Since the double recessive parent contributes only the recessive *ab* gametes to the back-cross progeny, each type of F_1 gamete can express its genotype directly in the progeny of the cross, as Fig. 9 shows, with the result that the four possible phenotypic varieties appear in approximately equal numbers. What can we expect to happen if assortment is *not* independent? Let us suppose that two parental genes are *linked* in such a way that they are always transmitted together (as *AB* and *ab* for example) to the gametes. This means that during formation of the F_1 gametes (Fig. 8) reassortment of *A* with *b* and of *a* with *B* is no longer possible, so that only two parental types of gamete, *AB* and *ab*, can arise. *The recombinant types of gamete are eliminated.* When the F_1 hybrid is crossed with the double recessive (Fig. 9), therefore, only the parental phenotypes will be found. Similarly if the original parents are of genotype *Ab*/*Ab* and *aB*/*aB*, and there is complete linkage between the determinants of the two characters, the F_1 hybrids will generate only *Ab* and *aB* gametes so that, once again, the parental phenotypes alone will appear on crossing to the double recessive.

Mendel had found that all of the seven pairs of characters which he investigated in the pea underwent independent assortment, which was, perhaps, a fortunate circumstance for the development of his theory. Mendel was lucky in this respect, for not long after the rediscovery of his work in 1900 many cases began to be reported where pairs of genes did not assort independently but tended to remain together in the gametes; that is, the genes showed *linkage* with one another. In nearly every such case, however, linkage was not complete since *some* progeny of recombinant phenotype appeared (Ab and aB in Fig. 9) when the F_1 hybrids were crossed to double recessives, but the proportion of recombinant to parental type progeny was only a fraction of the unity indicative of independent assortment. Furthermore, the proportion of recombinant to parental types, that is, the degree of linkage, turned out to be more or less constant from cross to cross so far as any particular pair of genes was concerned, but tended to vary widely for different pairs of genes.

We can illustrate and sum up these findings by imagining a cross between two individuals, one homozygous for the five dominant genes *A*, *B*, *C*, *D*, *E* and the other homozygous for the recessive alleles *a*, *b*, *c*, *d*, *e*. The F_1 hybrids are back-crossed, as before, to the homozygous-recessive parent. After scoring the numbers of each phenotype among the offspring, with respect to the five pairs of characters, we can then compare these pairwise for linkage between any two genes, just as in Fig. 9. All the offspring must possess one or the other allele of all the genes in order to be viable, so all we have to do is to look for the number of *recombinants* for any pair of genes and express this as a percentage of the total number of progeny. This gives the *recombination frequency* for the two genes under study and indicates the degree of linkage between them; the lower the frequency the closer is the linkage and *vice versa*. It is clear that *the recombination frequency cannot rise above* 50 *per cent*, since this is the level at which the number of recombinants equals the number of progeny having the parental combinations of the two genes when, as we have seen, assortment is independent and there is no linkage. There cannot be less linkage than none! The kind of result that might be found is given in Table 1. It will be seen that the five genes fall into two *linkage groups*. Genes *B* and *C* are both linked to *A* and to each other, the highest degree of linkage (5 per cent) being between *A* and *B*; similarly *D* and *E* are linked to about the same extent as *B* and *C*. On the other hand, the recombination frequencies for the pairs of genes *A* and *D*, *A* and *E*, and *B* and *E*, approximates to 50 per cent, showing that the two linkage groups *ABC* and *DE* assort independently and are not linked.

Table 1 illustrates the fact that genes are associated together in linkage groups which means that the genes belonging to a particular linkage group have a greater probability than chance of being transmitted to the gametes as a single block. In the hypothetical example we have given, the dominant alleles of all the genes were in one parent and the recessive alleles in the other. It is worth stressing, however,

TABLE 1

Demonstration of linkage by analysis of the inheritance of five pairs of characters.
Cross: F_1 hybrid x recessive parent.
Parental combinations of characters: *ABCDE*/*ABCDE* and *abcde*/*abcde*.

Genetic analysis of progeny						
Pairs of characters studied		Recombinant phenotypes observed: Type	number	Total number of progeny	Recombination frequency $\left(\frac{\text{Recombinants} \times 100}{\text{Total progeny}}\right)$	Linkage
A	B	Ab + aB	56	1120	5·0	++
A	C	Ac + aC	213	1120	19·0	+
B	C	Bc + bC	173	1120	15·5	+
A	D	Ad + aD	547	1120	48·8	None
A	E	Ae + aE	565	1120	50·4	None
B	E	Be + bE	573	1120	51·6	None
D	E	De + dE	206	1120	18·4	+

that the effect of linkage is simply to preserve the *parental* combinations of genes irrespective of the particular alleles which the parents carry. Thus if the parental genotypes are *AbC*/*AbC* and *aBc*/*aBc*, we shall find the same degree of linkage, as shown by the recombination frequency, between *A* and *b* (or *a* and *B*) as we found between *A* and *B* (or *a* and *b*) in the cross shown in Table 1.

LINEAR ARRANGEMENT OF GENES: THE ELEMENTS OF MAPPING

The next question is, can we infer anything about the relationships between the genes within a linkage group from their recombination frequencies with one another? Let us take the case of the three linked genes *A*, *B* and *C* (Table 1). Our reasoning has led us to the conclusion that the linkage between two genes is inversely proportional to the frequency with which they show recombination, since no recombination indicates complete linkage while 50 per cent recombination is the criterion of independent assortment, or no linkage. Thus *A* and *B*, with a recombination frequency (RF) of 5 per cent, are more closely linked than *A* and *C* (RF = 19·0 per cent). If we now look at the linkage between *B* and *C* (RF = 15·5 per cent) two kinds of relationship become apparent. The first is that *B* must lie somewhere between *A* and *C* because no other arrangement will give a closer linkage of *B* to *C* than of *A* to *C* (Fig. 10a, b). The second relationship is that, assuming the order *A—B—C*, the recombination frequency between *A* and *C* (19·0 per cent) approximates to the sum of the frequencies between *A* and *B*, and *B* and *C* (5+15·5=20·5) (Fig. 10b). This additivity is

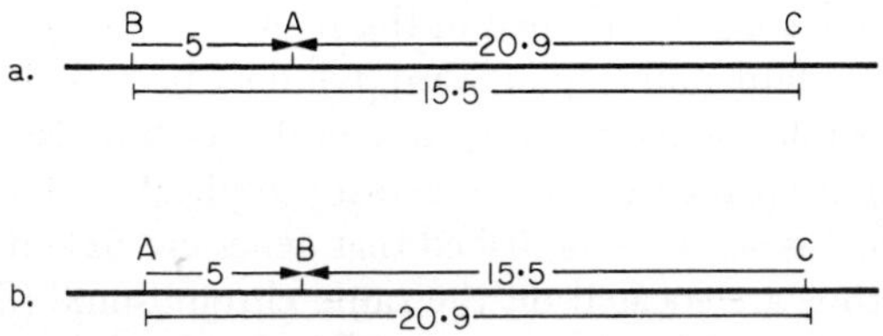

FIG. 10. The linear arrangement and order of genes on a linkage group, deduced from the pairwise recombination frequencies between them.

a. The order *B—A—C* is incompatible with the recombination frequencies which show that *B* and *C* are more closely linked (15·5 per cent) than are *A* and *C* (19·0 per cent).

b. The order *A—B—C* is not only compatible with the linkage data, but the recombination frequencies are approximately additive (i.e. *A—B* + *B—C* = *A—C*), indicating a linear arrangement.

(Based on data from Table 1.)

what would be expected if *A*, *B* and *C* were arranged in a line. Provided the genes to be mapped are not extremely close together (RF=approx. <0·5 per cent) or very far apart (RF>approx. 10 per cent) (see pp. 50, 51, 53), this relationship of additivity, indicating linear arrangement, has been found to hold throughout a wide range of plant and animal species. Thus, if we know the recombination frequencies between a series of linked genes, we can locate the position of these genes on the linkage group in their correct order and at their approximate relative distances apart. This process is known as *mapping*.

AN INTEGRATED MODEL OF INHERITANCE

We are now in a position to begin to equate the findings of Mendelian genetics, based on breeding experiments, with the morphological features of meiosis observed by the early microscopists. Up till now the situation has resembled that of two men, one of whom has been imprisoned in a motor car factory all his life and wondered what on earth all the systematic activity he saw could mean in terms of the world outside, while the other knew nothing of factories but owned a motor car and had taken it to pieces and wondered how it was built. The first person to suggest that the behaviour of the chromosomes at meiosis provided just the mechanism required for the segregation and re-assortment of Mendel's factors was W. S. Sutton in 1903. The subsequent intensive research by the American geneticist T. H. Morgan and his colleagues on inheritance in the fruit fly, *Drosophila melanogaster*, was a fundamental step forward, not only because it confirmed Mendel's principles in the most unequivocal way but also because it showed that these principles are universally applicable. These studies of Morgan and his school established that genes are linked in a linear way because they are located on the same chromosome (Sturtevant, 1913), and that recombination between linked genes is a consequence of crossing-over at the time of chiasma formation.

Beginning with its use by Morgan, *Drosophila* has proved an exceedingly useful tool for genetical research since it can raise a very large number of progeny in a ten-day cycle, numerous hereditary character differences are found, while its chromosomes are unusually well defined microscopically. A peculiarity of the fly is that chiasmata are not found at meiosis during sperm formation in the male, although they appear during oögenesis in the female. This is correlated with the absence of recombinant progeny when *male* hybrids are back-crossed to recessive females, although recombinants occur normally when *female* hybrids are crossed to recessive males. This absence of recombination during spermatogenesis makes an assessment of the number of linkage groups an easy matter since, in the hybrid male x recessive female cross, all genes belonging to the same linkage group show complete linkage while those of different linkage groups, of course, assort independently. Study of the inheritance of many hundreds of genes has revealed four linkage groups, three of which contain large numbers of genes while the fourth possesses only a few. From microscopic observation we know that *Drosophila melanogaster* has four pairs of chromosomes (haploid number=4), one of which is very small and dot-like.

To take another example, some 400 genes studied in maize (*Zea mays*) fall into ten linkage groups and there are also ten pairs of chromosomes; not only can each of the linkage groups be equated with one of the chromosome pairs, but the frequency of crossing-over (chiasma formation) seen at meiosis fits very well with the frequency of recombination found in breeding experiments. Many correlations of this kind are now known, while no case has arisen where the number of linkage groups exceeds the number of chromosomes. There are, of course, several organisms in which the number of linkage groups so far discovered is *less* than the number of chromosomes, but this is what we would expect if some of the chromosomes are small or the number of genes examined is low in proportion to the number of chromosomes. We may recall that the seven pairs of characters first studied by Mendel in the pea were unlinked to one another, although subsequent more extensive analysis revealed the presence of seven linkage groups in this plant, to match its seven chromosomes. We therefore see that the number of the chromosomes and their behaviour at meiosis corresponds in such a remarkable way to the picture inferred from the results of genetical crosses, as to leave little room for doubt that the chromosomes are indeed the carriers of the genes and constitute the material basis of inheritance. The main points of correspondence may be summarised as follows:

1. In general, in those species which have been adequately studied, the number of linkage groups turns out to be the same as the haploid number of chromosomes. There are, of course, exceptions to this rule, as in the case of tetraploids (p. 40) with four times the haploid number of chromosomes, in which there are only half as many linkage groups as gametic chromosomes, while 'empty' chromosomes, which carry no recognisable genes and are generally dispensible, are known in some species. Such cases, however, do not invalidate the correlation between the linkage group and haploid chromosome numbers.

2. The principle of segregation requires that the genetic contributions of the two parents segregate into separate gametes so that the gametes have only half the number of genes carried by the parent cells. Meiosis occurs only at gamete formation and its essential feature is that two consecutive cellular divisions are accompanied by only a single division of the chromosomes so that the diploid, somatic number of chromosomes is reduced to the haploid number in the gametes.

3. The principle of independent assortment implies that paternal and maternal sets of unlinked genes segregate independently so that, if *A* and *B* are unlinked genes, a hybrid of genotype *AB*/*ab* will segregate *AB*, *Ab*, *aB* and *ab* gametes. If we construe unlinked genes as separate chromosomes, this means that meiosis yields gametes possessing a haploid set of chromosomes among which the chromosomes of paternal and maternal origin are randomly distributed. Homologous chromosomes are normally identical in appearance, but nothing can be

seen to happen at the first metaphase of meiosis to suggest that the tetrads are so oriented on the equatorial plate that all the paternal pairs of strands face towards one pole of the cell and the maternal strands towards the other. Certain grasshoppers are peculiar in having some chromosome homologues which, although of different shape, pair regularly at meiosis; these homologues have, in fact, been observed to be randomly distributed to the gametes.

4. Breeding experiments show that reciprocal recombination occurs between the genes of homologous paternal and maternal linkage groups during gamete formation. Both the occurrence and the frequency of such exchanges is matched by crossing-over and chiasma formation at meiosis.

5. The frequencies with which pair-wise recombinations occur between three (or more) genes belonging to homologous linkage groups are usually additive, indicating that the genes are arranged linearly. This is entirely compatible with their location on a linear chromosome.

6. The maximum frequency of recombination between two genes on the same chromosome is 50 per cent, which is reached when the number of recombinant gametes issuing from a hybrid equals the number having the parental combinations of the two genes. This is just what chromosome behaviour would predict since, of the four chromosome strands involved in chiasma formation, only two show crossing-over, the other two remaining parental; thus a single cross-over at meiosis will give rise to two parental and two recombinant gametes.

There are other striking parallels between the morphology and behaviour of chromosomes on the one hand and, on the other, the theoretical predictions based on Mendel's laws, but these are concerned with such matters as the genetic basis of sex determination and various chromosomal aberrations (p. 40) which we do not propose to consider here because of their remote bearing on the genetics of bacteria in which we are primarily interested.

The relationship between meiosis and the genetic constitution of the gametes produced by the F_1 hybrid is shown diagrammatically in Fig. 11 which should make the mechanism of inheritance quite clear if it is studied closely in comparison with Fig. 8 (p. 29). In this diagram the behaviour of two pairs of chromosomes is portrayed, one of which carries the linked genes *A* and *B* (or their alleles *a* and *b*) and the other the gene *C* (or its allele *c*). The lower-left part of the diagram merely depicts how independent assortment of chromosomes occurs, and ignores

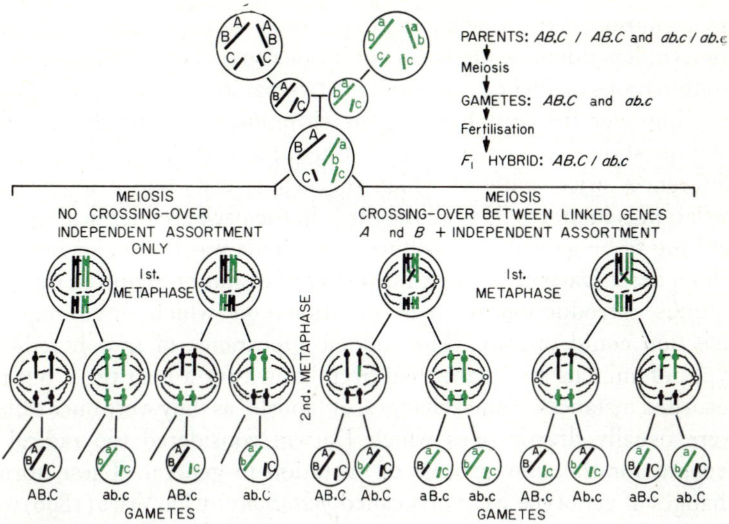

FIG. 11. The relationship between meiosis and the genotype of F_2 recombinants.
(For explanation, see text.)

crossing-over and recombination between *A* and *B* which should therefore be regarded as a single gene for purposes of comparison with Fig. 8. The lower-right part of Fig. 11 shows how eight different types of gamete, with respect to genes *A*, *B* and *C*, may result if the effect of a cross-over between *A* and *B* is imposed on that of independent assortment; to study recombination alone, gene *C* should be ignored altogether, thus reducing the products of meiosis to two parental and two recombinant gametes.

GENOTYPIC VARIATION AND MUTATION

We saw in Chapter 1 how the phenotype which an organism displays may depend on the interaction of its genotype with the environment. We have also discovered that we cannot infer the genotype of a diploid organism from its appearance, since the activity of recessive genes will be masked in hybrids by the presence of their dominant alleles and will only be revealed by breeding experiments which generate homozygous recessive progeny. Above all, we can now appreciate the unlimited possibilities for genotypic variation which are inherent in the sexual process, for not only has sexuality the ability to shuffle the pack of chromosomes from each parent and deal a new hand at every generation (independent assortment) but it can also try out innumerable new combinations of allelic genes from the two parents by means of random crossing-over between homologous chromosomes. But there is one vital question that we have not yet asked. How do the *initial* character differences arise, without which this whole elaborate machinery of variation is ineffectual? If all the cards in the pack are identical there is nothing to be gained by shuffling them. This was the major problem which beset Darwin, because his theory of evolution required the continuous introduction of new variations on which environmental selection could operate. The sudden appearance of new hereditary types of animals and plants had been known for a long time, but the recorded instances of such changes, or 'sports' as they were once called, were usually drastic ones which Darwin considered too radical to account for the slow process of evolution in general. These abrupt changes of genotype were first called *mutations* by De Vries (1890) who observed many striking examples of the phenomenon in the evening primrose (which, in the end, turned out to be not what we now call mutations at all!), but the fact that minute and barely perceptible changes could also arise in this way was not appreciated until the work of Morgan and his colleagues on mutation in *Drosophila* some twenty years later.

We can define a mutation as an abrupt, inheritable variation that is not due to segregation or recombination, but we now realise that such a definition is very imprecise since it may include many distinct types of alteration within the genetic constitution or *genome*. For example, inheritable changes may arise from alternations involving either the number or the structure of entire chromosomes (*chromosome mutations* or *aberrations*), or may be due to structural alterations within single genes (*gene mutations*). Among chromosome aberrations, several distinct kinds may be recognised:

1. The number of *sets* of chromosomes (or *ploidy*) may be changed. The somatic cells may come to have three sets (*triploid*) or four sets (*tetraploid*) of chromosomes, or more, instead of the two sets of the normal diploid. Increase in ploidy (*polyploidy*) is especially common in plants and has played an important role in agricultural development since it is often associated with large plant size due to increased cell volume.

2. The number of chromosomes in a set may be changed by the loss or gain of one or more chromosomes. A diploid cell which has lost a

homologue from one of its chromosome pairs (that is, has the diploid number of chromosomes *minus* one) is termed *monosomic*; a diploid cell which has gained a chromosome so that one chromosome of its set is represented three times (that is, it has the diploid number of chromosomes *plus* one) is termed *trisomic*.

3. There may be *translocation* or exchange of segments between two non-homogolous chromosomes, such that if the original chromosomes had the genes *ABCDEF* and *UVWXYZ*, the translocation might yield chromosomes of the constitution *ABCXYZ* and *UVWDEF*.

4. A single chromosome may suffer an *inversion* of a segment of its length so that the order of the genes changes from *ABCDEF* to *ABEDCF*.

5. A chromosome may lose one or more of its genes altogether; this defect is called a *deletion*.

In general, however, when we use the word 'mutation' we are referring specifically to *a change in the structure of a single gene* which results in either alteration or loss of its function; that is, the gene is changed to an allelic form.

Mutations may arise in any gene, in any cell, at any stage in the life cycle of an organism, but it is obvious that the fate of a particular mutation will be very dependent on the circumstances of its origin. If mutation occurs in a gamete, and the resulting allele is dominant, then a single one of the progeny of all the gametes will display the mutant phenotype. The occurrence of a mutation in a somatic cell of a sexual organism, however, will usually preclude its transmission to the gametes, since the gametes arise from a small minority of specialised cells, so that it will become extinct when the individual dies; if such an allele is recessive it will never express itself at all in diploid cells, but if it is dominant and arises *early* in the development of a multicellular organism, mitotic division will distribute it to a large population of cells so that a considerable portion of the mature individual may differ phenotypically from the rest. Individuals of this kind are called *mosaics*. A similar situation may be found in growing clones of bacteria or fungi, where an early mutation to a phenotypically distinct type may produce a visible *sector* of mutant cells in a colony on solid medium (p. 549), while no observable effect may follow a mutation in the last division or so. This relation between the time at which a mutation arises in a growing population of cells and the proportion of mutants among the final population, was used by Luria and Delbrück (p. 181) and by Newcombe (p. 185) to demonstrate the spontaneous origin of bacterial variants, and also forms a basis for measuring the rate at which mutation occurs (p. 195). Mutations arising in haploid cells, such as vegetatively dividing bacteria, will, of course, be expressed whether or not the allele is dominant or recessive.

Our aim here has been simply to introduce mutation in the form of a rather semantic explanation of the primary cause of the variation on which environmental selection and the mechanisms of sexual reproduction can then operate. In organisms where the vegetative form is predominant and diploid, the sexual process does much more than merely reshuffle mutations so as to present new combinations to the environment: it acts as a great store-house of recessive genes, transmitting them indefinitely in heterozygotes while periodically testing their environmental fitness in homozygous form. It has been estimated that, if all mutation were suddenly to cease, the sexual system has stored within it sufficient variation to carry evolution at least as far into the future as it has progressed from the past. In fact it is valid to assume that the greater part of evolution has resulted from selection acting on *this* type of variation, rather than on that due to new mutations which, in any case, tend to be deleterious and non-adaptive. The work of Mendel, far from demolishing Darwin's selectionist view that evolution had progressed by a series of virtually imperceptible steps by showing that heredity was discontinuous, turned out to provide the very answer that Darwin was looking for (see de Beer, 1964).

The mystery of mutagenesis and the new evidence concerning its nature which has been revealed by recent work on the physico-chemical basis of heredity and on the genetic analysis of fine chromosomal structure in bacteria and bacteriophage, will be studied in Chapter 13 (p. 297).

Part II
The Elements of Genetic Analysis

CHAPTER 3

Recombination Analysis and Chromosome Mapping

In the preceding Chapter we discussed briefly how the order of arrangement of a linear sequence of genes on a chromosome, and the relative distances between them, could be mapped by the analysis of recombination data derived from genetic crosses. The geographical map of a country can tell us far more than the shortest road to take in order to travel from one town to another. If, for example, we happen to know something of the products that are manufactured and consumed in different localities, then a map becomes really essential if we are to construct a model of how the country is organised. In the same way, the value of a chromosome map now goes far beyond its use in the design and control of breeding experiments. The development of biochemical genetics in recent years has enabled us to define individual genes in terms of their biochemical function and it turns out, as we might have expected, that what goes on at this fundamental level is a reflection of the organisation of the genetic material itself, which can be analysed by increasing refinements of the basic method of mapping which we have described (Chapter 7, p. 127). In fact, the scale of our maps can now be increased to reveal not only the lay-out of individual towns, but also the production lines within its factories. Before discussing how this can be done, however, it is necessary to understand something more of the techniques of map making and of the errors inherent in the instruments we use.

We have used the word 'gene' to mean the unit of inheritance which controls the expression of a single character. Mutation produces a change in the structure of a gene so that the function it determines is lost or altered. Two genes which control the expression of the same character in different ways are said to be alleles, and genetic analysis depends on observation of the quantitative way in which allelic genes in diploid cells segregate among progeny during sexual reproduction.

A basic assumption is that alleles of the same gene are located at precisely the same positions on homologous chromosomes, so that all the alleles of a particular gene will behave in an identical manner at recombination. We shall discover later on that this assumption is not true in the strictest sense but, like Euclidian geometry, it is sufficiently precise for all general purposes and only fails when the chromosomal distances to be measured become extremely small.

It is often convenient, when discussing mapping, to speak of the specific location, or *locus* (Latin: *locus*=a place), of a gene on the chromosome instead of referring to one or the other of several different alleles of that gene. Thus we can talk about the locus for flower colour in the pea without reference to whether the behaviour of the gene for purple flower colour, or that for white, is actually being observed; similarly, in bacteria, we may refer to the *lac* locus, determining lactose fermentation, irrespective of whether the gene involved determines the ability (lac^{+}gene) or inability (lac^{-}allele) to ferment this sugar. Thus the terms 'gene' and 'locus' are often used synonymously although they do not mean exactly the same thing.

THE RELATION BETWEEN RECOMBINATION AND CROSS-OVER FREQUENCIES

We have already noted that the degree of linkage between two genes on the same chromosome can be represented by the proportion of progeny in which these genes have become separated from one another so that they are no longer present in their parental couplings, that is, the proportion of progeny which are recombinant for the parental alleles of the two genes (Table 1, p. 34). This is called the recombination frequency (RF). We have also seen that this separation is effected by the occurrence of a cross-over between the two genes during the first meiotic division, resulting in a reciprocal exchange between two of the four chromatids (Fig. 11, p. 39). The closer together two genes are located on the chromosome, the less is the chance of a cross-over occurring between them and, therefore, the lower will be the recombination frequency. When the recombination frequencies between pairs of three or more linked genes (*A—B—C*) are estimated, they are found to be more or less additive; that is, the recombination frequency between *A—C* is approximately equal to the sum of the recombination frequencies between *A—B* and *B—C*. This indicates a linear arrangement, so that the three loci can now be placed in a linear order and at their approximate distances from one another on the chromosome. As the linkage relationships of more and more loci are investigated in the organism, the number which can be mapped on any particular chromosome will increase in proportion to the length of that chromosome; for if mutations are randomly distributed among all the chromosomes it is obvious that the longer the chromosome the greater is the number of genes it carries and, therefore, the greater is the chance that any one mutation will involve it. Now let us suppose that we have thus allotted a large number of loci to one chromosome; we can proceed to map them all in pairwise fashion just as we did with loci *A*, *B* and *C*. Since the number of loci is large, we can assume that the two extreme loci, *A* and *Z*, are fairly close to the extremities of the chromosome, and when we estimate directly the recombination frequency between them, by scoring the percentage of $Az+aZ$ recombinants, we find that this approaches, but does not exceed, the 50 per cent which indicates random assortment. This is what we expect to find because a single cross-over occurring *at any point* along the paired homologues must *always* lead to a recombination of the two extremities while at least one cross-over is likely to involve every long chromosome in every cell undergoing meiosis. There is clearly a second way of estimating the distance from *A* to *Z*, by adding together the sum of the recombination frequencies between all the intervening loci (A—B+B—C+C—D+ . . . +W—X+X—Y+Y—Z). Since we know that the arrangement is linear, we should expect this sum to add up to something less than 50 per cent. But it does not. On the contrary, it adds up to considerably *more* than 50 per cent and may exceed 100 per cent. Why is this? The model of recombination which we have visualised so far is a rather naïve one. According to this model the recombination frequency that we measure between any two points on the paired chromosomes was ascribed to a single cross-over between these points, and we assumed a random probability of crossing-over at any region along the chromosome length. We must now make our model more sophisticated by considering what happens if more than one cross-over occurs and if these cross-overs are not randomly distributed.

MULTIPLE CROSSING-OVER

The diagram below depicts three linked loci, *A*—*B*—*C*, *A* and *C* being equidistant from the intermediate *B* locus, and the recombination frequency between *A* and *B*, and between *B* and *C*, being 20 per cent.

If only a single cross-over can occur in the interval *A—C*, and its occurrence is random, it is obvious that the recombination frequency between *A* and *C* will be 40 per cent, since a cross-over will occur

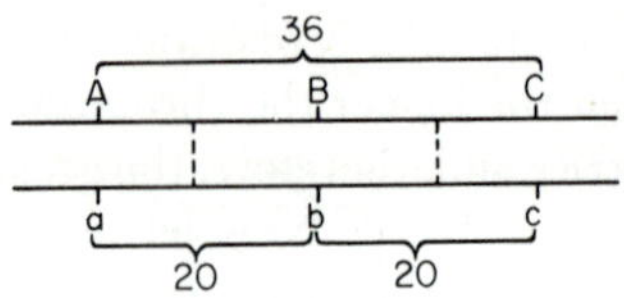

twice as often between these outside loci as between either *A* and *B* or *B* and *C*. If, on the other hand, more than one cross-over is possible, the probability of the coincidence of two cross-overs, one in the interval *A—B* and the other in the interval *B—C*, will be the *product* of the probabilities of single cross-overs in each interval separately multiplied by 2, that is, $20/100 \times 20/100 \times 2 = 8$ per cent. Now the genotypes of such double cross-overs will be *AbC* and *aBc*; but these are parental so far as *A* and *C* are concerned and will therefore not be scored as recombinants when the recombination frequency between *A* and *C* is estimated directly, so that the observed recombination frequency will be 40 per cent less 8 per cent, that is 32 per cent. In this way the recombination frequency between two outside loci is always less than the sum of the intervening frequencies when the distance between the outside loci is sufficiently long to permit double cross-overs. We know, of course, that multiple recombination does occur, both from the genotype of recombinants when the inheritance of a sequence of linked genes is scored (for example, *AbC* and *aBc* are obviously double recombinants although parental for *A* and *C*), as well as from the observation of multiple chiasmata. As we have seen, the occurrence of a single cross-over leads to recombination, while

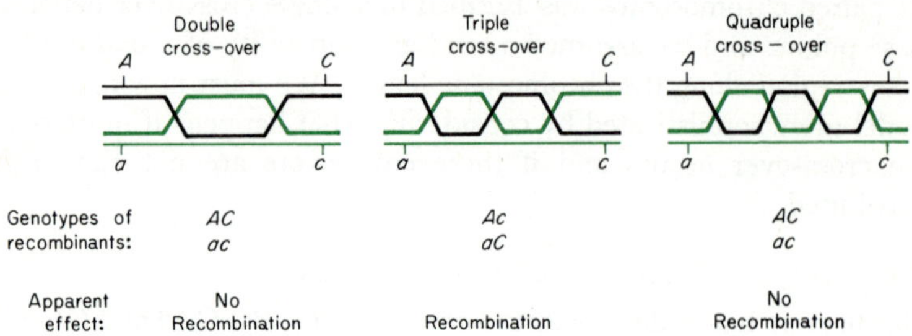

FIG. 12. Illustrating the effect on genotype of more than one cross-over between two loci.

that of a double cross-over in any region restores the parental combinations of genes situated on either side of the region. If a third cross-over occurs, this third exchange will once again lead to recombination of outside genes; similarly, a quadruple cross-over yields the parental couplings (see Fig. 12). In other words, an *odd* number of cross-overs results in the recombination of genes located at the extremities of the cross-over region, while an even number does not. This alternation means that genes separated by long distances, over which multiple crossing-over may be expected, have a probability of 0·5 of undergoing recombination so that the number of recombinants observed may be as low as one half that expected if only single cross-overs were assumed (see Fig. 13, p. 53).

By means of statistical adjustment it is possible to compensate for the effects of multiple recombination, so that a map showing the true positions of the genetic loci can be constructed, *provided* that the assumption of random crossing-over is retained. There are two ways of looking at this question of randomness. The first is to ask whether a single cross-over has an equal chance of occurring anywhere along a chromosome. Cytological observation of the frequency of chiasma formation has revealed a great variety of patterns. For instance in many species chiasmata are relatively rare in the vicinity of the centromere so that pairs of loci in this region show little recombination and so appear much closer together than they actually are; in other parts of the chromosome they may be frequent so that the loci appear spread out. In general, the distribution of chiasma formation may show wide variations not only between species but also between different chromosomes of the same species and even between the same chromosomes of the same type of cells under different physiological conditions. So far as visual observation can take us, then, there is good evidence that the mechanism of crossing-over does not arise in a chance manner.

INTERFERENCE

The second question we can ask about the randomness of crossing-over is whether the occurrence of one cross-over influences the chance of another arising close to it, or whether the two events are independent. The simplest case to consider is that of double recombination. The three loci shown in the diagram are ordered in the sequence *A*—*B*—*C*, the recombination frequency between *A* and *B* being 10 per cent, and between *B* and *C* 8 per cent. If crossing-over occurs randomly, the frequency of simultaneous events in intervals 1 and 2 will be

1.6 per cent, so that $AbC+aBc$ recombinants should arise at about this frequency. It is often found, however, that the actual number of double recombinants amounts to only a fraction, say 5 to 10 per cent, of the expected number. This phenomenon is known as *inter-*

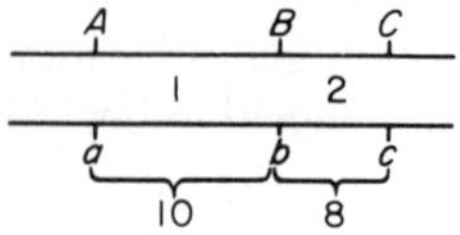

ference and indicates that although single recombination events occur in intervals 1 and 2, there is a strong tendency for double events to be suppressed or, to put it another way, that the occurrence of one event reduces the probability of another. The ratio of observed to expected number of double cross-overs is termed *coincidence* and is a measure of the degree of interference. It has been found experimentally that interference usually increases as the distance between loci becomes smaller until a point is reached when no double cross-overs are found, that is, the coincidence=0; conversely, above a certain distance interference disappears so that the coincidence=1. The classical explanation of this kind of interference is the mechanical difficulty of forming two chiasmata very close together when the chromosomes are greatly thickened and tightly coiled. The situation is somewhat analogous to trying to tie knots in a piece of thick rope; you can tie a single knot anywhere without much trouble, but it is difficult to tie two knots close together. This kind of interference is therefore called *chiasma interference.*

There are two other types of interference to be considered, both of which may influence apparent map distances in important ways. We have seen that the closer two genes are linked, the smaller is the proportion of recombinants which no longer carry parental combinations of these genes. This means that in order to study the relationships between very closely linked genes, very large numbers of progeny must be examined in order to find recombinants at all. This clearly sets a practical limit to the refinement of genetic analysis when recombinants must be scored by direct observation of character differences, as in the case of plants, animals and flies. As we shall presently see, a different situation exists in the case of many kinds of microorganism, where we can *select* for very rare recombinant types which have inherited certain combinations of nutritional and other characters of the two parents, which enable them to multiply and produce colonies on deficient culture media which repress parental growth. In this way recombinants arising in the proportion of one in several hundred thousand progeny can be scored.

Such highly refined analyses, carried out notably in the fungus *Aspergillus nidulans* and in bacteriophage, have unexpectedly revealed that over minute chromosomal regions the coincidence of recombination may rise to 20 or more, indicating *the occurrence of two cross-overs at a much greater than random frequency*. This phenomenon is just the opposite of chiasma interference and is known as *localised negative interference*. We saw earlier (p. 48) how the occurrence of double cross-overs reduces the observed recombination frequency between two genes, so that these genes appear to be closer together than they really are. The effect of localised negative interference is the same, since the occurrence of one cross-over greatly increases the probability of a second very close to it. Thus the observed frequency of recombinants for two very closely linked loci and, therefore, the distance between them, may be significantly underestimated. Negative interference is an important source of distortion in the genetic analysis of fine chromosomal structure, especially when it is necessary to express small chromosomal distances as a proportion of much longer distances over which this type of interference does not operate (see p. 144). It seems very unlikely that localised negative interference has anything to do with chiasma formation, so that it has important implications with regard to the mechanism of the recombination process which we will discuss later (Chapter 15, p. 379).

The third type of interference is called *chromatid interference* and although its significance, and even its existence in most cases, remains unknown, we will discuss it briefly for the sake of completeness. You will recollect that only two of the four chromatids, or strands, of the tetrad undergo exchange during any one recombination event, the other two strands remaining parental. Moreover, the genetic result of the exchange is the same irrespective of which of the two paternal and two maternal strands are involved, since the two strands of each pair arise from the division of a single chromosome and are identical (see Fig. 4, p. 19). On the other hand, the genetic outcome of a double cross-over will vary depending on whether the same or different strands participate in both events (Fig. 4). Thus crossing-over between the same two strands on both occasions (2-strand double) gives two doubly recombinant and two parental strands; if both strands exchanging at the second cross-over are different from those exchanging at the first

(4-strand double), we get four singly recombinant strands and no parental strands; finally, if only one of the two strands involved in the first event is also involved in the second (3-strand double), two singly recombinant strands, together with one doubly recombinant and one parental strand, are produced. Consideration of Fig. 4 will show that, *provided the involvement of any strand in the two events is entirely random*, 2-strand, 3-strand and 4-strand doubles will arise in the ratio 1:2:1. From this it may be deduced that, on the average, a double cross-over will yield double recombinant, single recombinant and parental strands in the ratio 1:2:1. If, however, the association of strands in neighbouring cross-overs is *not* random, that is, if there is chromatid interference, then this ratio will alter. Suppose, for example, that the involvement of a strand in one cross-over markedly reduces the likelihood of its participation in a second; the result will be to reduce the frequency of 2- and 3-strand doubles and increase that of 4-strand doubles which, as we have seen, yield four singly recombinant strands and no parental strands. In this way the proportion of gametes carrying singly recombinant chromosomes will rise at the expense of those carrying parental type chromosomes. This is the only theoretical mechanism, apart from pre-zygotic exclusion in bacterial crosses (p. 54 *et seq.*), whereby the frequency of recombination can rise above 50 per cent and cases are known where it occurs. In general, however, the problem of chromatid interference is a complicated one about which little is known although it can be analysed directly by means of *tetrad analysis* in organisms where the four products of meiosis in single zygotes can be isolated and examined (p. 68).

THE MAP UNIT

The main aim of this discussion has been to see how far we can relate the recombination frequencies obtained from genetic crosses to what appear to be the real events—the formation of chiasmata and crossing-over—which are observed at meiosis and which are more directly associated with chromosome length. We have noted that this relationship is affected by two major factors, the occurrence of double cross-overs and chiasma interference. In the absence of double cross-overs the recombination frequency and the cross-over frequency would exhibit a direct linear relationship (Fig. 13, curve *a*). The effect of double cross-overs, however, is to decrease the proportion of recombinant to parental progeny so that, with expanding distance, the frequency of recombination begins to decline as a function of cross-over frequency in *Drosophila*; the linear relationship no longer holds when the recombination frequency exceeds about 10 per cent (Fig. 13, curve *b*). Chiasma interference tends to compensate for this, however, as curve *c* (Fig. 13) shows, so that in practice the linear relationship may be extended up to 20–25 per cent recombination.

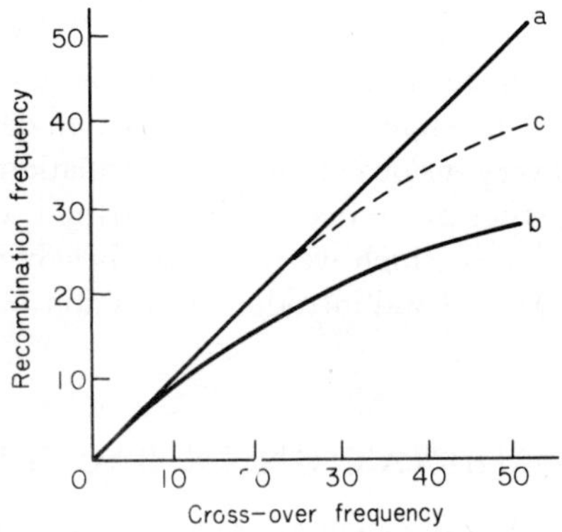

FIG. 13. The relation between recombination and cross-over frequencies.

a. Theoretical curve, assuming single cross-overs only, i.e. complete interference.

b. Theoretical curve, corrected for the effect of double cross-overs, assuming no interference.

c. Experimental curve (*Drosophila* data), showing how interference compensates for the effect of double cross-overs. (After Mather, 1938.)

The cross-over frequency can thus be obtained by correcting the observed recombination frequency for the effects of double crossing-over and chiasma interference although, thanks to the latter, this correction is mathematically rather complex. The unit which is used to measure chromosome lengths is called a *map unit*, or *cross-over unit* and is equal to the corrected recombination frequency, expressed as a percentage. For example, a corrected recombination frequency of five per cent equals five map units. In the case of all organisms which are amenable to genetic analysis of the classical type which we have been discussing so far, when the term 'map unit' is used it always accords with this definition irrespective of whether the organism is *Drosophila*, the maize plant, or fungi such as *Neurospora* or *Aspergillus*. While the specification of genetic maps in terms of map units is certainly more than a useful semantic convention and gives a truer picture of the disposition of genes than can be drawn in terms of simple recombination frequencies, it is not really very meaningful as an index for comparative studies since, as we have seen, the probability of crossing-over itself tends to be non-random and may differ widely between different species as well as varying with such factors as temperature, sex and the presence of aberrations, such as inversions, on other chromosomes. An assessment of the real distance between genes can only be made from cytological maps or their equivalent, which are derived independently of recombination data (see p. 64). It is not, therefore, a very serious matter that calculation of the map unit in bacteria is altogether precluded by the unorthodox nature of sexuality in these organisms which we must now briefly consider, in this context, in a very general and introductory fashion.

RECOMBINATION IN BACTERIA

In the classical systems an essential prelude to genetic recombination is the coming together within a single cell of the haploid nuclei from two gametic cells, and their subsequent fusion to form a diploid nucleus. How this nuclear transfer is achieved is a matter of detail which in no way affects the outcome of the cross, since the complete genomes of both gametes are involved. This is not so in the case of bacteria.

Bacteria are free-living haploid organisms and, among those species which display sexuality, are not differentiated, like many fungi, into somatic and gametic cells; every cell has the potentiality to behave as a gamete. Among the gametes, however, there is a functional differentiation into two types called *donors* (or males) and *recipients* (or females). This differentiation, as we shall shortly see, may be artificially imposed by the experimental method of mating in some systems, or it may result from a heritable physiological difference. The unique feature of sexuality in bacteria is that, as a general rule, *only a fraction of the genetic constitution of the donor cell is transferred to the recipient cell* which contributes not only its entire genome but also its cytoplasm. The recipient cell thus becomes a zygote which is incompletely diploid. The term *merozygote* (Greek: *meros*=part) has been applied to zygotes of this sort (Wollman, Jacob and Hayes, 1956) and it is obvious that, in them, the potentiality for recombination is limited to the diploid part of the genome. Those donor genes which are not transferred to the zygote clearly cannot be represented in its progeny. The genotype of the recipient bacteria therefore tends to predominate among recombinants. The extent of this predominance depends on the size of the donor fragment transferred to the recipient cell and this, in turn, is largely a function of the mechanism of transfer.

Three main mechanisms of genetic transfer have been described. In *transformation*, which operates principally in *Pneumococcus*, *Haemophilus*, *Bacillus* and *Neisseria*, virtually pure and highly polymerised deoxyribonucleic acid (DNA) extracted from the cells of one strain is able to induce permanent hereditary changes in the cells of a second strain to which it is added, with respect to those characters in which the two strains differ. The strain from which the DNA is extracted becomes the donor, while the second strain, to which donor characters are transferred by the DNA preparation, is the recipient. Transformation experiments yielded the first indication as to the chemical nature of genetic material (Chapter 20, p. 574; also p. 228).

Molecular weight determinations of transforming DNA preparations show that the size of the transforming fragments, which are of course of molecular dimensions, is about a hundredth part or less of the bacterial chromosome. It follows, as we might expect, that only one donor character is usually inherited by any one recipient cell as a result of a single transforming event. Thus if the donor strain differs from the recipient strain in two characters, such that the donor strain is of genotype *AB* and the recipient strain *ab*, among a population of recipient cells that have been treated with DNA from the donor strain there will be found some *Ab* and some *aB* cells. The frequency of occurrence of *AB* cells, however, will be small and will depend on the random probability of separate DNA molecules, one carrying gene *A* and the other gene *B*, independently encountering and being taken up by a single recipient cell at the same time; this probability will be equal to the *product* of the probabilities of the two independent events. In the case of certain pairs of characters, however, the frequency of double transformations is strikingly higher than could be expected by chance. This means that the genes determining these characters are so closely linked that, when the DNA is extracted, they are often left together on the *same* molecule of DNA. As we shall see when we come to study transformation in more detail, there is good evidence that the genes carried over in the transforming preparation are incorporated into the genome of the recipient cells, and replace their alleles there, by a process analogous to recombination. We are therefore justified in regarding transformation as a kind of sexual process which differs from sexual processes in more highly evolved organisms mainly in the somewhat artificial method of insemination andthe fractional genetic contribution of one of the parents.

The second mechanism of genetic transfer is called *transduction* (Chapter 21, p. 620). In this system, modified particles of a bacterial virus (bacteriophage) of low virulence act as vectors of bacterial genes. The occurrence of transduction has been described mainly in Gram-negative bacilli such as *Escherichia*, *Shigella*, *Salmonella* and *Pseudomonas* but is also found in gram-positive organsims such as *Bacillus* and *Staphylococcus*. A culture of the bacterial strain elected as donor is infected with the phage under conditions favouring its multiplication and lysis of the cells. The liberated phage particles are then freed from residual cells and used to infect a second, recipient strain, but this time under conditions encouraging bacterial survival. It is found that a small proportion of the progeny of the recipient cells have inherited donor characters. The frequency of such transduced cells is commonly about one per million particles of the transducing phage. The size of the fragment of genetic material transported in this way varies with the particular system but, in general, is rather larger than in the case of transformation. Transduction, because it restricts genetic analysis to small chromosomal regions, is of little use for mapping the relationships of genes in general; however this very defect can be put to good advantage, since the joint transduction of two or more genes indicates that they are very closely linked. Thus transduction offers a refined instrument for the analysis of genetic fine structure which, for example, first revealed the clustering together of genes concerned with different steps of a biosynthetic pathway.

The most highly developed form of sexuality in bacteria is the *conjugation* mechanism found in *Escherichia coli* (Chapter 22, p. 650). In conjugation, a physiologically differentiated donor bacterium first unites with a recipient bacterium and then transfers to the recipient a considerable part of its single chromosome. Very rarely, the whole of the donor chromosome appears to be transferred to form a complete zygote. Normally, however, the chromosome tends to break in transit to yield merozygotes containing, on the average, about ten to twenty per cent of the donor chromosome.

The essential peculiarity of bacterial crosses of all kinds, then, is the unequal genetic contribution of the two parents to the zygote which the various mechanisms of genetic transfer impose. How will this peculiarity reveal itself in recombination analysis? As Fig. 14B shows, two major effects are likely. The first is that reciprocal recombinants cannot arise from a single mating event for the simple reason that two complete chromosomes cannot be constructed from one and a bit, while a haploid cell is unlikely to be viable if an appreciable part of its chromosome is missing. The second effect is the probable need for a double cross-over (or, rather, any even number of cross-overs) if a viable recombinant is to emerge at all. This is obviously the case if the state of affairs depicted in Fig. 14B is a true, if diagrammatic,

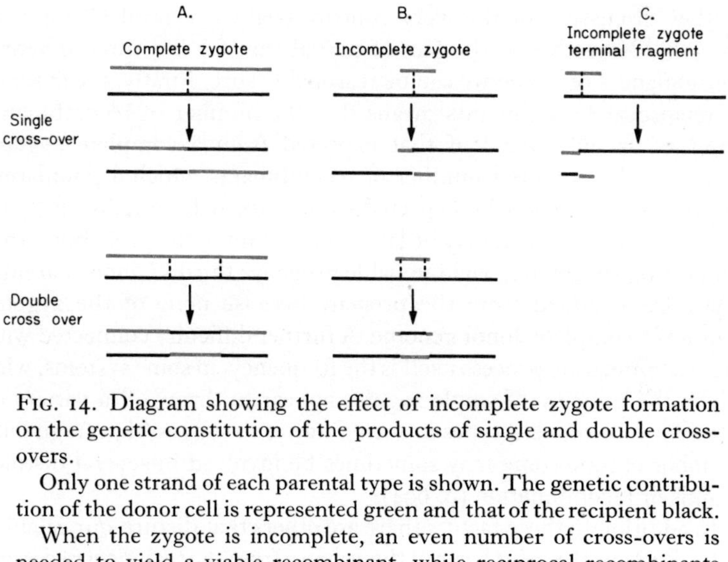

FIG. 14. Diagram showing the effect of incomplete zygote formation on the genetic constitution of the products of single and double cross-overs.

Only one strand of each parental type is shown. The genetic contribution of the donor cell is represented green and that of the recipient black.

When the zygote is incomplete, an even number of cross-overs is needed to yield a viable recombinant, while reciprocal recombinants cannot be formed.

representation of what actually happens, for if the recipient chromosome has two extremities and the donor fragment is homologous to, and pairs with, any region contained between these extremities, a single cross-over can yield only two incomplete chromosomes. It is usually assumed, in transformation and transduction studies, that two crossovers are required. Should the donor fragment include one of the *extremities* of the chromosome, however, then a complete recombinant chromosome, incorporating both extremities, is theoretically possible from a single cross-over as Fig. 14C shows. At first sight it might be thought that random fragmentation of the chromosome would so seldom involve an extremity that this possibility might be ignored in practice. On the other hand we now have rather compelling evidence for believing that the chromosome of recipient strains, at least of certain strains of *E. coli*, has a closed ring structure with no extremities (p. 550) which, theoretically, would permit formation of a viable recombinant from a single cross-over involving any fragment, *provided* certain specific assumptions are made about the mechanism of the recombination process. In fact we still know very little about how recombination is effected in higher organisms, much less in bacteria, so we will defer further discussion of this very controversial topic until Chapter 15 (p. 376). Nevertheless the fact of partial transfer has three inherent consequences of which we can be reasonably sure. Firstly, the absence of reciprocal recombinants means that the number of recombinants observed is only one half that expected from a complete zygote; secondly, the potential number of recombinants which a population of merozygotes might be expected to produce will be reduced by an unknown factor, which may be large, on account of the probability that single cross-overs may yield inviable progeny; thirdly, donor parental types are excluded from the progeny because none of the zygotes acquires a complete donor genome. A further difficulty connected with the recombination process itself is the frequency, in some systems, with which the progeny of single zygotes appear to throw off a variety of different recombinant types in such a way as to suggest that fragments of donor chromosome may sometimes be involved in several distinct rounds of recombination (p. 694).

In addition to these factors there are others that disturb our calculations. Perhaps the most basic of these arises from the technical necessity to identify recombinants by some *selective* process, because they are usually very rare, especially in transduction. This entails plating a large population of zygotes on a selective medium which, while repressing both parental types, will allow recombinant cells which have inherited a particular gene from the donor parent to multiply and form a colony. For example, the two parents may be auxotrophic and differ in their ability to synthesise two amino acids, A and B, the donor being of genotype *Ab* and the recipient *aB*. Neither parent can grow on a synthetic minimal medium lacking amino acids A and B, but those recipient bacteria which have received a fragment of donor chromosome which carries the *A* gene may incorporate it into their genomes by recombination, thus yielding *AB* recombinants which can multiply and form colonies. The adoption of a selective method therefore means that only a particular recombinant class can be counted, all other recombinant classes being excluded from consideration.

Another unavoidable complication in bacterial systems is the inability to distinguish zygotes from unfertilised recipient cells, especially in transformation and transduction where the proportion of zygotes is small. In general, we can only recognise zygotes posthumously by their recombinant descendants. But even if all the zygotes were complete and could be counted we would still be unable to assess the recombination frequency with accuracy because bacteria normally contain *several* haploid nuclei within a single cell (Plate 26, p. 544), only one of which usually participates in recombination. It thus turns out that we have no way of knowing how many, if any, of the recipient parental progeny issuing from a zygote emerged from a process analogous to meiosis, or from the simple segregation of those recipient nuclei which played no part in the process. In general, therefore, in bacterial systems, we are quite unable to assess, even on a statistical basis, the number of zygotes involved in a cross or the total number of parental and recombinant progeny which issue from them, so that the results we obtain cannot be expressed in terms of classical recombination frequencies, much less of map units.

The best we can do is to select recombinants inheriting a particular gene, or genes, from the donor parent, ignoring parental and other recombinant types, and then to score the percentage inheritance of other *unselected* donor genes among these recombinants. Provided the selected and unselected donor genes are situated sufficiently closely to one another as usually to be included together on the transferred fragment of donor chromosome, it is assumed that the frequency with which they are jointly inherited among recombinants is inversely proportional to the distance between them. For example, in the cross *ABC* (donor) × *abc* (recipient), in which inheritance of gene *A* from the donor is selected, if 90 per cent of the *A* recombinants have also inherited *B* (i.e. are *AB*) and 60 per cent are *AC*, the inference is that recombination events separate *A* and *B* in 10 per cent, and *A* and *C* in 40 per cent, of those merozygotes which received *A*—*B*—*C* fragments. Similarly the order and linearity of *A*,*B* and *C* can be assessed by estimating the linkage between *B* and *C*. If 70 per cent of *AB* recombinants are also *ABC*, then *B* is more closely linked to *C* (30 per cent) than *A* is (40 per cent), so that the order must be *A—B—C* and not *B—A—C*. Finally, since the distance *A—C* (40 per cent) equals the sum of the distances *A—B* (10 per cent) and *B—C* (30 per cent), the arrangement is a linear one (see Lederberg, 1947).

While this method of mapping is satisfactory enough in practice, its validity is necessarily bound up with the manner in which fragmentation of the donor chromosome occurs. If, in transduction for instance, the structure of the chromosome is such that the fragments result from breakage only at specific, predetermined points (p. 643), then all fragments which carry gene *A* will be the same and will also carry genes *B* and *C* so that exclusion of *B* or *C* from recombinants inheriting *A* will be a function of the process of recombination alone. On the other hand, if the donor fragments are formed by *random* breakage of the chromosome, as appears to be the case in transduction as well as conjugation in *E. coli*, then the absence of, say, gene *C* from recombinants selected for *A* may be due either to recombination between *A* and *C* (*post-zygotic exclusion*) or to fracture of the donor chromosome between *A* and *C* before transfer, so that *C* never entered the zygote (*pre-zygotic exclusion*).

An important consequence of pre-zygotic exclusion, by chromosome breakage prior to genetic transfer, is that apparent recombination frequencies greatly exceeding 50 per cent can occur between genes which are known to be linked. You will remember that the involvement of only two of the four chromosomal strands in each recombination event limits the proportion of recombinants to total progeny to 50 per cent, while the occurrence of multiple crossings-over will reduce this figure still further in the case of genes some distance apart on the chromosome (p. 49). What we observe in such a case is that the two genes preserve their parental couplings in not less than 50 per cent of recombinants. If, however, chromosome breakage in one of the parents was to exclude one of the genes from 80 per cent of zygotes, it is evident that, in at least 80 per cent of recombinants, these two genes would not be present in their parental couplings, thus mimicking a recombination frequency of over 80 per cent.

Enough has been said to indicate the unusual features of recombination in bacteria and the ways in which we must modify and adapt classical principles to fit these novel and useful systems. We will develop this topic in depth and detail when we come to consider the analysis of genetic fine structure (Chapter 7, p. 127) and to study individually the mechanisms of transformation (p. 574), transduction (p. 620) and conjugation (p. 650).

RECOMBINATION IN BACTERIOPHAGE

In their morphology and chemical constitution bacteriophages appear to be very simple creatures. The observed shape of most phage particles is delineated by a protein sheath or coat which takes the form of an hexagonal head, and a slender tail which usually terminates in a fibrillated end-plate whereby the particle attaches itself to the bacterial cell wall. The particle thus presents a tadpole-like appearance (Fig. 98, p. 418; Plates 13, 14, p. 416). The hexagonal head encloses a very long strand of deoxyribonucleic acid (DNA) which constitutes the genetic material of the organism. When a sensitive bacterium is infected by a virulent phage, the phage particle first attaches itself to the cell wall by its tail and then injects its genetic material into the cell. The protein coat remains outside. After injection, the phage genetic material behaves like a 'master-gene', taking control of the synthetic machinery of the bacterium and diverting it to the exclusive manufacture of new phage protein and genetic material, with the result that, 20–30 minutes later, the cell bursts to liberate several hundred progeny phage particles. This sequence is called the *lytic cycle.*

Recombination in phage is studied by the simultaneous infection of sensitive bacteria by two (or more) phage particles which differ in several characters as a result of mutation. The progeny particles emerging at the end of the lytic cycle are then examined for parental or recombinant phenotype. We need not here consider the nature of the characters employed and how they are scored (pp. 480–486), or the special features of different phage systems, but will merely note to what extent the broad and general properties of recombination in phage are consistent with classical principles.

In phage infection each particle injects its complete genome into the bacterial cell so that the problems of merozygosity which beset the interpretation of bacterial recombination do not arise. Shortly after infection new phage genetic material, consisting of DNA, begins to be synthesised in the 'DNA pool' of the cell. This synthesis, which can be followed chemically in the case of some phages, reflects the replication of the injected parental DNA which proceeds exponentially, with a generation time of 2–3 minutes, until the equivalent of about 50 new phage genomes have been produced. Up to this time no infective phage particles can be found when the cells are artificially ruptured. Phage genomes then begin to be irreversibly withdrawn from the pool at about the same rate as new DNA continues to be synthesised; these are fitted out with new protein coats so that mature, infective progeny particles begin to appear. Thus at the time the cell bursts there is about as much phage genetic material left in the DNA pool as there is in the liberated phage particles. If lysis is artificially delayed, the increase in the amount of DNA and the random withdrawal of genomes from it continues until each cell may contain more than 1000 mature phage particles. It is during multiplication of the phage genetic material in the DNA pool that recombination takes place. One might therefore anticipate that the larger the pool the greater would be the opportunity for contacts between the genetic material of the two parents and the more frequent the occurrence of recombination. By artificially rupturing infected cells early in the lytic cycle, the first particles to mature, by early withdrawal of the genomes from the pool, can be obtained and, among these, the frequency of recombinants is in fact found to be lower than normal.

The first and most vital point to grasp about phage recombination, therefore, is that we are no longer dealing with static pairs of chromosomes isolated in individual zygotes but with a dynamic system in which mating occurs randomly within a multiplying *population* of chromosomes, each member of which can indulge in a number of successive mating events, some of which may be incestuous. Under such circumstances we cannot hope to discover the outcome of any single mating event. The only reasonable approach is to regard phage recombination as a problem in population genetics which requires mathematical treatment for its analysis. This kind of analysis was first applied to bacteriophage crosses by Visconti and Delbrück (1953) who found that a good fit with the experimental data could be obtained if it was assumed: (1) that there is a complete mixing of the genetic structures of the parental phages in the DNA pool; (2) that each phage genome behaves as a single chromosome and does not break up into individual linkage groups which assort independently; (3) that there is repeated pairwise mating of these chromosomes which find their partners in a random manner. If the additional assumption was made that recombination in phage is similar to that in higher organisms and yields reciprocal recombinants, then it could be shown that five rounds of mating were needed to produce the observed results; that is, that each phage chromosome was involved, on the average, in five separate recombination events. The genetic effect of multiple rounds of mating mimics that of negative interference (p. 51), since each round increases the probability of recombination between two genes, so that the observed frequency of recombination is higher than expected.

This type of mathematical approach suffers from the defect that its success in predicting the right answer is no proof that the assumptions are true. In fact there is evidence that the assumption of pairwise mating is not necessarily correct and that so-called *group mating* can occur, in which more than two chromosomes participate in a single mating event. As an illustration of the kind of difficulty that may be encountered when *direct* evidence about phage recombination is sought, we will briefly discuss the important question of whether or not reciprocal recombinants are produced. It is technically possible to isolate all the progeny of a phage cross issuing from a *single* infected bacterium, and to score the genotype of each. This is called a *single burst experiment* (p. 415). The obvious approach is to decide upon a particular recombinant class and then to note whether the reciprocal class is also present among the progeny of the same burst. Two problems now arise, the first of which concerns a choice of loci. We may choose two loci which are situated far apart, on the ground that recombinants will then be sufficiently numerous to permit good statistical evaluation; but because recombination between such distant loci is frequent, they will certainly be involved in several further rounds of mating which will obliterate the initial reciprocity. The alternative is to pick a pair of loci so closely linked that their involvement in recombination during more than one round of mating is improbable; but in this case the number of recombinants may be too small for statistical analysis, especially when we remember that the genomes which appear in the phage progeny are drawn at random from the DNA pool and represent only about one half of those which participated in recombination.

The second problem is that even if reciprocal recombinants arise in equal numbers we do not know whether each pair of reciprocals has arisen from a single mating event or from separate events. This problem is not peculiar to phage genetics, of course, but is common to all the systems we have discussed so far in which the products of meiosis are randomly distributed, and can only be decisively solved in those rare organisms where tetrad analysis is possible (Chapter 4; p. 68). On the other hand, if it turned out that equivalent numbers of reciprocal recombinant types did *not* emerge from single bursts we would have good reason for doubting the orthodoxy of the recombination mechanism in phage provided, of course, that one of the types did not have a selective advantage. This, in fact, is what is found although we cannot yet be sure of its significance (see p. 494).

In many papers on phage genetics, chromosome measurements are recorded in 'map units' on the grounds that, since complete genomes are involved, everything goes according to rule and that the Visconti-Delbrück formulae compensate adequately for the effects of double cross-overs as well as of multiple rounds of mating. It now seems likely that some of the assumptions on which this mathematical treatment was based can no longer be regarded as valid ones; nor do the formulae take account of the high localised negative interference which is likely to operate when small regions of phage chromosome are being mapped (pp. 51 and 504).

CYTOLOGICAL MAPS

Linkage maps based on recombination frequencies may be likened to those useful diagrammatic maps of underground railway systems in which the stations are all arranged along straight lines and are equidistant from one another. If we assume that the trains run at a constant speed we can time the run between stations and correct the map so that the stations now lie at their true rail distances apart. But even this corrected map lacks geographical precision because the straight lines joining the stations may actually follow a circuitous course. To find the real relative positions of the stations we need a surface map on which the route of the railway is marked.

The construction of cytological maps, in which the true positions of genes on the chromosome may be directly observed, has proved possible in flies such as *Drosophila* and also in maize. One method is, briefly, as follows. Chromosome aberrations, such as deletions, translocations or inversions (p. 41), may readily be induced by X-irradiation and result in specific alterations of linkage relationships which can be precisely defined by genetic analysis. The location and limits of many of these aberrations can also be seen microscopically, especially at meiosis but sometimes also during mitotic division in somatic cells, so that the genetic and morphological findings can be correlated.

A second method depends on the fact that the salivary glands of fly larvae have the unusual property of growing to a large size without division of either cell or nucleus. At the same time the homologous chromosomes, which lie very closely together (as, indeed, seems to be common in somatic cells during the telophase of mitosis), completely

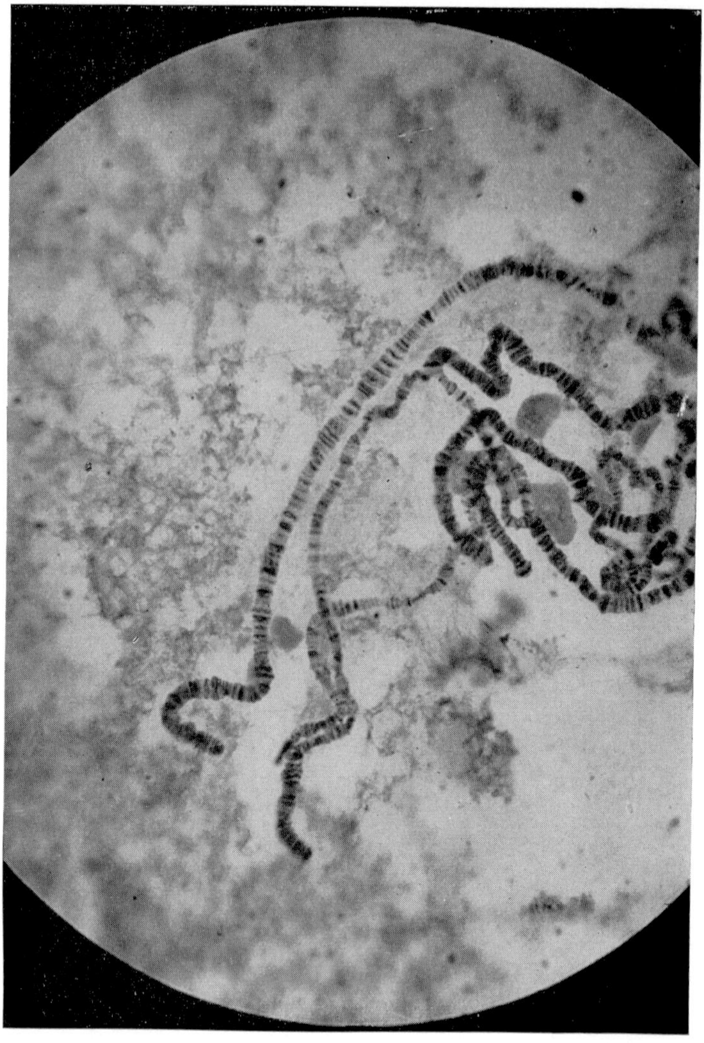

PLATE 2. Microphotograph of salivary gland chromosome of *Drosophila melanogaster*. (Kindly provided by Dr H. Slizynska)

facing p. 65

uncoil so that they are very much longer than normal, and then replicate many times. This longitudinal division is quite different from what happens at mitosis since there is no division of the nucleus and the replicas do not separate but stick together. The outcome is that long, large ribbon-like chromosomes are formed which are marked by stainable transverse discs or dots, rich in nucleic acid (DNA) (Plate 2). The number and position of these transverse striations are highly characteristic for each chromosome, although whether they actually represent the genes themselves is not known for certain; the point is that they serve as markers by means of which the site and extent of chromosome aberrations can readily be observed so that, once again, the genes inferred from recombination data can be located on a physical structure. For example, a deletion, whose position and extent is known from genetic analysis, may be correlated with the disappearance of a particular striation or group of striations.

When cytological maps constructed in this way are compared with their equivalent genetic maps, a complete correspondence is found so far as the linear order of the genes is concerned, but the distances between the genes may show wide relative variation. Close to the centromere, for example, genes which appear closely linked on the genetic map may actually lie quite far apart; elsewhere the converse may be found. In general, the spacing of genes is much more uniform on the cytological than on the genetic map. This is just what we might have anticipated from the non-random distribution of chiasma formation, and the effects of double crossing-over and interference. In fact, linkage maps are of the kind we would expect from a surveyor whose optical instruments had been fitted with an assortment of distorting lenses.

MAPPING BY INTERRUPTED MATING

Cytological maps of the chromosomes of higher organisms are very difficult and tiresome to construct. It turns out, however, that the same end can be achieved very simply, but by an entirely different method, in the bacterium *Escherichia coli*. This organism is normally haploid and has a single chromosome. It is also differentiated into two physiological types; the bacteria of one type (male) act as donors of their genetic material, while those of the other type (female) behave as recipients only. When young broth cultures of the two types are mixed, the male and female bacteria unite and are seen to be connected by a tube (Plates 32, 33, p. 672). The male cells then begin to inject their chromosomes into the females. The key feature of the injection process is that, in all the mating pairs, the same extremity of the male chromosome is always the first to penetrate the female parent while, thereafter, the rate of transfer of the chromosome is very constant under standard conditions. Moreover in this organism recombination can proceed even though the male has donated only a fraction of its chromosome to the zygote, though obviously only those genes that have actually entered the zygote can participate in recombination and be incorporated into the chromosome of a recombinant bacterium. Let

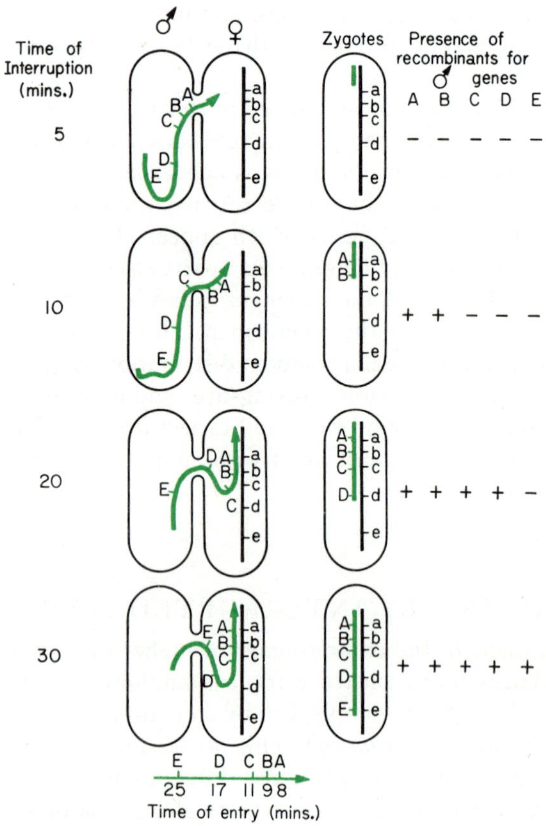

FIG. 15. Diagrammatic representation of the results of an interrupted mating experiment in *E. coli* (see text).

us suppose that the genes *A*, *B*, *C*, *D*, *E etc.* are arranged in that order on the male chromosome, gene *A* being the first to enter the zygote, while the chromosome of the female carries the alleles *a*, *b*, *c*, *d*, *e*, *f*, *g etc.* (Fig. 15). If samples of the mating cells are removed at intervals after mixing the parental bacteria, and subjected to violent agitation in a highspeed mixer or blendor, the cells are torn apart and the chromosome is broken in transit, although the viability of the cells themselves remains unaffected by this rather drastic treatment. Samples agitated before transfer of *A* will therefore yield no recombinants on subsequent plating, but as soon as *A* has begun to enter the zygotes the agitation will have no further effect on its inheritance so that *Abcd* . . . recombinants begin to appear; similarly the times of entry of genes *B* and *C* can be scored by noting the period required for the first appearance of recombinants inheriting them. In the example shown in Fig. 15, the genes *A*, *B*, *C*, *D*, *E* begin to enter the zygotes at 8, 9, 11, 17 and 25 minutes respectively after first mixing the parental cultures. This means that we can map the order of loci on the male chromosome, and the distances separating them, in terms of a time scale *which is related only to chromosomal transfer* and is thus independent of any aberrations inherent in the subsequent process of recombination. So long as the speed of chromosome transfer remains constant, therefore, the times of entry of the various loci are directly proportional to their true distances apart so that the equivalent of a cytological map is obtained (Chapter 22, p. 682). In fact we can go further than this. As we shall see, there is now indisputable evidence that the single chromosome of *E. coli* consists of only one, relatively enormously long, DNA polymer (Chapter 19, p. 550). By estimating the total amount of DNA per bacterium, and dividing this figure by the mean number of nuclei per bacterium, we obtain the amount of DNA per chromosome and, therefore, its length. Again, from interrupted mating experiments, we know that transfer of the whole chromosome would take about 90 minutes at constant speed. From this information it is possible, with some accuracy, to translate the time map of the distance between genes into molecular terms (p. 694).

CHAPTER 4
Specialised Genetic Systems in Fungi

TETRAD ANALYSIS

So far we have illustrated the fundamental principles of inheritance with examples derived mainly from classical systems in those plants and flies from whose study they emerged. The discovery of these principles and of their correlation with the observable behaviour of the chromosomes at meiosis represents a truly remarkable feat of imagination and intellect when one considers the complicated nature of the experimental material.

1. The characters studied are complex and their expression can only be observed at the end of a long chain of developmental events.

2. The individuals are diploid so that the genotype of their gametes can only be inferred from the phenotype of the next generation.

3. The gametes are numerous and randomly distributed, so that the relations between the four products of meiosis can only be deduced statistically from the proportions of various genotypes among large numbers of the progeny. For many years this led to reluctance by some geneticists to accept Mendel's principles on the ground, for example, that although the gametes from the total of all the F_1 plants yielded a ratio very close to 3 : 1 for each character, individual plants might (and did) give widely varying ratios.

Among microorganisms there are several species, especially of fungi, in which all these complications are neatly side-stepped. Character differences involving single biosynthetic steps may be studied in them; the vegetative individual consists of haploid cells, derived directly from mitotic division of the gametes or ascospores, in which recessive characters are directly expressed; finally, the four products of meiosis of every zygote are not only retained together in a single pod or ascus sac, but are precisely ordered so that each pair of haploid segregants from the second meiotic division lie next to one another. We will defer discussion of the important subject of biochemical genetics until the next chapter and, for the moment, will describe only those genetic features of the fungus *Neurospora* which bear on recombination analysis.

Neurospora is a haploid organism (with seven chromosomes), composed essentially of multinucleate hyphae forming a mass of mycelial filaments which bud off multinucleate macrospores and uninucleate microspores or conidia, both of which are asexual. Vegetative reproduction results from the mitotic division of either the hyphae or the germinated spores. Within the mycelial mass specialised hyphae arise which, in turn, yield long, slender filaments called *trichogynes* (Greek: *trichos*=hair; *gunē*=female) which form the female receptive organs. Fertilisation is effected by fusion of a trichogyne with a spore, or even with a mycelial fragment, which transfers its haploid nucleus to it. *Neurospora*, however, is heterothallic (p. 14) and has two mating types, *A* and *a*, so that although each individual is hermaphrodite, producing both spores and trichogynes, fertilisation demands union of the spores of one mating type with trichogynes of the other. This mating type difference is determined by a single gene, as we shall see. Following fertilisation, the pair of haploid nuclei at first remain separated and divide mitotically and in step, so that a number of new ascogenous (ascus-bearing) hyphae result, each having two nuclei, one descended from each parent. Ultimately the terminal cell of each ascogenous hypha elongates and bends over on itself to form a hook-shaped cell whose nuclei divide once mitotically to give four haploid nuclei (Fig. 16b). Two cell walls are now laid down so that the cell is divided into three cells, the first and last of which contain one nucleus while the middle one has two (Fig. 16 c). The first and last cells then fuse to form a single binucleate cell, while the middle one becomes the *initial cell of the ascus*. It is the subsequent behaviour of this initial cell that we must follow. Its two nuclei first fuse to form a single diploid nucleus (Fig. 16 d). The cell then elongates and the diploid nucleus goes through its first and second meiotic divisions (Fig. 16 e and f). The important point is that when the two haploid nuclei resulting from the first meiotic division themselves divide at the second meiotic division, each grand-daughter nucleus retains its position in the long axis of the cell so that the genetic consequences of each division are precisely reflected in the order of the nuclei. Each of the four nuclei then undergoes a single *mitotic* division to give four pairs of nuclei which continue to preserve their longitudinal order (Fig. 16 g); an

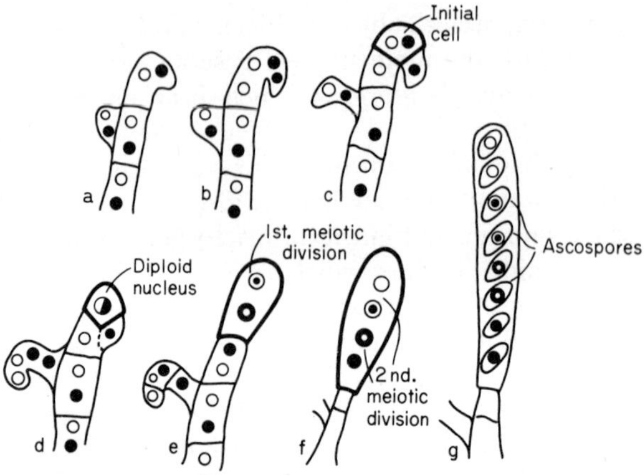

FIG. 16. Diagram showing the mode of development of ordered ascospores from ascogenous hyphae in *Neurospora*. (Redrawn from Catcheside, 1951.)

elliptical mass of cytoplasm condenses around each nucleus and is finally enclosed by a cell wall to form an *ascospore*. Thus each ascus sac contains eight ascospores (1,2,3,4,5,6,7,8) comprising four identical pairs (1,2:3,4:5,6:7,8) of which the upper two pairs (1,2:3,4) and the lower two pairs (5,6:7,8) represent segregation at the first meiotic division, while 1,2 and 3,4 on the one hand, and 5,6 and 7,8 on the other, are derived from second division segregations.

With the aid of a binocular dissecting microscope, the individual ascospores can be dissected out of the ascus sac and isolated in separate tubes of culture medium. It is usually necessary to stimulate germination by heating at 50°C. for 30 minutes, a procedure which has the additional advantage of killing conidia or mycelial fragments which may have been accidentally picked up during the dissection. In this way, separate cultures of each ascospore may be obtained, each of which can then be tested for inheritance of those characters in which the original parents differed. For example, the mating type of each culture may be determined by testing its fertility with each of the parent strains, *A* and *a*. When all the eight asci from a sufficient number of ascus sacs are tested in this way, three features become apparent.

1. Of the eight asci in each sac, four are always of one mating type and four of the other. This precise 1 : 1 ratio shows that mating type is determined by two genes which are strict alternatives or alleles; put another way, the 1 : 1 ratio indicates a difference involving a single gene.

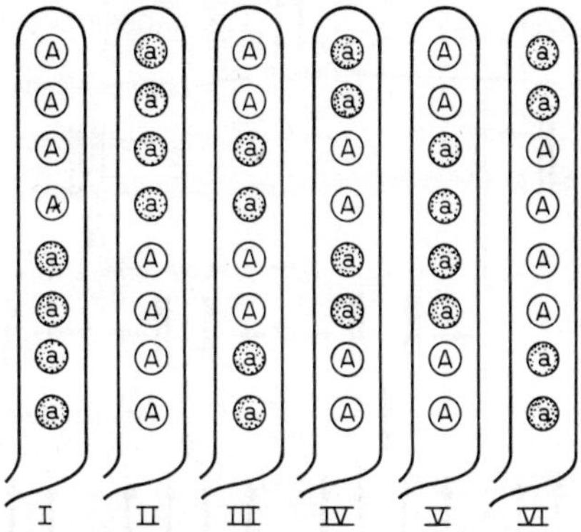

FIG. 17. Arrangements of ascospores, with respect to mating type, in *Neurospora*.

2. The arrangement of the asci in the sacs falls into the six different patterns shown in Fig. 17.

3. Patterns I and II arise in about equal numbers; similarly, patterns III, IV, V and VI occur with approximately the same frequency. On the other hand, the ratio (I + II)/(III + IV + V + VI) varies widely for different loci, but is more or less constant between crosses in the case of any given locus.

The relationship between these patterns and the events which occur during meiosis is interpreted in Fig. 18. Let us first consider what happens in the absence of crossing-over (Fig. 18A). You will recollect (p. 14) that at the first meiotic division the homologous chromosomes of the two parents synapse and then split longitudinally to form a tetrad of four chromatids, *but the centromeres of each homologue do not divide* so that, at anaphase, the pair of chromatids derived from each parent remain together and migrate to opposite poles. Thus, if there is no crossing-over, the allelic genes *A* and *a* will be separated at the first division. At the second meiotic division the centromeres do divide so that each chromatid segregates from its fellow into a separate grand-daughter cell. It follows that if the eight cells, emerging from these two meiotic divisions and the final mitotic division, preserve, in the ascus sac, the order imposed on them by the plane of the divisions, the two patterns I and II will result. Whether the four *A* ascospores or the four *a* ascospores lie uppermost in the sac is determined by the orientation of the two pairs of parental chromatids with respect to the two poles of the zygote cell during the first meiotic division; since this is normally

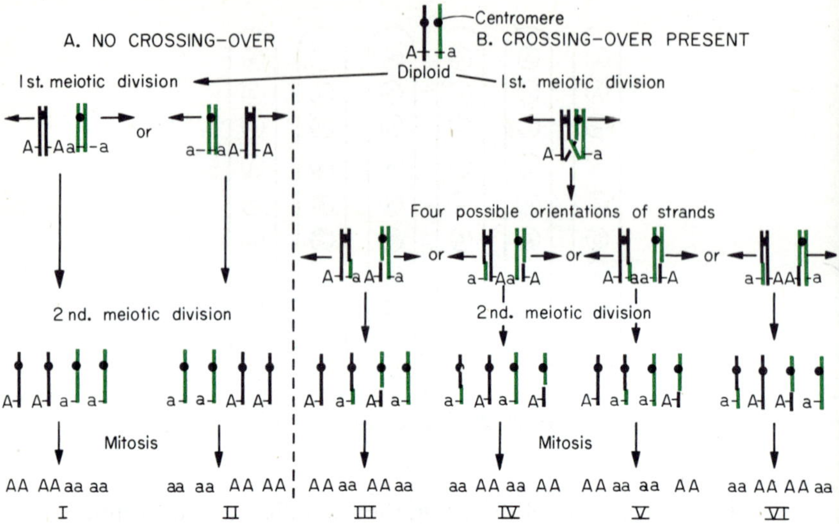

FIG. 18. Diagrammatic representation of how the six different ascospore arrangements arise in *Neurospora*, as a result of different events occurring at meiosis.

random in *Neurospora*, patterns I and II will arise with equal frequency. Thus patterns I and II, in which the parental alleles show a four by four arrangement, indicate *first division segregation* and no crossing-over between the centromere and the locus (see below).

Now let us see what happens when crossing-over occurs between two strands of the tetrad (Fig. 18B). Provided the cross-over arises between the centromere and the locus, its effect will be to render each pair of strands separated by the first meiotic division heterozygous for the gene, that is, *Aa* or *aA* instead of *AA* or *aa*. From this it follows that, in the ascus sac, an *A* pair of ascospores must always be adjacent to an *a* pair, thus giving rise to the four patterns III, IV, V, VI. Which of these four patterns is found depends, once again, on the orientation of the pairs of strands at the second meiotic metaphase; so long as this is random the four patterns will be randomly distributed. Patterns III, IV, V and VI therefore indicate *second division segregation*, that is, that separation of the allelic genes has been delayed until the second meiotic division due to the occurrence of crossing-over. It is important to understand clearly that *second division segregation will only be found in those cases where a cross-over has arisen between the centromere and the locus*. If the cross-over arises distal to the locus, as shown in Fig. 19A,

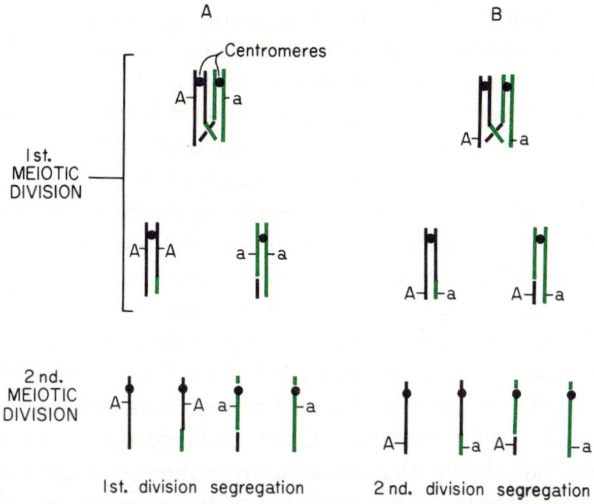

FIG. 19. Diagram showing the relation between the position of crossing-over, with respect to centromere and locus, and the segregation pattern revealed by tetrad analysis.

A. Crossing-over outside the centromere-locus intercept.

B. Crossing-over between the centromere and the locus.

it will not lead to heterozygosity of the pair of strands attached to each centromere so that first division segregation of the locus, yielding patterns I and II, will follow. Of course, if the segregation of a second pair of genes, *B* and *b*, situated on the same chromosome but distal to the *A* locus, is also being studied, a cross-over between *A* and *B* will yield recombinants for these pairs of genes, the ascospore pattern being I or II for locus *A* and III, IV, V or VI when the inheritance of *B* is scored.

Since the number of ascus sacs showing first division segregation (patterns I and II) represent the parental combinations while those showing second division segregation (patterns III–VI) arise from crossing-over between the centromere and the locus, the ratio of the second to the first is an index of the recombination frequency *between the centromere and the locus*, so that the position of the centromere can be mapped. Although the position of the centromere may be mapped in polyploids by a complicated method of genetic analysis, ascus analysis is the only method applicable to diploids which can distinguish between first and second division segregation, which is, of course, normally impossible when the products of meiosis are randomly distributed.

There are some mutations in *Neurospora*, as well as in other fungi where tetrad analysis is possible, which interfere with the pigmentation of thc ascospores themselves so that those ascospores that inherit the mutant gene appear pale while the wild type ascospores are dark. In crosses between such mutant and wild type strains, one can map the position of the locus in relation to the centromere without testing the individual spores, since the segregation patterns can be observed under the microscope and counted directly (Plate 3).

We have already noted, from studies of crosses in other organisms, that when recombination occurs between two pairs of linked genes (*AB* and *ab*), reciprocal recombinant types (*Ab* and *aB*) are produced in approximately equal numbers, but there was no proof that these arose from the reciprocity of every recombinational event although this seemed a valid inference. Careful microscopic examination of chaismata had shown that crossing-over involves only two of the four strands of the tetrad but, once again, direct *genetic* evidence of this was lacking. Tetrad analysis in *Neurospora* provided the proof that both inferences were correct, for here all the products of a single meiosis are assembled together in order in one ascus sac. It turns out that whenever a pair of recombinant ascospores is found, it is always associated with pairs of the reciprocal recombinant and both parental types. Actually, this statement is not *strictly* true since reciprocals may not be found in recombination between extremely closely linked genes, but that is another story (see p. 392).

Since tetrad analysis enables us to observe the segregation of all four strands in terms of their distribution at each meiotic division, it should permit an appraisal of how the various strands may participate in double cross-over events, and this is indeed the case. The consequences of 2-strand, 3-strand and 4-strand doubles are complicated and much easier to represent pictorially than to express in words; they are shown diagrammatically in Fig. 20. The centromere and three pairs of linked genes are depicted in a heterozygote of constitution *ABC*/*abc*. The positions of the cross-overs and the strands involved are shown in the left half of the Figure, and the constitution of the four emergent strands on the right. One cross-over arises between the centromere and the B locus, and the other between the B and A loci: the C locus is on the opposite side of the centromere and segregates at the first division in every case, but serves as a useful marker to define recombinant strands; for example, the 3-strand double yields one *cBA* strand, and the 4-strand double a *Cba* and a *cBA* strand, which would not be

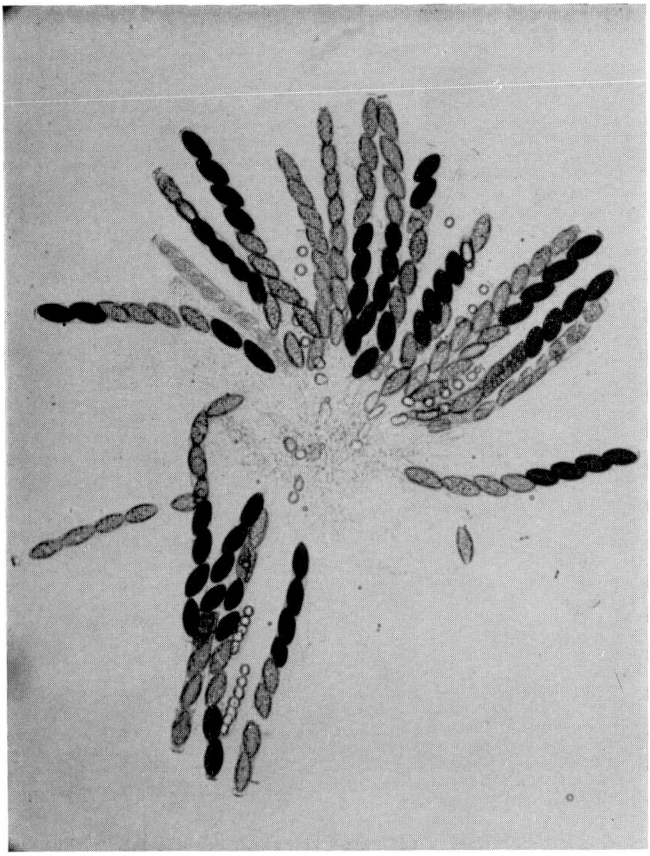

PLATE 3. Microphotograph showing the arrangement of ascospores from a cross between wild type *Neurospora* and a lysine-requiring mutant which exhibits delayed maturation and absence of spore pigmentation.

Of the asci showing segregation for this character, 9 show 1st division segregation (Fig. 19, patterns I & II) indicating absence of recombination, while 5 display 2nd division segregation, indicating recombination. Thus the recombination frequency between the centromere and the locus can be calculated. (Kindly provided by Dr D. R. Stadler)

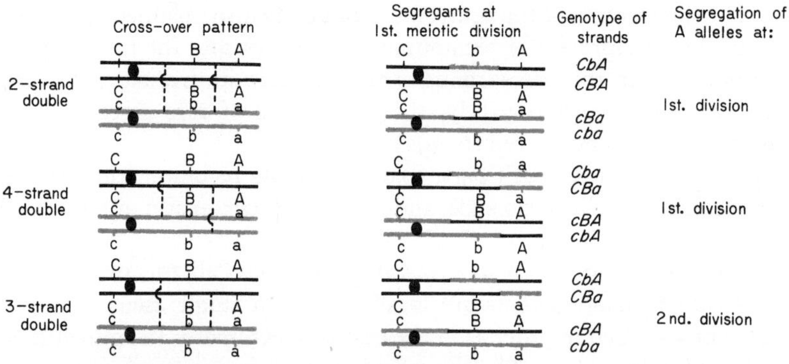

FIG. 20. Diagram showing the relationship between the strands involved in a double cross-over, in the centromere-*A* locus intercept, and the segregation pattern revealed by tetrad analysis.

Only 3-strand doubles lead to 2nd division segregation.

The existance of double crossing-over is, of course, inferred from the inheritance of an intermediate pair of alleles, *B* and *b*, among the progency (see text).

directly recognisable as recombinant strands but for the C alleles which they carry.

When we look at the genetic constitution of the four emergent strands in each case, it is evident that we will be able to recognise whether a 2-, 3- or 4-strand double cross-over has occurred from an examination of the genotype of the ascospores. The 2-strand double yields reciprocal double recombinants as well as both parental types of ascospore; the 4-strand double gives four singly recombinant types of ascospore and no parentals; the 3-strand double produces one double and two single (but non-reciprocal) recombinant types and one parental type (pp. 17, 19). Furthermore, if we look at segregation of the A locus we find that this occurs at the first division following both 2- and 4-strand double cross-overs arising between it and the centromere, while 3-strand doubles result in second division segregation. Tetrad analysis thus enables us to assess the existence of chromatid interference (p. 51). It appears that, in *Neurospora*, an excess of 4-strand doubles occurs between loci on the same side of the centromere, that is, chromatid interference is present, whereas when the loci lie on opposite sides of the centromere an excess of 2-strand doubles is found, indicating negative chromatid interference.

The unique advantages of tetrad analysis may be summarised as follows:

1. The location of the centromere can be readily mapped.

2. The reciprocity of recombinational events and the involvement of only two of the four strands of the tetrad in individual events can be established. Alternatively, the occurrence of rare cases of non-reciprocity can be proven (p. 392).

3. The strands which participate in double recombination can be identified and the existence and nature of chromatid interference verified.

In some fungi, such as *Aspergillus nidulans*, all the products of individual meiotic events are localised within single ascus sacs, but are not arranged in order as in the case of *Neurospora*. In spite of this, all the information yielded by tetrad analysis in *Neurospora* may still be obtained, although mapping of the centromere requires a more complicated technique.

The circumstances when special information of this kind is sought do not often arise, while the dissection of adequate numbers of individual ascospores is a tedious and lengthy business. In addition, the number of ascospores which must be dissected and examined is double that theoretically required, since each is represented twice in the sac; the tetrad analysis becomes an octad analysis in practice. Because of this it is usually simpler and quicker, for purposes of orthodox genetic analysis, to examine large numbers of randomly obtained ascospores on a statistical basis, just as one would the unordered gametes of other types of organism. Both fungi and bacteria have two decided advantages over diploid organisms in this respect. The first is that the gametes, on germination, directly reproduce the haploid vegetative organisms in which recessive as well as dominant characters are expressed. The second is the ability to employ nutritional or drug-resistant mutants which allow rare recombinants to be selected from a background of parental and other recombinant types. As a test of allelism, for example, this method is very simple and efficient. From the genetic point of view the aim here is to show that no recombination occurs between the two genes and for this large numbers of progeny must be examined, because the absence of recombinants among small numbers only implies close linkage and not the identity of the genetic loci. Suppose we have two auxotrophic strains of fungus or bacterium both of which differ from the wild type in being unable to grow on synthetic minimal medium, due to inability to synthesise a necessary growth factor. If the defects in the two strains are the result of mutations in different genes then, on crossing the strains, wild type prototrophic recombinants, capable of growing and producing visible colonies on the synthetic medium, will arise with a frequency proportional to the linkage between the genes, while progeny which are not recombinant for the two genes cannot grow. Thus, by merely plating large numbers of progeny on minimal medium, the presence of a minute proportion of prototrophic recombinants can be detected. In the following Chapters we will examine in more detail the value of nutritional mutants in revealing the sequential steps of biosynthetic pathways, and how their use in genetic studies has led to outstanding refinements in the analysis of chromosomal structure.

MITOTIC RECOMBINATION IN FUNGI

In order to inculcate the basis principles of genetic analysis we have, so far, dealt mainly with highly developed sexual organisms in which these principles are best displayed. However, we have not defined very dogmatically what is meant by sexuality. We could take the wide definition, for example, that the essence of sexuality is the formation of some kind of heterozygote which results in the production of recombinants; such a definition would include even transformation in bacteria as a sexual process. At the other extreme we might deny sex to those creatures which do not possess some morphological or, at least, physiological sexual differentiation, thus including *E. coli* as a sexual organism but excluding the yeasts. Alternatively we might degrade only the bacteria to a lower, sexless status by insisting on the formation of complete zygotes.

The most widely accepted criterion of sexuality is the existence of an alternating cycle of fusion and segregation of the genetic material from two separate lines of descent (Fig. 2, p. 13). This cycle must involve some fertilisation mechanism whereby the genetic complement of two haploid cells are brought together to form a diploid nucleus—a process known as *karyogamy*. The diploid cell, or certain of its descendants, ultimately passes through two meiotic divisions which lead to the re-emergence of haploid cells, some of which may be recombinants. The essential feature of sex, therefore, is meiosis which usually takes place in specialised cells and is, as we have seen, a process quite distinct from mitosis which mediates asexual or vegetative reproduction.

All somatic cells divide by mitosis although, as we shall see later (Chapter 19, p. 553), this is not true of the bacteria and a few other free-living organisms in which the segregation of daughter chromosomes is performed by a different mechanism. The somatic cells of all animals and most plants are diploid and, of course, have the same genetic constitution as the specialised cells in which meiosis occurs. We have already mentioned that the occurrence of mutation in a somatic cell early in the development of a multicellular organism may yield a clone of phenotypically altered cells which is large enough to be visible (p. 41). Thus a mutation affecting pigment production may result in the appearance of an unpigmented 'flash' of skin or hair in an otherwise uniformly pigmented animal.

Rare mosaics have been reported which display, not a single 'flash', but *adjacent* 'flashes' of different phenotype such as might be expected to arise from the segregation of reciprocally recombinant genotypes. Study of this phenomenon, especially by Stern (1936) in *Drosophila*, strongly suggested *mitotic recombination* as its basis.

G. Pontecorvo and his colleagues, working with the fungus *Aspergillus nidulans*, were the first to demonstrate directly that recombination can occur during mitotic division (Pontecorvo and Roper, 1952) and to show how it can be employed in the genetic analysis of fungi which lack a sexual cycle (review; Pontecorvo and Käfer, 1958). If asexual spores from two genetically dissimilar strains of a filamentous fungus are planted very close together on the surface of a solid medium, the emerging hyphae tend to fuse at points of contact. This fusion is followed by the transfer of nuclei from one hypha to the other so that *heterokaryons* are formed whose hyphae contain nuclei from both strains. As the heterokaryotic hyphae grow, the two types of nuclei multiply independently. At asexual spore formation, however, each initial sporogenous cell receives only a single nucleus of one type or the other. On subsequent multiplication of these cells, chains of spores are thus produced, each chain consisting of genotypically identical spores although the spores of adjacent chains may differ in genotype (Pontecorvo, 1946; see Fig. 25, p. 95). In this way heterokaryon formation, which is widespread among both sexual and asexual filamentous fungi, normally results in the clean segregation of the two types of parental, haploid nuclei in the asexual spores (microconidia). Very occasionally, however, two of the nuclei in the heterokaryotic hyphae may fuse to form a single *diploid* nucleus. This nucleus will multiply as such and be incorporated into spores in the same way as the haploid nuclei, so that rare chains of diploid spores will arise which, on subsequent germination, will yield diploid individuals. If the haploid parental strains differ in having complementary nutritional requirements, one being of genotype *Ab* (requiring B) and the other *aB* (requiring A) for example, then the diploid nucleus (*Ab*/*aB*) will possess genetic determinants for the synthesis of both substances needed for growth of the parents. The existence of diploid spores can therefore be recognised, and cultures from them obtained, by plating on media lacking the growth factors A and B (Roper, 1952).

Once diploid strains have been obtained which are heterozygous for a number of characters, involving, for example, colour or nutritional differences which are easy to recognise or select, the segregation of recombinant types among the asexual spores can be looked for. The fungus initially employed by Pontecorvo and his colleagues for studies of mitotic recombination was *Aspergillus nidulans* which has a sexual cycle very similar to that of *Neurospora crassa* and in which the haploid number of chromosomes (8) and the location on them of a considerable number of loci was already known from formal genetic analysis. This meant not only that the most suitably marked strains were already available to test for mitotic recombination, but also that the results of mapping by the two methods could be compared. It was found that recombinants did arise during mitosis in diploids, but with a frequency about 10,000 times lower than in the sexual cycle. Furthermore, linkage maps constructed from the mitotic recombination data revealed the same linkage groups, and the same arrangement of genes within each group, as those deduced from orthodox analysis.

In *Aspergillus nidulans* there is thus a cycle of events, called the *parasexual cycle* by Pontecorvo, leading to the production of recombinants which is quite distinct from, and much rarer than, the sexual cycle. We may summarise the essential sequence of events which comprises the parasexual cycle as follows:

1. The formation of a heterokaryon by hyphal fusion;
2. The rare fusion of two haploid nuclei within the heterokaryon to form a diploid nucleus;
3. The segregation of recombinant nuclei during mitotic division.

When these recombinant nuclei are haploid the cycle is complete but, as we shall see, this is by no means the rule in mitotic recombination. The method of genetic analysis by means of the parasexual cycle has since been applied successfully to a number of other fungi which do not possess a sexual cycle. In *Aspergillus niger*, for example, the linear order of 32 genes, distributed among six linkage groups, has been established exclusively by this method (Lhoas, 1961).

When recombination is mediated by the sexual cycle, we have seen that two meiotic divisions of the diploid zygote generate four haploid cells. Since mitosis is not a reductive process, however, what is the nature of mitotic recombination and what is the ploidy of recombinants issuing from it? The study of *Aspergillus nidulans* has shown that three distinct and independent processes may be involved, two of which give rise to diploid, and the third to haploid, segregants. We will deal with each of these processes in turn.

MITOTIC CROSSING-OVER

This is the only process of mitotic segregation whose outcome indicates that crossing-over has taken place between homologous chromosomes. You will remember that one of the essential observable differences between mitosis and meiosis is that, during mitosis in diploids, homologous chromosomes do not pair, but are arranged independently on the equatorial plate; after division of the centromeres, each pair of sister chromatids is then separately partitioned between the two daughter cells (Fig. 2, p. 13). However, since crossing-over must be preceded by pairing it is obvious that this must occasionally occur between a particular pair of homologous chromosomes during mitosis; in fact, the low frequency of mitotic crossing-over may well reflect the rarity of the event.

The data from which somatic crossing-over is inferred are of the following type. When several chromosomes of the diploid are heterozygous at a number of loci, diploid segregants may be isolated which are *homozygous* at all loci distal to a point on one of the arms of a particular chromosome; that is, they are homozygous diploid at these loci for the alleles derived from one parent. However, these recombinants are accompanied by a complementary group which are homozygous at the same loci for the alleles from the other parent. Both groups of recombinants normally remain heterozygous at all other loci. For example, if the normal diploid genotype is *A.BCDEF/a.bcdef*, where the *A* locus is on one chromosome and loci *B,C,D,E* and *F* are linked on another, the genotype of the segregants resulting from somatic crossing-over might be *A.BCDEF/a.bcdEF* and *A.BCDef/a.bcdef*, indicating crossing-over in the region between *D* and *E*.

The likely mechanism of crossing-over and segregation is shown diagrammatically in Fig. 21. The pair of homologous chromosomes synapse and, either during or after their division into bivalents, crossing-over occurs between two of the heterozygous chromatids in

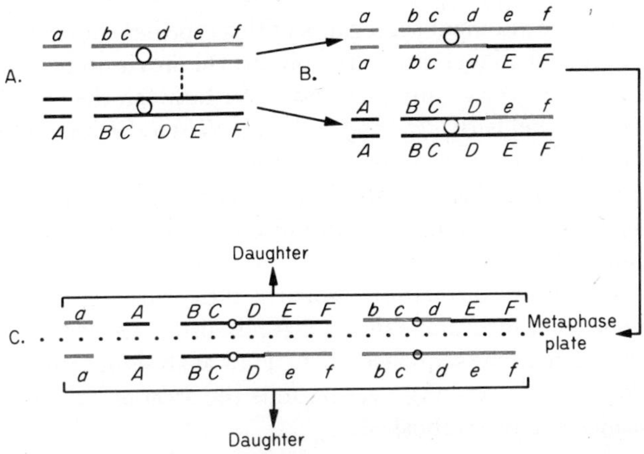

FIG. 21. Demonstrating the probable mechanism of somatic crossing-over and segregation in a heterozygous diploid cell.

A and *a* are alleles on one chromosome; *B* and *b*, and *C* and *c* are alleles on one arm of a second chromosome, while *D* and *d*, *E* and *e* and *F* and *f* are alleles on the other arm of the same chromosome. The chromatids carrying one set of alleles are drawn in green, those carrying the other set being black. The centromeres are indicated by circles. The vertical dotted line in A shows the position of crossing-over. The horizontal dotted line in C indicates the equatorial plate.

A and B show pairing, crossing-over and separation of the bivalents. C shows the orientation of the pairs of chromatids at metaphase and their distribution to the daughter cells at anaphase which lead to homozygosity of loci distal to the cross-over region (see text).

the interval between the *D* and *E* loci (Fig. 21A). The two chromatids of each bivalent remain joined by the undivided centromere, but one chromatid of each is now recombinant (Fig. 21B). The bivalents separate and, at metaphase, become arranged independently on the equatorial plate. The centromeres then divide and the chromatids are segregated into the two daughter cells (Fig. 21C). Observe that the segregation of two daughters which are reciprocally homozygous at the *E* and *F* loci depends on the relative orientation of the recombinant bivalents on the equatorial plate. For example, if one of those depicted in Fig 21C has been so oriented that the *BCDef* and *bcdEF* chromatids were segregated into the *same* daughter cell, then this daughter, although recombinant, would still be heterozygous at all the loci and indistinguishable phenotypically from the parental heterozygote. Thus only a *proportion* of the progeny of cells in which somatic crossing-over has occurred display homozygosity.

Analysis of recombinant segregants of this type has shown that they are homozygous at all loci situated on one arm of the chromosome distal to the cross-over. For example, a cross-over between the centromere and the *D* locus yields *BCdef*/*bcdef* and *bcDEF*/*BCDEF* recombinants, while recombinants resulting from a cross-over between *E* and *F* are homozygous for the *F* locus only. Put another way, if it is found that recombinants homozygous for *D* are also homozygous for *E* and *F*, while some are found which are homozygous for *E* and *F* but not for *D*, then *E* and *F* must be distal to *D*; similarly, the occurrence of recombinants homozygous for *F* alone shows that *F* is distal to *E*, so that the loci must be arranged in the sequence *DEF*. Thus analysis of recombinants arising from mitotic crossing-over permits the *order of loci on one arm of a chromosome* to be established.

MITOTIC NON-DISJUNCTION

From a diploid which is heterozygous for a number of loci on both arms of a particular chromosome, diploid segregants can be isolated which are homozygous for *all* these loci. For example, if the parental diploid has the genotype *A.BCDEF*/*a.bcdef*, the *A* locus being on one chromosome and the loci *B*,*C*,*D*,*E* and *F* on another, the segregants may be of either *A.BCDEF*/*a.BCDEF* or *A.bcdef*/*a.bcdef* genotype. When such segregants are found, any associated homozygosity (*P*/*P* or *p*/*p*) for an unmapped pair of alleles (*P*/*p*) indicates that the *P* locus lies on the same chromosome as *B*,*C*,*D*,*E* and *F*, so that this is a useful method for allocating genes to a particular chromosome although it gives no information as to the order of the genes.

The evidence suggests that this type of segregation may arise, in the first instance, from accidental loss of one of the pair of homologous chromosomes, so that the cell is *monosomic*, that is, is haploid (*A.BCDEF*/*a.*) for that chromosome (p. 41). If, at some subsequent mitotic division, the two chromatids of the remaining chromosome fail to segregate normally so that one daughter cell receives both, this daughter will become diploid again, but homozygous with respect to the chromosome. Since the other daughter receives no representative of this chromosome it will, of course, die (see Käfer, 1960).

HAPLOIDISATION

Haploidisation results in the segregation of cells which are haploid with respect to all the chromosomes, and represents the completion of the parasexual cycle. During this process there is no recombination between homologous chromosomes such as occurs at meiosis. On the other hand, which of the two parental chromosomes of the diploid is inherited by the haploid segregants is entirely random so that, although alleles from the two parents which are located on the same chromosome show no recombination, those located on different chromosomes show 50 per cent recombination. Haploidisation therefore offers a very efficient method of allocating genes to chromosomes, though it does not permit the mapping of genes on any one chromosome. Suppose that the diploid is heterozygous at three loci (*A.BC/a.bc*), of which the *A* locus is on one chromosome, and *B* and *C* on another. All the haploid segregants will be of either *BC* or *bc* genotype; half of the *BC* segregants will have the genotype *a*, and half the *bc* segregants the *A* genotype. Only a very small number of segregants need, therefore, be examined in order to establish linkage. Moreover the genotype of the segregants, being haploid, is directly expressed.

As in the case of mitotic non-disjunction, haploidisation appears to be initiated by the accidental loss of one partner of a pair of chromosomes. This introduces an unstable equilibrium which can only be balanced by restoration of the fully diploid state, as by mitotic non-disjunction (see above) or, alternatively, by the progressive loss of one of the members of other chromosome pairs during successive divisions until a fully haploid state is attained (Pontecorvo and Kafer, 1958). Haploidisation can be artificially induced in diploid fungal strains by treatment with *p*-fluorophenylalanine (Lhoas, 1961).

An intriguing corollary to the study of mitotic recombination in fungi is that the segregation mechanisms and methods of analysis which we have discussed may well be applicable to the genetic study of animal and human somatic cells growing in tissue culture. These cells are, of course, already diploid as well as heterozygous at many loci. The work of Pontecorvo suggests that segregation almost certainly occurs, or can be induced, in human somatic cells, but that the real difficulty is to find genetically determined character differences which are recognisable at the cellular level and which continue to be expressed in tissue culture (see Pontecorvo, 1962).

Part III
The Integration of Genetics and Biochemistry

CHAPTER 5

The Basis of Biochemical Genetics

We introduced the subject of genetics as the study of how character differences are inherited, and then went on to show how the genius of Mendel translated these differences into terms of alternative, or allelic, unitary factors which obeyed precise laws during their transmission from one generation to the next. These Mendelian factors then materialised as physical entities called genes, located at defined positions on the chromosomes; the structure of a gene could be changed in some mysterious way by mutation, leading to alteration in its function and, thus, in the expression of the character it determines. This evokes a picture of genes which are arranged like beads on a chromosomal string, each gene being capable of definition in terms of one or the other of its three properties of inheritance, mutation and function which are equivalent and coexistent. In this and the next few chapters we must take the analysis further by seeking an answer to some new questions. What is the function of the gene in the ultimate terms of biochemical processes? Does the gene have any spatial extension on the chromosome and, if so, can its limits be defined? Are the units of function, inheritance (recombination) and mutation really coterminous and, if not, how are they related to one another? We shall, first of all, discuss the *function* of the gene, not only on the grounds of historical precedence but also because the tools necessary for more refined genetic analysis of chromosomal structure are largely derived from functional concepts.

THEORETICAL PRINCIPLES

The fundamental ideas of biochemical genetics came from thinking about the function of genes in the light of a growing knowledge of the chemistry of intermediary metabolism, which supposed that the constituents of the body are built up from elementary precursors by a series of discrete steps, each of which is mediated by a specific enzyme.

Let us imagine the synthetic sequence A→B→C→D, shown in Fig. 22, which culminates in production of a required growth factor D. Three enzymes, a, b and c, are involved. Enzyme a acts on substrate A, converting it into B; enzyme b then converts B into C which, in turn,

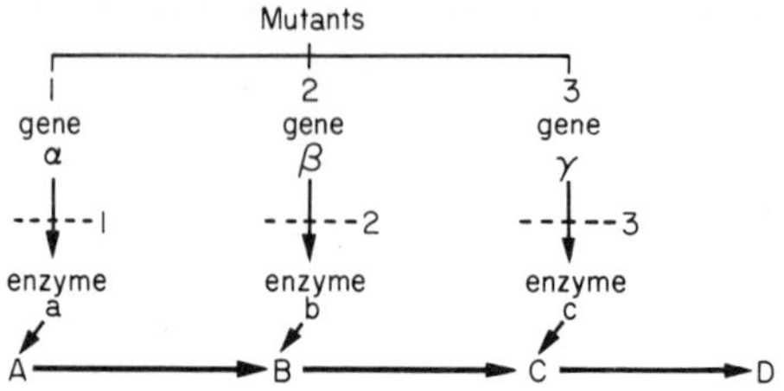

FIG. 22. Diagram representing the relationships between genes, enzymes, and an imaginary sequence of synthetic steps in intermediary metabolism (see text).

is acted upon by enzyme c to form D. Normally the cellular economy is controlled by the utilisation of the end-product D in such a way that the intermediates B and C are produced only as they are required and so are difficult to detect within the cell and are not excreted into the environment (see p. 705). If one now thinks about the intervention of genes in such a process, the simplest hypothesis is that they are responsible for deciding the specificity of the enzymes, so that we must postulate three genes, α, β and γ, which determine the synthesis of enzymes a, b and c respectively. The effect of mutation in any one of these genes will be either to prevent formation of the enzyme altogether or so to alter its specificity that its functional efficiency is greatly reduced or abolished.

Let us now examine the situation more closely. A mutation in gene γ (Fig. 22,3) will prevent formation of enzyme c so that substrate C can no longer be converted to D. Since D is an essential metabolite, metabolism will be arrested unless the organism is provided with an exogenous supply of substance D. On the other hand there is no block in the A→B→C chain of syntheses so that, if D is supplied in limiting amount so that metabolism continues, substance C will be synthesised. Since it cannot be converted to D, it will accumulate and may be poured into the environment in considerable amounts. If the mutation involves gene β (Fig. 22,2) the final outcome will be the same as before in so far as the synthesis of D is prevented and metabolism stops. In this case, however, the situation is different in two respects. Firstly, an exogenous supply of *either* C *or* D will permit metabolism to continue because there is no bar to the conversion of C into D; secondly, substance B, instead of C, will accumulate if a limiting amount of C or D is provided. In the same way the effects of a synthetic block from mutation in gene α (Fig. 22,1) can be overcome by supplying substances B, C or D, while substance A alone will accumulate.

The material with which the biochemist normally deals is like a cinematograph film of a synthetic pathway which he cannot arrest to examine 'stills' at any stage, or even observe in slow motion. All he is permitted to do is to mark the film at various places with isotopes or radioactive tracers so that he can briefly note the behaviour of certain elements as the film flashes on. The introduction of genetic blocks between the various steps of a synthetic pathway is analogous to a device for stopping the film and re-running short sections of it at any desired stage.

This hypothesis, and the kind of predictions that we can derive from it, impinges in important and very practical ways on both biochemistry and genetics. Let us assume that mutations are distributed more or less randomly among all the genes of the genome and that we can tell, by means of genetic analysis, whether independent mutations involving the same general phenotype, such as ability to synthesise substance D, are allelic or not. From the biochemical standpoint a number of possibilities present themselves.

1. If a sufficient number of D-dependent mutant strains are examined by genetic crosses, it should be found that although some of the mutations are allelic and block the biochemical pathway at the same position, others affect different genes determining other steps in the synthesis of D. Thus the mutations fall into a number of different allelic groups, each group corresponding to a distinct step in the biosynthetic pathway, so that an estimate of the probable number of steps in the sequence can be made.

2. It should be theoretically possible to align the mutants in a functional order corresponding to the sequence of the synthetic steps that are blocked in them. We saw from Fig. 22 that mutant 3, blocked at position 3 (gene γ), accumulated intermediate C behind the block, and that C might then diffuse into the environment. We also noted that mutants 2 and 1, blocked earlier in the pathway than mutant 3, could metabolise and grow if intermediate C, produced distal to the blocks, was exogenously supplied. Similarly mutant 2 accumulates intermediate B which permits growth of mutant 1 only. It follows that if we allow restricted metabolism of any mutant by offering it a limiting

amount of the end-product, D, it should be able to stimulate the growth of all those mutants which are blocked at a *preceding or proximal step*, but not of those mutants having a block in the same step, or one distal to its own in the synthetic pathway. The sequence of synthetic blocks in nutritionally defective mutants of some bacterial species, notably *Escherichia* and *Salmonella*, can be determined in this way by *cross-feeding* or *syntrophism* tests (Greek: *sūn*=together; *trophe*=

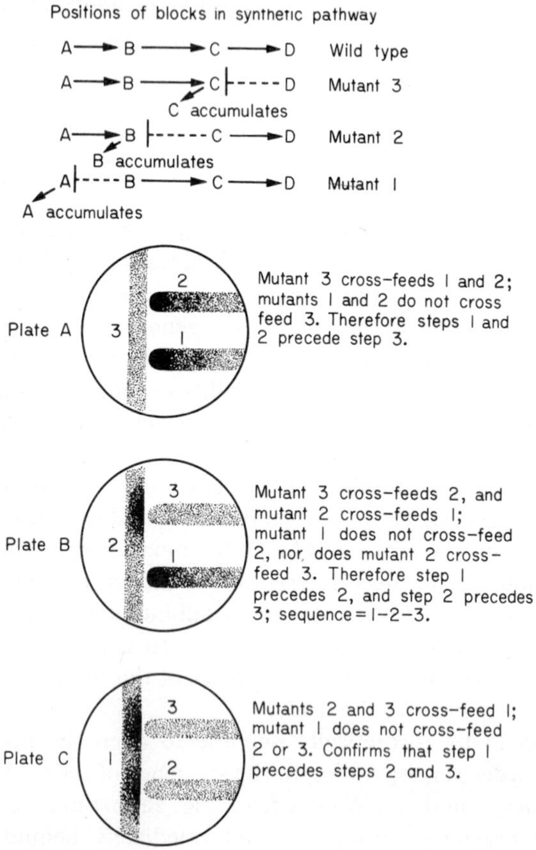

FIG. 23. Diagram illustrating the expected results of cross-feeding (syntrophism) tests between three mutant strains of a bacterial species, genetically blocked at different steps in the synthetic pathway of the required metabolite, D.

Washed cultures of the mutants are streaked as shown on plates of minimal agar, supplemented with limiting amounts of substance D.

The degree of shading indicates the luxuriance of growth (see text).

nourishment) (Fig. 23). The mutant strains are grown separately in a nutrient (complete) medium or in a synthetic minimal medium supplemented with an excess of the required end-product (D). After washing to remove excess nutrients, loopfuls of each mutant suspension are streaked on plates of minimal agar containing sufficient of the required end-product to permit slight growth. The streaks are arranged contiguously in such a way that the influence, on the growth of each strain, of metabolites diffusing from each of the other strains can be assessed, as shown in Fig. 23. The experimental conditions, such as the degree of supplementation of the minimal agar and the duration of incubation, must usually be rather precisely defined to ensure clearcut results. Over-supplementation or an extended period of incubation may lead to an obscure outcome from reciprocal cross-feeding; for example intermediate C, diffusing from mutant 3, will stimulate the growth of mutant 2 which may then produce sufficient surplus of the end-product, D, to allow further growth of mutant 3 so that, ultimately, the growth of the two mutants becomes indistinguishable. In those organisms which syntrophism tests are practicable, the functional sequence of different mutations can be found without having any knowledge of the chemical nature of the pathway involved.

3. If the end-product of the sequence (for example, substance D) is known and a number of mutants blocking its formation is available, the chemical nature of the intermediate steps may be elucidated. There are two possible approaches. The first involves divining what the intermediate metabolites are likely to be by means of 'inspired guesswork', and then testing whether these substances will, in fact, promote the growth of mutants blocked at earlier stages. For example, the chemist may conjecture that B and C are likely precursors in the synthesis of D. When the growth of a number of D-dependant mutants is tested in the presence of B and C, it is found that some respond to C only (mutant 2, Fig. 22), and others to the presence of either B or C (mutant 1); on the other hand, there is no response to analogues B′ or C′ which were considered as possible alternative intermediates. Thus the pathway B→C→D is confirmed. A second, corollary approach is the chemical analysis and identification of the intermediates which accumulate behind the genetic blocks and overflow into the culture medium. Paper chromatography is invaluable in studies of this sort which involve comparison of the products of growth of wild type and mutant strains in limiting media, and the isolation of unique mutant metabolites in as pure a form as possible.

4. When a metabolic pathway has been sketched in chemical terms by the various methods described above, it may turn out that a number of mutants that appeared to be blocked at the same step from the chemical point of view, nevertheless fall into more than one allelic group from the genetic standpoint. Put another way, mutations in, say, two genetically distinct genes are found to yield strains whose growth can be stimulated by substance C, but not by B, and which accumulate B in the medium (Fig. 22). The only interpretation, according to the theory, is that another hitherto unsuspected synthetic step intervenes between C and D in the pathway.

The hypothesis we have been discussing postulates that the function of genes is to determine the synthesis of all the enzymes which catalyse metabolic reactions, each gene being concerned with the synthesis of a single enzyme. This is the famous 'one gene–one enzyme hypothesis' of Beadle and Tatum whose work we shall shortly review. The theory is actually rather more comprehensive than its name implies. So far as we know all enzymes are proteins, but by no means all proteins are enzymes; moreover, many enzymes require coupling to a prosthetic group before they can manifest their activity. The theory allows for the fact that the synthesis of certain proteins which are not enzymes, such as haemoglobin, cytochrome *c*, bacterial flagella or components of the protein coat of bacteriophage, may be directly specified by genes, while the activity of enzymes which depend on a prosthetic group may require the operation of more than one gene. The one gene–one enzyme hypothesis is the focal point on which biochemistry and genetics converge. This convergence has illuminated not only the pathways of intermediary metabolism, but also two important aspects of theoretical genetics which have interacted to modify profoundly our conception of the gene. The first aspect concerns the ways in which genes may be analysed in terms of function rather than of inheritance; the second relates to the use of nutritionally deficient mutants in the analysis of genetic fine structure and this will be discussed in Chapter 7 and 8 (pp. 127, 158).

FUNCTIONAL ALLELISM AND COMPLEMENTATION TESTS

In the previous section we saw how one bacterial mutant may stimulate the growth of another different mutant lying adjacent to it, by supplying it with an essential metabolite which it could not synthesise for itself.

The example we took involved a rather special case of syntrophism, however, since each of the pair of mutants was blocked at a different step in the *same* pathway, imposing restrictions on the reciprocity of the process. But if two mutant strains having blocks in the synthesis of *different* required end-products are seeded together on minimal medium on which neither alone can multiply appreciably, it is usually found that mutual syntrophism occurs so that both mutants grow almost normally, provided a minute amount of nutrient is present, as a primer, to initiate metabolism. For example, two auxotrophic strains of *E. coli*, one requiring the amino acid arginine and the other leucine, will grow syntrophically in the absence of both requirements because the arginine-less strain will synthesise and excrete enough leucine, and the leucine-less strain enough arginine, into the medium to allow both to multiply more or less normally.

If mutual syntrophism can operate across the barriers which separate individual cells, the question arises whether it should not operate with greater efficiency and precision if the genomes of the two mutants are present together in the same cell. The answer is that it does. Diploid cells arising from the union of two mutant parental cells are phenotypically wild type *provided the mutations are in different functional loci.* In this case it doesn't matter whether the mutations block the synthesis of different growth factors, or different stages in the synthesis of the same growth factor; in either event the wild phenotype is equally

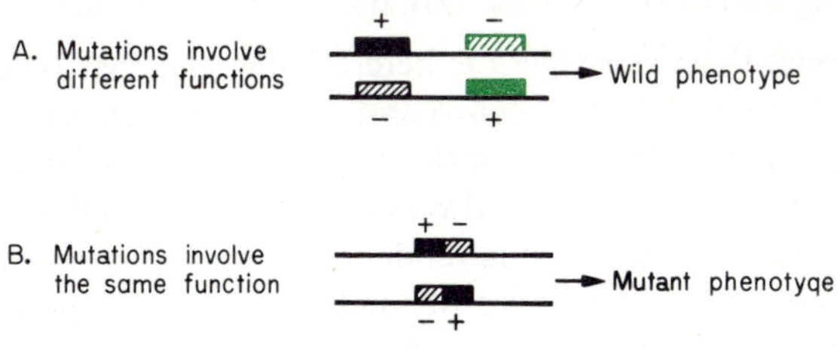

FIG. 24. Showing how complementation tests operate.

The black and green areas represent regions of the chromosome determining different functions (enzymes); the solid areas indicate wild type function, the hatched areas absence of function following mutation.

A. Two mutant genomes of complementary type are introduced into the same cell. The cell has one intact functional unit of each type, and can therefore synthesise both enzymes and display the wile phenotype.

B. Two mutant genomes having mutations in the same functional unit are introduced into the same cell. Neither can synthesise the enzyme, so that the cell displays the mutant phenotype.

expressed by the diploid. On the other hand, the diploid displays the mutant phenotype if the mutations in the two parents involve the same synthetic step, that is, if they are functionally identical (Fig. 24). We thus have a test of the functional similarity or dissimilarity of two genes, that is, of their *functional allelism*, which is independent of genetic recombination. This kind of test is called a *complementation test* since it tells us whether or not two mutant genomes are complementary with regard to each other's function. The situation has been likened (by R.D. Hotchkiss) to that of a man who wants to go somewhere and has two defective motor cars. If one of the cars has a flat tyre and the other a faulty carburettor, the man can fulfil his engagement by replacing either the flat tyre or the faulty carburettor with a good one from the other car; but if both cars have useless carburettors he can do nothing.

As in the case of syntrophism tests with bacteria (Fig. 23), complementation tests enable functional units to be defined genetically in the absence of any information about the chemical nature of the function. Complementation tests present little difficulty in diploid organisms with a sexual cycle, but these are just the organisms in which a genetic analysis of precisely defined biochemical characters is not usually feasible. On the other hand, in fungi or bacteria from which biochemical mutants can readily be isolated, either the vegetative individuals are haploid or a sexual cycle is lacking. Fortunately there are several ways of escape from this predicament.

HETEROKARYON FORMATION IN FUNGI

We have seen that *Neurospora* is heterothallic and that sexual reproduction will occur only between *A* and *a* strains which are of different mating type (p. 69). However, if the vegetative spores of two strains of the *same* mating type are planted very close together on the surface of a solid medium, there is a strong tendency for the emerging hyphae to fuse as they grow together. This fusion is accompanied by the transfer of nuclei from the hyphae of one strain to those of the other so that *heterokaryons* are formed whose hyphae contain nuclei of both strains. Under constant environmental conditions an equilibrium will be established between the proportion of both types of nucleus. Since the strains are of the same mating type there is normally no fusion of the nuclei to form a diploid nucleus (but see p. 78); instead, the nuclei continue their independent but balanced multiplication as the individual grows, until asexual spore formation segregates them again in the microconidia, each of which contains a single haploid nucleus of one or the other type (Fig. 25). Heterokaryon formation is very common among the *Fungi Imperfecti* which lack a sexual cycle (see Pontecorvo, 1946). Heterokaryons formed between two mutant strains having different functional defects grow like the wild type fungus, but if the

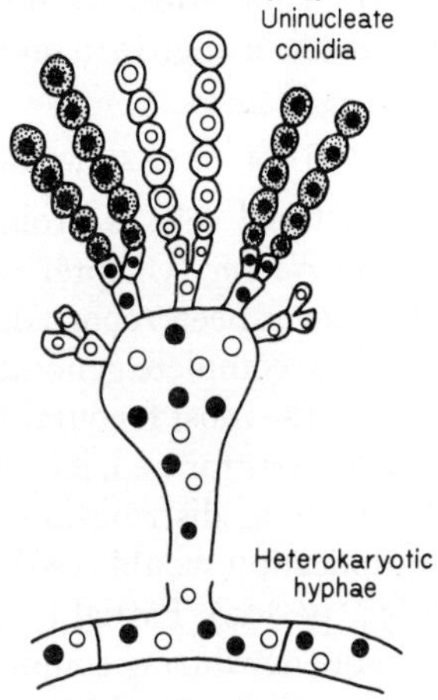

FIG. 25. Diagrammatic representation of the development of uninucleate conidia from a heterokaryon (*Aspergillus*).

In this case the pigmentation of the conidia is determined by the kind of nucleus which segregates in each conidium. (After Pontecorvo, 1946.)

mutants are functionally allelic the heterokaryon has the mutant phenotype (see Beadle and Coonradt, 1944). For example, if one mutant strain requires arginine and the other requires leucine, heterokaryons arising by hyphal fusion between the two strains develop normally on unsupplemented minimal medium on which neither parental strain will grow, since the two types of nuclei present in its hyphae are functionally complementary. Thus the development of wild type heterokaryons from functionally deficient mutant strains is proof that the mutations involve different functional loci. The converse is not necessarily true, however, unless it can be shown that heterokaryons have, in fact, arisen. This can be done by labelling the parental strains with a known pair of non-allelic mutations in addition to the pair under test. Suppose we have two tryptophanless *Neurospora* mutants of the same mating type which we wish to test for functional allelism. Attempts to isolate tryptophan-independent heterokaryons fail. If, now, a requirement for arginine is introduced into one strain and for leucine into the other (by crossing with known auxotrophs), we can isolate heterokaryons for the leucine and arginine loci by plating the strains together on minimal medium supplemented with tryptophan. These heterokaryons must also carry the pair of tryptophanless loci which, if non-allelic, will permit growth to occur on subculture to minimal agar without tryptophan; in such a case, absence of growth on minimal agar implies functional allelism.

COMPLEMENTATION TESTS IN BACTERIA

Although fusion between bacterial cells is probably quite common, anyway among the *Enterobacteriaceae* (Chapter 24), the formation of complete heterokaryons has never been reported. This may be partly due to the rarity with which the complete genome of one bacterium is transferred to another, even in the most favourable systems. However, even if complete heterokaryons *were* formed, it seems unlikely that they would be stable since the mode of distribution of nuclear bodies to daughter cells during multiplication would result in their segregation after a few division (Fig. 53, p. 202). Partial heterozygotes (merozygotes) are formed at high frequency during conjugation in *E. coli* K-12 but these are transient, segregating recombinants after a few generations, and are therefore not very suitable for complementation tests (but see p. 701). What is needed are homogeneous populations of bacteria, stably heterozygous for the region under study, whose phenotype can be examined in the usual way by plating on selective or differential media. Of the three types of complementation system now available in bacteria, two fulfil this criterion; however the one we will describe first, *abortive transduction*, is of a quite different and novel character.

Abortive transduction.

Transduction is a parasexual mechanism, especially prevalent among the *Enterobacteriaceae*, in which modified temperate bacteriophage particles act as vectors of small fragments of bacterial chromosome, transporting them from the cells of a donor strain on which the phage is grown to those of a recipient strain which the phage subsequently infects. It is characteristic of infection by temperate phages that a proportion of infected bacteria survive. When the two bacterial strains differ in genotype, donor genes transferred by the transducing phage particles may replace their alleles on the chromosome of the recipient bacteria, so that recombinant progeny arise. Because of the fragmentary nature of the donor contribution, recombinants are

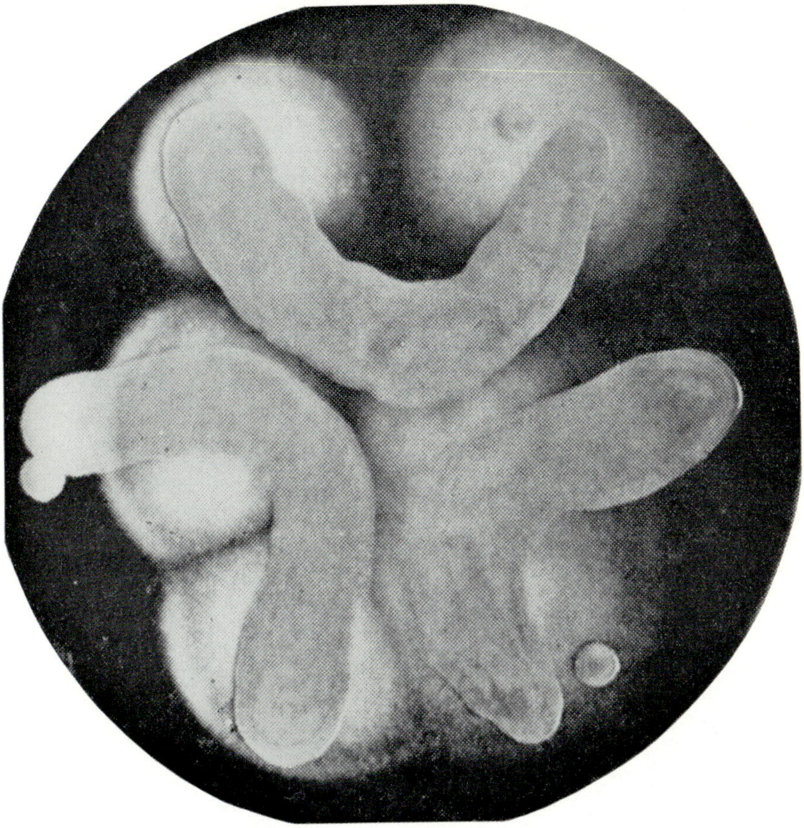

PLATE 4. Flares resulting from complete transduction of motility. Non-motile strain of *Salmonella typhimurium*, treated with phage lysate from a motile strain, is diluted 1 in 10 and spread on semi-solid gelatin-agar in the form of three U-shaped inocula. After primary incubation at 37° C., the visibility of swarms is enhanced by re-incubating at 23° C. at which growth continues but further spreading is prevented by solidification of the gelatin. The flares are seen as halos emanating from the dense growth of the inocula. Photographed by oblique, transmitted light. Magnification about × 1.

(Reproduced from Stocker, Zinder & Lederberg, 1953, by kind permission of the authors, and the editors of the *Journal of General Microbiology*)

facing p. 96

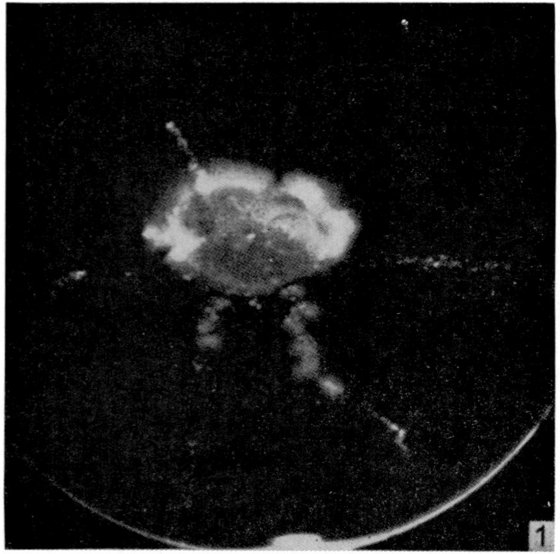

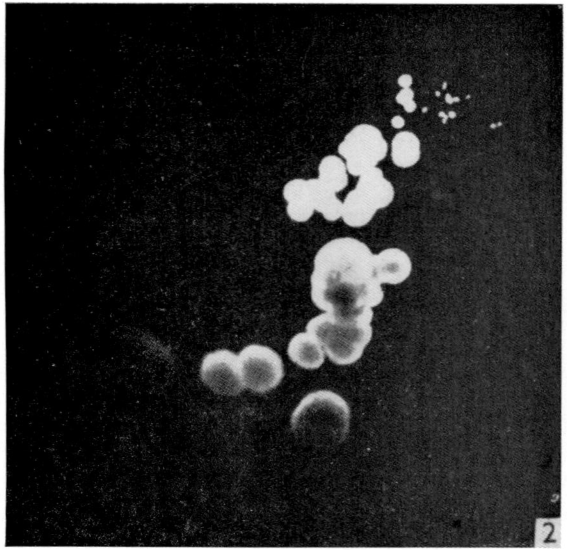

PLATE 5. Trails resulting from abortive transduction of motility. Non-motile strain of *Salmonella paratyphi* B is treated with phage lysate of motile *Salm. typhimurium* and spotted at the centre of a Petri dish of semi-solid gelatin-agar: incubated for 42 hours at 37°C. Trails of small, isolated colonies are seen emanating from the central growth of the inoculum.

1: magnification about ×1. 2: growing end of one of the trails from 1, magnified about ×8.

(Reproduced from Stocker, 1956, by kind permission of the author, and the editors of the *Journal of General Microbiology*)

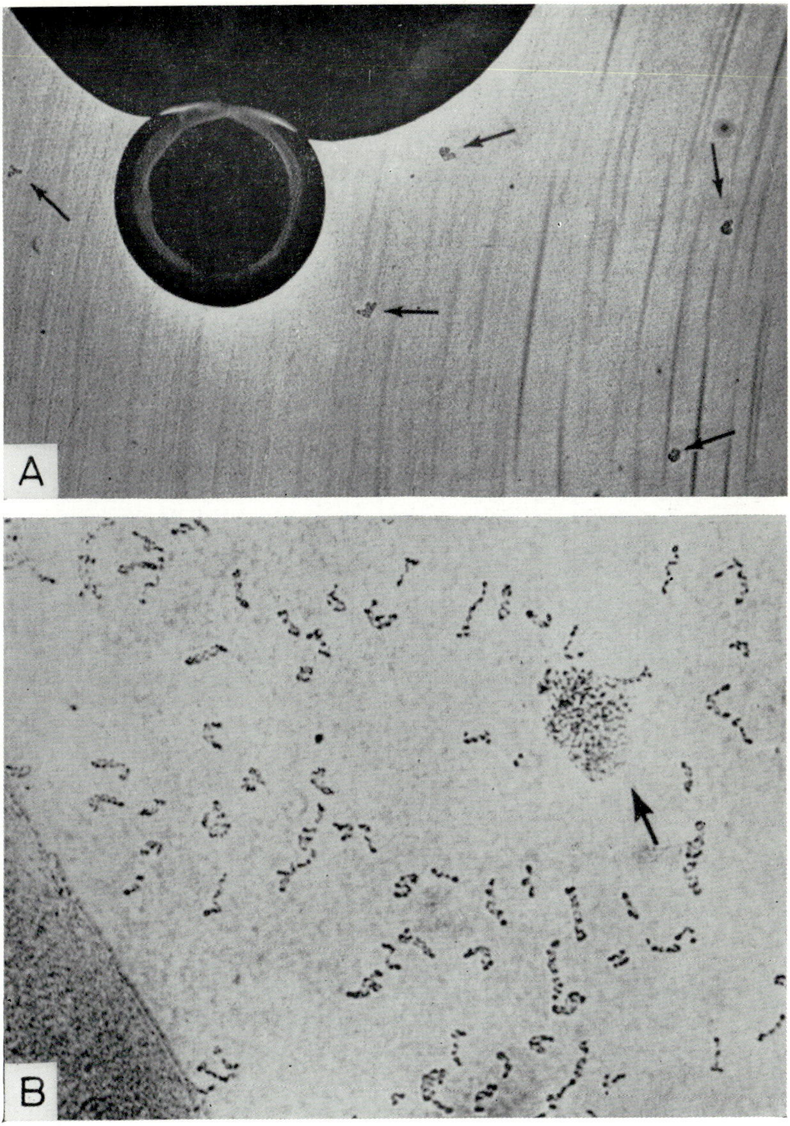

PLATE 6. The appearance of 'minute' colonies resulting from abortive transduction of a wild type nutritional marker (*Salmonella typhimurium*).

The 'minute' colonies, indicated by arrows, are microscopic in size and irregular in shape.

A: minimal agar surface after 72 hours at 37°C., enlarged about × 17.5. Part of a large colony derived from a complete transductant (wild type recombinant) is seen on the left.

B: minimal agar surface after 48 hours at 37°C., enlarged about × 175. The edge of a colony of wild type recombinant bacteria is seen at the

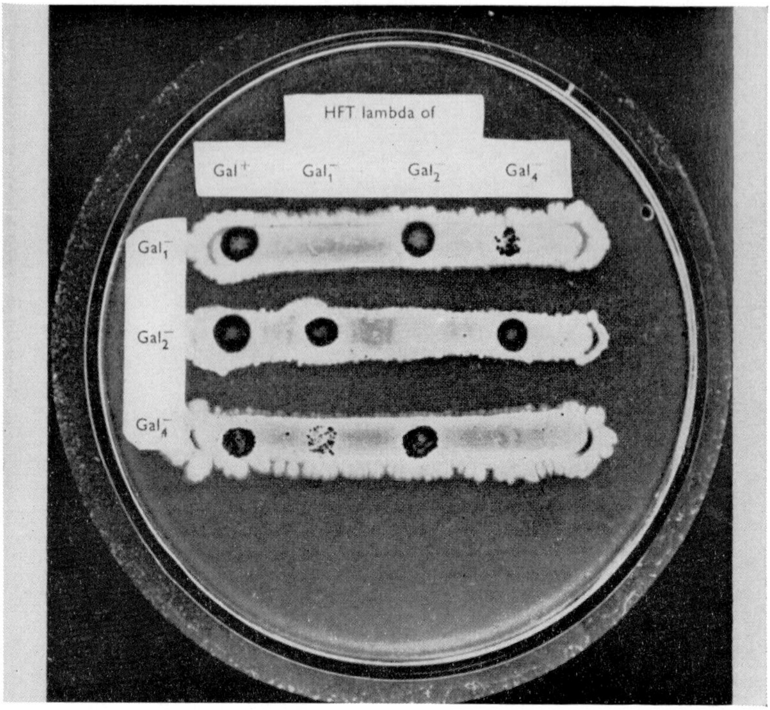

PLATE 7. 'Spot' tests of complementation between galactose mutants of *E. coli*, mediated by transductional heterogenotes.

The mutant recipient straints, indicated by the label at the left, are streaked on eosin-methylene blue medium containing galactose. An HFT preparation of phage λ, derived from wild type (*gal*+) bacteria and from each mutant (as indicated above), is spotted on the streak inocula and the plate is incubated. The areas of confluent pigmentation are due to the formation of lactose-fermenting heterogenotes and reveal complementation between mutants 1 and 2, and 2 and 4. In contrast, heterogenotes carrying gal_1^- and gal_4^- mutations show only flecks of pigmentation, due to rare *gal*+ bacteria arising from recombination between the two mutations which are in the same gene. The occasional flecks of pigmentation located outside the phage-infected zones are due to spontaneous mutational reversion in the recipient populations.

(Reproduced from Morse, Lederberg & Lederberg, 1956b, by kind permission of the authors and the editor of *Genetics*)

(PLATE 6 cont.) bottom-right corner. The 'minute' colony arising from abortive transduction is contrasted with the background of untransduced recipient bacteria which have divided not more than a few times on the minimal agar and appear as beaded chains.

(Reproduced from Hartman, Hartman & Šerman, 1960, by kind permission of the authors, and the editors of the *Journal of General Microbiology*) basically of recipient genotype, inheriting only single genes, or clusters of very closely linked genes, from the donor strain. Motility is one of many characters which are transducible in *Salmonella* by the phage P22. This transducing phage is grown on a motile donor strain; the liberated progeny phage particles are freed from residual bacteria and used to infect non-motile (usually non-flagellated) recipient bacteria.

Motile transductants are isolated by plating the infected recipient population in soft gelatin-agar in which the growth of non-motile cells remains circumscribed, while motile cells migrate outwards as they multiply to give an expanding 'flare' of growth (Plate 4). Stocker, Zinder and Lederberg (1953), who devised this technique, noticed that transduction plates showed, in addition to the flares, a number of linear 'trails' of isolated colonies leading out from the region of circumscribed growth (Plate 5). These trails were explained by supposing that the motility gene neither participates in recombination nor multiplies in a proportion of the cells receiving it, but is dominant and capable of determining flagellar synthesis. When such cells divide, only one daughter inherits the gene and is able to swim away from the parent culture. At the next division likewise, only one of the two progeny of this motile cell is motile and continues its outward progress; the other non-motile cell remains behind and produces a colony. This process of *uni-linear inheritance* of the transferred gene, termed *abortive transduction*, continues, leaving a trail of colonies of non-motile cells behind it, until the gene is lost for unknown reasons. This sequence of events is shown in Fig. 26A. In practice, a trail is never found to terminate in a flare, implying that if the gene fails to be integrated initially, the opportunity does not recur. This neat explanation of abortive transduction and its mechanism was confirmed in principle by single cell studies of transduced populations (Stocker, 1956; Lederberg, 1956a).

The same type of phenomenon was also observed by Ozeki (1956) in connection with the transduction of nutritional characters in *Salmonella*. He found that when an auxotrophic recipient strain, with a single growth requirement, is transduced by phage grown either on a wild type donor strain or on an auxotrophic strain having a different requirement, a considerable number of 'minute', slowly growing colonies arise on unsupplemented minimal medium, in addition to the large colonies of prototrophic recombinant cells (see Plate 6). Now if prototrophic colonies on a minimal agar plate are re-streaked by rubbing the surface of the agar with a spreader, every cell in the colony,

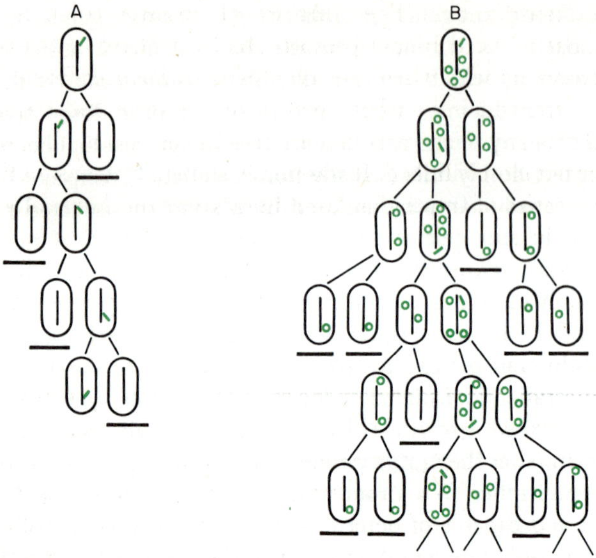

FIG. 26. Diagram showing the effect of linear inheritance of a functional unit of donor chromosome in abortive transduction.

The chromosome of the recipient parent is represented by a vertical, black line; the transduced fragment of donor chromosome by a short, green line.

Underlined cells are non-motile or, in the case of abortively transduced nutritional characters, are unable to divide.

A. Only daughters which inherit the abortively transduced donor fragment are shown as capable of motility (or division).

B. The abortively transduced fragment promotes synthesis of 'enzyme-forming systems' (green circles) which are distributed randomly among the progeny. Cells inheriting more than one 'enzyme-forming system' are motile (or can divide) whether or not the gene is also inherited (see text).

being nutritionally independent, will multiply to yield a colony, so that confluent growth results. When minute colonies are re-streaked in this way, however. although the bacteria contained in them are redistributed, the number of resulting minute colonies remains the same as before. The implication is that each minute colony contains only one cell which can generate another minute colony. This was substantiated experimentally by the analysis of individual colonies. In other words, the situation is analogous to the uni-linear inheritance of motility genes and provides another more general example of abortive transduction. The number of minute colonies usually exceeds that of recombinant prototrophs by a factor of ten or more. They always appear when the two parents have different growth requirements, and even when they have the same end-requirement provided this requirement is determined, in the two strains, by genes which are not closely linked. It was finally shown, by transduction tests between a series of tryptophanless mutants with biochemically defined functional defects, that minute colonies fail to arise only in those cases where the two parents suffer the same functional defect. Thus abortive transduction, as revealed by the development of minute colonies, offers a reliable complementation test although, so far, it has been demonstrated extensively only in *Salmonella* transduction.

The question arises as to how the minute colonies are formed if only a single cell in the entire colony possesses the complementing gene. In the case of a prototrophic bacterium the number of progeny increases exponentially at each division according to the series 2, 4, 8, 16, 32, 64 .... Assuming a negligible death rate, more than a million cells will have been produced by the end of the twentieth generation. In abortive transduction, if only that single cell which inherits the complementing gene at each generation is able to divide, the numbers of progeny will increase arithmetically according to the series 2, 3, 4, 5, 6, 7 ... (Fig. 26A), yielding only twenty cells at the end of twenty generations. Since minute colonies may contain as many as ten million cells after three days growth, this is clearly not what happens. It is necessary to suppose that the presence of the complementing gene in a cell stimulates the synthesis of a considerable number of non-replicating enzyme molecules, or 'enzyme forming systems', in excess of its requirements for one division; when the cell divides, sufficient of these particles will be distributed to the daughter which does *not* inherit the gene to enable it to continue to multiply, until further successive divisions dilute the number of particles below the operational threshold (Fig. 26B). Pedigree analysis of the abortive transduction of motility has afforded good evidence for this kind of inheritance (Stocker, 1956). (For further information on abortive transduction, see Chapter 21, p. 641).

Transductional heterogenotes.

Bacteria which survive infection by temperate phage usually become *lysogenic*. In lysogenic bacteria the chromosome of the phage, called *prophage*, is inserted into the bacterial chromosome and is inherited among progeny cells as if it were a natural part of it. This relationship, however, occasionally breaks down, with resulting development of the phage, so that the cell bursts and liberates mature virus particles. In the case of lysogenisation by what are called *inducible* phages, treatment of the lysogenic bacteria with small doses of ultraviolet (UV) light is followed, after an interval, by the bursting of virtually all the cells and the liberation from each of many mature, infective phage particles (see Chapter 17, p. 451).

The majority of temperate phages have fixed prophage locations on the bacterial chromosome which sometimes happen to be closely linked to known bacterial genes. For example, *E. coli* strain K-12 is normally lysogenic for an inducible phage called *lambda* (λ), the insertion site of the prophage being adjacent to a cluster of genes determining the fermentation of galactose (*gal*) (E. M. Lederberg and Lederberg, 1953; Wollman, 1953). Similarly, in this *E. coli* strain, the prophage of the inducible phage 80 is located very close to a group of genes mediating the synthesis of tryptophan (*try*). But let us discuss the case of phage λ since this was the first system to be discovered and is that most thoroughly investigated.

If a galactose-fermenting (gal^+) culture of *E. coli* K-12, lysogenic for phage λ, is induced by UV light, a very small proportion (10^{-6}) of the emerging phage particles carry the bacterial gal^+ gene, so that subsequent infection of a gal^- recipient strain results in the appearance of a small number of gal^+ transductant clones, the cells of which are also lysogenic. Unlike transduction of *Salmonella* by phage P22, however, only the *gal* region of the chromosome is transduced by λ phage. The gal^+ transductants show two unusual features. In the first place they are unstable and throw off gal^- segregants, so that they must still retain their original gal^- gene; this means that they are really diploid for the *gal* region (gal^-/gal^+). Secondly when they, in turn, are treated with UV light, as high as 50 per cent of the liberated phage particles can now transduce the gal^+ gene. Such a phage preparation is called HFT (for 'high frequency transduction') (Morse, Lederberg and Lederberg, 1956a, b).

The explanation is that, in the initial lysogenic population (Fig. 27A), a rare (10^{-6}) recombinational event occurs whereby a part of the bacterial chromosome, adjacent to the site of prophage insertion and including the gal^+ genes, becomes incorporated into the prophage chromosome (Fig. 27B,C,D). When the resulting transducing phage, liberated by induction, infects and lysogenises a gal^- recipient bacterium, the chromosome of the bacterium now carries not only the inserted prophage but also its associated gal^+ gene in addition to its own gal^- allele (Fig. 27E,F); that is, the bacterium is more or less stably diploid for the *gal* region. Since the gal^+ allele is dominant, the bacterium acquires the ability to ferment galactose. The prophage-bacterium complex, diploid for the *gal* region, is known as a *heterogenote* (Morse *et al.*, 1956b). Multiplication of a heterogenote will, of course, yield a population of heterogenotic cells, each of which carries a prophage bearing the gal^+ gene in its chromosome. Induction of this population, unlike that of the original culture, will therefore produce a very high yield of transducing phage.

In populations of $gal^-/\lambda\text{-}gal^+$ heterogenotes, not only do occasional cells lose their prophage to yield non-lysogenic, haploid gal^- segregants, but recombination also occurs between the homologous *gal* regions. The result is that a proportion of gal^- segregants are found to be still lysogenic and homozygous diploid for the gal^- allele. They are therefore *homogenotes*, $gal^-/\lambda\text{-}gal^-$, the prophage now carrying the gal^- allele of the recipient strain (p. 635). When phage from such a homogenotic gal^- culture is used to infect an independently isolated gal^- mutant strain of *E. coli*, new heterogenotes of the general type $gal_2^-/\lambda\text{-}gal_1^-$ are formed at high frequency. The ability of these heterogenotes to ferment galactose depends on whether the two mutations, gal_1^- and gal_2^-, are in the same or different functional units. Once a preparation of HFT phage carrying a particular gal^- allele has been made, the complementation test is easy to perform. The phage suspension is simply 'spotted' on to streak inocula of other gal^- bacterial strains on eosin-methylene blue (EMB) agar containing galactose. On this medium the growth of galactose-fermenting bacteria is darkly pigmented. Complementation is revealed, after incubation, by confluent pigmentation of the spotted area, from the formation and growth of large numbers of heterogenotes which can ferment the galactose (Plate 7). Occasional flecks of pigment may be found although the phage and recipient bacteria are known to carry non-complementing mutations; these are due to rare gal^+ *recombinants* arising from recombination between the two gal^- mutations (see E. M. Lederberg, 1960). This method suffers from the severe restriction that it can be applied only to those limited regions of chromosome which happen to lie adjacent to the sites of insertion of those rare temperate viruses which can mediate this type of transduction. The formation and structure of transductional heterogenotes will be considered in more detail in Chapter 21, p. 627 *et seq.*).

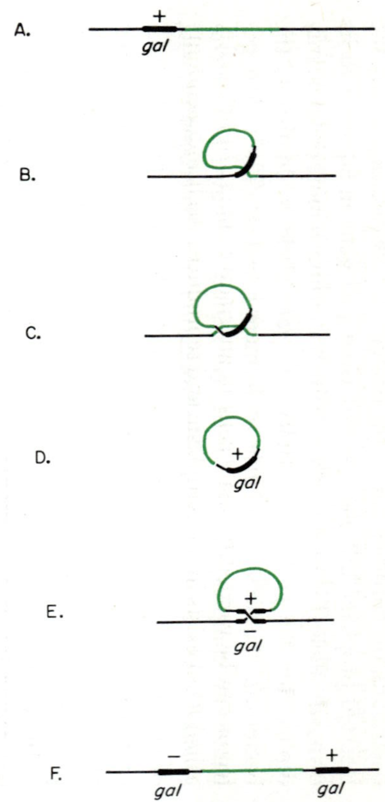

FIG. 27. Diagrammatic representation of the acquisition by λ phage of transducing ability for the *gal* region of the bacterial chromosome, and the formation of heterogenotes which are diploid for the *gal* region.

The black lines indicate the bacterial chromosome, the thickened segment representing the *gal* region. The green lines indicate the λ phage chromosome.

The evidence is that the chromosome of λ phage becomes circular in the cytoplasm of the host, and is inserted into the continuity of the bacterial chromosome by a recombination event and, on induction, is released again into the cytoplasm by a reversal of this sequence. The incorporation into the phage chromosome of an adjacent fragment of bacterial chromosome is due to its release by a rare recombination event between regions of phage and bacterial chromosome other than those which led to insertion, as shown.

A. The relationship of the *gal* region and the λ phage chromosome in gal^+ lysogenic bacteria.

B. Pairing of part of the phage chromosome with a region of bacterial chromosome distal to the *gal* region.

C. Release by recombination of the circular phage chromosome carrying the bacterial *gal* region (λ-gal^+) (D).

E and F. Infection of a gal^- recipient bacterium with λ-*gal* transducing phage. The phage chromosome becomes inserted into the recipient chromosome by a recombination event between allelic regions (not necessarily *gal* regions in the case of λ-*gal*). In this way, both *gal* alleles of the diploid heterogenote are present in tandem in the same chromosome.

Heterogenotes mediated by the sex factor of E. coli

In *E. coli* bacteria the character of maleness, which denotes their ability to conjugate with females, is conferred by the presence of a *sex factor* called *F* (for 'fertility'). Female bacteria lack this sex factor. The sex factor can be carried by male cells in alternative states. In one state, called F^+, it exists in the cytoplasm and reproduces itself independently of the bacterial chromosome; when F^+ bacteria conjugate with females they transfer their sex factors with high efficiency, so that the females are converted to F^+ males, but they are unable to transfer their chromosomes. The other state, called *Hfr* (for 'high frequency recombination'), arises with a low probability in F^+ populations; the sex factor leaves the cytoplasm and becomes more or less stably inserted, like prophage, at *one of many possible sites* on the bacterial chromosome. The progeny of a cell in which this $F^+ \rightarrow Hfr$ transition has occurred are able to transfer their chromosomes with high efficiency when they conjugate with females. For any given *Hfr* strain, a particular chromosomal locus, close to the site of sex factor insertion, is always the first to enter the female bacteria, to be followed by the other loci in the same sequence as their arrangement on the linkage map (p. 65; Fig. 15); the last genetic determinant to be transferred is the sex factor, *F*, itself.

We will review, later on, the intriguing problem of how this change in the state of the sex factor so dramatically affects the sexual behaviour of the male cell (Chapter 24, pp. 798–806). For the moment we will consider the sex factor only from the point of view of complementation tests. The sex factor of an *Hfr* strain tends, with rather a low probability, to leave the chromosome and revert to its cytoplasmic, F^+ state, probably as a result of another recombination event similar to that responsible for its insertion. Occasionally, however, the sex factor is released by a 'non-allelic' genetic exchange which incorporates a neighbouring region of the bacterial chromosome into its structure, just as in the case of transducing phage illustrated in Fig. 27. Unlike phage, however, the sex factor is a non-pathogenic agent whose replication in its cytoplasmic state is coordinated with that of the cell. After release, it continues to promote conjugation and, together with its incorporated fragment of bacterial chromosome, is transferred with high efficiency to female bacteria which propagate it just like a normal sex factor. The female bacteria and their descendants thus become diploid for any bacterial genes carried by the sex factor, as well as being converted into males. However, because the sex factor now carries a considerable region of good homology for the bacterial chromosome, pairing and recombination occur frequently between them so that the sex factor alternates rapidly between its chromosomal and cytoplasmic states (p. 794).

Sex factors which incorporate fragments of bacterial chromosome into their structure in this way are called *F-prime factors*, while the process of genetic transfer which they mediate has been given the somewhat dissonant name of *sexduction*. The heterogenotes they form behave genetically in precisely the same manner as transductional heterogenotes, so that recessive homogenotes and F-prime factors carrying recessive mutant genes can readily be obtained from them for complementation tests (for comparison of both types of heterogenotes for the *gal* region see E. M. Lederberg, 1960). The use of F-prime factors is more flexible and has a wider application than the transductional method, since not only may much longer lengths of chromosome be incorporated into the sex factor than into the phage chromosome, but F-prime factors incorporating many different regions of chromosome may be isolated (p. 677) from *Hfr* strains having different sites of sex factor insertion (Adelberg and Burns, 1959, 1960; Jacob, Schaeffer and Wollman, 1960; Adelberg and Pittard, 1965; see Scaife, 1967). Complementation tests and tests of dominance employing F-prime factors incorporating loci concerned with the fermentation of lactose (F-*lac* factors) have already contributed uniquely to our knowledge of the genetic regulation of biochemical pathways in bacteria (Jacob and Monod, 1961a,b; see Chapter 23, p. 712).

COMPLEMENTATION TESTS IN BACTERIOPHAGE SYSTEMS

In bacteriophage infection, the phage particles attach themselves to the bacterial cell wall by means of the tails of their protein coats (Plate 14, p. 416), and then inject their genetic material into the cell. In the case of virulent phage, the injected genetic material behaves like a 'master gene' which takes over control of the synthetic machinery of the cell, redirecting it to the manufacture of new phage protein and genetic material. Bacteriophage genetics is studied by infecting cells with particles of two (or more) genetic types of phage, and is complicated by the fact that replication and mating of the genetic material proceed together within the infected cell; there may be several rounds of mating (p. 62).

Although these unique features of phage crosses complicate the interpretation of genetic analyses, they do not affect the outcome of complementation tests between phage mutants. If the mutational defects in two types of phage particle involve different functional units, both of which are required for phage production by the infected bacterium, mixed infection will be followed by complementation so that both parental types can grow. When the cell lyses, both parental types of progeny, as well as recombinant particles, will be liberated. On the other hand, if the parental particles are mutant in the same function, no phage reproduction occurs and no progeny will be liberated.

Two types of complementation test have been described. The first is applicable to any type of mutation in phage which permits infection of the bacterial host but blocks some subsequent stage in phage development. As an example we will take the well known case of what are called r_{II} mutants of the virulent phage T4 (Benzer, 1955, 1957; see p. 169). The wild type phage, r^+, can infect and lyse two strains of *E. coli* called B and K (short for K-12). When grown on B it produces r_{II} mutants at quite high frequency. These are recognised in the first instance by the distinctive appearance of their plaques on strain B, but their most striking character is complete inability to grow and form plaques when plated on K bacteria; the bacteria are infected and killed, but do not lyse and liberate phage. When strain K is mixedly infected with pairs of independently isolated r_{II} mutants, however, it turns out that some pairs complement one another and produce plaques which contain the two mutant types of particle, while other pairs do not (Table 2). By examining pairwise combinations of a large number of r_{II} mutants in this way, Benzer (1957) found that they fell into one or the other of two functional groups, A and B. Any mutant belonging to group A is able to complement any group B mutant, while no pair of A mutants, or of B mutants, shows complementation. In Chapter 8 (p. 167) we shall see how the genetic mapping of these two functional groups of r_{II} mutants led to important new concepts about the relationship between genetic structure and function.

The second kind of complementation test concerns mutations which do not interfere with the intracellular multiplication of phage, but alter the specificity of the protein of the phage tail by which it attaches itself to its host. The test, devised by Streisinger (1956b; see also Brenner, 1957b), is based on a phenomenon known as *phenotypic mixing* (p. 511). If a bacterium is simultaneously infected with two closely related types

TABLE 2

The ability of wild-type (r^+) and r_{II} mutant strains of phage T4 to multiply in and lyse the cells of two strains of *E. coli*, B and K, following single and mixed infections.

Mutants r_{IIa} and r_{IIb} are not complementary and belong to the same functional group. Mutant r_{IIc} complements both r_{IIa} and r_{IIb} and so belongs to the other functional group.

Infection with	Ability to produce plaques on *E. coli* strain:	
	B	K
Wild-type (r^+) alone	+	+
Mutant r_{IIa} alone	+	—
Mutant r_{IIb} alone	+	—
Mutant r_{IIc} alone	+	—
Mutants $r_{IIa} + r_{IIb}$	...	—
Mutants $r_{IIa} + r_{IIc}$	...	+
Mutants $r_{IIb} + r_{IIc}$	...	+

of phage particle, P and Q, which differ in the specificity of their tail proteins, the cell synthesises both P and Q proteins as well as P and Q genomes. When the time comes for the proteins and genomes to be assembled into mature phage particles, however, it seems that the genomes have no cloakroom tickets to enable them to identify their own protein coats with the result that genomes and coats become randomly associated to yield progeny of constitution (P genome+Q protein) and (Q genome+P protein), as well as the parental types of phage. Such 'hybrid' particles can be recognised because those of constitution (P genome+Q protein), for example, are infective for bacteria to which only Q tail protein can adsorb. However, following single infection, the injected P genome can direct the synthesis of P protein only so that the 'second generation' progeny particles are no longer 'hybrid' but of pure (P genome+P protein) type. Therefore they are unable to infect the type of bacteria that produced them but have acquired infectivity for another bacterial host to which only P tail protein can adsorb.

Now let us return to the complementation test. From a wild type phage, characterised by infectivity for two bacterial hosts, A and B, a number of independent *host range mutants* are isolated which, as a result of alteration of their tail proteins, have lost the ability to attach to A bacteria while retaining infectivity for B. If the genetic defects in any pairs of these mutants involve different functional units, mixed infection of host B with such pairs should result in complementation and synthesis of wild type tail protein. Phenotypic mixing should therefore ensure that a significant proportion of the genetically mutant progeny should be fitted with wild type tail protein, and so display the wild phenotype by being able to infect and lyse A bacteria. Alternatively, the absence of such phenotypically wild type particles among the progeny of the mixed infection would imply that wild type protein had not been made and that the genetic defects were in the same functional unit. This latter result was what Streisinger (1956b) actually found. In order to demonstrate the principle of this type of complementation test, however, we have somewhat over-simplified and misrepresented the method used by him, which involves complexities we do not wish to discuss at this stage (but see pp. 482, 483, 511).

When we come to study the fine structure of the genetic material in Chapter 7, we shall see that one of the most fundamental discoveries to issue from complementation tests, not only in bacteriophage, bacteria and fungi but also in *Drosophila*, is that a single functional region of chromosome is divisible into many different mutational sites between which recombination can occur. This means that the genetic units of function and of recombination are not the same; the functional unit of the genetic material is linearly extended on the chromosome and can be subdivided further into smaller units of recombination and mutation. This elaboration of our conception of the gene has obvious semantic repercussions since words such as 'allele' and, indeed, 'gene' itself must either be abandoned or else be more precisely defined if they are to have real meaning.

CHAPTER 6

Biochemical Mutations in Various Genetic Systems

Nowadays it is widely thought that our modern notions of biochemical genetics, which have influenced our ideas on many important aspects of metabolic disorders in man, originated with the work of G. W. Beadle and E. L. Tatum and their colleagues (Beadle and Tatum, 1941; reviews, Beadle, 1945a, b; Catcheside, 1951) on nutritional mutants in *Neurospora*, and have since been developed extensively only in this and other fungi, and in bacteria. In fact there were a number of prior instances in other systems where genetic changes were found to be correlated with specific alterations in biochemical function.

HEREDITARY METABOLIC DISORDERS IN MAN

The earliest clear formulation of the concepts of biochemical genetics was made by a British physician, Sir Archibald Garrod, in 1908 (quoted by Yi-Yung Hsia, 1959), as a result of his studies on certain hereditary errors of human metabolism (Garrod, 1909, 1923).* Among the best known of these metabolic disorders is the rare defect known as alkaptonuria which is found in persons who are homozygous for a single recessive gene. The condition is due to a block in the degradation pathway of tyrosine and phenylalanine with the result that 2,5-dihydroxyphenylacetic (homogentisic) acid can no longer be broken down to maleyl-acetoacetate and ultimately to fumarate and aceto-acetate (Fig. 28, III). Homogentisic acid accumulates in the blood and is excreted in the urine where it turns black on exposure to air; when normal persons are administered homogentisic acid they do not excrete it, whereas alkaptonurics excrete it quantitatively. In 1914 it was shown that the blood of alkaptonurics was deficient in a specific

* For an appreciation of Garrod and his work, see Knox (1967).

enzyme, homogentisic oxidase, possessed by normal people, so that the story of this unusual condition was neatly concluded. In addition to alkaptonuria, three other genetic defects involving specific blocks in the tyrosine-phenylalanine pathway are now known and are summarised in Fig. 28. Phenylketonuria is important because it is a

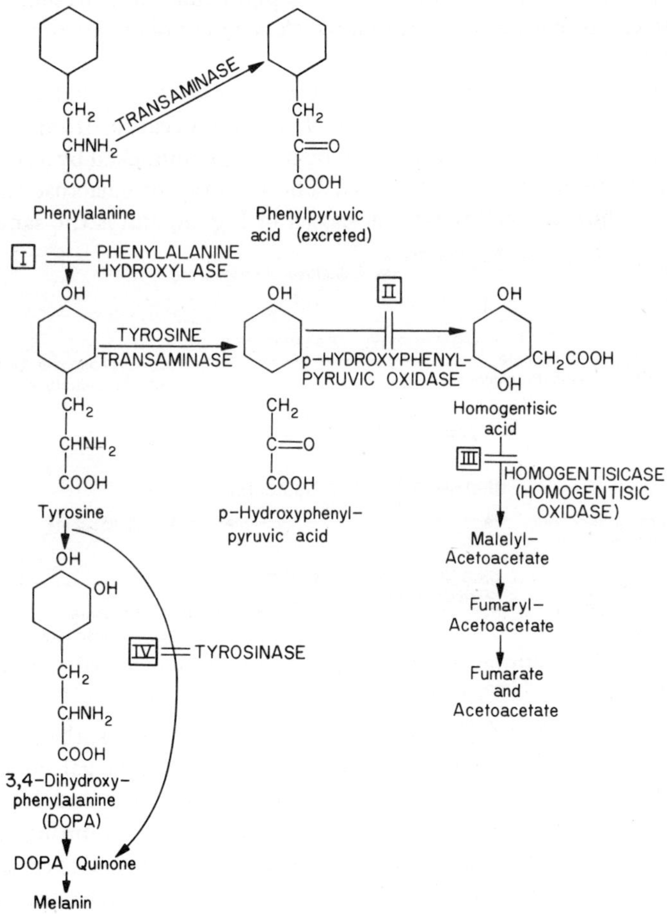

FIG. 28. Normal pathways for the degradation of phenylalanine and tyrosine in man.

The boxed, roman numerals indicate the positions of metabolic blocks in the following conditions:

I. Phenylketonuria.
II. Tyrosinosis.
III. Alkaptonuria.
IV. Albinism.

(Adapted from Yi-Yung Hsia, 1959.)

significant cause of mental retardation in children. The genetic block (Fig. 28, I) leads to excess of L-phenylalanine in the blood and spinal fluid; this is changed by the enzyme phenylalanine transaminase to phenylpyruvic acid which is excreted in the urine. The excess phenylalanine partially represses the normal metabolism of tyrosine to the black pigment melanin, so that reduced pigmentation is present. The disease can be successfully treated by exclusion of phenylalanine from the diet.

Another rare, hereditary metabolic disease of man called galactosaemia, first described in 1908, is of interest because, from a biochemical point of view, it appears to be precisely mimicked by a similar condition in the bacterium *E. coli*. The pathway of galactose metabolism in human erythrocytes is shown in Fig. 29 and is the same as

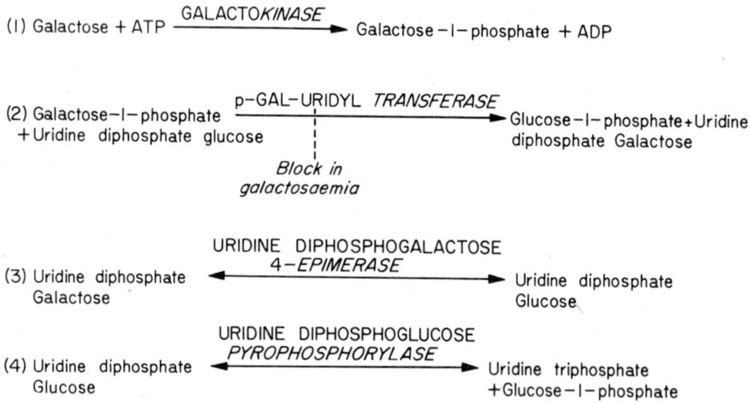

FIG. 29. The pathway of galactose metabolism in human erythrocytes. (From Yi-Yung Hsia, 1959.)

that utilised by *E. coli* (Kalckar, 1957). In galactosaemics, the genetic defect involves the transferase enzyme (step 2) so that galactose-1-phosphate accumulates (Schwarz, Golberg, Komrower and Holzel, 1956); ingestion of galactose leads to malnutrition, vomiting, and mental retardation if the infant survives, but the symptoms regress if milk is excluded from the diet. Mutants of *E. coli* have been isolated in which each of the first three steps is blocked (E. M. Lederberg, 1960). Those mutants which cannot synthesise transferase (step 2) accumulate galactose-1-phosphate as in man and in them, but not in the case of wild type or kinase (step 1) mutants, the addition of as little as 10 μg./ml. galactose to the synthetic medium depresses the growth rate (Kurahashi and Wahbe, 1958; Shapiro, 1967).

MUTATIONS IN METABOLIC PATHWAYS IN DROSOPHILA

Among the many other examples which are known of the influence of mutations on biochemical processes we shall recount that of eye pigmentation in *Drosophila* (Beadle and Ephrussi, 1936) since it demonstrates, in a totally different system, the same kind of syntrophism patterns shown by bacteria blocked at different stages of a synthetic pathway (Fig. 23, p. 90), and also because it led more or less directly to Beadle's *Neurospora* studies. The compound eyes of *Drosophila* and other dipterous (two-winged) insects develop from a bud or disc formed during the early development of the larval stage. The eye disc can be surgically removed at this stage and transplanted to the body cavity of another larva in which it continues its development. When the host larva undergoes metamorphosis and becomes an adult fly, the implanted parasitic eye can be dissected out, or viewed through the abdominal cuticle, and its colour observed. Many mutations are known in which the wild type eye colour is altered from brown to white, pink, peach, carmine, vermillion, cinnabar and so on. It was found that most of these mutant eye discs preserve the colour determined by their own genotype when transplanted to wild type larvae, and *vice versa*. In the case of two mutants, vermillion (*v*) and cinnabar (*cn*), however, the mutant discs develop into wild type eyes when grown in the wild type larvae. The body fluids of the wild type larva therefore possess some substance or substances which the mutants are unable to synthesise. When *v* discs are transplanted to *cn* larvae they develop wild type pigmentation so that *cn* larvae produce something which enables *v* discs, but not *cn* discs, to develop normally. The converse

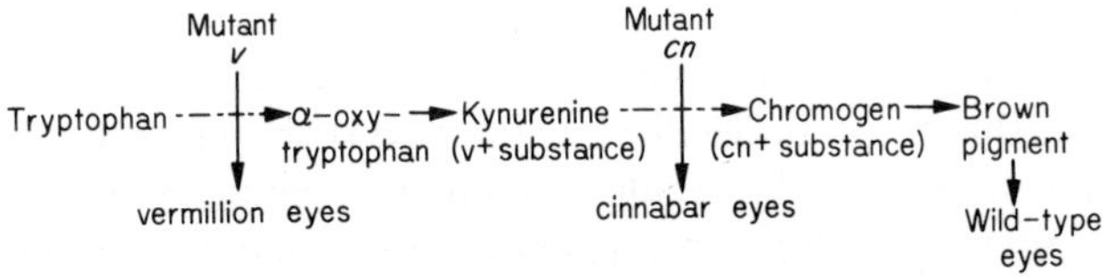

FIG. 30. Steps in the synthesis of brown eye pigment in *Drosophila* which are blocked in vermillion and cinnabar eye-colour mutants. The various shades of red eye colour shown by these mutants is due to a second, competitive chain of reactions (not shown here), leading to formation of a red pigment, which becomes dominant when chromogen is not made.

does not hold, however, since *cn* discs retain their mutant phenotype in *v* larvae. The inference is that two substances, v^+ and cn^+, are consecutively produced in the synthesis of wild type pigment, wild type larvae being able to synthesise both and *v* mutants neither, while *cn* mutants can make cn^+ but not v^+ substance. The v^+ substance was later identified with kynurenine, the *v* mutation blocking the formation of its precursor, α-oxytryptophan, from tryptophan; the cn^+ substance is a chromogen precursor of wild type pigment produced from kynurenine and it is this step that the *cn* mutation blocks (Fig. 30).

BIOCHEMICAL MUTATIONS IN NEUROSPORA

The biochemical mutations which we have so far reviewed were those that had presented themselves by chance to several astute observers. 'Fortune favours only the prepared mind', however, and already the idea had taken root that there exists a one-to-one relation between genes and specific chemical reactions. Beadle and Tatum (1941; Beadle, 1945a, b) decided to approach this general problem from the other end by attempting deliberately to induce mutations to inability to synthesise specific chemical substances. They selected *Neurospora* as the best organism to use on account of the suitability of its genetic system (p. 68), but especially because this fungus can grow on a defined minimal medium, synthesising all its amino acids, vitamins (except biotin) and nucleic acid bases from sugar, an inorganic source of nitrogen, some salts and 'trace elements'. Thus biochemical mutants can be recognised by their inability to grow in minimal medium, and the defective pathways identified by finding a particular amino acid, vitamin or nucleotide which, on adding to the medium, remedies the defect.

Asexual spores (conidia) were irradiated with X-rays or ultraviolet light and then used to fertilise a wild type strain of opposite mating type. From each of the resulting ascus sacs single ascospores were taken (to avoid picking the same mutant twice) and grown individually in small tubes of nutrient medium containing as complete a supplement as possible of amino acids, vitamins and other growth factors. This seemingly complicated procedure of imposing meiosis on the irradiated nuclei was adopted in order to ensure the purity of emergent mutant clones. It is possible that the chromosomes of haploid nuclei may be effectively double so that mutation in one of a pair of chromosomes would be masked by the subsequent growth of the wild type progeny inheriting the other chromosome. Thus there was no guarantee that the vegetative descendants of even single irradiated haploid microconidia would be genetically homogeneous. When the single-ascus cultures became established on complete medium, vegetative conidia from them were subcultured to minimal medium. Those that failed to grow were assumed to have a mutation blocking an essential biochemical pathway and were then systematically tested in the presence of defined supplements until one which promoted growth was found.

The next step was to determine how the mutants isolated in this way differed genetically from the parent strain. This was done by crossing them with wild type strains of opposite mating type and scoring all the eight ascospores from a considerable number of ascus sacs for inheritance of the mutant or wild type character. If, from each set of eight ascospores, four mutant and four wild type strains were obtained, and these were found to be arranged in orthodox order (Fig. 17, p. 71), the conclusion was drawn that the mutant phenotype had resulted from the alteration of a single gene (p. 70). In a number of cases the functional similarity or dissimilarity of pairs of mutants were tested by means of complementation tests in heterokaryons.

In this work some 380 mutant strains were isolated from over 68,000 single ascospores examined, the great majority of which required an amino acid, a B-group vitamin, or a purine or pyrimidine nucleic acid base (p. 229). An analytical classification of the mutants presents some difficulty since various kinds of selective bias are unavoidably associated with the methods used for isolating and characterising them. In many cases these sources of bias apply equally to the isolation of bacterial mutants and should be borne in mind.

1. Complete media, especially if yeast extract is present, may contain substances known to be inhibitory to certain mutants which will therefore be selected against.

2. Even the best of complete media may not promote the growth of all deficient mutants, either because it lacks an unstable or unusual metabolite which the mutant requires, or because the mutant fails to synthesise an enzyme (permease) without which the metabolite cannot penetrate the cell wall.

3. Some mutants have a different pH optimum from that of the wild type and may not be detected unless a medium of correct pH is employed.

4. Strains mutant with respect to the ability to utilise such substances

as fats, fatty acids or specific carbohydrates will not appear as mutants unless these substances are substituted as the sole source of carbon in the minimal medium.

5. Mutants unable to reduce nitrates to nitrites, or nitrites to ammonia, will appear as nutritional mutants in minimal medium in which nitrate is the only nitrogen source, but will not respond to the usual supplements.

6. There is a strong tendency to concentrate on those groups of mutants which can readily be classified which means, in effect, those showing a response to growth factors or intermediates which are available in chemically pure form.

7. In the absence of reliable complementation tests one cannot be sure whether independently isolated mutants involving the same pathway are due to different mutations or to recurrences of the same one.

Despite these sources of statistical inexactitude, it was possible to estimate that the 380 mutants isolated by Beadle and Tatum represented mutations in about 100 separate genes responsible for vital biochemical reactions. It is significant that most of the clearcut mutants concerned the synthesis of known compounds; moreover, *Neurospora* mutants deficient for each of the seven known B-group vitamins and for nearly all of the naturally occurring amino acids have been isolated by various workers. So far as the B-group vitamins are concerned, it therefore seems unlikely that many more remain to be discovered. Before going on to more recent studies of biochemical mutations in bacteria and the important genetical ideas that have sprung from them, we will illustrate the value of biochemical genetics in elucidating the steps of intermediary metabolism by means of a few representative examples concerning amino acid synthesis in *Neurospora* and other fungi, as well as in bacteria.

AMINO ACID SYNTHESIS IN FUNGI AND BACTERIA

Arginine synthesis

This was the first metabolic pathway to be analysed by means of the new genetic tools (Srb and Horowitz, 1944). Fifteen arginine-requiring *Neurospora* mutants were analysed genetically, by crosses as well as by complementation tests in heterokaryons, and were found to comprise seven distinct allelic groups, indicating at least seven steps in the synthetic pathway. Previous studies of arginine metabolism in mammalian liver had not only shown that the enzyme, arginase, catalyses the splitting of arginine into ornithine and urea, but had also suggested that ornithine and citrulline are precursors in the synthesis of arginine. When the ability of these substances to promote the growth of a representative from each of the seven groups of arginine-requiring mutants was investigated, one was found to respond only to arginine, two to either arginine or citrulline, and four to ornithine, citrulline or arginine. It was therefore clear that, in *Neurospora* as well as in mammalian liver, ornithine and citrulline are arginine precursors and that the synthetic sequence must be ornithine→citrulline→arginine, for if it were citrulline→ornithine→arginine the existence of two mutant groups responding to citrulline but not to ornithine would not make sense (pp. 88, 89). The fact that two genes intervene between ornithine and citrulline (Fig. 31; 2, 3) suggested that there is an intermediate

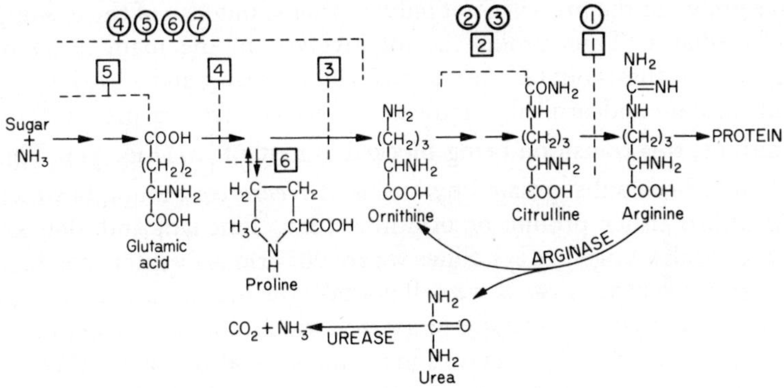

FIG. 31. Synthetic steps in the synthesis of arginine. The numbers indicate genetic blocks, inferred from
{ ○ = *Neurospora* mutants studied by Srb and Horowitz (1944).
{ □ = *Penicillium notatum* mutants (Bonner, 1946).

step, probably involving carbamino ornithine, in this transformation. Analysis of this chain of reactions was extended by Bonner (1946a; review, 1946b) using over 50 independent isolations of arginineless mutants of *Penicillium notatum*. Since this fungus lacks a sexual cycle, while complementation tests in heterokaryons proved ineffective, no direct evidence of genetic control was obtained; the mutants were simply tested for growth response to an extended range of presumptive arginine precursors and were found to fall into five groups as follows:

Group	Growth in presence of: Arginine	Citrulline	Ornithine	Proline	Glutamic acid
1	+	—	—	—	—
2	+	+	(±)	—	—
3	+	+	+	—	—
4	+	+	+	+	—
5	+	+	+	+	+

Assuming, as before, that the inability of an intermediate metabolite to promote growth is symptomatic of a genetic block further along the pathway, these patterns of response would seem to indicate the synthetic sequence, →glutamic acid→proline→ornithine→citrulline→arginine (Fig. 31; 5, 4, 3, 2, 1). However, another mutant was found (Fig. 31; 6) which responded to proline alone, and not to arginine or any other of its precursors. This made it clear that proline cannot precede arginine directly since, if it did, this prolineless mutant should have grown in the presence not only of proline but also of the ensuing intermediates. Thus proline is not involved in the main chain of arginine synthesis but (as the responses of mutants 3 and 4 to glutamic acid, proline and ornithine indicate) is derived from a precursor of ornithine, this transition being blocked in mutant [6] (Fig. 31). This has since been substantiated by means of *Neurospora* mutants which can utilise either proline or ornithine (Srb, Fincham and Bonner, 1950; see also Vogel, 1955). Since we are not primarily interested here in arginine synthesis *per se* we will not take the matter further, except to mention that the enzyme arginase, which converts arginine to ornithine with the release of urea in mammals, is also found in *Neurospora* so that the same 'arginine cycle' operates in both. In mammals the urea is excreted as such, but *Neurospora* makes another enzyme, urease, which finally converts it to carbon dioxide and ammonia (Fig. 31).

Tryptophan synthesis

A brief account of this pathway is included for two reasons. Firstly, the study of *Neurospora* mutants provided the first demonstration of how tryptophan is synthesised in living cells, as well as of the absence of an enzyme from a mutant strain; secondly, tryptophan-requiring mutants have been an important tool for exploring genetic fine structure as well as the complexities of enzyme structure and function in bacteria, so that a preview of the pathway is appropriate here.

An initial survey of some 30 tryptophanless *Neurospora* mutants revealed two groups which yielded wild type recombinants when crossed, and showed functional complementation in heterokaryons (Tatum, Bonner and Beadle, 1944). At least two genes and two biochemical steps are therefore implicated. In both types of mutant the requirement for tryptophan could be satisfied by indole (Tatum and Bonner, 1944). The next step was the discovery that the uptake of indole, and its disappearance from the medium, is greatly accelerated if serine is present and that, under these circumstances, tryptophan appears in the medium. Since serine is used stoichiometrically in the process it was concluded that tryptophan is formed by the condensation of indole and serine. This was subsequently confirmed by the isolation of a mutant in which this condensation is blocked (Mitchell and Lein 1948); cell-free extracts of this mutant, unlike similar extracts prepared from wild type strains, lack the enzyme, tryptophan synthetase, and are unable to synthesise tryptophan in the presence of indole and serine (review: Yanofsky, 1960). When one of the two original types of mutant (Fig. 32, gene 1), responsive to either indole or tryptophan, was grown in the presence of very small amounts of these substances, just sufficient to allow slight growth, it was found that something is produced in the medium which stimulates the growth of the second genetic type (Fig. 32, gene 2). This substance was isolated in pure form

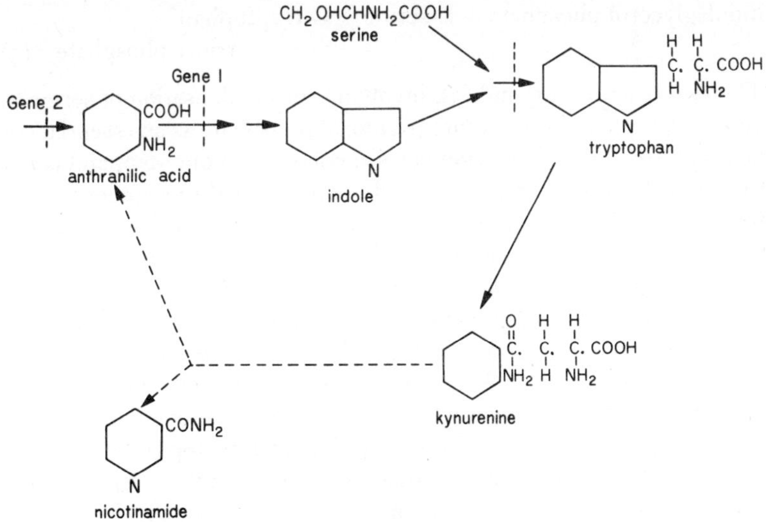

FIG. 32. Tryptophan synthesis in *Neurospora*.

and identified as anthranilic acid. Anthranilic acid is thus implicated as a precursor of indole which accumulates when the synthesis of indole is blocked. Thus of the two mutant types first isolated, one responds only to indole and tryptophan and the other to anthranilic acid, indole and tryptophan. In this way the sequence

$$\rightarrow \text{anthranilic acid} \rightarrow \text{indole} + \text{serine} \rightarrow \text{tryptophan}$$

was established (Fig. 32). Recent work has expanded this picture considerably and has disclosed the existence of a tryptophan cycle. For example, indole-3-glycerol phosphate is now known to intervene between anthranilic acid and indole, while tryptophan is itself a precursor of kynurenine (see p. 112 and Fig. 30) which is an intermediate in a pathway leading back to anthranilic acid and on to the B-group vitamin, nicotinamide, at least in the case of fungi.

The enzyme catalysing the reaction

$$\text{indole} + \text{L-serine} \rightarrow \text{L-tryptophan} \qquad (1)$$

is known as tryptophan synthetase. *In vitro* studies of this enzyme from *Neurospora*, however, have recently shown that it can also mediate two further reactions, each implicating the precursor of indole, indoleglycerol phosphate, thus:

$$\text{indoleglycerol phosphate} \rightleftharpoons \text{indole} + \text{triose phosphate} \qquad (2)$$

$$\text{indoleglycerol phosphate} + \text{L-serine} \rightarrow \text{L-tryptophan} + \text{triose phosphate} \qquad (3)$$

The two reactions (1) and (3), involving the condensation of serine to yield tryptophan, both require pyridoxal phosphate as an essential cofactor. Furthermore, reaction (3) is carried out in one step, and is not the summation of reactions (2) and (1) performed in sequence, since free indole is not found as an intermediate and since the reaction is carried out at a much faster rate than is reaction (2) (Crawford and Yanofsky, 1958; Bonner *et al.*, 1960). The enzyme appears to be genetically determined by a single gene, the particular enzymic activity affected depending on the region of the gene in which the mutation occurs (Bonner *et al.*, 1960). Thus, in this case, independent mutations occurring in a single gene which determines only one enzyme may nevertheless yield blocks in different biochemical steps of a synthetic pathway. The tryptophan synthetase produced by *E. coli* shows a similar range of enzymic functions but differs from the *Neurospora* enzyme in being constructed of two distinct protein sub-units, A and B, which are genetically determined by different, though adjacent, genes. The evidence so far available suggests that component A acts primarily on indoleglycerol phosphate so that, in mutant strains where it is defective, reactions (2) and (3) are blocked and indoleglycerol phosphate accumulates; component B is mainly concerned with the conversions involving serine [reactions (1) and (3)]. The efficient performance of any reaction by the purified components *in vitro*, however, requires the presence of both in association (Crawford and Yanofsky, 1958; review, Yanofsky, 1960).* We shall return to study the genetical and functional problems posed by such complex enzymes as tryptophan synthetase when we come to discuss how the results of complementation tests may be related to genetic fine structure in the next Chapter (see p. 155; Fig. 43).

MUTANTS WITH MULTIPLE REQUIREMENTS

Most of the nutritional mutants isolated from *Neurospora*, like those blocking various steps in the synthesis of arginine and tryptophan which we have described, were found to have a requirement for a single growth factor and clearly supported the idea that each gene controls the production of only one enzyme. In contrast, as the survey extended, several groups of mutants were isolated which required more than one amino acid for growth although genetic analysis showed that the mutations involved only single genes. The intrusion of this kind of complication into the one gene–one enzyme hypothesis may be exemplified by two mutant groups, one having a requirement for more than one amino acid of the aromatic series, and the other for isoleucine and valine.

The aromatic amino acids

Aromatic amino acids are characterised by the presence of the benzene ring in their structure. The most important are tyrosine, phenylalanine, tryptophan and *p*-aminobenzoic acid. Several mutants of *Neurospora* (see Tatum *et al.*, 1954) and many *E. coli* mutants (see Davis, 1955) have been isolated which require more than one, and even all, of these amino acids for growth. The reason for this became apparent when it was found that a single rare plant acid, shikimic acid, would alone promote the growth of a number of these polyauxotrophic mutants, the response being proportional to the concentration of the acid. The inference was that shikimic acid is a common precursor of all the aromatic

* A summary of present knowledge of the enzymes of the tryptophan pathway in *E. coli* is by Imamoto, Ito and Yanofsky, 1966.

acids and therefore makes good the requirements of those poly-auxotrophic mutants in which the genetic block occurs earlier in the synthetic pathway (Fig. 33). Subsequently, bacterial mutants were found

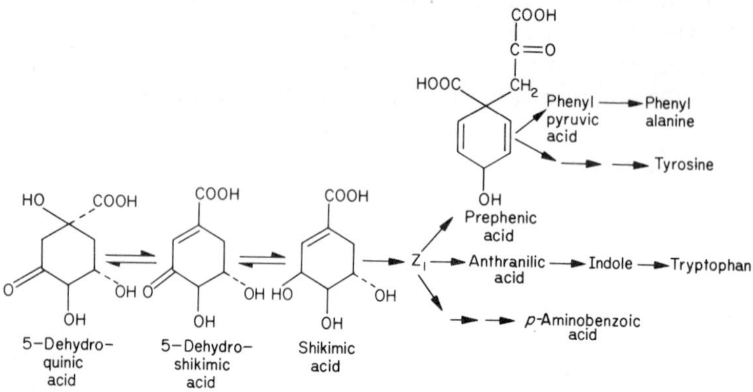

FIG. 33. The synthetic pathway of the aromatic amino acids. (After Davis, 1955.)

which were blocked later in the sequence and accumulated shikimic acid in the culture medium, thus confirming this substance as a genuine intermediate. The later intermediate compounds, Z_1 and prephenic acid (Fig. 33), are peculiar because, although they are accumulated in large quantities by appropriate mutants, they fail to promote the growth of any mutant. Compound Z_1 however, readily yields shikimic acid on heating at low pHs and its position on the pathway was allocated on the grounds that those mutants which accumulate it also accumulate traces of shikimic acid, while no Z_1 was found in the case of those mutants which accumulate appreciable amounts of shikimic acid. Accumulation of prephenic acid has been found mainly in mutants requiring phenylalanine or phenylpyruvic acid for growth. Prephenic acid is extremely acid-labile, having a half-life at room temperature and pH 6 of only 13 hours, and decays to phenylpyruvic acid. This leads to the unusual situation that *E. coli* mutants which are blocked between prephenic acid and phenylpyruvic acid can build up their own required growth factor again and show delayed growth unless the culture medium is kept alkaline.

Threonine and methionine

Many similar mutants which have two or more growth factor requirements as a result of a genetic block in a common synthetic pathway have been recorded both in *Neurospora* and in bacteria. A good example of this is a dual requirement for threonine and methionine in mutants blocked in the pathway leading to synthesis of homoserine which is an intermediate common to both these amino acids (Fig. 34; for *Neurospora* data see Teas, Horowitz and Fling, 1948; for *E. coli* data see Cohen and Hirsch, 1954).

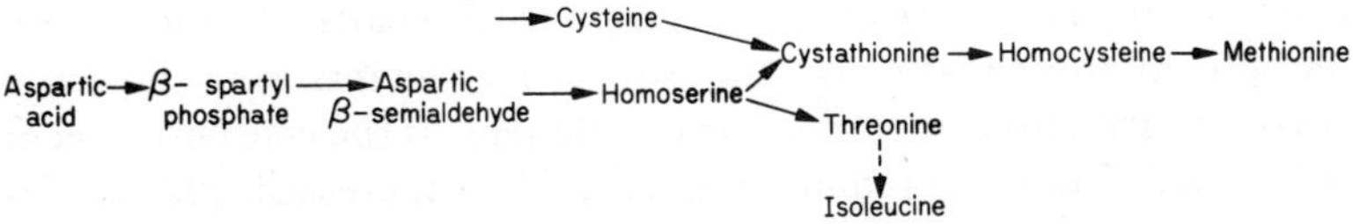

FIG. 34. The synthesis of methionine and threonine.

Isoleucine and valine

The syntheses of the branched-chain amino acids isoleucine and valine proceed, for the final four steps of their course, along different but parallel pathways as Fig. 35 shows, although valine originates from pyruvic acid while isoleucine is derived from a condensation of

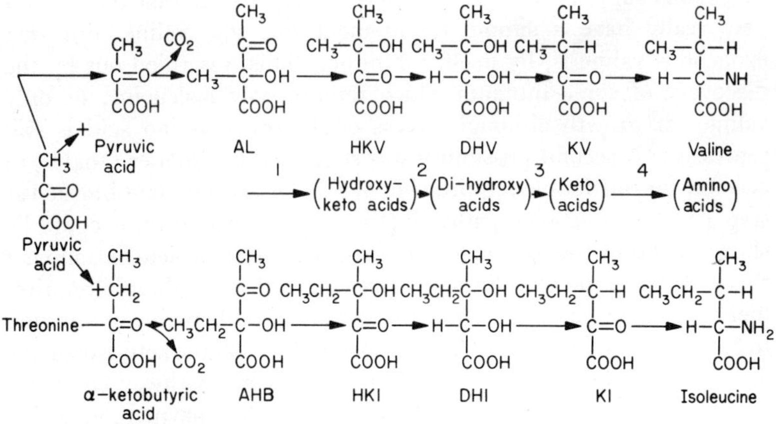

FIG. 35. Synthetic pathways of valine and isoleucine.
AL = α-acetolatic acid; HKV = α-keto-β-hydroxyisovaleric acid;
DHV = α, β-dihydroxyisovaleric acid; KV = keto-valine.
AHB = α-aceto-α-hydroxybutyric acid; HKI = α-keto-β-hydroxy-β-methylvaleric acid; DHI = α,β-dihydroxy-β-methylvaleric acid;
KI = keto-isoleucine.

The analogous, parallel steps of the two pathways are designated 1, 2, 3 and 4. (after Wagner and Bergquist, 1960.)

deaminated threonine (α-ketobutyric acid) and pyruvic acid (reviews: Adelberg, 1955; Knox and Behrman, 1959). These sequences were primarily worked out by observing the growth-stimulating intermediates, accumulations and enzyme deficiencies of a number of *Neurospora* and *E. coli* mutants in the way we have described for other amino acids, as well as by isotopic tracer studies. A peculiarity of most of these mutants, however, is the fact that they have a joint requirement for both isoleucine and valine, supplied in a particular ratio (3:7), for optimal growth, although the *Neurospora* mutants were shown by crosses to arise from alterations of a single gene. If the concentration of either one of the two required amino acids is increased, growth decreases until complete suppression is reached at a ratio of 10:1. The further addition of the other amino acid to the medium, however, reverses the inhibition until optimal growth is again achieved at an isoleucine: valine ratio of 3:7. Both amino acids are therefore necessary for growth but display *competitive inhibition* when either is present in excess.

One possible explanation of the double requirement is that it could be a direct consequence of competitive inhibition between exogenous supplies of isoleucine and valine; that is, that the mutant may really have a simple requirement for, say, valine, but that exogenous valine is, by itself, inhibitory. This was ruled out by the discovery of some mutants which require only isoleucine, or only valine, for growth although excess of the other amino acid is still repressive. A second possibility was suggested by Bonner (1946b) to account for the double requirement of a *Neurospora* mutant blocked at step 4 in the isoleucine pathway (Fig. 35). In this strain it could be shown that only the conversion of keto-isoleucine to isoleucine, and not that of keto-valine to valine, was arrested so that the double requirement could not be due to a genetic block in a single pathway common to both amino acids. It was therefore postulated that the keto-isoleucine accumulating behind the block in the isoleucine pathway competes with its keto-acid analogue, keto-valine, for the enzyme converting keto-valine to valine thus effectively repressing the analogous step in the valine pathway.

Although such a mechanism may well operate in some other pathways, it has been shown by Adelberg (1955) that it is not true for mutants requiring isoleucine and valine, at least in the case of *E. coli*. The final conversion of both keto-acids to their respective amino acids (Fig. 35, step 4) is due to transamination. A transaminase can be extracted from wild type *E. coli* which catalyses amino transfer between any two of the amino acids isoleucine, valine, leucine, norleucine, norvaline and glutamic acid, but this enzyme is completely absent from mutants blocked at step 4. It seems likely that this is a single enzyme which operates in the formation of both amino acids so that its absence explains how the double requirement stems from a single mutation. However, both wild type and mutant strains possess another enzyme transaminating valine alone, with alanine or α-amino-butyric acid, so that a limited amount of valine synthesis is permitted. Adelberg and Umbarger (1953) obtained a secondary mutant from one of the isoleucine-valine-requiring strains which yielded a 4–5 fold increase in its valine-alanine transaminase and this was accompanied by the capacity to grow maximally on isoleucine alone. A comparable state of affairs has recently been reported in *Salmonella typhimurium* (Wagner and Bergquist, 1960). There is good biochemical evidence that a single enzyme promotes both the AL→HKV as well as the AHB→HKI conversion (Fig. 35, step 1), while the genetic evidence, coupled with enzyme extractions, strongly suggests that single enzymes act on both pathways at steps 3 and 4, so that the double requirement of mutants blocked at each of these steps may be similarly explained. Once again we find that the apparent exceptions to the one gene-one enzyme hypothesis serve only to confirm it when they are investigated in depth.

THE NATURE OF COMPETITIVE INHIBITION

The nature of the competitive inhibition between isoleucine and valine in these mutants is not known. In fact the situation here is far from being unique for many other examples of the same phenomenon, but involving other pairs of amino acids, have been reported in auxotrophs of both bacteria and fungi. A case in point is the *Neurospora* mutant, mentioned above, which had a double requirement for methionine and threonine as a consequence of a genetic block in the common pathway leading to homoserine synthesis; a rather precise ratio between the concentrations of exogenous methionine and threonine is needed for growth, excess of either amino acid suppressing metabolism. The phenomenon cannot glibly be explained on the basis either of competition for a common enzyme mediating the blocked pathway, or of the operation of a normal 'feedback' mechanism controlling balanced synthesis within the cell, whereby excess of the end-product of a pathway represses the activity of earlier steps (see pp. 706–712); for not only do the positions of the blocks in the auxotrophs, and the particular pairs of amino acids concerned, usually not fit such explanations, but the amino acids fail to inhibit growth of the wild type strains. This suggests the involvement of some supernumerary transport system. In other words, the competition does not implicate the production line which *normally* manufactures an essential component of the cell from raw materials. It is found only when the production line no longer operates, so that the finished component has to be imported into the factory from outside; we can imagine that a special conveyor belt exists for this purpose, from which the required component can be displaced competitively if certain other components are fed on to it.

An interesting and important analysis of this problem, which shows how genetical techniques may be used to study the topography of biochemical reactions within the cell, was made by Pontecorvo (1952a). He found competitive inhibition between lysine and arginine in certain mutant strains of *Aspergillus nidulans* such that exogenous arginine completely suppresses the growth of lysine-requiring strains, and exogenous lysine that of arginine-requiring strains, when the molar ratio of the inhibiting to the required amino acid reaches 2:1. A doubly mutant haploid strain, made by recombination, grew only within a narrow range of molar ratios of the two amino acids. Using a special technique (Roper, 1952; see p. 78), heterozygous *diploid* strains were then made in which the two mutations were introduced into the diploid nucleus either by one of the two parents ($arg^- \ lys^-/++$) or by both ($arg^-.+/+.lys^-$). Not only were the diploids of both kinds independent of growth factor requirements, as might be expected since the two mutations concerned different functions, but their growth was unaffected by the presence of exogenous arginine or lysine. In contradistinction, *heterokaryons* (pp. 78, 94), carrying the two mutations *in different nuclei* of the same cell, were highly sensitive to exogenous amino acids. This sensitivity of heterokaryons, coupled with the resistance of the diploids, showed that the site of amino acid competition must be within the cell but outside the nucleus.

Part IV
The Analysis of Genetic Fine Structure in Microorganisms

CHAPTER 7

Some Theoretical Aspects of Fine Structure Analysis

HISTORICAL INTRODUCTION

Two of the most important words in genetics are 'gene' and 'allele', if only because they have for a long time provoked the question, 'What do they mean?' Much of the history of genetics may be written in terms of attempts to answer this question. This Chapter is. We first defined a gene as a particulate unit of genetic material which obeyed Mendel's laws of inheritance and determined the expression of a particular character. An allele was initially presented as one of the two alternative states, dominant and recessive, in which a gene could exist but it later became evident that a gene could be modified by mutation in many more ways than two, each allele determining a different phenotype without necessarily being recessive to wild type. Then came the complementation test (p. 92 *et seq.*) with its ability to discriminate between mutations involving the same or different basic functions, the cooperative action of a number of such functional units usually being required for the expression of the observed character. One of the most interesting facts to emerge from complementation tests was, as we shall see, that a number of distinct mutational sites, separable by recombination, might be found within the same functional region of chromosome. We are therefore confronted by the dilemma that the gene cannot at the same time be both the determinant of a character and the unit of inheritance since these two components of the original definition are clearly quite different things.

Although, in this and the next Chapter, we will be dealing principally with the methods and findings of transduction experiments with bacteria and of genetic crosses with bacteriophage, which have greatly illuminated the finer details of the relationship between genetic structure and function, it should be noted that these modern studies are the direct and logical culmination of earlier work with *Drosophila*, initiated over 30 years ago by Dubinin and other Russian geneticists. These early investigations concerned a series of alleles of a gene called 'achaete-scute' which determines the number and distribution of bristles on the fly. Each allele was characterised by the absence of specific individual bristles and so caused a distinctive bristle pattern. Two striking findings emerged from study of these alleles. The first was that the alleles could be arranged in a definite series on the basis of the bristle patterns. Secondly, when two alleles which showed a different but overlapping pattern in homozygotes, were combined in heterozygotes, only those bristles which both alleles lacked in common failed to develop. It was deduced that the gene was compounded of a linearly arranged series of separate functional units. Finally, four cross-overs within the gene were found among some 75,600 flies examined (Dubinin, 1932, 1933; reviewed Demerec and Hartman, 1959).

Later studies of other genes in *Drosophila* showed clearly that alleles displaying the same phenotypic effect were often separable by recombination and that, among such alleles, some pairs showed complementation while others did not. In fact it was in *Drosophila* that the complementation test was first developed by Lewis (1951) who called it the *cis-trans test* because the phenotypic effects of two opposing arrangements of a pair of mutant genes are compared. In the *trans* arrangement (Latin = on the other side of), each genome of the diploid nucleus has one wild type gene and one mutant gene (*Ab*/*aB*). When the mutations are in different functional units, the *trans* arrangement yields the wild phenotype since, as we have already seen (Fig. 24A, p. 93), each cell possesses one fully functional unit of each type. But if both mutations are in the same functional unit then, in the *trans* arrangement, both representatives of this functional unit are defective so that the mutant phenotype is displayed (Fig. 24B). In the *cis* arrangement (Latin=on this side of) the two wild type genes are on one genome of the diploid and their two mutant alleles are together on the other (*AB*/*ab*) so that, whether the mutations involve the same or different functional units, every cell has one normal, wild type genome and should (and does) always display the wild phenotype. The *cis* arrangement is therefore merely a formal control for the *trans* test which constitutes the complementation test proper and may, in practice, be omitted. The real significance of the *cis* test lay in the fact that it could be performed at all in *Drosophila*, by obtaining recombinants in which two mutations, which were non-complementary in the *trans* arrangement, were present on the same chromosome. Thus the occurrence of recombination between functionally allelic mutations was demonstrated. In such cases the phenotypic distinction between the *cis* and *trans* arrangement is often referred to as a *position effect*. This should not be confused with alterations in function which may accompany other kinds of genetic rearrangement such as inversion (p. 41) which are called by the same name.

An investigation of mutational pattern in two well defined genes in maize (Stadler, 1951) also indicated that genetic loci were divisible and composite structures. Such loci became known as 'complex loci' and their alleles, which were separable by recombination, as 'pseudo-alleles'. A proper investigation of pseudo-allelism clearly demanded definitive analysis of the relationships between the three properties ascribed to the gene—the ability to mutate to allelic forms, to determine a specific function, and to undergo recombination with other genes. Technical difficulties presented a formidable deterrent to the extension of this kind of analysis in an organism such as *Drosophila*, the limiting factor being the immense labour of searching for very rare recombinants by simple inspection of tens of thousands of progeny, since the distance between pseudo-allelic sites proved to be of the order of only a few thousandths of a map unit. What was needed was an organism in which mutations affecting known biochemical reactions could be isolated, complementation tests performed, and exceedingly rare wild type recombinants recovered quantitatively by a selective method.

The obvious solution was to employ the sexual fungi, and it is an interesting example of the way in which the evolution of new ideas in science is so often accompanied by the independent development of new methods for their exploitation, that the decade which saw a clear definition of the problems of pseudo-allelism in *Drosophila* was also distinguished by the birth of biochemical genetics in microoganisms. The earliest contributions came from studies in *Neurospora*. One was by Bonner (1951) who set out to find whether genetic complexity could be revealed in genes known to determine a specific, chemically defined reaction; he demonstrated the formation of wild type recombinants between three groups of mutants, each of which blocked the conversion of 3-hydroxyanthranilic acid to niacin and accumulated quinolinic acib, and none of which showed any complementation in heterokaryons with any one of the others. Another rather different type of study by Giles (1951) showed that a series of eight independent mutations to inadility to utilise inositol, all different as judged by their rates of reverse mutation to wild type, nevertheless belonged to the same genetic locus and blocked the same function.

Similar findings emerged from studies of biochemical mutations in *Aspergillus*. Strangely enough, the first of these investigations was devised to test a quite independent (and incorrect) theory that genes specifying distinct but closely related functions might be expected to be closely linked in order to facilitate the participation of their products in a sequence of associated biochemical reactions (Pontecorvo, 1950). Three biotin-requiring mutants were isolated whose mutational sites were found to recombine with a frequency of about 10^{-3}, showing that they were very closely linked; it then turned out that none of these mutants showed complementation with any other so that they had to be considered as alleles of the same gene (Roper, 1952). These findings led Pontecorvo (1952b) to postulate that a unit of function contained many mutational sites which were separable by recombination, and was therefore much larger than the unit of recombination. Subsequently, a number of well separated adenine loci were mapped in *Aspergillus*; within one of these loci (*ade*-8) four linearly arranged but non-complementary mutational sites were identified, as were five sites within another locus (*ade*-9) (Pritchard, 1955; Calef, 1957). Similar observations were also made with respect to a number of adenine loci in two varieties of yeast (Roman, 1956; Leupold, 1958). This analysis of the gene was then extended to bacteria and bacteriophage, and culminated in a brilliant and revealing series of investigations on *Salmonella typhimurium* by M. Demerec and his colleagues at Cold Spring Harbor, and on bacteriophage T4 by S. Benzer.

Demerec and his group were initially interested in studying patterns of mutagenesis in bacteria and for this purpose had accumulated several hundred auxotrophic mutants of *Salm. typhimurium*, the wild type of which is nutritionally non-exacting and grows normally on unsupplemented minimal medium. Their first aim was to cross pairs of mutants of similar phenotype, as judged by the biosynthetic steps blocked in them, by means of transduction (p. 56; Chapter 21, p. 620) and selection for wild type recombinants, in order to ascertain which mutants were 'allelic' according to the then prevalent assumption that recombination could occur only between mutational sites involving non-identical functions. They found, in fact, that nearly all pairs of mutants, blocked in the same biochemical step, *did* yield recombinants though at a much lower frequency than in crosses between mutants of different phenotype or in wild type × mutant crosses (Table 3).

TABLE 3
Comparison of the number of wild type (prototrophic) recombinants arising from transductional crosses between various auxotrophic mutants of *Salmonella syphimurium*.
1. Wild type × mutant cross.
2. Crosses between phenotypically different (non-allelic) mutants.
3. Crosses between phenotypically identical (allelic) mutants.

	Recipient	Donor	No. of wild type recombinants
1.	*try*-D10	*try*-D10	0
	try-D10	wild type	1822
	try-D10	*his*-22	1456
	try-D10	*met*-15	1617
	try-D10	*try*-A8	208
2.	*try*-D10	*try*-B4	602
	try-D10	*try*-C3	270
	try-C3	*try*-D10	88
	try-D10	*try*-D1	4
	try-D10	*try*-D6	2
3.	*try*-D10	*try*-D7	7
	try-D10	*try*-D9	12

8×10^7 cells of the recipient were infected with 4×10^8 phage particles from the donor and plated on minimal medium. Prototrophic colonies were counted after 48 hours at 37°C.
try, his, met = mutants requiring tryptophan, histidine and methionine respectively.
The capital letters, A, B, C, D, indicate loci determining different biochemical functions: the figures refer to individual mutant isolates. Thus *try*-D10, *try*-A8 and *try*-C3 are mutants blocked in different steps in tryptophan synthesis; *try*-D1, *try*-D6, *try*-D7 &c. are different mutants blocks in the same step. (From Demerec and Hartman, 1956.)

Subsequent investigation of some 1200 auxotrophic and sugar fermentation mutants, by means of chemical specification of biosynthetic blocks and complementation tests (abortive transduction, p. 96), coupled with transductional analysis, revealed more than 60 biochemically defined genetic loci. Within nearly all of these loci more than one mutational site was found, nearly all the sites within each locus showing recombination with one another. Following up these observations, a most extensive and striking study was made by Hartman *et al.* (1966b) of a series of over 200 *Salm. typhimurium* mutants, all of which required histidine for growth. Analysis by biochemical methods and complementation tests disclosed seven functional loci, each of which determined a specific step in histidine synthesis; within two of these loci, 31 out of 34, and 33 out of 35, mutational sites gave recombinants with one another (reviewed, Demerec and Hartman, 1959). We shall enlarge upon this work later in the Chapter. The point we wish to make now is that not only did it confirm the existence of complex loci in bacteria and show that they were the rule rather than the exception, but it also made it clearly manifest that the genetic unit of function must be constructed of a large number of mutational sites separable by recombination, since recurrent mutations at the same site were rarely found. Demerec therefore abandoned the term 'pseudo-allele'. Instead, he suggested that the word *gene locus* should connote a region of chromosome controlling a single function, any mutational alterations within such a locus being termed alleles; pairs of alleles which yielded no recombinants he called *identical alleles*, implying that the mutations arose at the same site, while those that did show recombination were *non-identical alleles*. These two types of allele have also been referred to as *homoalleles* and *heteroalleles* respectively (Roman, 1956).

We began this Chapter by looking at analysis of the gene from the historical point of view in order to put the evolution of our knowledge in proper perspective against the background of those classical concepts which, up to now, we have concentrated upon. We have seen that while the idea of complex genes first became explicit in higher organisms, insight into the nature of the complexity could only be revealed by the greatly increased magnifying power which the use of simple, biochemically defined characters and the ability to select very rare recombinant types bestow on genetic analysis in fungi and bacteria. In fact the most refined analysis of the gene yet attained (or even, perhaps, attainable) is that performed by Benzer (1955, 1957) on a small chromosomal region of bacteriophage T4. Before looking at individual systems, however, we will first consider some general principles which are involved in all analyses of this sort.

THE THEORY OF FINE STRUCTURE ANALYSIS

Any type of recombinational analysis requires marker points on each of the two parental chromosomes, whose distance apart can then be measured by the frequency with which recombinations of them are found among the progeny of the cross. These points are the sites on the chromosome at which mutations occur which result in observable alterations of phenotype. If we assume that the probability of a mutation occurring at any site is random, then the smaller the region of chromosome the less is the likelihood that we shall find it marked by two mutations; or, to put it another way, the larger is the number of mutant strains we must examine before we are likely to discover two mutational sites within the region. For example, if a chromosome possesses 100,000 sites at which mutation can occur (and this is probably an underestimate in the case of most organisms), the random probability of two mutations arising independently at adjacent sites is 1 in 50,000. The number of mutants which would have to be isolated before we could begin to estimate the probable distance between adjacent sites is, therefore, very large. In order to carry out the genetic analysis, all possible combinations of mutant pairs must be crossed to find the pair with the smallest recombination frequency. If there are n mutants, the number of crosses is $n(n-1)/2$, or approximately $n^2/2$ when n is large. Thus the analysis of even 10,000 mutants would involve 50 million crosses! Fortunately there are a number of tricks we can employ to transform this Herculean task into a practicable one, by restricting the examination to only a minute chromosomal region.

THE PRELIMINARY GROUPING OF MUTATIONS WITHIN SMALL REGIONS

The use of phenotypically similar mutants.

It is assumed (though this is far from being necessarily true) that a particular function, or related series of functions, is determined by only one gene, or by a closely linked cluster of genes; only those mutations which involve this function are then selected and mapped. For example, there is a semi-quantitative method for isolating bacterial mutants which have a specific nutritional requirement for, say, tryptophan or histidine (Chapter 10, p. 212), so that large numbers of independent, but phenotypically similar, mutations can easily be obtained which, it is hoped, will lie together within a minute region of chromosome. Similarly, mutants of bacteriophage which display a specific alteration of host-range or of plaque type are easy to isolate (Chapter 18, p. 480).

Mapping by complementation tests.

We have seen (Fig. 24, p. 93) that two mutations in the *trans* configuration can often complement one another to produce the wild phenotype when they involve different functional units of the chromosome, whereas the mutant phenotype continues to be expressed when the same functional unit is affected. Thus systems where two mutant genomes can readily be introduced into the same cell offer a simple method of testing whether or not two mutations lie within the same locus (Benzer, 1957; Hartman, Loper and Šerman, 1960). Since the same phenotype, upon which we must rely for the preliminary selection of mutant strains, often results from mutation in different, spatially separated functional units, complementation tests permit an initial selection of mutants which are defective at the same locus and, therefore, are suitable for fine structure analysis. This is especially the case where no knowledge of the biochemical basis of the phenotype is available. The method, however, is a much less refined one than the use of multisite mutants which operate at the intra-genic level.

Mapping by deletion or 'multisite' mutants.

These mutants which, for many regions, constitute a significant proportion of all mutants isolated, are characterised by two properties; they are completely stable, being unable to revert to wild type by back-mutation, and they cannot recombine with two or more contiguous mutational sites which can themselves recombine to yield wild type recombinants. They therefore behave as if they were completely lacking a part of the chromosome (p. 41). The extent of the deletion can be defined by crossing the multisite mutant with a series of single-site, revertable mutants whose mutational sites on the chromosome have already been mapped. For example, in Fig. 36, if it is found that a stable, presumably multisite, mutant yields wild type recombinants in crosses with mutants 1, 2, 3, 9, 10, 11, 12, 13, 14, 15 and 16, but no recombinants with mutants 4, 5, 6, 7, and 8, then the deletion extends over region E; similarly, a deletion covering region F is shown by the absence of recombination with mutants 8 to 13, but not with any other mutants. Two multisite mutants whose deletions overlap, such as A and E, or C and F (Fig. 36) are clearly unable to produce between them a wild type recombinant; on the other hand, if there is no overlap, as in the case of mutants A and B, then recombination in the region intervening between the deletions can reconstitute a wild type chromosome.

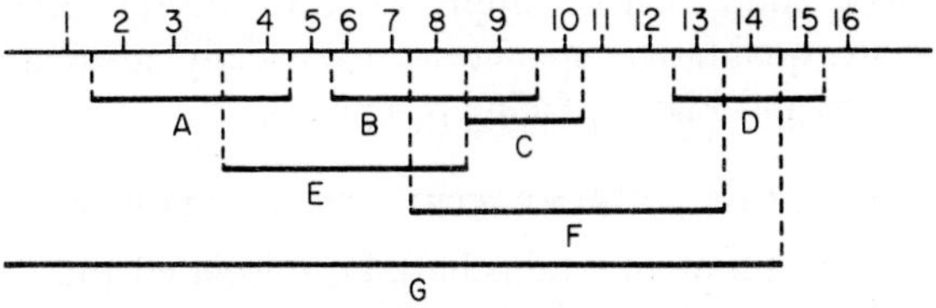

FIG. 36. The use of deletion or 'multisite' mutations in mapping. 1–16 represent single-site 'revertable' mutations which have been mapped on the chromosome as shown; each of these sites can undergo recombination with any other site, yielding wild type recombinants.

Lines A–G represent the extent of chromosome deleted in each of a series of seven multisite mutants. These mutants do not revert to wild type, and show no wild type recombinants when crossed with those single-site mutants whose sites lie within the deleted region; they do, however, yield recombinants with all other single-site mutants and with other multisite mutants whose extent does not overlap their own.

A range of overlapping, multisite mutants can enormously reduce the work involved in mapping single-site mutations which, by their use, can be rapidly and accurately allocated to a small region; only those mutations within the same small region need then be ordered, and the distance between them measured, by pair-wise crosses between the mutant strains. The mutants to be mapped are crossed with each of the multisite mutants and the presence or absence of recombination noted. Some examples from Fig. 36 should make the method clear. Suppose that a mutant yields wild type recombinants with multisite mutants A, B, C, D and F, but none with E or G. Since it does not cross with E the mutational site must be within the region covered by this deletion; the presence of recombination with A and B, however, shows that it is outside these regions so that it must be located between sites 4 and 6. Again, a mutation which shows recombination with A, D and E, but none with B, C, F or G must be situated within a region where these last four deletions overlap, that is, between sites 8 and 10. Similarly any mutant which shows recombination with G must have its site outside this region, somewhere to the right of site 14.

The usefulness of this method can be gauged from the fact that the mapping of 100 mutants by pairwise recombination would require just under 5000 quantitatively precise crosses, as against 700 crosses if seven overlapping multisite mutants are available for the region. Moreover, since the information required from these latter crosses is of the 'all or none' kind, that is, simply whether recombination does or does not occur, qualitative 'spot' tests are usually adequate (Benzer, 1957; see p. 170 and Plate 8, facing p. 171).

METHODS OF FINE STRUCTURE MAPPING

When a number of mutant strains have been isolated and their mutational sites located within a single locus, or part of a locus, the next operation is to map their relative positions and the distances between them by means of pair-wise crosses. In the case of bacteria and bacteriophages we are faced with a number of diverse genetic systems, each of which requires a somewhat different conceptual and experimental approach with respect to the details of recombinational analysis. For instance recombination in phage involves the complete genomes of both parents but is complicated by its occurrence in a multiplying 'pool' of chromosomes which are randomly withdrawn during the growth cycle to be incorporated into progeny particles (see p. 61). In the case of transformation and transduction, only small fragments of donor chromosome are transferred to the recipient bacteria so that an even number of cross-overs is needed to produce viable recombinants, while linkage analysis is restricted to small chromosomal regions (p. 56, Fig. 14). In conjugation in *E. coli*, on the other hand, relatively large segments of donor chromosome may be transferred but the orientation of the transfer process, and the occurrence of random chromosomal breakage as it proceeds, introduces a polarity with respect to genetic analysis, that is, the validity of the analysis depends upon whether a particular locus enters the recipient bacteria before or after the selected locus (p. 690; Fig. 15, p. 66). The methodology best suited to each system will be discussed later when we come to consider them in detail. For the moment let us examine the general theoretical aspects of fine structure mapping which are common to them all.

Mapping by comparison of recombination frequencies in '2-factor' crosses.

If we know that a number of points are arranged in a straight line, and if we also know the distance between each pair of them, it is obvious that we can arrange them in a specific order in relation to one another. Each pair of mutants is crossed under standard conditions and plated on a selective medium on which only wild type recombinants can grow. The number of recombinant colonies (or phage plaques) is then counted and compared for each pair of mutants. It is assumed that the closer the two mutant sites in each parent, the lower is the probability of recombination between them and, therefore, the smaller the number of recombinants. In the case of phage, where two complete chromosomes are involved, only a single recombination event or cross-over is postulated (Fig 37A, cross-over in region 2). In merozygotic systems,

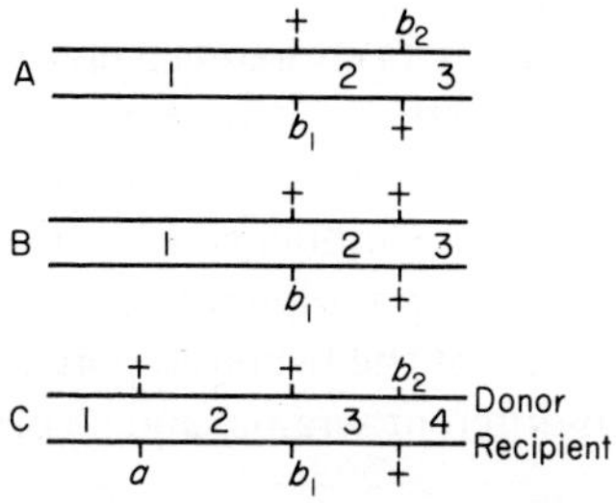

FIG. 37. Mapping by comparison of recombination frequencies. b_1 and b_2 are different mutational sites within a single functional locus. *a* is a mutational site in a different locus, closely linked to the *b* locus. The numerals indicate various cross-over regions between the sites.

A: Mutant × mutant cross; wild type recombinants are produced by a single cross-over in region 2 in bacteriophage, or by two cross-overs, in regions 1 and 2, in merozygotic systems.

B: Wild type × mutant cross; wild type recombinants are produced by a cross-over in either regions 2 or 3 in bacteriophage; in merozygotic systems two cross-overs, in regions 1 and 2 or 1 and 3 are required.

C: Ratio test: in merozygotic systems, the number of recombinants for the *b* locus can be standardised, from cross to cross, by comparison with the total number of recombinants for the linked *a* locus (see text).

where the genetic contribution of the donor parent is incomplete, an even number of cross-overs is necessary, as in regions 1 and 2 in Fig. 37A where the upper line represents the fragment of donor chromosome and the lower line the complete chromosome of a recipient bacterium.

The number of wild type recombinants can either be compared directly or related to some standard which is common to all the crosses. Thus in phage recombination, the number of wild type recombinants is usually expressed as a percentage of the total number of progeny particles emerging from the cross, that is, as a true recombination frequency. In merozygotic systems, where the total progeny is unknown (p. 59), the number of wild type recombinants arising from crosses between two mutant strains (Fig. 37A) can be related to the number emerging from wild type × mutant crosses (Fig. 37B); for example, in tranductional analysis of the region controlling histidine synthesis in

Salmonella, when transducing phage grown on one *his* mutant is applied to another *his* recipient strain, the number of prototrophic recombinants may range from 60 per cent to as low as 0·1 per cent of that obtained when transducing phage grown on a wild type strain is used with the same recipient (Hartman, Hartman and Sĕrman, 1960; see Table 3, p. 131).

Although this method has been extensively used in mapping, and may be the only possible method in some circumstances, it embodies many inherent sources of error since the results of *different* crosses are compared, and these may be influenced by all kinds of extraneous factors such as variations of temperature, medium, conditions of plating, physiological state of the bacteria, potency of the transducing phage when grown on different strains, and so on.

The ratio test.

The effects of external variables are here controlled by relating the number of recombinants that are wild type for the region being studied to some other class of recombinants *emerging from the same cross*. This can be accomplished in two ways depending on whether the mutations to be mapped belong to the same or different functional loci. If they belong to the same functional locus, then one of the parental strains is marked by a mutation at another, closely linked locus. This is illustrated in Fig. 37C where the recombination frequency between mutational sites b_1 and b_2, involving a specific nutritional requirement for a growth factor, B, is to be measured by transduction. The recipient strain is additionally marked by a mutation in the closely linked locus *a*, so that it requires substance A for growth. Samples of the recipient bacteria, after infection by transducing phage grown on the donor strain, are plated on two different selective media, one unsupplemented and the other supplemented with substance B. The number of colonies arising on the unsupplemented medium indicates the frequency of recombination in region 1 and region 3, in which we are interested, while that on the B-supplemented medium, which allows the growth of all *b* mutants, gives the frequency in region 1 and in the much longer region 2+3+4. The ratio

Colonies on unsupplemented medium/colonies on B-supplemented medium

remains very constant between crosses and independent of the experimental conditions. Crosses of this type, however, are actually 3-factor crosses since three mutational sites are involved and, as we shall shortly see, the number of recombinants obtained depends on the order in which the markers are arranged as well as on which of the two parents is used as donor. In fact, it is not necessary that the locus employed to standardise the results of a cross be linked to the region being measured. The frequency of recombinants for *any* outside pair of alleles suffices to control the efficiency of the crossing procedure *as a whole* even though, as is the case with unlinked loci in transduction, the controlling locus and that under study are carried on different chromosomal fragments so that the two classes of recombination event take place in different bacteria.

If the two mutations to be mapped happen to belong to different functional loci, so that the mutant strains can be distinguished by their growth response to different supplements, then a standardised recombination ratio can often be obtained in transductional analysis in the absence of a third marker (Hartman, Hartman and Šerman, 1960). Let us suppose, for example, that the two mutations b_1 and b_2, depicted in Fig. 37A are such that the donor strain carrying b_2 ($+b_2$) can grow in the presence of substance B_2 while the recipient strain, carrying b_1 (b_1+), cannot do so. If the cross is made on unsupplemented medium only wild type recombinants ($++$) arising from crossing-over in regions 1 and 2 can grow; on medium supplemented with B_2, on the other hand, recombinants of donor genotype ($+b_2$), due to recombination in regions 1 and 3, can grow in addition to wild type recombinants. Thus the ratio wild type/donor type recombinants can be obtained. Moreover the use of a limiting concentration of substance B_2 will restrict the size of colonies of the donor type without affecting that of wild type colonies, so that the two types of recombinant can be distinguished and counted on the same plate, thus providing a further degree of internal control. It should be stressed, however, that this type of test is only applicable to transduction and transformation systems in which the donor bacteria themselves are excluded from the plating medium, so that all colonies of donor type cells must have arisen through recombination.

Mapping by 3-factor reciprocal crosses.

Although mapping by means of comparative recombination frequencies has been extensively used in the analysis of genetic fine structure, and may be the only practicable method (see below), its sensitivity to extraneous sources of error as well as its inability to compensate for intrinsic variability in the recombination mechanism itself, such as that introduced by localised negative interference (pp. 51, 64), renders it suspect when the distances between mutational sites become extremely small. The only unambiguous method for determining the order of sites is by means of 3-factor reciprocal crosses. The principle of the method, as applied to merozygotic systems such as transduction, is set out in Fig. 38. It is assumed that we know, from comparative

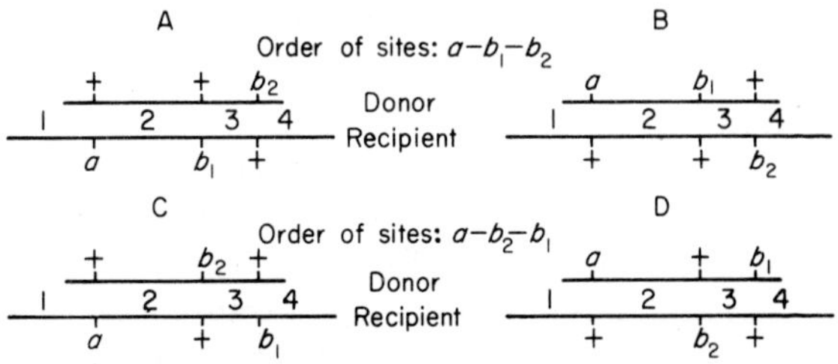

FIG. 38. The mapping of sites by reciprocal, three-factor crosses. (Transduction).

A and B are reciprocal crosses in which each parent is alternately used as donor and recipient. The order of sites is assumed to be a—b_1—b_2.

Similarly, C and D represent reciprocal crosses, but the order of the sites is here assumed to be a—b_2—b_1.

To obtain wild type (+ + +) recombinants, cross-overs must occur in the following regions in each cross:

Order a—b_1—b_2 {A: 1 and 3; B: 3 and 4} Ratio A/B recombinants = about 1.

Order a—b_2—b_1 {C: 1, 2, 3 and 4; D: 2 and 3} Ratio C/D recombinants = low.

On the assumption that four cross-overs occur much less frequently than two, the ratio of the number of recombinants from reciprocal crosses indicates the order of the three sites.

recombination frequencies, that site *a* lies to the left of the *b* region but do not know whether the order of the three sites is a—b_1—b_2, as in Fig. 38A, or a—b_2—b_1, as in Fig. 38C. Reciprocal crosses, in which each parent is alternately used as donor and recipient, are set up by growing the transducing phage on each mutant strain and then using it to transduce the other strain; the two crosses are represented in Fig. 38A and B. You will see that in both these crosses, where the order is assumed to be a—b_1—b_2, two cross-overs are required to yield wild type (+++) recombinants; in regions 1 and 3 for cross A, and in regions 3 and 4 for cross B. This means that if the order a—b_1—b_2 is the correct one, the numbers of wild type recombinants issuing from the reciprocal crosses will not be very different. But suppose the order is really a—b_2—b_1; what results may we expect in this case? In Fig. 38C and D this order is assumed; you will see that, in cross C, *four* cross-overs (in regions 1, 2, 3 and 4) are now needed to produce +++ recombinants while the reciprocal cross, D, requires only two cross-overs so that the ratio of the number of wild type recombinants issuing from cross C to that from cross D will be low.

In merozygotic systems where the genetic contribution of the donor to the zygote is only fractional, reciprocal crosses of the kind shown in Fig. 38 can be made by merely reversing the donor-recipient relationship. In the case of transformation and transduction this is a simple procedure since this relationship is arbitrarily determined by selecting one or the other of the two parental strains as the source of transforming principle (DNA) or of transducing phage respectively. In the case of conjugation in *E. coli*, however, a reversal of the donor-recipient relationship is much more difficult since the donor state is determined genetically and not by the experimental procedure (see Chapter 22).

In systems where the complete genomes of the two parents participate in recombination, as with bacteriophage, reciprocal crosses can

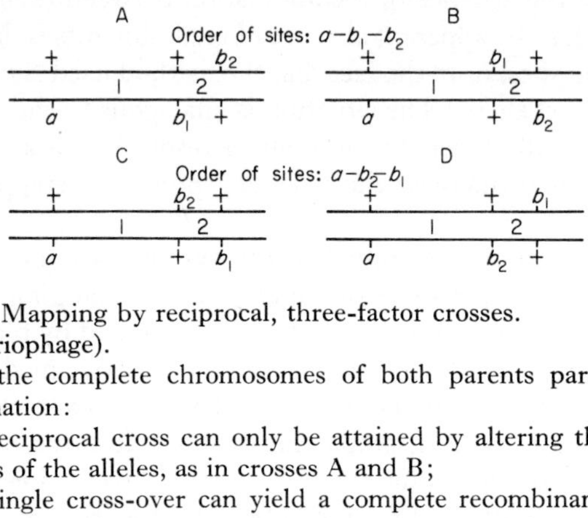

FIG. 39. Mapping by reciprocal, three-factor crosses. (Bacteriophage).

Since the complete chromosomes of both parents participate in recombination:

1. A reciprocal cross can only be attained by altering the parental couplings of the alleles, as in crosses A and B;

2. A single cross-over can yield a complete recombinant chromosome.

To obtain wild type (+ + +) recombinants, cross-overs must occur in the following regions in each cross:

Order a—b_1—b_2 {A: 2; B: 1 and 2} Ratio A/B recombinants = high.

Order a—b_2—b_1 {C: 1 and 2; D: 2} Ratio C/D recombinants = low.

be made only by having *different associations of the alleles in each parent* in the two crosses. This is shown in Fig. 39. In cross A the genotypes of the parents are $++b_2$ and $a\ b_1+$, while in cross B they are $+b_1+$ and $a+b_2$; that is, the parental couplings of the alleles have been switched in the two crosses. This means that reciprocal crosses require the synthesis of two new strains by means of preliminary recombination experiments, and it is obvious that the amount of work involved is prohibitive when a large number of closely linked sites must be mapped. Hence, fine structure mapping of a large number of mutational sites in bacteriophage must rely on comparative recombination frequencies.

THE INTERPRETATION OF FINE STRUCTURE ANALYSES

The resolving power of genetic systems (Pontecorvo, 1958).

The ultimate limit of genetic analysis is reached when it is possible to define, by experiment, the closest mutational sites which are capable of separation by recombination. If the number of mutations at our disposal within a single locus is large, we may anticipate that some of them will have occurred at identical sites, that is, are identical alleles as defined by the absence of recombination between them. But how do we know that the apparent absence of recombinants is due to actual identity of the sites, or to the fact that the method used is not sensitive enough to detect them? The situation is analogous to the discrimination of very small objects at the limit of resolution of a microscope. The solution is to increase the resolving power by stepping up the magnification.

The magnification of microbial genetic systems is a function of their ability to select very rare recombinant classes emerging from large populations of zygotes. It can therefore be increased by plating larger numbers of zygotes, or their progeny, on the selective medium so that the probability of obtaining the desired recombinant class is proportionately increased. If, when this is done, it turns out that the number of recombinants for the most closely linked sites within the locus rises markedly, while recombinants for the presumptively identical sites continue absent, then it is usually assumed that the sites are, in fact, identical. An example from transduction experiments involving histidine-requiring (*his*) mutants of *Salm. typhimurium* will make the point clear. The majority of crosses between various *his* mutants, where 2×10^7 recipient bacteria were plated after infection by the transducing phage at a multiplicity of 2·5, yielded at least ten prototrophic recombinant colonies. A small number of crosses, however, failed to give any recombinants. When the number of recipient bacteria plated was increased to 2×10^8, and the multiplicity of infection to 10, the minimum number of recombinant colonies emerging from the previously fertile crosses rose to 100, while the unproductive crosses remained unproductive (Hartman, 1956). In fact, a 100-fold increase in the resolving power of the system still fails to yield recombinants for *his* mutants which are unproductive under standard conditions. Either their mutational sites are truly identical as judged by recombination, or else the efficiency of recombination falls dramatically when the distance between sites drops below a certain critical value. On the other hand, the majority of sites found to be identical by recombination tests, both in bacteria and bacteriophages, can be distinguished by other criteria such as their rates of mutational reversion to wild type, suggesting that several types of mutational alteration may occur within a single site. We will discuss this problem further when we come to consider the molecular basis of mutation (Chapter 13, p. 301).

An important limitation to the degree of resolution obtainable by recombinational analysis is the frequency with which the parental mutants revert spontaneously to wild type, for it is impossible to distinguish, in any practicable way, between wild type clones resulting from recombination and those due to mutational reversion. Thus at high resolution the number of recombinants may have to be scored against a background, or 'noise level', of wild type mutants. As the resolution is increased the noise level will rise until the meaningful information can no longer be heard. We find ourselves here confronted with a problem somewhat similar to that of 'indeterminacy' in physics, whereby velocity and position become mutually exclusive properties of sub-atomic particles when we try to measure them. For example, in attempting to estimate the distance between adjacent sites, we can begin by making sure that the mutations to be mapped really involve only single sites; but the criterion that this is so is that the mutants should be unstable and revert to wild type with demonstrable frequency. Alternatively we can attempt to endow our system with the maximum resolution by working with stable mutants; but these are likely to be multisite mutants which have suffered deletions covering a variable number of adjacent sites so that the resolution of the analysis again becomes self-limiting. In practice, of course, a compromise must be adopted by utilising only the minority of mutants that revert with a demonstrable but low frequency.

Measurement of the gene and its components.

The ultimate goal of fine structure analysis is to achieve a resolving power which can discriminate between components of the gene at the molecular level so that their behaviour can be interpreted in terms of the molecular architecture of the genetic material. The first step in such an interpretation is to find a way of converting the abstract measurements, in recombination frequencies or units, which genetic analysis provides into real measurements of absolute length. In those cases where a large number of genetic loci have been mapped in bacteria or bacteriophages, only a single linkage group has been found; that is, there is only one chromosome per bacterial nucleus or phage particle. We shall see in Chapter 11 (p. 227) that it is now possible to define the genetic material as deoxyribonucleic acid (DNA) and to estimate the amount of it in each bacterial nucleus or phage particle by chemical or other means. Thus measurement of the dimensions of the gene, or of the distance between mutational sites, in terms of DNA structure becomes possible if we can express the small recombination frequencies obtained in fine structure analysis as a reasonably true proportion of the length of the chromosome as a whole. Several complications arise when we attempt to do this.

We have seen in Chapter 3 (p. 46 *et seq.*) that the length of the genetic map of a chromosome is determined, in effect, by adding together the recombination frequencies obtained between pairs of adjacent loci distributed along the length of the chromosome. If, however, pairs of loci happen to be rather far apart, the occurrence of double exchanges between them will reduce the recombination frequency so that they appear closer than they really are; that is, the length of the map will be under-estimated. For this reason, as well as to ensure that some loci are located near the extremities of the chromosome, the number of loci mapped must be large. This is especially so if, as appears to be the case in *E. coli* and probably also in other bacteria, recombination events occur very frequently between pairs of chromosomes, since the distances over which double exchanges are unlikely will be proportionately reduced. In fact, the map length of a chromosome is directly proportional to the probability of recombination per unit of its length. Thus formally speaking, the chromosomes of *Drosophila* males, which show no recombination at meiosis, have a map length of zero; at the other extreme, the map length of the single chromosome of *E. coli* has been estimated to be about 2000 recombination units (Jacob and Wollman, 1961a; see p. 695) while that of bacteriophage T4 appears to be about 800 units (Benzer, 1957; but see also Stahl, Edgar and Steinberg, 1964).

Assuming that the total map length is known with some accuracy, the further difficulty of localised negative interference arises when we try to relate it to the very small distances inherent in fine structure analysis. At these distances it has been shown, in *Aspergillus* (Pritchard, 1955), *E. coli* (Jacob and Wollman, 1961a, p. 229) and bacteriophage (Chase and Doermann, 1958), that the frequency of recombination may greatly increase so that the regions involved will apparently shrink in proportion to the rest of the map. Little is known about the degree of such negative interference and the distances over which it may operate in bacteria, so that it cannot be properly corrected for (see Maccacaro and Hayes, 1961; also Chapter 15, p. 379 *et seq.*).

In addition to all these factors, other imponderables present themselves. For example, we do not know whether, apart from interference, the probability of recombination remains constant per unit length of chromosome; nor are we sure, when attempting to interpret genetic findings in terms of DNA structure, whether all the DNA of the chromosome is genetically functional. Our calculations are therefore dependent on a number of assumptions, based largely on good guess-work, so that although recombinational analysis may indicate the scale of the genetic fine structure, at the present time it falls far short of being a precision instrument.

In the light of what has been said, let us look briefly at the three systems which have mainly been used for fine structure analysis, namely, bacteriophage T4, transduction in *Salmonella* and conjugation in *E. coli*. The bacteriophage system has undoubtedly yielded the greatest amount of detailed information about the fine structure of the gene, due mainly to the large number of mutants which have been isolated, the great sensitivity of the method for selecting recombinants, and the simplicity of the crossing techniques (see p. 169). Moreover since it is the only system in which the entire chromosomes of both parents participate in recombination, the total map length can readily be estimated if it is assumed that the complication of successive rounds of random mating is adequately compensated for by the formula of Visconti and Delbrück (1953; see p. 62). This formula, however, does not take account of localised negative interference, while the functions of those phage genes which are best adapted to fine structure analysis and complementation tests are, unfortunately, not understood at the biochemical level.

In transduction, on the other hand, the fragmentary nature of the genetic contribution from the donor bacteria precludes the possibility of establishing the distances between any loci that are not closely linked, since these are never transferred together to the same zygote; only the linkage relationships of loci which are not more than about one hundredth the total length of the chromosome apart can be effectively mapped in relation to one another. It follows that the total map length cannot be estimated by genetic means. The great advantage of transduction in bacteria is that the functions of the particular genes under study may be precisely specified by both biochemical and complementation tests, so that the relationship of genetic structure to biochemical behaviour can be studied (see p. 158 *et seq.*).

Conjugation in *E. coli* is also a merozygotic system and might be expected to possess all the advantages and disadvantages of transduction as an analytical tool, except that much longer regions of the chromosome can be mapped. It turns out, however, that the unique and peculiar features of genetic transfer in conjugation permit the chromosome to be mapped and its entire length measured in *absolute units which are independent of recombination.* Conjugation therefore appears to be an ideal instrument for fine structure analysis from the genetic point of view. Formerly it suffered the disadvantage that it lacked a system for performing complementation tests but, with the advent of F-prime factors and sexduction which serve this purpose admirably, the full possibilities of conjugation are beginning to be realised. We will defer a fuller discussion of this system to Chapter 22, p. 674.

Estimation of the number of sites per locus.

A gene locus is here defined as a region of chromosome which determines a single function at the biochemical level. We will discuss what is meant by this more precisely in the next section. A mutation affects the gene in such a way that its function is altered or destroyed. Multisite mutations may exceed an entire locus in length and are usually ascribed to actual deletions of the genetic material; on the other hand, the so-called 'single-site' or 'point' mutations, which are reversible, are limited to very small fractions of the locus called sites. The main achievement of fine structure analysis has been to show that mutation may result from alterations at any one of a number of sites within the locus.

How many sites are there per locus? The ability to answer this question is important since it provides an experimental basis for relating molecular structure to function. One approach is based on the ratio of the recombination frequencies between the closest and the most widely separated sites within the locus and is subject to the various sources of quantitative error mentioned in the previous section. Thus if recombinants for the most widely separated sites arise with a frequency of 1 per cent, and those for the closest sites with a frequency of 0·01 per cent, the number of sites in the locus would be about 100. Such estimates for a number of different loci, in organisms ranging from *Drosophila* to bacteriophage, are consistent in suggesting that the number of sites per locus may be of the order of many hundreds (Pontecorvo, 1958).

Other methods, which do not rely on recombination frequencies, are based on statistical treatment of the actual numbers of non-identical and identical sites found within a locus, and tend to give lower estimates although these are still of the order of 100 or more (Demerec and Hartman, 1959; Benzer, 1957, 1961). None of these methods allows for the likelihood that there are many sites at which mutation may occur without producing any alteration in phenotype. For this and other reasons, therefore, the estimates are certainly low. Moreover there are other phenomena which must be taken into account in making estimates of this sort. Among the more important of these is the tendency of mutations to pile up at particular sites or 'hot spots' (Benzer, 1957, 1961; Fig. 49, p. 175), and the fact that mutations induced by different classes of mutagenic agent may involve quite different sites (Brenner *et al.*, 1958; Benzer, 1961; see Chapter 13, p. 302).

GENETIC STRUCTURE AND FUNCTION: QUALITATIVE ASPECTS OF FINE STRUCTURE ANALYSIS

The 'one gene–one enzyme' hypothesis of Beadle and Tatum (Chapters 5 and 6, pp. 87 and 112 *et seq.*) postulates that the function of a gene is to determine the structure of a protein, that is (usually), of a single enzyme which mediates a particular step in biosynthesis. We should, perhaps, mention here that we now know that some genes and or, at least, chromosomal regions, do not specify protein structure, but, instead, the structure of certain ribonucleic (RNA) components of the cell (p. 289), but this in no way affects the validity of the hypothesis from our present standpoint.

Bacterial mutants suitable for fine structure analysis are initially chosen on the basis of inability to synthesise a particular amino acid such as tryptophan or histidine. As explained in Chapter 5, these mutants may then be sub-divided into groups, each of which is blocked at a specific step in the synthesis, by cross-feeding experiments and biochemical tests. If detailed knowledge of the pathway is available, the biochemical criteria for involvement of a single, specific step can be made very rigorous. For example, it can be shown that the mutants of a particular group are identical in their growth response to certain chemical intermediates but not to others, in the intermediates they accumulate and in their lack of a single enzyme which, when extracted from wild type bacteria, can carry out, *in vitro*, the conversion of which the mutants are incapable. When the mutational sites of such a group of mutants are mapped they are invariably found to be clustered together in a small region of chromosome (locus) which is distinct from that occupied by the mutational sites of other functional groups serving the same synthetic pathway. The sites of mutations affecting different synthetic steps are never found to overlap or to be interspersed. Where all these criteria are satisfied, therefore, we may be very sure that all the mutations within the group affect the structure of one enzyme only, although each non-identical mutation probably does so in a different way.

An independent method of ascertaining whether two mutations belong to the same or different functional groups is the complementation test in which the mutations are introduced into the same cell on opposite chromosomes (*trans* position) (see p. 93; Fig. 24). If the cell displays the wild phenotype the inference is that the two defective genomes can, between them, provide all the requirements for cellular growth, so that the mutations must involve different functions. Conversely, continuance of the mutant phenotype indicates identity of function. We might reasonably expect the results of complementation tests and those based on the biochemical characterisation and mapping of mutants to coincide. It turns out, however, that this is not always so. Many cases are now known, especially in *Neurospora* and yeast but also in bacteria, where, among a series of mutational sites all blocking synthesis of a single enzyme and all mapping together within a single locus, some complement one another and form active enzyme, while others do not. This phenomenon is known as *inter-allelic* or *intra-genic complementation*.

The features of interallelic complementation.

We do not propose to examine this rather complex phenomenon in any detail here, but only to say enough to reveal its nature and to indicate how it affects our concepts of the relation between genes and enzymes.

We have seen how deletions may be represented on a genetic map as lines overlapping those sites or other deletions with which they do not yield recombinants (Fig. 36, p. 135). In the same way, complementation maps of genetic regions or loci may be drawn as a series of lines, each line representing groups of mutants which do not complement one another but behave identically in complementation tests with other mutants. For example, in Fig. 40, the three lines A, B and C represent

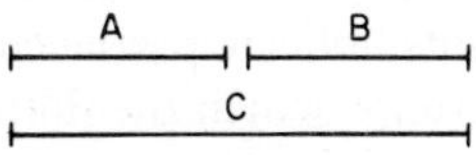

FIG. 40. Linear representation of complementation relations between three groups of mutants, A, B and C.

Pairs of mutants within any group fail to show complementation.

Any mutant of Group A complements any mutant of Group B.

Mutants of Group C do not complement any mutant of either Groups A or B.

three groups of mutants. No pair of mutants within any one of these groups are complementary; lines A and B do not overlap, indicating that any group A mutant is complemented by any group B mutant. Group C mutants belong to a different class in that, not only do they not complement one another but, in addition, they fail to show complementation with any group A or group B mutant; hence, line C overlaps both A and B. It is important to understand clearly that complementation maps indicate only the complementation *patterns* displayed by groups of mutants and are not necessarily related in an overt way to genetic maps of the mutational sites of these mutants; nor do the lengths of the lines in complementation maps indicate any numerical relationship to the number of mutants. In Fig. 40, for example, line C could equally well represent a single mutant or twenty mutants which failed to complement A and B mutants.

Now turn to Fig. 41 in which are compared the genetic and complementation maps of a region of *Salm. typhimurium* chromosome controlling the synthesis of histidine (Hartman *et al.*, 1960a, b), ignoring for the moment the adjacent locations of the various loci, H, B, C, D and G (see p. 160). It has been shown by stringent biochemical analysis that all the mutants which map together in each of the loci B, C and D fail to synthesis a particular enzyme (Ames *et al.*, 1960). When the biochemically and genetically homogeneous mutants of each locus are tested in pairs (by abortive transduction, p. 96) for complementation with one another, as well as with the mutants of other loci, they are found to yield quite different patterns as the complementation map reveals (Fig. 41, lower map). For instance, the thirteen mutants of locus C behave in a reasonable way in that none complement one another while all show complementation with mutants from other loci; in contrast, the mutants of locus D fall into three complementation groups analogous to those of Fig. 40, while four clearcut groups can be defined in locus B. The dotted lines on the complementation maps of some of the locus B mutants indicate partial complementation with the mutants of those other groups which the dotted lines overlap; that is, 'minute' colonies indicating abortive transduction were found, but these were much smaller than those produced when wild type donors were employed.

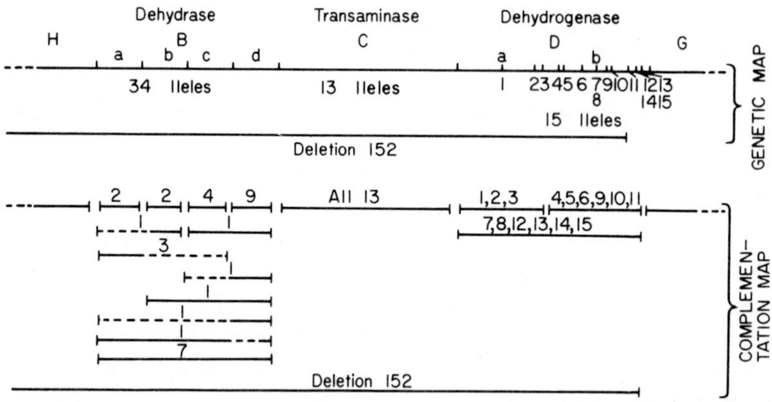

FIG. 41. Genetic and complementation maps of loci determining various enzymes in the pathway controlling histidine synthesis in *Salmonella*.

The figures of the complementation maps of the B locus indicate the total number of mutants belonging to each complementation class.

The italicised figures on the D locus complementation maps indicate individual mutants, the location of whose sites are shown on the genetic map. Data from Hartman *et al.*, 1960a, b.

In these studies the mutational sites of mutants belonging to loci B and C were not accurately mapped so that the genetic and complementation maps cannot be correlated. In the case of locus D, however, the relative positions of a number of sites, and the distances between them, were assessed by ratio tests (p. 138) and are approximately indicated in Fig. 41 where, for this locus, the italicised numbers refer to individual mutant and their sites. It will be seen that whereas the sites of mutants of complementation groups a and b are located in separate regions of the genetic map of the locus, so that the genetic and complementation maps coincide, the sites of two of the six mutants (*7* and *8*) which do not complement groups a or b fall in the Db region among the b mutant sites. Finally the similarity of the genetic and complementation map of mutant 152 may be noted; this mutant has a large deletion covering loci H, B, C and D, produces none of the three defined enzymes, but does yield recombinants with mutants whose sites lie in locus G.

The nature of interallelic complementation.

An explanation of interallelic complementation is best sought among studies of this phenomenon in *Neurospora* where the availability of stable, complementing heterokaryons (p. 94) permits a fuller investigation of the nature of the enzymes produced. In general, four characteristic features of interallelic complementation are becoming apparent (Fincham, 1962a).

1. The majority of mutants are non-complementary with all others at the same locus and, of these, most are point mutations so that their behaviour cannot be ascribed to deletions covering the locus. This suggests that the gene is not divided into sub-sections determining discrete, non-overlapping functions for, if this were so, the genetic and complementation maps would coincide as, in fact, they do for *different* genes; the alternative is that the function of the gene is unitary but that mutation at a number of different sites can affect it in different ways.

2. For interallelic complementation to occur, the derangement of synthesis of the protein in each of the two mutants must not be severe. Thus in the case of glutamic dehydrogenase and tryptophan synthetase in *Neurospora*, complementing mutants either continue to produce an enzyme of very low activity or else yield an inactive protein which, nevertheless, reacts serologically with antiserum prepared against the enzyme.

3. The level of enzyme activity shown by heterokaryons in interallelic complementation is always low and rarely exceeds 25 per cent of the activity found in wild type strains or as a resultof complementation between *different* genes.

4. The enzyme produced by interallelic complementation can often be shown to be, and perhaps always is, *qualitatively* different from wild type enzyme in such properties as temperature lability, and may also be distinguishable from the relatively inactive enzyme synthesised by the complementing mutants.

The specificity of proteins is undoubtedly determined at the genetic level by a linear arrangement of sites within the genes. The proteins themselves, however, are three-dimensional structures whose configurations and unique activities are governed by the highly specific ways in which their primary polypeptide chains are folded. This specific folding of the molecule is exclusively layed down by the sequence of amino acids along the polypeptide chain of which it is composed; that is, the precise folding of the primary polypeptide is determined by such factors as cross-linkages between the side-chains of particular amino acids located at specific positions in the chain. On this basis the effect of mutation is to change the linear sequence of amino acids in the polypeptide so that either the architecture of the protein molecule as a whole, or the molecular groups on which its specific activity depends, is changed (Chapter 12, p. 259 *et seq.*).

It has recently been found that many large protein molecules are not unitary structures but are built up of subunits which may or may not be identical. The most striking and best worked out example of this is the haemoglobin molecule, which has been shown by X-ray diffraction analysis to comprise four subunits which are fitted together in a simple and very symmetrical way (Perutz *et al.*, 1960). The subunits consist of two chemically different types of polypeptide chain, called α and β, each of which closely resembles a myoglobin molecule in its three-dimensional structure (Kendrew, 1960; see Plate 11, p. 262) and which are assembled in identical pairs. The molecule of haemoglobin can thus be written in the form $\alpha_2\beta_2$. Various abnormal types of haemoglobin may arise in man as a result of mutation (p. 27, 174) and can be shown by physico-chemical analysis to be due to alteration in either the α or the β chain.

Familial data have provided genetic evidence that synthesis of the α and β chains is determined at different loci which are probably not closely linked so that assembly of the polypeptide chains to form haemoglobin molecules must occur *after* their synthesis and is not directly determined at the genetic level. In support of this, cases have been described of individuals who appear to be heterozygous for alleles at each locus. For example, such individuals may inherit a defect in the α chain ($\alpha_2^x\beta_2$) from one parent and a defect in the β chain ($\alpha_2\beta_2^x$) from the other. The red blood cells of such individuals contain *four* different types of haemoglobin molecule, sometimes in approximately equal amounts (Itano and Robinson, 1960; Baglioni and Ingram, 1961). These four types represent the four possible combinations of normal and abnormal chains; the two parental types $\alpha_2^x\beta_2$ and $\alpha_2\beta_2^x$, and the two 'recombinant' types $\alpha_2\beta_2$ (normal haemoglobin) and $\alpha_2^x\beta_2^x$ (haemoglobin defective in both chains). The conclusion is that at the level of protein structure new types of molecule, which do not reflect the genotype, can arise from the random reassortment of polypeptide subunits.

Arrogant microbial geneticists should note that this is not the first time that the biochemical and genetic study of man has helped to illuminate their problems! The formation of functional protein by the aggregation of polypeptide subunits offers a plausible and satisfying explanation of the general facts of interallelic complementation although, of course, we may expect the details to differ from case to case with the vagaries of protein structure; whether, for instance, there are two or more subunits (Catcheside and Overton, 1958; Fincham, 1959).

Let us suppose, by way of illustration, that a particular enzyme is composed of two identical subunits which, alone, do not have appreciable enzyme activity. Fig. 42A shows, schematically, a number of different ways in which the subunit could be rendered defective by mutation in the locus determining its formation. The defects may be visualised as interfering with the aggregation of two subunits or, if they do aggregate, as altering the configuration of the resulting protein so that its function is impaired or absent. When, in complementation tests, subunits altered in different ways are synthesised in the same cell, some pairs, represented as having non-overlapping defects in Fig. 42A, may be able to compensate one another's defects and so form a 'hybrid' structure which is stable and possesses some enzyme activity but which, nevertheless, is different from wild type enzyme. Other pairs, shown in Fig. 42A as having overlapping defects, cannot compensate each other in this way so that stable, hybrid molecules are not formed (see Fincham, 1960; 1962a). It will be seen from Fig. 42A, B, C

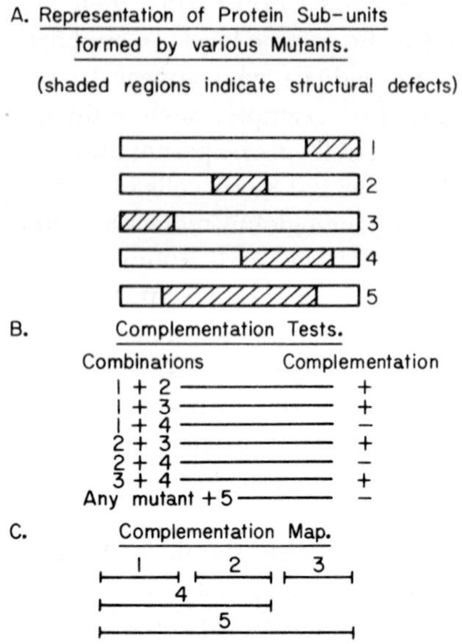

FIG. 42. Inter-allelic complementation interpreted in terms of alterations in the structure of protein subunits (polypeptide chains). In this illustration, active enzyme is formed by the aggregation of two *identical* subunits.

that more than two complementing groups may occur within a locus in systems involving only two identical sub-units, so that the number of subunits cannot be inferred from the complexity of the complementation map. On the other hand, the absence of interallelic complementation (Fig. 41, locus C) would, on this view, suggest that the enzyme concerned consists of only a single polypeptide chain. Surprisingly, however, this turns out not to be the case since the transaminase determined by gene C can be dissociated into what appear to be two identical subunits (Loper *et al.*, 1964).

Another feature of this attractive concept is that, as well as explaining the general characteristics of interallelic complementation (p. 149), it also allows for the fact that the mutational sites of non-complementing mutants may be scattered anywhere in the locus (see Fig. 41, locus D, sites 7 and 8). The absence of complementation merely implies that the mutation has so altered the subunit that it is incapable of participating in the formation of a hybrid molecule; it is probable that there are many critical points along a polypeptide chain where a mutation will wreck the structure. It is known, for example, that a high proportion of revertable, spontaneous mutations may result in gross derangements of protein synthesis.

Experimental evidence is accumulating that some, at least, of the enzymes involved in interallelic complementation owe their activity to the interaction of subunits. For example, we have seen that the *his*B locus of *Salm. typhimurium*, which determines the enzyme imidazoleglycerol phosphate dehydrase, accommodates four complementary groups of mutants (Fig. 41, p. 150). It turns out that when *extracts* of pairs of complementary mutants from these groups are mixed *in vitro*, enzymic activity is restored. Mutants of the Bc group are unique in lacking the enzyme histidinol phosphate phosphatase in addition to the dehydrase, but immunological and physico-chemical tests show that the activities of the two enzymes, as well as that of normal dehydrase, reside in closely similar molecules. It is therefore likely that the *his* B locus controls the formation of a single, bifunctional protein formed by the interaction of similar subunits (Loper, 1961).

Another example of enzyme complexity is that of tryptophan synthetase in *E. coli*. This enzyme catalyses the last three steps in tryptophan synthesis and its activity has been shown to result from the association of two chromatographically different protein sub-units, A and B. These two components are genetically determined by distinct but adjacent loci in the tryptophan region. Purified A and B components appear to be responsible for different aspects of synthesis, but separately display only feeble activity. Normal synthetase activity is only attained by mixtures of the two components. It therefore seems probable that all three reactions are catalysed on a single surface formed by the union of the A and B sub-units (Fig. 43) (Crawford and Yanofsky, 1958; Yanofsky, 1960, 1964; see p. 118).

Neurospora tryptophan synthetase, on the other hand, is genetically determined by a single locus, *try*-3, and does not appear to be separable into sub-units, although it catalyses the same three reactions as does the *E. coli* enzyme. The majority of mutations within the *try*-3 locus yields protein which retains the immunological specificity of the enzyme but has lost one or more of its catalytic functions. It turns out that the mutational sites involving different enzymic activities of the protein are clustered towards different extremities of the *try*-3 locus, while those resulting in complete functional deficiency are located in the central region (Bonner *et al.*, 1960). This again suggests a single molecule but one having functionally differentiated regions which,

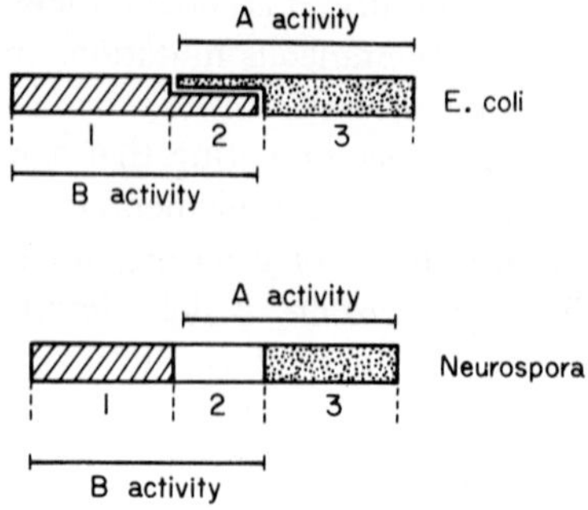

FIG. 43. A schematic representation of the relation between molecular structure and function in tryptophan synthetase from *E. coli* and *Neurospora*.

In both cases the structural integrity of region 2 is necessary for proper manifestation of the different functions of regions 1 and 3.

nevertheless, share a common structural component. In Fig. 43, the behaviour of tryptophan synthetase from *E. coli* and *Neurospora* is compared in a schematic way. In addition to the enzymes imidazoleglycerol phosphate dehydrase and tryptophan synthetase, there is evidence that the molecule of *Neurospora* glutamic dehydrogenase, as well as that of β-galactosidase from *E. coli*, consists of a number of identical subunits (see Fincham, 1962b; Jacob and Monod, 1961b).

Recent studies of *in vitro* complementation, using highly purified, antigenically-related proteins extracted from strains showing interallelic complementation, have provided proof of the hypothesis that restoration of enzyme activity is due to the union of defective polypeptide subunits of each type to form hybrid protein molecules. The first and most clearcut demonstration was provided by the enzyme alkaline phosphatase in *E. coli* (Schlesinger and Levinthal, 1963, 1965; Schlesinger, 1964). The wild type enzyme is known to be composed of two polypeptide subunits which can be reversibly dissociated at low pH to yield inactive monomers. Highly purified mutant proteins can similarly be dissociated into monomer subunits. When monomeric preparations of two such complementing, mutant proteins were allowed to interact at neutral pH, a 25-fold increase in phosphatase activity was observed. This active protein was shown to have twice the molecular weight of the monomer, as well as an electrophoretic mobility midway between those of the two mutant proteins, and to be distinct from the wild type enzyme in its heat lability, while its formation obeyed bimolecular kinetics. Finally, the hybrid nature of the active protein was directly demonstrated by incorporating heavy isotopes of hydrogen (^{2}H) and of nitrogen (^{15}N), as well as a radioactive sulphur (^{35}S) label, into one of the proteins so that it was denser than normal. When a monomeric preparation of this dense protein was mixed with an excess of the complementing protein of normal density, the resulting active enzyme was shown, by density gradient centrifugation (p. 243), to lie midway between the two parental proteins in density (Schlesinger and Levinthal, 1965).

Essentially similar findings have been obtained in the case of glutamic dehydrogenase of *Neurospora crassa*, in which the active enzyme appears to comprise 6–8 polypeptide subunits (Fincham and Coddington, 1963; Coddington and Fincham, 1965). The pairs of mutant proteins were here isolated in purified form by column fractionation, and monomeric preparations made by acidification. It was shown that a radioactive label (^{35}S) incorporated into either mutant protein always appeared in the active enzyme formed by *in vitro* complementation, thus proving the participation of both proteins in its formation.

The subject of interallelic complementation is reviewed by Catcheside (1960, 1964), by Fincham and Day (1963), by Schlesinger and Levinthal (1965) and by Siddiqi (1965). In addition, the Brookhaven Symposium in Biology, No. 17, 1964 is devoted to 'subunit structure of proteins: biochemical and genetic aspects'. A recent monograph is by Fincham (1966).

CHAPTER 8

Fine Structure Analysis in Practice

The new concepts of genetic organisation which have come from high resolution analysis provide a focal point at which genetics and molecular biology meet. They are so important that we need no excuse for looking more closely at some of the results actually obtained with *Salmonella* and bacteriophage.

TRYPTOPHAN SYNTHESIS IN SALMONELLA

Ten independent, tryptophan-requiring auxotrophic mutants of *Salm. typhimurium* were isolated and found to fall into four clearcut groups according to the steps in the pathway of tryptophan synthesis which are blocked in them (Brenner, 1955). The biochemical characteristics of these groups are set out in Table 4, from which it will be seen that the position of the block in each is decisively pin-pointed by *in vitro* conversion studies as well as by the substances it accumulates, the intermediates which stimulate its growth and its ability to cross-feed the mutants of other groups (p. 87 *et seq.*; Figs. 22, 23). The groups can thus be arranged in the functional sequence A, B, C, D, corresponding to the successive steps in tryptophan synthesis which they are unable to perform (Fig. 44, 1).

The order of the four loci on the chromosome was then mapped by means of transduction, using the transducing phage PLT22 as vector (Demerec and Z. Hartman, 1956). It so happens that a particular locus controlling one of the steps in cysteine synthesis (*cys* B) is closely linked to each of the *try* loci, so that *cys*B and the whole of the *try* region are often transferred together to recipient bacteria on the same fragment of donor chromosome; for example, doubly auxotrophic recipients of genotype *cys*B*try*A, *cys*B*try*B, *cys*B*try*C or *cys*B*try*D all readily yield

TABLE 4

Biochemical characteristics of different groups of tryptophan-requiring mutants of *Salm. typhimurium.*

Mutant group	Conversion blocked	Substances accumulated	Growth stimulants	Cross-feeds mutant of groups:
A	→anthranilic acid	...	anthranilic acid indolegylcerol phosphate indole tryptophan	None
B	anthranilic acid→ indoleglycerol phosphate	anthranilic acid	indoleglycerol phosphate indole tryptophan	A
C	indoleglycerol phosphate→ indole	indoleglycerol phosphate	indole tryptophan	A, B
D	indole→ tryptophan	indole and indoleglycerol phosphate	tryptophan	A, B, C,

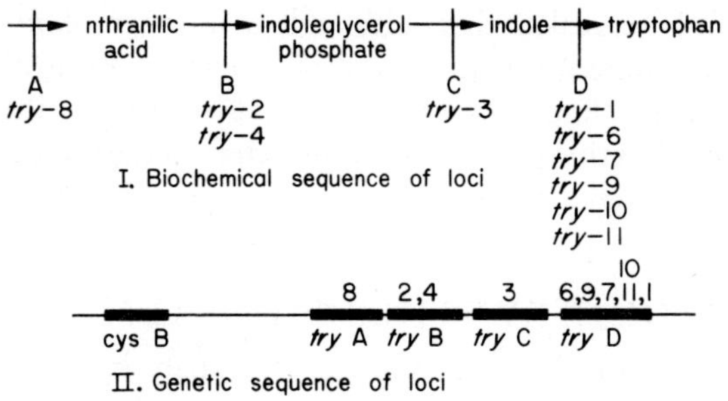

FIG. 44. The sequence of loci determining steps in tryptophan synthesis in *Salmonella*, as determined by biochemical and genetic analysis.

The capital letters represent groups of mutants which are blocked in the same synthetic step.

The arabic numerals indicate independently isolated mutants.

prototrophic recombinants when transduced by phage grown on wild type (++) donors. It was therefore possible to establish an unambiguous order for any pair of *try* loci with respect to the *cys*B locus by means of reciprocal 3-point tests (p. 139; Fig. 38). Such tests, involving all possible combinations of pairs of *try* loci with *cys*B, gave the following relative arrangements:

(1) *cys*B—*try*B—*try*D
(2) *cys*B—*try*A—*try*D
(3) *cys*B—*try*A—*try*B
(4) *cys*B—*try*C—*try*D
(5) *cys*B—*try*A—*try*C
(6) *cys*B—*try*B—*try*C

It is clear from (1) and (3) that the order *try*A—*try*B—*try*D is the only one compatible with the results, while (4) and (6) show that *try*C lies between *try*B and *try*D. Thus all the *try* loci are located close together on the same side of *cys*B, in the order *cys*B—*try*A—*try*B—*try*C—*try*D (Fig. 44, II). This means that the order of arrangement of the loci on the chromosome corresponds to the sequence of the biochemical steps they determine. This striking 'assembly line' arrangement, or *sequential order of genes* as it was called, caused much excitement when it was first discovered since it seemed reasonable to suppose that it had significance not only from the evolutionary but also from the functional point of view. Early results of the genetic analysis of the histidine pathway at first appeared to confirm this but, as we shall see, more detailed study revealed that, although a number of *his* genes *are* sequentially ordered, others are out of sequence. From the study of many loci we now know that the significant point is not the sequential order of genes but that genes determining steps of the same biosynthetic pathway are *clustered* within the same chromosomal region (p. 163).

HISTIDINE SYNTHESIS IN SALMONELLA

The combined biochemical and genetic analysis of the histidine pathway by B. N. Ames and P. E. Hartman, and their colleagues, is by far the most thorough and extensive investigation of a biochemical pathway so far undertaken with bacteria—or, for that matter, with any organism (Hartman *et al.*, 1960a, b; Ames *et al.*, 1960; Loper *et al.* 1964; Smith and Ames, 1964, 1965). These studies include the isolation of over 900 independent, histidine-requiring (*his*) mutants and the mapping of 540 of these by means of transduction; the genetic and biochemical recognition of nine genes serving the pathway, and the isolation and analysis of the physical nature of the various enzymes involved; and, finally, analysis of the results of many intragenic complementation tests (including more than 2,500 in locus B alone!) in terms of the molecular structure of the enzyme concerned (Loper *et al.*, 1964). We have already referred to some of the findings of this study in connection with interallelic complementation (p. 149) and will here outline only its more general features.

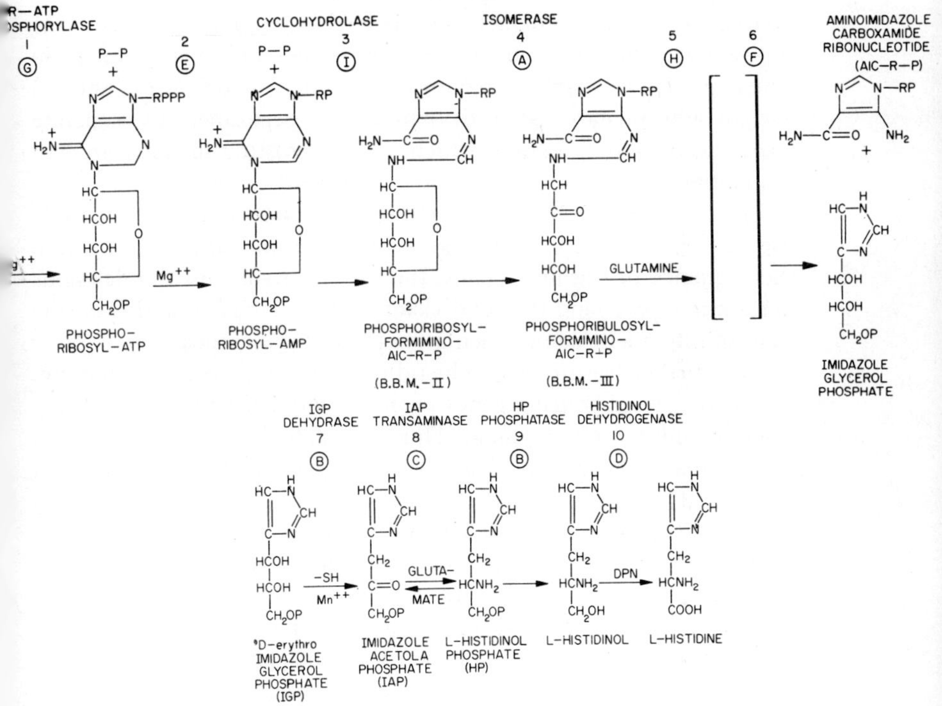

FIG. 45. The biochemical steps in the synthesis of histidine.

The structure of the compound in brackets is not certain. The sequence of the F and H steps could be reversed. P = phosphate, R = ribose, AIC = aminoimidazole carboxamide. The data are derived from the work and papers of B. N. Ames and his colleagues.

The mutants were mapped by crossing them with an extensive range of deletion mutants which enabled them to be allocated to one or another of nearly 40 chromosomal segments, as well as by 2-factor crosses (p. 136) and, in a few cases, by ratio tests (p. 138). Although a number of 3-factor crosses were performed, the absence of an 'outside marker' closely linked to the *his* region, as *cys*B is to the *try* region, precluded the general use of this accurate mapping procedure.

All the mutations were found to map together in a small region, the *his* region, comprising about 0·26 per cent of the total length of the *Salmonella* chromosome. This region is subdivided into nine sub-regions, A–I, each of which is concerned with the synthesis of a different enzyme of the histidine pathway; that is, the pathway is served by nine genes. As a matter of fact there are ten biochemical steps mediating this pathway, but one of the genes (*B*) specifies a bifunctional protein which plays two distinct enzymic roles—that of dehydrase (step 7) and of phosphatase (step 9) which are performed by separate enzymes in *Neurospora* and which, it is interesting to note, are not sequential.

The biochemical steps of the pathway are shown in Fig. 45, while Fig. 46 represents a simplified genetic map of the *his* region showing the sequence of the nine genes, the extent of some of the deletions used in mapping, and the subdivisions of genes *E*, *I*, *B* and *D* on the basis of interallelic complementation (Loper *et al.*, 1964).

Two further features of the histidine region are worth mentioning. One is that neighbouring genes of the region, although adjacent, are discrete and non-overlapping. Thus a mutation which maps within a particular gene always displays the expected biochemical defect characteristic of mutation in that gene; conversely, mutations which are first characterised by biochemical analysis are always found to map in the anticipated gene. Moreover, when large numbers of re-combining sites have been found in adjacent loci, such as *his* C(*c*.107) and *his* D(*c*.125), the low recombination frequencies between the nearest sites in the two loci, together with the clarity with which they can be allotted to their respective genes by complementation tests, implies that the genes are contiguous but do not overlap.

Secondly, as the simplified map in Fig. 46 suggests, by combining the three criteria of biochemical phenotype, interallelic complementation and fertility with the series of deletion mutants, the *his* region can be divided into some 50 sub-regions to one of which every single-site mutant can be clearly allocated. Such a refinement of methodology has

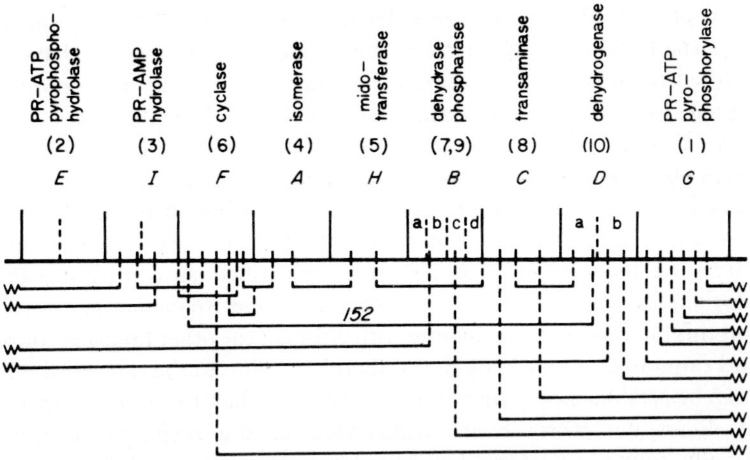

FIG. 46. Genetic map of the histidine (*his*) region of the *Salm. typhimurium* chromosome.

The thick horizontal line represents the chromosome region, divided into 9 biochemically and genetically defined functional loci, marked *E, I, F, A, H, B, C, D, G* in sequence, by the uninterrupted vertical lines.

The dotted lines above the chromosome indicate further subdivisions of loci *E, I, B* and *D* on the basis of interallelic complementation tests (see Fig. 41).

The bracketed figures (1)–(10) indicate the step number in the synthetic pathway of the enzyme determined by each gene (cp. Fig. 45).

The series of horizontal lines below the chromosome delineate the extent of a number of multisite mutations used for mapping. The wave termination of some of these lines at the extremities of the figure show that this end of the deletion has not been defined.

(Adapted from Loper *et al.*, 1964)

all the attributes of an automatic sorting machine in mitigating the labour of fine structure analysis.

THE CLUSTERING OF FUNCTIONALLY RELATED GENES

When the loci determining the various steps of the biochemical pathways concerned in the synthesis of tryptophan and histidine were first mapped, it seemed that these loci were arranged in the same order as the biochemical sequence. Subsequently, many other examples of apparent sequential order were reported. However, as we have just seen in the case of histidine synthesis, more refined analysis has generally shown that, despite some degree of correlation between the sequences of genes and biochemical steps, this correlation is far from perfect. What has been substantiated beyond question, and is important, is that, in bacteria, functionally related genes have a marked tendency to be clustered together in the same chromosomal region. For example, one serine (*ser*B) and four threonine loci (*thr*A, B, C, D) have been found to be closely linked; similarly, an isoleucine locus (*ile*A) and four isoleucine-valine loci (*ilva*A, B, C, D) are all linked, probably in the same sequence as their biochemical reactions (Glanville and Demerec, 1960; see p. 121, Figs. 34, 35). Another threonine locus (*thr*E) has also been identified as controlling the biochemical step which links the threonine and isoleucine pathways, but this locus is not transducible with either of the other linkage groups and cannot be mapped; no doubt its relationship to the other loci will soon be established by means of conjugation whereby larger segments of donor chromosome are transferred to the recipient bacteria. Finally, four proline (*pro*) loci have been recognised in *Salmonella*, at least three of which have been shown to be contiguous (Miyake, in Demerec *et al.*, 1958). Demerec (1964) states that, of eighty-seven loci mediating eighteen pathways in *Salmonella*, over 70 per cent are arranged in clusters of two or more on the chromosome. For a recent summary see Sanderson (1967).

So far as species other than *Salmonella* are concerned, a clustering of genes has been established in *E. coli* for the same four tryptophan loci as in *Salmonella* (Yanofsky and Lennox, 1959) as well as for three loci controlling arabinose fermentation (Gross and Englesberg, 1959), transduction by phage P1 being used in both investigations. In addition, four out of five genes controlling pyrimidine synthesis have been shown to be clustered (Beckwith *et al.*, 1962). Clustering of the genes of nine pathways has been reported in *E. coli*. (Demerec, 1964). For a recent summary see Taylor and Trotter (1967) and Fig. 127 (p. 666). Analysis of *B. subtilis* by transformation has so far revealed four clusters of phenotypically related genes (Ephrati-Elizur, Srinivasan and Zamenhof, 1961; Nester, Schafer and Lederberg, 1963; Anagnostopoulos, Barat and Schneider, 1964). Finally, gene clustering has also been reported in the Actinomycete, *Streptomyces coelicolor* which is amenable to mapping by conjugation (Hopwood, 1965a, b).

Some cases are known, however, where loci serving the same general phenotype are widely separated on the chromosome, as exemplified by arginine-requiring mutants of *E. coli*, mapped by means of conjugation (p. 731). Nevertheless, instances of the clustering of functionally related genes are now so numerous and striking as to constitute the rule rather than the exception in *Salmonella*, *E. coli* and *B. subtilis*, although in *Pseudomonas aeruginosa* it appears to occur rarely, if, indeed, with any significance at all (Fargie and Holloway, 1965). In other organisms, from *Drosophila* to fungi, a very different situation exists, although the distribution of functionally related genes on their chromosomes is apparently not random. Thus in *Neurospora* various tryptophan and histidine loci, which appear to be functionally analogous to those of *Salmonella*, are located in different linkage groups and probably in different chromosomes (Barratt *et al.*, 1954).

How can we account for this clustering of genes in bacteria? Perhaps it would be more meaningful to ask how we can explain the difference between bacteria and other cells in this respect. It is obvious that the arrangement must be an advantageous one, for otherwise it would not have survived the selective pressures of evolution. First of all, it seems reasonable to discard all theories based on differences in the mechanism of gene expression. One could suppose, for example, that if gene products were extremely unstable, their cooperation synthesis might require that the genes be very close together (Pontecorvo, 1950, 1952b); alternatively, it has been suggested that biochemical functions which are implemented in cytoplasmic organelles in other organisms may be performed at the gene surface in bacteria (Demerec and Demerec, 1956). In fact it is now known that, in bacteria, the synthesis of protein takes place in cytoplasmic ribosomal particles, while the enzymes and intermediates they utilise do not differ materially from those of other organisms. More plausible explanations may be found at the level of genetic organisation rather than of function. For example, a striking feature of all parasexual systems in bacteria is the fragmentary nature of genetic transfer; it has been suggested that, in these circumstances, the grouping of genes serving coordinated synthetic pathways would prevent their disruption during transfer and thus enhance the efficiency of sexuality in restoring mutational damage (Clowes, 1960).

Recent studies of one of the mechanisms for the regulation of enzyme synthesis in bacteria appear to have a crucial bearing on the problem. In the case of many synthetic pathways, including those for tryptophan and histidine, the end-product of the chain of reactions (e.g. tryptophan or histidine) has the effect of repressing synthesis of all the enzymes of the pathway. This is known as *coordinated enzyme repression*. The evidence suggests that the repressor is not the end-product itself, but a cytoplasmic entity formed by interaction of the end-product with a substance produced by an independent 'regulator' locus (*R*). The repressor acts on a small, specific region of chromosome known as the *operator* which is very closely linked to the sequence of genes determining the structures of the repressed enzymes. Thus the operator, controlled by the repressor, behaves like a 'master switch' which can turn on, or turn off, the activity of a series of structural genes which are linked to it. Moreover the operator can only influence the activity of those genes which are located on the same chromosome as itself; that is, it cannot switch off the activity of an identical sequence of genes if these are on a second homologous chromosome in a diploid cell. The integrated unit of control, comprising the operator together with its chain-gang of subservient genes, is called an *operon* (Jacob and Monod, 1961a, b; see Chapter 23, p. 716). The structural integrity of the operon is probably essential for the proper functioning of this regulatory mechanism, so that the clustering of genes under the control of a single operator would confer evolutionary advantage. Apart from two cases in *Neurospora*, coordinated enzyme repression has not yet been described in organisms other than bacteria; in higher organisms it may have been superseded by more refined mechanisms of control.

THE FUNCTIONAL GROUPING OF ALLELIC SITES WITHIN LOCI

All mutations at the same locus do not necessarily produce mutants of identical phenotype as we have already seen from the occurrence of interallelic complementation (Fig. 41, p. 150). The resolution of fine structure analysis enables the order of mutational sites within a single locus to be mapped with some confidence, especially in cases where the proximity of two different loci permits the use of 3-factor crosses. The question arises whether there is any relation between the position of a mutational site within a locus and the phenotype of the mutant. A good example of a study of this point is a transductional analysis of the distribution of nine sites within a particular adenine-thiamine locus (*adth*A) of *Salmonella* (Ozeki, in Demerec *et al.*, 1956). Mutants at this locus differ from one another in their growth response to pantothenate as a substitute for thiamine in their double requirement; on adenine-pantothenate medium two mutants did not grow at all (I), five showed heterogeneous growth (II), while two grew normally (III). The sequence of these different classes of mutation was found to be III—II—II—I—I—II—III. A comparable non-random distribution of sites within the *cys*B locus of *Salmonella* has been reported by Clowes (1958a, b; review: Demerec and Hartman, 1959) and we have already noted the differentiation of sites observed by Bonner *et* *al.* (1960) within the *Neurospora* locus, *try*-3, determining production of tryptophan synthetase (p. 155).

As in the case of interallelic complementation, results of this sort can be interpreted at the level of protein structure. If the sequence of amino acids in a polypeptide is laid down by the sequence of sites within the locus, then we may expect alterations at different sites or regions within the locus to alter the protein in different ways. Thus we might deduce from Ozeki's study of the *adth*A locus that the extremities of the particular polypeptide chain which the locus determines can be modified with relative impunity, whereas the protein structure is very sensitive to changes affecting its middle region (see p. 265).

THE GENETIC FINE STRUCTURE OF BACTERIOPHAGE T4

S. Benzer's brilliant and remarkable studies of a small region of the chromosome of bacteriophage T4 approach the ultimate in fine structure analysis, so we have left a description of them to the last (Benzer, 1957, 1961).

When wild type phage T4 is plated on its host, *E. coli* B, it produces rather small plaques with fuzzy edges, designated 'r^+'. If a large number of plaques are looked at, about 1 in 10,000 can be discerned which are easily distinguishable from the r^+ type in being larger and having sharper edges; these are designated '*r*' and are produced by mutant phage particles (see Plate 22, p. 489). Mutants which display the *r* phenotype when plated on *E. coli* B may fail to yield plaques, or may give rise to plaques of either *r* or r^+ type, when plated on two other strains of *E. coli* called S and K, as shown in Table 5. Thus *r*

TABLE 5.

The phenotype (plaque morphology) of *r* mutants of phage T4, isolated on *E. coli* B, when plated on *E. coli* strains S and K. (From Benzer, 1957).

Strain of phage	Strain of *E. coli* B	S*	K*
wild	wild	wild	wild
r_{I}	*r*	*r*	*r*
r_{II}	*r*	wild	...
r_{III}	*r*	wild	wild

* K = *E. coli* K-12, lysogenised by phage λ.
S = non-lysogenic (sensitive) strain of *E. coli* K-12.

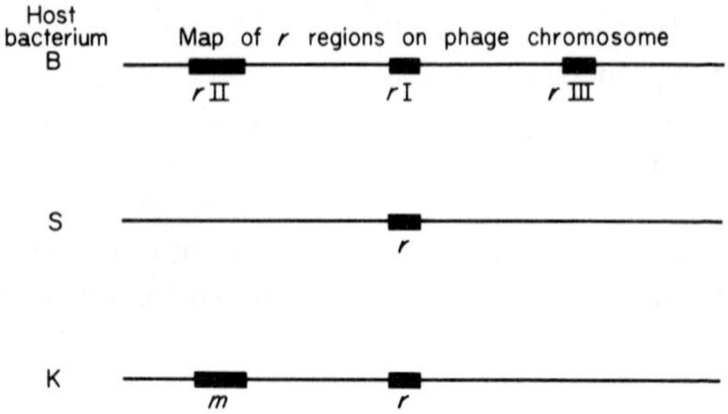

FIG. 47. Representation of how *r* mutations might map on the chromosome of phage T4, when each of the three strains of *E. coli*, B, S and K, is used as host. (After Benzer, 1957.)

mutants can be sub-divided into three phenotypic varieties called r_{I}, r_{II} and r_{III}. If pairs of *r* mutants are crossed by mixedly infecting *E. coli* B, wild type (r^{+}) recombinant phage particles are produced so that the sites of the *r* mutations can be mapped by observing the

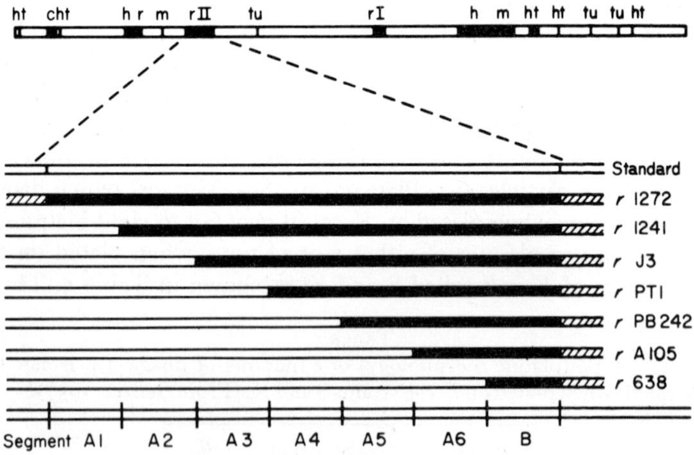

FIG. 48. The sub-division of the r_{II} region of bacteriophage T4 into seven segments by means of deletion mutants.

At the top, the r_{II} region is shown compared with the entire genetic map of the phage. The map is a composite of markers mapped in T4 and the related phage T2. Seven segments of the r_{II} region are defined by a set of deletions beginning at different points and extending to the right-hand end (and possibly beyond as indicated by shading). (Reproduced from Benzer, 1961 by kind permission of the author and the National Academy of Sciences, Washington).

recombination frequencies between them as well as their relation to other markers. It is found that r_{I}, r_{II} and r_{III} mutations map at well separated loci on the phage chromosome (Fig. 47, host B; see also Fig. 48).

It is worth digressing here for a moment to look at an important feature of genetic maps in general which is emphasised by the three varieties of *r* mutant; namely, that they are constructed from a highly selected group of mutations. Those mutations which produce no visible change of phenotype or which, in haploid organisms, are lethal, do not appear on the map. However, the expression of a mutation may depend on environmental factors so that the genetic map may vary as the environment changes. This is well exemplified by Fig. 47 which shows how *r* mutations might be represented if they were isolated and mapped on each of the three *E. coli* strains, B, S and K (Benzer, 1957). Only r_{I} mutants would be isolated on S. On K, r_{III} mutants would appear as wild type and remain unisolated; some r_{II} mutants, however, are capable of growing very poorly in this strain, producing tiny plaques which would not be classified as *r* mutants at all, but as a different type of mutant called 'minute' (*m*).

THE r_{II} SYSTEM AND ITS METHODOLOGY

The system of fine structure analysis in phage T4 was the first to be evolved (Benzer, 1955) and resulted from Benzer's discovery that r_{II} mutants, in general, do not produce plaques on *E. coli* K although they infect and kill this host strain. This discovery was cleverly exploited by Benzer in two important ways. The first was the development of a *complementation* or *cis-trans* test (Table 2; p. 106) for determining whether two r_{II} mutants belong to the same or different functional groups. Mixed infection of K by wild type phage (++), together with phage carrying two r_{II} mutations ($r_{\mathrm{IIa}}r_{\mathrm{IIb}}$), always results in the formation of plaques containing both wild type and doubly mutant particles; thus the wild type phage promotes the growth of r_{II} mutants by making good their functional defects. On the other hand, when K is mixedly infected with various combinations of *pairs* of r_{II} mutants (*trans* arrangement: $r_{\mathrm{IIa}}+/+r_{\mathrm{IIb}}$), neither of which yield plaques on this host when plated alone, either no plaques are produced or normal *r* plaques develop. These plaques arise from complementation and not from reconstitution of the wild type by recombination, since the particles in them usually comprise the two

r_{II} mutant types and not the wild type. On the basis of this test r_{II} mutants can be classified into two clearcut functional groups, A and B.

The second important feature of the r_{II} system is the availability of a highly selective method for identifying and counting recombinants. Crosses between pairs of r_{II} mutants are made in host B on which both mutant and wild type phages plate with equal efficiency. Provided the mutants are not identical, the progeny of the cross will comprise an excess of the two mutant parental types together with a small proportion of wild type recombinant particles whose number will depend on the distance between the r_{II} mutational sites. The total number of progeny is estimated by plating appropriate dilutions of the lysate on B and counting the resulting plaques. If the progeny particles are plated on K, however, only the wild type recombinants can produce plaques so that the ratio of the number of plaques on K to that on B indicates the recombination frequency. When the recombination frequency is fairly high the method can be checked by comparing the number of plaques on K with the number of wild type plaques observed directly by plating on B; the two are found to be the same.

The sensitivity of this method is high enough to detect a recombination frequency as low as 10^{-8}(0·000001 per cent) so that, in practice, it is limited only by the rate at which reversion of the mutants to wild type occurs. Since only mutants which show observable but very low reversion rates (that is, mutants which are not due to deletions) can be mapped with precision (p. 143), an estimation of reversion frequency, by plating known numbers of particles on K, is an essential first step in mapping. The reversion frequency of r_{II} mutants may vary enormously, ranging from less than 10^{-8} for stable mutants to as high as several per cent.

Benzer was the first to utilise deletion mutants for the preliminary mapping of single-site mutants in fine structure analysis (p. 134), and also developed ingenious 'spot' tests which greatly simplified their use. Deletion mutants in the r_{II} region are quite easy to isolate since they can be presumptively recognised by their stability, that is, by the absence of plaques when more than about 10^8 particles are plated on K; the extent of the deletion is then defined by the range of point mutations with which no recombination is obtained.

To test for recombination, two mutants must be grown together on B and the progeny then tested for ability to plaque on K. A simple and semi-quantitative way of doing this in a single operation is to add a drop of a suspension of each of the mutants to a few drops of a culture of *E.*

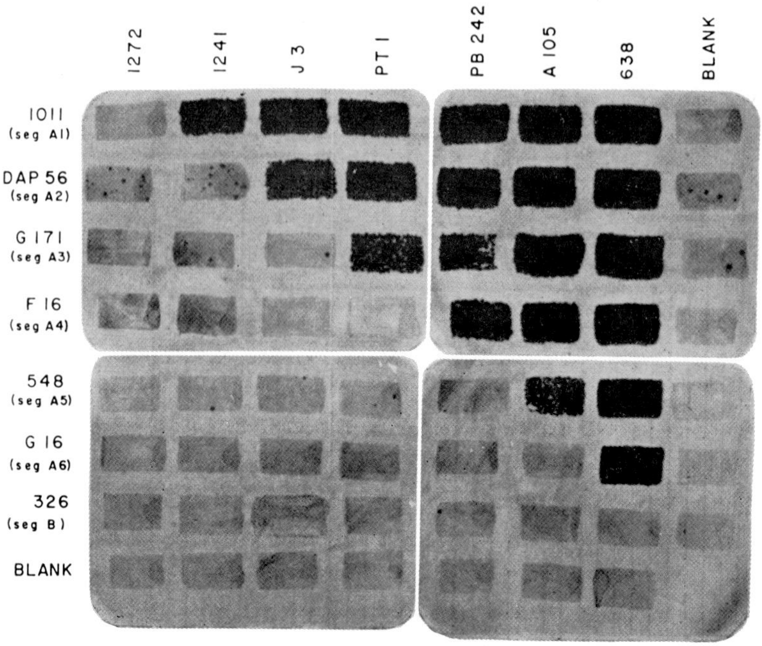

PLATE 8. Crosses for mapping r_{II} mutations. The photograph shows a series of 'spot' tests in which a number of mutants, indicated on the left, are crossed with each of the 7 deletion mutants (indicated above) of Fig. 48. The dark areas of confluent, or almost confluent, lysis are due to r^+ recombinant particles. The isolated plaques seen in the blanks and elsewhere are caused by r^+ revertants present in the mutant stocks. The results show that all the mutations are located in different segments.

(Reproduced from Benzer, 1961, by kind permission of the author, and the National Academy of Sciences, Washington) *coli* B in a small tube; after standing for a few minutes to allow adsorption, a droplet of the mixture is spotted on an agar surface previously seeded with a culture of *E. coli* K. The plate is incubated. When the infected B cells burst, any bacterium that liberates one or more wild type recombinant particles can initiate a spreading infection through the K cells and thus produce a plaque; no B cells are left on which r_{II} particles can plaque since the initial infection has destroyed them. In practice, spot tests of this kind are most useful in preliminary mapping by means of deletion mutants where only a qualitative test is required, indicating whether recombination does or does not occur. Plate 8 shows the results of a test of this kind in which seven single-site mutants are allocated to different segments of the r_{II} region by crossing each with the series of seven deletion mutants represented in Fig. 48. Observe that some of the mutants yield plaques in the singly infected control platings (blanks) due to reverse mutation.

For the accurate measurement of recombination frequencies, of course, more conventional, standard methods are needed. B bacteria are doubly infected with the pairs of mutants and allowed to burst in fluid medium. Dilutions of this lysate are then plated separately with cultures of B and K, and the numbers of plaques produced on each is counted. Since statistically equal numbers of wild type (++) and doubly mutant ($r_{IIa}r_{IIb}$) reciprocal recombinants emerge from the cross, but since the latter do not form plaques on K, the number of plaques on K is doubled for the purpose of estimating the recombination frequency.

RESULTS OF THE ANALYSIS

Over 2,400 spontaneous and induced r_{II} mutants, all having a low reversion rate, have now been isolated and mapped (Benzer, 1961). The sites of all of them are located within a small region of the chromosome, the limits of which recombine with a frequency of about 8 per cent. Assuming a total map length corresponding to about 800 per cent recombination (recombination units), this means that the r_{II} region is about 1/100 the length of the entire phage chromosome.

Mapping was carried out by means of two sets of spot tests against deletion mutants. Each mutant was first allocated to one of seven segments by means of the set of deletion mutants shown in Fig. 48. As will be seen from Plate 8, the number of the segment into which the mutant falls is found by simply counting the number of zero readings from the left. Then, by using another series of deletion mutants intruding into each of these segments, the mutant could be further localised into one or another of 47 segments into which the r_{II} region can thus be divided. The order of these segments, as inferred from the presence or absence of recombination among the deletion mutants, was independently checked by the conventional measurement of recombination frequencies between representative mutants from each; both methods gave the same order. In general, however, the mutants falling within each small segment were not mapped by recombination frequencies but merely tested for the presence of recombination between them to determine whether or not their sites were identical.

The most illuminating features of chromosomal architecture revealed by Benzer's studies can be summarised as follows.

Structure and function: the cistron.

We have shown how r_{II} mutants can be subdivided into two functional groups, A and B, by means of complementation tests. When the mutational sites of the two groups are mapped on the chromosome, it is found that all the A mutants lie together in one region and all the B mutants in another, adjacent region. This coincidence of map region and function is complete in that no A group site is located between two B sites, and *vice versa*; nevertheless the frequency of recombination between the nearest A and B sites is of the same order as that between adjacent sites in the same functional unit, suggesting that the two regions are contiguous. Benzer proposed the name *cistron* to denote the functional unit of a chromosome, that is, that length of chromosome which determines a single function as defined by complementation (*cis-trans*) tests. From the recombination frequencies between the most widely separated mutations in the A and B cistrons, their relative lengths can be estimated. The A cistron turns out to be about five units long, while the B cistron is three units; the same relative measurements are given by the ratio of the number of sites which have been identified in each, 200 being found in the A cistron as against 108 in the B cistron (see p. 146).

Components of the cistron and their organisation.

The cistron is divisible into a number of mutational sites which are recognised because they have produced a change in phenotype and can be separated by recombination. Thus in addition to the unit of function, the cistron, we can postulate two other kinds of smaller genetic unit, namely, a unit of mutation and a unit of recombination

(Pontecorvo, 1952b), for which Benzer (1957) proposed the names *muton* and *recon* respectively. The muton was defined as the smallest element that, when altered, can produce a mutation: similarly the recon was the smallest element that can be exchanged, but not divided, in recombination. As we shall see when we come to look at the gene in physico-chemical terms, the distinction between the muton and the recon is not very meaningful. Very approximate estimations made by Benzer (1957) on the basis of recombination frequencies, show that their sizes cannot be very disparate but, within the confines of Benzer's analytical system, we can set only an upper limit to these unless the region of the map we are looking at is saturated with mutational sites; that is, unless no new sites are discovered as more mutants are analysed. However, as we shall see later, in other systems where the results of genetic analysis of mutational sites within a cistron can be correlated with specific amino acid alterations in its protein product, it becomes possible to define both the muton and the recon precisely (p. 300).

Experimentally, the smallest recombination frequency so far observed between two r_{II} mutants sites is about 0·02 per cent. If it is assumed that the probability of recombination within the r_{II} region remains constant, this represents a distance of 0·02/8 or approximately 1/400 part of the r_{II} region. Within this region a total of 308 sites have been discovered by the examination of over 2,400 mutants and it can be estimated that, if new sites continue to be found in the same proportion, another 120 sites remain unaccounted for (Benzer, 1961). Thus the total number of identifiable sites works out at 428 which fits very well with the calculation that the nearest sites are 1/400 of the r_{II} region apart. Such considerations, of course, take no account of the possibility that there are a large number, maybe a majority, of sites at which mutation produces no observable change.

How are these sites arranged? Map distances derived from recombination frequencies within the r_{II} region are approximately additive so that it is reasonable to assume that the sites are arranged linearly. Moreover the complete absence of ambiguity in the mapping of thousands of mutational sites by means of the deletion mutants, and the conformity of the results with those obtained from recombination frequencies, shows that the segments defined by the deletion mutants are also linearly ordered. Although branching of the chromosome cannot be absolutely excluded, no branch can contain more than one of the forty-seven segments of the r_{II} region (Benzer, 1961).

The distribution of mutations on the chromosome.

One of the most surprising features of what Benzer calls the 'topography' of the r_{II} region, is that the distribution of point mutations on the chomosome is far from random. Of the sites discovered, some are represented by only a single mutation and others by large numbers. For example, out of 1,612 spontaneous mutations mapping at 251 sites, more than 500 occur at a single site or 'hot spot' in the B cistron and nearly 300 at another site in the A cistron. The topography of these spontaneous mutations is shown in Fig. 49, in which each small square represents one occurrence of a mutation at a site. The meaning of these sites of high mutability is not understood. We shall discuss their possible significance later when we come to consider the nature of mutation in terms of the physico-chemical structure of the genetic material (p. 325).

CONCLUSIONS

We embarked upon this account of the fine structure of genetic material with the intention of defining more precisely what we mean when we talk about genes. We have now travelled about as far as purely genetic analysis is able to carry us. What we have found? In some ways the situation has become more complicated because instead of thinking of the gene as a simple unit of inheritance we now have to visualise a linear segment of the chromosome, the cistron, which defines a specific function and which is divisible by recombination into many hundreds of sites, the alteration of any one of which by mutation may disrupt the expression of the function. On the other hand, this more detailed picture of the gene begins to become really meaningful in terms of what the gene does, since we can begin to equate the linear array of sites with the sequence of amino acids in the protein which the gene specifies. For example, Ingram (1957) has shown that the effect of a single mutation involving the structure of human haemoglobin is the substitution of a single, specific amino acid among approximately 300 amino acids which the protein contains.

The need for an analytical definition of the gene was revealed by the popularity accorded to Benzer's introduction of the word 'cistron'. Useful as this word has been, however, it has engendered some confusion on account of the disparities which have become apparent between the two ways in which a function may be defined. You will

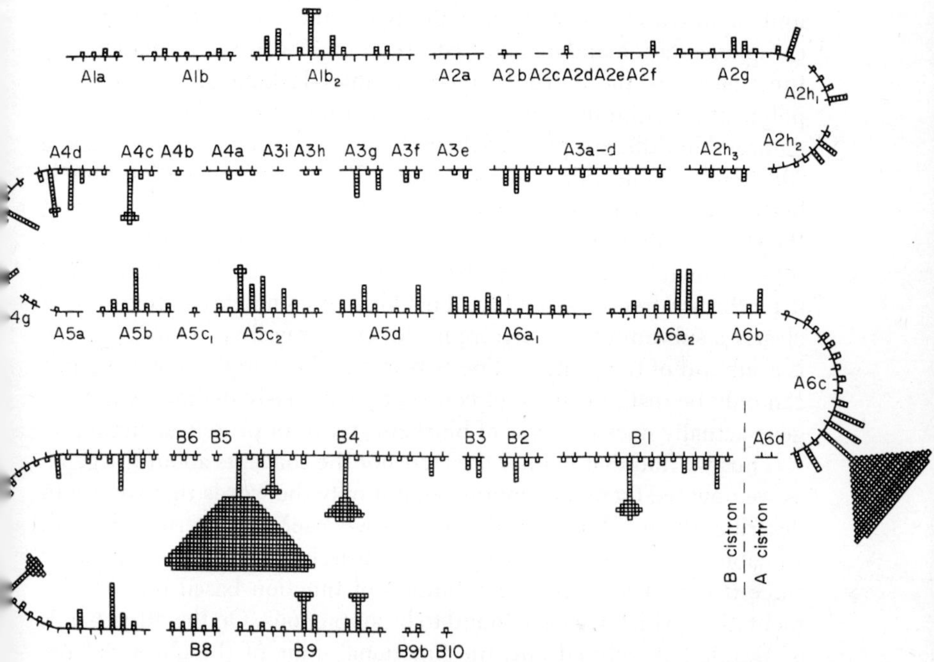

FIG. 49. The distribution of spontaneous, r_{II} mutational sites on the A and B cistrons of bacteriophage T4.

Each small square represents the site of an independently isolated mutation; the designations A1a, A1b, A2, B1, B2 etc., refer to the different segments into which the r_{II} region can be divided by means of deletion mutants. (Reproduced from Benzer, 1961).

recollect that the functions of the A and B cistrons of the r_{II} region were, of necessity, defined by complementation tests alone and that, in this system, the genetic and functional maps agree completely. However, when we define functions by biochemical methods, in terms of enzyme activity, we often find that functional regions can be subdivided by interallelic complementation, and that the genetic and complementation maps do not coincide. It turns out that this paradox is more semantic than real, since many proteins are built up of polypeptide subunits which aggregate, after their formation, to produce functionally active complexes. In theory, therefore, the complementation test can define functional units at two distinct levels. In those cases where the active enzyme consists of a single polypeptide chain, the unit of function, the cistron, will appear the same by biochemical, complementation and genetic criteria. If, however, the formation of a functional enzyme depends upon the inter-locking of two identical polypeptide subunits, then a new situation arises where subunits, damaged in different parts of their chains, can still combine to yield a sufficiently stable and active structure. The complementation here is between units that possess precisely the same function, but it occurs at the structural rather than at the operational level; it is a matter of the way in which the machine can be put together rather than of the duties it performs after its assembly. This kind of complementation unit is clearly a subunit of the cistron, in the same way that the polypeptide is a subunit of the protein. The cistron and the complementation unit can only be distinguished, of course, by rigorously defining what the gene actually does in terms of biochemistry or of protein structure.

Thus we find that neither the facts nor the concepts about the gene, as we now see them, are confusing, but only the words that we use to describe them. What words should we use? The term 'cistron', valuable as it has been in clarifying our ideas, is not entirely satisfactory since it is tied to a specific definition of function based on complementation, which we have found to be ambiguous. On the other hand, as Beadle has pointed out, the functional units of the chromosome, as revealed by the joint assaults of chemical, complementation and high resolution genetic analyses, '. . . clearly correspond to the units that geneticists have long called genes. That these units can be resolved into subunits capable of separate mutation and of reassociation by some mechanism of intragenic recombination seems an inadequate reason to discontinue calling them genes. . . . In the meantime it is far more important that we make use of all possible methods of learning more about the nature of the genetic material than that attempts be made to answer with finality questions of terminology' (Beadle, 1957). The word 'gene' is a particularly suitable one to preserve at a time when ideas are undergoing rapid change, for it has survived the evolution of so many different concepts that it is associated with none. The present preoccupation of genetics is to seek explanations in terms of the structure and behaviour of macromolecules which are the basis of all vital phenomena. From the point of view of a macromolecule the face of the gene wears a new look; the idea of 'one gene–one enzyme' has been transmuted to the new concept of 'one gene–one polypeptide chain'.

Part V
Mutation in Bacteria

CHAPTER 9

The Nature of Bacterial Variation

Although we have followed the development of the science of genetics to a stage where some very sophisticated ideas are beginning to emerge, we have so far refrained from any discussion of the nature of mutation itself which, after all, appears to be the prime source of variation in all living things. In fact, if evolutionary theory is correct, mutation is the central fact of life, for without it there could have been no evolution and life as we know it could not have achieved any significant complexity. Our reason for not discussing mutation up till now is a good one—there has been nothing much to discuss. The word 'mutation' means 'change' and, from what we have learnt so far, this is nearly all that we can say about it, except that we now know that the change can result from alteration of a very small part of the gene. The words 'gene' and 'mutation' only begin to take on real operational significance when, in Chapter 13, we begin to interpret them in terms of the arrangement of atoms and molecules that make up the genetic material. Before taking this step, however, we must look at some of the extrinsic factors which influence the occurrence of mutation and its expression and, especially, those novel features of bacteria which must be taken into account when they are used as genetic tools. Except where specifically indicated, the word 'mutation' will refer only to alterations of single sites within a gene.

THE NATURE OF BACTERIAL VARIATION

Until about twentyfive years ago bacteria were almost universally regarded as manifestations of a form of life quite different from that of other creatures, in spite of the half century of intensive study of their behaviour which their importance to medicine and agriculture had stimulated. The basis of this belief in the uniqueness of bacteria was two-fold. In the first place they are extremely small and pleomorphic, and did not seem to possess a nuclear apparatus like other cells (see

Chapter 19, p. 546). Secondly, they display an incomparable aptitude for adapting themselves to new environmental conditions; once an adaptation has been accomplished, however, it becomes an inheritable property among the descendants of the altered cells.

A classical example of this kind of adaptation is the case of the coliform bacillus, *E. coli mutabile*. Cells of this organism are unable to ferment lactose, and retain this inability indefinitely when cultivated in media devoid of the sugar. However, when they are transferred to a medium containing lactose, fermentation begins after a day or so; if, at this stage, sub-cultures are made to further tubes of lactose-medium, fermentation starts at once. Moreover this ability to ferment lactose without delay, acquired as a result of exposure to the sugar, is now retained, even though the bacteria are subsequently subcultured through many transfers in sugar-free medium. This kind of behaviour, typical of a wide range of bacterial characters such as resistance to drugs and ability to dispense with specific nutritional requirements, which lend themselves to ready demonstration, promoted the theory that bacterial variation resulted from *adaptation* rather than from mutation; that is, that the *environment* provoked an inheritable alteration in the bacteria exposed to it. Thus bacteriology became 'the last stronghold of Lamarckism' (Luria 1947a), which is the doctrine of the inheritance of acquired characteristics.

On the other hand, there were some bacteriologists who saw in the vigorous adaptibility of bacteria a clear expression of mutation and selection, operating on very large populations of rapidly growing, individual cells. It is perhaps difficult for us now, looking back, to understand why the interminable controversy between these two schools of thought remained inconclusive for so long. It must be remembered, however, that it was conducted, not by geneticists, but by bacteriologists who were accustomed to thinking in terms of the activity of *cultures* containing thousands of millions of cells, rather than of the behaviour of individual bacteria and their descendants, that is, in terms of *clones*. Thus an analytical way of thinking about bacterial populations, as well as the techniques necessary for decisive experiments, were slow to evolve.

I. M. Lewis (1934) was one of the first to devise an experiment to decide whether all, or only a small proportion, of the bacteria in a culture become adapted to growth in a new environment. He plated suspensions of single colonies of *E. coli mutabile*, which readily acquires the ability to ferment lactose in the presence of this sugar, on plates of synthetic medium containing either glucose or lactose as the sole source of carbon, as well as on nutrient agar. He found that only about 1 in 10^5 of the bacteria growing on glucose-synthetic or nutrient agar were able to form colonies on lactose-synthetic agar; however, all the bacteria in those colonies which did arise on the lactose-synthetic medium were lactose fermenters and, unlike the cells of the original strain, grew equally well on all three media. This experiment seemed to demonstrate that the adaptation of this organism was due to the presence of a minority of cells in the population which were initially able to ferment lactose and so had a selective advantage enabling them to outstrip the growth of the original type in the presence of the sugar. The proponents of the adaptation theory countered this sort of experiment by pointing out that it had not, in fact, been shown that the minority cells were initial fermenters since they could only be identified by exposing them to lactose. Thus evidence that spontaneous mutation was responsible for any particular case of adaptation required the demonstration that mutant bacteria are present in the population *before exposure to the selective agent* necessary to reveal them. The first line of evidence fulfilling this requirement was the *fluctuation test* devised by Luria and Delbrück (1943), who were the first to approach bacterial inheritance as a problem in population genetics, amenable to statistical analysis.

THE FLUCTUATION TEST

The system studied by Luria and Delbrück was the acquisition of resistance to the virulent bacteriophage T1 by *E. coli*. When a culture of sensitive *E. coli* is plated in the presence of an excess of the phage, although virtually all the bacteria are killed and lysed, a small number survive and give rise to colonies on the plate. These colonies consist of cells which are resistant because they no longer adsorb the phage, and which pass on this property indefinitely to their descendants. This behaviour is found even with cultures started from single bacteria so that the initial introduction of resistant cells may be excluded.

Such a bacterial population, started from a single cell at zero time (t_o), may be visualised at the time of plating (t_x) as possessing two dimensions at right angles to one another. One dimension is the *number of bacteria* in the population at time t_x (Fig. 50, horizontal line); the other is a time dimension, expressed as the number of generations which have arisen, between zero time (t_0) and the time of observation (t_x), to produce this population (Fig. 50, vertical line). We can now

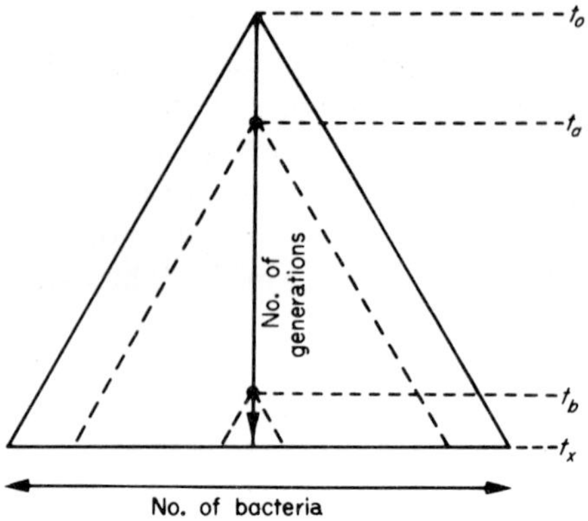

FIG. 50. Diagrammatic representation of the growth of a bacterial population started, at the apex, from a single cell, or a small number of cells, at time t_0.

The length of the base of the large triangle indicates the number of bacteria present in the population at time t_x.

The vertical line is a time axis on which the two dots represent the occurrence of mutations, one early at time t_a and the other late at time t_b. The lengths of the bases of the smaller triangles, having these points at their apices, indicate the proportions of the total population which the descendants of these mutants will comprise.

look at the adaptation and mutation hypotheses in terms of these dimensions. The adaptation hypothesis says that every bacterium in the population at time t_x has a small, but constant, random probability of acquiring resistance in response to the stimulus of contact with the phage, so that the number of resistant colonies observed will depend only on the number of bacteria plated and be independent of the time dimension. According to the mutation hypothesis, on the other hand, each bacterium has a small, but constant, random probability *per generation* of undergoing mutation to resistance; once the mutation has occurred, all the descendants of the resistant mutant will be resistant. In this case the probability is distributed uniformly over the time dimension. If the mutation occurs early, as at time t_a (Fig. 50), it will have generated a large *clone* of resistant cells by the time (t_x) at which the population is exposed to the phage; but if the mutation occurs late on the time scale (Fig. 50, t_b) then the population will contain only a small number of resistant cells.

Instead of looking at a single culture, let us now see what happens if a number of *independent* cultures, each started from a few cells and under identical conditions, is set up. At time t_x, each of these cultures will contain approximately the same total number of bacteria. A sample from each is now plated with phage and the number of resistant colonies developing is counted. If the adaptation hypothesis is correct, each culture should contain approximately the same number of resistant cells since each cell, at time t_x, has the same probability of becoming resistant. The mutation hypothesis leads to a quite different prediction. In this case, although the independent cultures are the same with respect to the total size of population, some may have experienced a mutation at an early generation and thus yield a large number of resistant colonies, while with others a mutation may have occurred just prior to the time of plating so that only a few colonies arise. This means

TABLE 6

Comparison of the number of phage-resistant bacteria in different samples of the same culture, and in series of different cultures. (Data from Luria and Delbrück, 1943.)

Expt. No.	11		17
	Samples from same culture	Samples from independent cultures	Samples from independent cultures
Volume of cultures	...	10 ml	0·2 ml
Volume of samples	0·05 ml	0·05 ml	0·08 ml
Sample No.			
1	46	30	1
2	56	10	0
3	52	40	0
4	48	45	7
5	65	183	0
6	44	12	303
7	49	173	0
8	51	23	0
9	56	57	3
10	47	51	48
11	...	...	1
12	...	...	4
Average per sample	51·4	62	30
Variance	27	3498	6620

that the independent cultures will show a wide *fluctuation* in their content of resistant bacteria, indicating that these bacteria are distributed clonally in the population. Put another way, a wide fluctuation in the content of resistant bacteria between independent cultures indicates the presence of clones of resistant cells whose progenitors must therefore have arisen at some time prior to exposure of the population to the phage.

From the statistical point of view, the significance of the results of a fluctuation test is assessed by comparing the variance between the number of resistant colonies emerging from independent cultures with

TABLE 7

Distribution of the numbers of phage-resistant bacteria in series of similar cultures. (Data from Luria and Delbrück, 1943.)

Number of cultures = 87
Volume of cultures = 0·2 ml
Volume of samples = 0·2 ml

Number of resistant bacteria	Number of cultures
0	29
1	17
2	4
3	3
4	3
5	2
6–10	5
11–20	6
21–50	7
51–100	5
101–200	2
201–500	4
501–1000	0

Average per sample	28·6
Variance	6431
Average per culture	28·6
Bacteria per culture	$2{\cdot}4 \times 10^{8}$
Mutation Rate	
Method 1	$0{\cdot}32 \times 10^{-8}$ } see p. 195
Method 2	$2{\cdot}37 \times 10^{-8}$ }

that given by different samples of a single culture. Examples from some of the results of the original experiments of Luria and Delbrück (1943) are given in Tables 6 and 7. The greatest degree of fluctuation is observed when the size of the population at the time of plating is such that the probability of mutation has been rather low, as a comparison of Experiments 11 and 17 (Table 6) shows. In large populations or samples, the occurrence of a considerable number of mutations distributed over the time dimension will tend to iron out the differences between independent cultures.

The mechanism of variation with respect to many types of character (resistance to bacteriophages, streptomycin, penicillin, sulphonamide, radiation; independence of required growth factors; capacity to ferment various carbohydrates) in a number of different bacterial species (*E. coli*, *Salmonella*, *Staphylococcus pyogenes*) has now been investigated by means of the fluctuation test. In every case the variants have exhibited the clonal distribution indicative of mutation

THE NEWCOMBE EXPERIMENT

So great was the inertia of old concepts that six years elapsed before a new and more direct method of investigating the origin of bacterial variants was devised (Newcombe, 1949). This method, like the fluctuation test, was initially applied to the development of resistance by *E. coli* to virulent bacteriophage, and has an appealing simplicity. Nutrient agar plates are spread with similar samples of a broth culture of sensitive bacteria. The plates are incubated. After some hours, when each bacterium has multiplied to produce a clone, the surfaces of half the plates are rubbed with a spreader so as to re-distribute the bacteria. The bacteria on the remaining plates are left undisturbed. All the plates are then sprayed with a suspension of the phage so that sensitive bacteria are destroyed and only resistant variants can grow to produce colonies. If resistance is acquired by a small and constant proportion of the bacteria in response to the stimulus of contact with the phage, the number of resistant colonies arising on the undisturbed and respread plates should be approximately the same, since respreading does not influence the *number* of bacteria on the plates but only their distribution. On the other hand, if the original inoculum contained resistant mutants, or if these developed on the plates during the initial period of incubation, each mutant would yield a localised clone of resistant cells. On the undisturbed plates each of these clones would give rise to only one colony. On the respread plates, however, the bacteria within every resistant clone would be separated and redistributed on the agar surface so that each bacterium would now produce a resistant colony. Thus if the mean number of bacteria per resistant clone was 100, the respread plates would be expected to show about 100 times more resistant colonies than the undisturbed plates after spraying with phage. This was the type of result found experimentally, as Table 8 shows. A possible objection to the method is that the phage might be less able to evoke an adaptive response when the bacteria are crowded in colonies than when they are isolated from one another by respreading. If this were the case, the proportion of resistant colonies on the undisturbed plates should decrease as the clone size increases. Table 8 demonstrates that, in fact, the opposite effect is found; between the fifth and sixth hour of incubation the increase in the total number of bacteria (that is, the clone size) is about ten-fold while the number of resistant colonies on the unspread plates rises proportionately.

Newcombe's method may be used to assess the nature of any type of variation, provided that this permits selection to be imposed on the bacterial populations, after an initial period of unrestricted multiplication, without disturbing the positions of clones on the control (unspread) plates. It is therefore particularly applicable to variations involving resistance to biological and chemical anti-bacterial agents.

TABLE 8

The effect of redistribution of the cells of a culture of *E. coli*, growing on the surface of nutrient agar, on the yield of phage-resistant colonies. (Data from Newcombe, 1949.)

Time of incubation after initial seeding	5 hours		6 hours	
Number of bacteria seeded	$5{\cdot}1 \times 10^4$			
Number of bacteria after incubation	$2{\cdot}6 \times 10^8$		$2{\cdot}8 \times 10^9$	
	Plates		Plates	
	Undisturbed	Spread	Undisturbed	Spread
Number of colonies of resistant bacteria isolated (total of 6 plates)	28	353	240	12,638

THE INDIRECT SELECTION OF BACTERIAL MUTANTS: REPLICA PLATING

The technique of replica plating was devised by Lederberg and Lederberg (1952) to alleviate the irksome task, inherent in genetic analysis, of scoring large numbers of colonies of recombinant bacteria for inheritance of various characters. Before the advent of replica plating this entailed the individual transplantation of a sample of each colony, by means of a sterile wire needle or loop, to a series of plates of differential or selective media on which the nutritional requirements, fermentative capacities, resistance patterns and other characters of the bacteria could be assessed. Since the examination of, say, 500 recombinants for the inheritance of five characters requires 2,500 such manual transfers, even a limited genetic analysis was a formidable chore.

The idea behind replica plating was to substitute many operations of one needle by a large multiplicity of needles attached to a base plate, which could sample the entire pattern of bacterial growth on the agar surface of an 'initial' or 'master' plate in a single operation, and then transfer this pattern as a whole, and in turn, to the surfaces of the series of secondary (replica) plates used for analysis. The theoretical requirements of replica plating are admirably met by pile fabrics such as velvet or velveteen, the pile of which serves as a close-set and orderly array of short, flexible needles. A square of sterile velveteen is placed, pile upwards, over the flat end of a wooden or metal cylindrical block of diameter slightly less than that of a standard plate, the material being held in position by a metal collar (Fig. 51). The initial plate bearing the colonies or bacterial growth to be replicated is inverted over the fabric and the agar surface pressed gently against the pile. The plate is then carefully removed. In those positions where the pile penetrates the bacterial growth, samples of the cells adhere to it so that the pile carries a precisely positioned print of the bacterial growth on the initial plate. The various replica plates, marked with a reference point for subsequent orientation, are now inverted in turn over the fabric, pressed gently against it so as to pick up a sample of the bacteria adhering to the pile, and then removed and incubated. In this way as many as six to eight replicas may be printed from a single pad. In practice, about 10 to 30 per cent. of the bacteria on the initial plate are transferred to the fabric, and about the same proportion of these is deposited on the replica plates (Lederberg and Lederberg, 1952).

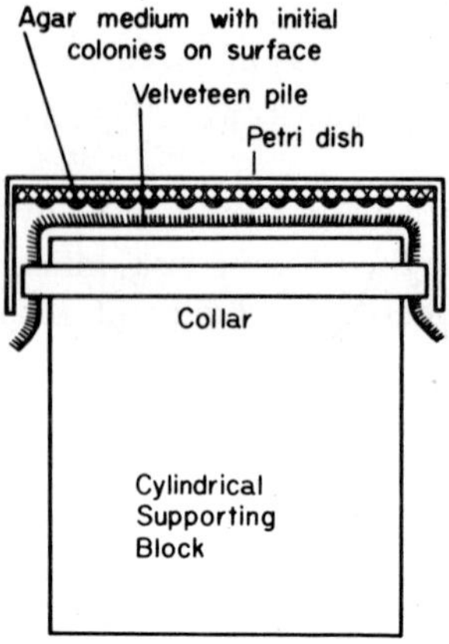

FIG. 51. Diagram showing the way in which velveteen squares are secured, and the replica of the 'initial' plate made, in the replica plating technique.

We will not emphasise further the many and valuable practicable applications of replica plating, but will pass on to the way the Lederbergs (1952) used the technique for the direct demonstration of pre-existing mutants in bacterial cultures. They first studied the acquisition by *E. coli* of resistance to phage T1 in the following way.

Initial plates of nutrient agar are seeded with a small volume of a dense broth culture of sensitive bacteria, derived from a single colony. The plates are incubated for several hours to allow each bacterium to grow into a sizable clone. Replicas of each initial plate are then transferred to two or more nutrient agar plates *impregnated with phage*, all the plates being marked with reference to their orientation on the replicating pad. Following incubation, only colonies of resistant cells appear on the replica plates, sensitive bacteria having been destroyed by the phage. When the positions of resistant colonies on two or more replicas of the same initial plate are compared, many are found to be superimposable: that is, they occupy the same positions on the plates with respect to the marker point (Fig. 52, top row). This reiteration of resistant colonies means that the velveteen pad must have picked up a *number* of resistant bacteria at equivalent positions on the initial plate and is good evidence that clones of resistant bacteria were located there at the time of replication.

The next step is to isolate these clones on the initial plate, and demonstrate that they really are resistant although they have never been in

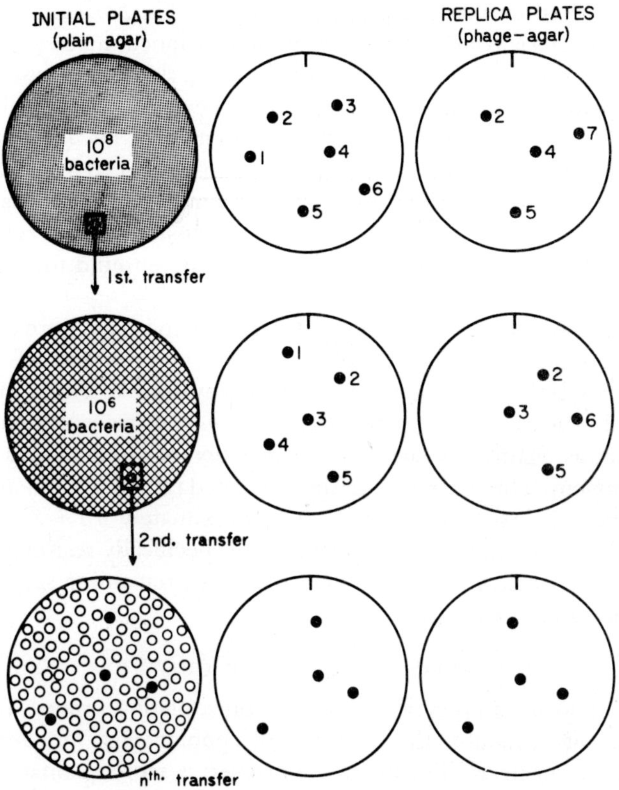

FIG. 52. Illustrating how the technique of replica plating can be used for the isolation of mutant clones by indirect selection.
For explanation, see text.

contact with the phage. The initial plate, of course, being heavily seeded with sensitive bacteria, presents a confluent growth after incubation so that, although the sites of resistant clones on it can be inferred from the positions of congruent colonies on the replicas, these sites are not marked by isolated colonies. A small area of agar enclosing one of these presumptive clones is cut out and a suspension made of the bacteria on it. The resistant bacteria in this suspension should be concentrated to the extent that the area of the whole plate is greater than the small area removed—usually about 100-fold. A second initial plate of plain nutrient agar is now seeded, from a dilution of this suspension, with about one hundredth of the inoculum used for the first series of initial plates (Fig. 52, 2nd transfer). Following a period of growth, replicas are transferred to phage-agar plates as before. The enrichment of the resistant bacteria compensates for the smaller inoculum so that about the same number of resistant colonies as formerly should arise on the replica plates. The whole process is then repeated. If the initial plates at the first transfer were seeded with about 10^8 bacteria and the inoculum size is reduced 100-fold at each subsequent transfer, then, at the 4th transfer, the initial plate should contain only 100 bacteria, yielding isolated colonies (Fig. 52, n^{th} transfer). Some of these colonies, on replica plating to phage agar, will be found to consist of resistant bacteria.

In this way the Lederbergs (1952) were able to isolate pure cultures of phage-resistant and of streptomycin-resistant variants of *E. coli* without directly exposing the bacteria at any time to either of these agents. In the case of phage-resistant mutants four transfer steps were required, as against six steps for the isolation of mutants resistant to streptomycin. This accords with the calculated rates of mutation to the two types of resistance, which are approximately 10^{-7} and 10^{-10} respectively. In each case the isolates were specifically resistant, those indirectly selected for streptomycin resistance remaining sensitive to the phage and *vice versa*.

SIB SELECTION IN FLUID CULTURES

The method of indirect selection by replica plating proved, beyond any reasonable doubt, the existence of spontaneous, 'pre-adaptive' mutation in bacteria. The method, however, is more qualitative than quantitative and fails to show that *all* the variants of any particular type that arise in a culture do so by mutation. Due to the low rate at which variation occurs it is obviously not feasible to attempt a more direct test, even with the help of replica plating, since this would necessarily involve the scoring of many millions of isolated colonies.

A way of escape from this predicament was found by Cavalli-Sforza and Lederberg (1956) who devised a method of indirect selection in fluid media, based on the fluctuation test, whereby the observed and predicted degrees of enrichment can be compared at every step; agreement between the two would show that fluctuation is indeed a result of the random occurrence of mutation, and not of environmental variation as supporters of the adaptation hypothesis suggested, and would expose mutation as the sole source of variation.

A fluctuation test is set up by inoculating each of a large number of broth tubes with a small number of sensitive bacteria (see p. 182;

Table 6, experiment 17). When the cultures are fully grown, samples are plated on selective medium to determine the number of resistant bacteria in each, as well as on plain agar to assess the total number of bacteria; the broth cultures are stored in the refrigerator until the results of the assay are known. The proportion of resistant to sensitive bacteria in the culture yielding the greatest number of resistants (the 'best' tube) is estimated. Suppose this proportion is 1 in 10^6, the culture containing 10^9 sensitive bacteria and 10^3 resistants per ml. If this culture is now diluted 10^{-4}, each ml will contain 10^5 sensitives and 0·1 resistant bacteria. In practical terms this means that if each of another series of broth tubes is inoculated with one ml of this dilution, the probability is that *one tube in every ten will receive one resistant and* 10^5 *sensitive bacteria*; in such a tube, therefore, the proportion of resistant to sensitive bacteria is 1 in 10^5, that is, the proportion of resistants has been increased ten-fold. This second series of cultures is grown up and samples plated for resistants as before. Again, the proportion of resistants to sensitives in the 'best' tube is estimated; the culture is diluted so that each ml has a probability 0·1 of containing a resistant cell, and one ml samples are then transferred to a third series of broth tubes. Thus, at each transfer, a calculable degree of enrichment of resistants to sensitives is obtained without exposing the bacteria to the selective agent so that, ultimately, a culture is procured which yields isolated colonies of resistant bacteria on plating. The reason for choosing the 'best' tube, having the highest proportion of resistant bacteria, at each step of enrichment is to ensure that descendants (sibs) of the initial resistant clone continue to be selected, and not those of newly arisen mutants in the 90 per cent of cultures to which a resistant cell was not transferred; if one of these latter mutants were picked, the whole process would, of course, automatically be returned to step 1. It is obvious that the 10 per cent of tubes which receive a resistant bacterium at the outset are likely to yield more resistant cells than those other tubes in which a mutation to resistance occurs randomly at some later time during growth.

This method was successfully employed for the indirect selection of streptomycin-resistant and of chloramphenicol-resistant mutants. In both cases the observed rate of enrichment, after certain corrections had been made (see below), was comparable with that predicted for the experimental conditions used, and agreed with the hypothesis that all the resistant variants isolated on the selective media had resulted from spontaneous mutation (Cavalli-Sforza and Lederberg, 1956).

In practice the calculations may prove more complicated than in the simple case we have described, where we assumed that mutant and wild type bacteria are identical in every respect but resistance. This is not necessarily so. For example, streptomycin-resistant mutants frequently grow more slowly than the wild type and are at a selective disadvantage in mixed cultures, so that the number of cycles of indirect selection required for their isolation will be greater than that predicted. This was found in the experiments of Cavalli-Sforza and Lederberg (1956); by calculating back from the ratio of observed to predicted enrichment, they found that the discrepancy could be accounted for if the relative growth rate of the mutant was about 0·88 that of the sensitive strain. When the mutant was isolated and then tested directly in mixed culture with the sensitive strain, its relative growth rate turned out to be 0·856.

Despite the formidable array of evidence that bacterial variation is exclusively due to spontaneous mutation, and the absence of convincing experimental testimony for the contrary view, the controversy is not yet dead although the adaptation hypothesis has now shifted far from its original Lamarckian stand-point. The current concept of adaptation is mainly derived from Hinshelwood's (1946) theoretical approach to the chemical kinetics of bacteria. It postulates that contact of sensitive bacteria with a low concentration of some anti-bacterial drug may shift the normal equilibrium of cellular chemical reactions to a new equilibrium which is less susceptible to interference by the drug. Since bacterial inheritance is normally mediated by binary fission, in which the cytoplasm of parent bacteria is distributed between their daughters, it is certainly possible that such a new equilibrium, if attained, might prove stable in the absence of the drug, and thus be perpetuated cytoplasmically for many generations. The proponents of this view neither deny the occurrence of mutation and selection in those cases where this has been clearly demonstrated, nor do they assert that adaptations are *genetically* inherited (Dean and Hinshelwood, 1957) but, in our opinion, they have yet to prove their point. If adaptations of this kind do occur, one might expect them to involve rather small differences of phenotype which are difficult to analyse genetically. On the contrary, some of the reported adaptive responses (for example, to chloramphenicol) have been strikingly large and are claimed to have occurred under conditions precluding mutant selection (Woof and Hinshelwood, 1960; Dean, 1960; Dean and Broadbridge, 1963). Unfortunately the organism used in all these experiments (*Aerobacter aerogenes*) lacks a sexual system so that a direct experimental distinction between genetic and cytoplasmic inheritance of the reported adaptations has not yet been made. For discussions of various aspects of the adaptation hypothesis, see Dean and Hinshelwood, 1957, 1963, 1964a, b, 1965: see also Sevag, 1946).

THE RATE OF MUTATION

We have seen that mutations arise from events distributed randomly along the time axis during the growth of bacterial populations (Fig. 50). It follows that the rate of mutation must be expressed as the probability of the event in terms of time units and *not* in terms of population size. What kind of time unit should be employed? The choice is between 'clock time' measured in minutes or hours, and 'biological time', the unit of which would be the mean period required for the reproductive cycle, that is, the *generation time*, which of course may vary enormously with the physiological and environmental state of the bacteria. The decision rests primarily on what we imagine the nature of mutation to be. *A priori* it is reasonable to assume that mutation results from interference with some metabolic process so that copying errors are made during gene replication which occurs at each division cycle. It has been shown experimentally that once cultures of *E. coli* in broth are fully grown, the proportion of resistant to total bacteria remains constant over several days, even when the bacteria begin to die off as the cultures age (Luria and Delbrück, 1943). Furthermore, studies of mutation in *E. coli* growing in a 'chemostat', whereby controlled multiplication is maintained indefinitely by a continuous flow of medium having a limiting concentration of a required nutrient, have shown that at high growth rates the mutation rate per generation is constant (Fox, 1955). The opposite result may be found, however, when the generation time is greatly extended (Novick and Szilard, 1950). The mutation rate is therefore now generally defined as the *probability of mutation per cell per generation*.

A number of methods have been devised for estimating mutation rates. We do not intend to describe here the statistical or experimental techniques which these methods involve, which may be found in the original papers, but only to outline the general principles on which the methods are based.

Since the mutation rate is expressed in terms of the number of mutations per bacterium in each division cycle, the essential data are the number of mutations arising, the number of generations that have occurred and the mean number of bacteria present, over a given time interval. Since we are dealing with unsynchronised, exponentially growing populations, we at once become involved in a statistical consideration of population dynamics. Imagine a population of 10^8 bacteria which doubles over a thirty minute period which is thus the generation time. Clearly 10^8 divisions have taken place during this time so that, if one mutation was observed to arise, the mutation rate might be thought to be 1×10^{-8}. However the population has also been increasing over the thirty minute period, at any time during which the mutation may have arisen, so that the mutation rate as a function of mutation *per bacterium* per division must actually be *less* than this. The average number of cells in an exponentially growing population throughout the mean generation time can be calculated by dividing the initial number by the *exponential function*, $\log_e 2$(0·6931). The mutation rate is therefore $0{\cdot}693 \times 10^{-8}$, and not 1×10^{-8}. Thus the term $\log_e 2$ appears in all methods for estimating mutation rate, as a factor which corrects for the average number of cells present in the population.

The number of bacterial generations which has occurred over a given period is usually reckoned to be the same as the total number of bacteria at the end of the period—or, more accurately, this number divided by $\log_e 2$—provided that the ratio of the final number to the initial number is large. To understand why this should be so, we must clearly distinguish between the number of generation (division) cycles and the total number of bacterial generations. As an example, Table 9 shows

TABLE 9

The relation between the final number of bacteria in a population and the total number of preceding bacterial generations.

Generation (division) cycles		1	2	3	4	5
No. bacteria	1	2	4	8	16	32
Total No. bacterial generations		1	1 + 2 = 3	1 + 2 + 4 = 7	1 + 2 + 4 + 8 = 15	1 + 2 + 4 + 8 + 16 = 31

what happens during the multiplication of a single bacterium. At the end of five generation cycles there are thirty-two bacteria, but these have emerged from a total of thirty-one bacterial generations, any one of which had an equal probability of sustaining a mutation. At the end of the 1st division cycle there had been only one generation to produce two bacteria; at the end of the second cycle, however, each of these two bacteria had gone through a generation, so that a total of three generations had yielded four bacteria at the end of the second division cycle. Similarly, at the end of the third cycle, the four preceding bacteria had each gone through a division so that the total number of bacterial generations had risen to seven. It is obvious that when the final number of bacteria is very large compared to the initial number, the number of intervening bacterial generations approximates so closely to it that any difference between them lies well within the experimental error. Thus if the initial number of bacteria is N_1, and the final number after time t is N_2, the total number of generations during time t is given by the formula $(N_2—N_1)/\log_e 2$.

We may now proceed to consider four actual methods, representing various conceptual approaches to the problem. Two of these were developed by Luria and Delbrück (1943) and are based on statistical analyses of the results of fluctuation tests. In the first, a large number of independent cultures are started from small inocula, the volume of the cultures being such that only a proportion of them yield mutants when growth is complete. On the assumption that the number of mutations is distributed randomly among the cultures, the average number of mutations per culture may be calculated, by means of Poisson's formula, from the proportion of tubes which show no mutants. For example, in Table 7 (p. 184), twenty-nine out of eighty-seven cultures yielded no phage-resistant mutants, a proportion of 0·33. Assuming a Poisson distribution of mutants among the independent cultures, this means that the average number of mutants per culture was 1·10. The total number of bacteria per culture was $2{\cdot}4 \times 10^8$ so that the total number of bacterial generations $= 2{\cdot}4 \times 10^8/\log_e 2(0{\cdot}693)$ $= 3{\cdot}4 \times 10^8$. Therefore, the number of mutations per bacterium per generation $= {}^{1.1}/_{3.4} \times 10^{-8} = 0{\cdot}32 \times 10^{-8}$. This method has two disadvantages; a large number of cultures must be employed for a significant estimate of the proportion which yield no mutants, while inefficient use is made of the data which the fluctuation test provides.

Luria and Delbrück's (1943) second method does make effective use of these data by taking into account the average number of mutants per culture, but, from the statistical point of view, is very sophisticated. When a mutation occurs, the number of mutant bacteria stemming from it will increase exponentially at the same rate as the population as a whole, provided the growth rates of mutant and wild type bacteria are the same. Since, however, the number of bacterial generations increases at the same rate as the population (Table 9), the probability of the occurrence of new mutations also increases exponentially. From this it follows that, *when the population becomes large enough for mutation to be a likely event*, the average number of mutants per culture will be approximately the same irrespective of whether they arose early or late; before this time, however, the occurrence of an early mutation will increase the number of mutants above the average.

We can take as a critical time, after which the average number of mutants remains constant, the time (t_o) at which just one mutation occurs *throughout the whole series of cultures* in a fluctuation test, that is, when the aggregate population reaches the mutation frequency. If the average number of mutants is known, and it is assumed that they have all arisen between time t_o and the time of observation, t, two equations can be drawn up. In one the average number of mutants is related to the actual mutation rate and the average number of bacteria per culture at time t, expressed as a function of the time interval ($t—t_o$); in the other, ($t—t_o$) is related to the mutation rate, the average number of bacteria at time t and the total number of cultures observed. These two equations can be combined to eliminate the term ($t—t_o$), so that the mutation rate can be estimated in terms of the average number of mutants per culture, the average number of bacteria at time t and the total number of cultures.

Experimentally, mutation rates estimated by this method are found to be independent of the volume of the cultures or the growth rate of the bacteria. They are, however, much higher than those given by the first method (Table 7) due to the fact that early mutations, prior to time t_o, do arise and inflate the average number of mutants. Moreover the second method, unlike the first, is affected by any difference between the growth rates of mutant and wild type bacteria (Luria and Delbrück, 1943).

The third method is much more obvious and direct. Two series of agar plates are seeded with the same inoculum of sensitive bacteria and incubated. At intervals thereafter, sample plates of one series are removed, the growth on each is washed off into suspension and the average number of viable bacteria per plate estimated by counting the number of colonies arising from appropriate dilutions; at the same time equivalent plates of the other series are sprayed with phage (or other anti-bacterial substance) and re-incubated to assess the average number of *clones* of resistant mutants present per plate, as in the case of the undisturbed plates in the Newcombe experiment (p. 185; Table 8). In this way the increase in the number of mutant clones, and hence of mutations, that has arisen over a given time can be related to the increase in the total number of bacteria and, therefore, to the number of bacterial generations that have occurred over the same time. The mutation rate, as we have seen, may then be derived from the formula

$$m = \log_e 2 (M2 - M1)/(N2 - N1)$$

where M1 and M2 are the number of mutant clones (resistant colonies) arising from plates sprayed at times 1 and 2, and N1 and N2 are the corresponding total bacterial counts (Beale, 1948; Newcombe, 1948).

The information required for estimating the mutation rate to phage resistance by this method is provided by the data relating to the control half of the Newcombe experiment in Table 8 (p. 186). Between five and six hours after incubation, the total number of bacteria rose from $2{\cdot}6 \times 10^8$ to $2{\cdot}8 \times 10^9$, so that

$$(N2 - N1) = 25{\cdot}4 \times 10^8;$$

over the same interval, the number of resistant colonies on the undisturbed plates rose from 28 to 240, so that

$$(M2 - M1) = 212.$$

The mutation rate, therefore, is given by

$$\begin{aligned} m &= 0{\cdot}693 \times 212/25{\cdot}4 \times 10^8 \\ &= 5{\cdot}8 \times 10^{-8} \text{ approximately.} \end{aligned}$$

Given careful technique, this method is accurate and is unaffected by differences in growth rate between mutant and wild type bacteria; on the other hand it can be applied only to mutants which allow selection to be imposed on plate cultures without disturbing the growing clones.

The fourth and last method we shall describe is based on the theoretical prediction of Luria and Delbrück (1943) that the *proportion* of mutants in a growing culture should increase with time, if the growth rate of mutant and wild type bacteria is the same, since new mutants are constantly being added to the exponentially growing population of old ones. In fact it was during an attempt to check this prediction experimentally over long periods of exponential growth, involving frequent subculture to fresh broth, that they noticed great variations in the proportion of mutants from culture to culture, and eventually realised that this fluctuation was an inherent property of the mutation hypothesis itself.

The way in which the proportion of mutants in a growing culture may be expected to increase is illustrated in Table 10. Imagine a large

TABLE 10

The proportion of mutants to total cells in a growing population of bacteria. It is assumed that:

1. the growth rates of mutant and wild type bacteria are identical;
2. the mutation rate (m) is very low so that the number of wild type bacteria available to generate new mutants is virtually the same as the total number;
3. the probability of back-mutation of mutant bacteria to wild type is negligible over the experimental period.

Generation (division) cycles		1	2	3	4
Total No. bacteria	N →	2N →	4N →	8N →	16N
No. old mutants			$4mN$	$16mN$	$48mN$
No. new mutants		$2mN$	$4mN$	$8mN$	$16mN$
			$4mN + 4mN = 8mN$	$16mN + 8mN = 24mN$	$48mN + 16mN = 64mN$
Proportion mutant/total bacteria		m	$2m$	$3m$	$4m$

population of wild type bacteria, N, which mutates at a rate m, but initially contains no mutants. At the end of the first generation cycle there will be a total of 2N bacteria of which $2m$N are likely to be mutants; the ratio of mutant to wild type bacteria is therefore $2mN/2N = m$. If the mutation rate is very small, as is usually the case, we can regard the total number of bacteria, 2N, as being all wild type. If we make the further assumption that mutant and wild type bacteria multiply at the same rate, then at the end of the second generation cycle there will be a total of 4N bacteria, of which $4m$N will be mutants derived from division of the $2m$N mutants present at the end of the first cycle; to these mutants, however, must be added $4m$N new mutants which arise *de novo* among the 4N wild type bacteria, so that the number of mutants at the end of the second generation cycle is $8mN$, and the proportion of mutants $8mN/4N=2m$. Similarly the proportion of mutants at the end of the third generation cycle is $3m$, at the end of the fourth cycle is $4m$, and at the end of g cycles is gm. Thus the proportion of mutants increases linearly with time, the increase per generation cycle being proportional to the mutation rate. For example, if the mutation rate to phage resistance is 5×10^{-8} then, at each generation cycle, the number of resistant colonies arising will increase by 5 for every 10^8 bacteria present.

To obtain the mutation rate, the culture must be grown exponentially over a long period, the proportion of mutants being estimated at intervals and plotted as a function of the number of generations; the mutation rate is then calculated from the slope of the curve (Stocker, 1949). Alternatively, if the number of mutants and the total number of bacteria at two time intervals, together with the number of intervening generations, is known, the mutation rate may be assessed from the formula

$$m=2\log_e 2\left(\frac{M2}{N2}-\frac{M1}{N1}\right)/g$$

where M1 and M2, and N1 and N2, are the number of mutants and total bacteria respectively at times 1 and 2, and g is the number of generations (Catcheside, 1951, p. 161).

In this method, cultures must be started from large inocula, to prevent errors arising from statistical fluctuations in the number of mutants developing during the early generations, and must be maintained in exponential growth over a large number of generations. This means, in practice, that the experiment must be followed during many subcultures (Stocker, 1949), or else must be carried out in a 'chemostat' in which exponential growth continues indefinitely. The method is statistically sound and widely applicable to all sorts of mutants, but is obviously dependent on the absence of any selective advantage for either wild type or mutant in the growth medium.

DRIFT TO MUTATIONAL EQUILIBRIUM

If a population is maintained in continuous exponential growth, the rate of increase of mutants will ultimately tail off and cease to be linear. There are two reasons for this. The first is that when the rising number of mutants becomes a significant fraction of the total population, the fraction of wild type bacteria falls proportionately so that fewer new mutants are produced at each generation cycle. Secondly, when the number of mutants becomes large, the probability of *back-mutation* (*reversion*) to the wild type comes into play. Since the reversion rate is usually of the same order of magnitude as the original mutation rate, a stage will eventually be reached at which the number of new mutations produced is balanced by the number of reversions. When this state of mutational equilibrium is achieved, although the culture continues to multiply exponentially, the number of mutants remains constant at a level determined by the relative rates of mutation in each direction (Stocker, 1949).

Under actual experimental conditions many factors may intervene to prevent the attainment of mutational equilibrium, especially when synthetic media are employed in the study of mutations involving nutritional requirements. Among the most important of these factors is the appearance of a class of mutants, quite unrelated to those under study, which have a higher growth rate in the medium used. These mutants will rapidly outstrip and replace both the original wild type and the mutant bacteria so that a new wild type population will be established in which there are virtually none of the mutants under study. These latter will begin to arise again, and to increase, in the new wild type population until cut short by the occurrence of another new, unrelated mutation better adapted than the first to growth in the medium. This process may continue indefinitely. Thus the mutant under study may be observed to undergo a cyclical rise and fall as a result of the *periodic selection* by the environment of independently arising mutants (Atwood, Schneider and Ryan, 1951). The mutants responsible for these 'adaptive leaps' tend, of course, to stem from those bacteria which constitute the overwhelming majority of the total population, that is, the wild type, so that it is the observed mutant which is periodically eliminated. If by chance, however, the adaptive mutation happened to arise in one of the mutant cells, then the experiment would be brought to an end by the rapid emergence of an exclusively mutant culture. Another possibility, where the observed mutation rate is very high, is that mutant and wild type bacteria may have attained a 50 per cent equilibrium by the time an adaptive mutation is likely to occur. In such a case either the original wild type or the mutant type would be eliminated with equal probability.

CHAPTER 10

The Expression of Mutations

SEGREGATION AND PHENOTYPIC LAG

So far we have treated the kinetics of mutation in growing cultures as if the event actually observed at a particular time, such as the appearance and subsequent multiplication of a resistant bacterium, really reflects the mutational event itself. Such a simple relationship rarely holds in practice because of certain peculiarities inherent in bacteria, or in the nature of the characters used in mutational studies. In the first place each bacterium, whether defined as a morphological or as a viable unit, usually contains a number of identical nuclei, especially during the phase of exponential growth (see Chapter 19, p. 546;) nuclear division precedes division of the cell, when the nuclei are distributed between the two daughters in a precise and orderly way. A bacterium, therefore, which contains one mutant nucleus is really a heterokaryon (p. 78) so that, although each nucleus is haploid, the question of dominance arises in the expression of mutations.

Consider the bacterium represented in Fig. 53 which has experienced a mutation in one of its four haploid nuclei. If the wild type bacterium is auxotrophic, requiring, say, histidine for growth (*his*$^-$), and the mutation overcomes this defect (*his*$^-$$\rightarrow$*his*$^+$), then the mutant will be dominant and will express itself rapidly, because histidine will be synthesised and the cell as a whole enabled to divide in the absence of added amino acid. Because of the mode of nuclear segregation, however, only one of the two daughter cells arising at each of the first two divisions will inherit mutant nuclei, while two divisions are needed to yield a homokaryotic bacterium in which all the nuclei are mutant. An increase in the number of mutant bacteria will therefore be delayed until the third division cycle after the occurrence of the mutation (Fig. 53). This delay in the initiation of mutant clones is called *segregation lag* and has been shown to account for discrepancies between the observed rate of increase in the proportion of mutants and that expected from reconstruction experiments demonstrating an equal growth rate for

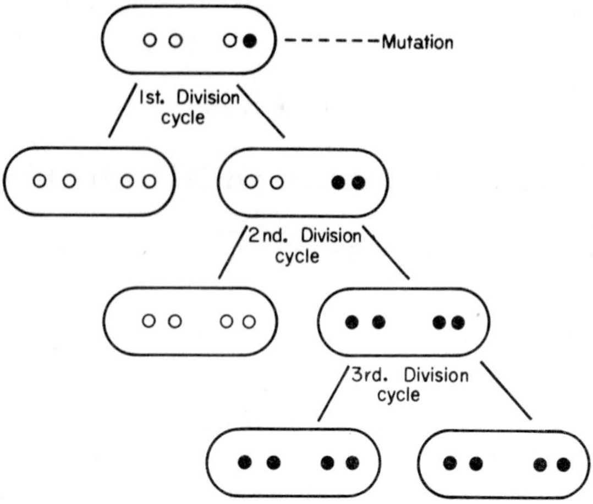

FIG. 53. The segregation of mutant nuclei in multinucleate bacteria. Wild type nuclei are represented by white circles, and mutant nuclei by black circles.

mutant and wild type bacteria in mixed culture (Ryan and Wainwright, 1954). In assessing mutation rates by methods involving the total number of mutant bacteria observed at a given time, segregation lag will mimic a selective disadvantage for the mutant and lead to underestimates. Moreover, the multinucleate nature of bacteria implies that mutation rates expressed as the probability of mutation *per bacterium* per generation will vary according to the mean number of nuclei per cell and will be different from the probability *per multable unit* or nucleus (Ryan and Wainwright, 1954).

If the mutation arising in one of the nuclei of a quadrinucleate cell is recessive, a rather different state of affairs is found (Fig. 53). Segregation lag operates as before but, in addition, the *mutation remains unexpressed* until the end of the second division when the dominant, wild type nuclei are segregated out and the cell becomes homokaryotic recessive. This type of delay in the expression of a newly acquired character is called *phenotypic lag*. In assessing mutation rates to recessive alleles by any method which involves counting the number of mutant bacteria present by plating samples of the population on a selective medium, the true number of mutants will again be underestimated on account of phenotypic lag, to the extent that some of them are not expressed at the time of plating and so will not multiply to produce colonies. Since the number of mutations arising in a growing population increases exponentially, an appreciable proportion at any time tends to be of recent origin, so that phenotypic lag can lead to a large under-estimate of mutation rate over and above that due to segregation lag alone.

The study of diploid bacteria, formed by conjugation in *E. coli*, has shown that resistance both to streptomycin (J. Lederberg, 1951a) and to phage T1 (Lederberg, 1949; Hayes, 1957a) is recessive to sensitivity in this organism. In both cases the selective agent for scoring resistant mutants is rapidly lethal to phenotypically sensitive bacteria, so that the loss of large numbers of potential mutants from phenotypic lag is likely. This is especially so where the expression of phage resistance is concerned since this does not depend on segregation lag alone. Resistance to bacteriophage results from specific alterations of cell wall structure so that the phage is no longer able to adsorb to the cell and so cannot infect it. When a mutation to resistance occurs in a sensitive cell, the expression of resistance not only requires the synthesis of new cell wall substance which lacks specific phage receptors, but also the elmination of pre-existing receptors. When the cell divides, although no new receptors are synthesised, the old receptors remain and are shared out among the cell walls of its descendants. Several generations may therefore elapse before cells begin to appear in which all the old receptors have been diluted out and replaced by new material, so that resistance is complete (Fig. 134, p. 702).

Apart from these inherent properties of bacteria, purely environmental influences may play a major role in determining the kinetic relationship between wild type and mutant cells in growing bacterial cultures. This is well exemplified by the behaviour of histidine-independent (*his*$^+$) mutants arising in a culture of histidine-requiring (*his*$^-$) *E. coli* in synthetic medium containing a strictly limited amount of histidine (Ryan and Schneider, 1949). One would naturally predict that, under these conditions, the *his*$^+$ mutants would have a great selective advantage and would rapidly replace the majority of *his*$^-$ bacteria. But this is not what happens. The limited growth of the *his*$^-$ cells represses that of the *his*$^+$ mutants by using up the energy source when cultivated aerobically and, under anaerobic conditions, by producing acid and other deleterious metabolic products. Apart altogether from the statistical and technical advantages and disadvantages of the various methods for assessing mutation rates, the nature of bacterial systems introduces so many uncontrollable complications that it must be admitted that no really satisfactory method exists.

THE INDUCTION OF MUTATIONS

There are many reasons why bacteria (and bacteriophages) are ideally suited to the study of mutagenesis. They have a short generation time, enormous populations can easily be obtained and manipulated under controlled conditions, while very small numbers of mutants can be detected and counted by selective methods. Their haploid nature greatly facilitates the recognition of recessive mutations and permits comparative studies of mutation and reverse-mutation (*forward* and *backward mutation*) to be made. Finally, the advent of methods of high resolution genetic analysis enables the sites of mutation to be pin-pointed on the chromosome. However, the randomness and rarity of spontaneous mutation has so far forced us to look at the phenomenon rather indirectly, in terms of probabilities inferred from the kinetics of growing populations. To study the nature of the mutational event itself we need a method of *inducing* the synchronous onset of mutation so that the metabolic and genetic behaviour of populations of cells in which it has occurred may be followed. The number of induced mutants must greatly exceed the background of spontaneously arising mutants so that any changes found after the induction can be assumed to have been due to it and to have started at the same time.

The artificial induction of mutation was first accomplished by means of X-rays in *Drosophila* (Muller, 1928) and was followed by the discovery that ionising radiations in general (γ-rays, neutrons, *a*-rays) were also mutagenic. The induction of mutation in *Drosophila* by ultraviolet (UV) light was then reported (Promptov, 1932). A detailed study of the induction of mutations in bacteria (*E. coli*) by X-rays and UV light was first undertaken by Demerec and Latarjet (1946) and since then bacteria have been increasingly used for both irradiation and mutational research. The mutagenicity of certain chemical substances, the nitrogen mustards, emerged during the Second World War, from studies of the genetic effects of mustard gas on *Drosophila* (Auerbach and Robson, 1946; Auerbach, Robson and Carr, 1948). Since then a vast range of mutagenic chemical compounds have been described; these include inorganic acids, alkalis, ammonia, peroxides, metal salts, organic acids, formaldehyde, phenols, carbamates, dyes, various carcinogenic substances, epioxides, ethyleneimines, alkylating agents, acridines, and purine and pyrimidine analogues. (A comprehensive list of chemical mutagens is given by Szybalski, 1958.)

The great diversity of these mutagenic substances discouraged the hope that some simple chemical configuration might appear as a mutagenic principle which would clarify the chemical basis of mutation. However this was not the only complication. Different physical and chemical agents were found to vary greatly in their toxicity as compared with their mutagenic activity; for example most mutagens kill bacteria rapidly, but ethyl methane sulphonate (EMS), although highly mutagenic, has been reported to have little effect on viability (Loveless and Howarth, 1959). Some chemical mutagens induce mutations at certain genetic sites but not at others within the same locus; on the other hand, some mutations which spontaneously revert at a normal rate are stable to the action of artificial mutagens (Demerec, 1953; Glover, 1956).

The induction of mutation is often reversible by environmental factors. Thus the action of X-rays is reduced by catalase (presumably by destruction of organic peroxides produced by the irradiation) and in the absence of oxygen, while the mutagenic effect of UV light can be reversed by subsequent exposure of the irradiated cells to visible light (*photoreactivation* or *photoreversal*) (Kelner, 1949; see p. 331); again, the presence of purine ribonucleosides can completely suppress the induction of mutations by purine analogues, and even the occurrence of a proportion of spontaneous mutations (Novick, 1956).

The number, variety and complexity of the observations concerning induced mutagenesis that have accumulated during recent years make the writing of a concise and digestible review very difficult (but see Westergaard, 1957). We do not propose to attempt it since much of the information relates more to the mode of action of radiations than to genetics, while little of it fits together to form a rational picture of mutagenesis. On the other hand we will see in Chapter 13 that a very plausible model of how mutants are formed, at the molecular level, can be constructed from our knowledge of the physico-chemical structure of the genetic material, coupled with genetic analysis of bacteriophage mutants induced by chemicals which act specifically on this structure. For the moment we will pave the way to these new ideas by discussing some biochemical requirements for the establishment of induced mutations.

THE KINETICS OF EXPRESSION OF INDUCED MUTATIONS

Early studies of the induction of many types of bacterial mutant by UV light showed that the maximum yield is obtained only when the irradiated bacteria are allowed to multiply for many generations before plating them on selective medium. In the case of mutation to phage resistance, for example, the delay in the appearance of mutants extended to as long as twelve division cycles (Demerec, 1946). A number of theories were formulated to explain this delay. Apart from phenotypic and segregation lag, the more obvious of these were that the onset of division is delayed in those cells destined to become mutants, or that the mutational event itself is a delayed effect of the induction. There is little doubt, as we have seen, that both phenotypic lag and segregation lag can operate in delaying the expression of mutations but this was not the important thing, for it turned out that the relevance of all these theories was largely by-passed by the outcome of a brilliant series of experiments by Evelyn Witkin (1956), originally designed to test their validity.

In order to determine the extent to which phenotypic and segregation lag might account for the delay, Witkin (1956) began by comparing the kinetics of the appearance of UV-induced mutants on the one hand, and of transductional recombinants of the same type on the other, under identical conditions. A suspension of tryptophan-requiring (try^-) *Salm. typhimurium* was irradiated and divided into two parts, A and B, each of which was treated with a preparation of transducing phage as follows:

A. try^-/UV recipient+phage grown on a UV-induced try^+ mutant strain as donor.

B. try^-/UV recipient+phage grown on the same (recipient) try^- strain as donor.

Both suspensions are thus identical with respect to previous history, UV-treatment and phage infection. Suspension A, however, should yield a predominance of try^+ *recombinants* since the frequency of transduction is much higher than that of induced mutation; suspension B cannot yield recombinants but only induced try^+ *mutants* because the transducing phage was grown on the same try^- strain. Both suspensions were spread on a number of plates of minimal medium containing various limiting concentrations of nutrient broth so that, on each, the number of residual generations through which the try^- bacteria could pass was restricted. After incubation, the number of try^+ colonies developing, expressed as a proportion of the maximum, was correlated with the number of residual generations which had elapsed after plating. The result was that the maximum number of *try*$^+$ transductants is achieved after a single division cycle while that of induced *try*$^+$ mutations requires six division cycles. This finding was confirmed in similar experiments with lysine- and adenine-requiring *Salmonella* auxotrophs. It was plain that neither phenotypic nor segregation lag, nor any non-genetic physiological effect of the irradiation, could account for the UV-induction delay since these factors are common to both sets of experiments.

If the late appearance of mutants is due either to delay in the division of those bacteria destined to become mutants, or to a lag in the mutagenic action of the inducer, this should be revealed by a pronounced diminution in the rate of increase of newly induced mutant clones as compared with that of clones of an equivalent number of *already established* mutant bacteria, similarly irradiated. Auxotrophic (*try*$^-$) bacteria on the one hand, and small numbers of prototrophic (*try*$^+$) bacteria on the other, were irradiated and plated on partially enriched minimal agar under identical conditions. At intervals after incubation the clonal increase of prototrophs was assessed by the respreading technique (p. 185). It turned out that although the rate of increase of newly induced mutant clones is a little slower than that of clones of irradiated, established mutants, the difference between them is far too small to account for the delay previously found (Witkin, 1956). Thus the facts are at variance with all the hypotheses.

Witkin conceived that this paradox might be resolved if the operative factor determining the appearance of mutants is not the number of post-irradiation divisions *per se* but the degree of enrichment of the medium used to control these divisions. For example, a concentration of nutrient broth which permitted ten residual divisions might allow much more active metabolism *during the first division cycle* than a concentration sufficient only for two divisions. Thus what appeared to be a lag might, in reality, be a quantitative nutritional effect deciding the irreversible establishment of mutation during a short, early period after irradiation. This possibility was investigated by transplanting the irradiated bacteria from unenriched to enriched minimal agar, and *vice versa*, at various times after plating on the initial medium, and then counting the number of prototrophic mutant colonies that arose. The results of a sample experiment are shown in Table 11. You will see that the crucial factor determining mutant yield is whether or not the bacteria are supplied with a source of nutriment during their first hour after irradiation, and that the ultimate yield is independent of the number of residual divisions. For example, in Experiment 3 (Table 11),

TABLE 11

The effects of the addition or deprivation of nutriment, during the 1st. hour after UV-irradiation, on the number of induced, prototrophic (try^+) mutants. $4{\cdot}3 \times 10^7$ *Salm. typhimurium* (try^-) bacteria, surviving treatment with UV light, were plated on the initial medium. After one hour at 37°C. they were washed from the surface, concentrated and replated on the final plate. After 48 hours incubation, the number of prototrophic colonies on the final plates were counted. (Data from Witkin, 1956.)

Enrichment { − = minimal medium alone. + = minimal medium enriched with 2·5 per cent. nutrient broth.

Experiment No.	Enrichment Initial → medium (one hour)	Final medium	No. viable cells replated	Average no. induced prototrophs per plate	No. residual divisions
1	−	−	$4{\cdot}3 \times 10^7$	4	0·4
2	+	+	$4{\cdot}1 \times 10^7$	127·5	6·5
3	−	+	$4{\cdot}0 \times 10^7$	2·5	6·1
4	+	−	$3{\cdot}9 \times 10^7$	127	1·5

where the irradiated bacteria are initially plated on minimal medium and then transplanted to enriched medium, the number of mutants is as low as when they are grown throughout on minimal medium (Experiment 1), although they go through 6·1 divisions on the final, enriched medium; alternatively, in Experiment 4, the mutant yield is the same as that of Experiment 2, where both initial and final plates are enriched, although the number of residual divisions is more than 4-fold lower. Other experiments established that the length of the 'sensitive' period after irradiation, during which nutrient broth must be present if the maximum yield of mutants is to be obtained, is about one hour.

What specific metabolic activity is promoted by the nutrient broth? The results of a number of experiments point unambiguously to protein synthesis. For example, a mixture of all the biological amino acids can substitute fully for broth, although the single amino acid required for growth (tryptophan) is ineffective alone. The appearance of mutants is dramatically suppressed if chloramphenicol, which specifically inhibits protein synthesis (Wisseman *et al.*, 1954; Hahn *et al.*, 1954), is added to the enriched initial medium for the first hour, but not if it is added later. Furthermore, reduction of the concentration of amino acids reduces the number of mutants and the growth rate proportionately, reduction of the latter presumably being proportional to the rate of protein synthesis when this is growth limiting. Finally, if the auxotrophic strain is grown, *before* irradiation, in minimal medium supplemented with the required amino acid, so that it is 'pre-adapted' to protein synthesis in this medium, the subsequent yield of mutants is largely independent of exogenous amino acids although still sensitive to chloramphenicol suppression. On the other hand, a number of auxotrophic strains of *E. coli* were isolated which, when grown in broth and then irradiated, gave a maximum yield of prototrophic mutants, with no lag, on unenriched minimal agar; nevertheless the appearance of these mutants could be suppressed by chloramphenicol (Witkin, 1956).

We have seen that the sensitive period, during which exogenous amino acids must be present if mutants are to develop, is about one hour at 37°C., or about the first third of the first division cycle on minimal agar supplemented with the required amino acid. If the irradiated bacteria are left on unenriched medium, or are exposed to chloramphenicol on an enriched medium, for any appreciable part of this period, the yield of mutants is irreversibly depressed. On the other hand, if *all* the metabolic activities of the irradiated bacteria are arrested by incubating them in buffer, they then preserve their capacity to produce mutants for a much longer time. It follows that, on minimal agar, some metabolic process *other than protein synthesis* is initiated which, when completed early in the first division cycle, shuts the door on mutational alteration. This is confirmed by the fact that when the duration of the post-irradiation division cycle is altered by changing the overall rate of metabolism as, for example, by incubation at different temperatures, the sensitive period is changed proportionately so as to comprise about one third of the lag period; on the other hand, when the lag is extended by limiting or temporarily arresting protein synthesis at 37°C., the sensitive period is not changed but remains constant at one hour (Witkin, 1953, 1956).

What is the metabolic process that controls the duration of the sensitive period? A plausible candidate is synthesis of the genetic material itself, DNA. For instance, the amount of DNA just doubles during the sensitive period (Kanazir, in Witkin, 1956), while prolongation of the lag period by increasing the dose of UV light or by post-treatment with varying concentrations of the purine analogue caffeine, both of which specifically depress DNA synthesis (Kelner, 1953; Witkin, 1958), affect the sensitive period proportionately.

We have dealt with this line of research at some length in order to illustrate a new and promising approach to the study of mutagenesis, and not to explain how it occurs. It appears, in fact, that the phenomena described tend to be restricted to the induction by UV light of mutations which determine the synthesis of new enzymes and, therefore, are dominant. Thus while mutations to lactose fermentation ($lac^- \rightarrow lac^+$) behave like mutations to prototrophy in their susceptibility to irreversible repression by chloramphenicol, mutations from streptomycin sensitivity to resistance (which happens to be recessive; Lederberg, 1951a) in the same bacteria display no sensitive period (Witkin and Theil, 1960). The behaviour of induced mutations to auxotrophy is not known since these cannot be selected for directly (see next section). Furthermore, mutations to prototrophy induced by alkylating agents not only are not repressed by post-treatment with chloramphenicol but are actually stimulated by the drug (Strauss and Okubo, 1960).

THE ISOLATION OF AUXOTROPHIC MUTANTS

Mutants which have acquired resistance to bacteriophages or to drugs are easy to isolate since they can be selected by plating large populations of sensitive cells on nutrient agar containing the anti-bacterial agent; the sensitive cells are killed, or their growth suppressed, so that only the resistant mutants can multiply to produce colonies. Similarly, as we have seen, prototrophic mutants are readily selected from cultures of auxotrophic strains by plating the washed culture on minimal medium which lacks the required growth factor. Not so with auxotrophic mutants whose distinguishing feature is that they will *not* grow on media which support growth of the wild type. Since auxotrophic mutants of bacteria are so widely used in biochemical as well as in genetic research we will say a little about the general methods used to obtain them.

The initial step is usually to treat a culture of the wild type strain with some mutagenic agent in order to increase the proportion of mutants among the survivors. Although highly efficient chemical mutagens such as nitrosoguanidine (p. 213) are now usually employed, until recently ultraviolet (UV) light was the most widely used agent, a washed suspension of the bacteria being irradiated to about 1·0–0·1 per cent survival. (A brief account of irradiation 'target' theory, and of the relationship between dose and survival, will be found on p. 525). As a rule the irradiated or otherwise mutagenised bacteria should first be suspended in a complete medium, such as nutrient broth, and incubated to allow the recessive auxotrophic mutations to segregate and express themselves. The small proportion of auxotrophic mutants must now be isolated and recognised, and their specific nutritional requirements identified. There are several ways of doing this. In one early method, called the *delayed enrichment* method, the culture containing the mutants is diluted and plated on minimal agar, so as to yield isolated colonies after incubation; a small volume of molten minimal agar is then poured over the spread surface and allowed to set, so that the bacteria are sandwiched between two layers of minimal agar. On incubation, only the non-mutant, prototrophic bacteria give rise to colonies; the position of these colonies is marked. A layer of *nutrient* agar is then poured over the minimal agar surface, and the plate re-incubated. The nutrients diffuse into the minimal agar, allowing the auxotrophic mutants to produce colonies which are subsequently identified by their delayed appearance. In another method, that of *limited enrichment*, the irradiated culture is spread on a minimal agar plate containing a limiting concentration of nutrient, so that presumptively auxotrophic bacteria are recognised directly by the small size of their colonies.

These methods have now been largely superseded by *replica plating* (p. 187). The irradiated culture is diluted and spread on initial plates of nutrient agar; the resulting colonies are then sampled with a velveteen pad and printed to replica plates containing minimal agar. When the initial and replica plates are compared, colonies of auxotrophic mutants growing on the initial plates are recognised by their absence from the replica plates when the two are superimposed. The auxotroph colonies must then be purified by replating, checked for auxotrophy, and their specific growth requirements finally determined by streaking (or replica plating) to minimal agar supplemented with various combinations of amino acids, B group vitamins, and purine and pyrimidine bases (see Holliday, 1956). It should be noted that, as a rule, only a proportion of the presumptive auxotrophic isolates, identified by their initial failure to grow on minimal agar, turn out to be stable mutants.

The isolation of auxotrophic mutants was greatly facilitated by the introduction of a semi-selective method based on the mode of action of penicillin (Davis, 1948; Lederberg, 1950). Penicillin (and, for that matter, streptomycin) is an antibiotic which rapidly kills growing bacteria but is innocuous to those which are not metabolising. If, therefore, a mixture of prototrophic and auxotrophic bacteria is well washed to remove all nutrient material and then incubated in minimal medium containing a lethal concentration of penicillin, the prototrophs will be killed as soon as they begin to metabolise, but the auxotrophs, being unable to grow, will remain viable. Thus a proportionate increase in the number of auxotroph colonies will be found on subsequent plating, Although penicillin is primarily effective against Gram-positive organisms, it is also lethal for a wide range of Gram-negative bacilli, such as *Escherichia*, *Salmonella*, *Shigella* and *Proteus*, if used in fairly high concentration, and especially in synthetic media.

The method works well in practice and enables spontaneous as well as induced mutants to be recovered with ease, but is it not sufficiently controllable to permit quantitative studies of mutation to auxotrophy. Moreover, reliable results are obtained only by rigorous attention to detail. It is essential, for example, that penicillin treatment should be preceded by adequate growth of induced cultures in nutrient broth to allow full segregation and expression of the mutants; subsequent starvation of the bacteria in a nitrogen-free medium is desirable, in addition to washing, to ensure the metabolic inertness of auxotrophs. The primary action of penicillin on sensitive bacteria is to induce protoplast formation by interfering with cell wall synthesis (Lederberg, 1956b); the protoplasts then burst osmotically, releasing nutrients into the minimal medium which promote growth of the auxotrophic mutants, so that they, in turn, become susceptible to the penicillin. For this reason the treated population should be restricted to not more than about 10^7 bacteria per ml, the concentration of penicillin should be high and the duration of treatment as short as possible. The use of hypertonic minimal medium, containing 20 per cent sucrose, to prevent the bursting of protoplasts has been reported to increase greatly the auxotroph yield (Gorini and Kaufman, 1960).

The methods we have described, and particularly the penicillin method, can be adapted to the selection of auxotrophs with specific growth requirements. Suppose we wish to isolate an arginine-requiring mutant without the trouble of searching for one among a wide range of other auxotrophs. Selection can first be applied during the stage of post-irradiation growth to allow mutational segregation and expression. If the irradiated culture is incubated in minimal medium containing arginine instead of in broth, *arg*⁻ mutants, but no other types of auxotroph, can grow and keep pace with the wild type population. Secondly, if the penicillin-minimal medium is supplemented with all the amino acids *except* arginine, only *arg*$^-$ mutants will be unable to metabolise and will be spared by the penicillin; in this case, of course, highly purified amino acid preparations must be employed. In the same way, double auxotrophs, having two requirements, may be obtained by starting with a strain having one of the requirements, such as for methionine; in this case the initial type is selected against by adding methionine to the penicillin-minimal medium.

Recently a new chemical mutagen of great potency, N-methyl-N^1-nitro-nitrosoguanidine (NTG), has come into wide use for obtaining bacterial mutants, although its mode of action at the molecular level remains unknown. This substance is so efficient that it has been shown to induce at least one mutation per treated cell under conditions permitting over 50 per cent survival, while auxotroph yields well in excess of 10 per cent of the treated bacterial population can be achieved (Adelberg, Mandel and Chein Ching Chen, 1965). This order of effectiveness may be a positive disadvantage since every mutant is likely to have one or more other mutations at sites different from that selected. For a description of optimal conditions for mutagenesis by nitrosoguanidine, see Adelberg *et al.* (1965).

SUPPRESSOR MUTATIONS

When a mutant reverts to wild type as a result of another mutation at a different locus or site, the second mutation is called a *suppressor* of the first. If the original mutation imposed a requirement for, say, tryptophan (*trp*), a suppressor of this mutation is designated *su-trp*; the genotype of the original wild type would thus be written as *trp*$^+$*su-trp*$^+$, of the mutant as *trp.su-trp*$^+$, and of the suppressed mutant as *trp.su-trp*. Revertants due to suppressor mutations are usually phenotypically distinguishable from true wild type, as well as from revertants due to back-mutation at the original site, particularly in displaying a lower grade of function; that is, they are *pseudo-wild*. For example, suppressors of an auxotrophic mutation yield revertants which grow more slowly and produce smaller colonies on minimal agar than does the wild type; when the enzyme itself can be estimated, this may be found to be synthesised at a lower rate than normal.

The occurrence of a suppressor mutation can only be demonstrated by genetic crosses which reveal the non-identity of the sites of the original and reverse mutations. By definition, reversion by a suppressor mutation leaves the original mutation intact, so that the original mutant type should be recovered among the progeny of a cross between wild type and revertant, as Fig. 54 shows. The frequency with which such mutant recombinants are found depends, of course, on the distance between the suppressor site and that of the original mutation; this distance may in some cases be very small, as we shall see, so that a selective method may be required to detect the recombinants. By means of appropriate crosses, a suppressor mutation isolated in a particular mutant can be introduced into other mutants so that its effect on other alleles and loci may be studied.

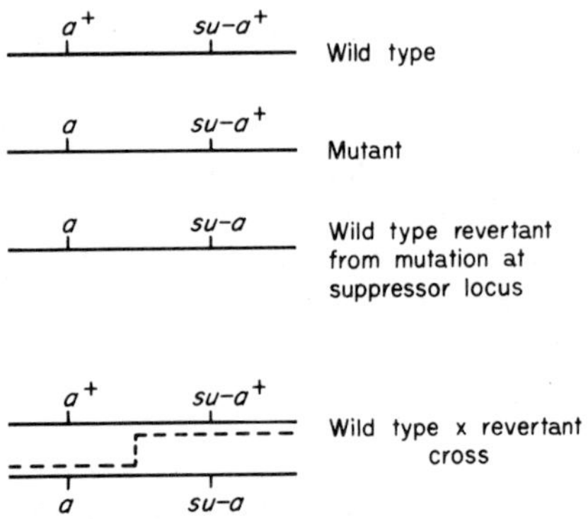

FIG. 54. The genetic identification of suppressor mutations.
In the wild type × revertant cross, recombination in the interval between the sites of the original and of the suppressor mutation will yield a mutant recombinant, as indicated by the interrupted line.

The use of the single word 'suppressor' to connote all reversions due to mutations at secondary sites is misleading, since it implies a unity of mechanism which certainly does not exist. In fact, a great variety of different spatial and functional relationships have been found between suppressors and the alleles they suppress. Thus suppressor loci may be either unlinked or closely linked to the genes they control; the sites of suppressors of spontaneous and acridine-induced r_{II} mutations in bacteriophage T4 (p. 323) are actually located within the same cistron as the original sites and exceedingly close to them (Crick *et al.*, 1961; see p. 346). From the functional point of view, the action of a suppressor mutation may involve a number of loci having different and even, apparently, unrelated functions, or may be restricted to the alleles of a single locus; alternatively, a number of suppressors have been described which are allele-specific, that is, they only suppress mutations which arise at particular sites within a locus.

A number of different mechanisms have been postulated to explain how suppressors operate. For example, suppressors of mutant loci, or groups of loci, which block a biochemical pathway could act by opening up an alternative pathway. Again, it has been suggested that the mode of action of some allele-specific suppressors may be to restrain the inactivation of specifically altered enzyme by some metabolite. For example, a specific allele of the tryptophan synthetase locus in *Neurospora* (td_{24}) yields an enzyme which is inhibited by zinc so that only highly purified preparations are active; a specific, unlinked suppressor for this allele is known and may operate by preventing zinc from uniting with the altered enzyme (Suskind and Kurek, 1959).

Another postulated mechanism, for which considerable experimental evidence now exists, involves the structure of what is called *amino acid-transfer ribonucleic acid* or, for short, *transfer* RNA which we will discuss in more detail later (p. 280). For the moment it is enough to know that the function of this low molecular weight RNA fraction is to transport amino acids to the ribosomal particles where proteins are made, and there to align them on a template, based on the information carried by the gene, in the sequence characteristic of the particular polypeptide to be synthesised. For this purpose there exist a number of different species of transfer RNA, each of which attaches specifically to only one of the 20 biological amino acids when these have been enzymatically 'activated'. When a mutation occurs which leads to formation of a faulty protein, the indirect result is that the ribosomal template is altered in such a way that the wrong transfer RNA species (*b* instead of *a*), carrying a different amino acid (B instead of A), is lined up at a particular place in the sequence. It is clear that if another mutation could change a proportion of these 'wrong' RNA molecules so that they now transported the 'right' amino acid (A), then good protein would be synthesised. This would mean, of course, that amino acid A would be substituted for B at other positions in the same protein, as well as in other proteins, so that the suppressor might be expected to restore only a proportion of the activity of normal wild type enzyme and to have a widespread effect. A mutation involving the specificity of one of the activating enzymes responsible for attaching the 'right' amino acid to the 'right' RNA molecule might be assumed to exert the same suppressor action.

The fact that a proportion of r_{II} mutants of bacteriophage T4 temporarily display the wild phenotype in the presence of the base analogue 5-fluorouracil (5FU), which can substitute for uracil in RNA, and at the same time can grow on certain mutant derivatives of *E. coli* K, originally suggested that this kind of suppression might occur, especially since certain bacterial mutations involving enzymes such as β-galactosidase and alkaline phosphatase also proved responsive to 5FU by showing the wild phenotype (see p. 370; Benzer and Champe, 1961). It was then found that a number of classes of suppressors of this type exist in *E. coli*, mapping at different parts of the chromosome, which reverse equivalent groups of mutations irrespective of whether these involve the r_{II} region of phage T4 or the lactose pathway of *E. coli*, by correcting certain errors in the genetic code of the gene which result in failure to synthesise the derivative polypeptide chain beyond a certain point (p. 361; Benzer and Champe, 1961; Brenner and Beckwith, 1965). Moreover, it is now known that certain regions of the bacterial chromosome in fact determine the structure of transfer RNA molecules, which can thus be altered by mutation in the postulated way (p. 289).

A mechanism of suppression about which there can be no doubt is that described by Crick *et al.* (1961) whereby reversion results from a mutation in the same cistron as the original mutational site and very close to it. Since discussion of this mechanism involves the nature of the genetic code, we will relegate it to Chapter 14 (p. 346). Another type of suppression, which results from restoration of a stable, active protein structure by mutation at a secondary site, is described on p. 301.

Most of the suppressor systems reported in bacteria have arisen in *Salm.typhimurium*, simply because this organism has been so extensively used in transduction studies. Among them are allele-specific suppressors for adenine-thiamine (*adth*) (Yura, 1956a) and leucine (*leu*) (Smith-Keary, 1960) mutants, and non-specific suppressors for several cystine (*cys*) loci (Howarth, 1958). An interesting example of non-specific suppressor activity which does not involve auxotrophic mutants has been described by Hashimoto (1960) who found, by transduction analysis, that mutations to either streptomycin-resistance or dependence in *E. coli* are clustered within a very small region of the chromosome; sixteen different suppressor alleles were identified and mapped within a closely linked locus, any one of which can suppress all of the resistance and dependence mutations tested.

DELETIONS AND REARRANGEMENTS

We have already noted that multisite mutants, or deletions, occur both in bacteria and in bacteriophages and have proved of great value in fine structure mapping. Their characteristic feature is that they cannot be represented as points, but behave as if they covered segments of the chromosome; thus they do not revert to wild type and do not show recombination with any mutational site within the regions over which they extend, although these sites can recombine with one another to yield wild type recombinants.

In *Salmonella*, the frequency with which deletions arise varies for different regions of the chromosome, but is generally low. Thus eight of 279 histidine mutants, six of 38 proline mutants, one of seven cysteine A mutants and one of 65 tryptophan mutants were deletions, while none were found among 120 leucine, 37 threonine, 16 isoleucine-valine, 12 isoleucine and 14 aromatic compound mutants (Demerec and Hartman, 1959). On the other hand, about 40 per cent of mutations of the CYS-C region are deletions. Over 500 CYS-C mutants have been isolated and analysed by transduction, so that the region is unusually well mapped. These mutants fall into five complementing groups, representing five gene loci (*cys*C, D, H, I and J). Of the 93 deletions studied by Demerec, 81 are very similar; they start at a particular location between the D and H loci and then 'run off the map' whose extent, of course, is strictly limited in transductional analysis. This location is peculiar for another reason, because the closest mutations on either side of it show a recombination frequency about twice that of other adjacent mutations in the region. This implies that the location where the deletions originate is actually a small 'silent' region of the chromosome which is not subject to mutation and, therefore, specifies no genetic information. It has been equated with regions of *heterochromatin* which can be observed on the chromosomes of higher organisms and may carry a special liability to breakage (Demerec, Gillespie and Mizobuchi, 1963). Comparable deletions have been described in the lactose region of *E. coli*, where as many as 30 per cent of mutations may be long deletions (Cook and Lederberg, 1962).

In general, deletions, as well as other chromosomal aberrations (p. 41), are particularly prone to follow treatment with ionising radiations which are known, from morphological studies, to produce chromosomal breaks in higher organisms. In the CYS-C region of *Salmonella*, however, approximately the same proportion of deletions are found among mutants whether these are spontaneous or induced by X-rays, UV light or nitrous acid (p. 304); on the other hand, no deletions are found among mutants induced by the nucleic acid base analogue, 2-amino purine (Demerec, 1960) (see p. 303). Deletions amounting to one per cent of the chromosome have been found in *E. coli* (Jacob and Wollman, 1961a, p. 167).

It is usually assumed that deletion mutants have suffered an actual excision of a fragment of genetic material. Such an excision should lead to a reduction of the distance between outside markers close to the two extremities of the deleted region. If two strains carrying the same deletion, but differing with respect to the outside markers, are crossed, the normal recombination frequency between the markers should be reduced in proportion to the extent of the deletion. This result has been found in the case of deletions in the r_{II} region of phage T4 (p. 170). It might be expected that a physical loss of genetic material (DNA) as the result of a deletion would be reflected by a reduction in the density of the phage particles, since less DNA would be packed into the phage head. The densities of phage particles can readily be compared by centrifugation in a density gradient of caesium chloride having a mean density approximating that of the phage (p. 394). In fact a density reduction was not found to accompany the decrease in recombination frequency. However, as we have seen, genetic analysis suggests that the entire r_{II} region constitutes only about one per cent of the phage genome; a loss of this extent would be scarcely detectable as a density difference by the method employed (Nomura and Benzer, 1961). On the other hand, viable mutant strains of the *E. coli* phage, λ, have been isolated which, by density measurements in a caesium chloride density gradient, lack as much as eighteen per cent of their genetic material (Weigle, 1960; Kellenberger, Zichichi and Weigle, 1961b, see p. 470).

In addition to deletions, chromosomal transpositions have been reported following treatment of *E. coli* with X-rays and nitrogen mustard; appreciable segments of chromosome have been removed from their proper regions and inserted elsewhere (Jacob and Wollman, 1961a, p. 167). Aberrations of this sort are easy to recognise when they occur in *Hfr* male strains of *E. coli* K-12 for, on conjugation with females, such strains transfer their chromosomes linearly from one end, so that alterations in the sequence of the genes can be established by interrupted mating experiments (p. 65, Fig. 15; Chapter 22, p. 682 *et seq.*).

More recently, inversions have been described in *E. coli* K12 in which segments of bacterial chromosome have become inserted the wrong way round, usually at some location different from the normal one; that is, the inversion is associated with a transposition. In most of these cases the segment of bacterial chromosome has first become integrated into the sex factor chromosome to form an F-prime factor (pp. 674, 794) which can then insert itself, together with its bacterial component, either way round at different chromosomal sites (see p. 374 ; Cuzin and Jacob, 1964; Berg, 1966; Berg and Curtis, 1967). The occurrence of inversions has important implications for the way in which the genetic information carried by the chromosome is translated into functional terms, and we will consider it again when we discuss the genetic code in Chapter 14 (p. 374).

HIGH MUTABILITY GENES

Strains of *E. coli* and *Salmonella* have been isolated which are characterised by an unusually high spontaneous mutation rate. In these strains the rate of mutation of a wide range of genes may be raised by as much as 1000-fold above the normal. Genetic analysis has shown that the character of high mutability is determined at one or another specific chromosomal locus (*mut*) (Treffers *et al.*, 1954; Miyake, 1960).

There is some evidence for the operation of one possible mechanism for this phenomenon, namely, that mutation in a gene determining the synthesis of a normal nucleic acid base may result in the formation of a mutagenic base analogue. It was observed that virtually all histidine (*his*) mutations arising spontaneously in a high mutability strain of *Salm. typhimurium* are of the same type as those induced by analogues of the nucleic acid bases (p. 309; Kirchner, 1960). Chromatographic analysis of the bases of this strain has revealed that, in addition to the normal bases, adenine, guanine, thymine, and cytosine it possesses an extra base not present in strains which mutate at the normal rate. This base, analogous to guanine, appears to be very similar to 2-dimethylamino 6-hydroxy-purine. Moreover, the mutator gene in *Salm. typhimurium* seems to be closely linked to markers involved in purine or pyrimidine synthesis (Kirchner, 1962; personal communication; Kirchner and Rudden, 1966).

More recently, a quite different mechanism has been found in phage T4. Certain mutations in a gene specifying the structure of an enzyme (polymerase) which is responsible for the replication of the phage chromosomal DNA, lead to the production of a high proportion of mutant progeny phages by making mistakes in copying (Speyer, 1965; see p. 256).

A recent analysis of the *E. coli* mutator stock isolated by Treffers *et al.* (1954) has shown that its behaviour differs fundamentally from that of the *Salmonella* stock investigated by Kirchner (above), in that all the mutations produced are of a very specific type (transversions; p. 309) distinct from those induced by base analogues and other chemical mutagens, and are one-way; that is, the mutations resulting from mutator action are not reverted by it (Yanofsky, Cox and Horn, 1966).

CONTROL OF MUTATION BY NON-CHROMOSOMAL GENETIC ELEMENTS

Certain slowly growing prototrophic revertants of a particular leucine-requiring auxotroph (*leu-151*) of *Salm. tyhpimurium*, due to mutation at a closely linked suppressor locus (p. 213), were found to be very unstable. Thus, a variable and usually high proportion of the bacteria of a revertant colony turned out to be auxotrophs which were themselves unstable. From these unstable lines, stable, slowly growing prototrophs, as well as stable auxotrophs, could be isolated. The property of instability proved to be transducible and was initially restricted to a particular suppressor locus.

On this evidence alone the instability might be ascribable to some inherent property of the suppressor gene, influenced perhaps by the genetic background of the cell. However, two further findings showed that this was not so. Firstly, when the suppressor-*leu-151* region of chromosome was transduced from a stable derivative to another *Salm. typhimurium* strain (*ara B-9*), many of the slowly growing, leucine-independent transductants were found now to be unstable, and this instability could be transferred back to a stable *leu-151* line by transduction. Secondly, when another mutation determining a proline requirement (*pro-401*) was introduced into the *leu-151* strain, some proline-independent reversions derived from it were unstable, and this new instability was not only now co-transducible with the *pro* locus but was highly correlated with stabilisation of the *su. leu-151* region; in other words, the factor responsible for instability was capable of shifting from one locus to another (Dawson and Smith-Keary, 1963).

About two years after the initial isolation and investigation of the *pro-401* mutation just described, this strain was again studied and was found to have undergone a remarkable change in behaviour. Unlike the original auxotroph it no longer yielded any wild type reversions. However, when plated on minimal agar lacking proline, a considerable number of apparently prototrophic colonies arose; when diluted suspensions of these colonies were plated on the same medium, no colonies at all emerged—the original colonies consisted entirely of *pro*$^-$ auxotrophs! On the other hand if one of the pseudo-prototrophic (or 'auxotrophic revertant') colonies was streaked out on the medium, growth occurred where the inoculum was heavy but declined rapidly with decrease in cell density. However, the most striking phenomenon to emerge from this study was that the *pro-401* strain is self-transducible; when transducing phage grown on one culture of this strain is applied to another, colonies of stable, *pro*$^+$ transductants arise (Smith-Keary and Dawson, 1964).

These facts are explicable on the hypothesis, proposed by Smith-Keary and Dawson (1964; Dawson and Smith-Keary, 1963), that the *pro-401* auxotroph did not arise from a true mutation but from attachment of a 'controlling episome' (previously invoked as the cause of instability at the *su. leu-151* locus) to a site within the *pro* region of chromosome, thus blocking the action of the gene involved. If this element is assumed to shift from one site to another within the region, then a *pro-401* population will contain bacteria blocked at different genetic sites so that transduction between them can yield prototrophic recombinants. Moreover if the blocking element can move to genes mediating other steps in the pathway of proline synthesis, then the growth, on proline-deficient media, of colonies which contain only auxotrophic bacteria can be explained by the occurrence of cross-feeding (p. 90). The existence of such complementing mutant bacteria in cultures of strain *pro-401* was verified by the demonstration that abortive transductants arise in self-transduction experiments. In addition, transductions between clones that have been separately subcultured for many generations may yield as many as one hundred times more prototrophs than self-transduction within the same clone showing that the blocking element may spread to relatively distant proline sites (Smith-Keary and Dawson, 1964; see also Smith-Keary, 1966).

In their first paper Dawson and Smith-Keary (1963) drew attention to the close similarity between the behaviour of their system and that of the 'controlling elements' described by Barbara McClintock in maize, which jump from one chromosomal locus to another, modifying the stability and expression of those genes to which they are temporarily attached (review: McClintock, 1956). We think the phenomenon is both interesting and important, but would warn readers against the possible confusion that could arise from use of the word 'episome' to describe the controlling element. This word was coined to define a group of genetic elements whose principal feature is the ability to adopt alternative cytoplasmic or chromosomal states in the cell (Jacob and Wollman, 1956c). The definition was specifically based on two elements which display this feature—the sex factor of *E. coli* and the genetic material of the temperate bacteriophage λ, both consisting of DNA equivalent to about one per cent of the DNA of the bacterial chromosome. There is as yet no evidence concerning the physical constitution or size of the controlling elements in *Salm. typhimurium*, while that for an independent cytoplasmic existence is far from compelling. The controlling element may turn out to be of a fundamentally different nature from those other elements which the word 'episome' now brings to mind (p. 747).

On the other hand, a precedent already exists for the induction of mutations in *E. coli* by infection with a temperate bacteriophage (Mu 1) (Taylor, 1963). This mutagenic action, which induces single auxotrophic and other mutations at many different loci with a frequency of about 2 per cent, was found to be associated only with the process of lysogenisation; the mutation rate of established lysogens was no higher than that of the non-lysogenic strain. Finally, genetic mapping of the chromosomal locations of the prophage in various mutant strains, by means of conjugation, revealed that these always coincide with, or are closely linked to, the sites of mutation. Recent evidence strongly suggests that lysogeny results from a single, reciprocal recombination event between circular bacterial and viral chromosomes which unites the two into a single, continuous structure; at the site of insertion, of course, the continuity of the bacterial chromosome will be interrupted by the prophage (p. 458; Fig. 102). The most likely explanation of the mutagenic effect of lysogenisation by phage Mu 1 is that the prophage has pairing affinities within many genes whose function is disrupted by insertion, since the two parts of the gene are now separated by the prophage (Taylor, 1963). What may be a similar mutagenic effect has been reported to follow infection of *Actinomyces olivaceus* by actinophage (Alikhanian and Iljina, 1960).

A phenomenon superficially similar to the self-transduction of *Salm. typhimurium* strain *pro-401* (above) was described at about the same time by Demerec (1960). He discovered that certain mutant strains of *Salm. typhimurium* gave rise to a significantly larger number of prototrophic clones when treated with phage grown on the same strain, than occurs by spontaneous reversion in control cultures. These strains, which he called 'selfers', comprised about 40 per cent of 201 auxotrophic mutants tested, representing 22 gene loci. They are not distributed randomly either within or among the loci with which they are associated, and are singularly absent from some loci such as the four *his* loci investigated. A most unexpected finding was that prototrophs continue to arise when transducing phage grown on a donor strain having a deletion covering the mutant site of the selfer recipient is used. Demerec (1963) concluded that selfing could not be due to transduction as normally understood, and suggested that the mutability of the mutant site of a selfer gene is stimulated by the pairing of an immigrant chromosome fragment near to it. Alternatively the mutational defect in the recipient chromosome might be repaired by substitution of a fragment of donor chromosome from some non-allelic region, which happens to have the proper sequence of genetic sites. In molecular terms, such non-allelic homologies would be due to chance reiterations of nucleotide sequences in the chromosomal DNA (Clark, 1964; see p. 798). Whatever its cause, the phenomenon of selfing is clearly quite different from self-transduction in *Salm. typhimurium* strain *pro-401*, which is associated with production of auxotrophic revertants and the occurrence of abortive transductants (Smith-Keary and Dawson, 1964).

Part VI
The Physico-Chemical Mechanisms of Heredity

CHAPTER 11

Deoxyribonucleic Acid

Biologists have for many years been interested in the group of chemical substances known as the nucleic acids which, together with the proteins, comprise the principal macromolecular constituents of cells. The story of the nucleic acids begins as long ago as 1869 when Friedrich Miescher, a Swiss biochemist working at Tübingen, isolated a substance which he called 'nuclein' from the nuclei of pus cells. Miescher later returned to Basle and continued his work on 'nuclein', extracting it from the sperm of salmon which at that time came up the Rhine as far as Basle to spawn. It was later realised that 'nuclein' had the properties of an acid and so became known as nucleic acid. Because of the particular association of nucleic acids with the nuclei of cells, their nature and function became of particular interest to geneticists.

There are two types of nucleic acid, deoxyribonucleic acid (DNA) and ribonucleic acid (RNA) whose chemistry we will shortly discuss. Chemical analysis and the development of histochemical staining techniques revealed that the DNA of cells is restricted to the nuclei and appears to be localised in the chromosomes, while the RNA is found elsewhere in the nuclei as well as in the cytoplasm. In addition, much indirect evidence pointed to an intimate and universal association of DNA with the genetic material. First, DNA is present in all cells, whether animal, plant or bacterial, and, in them, is virtually restricted to the nucleus. Second, all the somatic cells of any given species contain a very constant amount of DNA, irrespective of the functional differentiation or metabolic state of the cells. The amount of RNA, on the contrary, varies markedly with metabolic activity. Third, the amount of DNA present in haploid germ cells, which possess only one set of chromosomes, is just one half that found in diploid somatic cells containing two sets of chromosomes. Similarly when the ploidy of somatic cells is further increased, the amount of DNA per cell rises proportionately. Finally, the mutagenic action of UV light has its peak at that part of the spectrum (2537 Å) which is most completely absorbed by nucleic acids.

However DNA is usually found in the nucleus in the form of nucleoprotein and, in view of the complexity of the instructions which the genetic material has to impart to the cell, it seemed more probable, at one time, that it was the protein moiety that conferred specificity because of the great number of unique configurations that protein can adopt.

EVIDENCE FOR NUCLEIC ACIDS AS GENETIC MATERIAL

General acceptance of the unique role of the nucleic acids in the transmission of hereditary characters awaited the evidence provided by three kinds of definitive experiments in bacterial and virus genetics. The first of these was the demonstration by Avery, Macleod and McCarty (1944) that the active principle mediating genetic transformation in pneumococci is highly polymerised DNA. The ability of the DNA to transmit genetic characters in this way is completely and specifically destroyed by the enzyme deoxyribonuclease; moreover, the higher the degree of purification of the DNA the greater becomes its efficiency in transformation (Hotchkiss, 1949; see Chapter 20, p. 575).

The second type of experiment was concerned with the problem of how virulent bacteriophages infect their host cells. Generally speaking, bacteriophages are simple particles consisting of a tadpole-shaped protein sheath, the head of which encloses only one type of nucleic acid—DNA. By specifically labelling the phage protein with radioactive sulphur (^{35}S) and the phage DNA with radioactive phosphorus (^{32}P), Hershey and Chase (1952) were able to show that, on infection, the phage DNA enters the bacterium whereas nearly all the phage protein remains outside. Once DNA injection has occurred the protein sheath of the phage, adsorbed to the cell wall, can be stripped off by violent agitation in a blendor without affecting the subsequent course of the infection (see Chapter 16, p. 422). Thus it is the phage DNA which 'programmes' the infected cell so that new phage DNA and protein are synthesised and finally incorporated in progeny phage particles.

Many animal and plant viruses, and several types of bacterial virus, are composed of protein and RNA, and are devoid of DNA. The third type of study we will mention established the fact that, in these viruses, the specifications for new virus synthesis reside in the RNA which, in them, supplants DNA as the genetic material. Tobacco mosaic virus, for example, consists of a protein rod with an axial hole running through it, the RNA being wound helically around the hole throughout the length of the rod. When a preparation of the virus is fractionated chemically, so that the RNA and the protein are separated, it turns out that the RNA *by itself* is infective, though to a much lower titre than its equivalent in intact virus, while the protein is inert (Gierer and Schramm, 1956; Fraenkel-Conrat *et al.*, 1957). The infectivity of the RNA is rapidly destroyed by ribonuclease but is unaffected by antiserum against the virus protein. A significant corollary is the discovery by Fraenkel-Conrat and Williams (1955) that when the fractionated and purified virus protein and RNA are mixed together in a test-tube, the two fractions combine automatically, by a kind of crystallisation process, to reconstitute particles which are morphologically mature and fully infective. Various strains of tobacco mosaic virus exist whose protein components can be distinguished both serologically and by their amino acid content. If two such strains of virus are fractionated, and the protein from one strain is then mixed with RNA from the other, virus rods are reconstituted as before. When the tobacco plant is infected with these synthetic, 'hybrid' virus particles, the progeny particles which are subsequently released are found to have protein of the same type as that of the original particles from which the RNA was extracted, and different from that of the infecting particles. This means that the specificity of the protein of the progeny particles, issuing from the infection, is determined by the RNA and not the protein of the infecting particles (Fraenkel-Conrat and Singer, 1957; review, Gierer, 1960).

These, then, were the revelations which spotlighted the nucleic acids as the repositories of genetic information, and paved the way for the elucidation of the physico-chemical structure of DNA which, beyond doubt, has been the most important and provocative biological discovery of this century.

THE PHYSICO-CHEMICAL STRUCTURE OF DNA

Hydrolytic degradation of DNA, by means of acids or enzymes, has shown that it is constructed of three essential components; these are the pentose sugar 2-deoxy-D-ribose, phosphoric acid which confers on DNA its acid properties, and nitrogenous bases. The DNA from all types of cell contains two kinds of nitrogenous base—the purines which have a double ring structure, and the pyrimidines with only a single ring. In general, and especially in bacteria, four bases predominate;

the two purines, adenine (A) and guanine (G), and the two pyrimidines, thymine (T) and cytosine (C). The chemical structure of these bases is shown in Fig. 55A, and that of the sugar, 2-deoxy-D-ribose, in Fig. 55B (bottom left). The sugar molecule can combine with any one of the

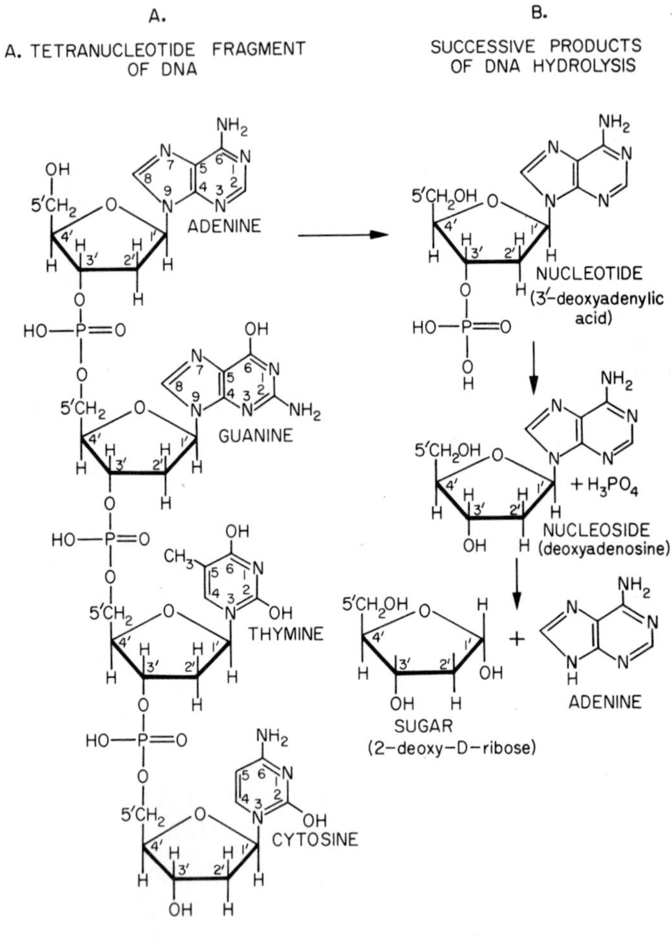

FIG. 55. The chemical components of deoxyribonucleic acid (DNA).
A. An intact tetranucleotide fragment, showing how the individual nucleosides are joined together by phosphate molecules, which link the 3′-position of one sugar residue to the 5′-position of the next.
B. The products of progressive DNA hydrolysis (see text).

four bases to form a glycosidic compound known as a *nucleoside* (Fig. 55B). Attachment of a phosphate group to the sugar of the nucleoside yields a *nucleotide* (Fig. 55B, top). There are thus four general types of nucleotide, depending on which of the bases is attached to the sugar-phosphate ester; deoxyadenylic acid, deoxyguanidylic acid, deoxythymidylic acid and deoxycytidylic acid.

Chemical analysis has shown that DNA is a polymer built up of a long chain of nucleotides, that is, a *polynucleotide*. To form this polynucleotide, the sugar residues of the nucleotides are joined together by their phosphate molecules in such a way that the C3′ position of one deoxyribose is linked to the C5′ position of the next (Fig. 55A) (review: Brown and Todd, 1955). This 3′–5′–3′–5′ linkage is important because, as we shall see, it introduces a polarity into the structure of the polynucleotide strand which is relevant to the genetic function of the molecule. Three further fundamental features of DNA were revealed by chemical analysis.

1. Whatever the source of the DNA, it contains an equivalent number of purine and pyrimidine bases, that is, A+G=T+C.

2. There is also equivalence between the amounts of adenine and thymine, and between the amounts of guanine and cytosine; that is, A=T and G=C.

3. In DNA isolated from different sources, the ratio A+T/G+C, called the *base ratio*, may vary widely although it remains constant for any one species. Among different bacterial species, for example, base ratios varying from 0·35 to 2·7 are found (see Chargaff, 1955: Sueoka, 1961a).

This was about as far as purely chemical investigation could go in showing how the DNA molecule is built up. The arrangement of the components of the polymer in three dimensions required a quite different approach to molecular structure which was provided by X-ray diffraction analysis. There is nothing inherently difficult in the principle of this method, although the interpretation of the results is, of course, a highly complicated and esoteric business. If a crystalline structure such as a diamond is rotated, in one plane, in a beam of light, the light will be reflected by the regularly repeating facets of the crystal to form a pattern, the intensity of the reflected light at any point depending on the degree to which the emergent rays augment or cancel one another through interference. If this pattern is recorded photographically, the architecture of the crystal can be deduced from it. However the degree of resolution of the method, as that of microscopic resolution, is limited by the wavelength of light. The wavelength of X-rays is sufficiently short to be capable of being reflected by organic molecules in which the average distance between atoms is about 2Å, so that the same principle can be applied to those molecules which are like crystals in possessing regularly repeating atomic patterns. To obtain a diffraction pattern, a fibre of DNA is drawn out so that all its molecules are oriented the same way, and rotated in a narrow beam of monochromatic X-rays. The pattern of the reflected rays is recorded on a photographic plate (Plate 9); alternatively, the angle of incidence of the beam may be altered rhythmically, the varying intensity of the reflected rays being converted, by a counting device, into electric impulses which are then analysed and recorded on tape by an electronic computer. This latter method is especially employed in the diffraction analysis of proteins where an immense amount of information must be obtained and collated (see p. 261). For further information about the methods and techniques of X-ray diffraction analysis, see Crick and Kendrew (1958), Rich and Green (1961), Perutz (1962), Haggis *et al.* (1964).

All the fundamental data which X-ray diffraction analysis could provide about DNA were acquired, and elegantly interpreted, by M. F. H. Wilkins and his colleagues. Not only did the diffraction patterns given by DNA from different sources prove to be similar, but the same patterns were obtained irrespective of whether isolated DNA fibres or intact material, such as sperm heads or bacteriophage particles, were employed (Wilkins and Randall, 1953). The essential structural features revealed by these patterns were as follows (Wilkins, Stokes and Wilson, 1953; Franklin and Gosling, 1953).

1. The purine and pyrimidine bases, which are flat structures, are arranged at right angles to the long axis of the polynucleotide chain, and are stacked one above the other like 'a pile of pennies', the central planes of neigbouring bases being 3·4Å.

2. The chain is not straight but is wound helically around a central axis, one full turn of the helix extending 34 Å.

3. A comparison of the measured density of DNA with that calculated on the basis of the atomic spacing showed that DNA consists of more than one helically arranged polynucleotide chain.

The final step in this series of diverse investigations was taken by J. D. Watson and F. H. C. Crick (1953a; Crick and Watson, 1953) who, by a brilliant synthesis, fitted the chemical and X-ray diffraction data together into a symmetrical structure which was not only

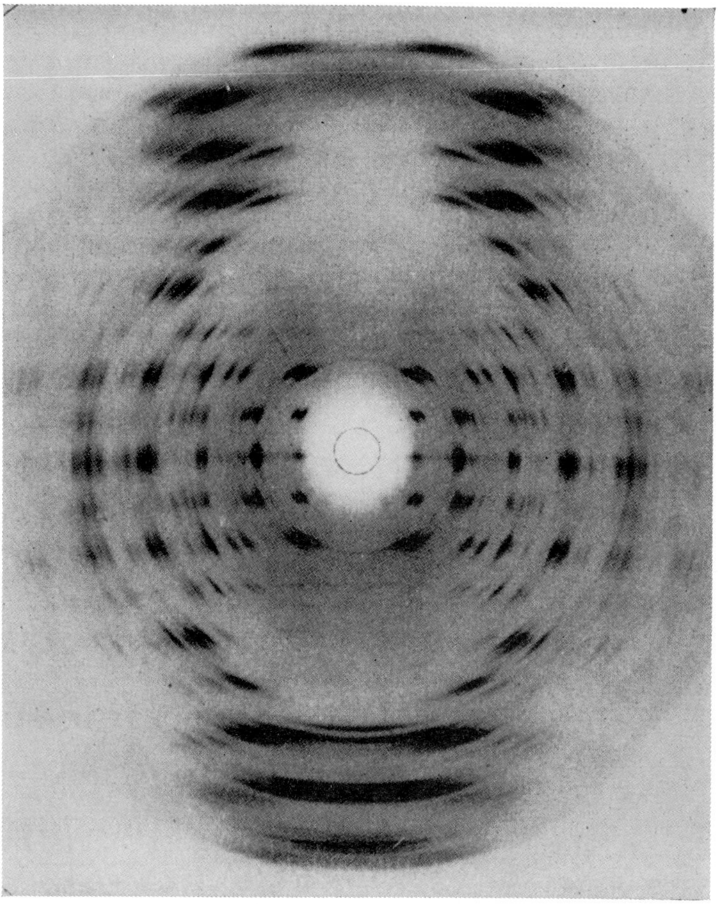

PLATE 9. The X-ray diffraction pattern of DNA. This pattern is produced by the relatively dehydrated, crystalline form of DNA. In natural DNA, which is more highly hydrated, much of the detail of the diffraction pattern is lost.

(Reproduced from Wilkins, 1956, by kind permission of the author, and the Long Island Biological Association.)

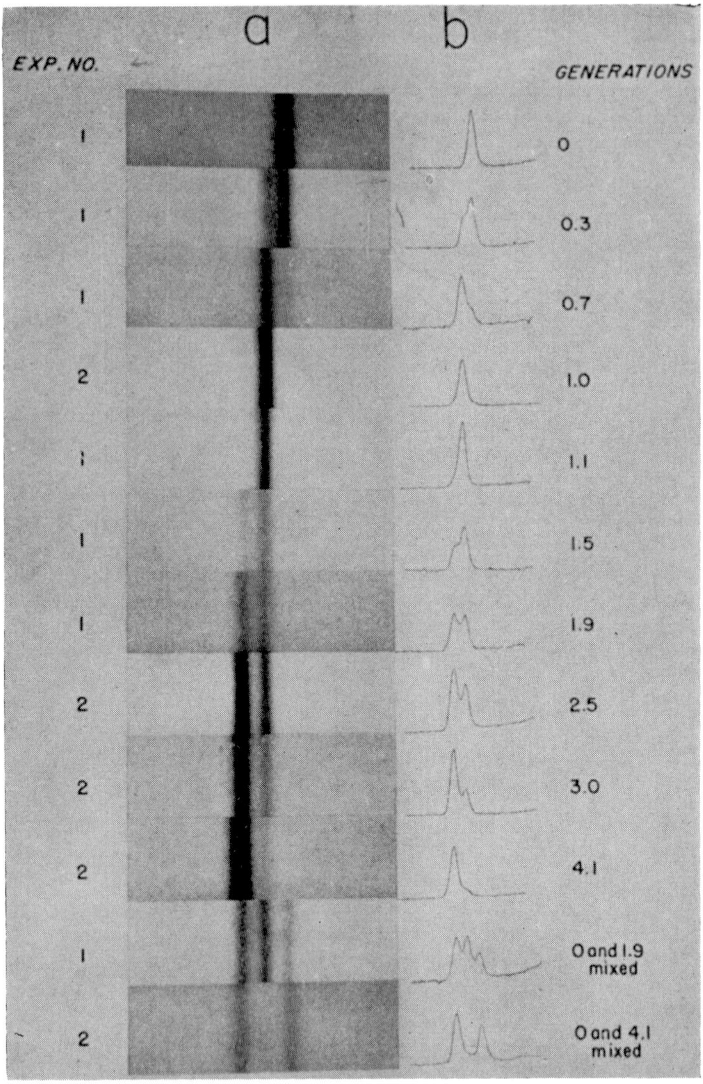

PLATE 10. Results of the Meselson-Stahl experiment, demonstrating the semi-conservative replication of *E. coli* DNA.

The DNA is extracted from multiplying, ^{15}N-labelled bacteria at intervals (expressed in generation times) following addition of an excess of ^{14}N substrates, and then spun for 20 hours, in a caesium chloride density gradient, in an analytical ultra-centrifuge. (*continued on foot of facing page*)

compatible with all the facts but possessed the inherent properties one would expect of genetic material. The Watson-Crick model of DNA structure comprises two polynucleotide chains which are interwoven

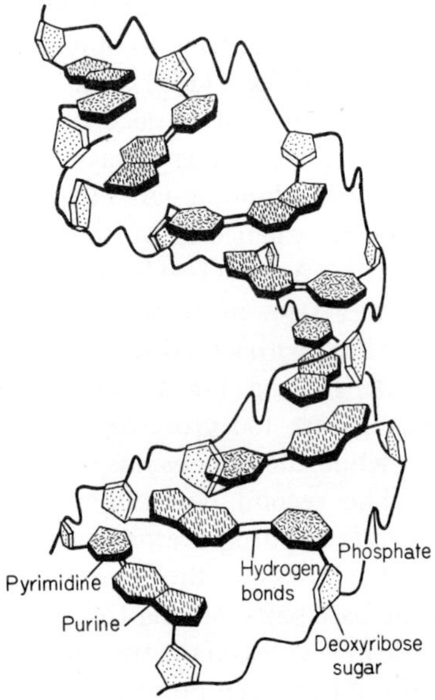

Fig. 56. Schematic drawing of part of a molecule of deoxyribonucleic acid, showing the general shape and arrangement of its components according to the Watson–Crick model.

The white, dotted pentagons, connected by the wavy lines, represent the molecules of deoxyribose sugar; the kinks in the wavy lines are the phosphate molecules joining the sugars.

The bases are stippled and outlined by black bands; the pyrimidines are simple hexagons, and the purines the larger compounds of hexagon and pentagon.

The short, straight lines connecting the opposing bases of the two strands indicate hydrogen bonds. (After Crick, 1954; redrawn from a photograph by kind permission of *Scientific American*.)

---

a. Ultraviolet absorption photographs showing the positions of the DNA bands in the gradient and, therefore, the density of the DNA at the time of sampling. The density increases to the right.

b. Micro-densitometer tracings of the DNA bands shown in the photographs. The height of each curve above the base line is directly proportional to the concentration of DNA at the corresponding density. The lowermost photograph and tracing serve as a density reference for pure ^{14}N- and ^{15}N-DNA. The photograph and tracing above it (mixture of DNAs extracted at generations 0 and 1·9) show the relative position of the DNA of intermediate density, whose peak is found to be centered at 50 ± 2 per cent. of the distance between the ^{14}N and ^{15}N peaks, when allowance is made for the relative amounts of DNA in the three peaks.

(Reproduced from Meselson & Stahl, 1958a, by kind permission of the authors, and the National Academy of Sciences, Washington.)

to form a double helix. The two chains are joined together by hydrogen bonding between the bases which face inwards towards one another (Fig. 56). In fact the double helix can be visualised schematically as a ladder, the uprights of which represent the sugar-phosphate 'backbones' of each chain, the rungs being the pairs of hydrogen-bonded bases; the two ends of the ladder are then twisted in opposite directions so that the uprights wind round one another (Fig. 59B).

The most important feature of the model is that it necessitates specific pairing between the bases. Not only must purines always pair with pyrimidines, but adenine must pair with thymine (A—T) and guanine with cytosine (G—C). Thus the equivalence of purine and pyrimidine bases, as well as the unity of the A+T/G+C ratio, are explained. There are two reasons why the bases must pair in this specific way. The most obvious is that purines, with a double ring, are larger structures than pyrimidines, with a single ring (Fig. 55A), so that if two purines are paired their dimensions are too great to fit the constant diameter of the double helix, while the dimensions of two pyrimidines are too small (Fig. 56). The second determinant of specificity is the positions, on the bases, of the hydrogen atoms which can participate in bonding. Fig. 57 (a and b) shows how the orientation of adenine-thymine and guanine-cytosine base pairs, arising from the attachment of the bases to the deoxyribose molecules on the two chains of the double helix, fits the positions of the hydrogen bonds. On the other hand, the relative positions of the hydrogen atoms on adenine and cytosine are normally incompatible with bonding (Fig. 57c) without distorting the symmetry of the double helix. Occasionally, however, hydrogen atoms undergo tautomeric shifts to other positions and, when this occurs, new pairing interactions may be possible. For example, if the hydrogen atom normally present at the 6-amino position in adenine shifts to the N1 position, the adenine will pair with cytosine instead of with thymine (Fig. 57d). For this reason the Watson–Crick structure is based on the most stable tautomeric forms of the bases and does not exclude occasional paradoxical base pairing (see pp. 238, 302).

An important feature of the model is that, although the specificity of base pairing means that every base on one polynucleotide chain determines the equivalent base on the other chain, that is, that the sequences of the bases on the two chains are complementary, *there is no theoretical restriction whatsoever on what this sequence will be*. So far as any one chain is concerned, any sequence of bases is allowed. Another

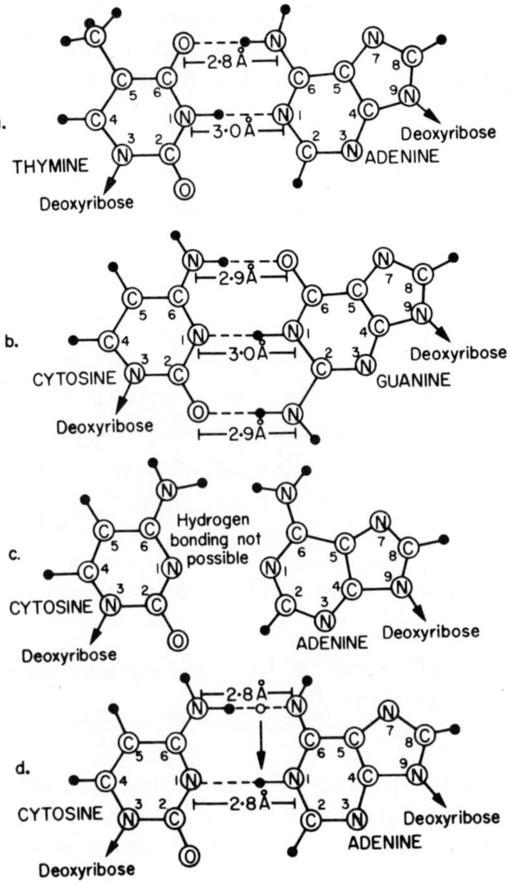

FIG. 57. Demonstrating how the specificity of base pairing in DNA is determined by hydrogen bonding.

The hydrogen atoms are indicated by solid circles, and the hydrogen bonds by interrupted lines.

a. The pairing of thymine and adenine.

b. The pairing of cytosine and guanine.

c. Shows why the pairing of cytosine with an adenine molecule, having the most stable distribution of its hydrogen atoms, cannot lead to hydrogen bonding.

d. Shows how the shift of a hydrogen atom in an adenine molecule, from the 6-amino to the N1 position, permits hydrogen bonding with cytosine. The normal position of the hydrogen atom is indicated by the small, open circle.

The dimensions shown are only approximate.

property of the model is that the two chains run in opposite directions in terms of the 3′–5′ phosphate-deoxyribose linkages. Both these features of the double helix are demonstrated in Fig. 58. A scale

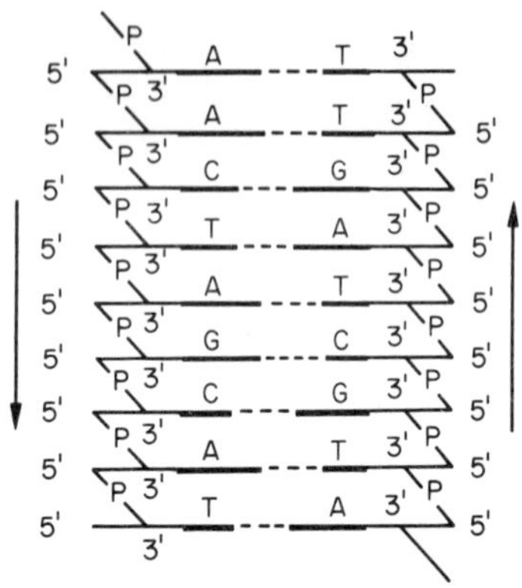

FIG. 58. Diagram of the DNA double helix, showing:
a. The specific pairing of the bases.
b. The unrestricted sequence of bases along any one chain.
c. The reversed direction of the 3′–5′ phospho-diester linkages of the two chains.
The heavy lines indicate the bases, and the interrupted lines the hydrogen bondings between base pairs.
A = adenine; G = guanine; T = thymine; C = cytosine.

model of part of the DNA molecule, comprising 1½ turns of the double helix, is shown in Fig. 59A, in comparison with a schematic representation of certain dimensions of the molecule drawn to the same scale (Fig. 59B).

Since the Watson–Crick structure was first elaborated, its dimensions and general architecture have been checked by much more refined diffraction analysis made possible by use of the lithium salt of DNA, which forms a far more perfect crystalline structure than does the sodium salt (Langridge *et al.*, 1960a, b; Marvin *et al.*, 1961; Wilkins, 1961). In addition, all the essential features of the model, such as the hydrogen bonding of the bases, the specificity of base pairing, the opposite polarity of the two chains and the arbitrary sequence of bases along any one chain, have been confirmed by a wide variety of physical and chemical methods. We shall describe some of these methods in due course.

For the moment, let us glance rather superficially at the model to see how it matches up to the functional requirements of genetic material which, after all, is the main reason for our interest in it. These requirements are three-fold.

1. *The genetic material is the blueprint for the cell and carries the information* necessary to direct all its specific activities, presumably in

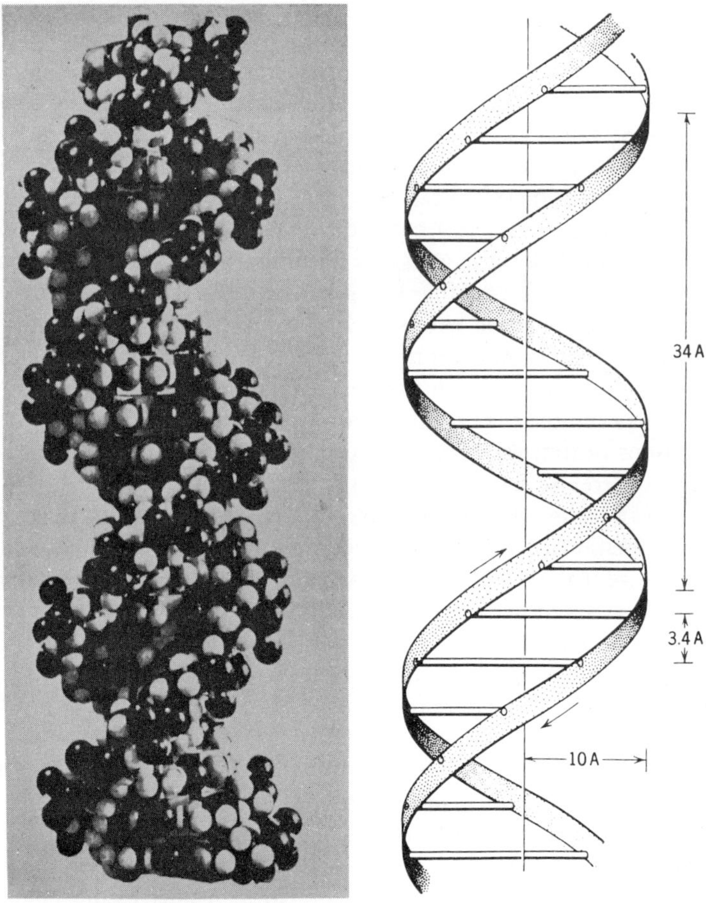

A B

FIG. 59. A molecular model of deoxyribonucleic acid (A), compared with a schematic representation of its structure drawn to the same scale and showing some dimensions of the molecule (B).

(By courtesy of Dr M. F. H. Wilkins)

some kind of chemical code. Can we find, in the Watson–Crick structure, a clue to the nature of this code? The answer is that the only possible irregularity in the structure which could be used as a code is the sequence of the four bases, adenine, guanine, thymine and cytosine, or of pairs of these bases, along the long axis of the duplex. The code would therefore consist of four letters. What kind of information do we expect the code to provide? We can answer this in a simple way. The metabolism of living cells is mediated by a very large number of enzymes which are protein in nature. The most plausible, current view, known as the *sequence hypothesis*, assumes that the specific activity of these proteins is a function of the sequence of twenty amino acids within the polypeptide chains of which they are composed (p. 259). The coding problem therefore resolves itself into how a linear sequence o four bases, or pairs of bases, can code for a linear array of twenty amino acids. Theoretically, this presents no difficulty, as we shall later see, since from an alphabet of four letters can be made as many as sixty-four distinct words of three letters each. However, the code can only be considered fully in relation to the mechanism of protein synthesis as a whole, just as it is difficult to discuss the principles of tape recording if we are permitted to talk only about the tape itself, and must omit the operations of transcription and reproduction of the sounds it carries. We will therefore return again to this problem in Chapter 14, p. 339.

2. *The genetic material replicates*, that is, reproduces itself, so that the information it carries is inherited, in a very precise way, by daughter cells. As we shall show in the next Section, the Watson–Crick structure is ideally adapted to self-reproduction since the complementarity of the bases on the two chains means that, on separation, each can act as a template for the synthesis of the other with the result that two new, identical double helices are formed.

3. *The genetic material can undergo mutation* so that the message is altered in a specific and heritable way. It has been shown, for such diverse material as human haemoglobin (Ingram, 1957) and the protein of tobacco mosaic virus (Tsugita and Fraenkel-Conrat, 1960, 1962), that a single mutation results in the substitution of a single amino acid by another in the protein. If each amino acid is specified by the sequence of a small number of bases in the DNA (or RNA), the substitution of one base by another might clearly alter the sequence in such a way that it now codes for a different amino acid. Watson and Crick (1953a) suggested, for example, that natural mutation could be explained by assuming occasional unstable tautomeric shifts in the positions of

hydrogen atoms on the purine and pyrimidine bases. Thus if, during DNA replication, the hydrogen atom at the 6-amino position of adenine should shift to the N1 position (Fig. 57, c and d), then the adenine will pair with cytosine instead of with thymine, so that one of the chains will now have cytosine instead of thymine in its base sequence at this point. At the next replication this cytosine will pair normally with guanine, so that the original T—A base pair is replaced by a C—G pair and the code will be permanently altered. We will discuss the molecular basis of mutation in detail in Chapter 13, p. 302.

THE REPLICATION OF DNA

Earlier hypotheses of replication assumed the formation of a template which then turned out copies of the genetic material. The beauty of the Watson–Crick structure is that it embodies a built-in template system for self-replication. Due to the specificity of base-pairing, whatever may be the sequence of bases along one chain, the sequence along the other is automatically determined. Thus each chain of the double helix can serve as a template for the synthesis of the other. This suggests the scheme outlined in Fig. 60. All that is required is that the hydrogen

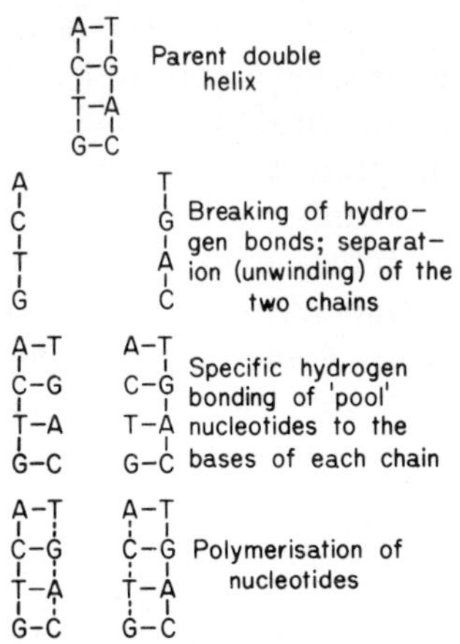

FIG. 60. Schematic representation of the replication of DNA.

A, G, T and C are adenine, guanine, thymine and cytosine nucleotides, respectively.

The horizontal lines indicate hydrogen bonds between the bases; the vertical lines indicate the phospho-diester linkages between the nucleotides of each chain.

bonds joining the bases should break and that the two chains should separate. If this occurs in a 'pool' of the four nucleotides, which have previously been synthesised by the cell, one can imagine the bases of these nucleotides jostling to form hydrogen bonds with the bases of the two chains. When adjacent nucleotides pair correctly with their partners on the chains, they find themselves suitably oriented for their sugar-phosphate molecules to knit together, so that new chains begin to form along the old ones. Nucleotides which pair incorrectly will be improperly oriented for incorporation into these new chains and so will be supplanted by others until the right partner is found (but see p. 253). In this way two new polynucleotide chains are assembled, the bases of which are exactly complementary to the bases of the old chains, so that two new double helices of DNA, identical with one another and with the original double helix, emerge.

The main problem raised by this simple and elegant mode of replication, which was originally proposed by Watson and Crick (1953a, b), is that it requires unwinding of the two chains. The uncoupling of the hydrogen bonds involves no difficulty since these are very weak compared with the other, covalent links holding the molecule together. The separation of the polynucleotide chains is a different matter. These are interwoven like a braid (plectonemic coiling) and can only be separated either by repeated breakage of the chains, or by untwisting them in much the same way that a length of double-stranded electrical 'flex' may be unravelled by holding one end and rotating the other. This untwisting presents a formidable theoretical problem when one considers the number of turns involved. For example, the average amount of DNA in the chromosome of a bacteriophage, which is only about 1/100 the amount contained in a bacterial chromosome, represents a double helix of the order 34μ long comprising some 100,000 base pairs and 10,000 turns; nevertheless this structure duplicates itself about every two minutes in an infected bacterium. Although one may boggle at the mechanics of the problem, it has been calculated that the chemical energy required for unwinding is only a fraction of that needed for polymerisation of the 'backbones' of the new chains (Discussion, Delbrück and Stent, 1957).

To mitigate the unwinding difficulty, Watson and Crick proposed that the two chains may not be unwound completely before synthesis of the new chains begins, but that the process of unwinding and synthesis may proceed hand in hand from one end of the molecule, as shown in Fig. 61. Recently, the unwinding problem *per se* has become submerged in a flood of evidence supporting the basic Watson–Crick replication mechanism, some of which we will now review.

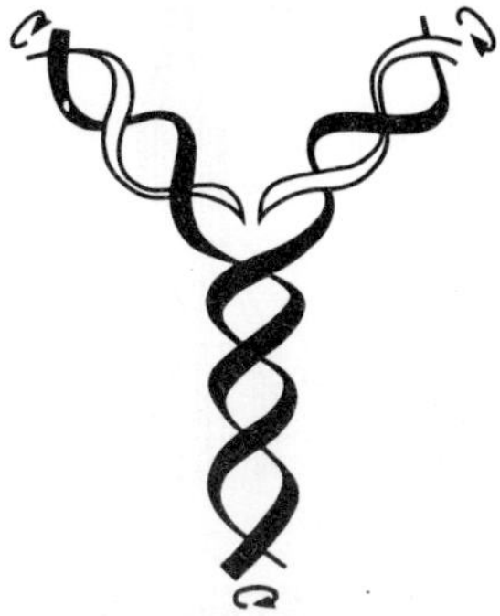

FIG. 61. Drawing representing how unwinding and replication of the DNA double helix may proceed together.

The directions of rotation of the old and new molecules are indicated by the arrows.

The old chains are shown black, and the new chains white. (Modified from Delbrück and Stent, 1957.)

THE MESELSON–STAHL EXPERIMENT

An obvious experimental approach to the problem of DNA replication is to introduce isotopically labelled atoms into the DNA of living bacteria, or bacteriophage, and then to observe how the labelled atoms are distributed in the DNA of the progeny, when the organisms divide in an unlabelled medium. There are three possible patterns of distribution which depend on the mechanism of replication (see Delbrück and Stent, 1957).

1. The parental DNA duplex does not unwind but serves as an intact template for the synthesis of an entirely new daughter duplex. Thus the DNA of the first generation progeny consists of duplexes which are either fully labelled or not labelled at all, the labelled duplexes being transmitted intact to the next generation (Fig. 62, scheme 1). This is called *conservative replication* since the parental duplex is conserved as a whole.

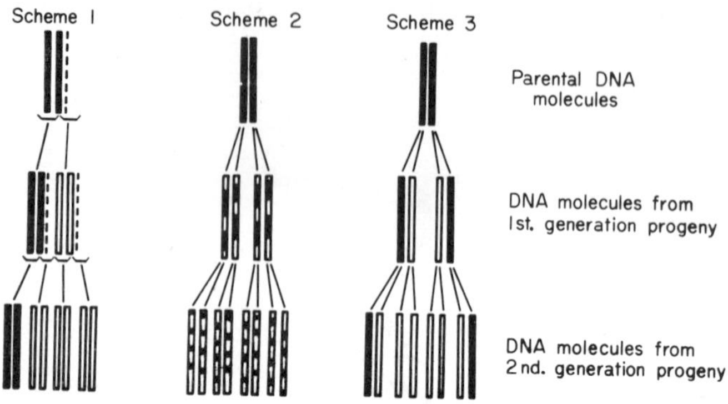

FIG. 62. Demonstrating the distribution of labelled atoms, in a parental DNA duplex, among daughter and grand-daughter duplexes, according to three possible schemes of DNA replication. The isotopic label is shown black (see text).

2. The parental duplex unravels by fragmentation of the chains at every half turn. New synthesis occurs along the fragments, which then reunite crosswise to form two duplexes. The outcome of this scheme is that both chains of each daughter duplex are composed of alternating fragments of old and new chains. Similarly, at the next replication, the labelled atoms will be shared out again among both chains of both grand-daughter duplexes; no unlabelled duplexes appear (Fig. 62, scheme 2). Since the atoms of the parental duplex are dispersed uniformly among the daughter and grand-daughter duplexes, this is termed *dispersive replication*.

3. The parental duplex unwinds, each chain serving as a template for the synthesis of a new chain as proposed by Watson and Crick. Here, each daughter duplex consists of one fully labelled chain and one unlabelled new one; that is, each daughter duplex is 'hybrid' with respect to the labelling of its chains. At the next generation each labelled chain will generate a new 'hybrid' duplex, and each unlabelled chain a completely unlabelled duplex (Fig. 62, scheme 3). This is called *semi-conservative replication* because, although the parental duplex disappears, each of its two chains is conserved among the descendant duplexes.

The way in which the atoms of the parental duplex are distributed among the polynucleotide chains of the daughter and grand-daughter duplexes, in each of these three types of replication mechanism, is shown in Fig. 62, in which the parental, labelled chains are black, and newly synthesised, unlabelled chains are white. Observe that the three possible systems of replication, based on the Watson–Crick structure, give quite different results. In particular, semi-conservative differs from conservative replication in that the fully labelled parental duplex disappears after the first generation, being replaced by two 'hybrid' duplexes; it also differs from dispersive replication in the emergence of completely unlabelled duplexes during the second replication.

The first experiment of this kind to give definitive results was done by Meselson and Stahl (1958a, b) using an ingenious analytical method called *density gradient centrifugation*, devised by Vinograd and themselves. If a solution of caesium chloride, at about 6M concentration, is centrifuged for many hours at a speed sufficient to exert a centrifugal force of more than 100,000***g*** on the solution, an equilibrium is reached in which the centrifugal force tending to deposit the caesium ions is counteracted by diffusion which tends to disperse them. The situation is analogous to the action of gravity on the atmosphere and the result is the same; a density gradient is set up, the greatest density being at the bottom of the centrifuge tube. It so happens that the density of DNA lies within this gradient, so that if DNA is added to the caesium chloride solution and the mixture is centrifuged, the DNA will float at a level in the gradient determined by its own buoyant density, like a balloon in air, and will thus be concentrated into a discrete band. The method is so sensitive that if normal DNA is mixed with DNA in which the nitrogen (^{14}N) of the bases has been replaced by the heavy isotope of nitrogen (^{15}N), the two types of DNA molecule will separate in the gradient into two well-defined bands, even though the density difference is only 0·8 per cent, just as balloons filled with either hydrogen or helium will rise to different levels in the atmosphere. When an analytical ultracentrifuge is used the positions of the bands may be recorded photographically by UV light as in the experiment to be recounted. However, a simple alternative method is available for use with a preparative type of centrifuge (p. 394).

The experiment was as follows. *Escherichia coli* was grown for many generations in a medium containing ^{15}N as the only source of nitrogen so that all the ^{14}N of the cells, including that in the DNA bases, was replaced by the heavy ^{15}N. A great excess of ^{14}N was then added so that, during all subsequent growth, virtually only ^{14}N atoms would be incorporated. Just before the addition of the ^{14}N, and at intervals thereafter, samples of the culture were removed, and the DNA extracted from the bacteria and analysed in a caesium chloride density gradient. The results obtained are reproduced in Plate 10 (opposite p. 233). The sample removed just prior to adding the ^{14}N shows a single band of heavy DNA. One generation later this heavy DNA band has disappeared and been replaced by a single band of new DNA having a density half way between that of heavy and light DNA, indicating that the DNA is composed of equal numbers of heavy and light molecules. At the end of the 2nd generation this intermediate band is conserved, but a new band of wholly light DNA has appeared which thereafter increases in amount. Reference to Fig. 62 will show that these results are entirely compatible with the semi-conservative mechanism of Watson and Crick, and incompatible with either of the other schemes, *provided* it can be shown that the DNA molecules examined are really in the form of double helices and are not, for example, aggregates of duplexes.

One of the properties of DNA is that, on heating to 80–100°C. in aqueous solution, the hydrogen bonds break and the two chains of the duplex come apart, and do not reform again into duplexes unless the solution is cooled extremely slowly (see Marmur, Rownd and Schildkraut, 1963). Meselson and Stahl (1958b) therefore isolated the intermediate density DNA from the gradient and heated it to about 100°C.; the heated DNA was then re-centrifuged in a caesium chloride gradient. Instead of the single intermediate band two bands now appeared, one heavy and one light. Thus the DNA of intermediate density must have been truly 'hybrid', consisting of molecules with one heavy and one light polynucleotide chain.

This method has since revealed the appearance of 'hybrid' DNA molecules during the replication of other organisms, including *Chlamydamonas* at one extreme and several types of bacteriophage at the other (p. 545), so that the evidence for semi-conservative DNA replication is very convincing. An interesting sidelight on the semi-conservative replication of DNA is provided by the case of phage ϕX174, whose genetic material consists of only a single DNA strand. As soon as this DNA infects a host cell a complementary polynucleotide chain is formed along it, and it is by means of this double-stranded structure that the phage DNA is replicated (see p. 437). In fact, the single-stranded RNA of RNA viruses is also replicated by means of a double-stranded form (p. 443)

Experiments of this sort only tell us how DNA *molecules* divide and do not necessarily provide any information about the division of *chromosomes* unless, of course, the chromosome and the DNA double helix are synonymous (see pp. 555, 564). This, in fact, seems certain in the case of bacteriophage (p. 535) and of bacteria (p. 550), but we should be cautious about extrapolation to organisms whose chromosomes are clearly more highly organised from the point of view of morphology and behaviour. Attempts have been made to follow the distribution of atoms of labelled *chromosomes* among daughter and grand-daughter chromosomes *during mitosis*, using root cells of the broad bean, *Vicia faba* (Taylor *et al.*, 1957). Seedlings were grown in a medium containing the thymine deoxynucleoside, thymidine, whose hydrogen atoms had been replaced by the radioactive isotope tritium (^{3}H). Thymidine is incorporated exclusively into DNA. The roots were then transferred to medium containing non-radioactive thymidine, together with colchicine which prevents the segregation of chromosomes to daughter cells without affecting their duplication, so that the number of chromosome divisions can be inferred by counting the number of chromosomes in the cells. At intervals after transfer autoradiographs of the material were made so that the positions of the tritium label could be correlated with those of the dividing chromosomes. The result was that, after one division, only one of each pair of sister chromatids was labelled; after two divisions about half the chromosomes contained one labelled and one unlabelled chromatid, while both chromatids of the remainder were completely unlabelled. This is a precise reflection, at the chromosome level, of the results of the Meselson–Stahl experiment. However an independent repetition of Taylor's experiment has cast doubt on the generality of this result. It was found that if colchicine was omitted from the medium, the tritium label was distributed between the two sister chromatids (LaCour and Pelc, 1958, 1959). Attempts to elaborate models of chromosome replication in molecular terms does not therefore appear to be a profitable venture at this stage. The replication of the bacterial chromosome is discussed in Chapter 19 (p. 553).

THE *IN VITRO* SYNTHESIS OF DNA

We have so far looked at DNA synthesis in a mechanical sort of way, from the viewpoint of structural models, and this has led us to talk rather vaguely of nucleotides 'knitting together', once they have been properly oriented by bonding to their complementary bases on the 'template' polynucleotide chain. By what sort of biochemical processes is this accomplished in the living cell? Most of what we know of these processes derives from the work of A. Kornberg and his colleagues which has culminated in the *in vitro* synthesis of specific DNA under quite precisely defined conditions (reviews; Hayes, D., 1967; Kornberg, 1957, 1960, 1961; Smellie, 1965; Davidson and Cohn, 1963a, b, 1964, 1965, 1966; also Cold Spring Harbour Symposium on Quantitative Biology, 'Synthesis and structure of macromolecules', Vol. 28, 1963).

When DNA is enzymatically degraded by certain types of deoxyribonuclease (DNase), the primary breakdown products are nucleotides in which a single phosphate molecule is attached to the 3′ position of the deoxyribose (Fig. 55, p. 230; Fig. 63). Such nucleotides may be descriptively called deoxyribonucleoside 3′-phosphates. The pathways of purine and pyrimidine synthesis, on the other hand, lead to the formation of deoxyribonucleoside 5′-phosphates, that is, of nucleotides in which a phosphate molecule is attached to the 5′ instead of to the 3′ position of the sugar, as shown in Fig. 63A. However, the basic building blocks for the synthesis of DNA are not nucleoside *mono*phosphates, but nucleoside *tri*phosphates. These triphosphates are formed by the interaction of the monophosphates with adenosine triphosphate (ATP), by means of various kinases. The deoxyribonucleoside 5′-triphosphates then interact with the 3′-hydroxyl group of the deoxyribose at the growing end of a polynucleotide chain, to which they become attached with liberation of inorganic pyrophosphates according to the scheme.

$$\left.\begin{array}{l}\text{A-p-p-p}\\[1ex]\text{G-p-p-p}\\[1ex]\text{T-p-p-p}\\[1ex]\text{C-p-p-p}\end{array}\right\}\rightarrow\;\downarrow\;\begin{array}{l}\text{A-p}\\|\\\text{G-p}\\|\\\text{T-p}\\|\\\text{C-p}\end{array}\;+4\text{ p-p (inorganic pyrophosphate) (see Fig. 63).}$$

where A, G, T and C represent adenosine, guanosine, thymine and cytosine respectively, and p represents a single phosphate group. Thus the polynucleotide chain grows from one end by the step-wise addition of nucleotides, the direction of the synthesis being determined by the fact that the nucleoside 5′-phosphates must bond to the 3′-hydroxyl group of the preceding deoxyribose.

The evidence for this comes from the requirements for *in vitro* synthesis of DNA. These are:

1. A mixture of the nucleoside 5′-triphosphates of all four bases.

2. A purified enzyme, DNA polymerase, obtained from cell-free extracts of *E. coli*.

3. High molecular weight DNA, which acts either as a 'template' or as a 'primer' for the synthesis in a way we shall shortly describe.

When all these are added together in a test-tube in the presence of Mg^{++} ions, a net increase in the amount of DNA occurs which may be more than twenty times the amount added as 'primer'. The DNA accumulates until one of the nucleoside triphosphates becomes exhausted, so that at least 95 per cent of it is constructed from the added nucleosides, while equimolar quantities of inorganic pyrophosphate are released in the process.

The reaction does not take place with nucleoside 5′-*di*phosphates, nor if one of the four nucleoside 5′-triphosphates is omitted from the mixture. The key to the synthesis is, of course, the DNA polymerase and, in the first instance, the achievement of a significant reaction depended on its partial purification. The reason for this is that crude cell extracts contain a considerable number of nucleases and diesterases which either attack the subtrates or break down the product as soon as it is formed. The polymerase is assumed to promote synthesis by effecting the 3′–5′ phospho-diester linkages.

With regard to the requirement for pre-existing DNA to 'prime' the reaction, it turns out that DNA from any source is effective and that the introduction of a limited number of single-strand breaks by controlled DNase activity or by sonication does not reduce this effectiveness. On the other hand, it appears that polymerases from different sources may differ markedly in the efficiency with which they can employ native DNA as primer. The evidence suggests that while bacterial polymerases, isolated from *E. coli* and *B. subtilis*, may operate equally well on either native (double-stranded) or heat-denatured (single-stranded) DNA (Richardson *et al.*, 1964; Okazaki and Kornberg, 1964), polymerases extracted from various animal cells (Keir, Binnie and Smellie, 1962; Bollum, 1963) and from *E. coli* following infection with phage T2 (Aphosian and Kornberg, 1962) are efficient only in the presence of denatured DNA or, particularly, of the naturally occurring, single-stranded DNA of phage ϕX174 (Keir *et al.*, 1962).

Moreover in the case of mammalian polymerases the net synthesis of DNA does not normally exceed the amount initially present as primer. In the present state of our knowledge it is impossible to decide to what extent these differences in polymerase activity are more apparent than real because, as we will discuss in the last part of this

Chapter, not only do cells probably possess more than one type of polymerase to cope with the different demands of chromosome replication and the repair of DNA damage, but DNA synthesis is itself a complex operation which may need the cooperation of other enzymes as initiators, to open up the double-stranded molecule, for example, so that the base sequence on the two strands can be copied.

What is the role of the DNA 'primer'? Does it simply serve as a growing point for the terminal, random addition of nucleotides, (as the word 'primer' implies) or does it serve as a template for the building up of molecules identical with itself? The requirement for all four bases, together with the suggestion that the true primer is really a single polynucleotide chain, support the template hypothesis. But there is much more compelling evidence than this. First of all, the structure of natural DNA is not rigidly circumscribed by the four bases, adenine, guanine, thymine and cytosine. For example, DNA from grasses contains large amounts of 5-methylcytosine which is an analogue of cytosine, while in DNA from *E. coli* bacteriophages T2, T4 and T6 cytosine is entirely replaced by its analogue 5-hydroxymethylcytosine (HMC) (see p. 426). Similarly, DNA phages of *B. subtilis* are known in which thymine is totally replaced by uracil or by 5-hydroxymethyl uracil (Kallen, Simon and Marmur, 1962). Moreover, synthetic purine and pyrimidine analogues can be incorporated into the DNA of living bacteria and bacteriophages although, as we shall see, they are usually mutagenic (Chapter 13). Indeed, Nature is very tolerant of variations in the structure of DNA bases, *provided* that these conform to one rule—they must permit proper pairing and hydrogen bonding with one of the natural bases. According to this rule, for instance, 5-bromouracil is an analogue of thymine and pairs only with adenine; 5-bromocytosine is an analogue of cytosine and pairs only with guanine; hypoxanthine is an analogue of guanine and pairs only with cytosine.

When deoxyribonucleoside 5′-triphosphates of various of these unnatural bases are substituted for each of the usual bases in turn, in the *in vitro* synthesis of DNA, no new DNA is formed unless the normal base is replaced by its analogue. Thus there is a normal yield when 5-bromouracil is substituted for thymine, but none when it is added instead of adenine, guanine or cytosine. From this it follows that specific hydrogen bonding must be involved in the synthesis.

We have already referred to the fact that DNAS from different sources may vary widely in their base ratios; that is, in the ratio, adenine+thymine/guanine+cytosine. Various natural DNAs, having diverse base ratios, may be used as 'primers' in the *in vitro* system; the base ratio of the newly synthesised DNA can then be determined and compared with that of the 'primer'. It turns out that the two are the same, as Table 12 shows. It will be seen, in addition, that the ratio of total purines to pyrimidines is unity, as it is in natural DNA. An interesting example of the specificity of the newly synthesised DNA is revealed by the bottom lines of the two parts of Table 12. If *E. coli* DNA

TABLE 12

A comparison of base ratios of various DNA 'primers' with those of the derivative DNA synthesised *in vitro*. (Data from Kornberg, 1960.)

Ratio	Primer DNA	Derivative DNA	Source of primer DNA
$\frac{A+T}{G+C}$	0·49	0·48	*Mycobacterium phlei*
	0·97	1·02	*Escherichia coli*
	1·25	1·29	Calf thymus
	1·92	1·90	Bacteriophage T2
	...	>40	A–T co-polymer
$\frac{A+G}{T+C}$	1·01	0·99	*Mycobacterium phlei*
	0·98	1·01	*Escherichia coli*
	1·05	1·02	Calf thymus
	0·98	1·02	Bacteriophage T2
	1·00	1·03	A–T co-polymer

A, G, T and C represent adenine, guanine, thymine and cytosine respectively.
A–T co-polymer = a synthetic DNA containing adenine and thymine deoxyribonucleotides exclusively, in alternating sequence in each strand.

polymerase is incubated with the deoxyribonucleoside 5′-triphosphates of adenine (dATP) and thymine (dTTP) only, and *priming DNA is omitted from the mixture*, high molecular weight DNA begins to be formed after a long lag. This DNA is found to consist of a double helix which dissociates into two strands on heating, each strand having alternating adenine and thymine bases in strict sequence (Kornberg, 1961). When this synthetic DNA is used as primer in a normal *in vitro* synthesis experiment, the derivative DNA contains only adenine and thymine; neither guanine nor cytosine are incorporated although the nucleoside triphosphates of all four bases are present.

Natural DNAs from diverse sources may be expected to vary not only in their base ratios, which may reflect fundamental evolutionary differences in protein synthesis or in the variant of the genetic code which determines it, but also in the details of the information carried by the code. This should be reflected by differences in the sequence of bases in the long axis of the polynucleotide chain. Analysis of base sequence over any appreciable length of DNA is not yet technically practicable. However, the *in vitro* system of DNA synthesis has provided a beautifully ingenious method, called *nearest-neighbour sequence analysis*, for estimating the relative frequencies with which pairs of each of the four bases lie next to one another (Josse *et al.*, 1961; review, Kornberg, 1960).

We have seen that DNA is synthesised by the bonding of the phosphate molecules attached to the 5′ position of a deoxyribonucleoside triphosphate to the 3′ position of the nucleotide immediately preceding it in the polynucleotide chain, the particular nucleoside triphosphate selected for coupling being determined by its ability to hydrogen bond to its partner on the 'primer' (template) strand (Fig. 63). On the other hand, when the DNA so formed is degraded by DNase, the phospho-diester bond is cut at its attachment to the 5′ position of the sugar below, so that nucleoside 3′-phosphates are liberated (Fig. 63B). This means that the phosphorus atoms which were attached to the 5′ positions of the precursor nucleotides are transferred, at degradation, to the adjoining nucleotides which precede them in the sequence. DNA synthesis is carried out with the four nucleoside 5′-triphosphates, *the phosphorous atoms of one of which* (adenine in Fig. 63) *are labelled with the radioactive isotope*, ^{32}P. The newly synthesised DNA is then hydrolysed by DNase; each of the liberated nucleotides is isolated by paper electrophoresis, and its content of ^{32}P estimated. In the theoretical example given in Fig. 63, the nucleotides of guanine and cytosine will be equally labelled, showing that these two nucleotides are equally frequent neighbours of adenine; if the thymine nucleotides turned out to have ten per cent. of the ^{32}P content of the guanine and cytosine nucleotides, then it would follow that thymine precedes adenine in the chain with only ten per cent. their frequency. By carrying out this procedure four times, each of the four nucleoside 5′-triphosphates being labelled in turn, it is

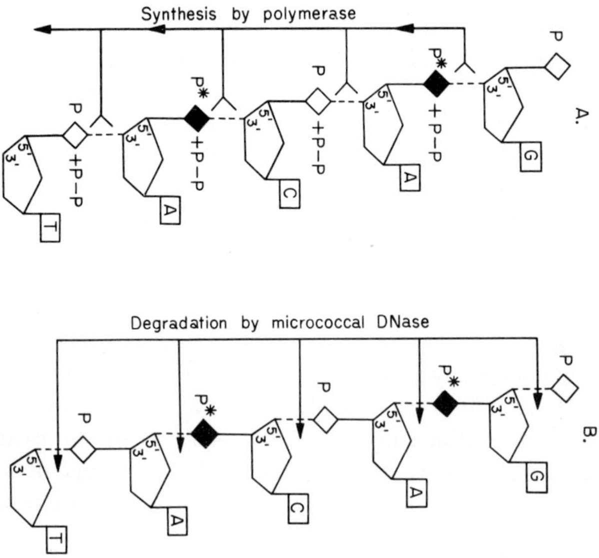

FIG. 63. To demonstrate the operation of nearest-neighbour sequence analysis.

A shows how deoxyribonucleoside 5′-triphosphates are enzymatically coupled by polymerase to the 3′ position of the preceding nucleotide. The arrows show the direction of synthesis.

B shows how the phosphate molecule carried by each precursor nucleotide is transferred to the neighbour nucleotide preceding it in the sequence, as a result of DNase degradation.

Flags marked A, G, T and C indicate the bases adenine, guanine, thymine and cytosine respectively. The diamonds marked 'P' indicate phosphate molecules; black diamonds marked 'P*' are labelled with ^{32}P atoms.

possible to estimate the relative frequencies with which each of the sixteen possible dinucleotide combinations arises.

The results of this type of analysis reveal that DNAs from different sources have specific, reproducible patterns of nearest neighbour sequences which are not predictable from the base ratios (Table 12), but in all the patterns every one of the sixteen possible dinucleotide pairs occurs. Most important from the point of view of DNA structure, the analyses make it quite clear not only that enzymatic synthesis involves specific base pairing, but that the two strands of the product of the synthesis, like the primer, have opposite polarities; for when the frequencies of the sixteen pairs of neighbouring nucleotides are compared it is found, for example, that AG and CT, and GT and AC, have equivalent frequencies. Since A always pairs with T, and G with C, in the double-stranded DNA, the interpretation is that copying by the polymerase of the pair sequence A-G on one strand is correlated with copying the paired bases T and C, respectively, on the other strand in the opposite direction, thus:

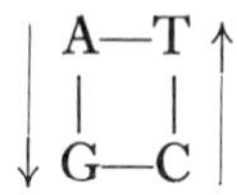

Again, sequence analysis shows that the dinucleotide frequencies of AA and TT, as well as of GG andCC, are the same, implying that the synthesis involves the copying of both strands of the primer (see Bessman, 1963; review, Smellie, 1965). An independent chemical method of studying the various sequences of pyrimidine nucleotides which occur between two purine nucleotides in DNA has been evolved by Burton and Peterson (1960). The two methods agree in their assessment of the frequencies of adjacent purines, which is the extent to which the information given by them is comparable.

We may conclude, therefore, that the physico-chemical structure of DNA proposed by Watson and Crick may be regarded as proven, while the mechanisms they postulated for its replication and the transport of genetic information now have a substantial experimental basis. The implications of this structure for genetics have been so revolutionary that, nowadays, it is almost impossible to discuss basic genetic phenomena in a meaningful way without thinking in terms of it.

THE SIGNIFICANCE OF *IN VITRO* DNA SYNTHESIS

The achievement of DNA synthesis in a test tube at once prompts the question, "Is this the mechanism whereby the chromosome is replicated in the living cell?' We have already reviewed several apparent differences in the *in vitro* behaviour of bacterial and mammalian DNA polymerases, and especially the low net yield of DNA by the latter and their requirement for single-stranded DNA as primer. Although, perhaps, we might expect there to be differences in the specific activity of enzymes from such diverse types of cell, it turns out that the differences may not be so significant as they seem. For example, although polymerase from ascites tumour cells is seven times more active on denatured than on native DNA as primer, and shows a further marked enhancement on single-stranded DNA from phage ϕX174, it is as active on native as on denatured phage T2 DNA which is double-stranded (Keir *et al.*, 1962). Again, the activity of the most highly purified preparations of *E. coli* DNA polymerase is feeble compared with that of crude preparations, almost certainly due to the removal of nucleases which activate the priming capacity of double-stranded DNA. In fact it has so far proved impossible to dissociate one nuclease (exonuclease II) from the polymerase (Richardson *et al.*, 1964).

When we come to match the product of *in vitro* DNA synthesis against the native DNA primer we have already seen some of the striking evidence, especially from a comparison of base ratios and neighbouring nucleotide frequencies, that product and primer are the same. Nevertheless the *in vitro* system falls short of the requirements of chromosome replication in several important ways:

1. The rate of synthesis is far too low to permit replication of the whole chromosome within the normal generation time;

2. Electron microscopy of the synthesised product shows that, unlike the native DNA used as primer, it consists of a highly branched filament (Richardson, Inman and Kornberg, 1964).

3. If the product is heated to denaturation temperature and rapidly cooled, it quickly re-establishes the native, double-stranded form; this implies that the two strands remain held together in such a way that, on cooling, each can readily find and pair again with its complementary strand (Schildkraut, Richardson and Kornberg, 1964).

4. The chain has so far been shown to grow from the 3′-hydroxyl end only. Because of the opposite polarity of the two DNA strands, this means that a polymerase of this type can replicate both strands only by starting along each from opposite ends of the molecule.

This polarity of the polymerase is the basis of a plausible model to explain both the branching and the resistance to denaturation of synthetic DNA. This is shown in Fig. 64 where the black lines represent the strands of the priming DNA and the green lines the newly synthesised, growing strand. Because the polymerase is polarised, the sister of the priming strand cannot be copied from the same end and so is loose (Fig. 64A); after a time, however, this loose strand may attract the

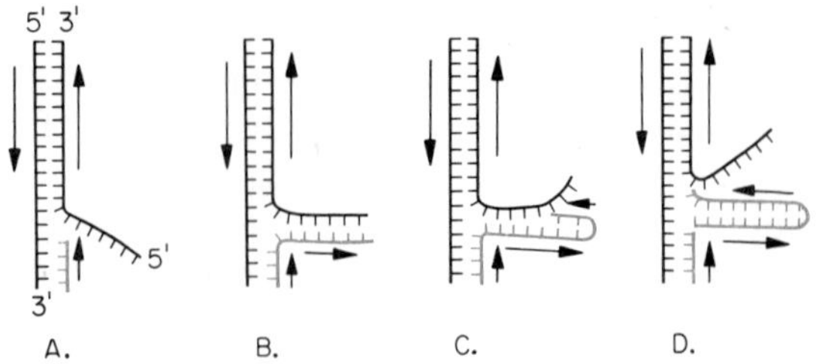

FIG. 64. A hypothetical model to explain the branching and resistance to denaturation of DNA synthesised *in vitro* on a double-stranded primer.

The two strands of the priming DNA are black, while the newly synthesised, growing strand is green. The bases of each strand are represented by the short lines projecting at right angles from the strands. The arrows indicate the polarity of synthesis of each strand.

A. Copying of the template strand has begun at its 3′-hydroxyl end. The other strand of the primer DNA cannot be copied from this end and remains loose.

B. The loose strand attracts the polymerase which switches to copy it with the same polarity as before, i.e. from the 3′ towards the 5′ end.

C. When the extremity is reached, synthesis doubles back to copy the newly synthesised strand, again displacing the original 5′ strand.

D. The branch is complete and synthesis is resumed along the original 3′ strand. The position is now similar to that shown at A., so that another branch can be formed.

After Schildkraut *et al.*, 1964.

polymerase which proceeds to run back along it, forming a complementary strand (B). When the extremity is reached, the newly synthesised strand may then double back to copy itself, again displacing the original 5′ strand (C, D). Thus repeated branches may be formed, the two strands of which are continuous at their ends and so cannot be irreversably separated by heating.

When we come to consider the nature of the bacterial chromosome in more detail (pp. 546–568) we will find indisputable evidence from autoradiographic studies not only that the chromosome of *E. coli* consists of a single, continuous, double-stranded molecule of DNA about 1,100μ long, but that it is replicated from a single starting point and in the same direction along both polynucleotide strands (Cairns, 1963a,b). This suggests either that chromosome replication *in vivo* is mediated by two polymerases of opposite polarity, only one of which has so far been isolated and demonstrated in the test tube, or else that the Kornberg DNA polymerase is not that responsible for normal chromosome replication. We have already noted some aberrant features of the Kornberg system which suggest that this latter alternative may be the correct one, although we must remember that they could also result from failure to provide *in vitro* some essential feature of the complex intracellular environment. It must be pointed out, however, that mutations in gene 43 of phage T4 (p. 491) lack the Kornberg polymerase and do not make any phage DNA (de Waard, Paul and Lehman, 1965).

If the Kornberg polymerase is not the enzyme responsible for chromosome replication, what is its normal role in the cellular economy? One of the main types of damage inflicted on DNA by ultraviolet light is the formation of thymine dimers by covalent bonding between adjacent thymine residues (p. 328). It has recently been found that a natural mechanism exists for repair of this damage since the thymine dimers, together with a considerable number of neighbouring nucleotides, are excised from the damaged strand and appear free in the medium (Setlow and Carrier, 1964; Boyce and Howard-Flanders, 1964a). Presumably repair is completed by re-synthesising the excised single-stranded fragments, using the undamaged complementary

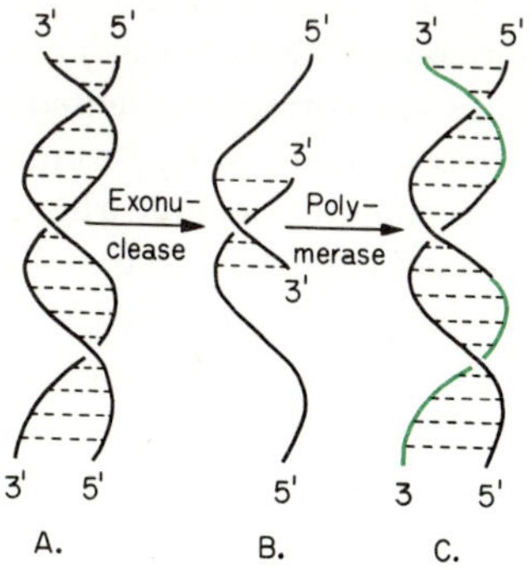

FIG. 65. Scheme for repair, by the Kornberg DNA polymerase, of DNA degraded by exonuclease III.

Exonuclease III nibbles away the DNA strands from the 3′-hydroxyl ends only, resulting in molecules which are double-stranded in the middle, with single-stranded extremities (B). In the presence of polymerase, the original double-stranded structure is restored by synthesis of new strands (green), using the 3′ ends as true primers to which they are covalently bonded, and the single-strand remnants as templates.

After Smellie, 1965.

strand as template (p. 333). It has been suggested that this may be the real function of the Kornberg polymerase (Howard–Flanders, quoted by Stacey, 1965a) and, indeed, there is some good, direct evidence for this. An enzyme, exonuclease III, has been isolated which attacks double-stranded DNA from the 3′-hydroxyl ends only, progressively removing nucleotide residues from both extremities so that molecules with a double-stranded central region and single-stranded ends are produced, as shown in Fig. 65B. When this partially degraded DNA is added to the polymerase system, it is repaired by copying the remaining strand, while the newly synthesised DNA is covalently bonded to the 3′ ends of the degraded strands (Fig. 65C). Not only does this repair synthesis proceed at a rate about five times greater than when native DNA is used as primer, but the double-stranded product is unbranched, can be normally denatured and, in transformation experiments, shows restoration of genetic function (Richardson *et al.*, 1964b).

We will discuss some further aspects of DNA synthesis when we describe the replication of single-stranded DNA (p. 436) and of the bacterial chromosome (p. 553). But before we conclude this chapter there are two final points which may be made, since they have a direct bearing on the nature of DNA polymerase. The first concerns what the polymerase actually does. Earlier on we drew a rather naive picture of the various bases of the nucleotide pool jostling for position to hydrogen bond to their complementary bases on the template strand; when all the nucleotides are correctly oriented by this process, the polymerase simply connects them up by forming covalent phosphodiester linkages. It now seems questionable whether this sort of process is efficient enough to account for the speed and accuracy of chromosome replication. An interesting alternative is that the specificity of base pairing is enzymatically determined, and some evidence implicating DNA polymerase in such a role may be found in the recent report of a mutant polymerase which generates a high mutation rate, presumably by introducing copying mistakes during replication (Speyer, 1965).

The second point concerns the site of polymerase activity. There is now increasing experimental support for the hypothesis that the bacterial chromosome has a specific attachment to the cell membrane, and that the attachment site may play an important part in initiating and controlling replication of the chromosome as well as in segregation of the daughter replicas (Jacob, Brenner and Cuzin, 1963; Jacob, Ryter and Cuzin, 1966). In the case of *B. subtilis*, DNA polymerase activity, precipitable with DNA, has recently been described as occurring in rather constant amounts in the cell membrane fractions of both bacteria and spores. It is suggested that this membrane-bound polymerase may be the one responsible for chromosome replication, while the supernatant enzyme may be involved in DNA repair (Yoshikawa, 1965) (see p. 568 *et seq.*).

Since revision of this chapter, the *in vitro* synthesis of infectious phage φX174 DNA has been reported by Goulian, Kornberg and Sinsheimer (1967). The DNA of phage φX174 is circular and normally occurs as a single strand, from which a double-stranded 'replicative form' is synthesised following infection (p. 437). In the *in vitro* synthesis, the single-stranded DNA was used as a template on which a complementary strand was synthesised by *E. coli* DNA polymerase (the Kornberg enzyme). This yields a double-stranded circular molecule, the newly synthesised strand of which remains unclosed. The secret of success was to add another enzyme, *polynucleotide ligase*, which covalently bonds the extremities of this open strand, to form a completely closed molecule of double-stranded DNA. The newly synthesised strand can then be separated and used, in turn, as a template for making a completely synthetic, circular duplex. Both of these synthetic strands, as well as the duplex itself, proved infectious (Goulian *et al.*, 1967).

CHAPTER 12

Ribonucleic Acid and Protein Synthesis

We are now approaching the focal point of our story, where we can begin to interpret the inferences of genetic analysis in the strict and definitive terms of physics and chemistry. It is not very long since genetics and biochemistry were quite separate sciences, each seeking some fundamental functional unit, which would serve as a key to unlock the mystery of life. The biochemists found enzymes, the geneticists genes. These two basic units were equated in the one gene–one enzyme hypothesis and from then on began a partnership which led to a rapidity of progress previously unparalleled in biological research. From the point of view of genetic analysis this partnership reached its zenith when, as we have seen, the fine structure of the gene was correlated with the polypeptide composition of the derivative protein. It is difficult to see how genetic analysis could have progressed much further, had it not been for the recent dramatic developments in the chemistry and physical chemistry of the nucleic acids.

The result of these developments has been that biochemistry and genetics are now welded together as part of a single, fundamental science known as *molecular biology*, in which genetic analysis is only one of many basic tools. Molecular biology now has so many aspects, and impinges on such diverse problems as the operation of genes, membranes or muscle fibres, that it is hard, if not impossible to define it to everyone's satisfaction. A broad working definition is that molecular biology seeks to explain cellular activity at the molecular level. This definition, of course, embraces much pure biochemistry such as the mechanism of energy production and of amino acid synthesis, and there is no reason why it should not. In practice, however, the molecules which we will be discussing here, nucleic acids and proteins, are all long-chain polymers or macromolecules, and our main aim will be to analyse their functions in terms of their molecular architecture. We have already seen, in the case of DNA, the elegant way in which function can be related to macromolecular structure.

In the following account we do not intend to offer a detailed description of the chemistry of proteins and ribonucleic acid (even if we were competent to do so), but only to outline the picture that is beginning to form as the pieces of the jig-saw puzzle fit into place. To do this in a concise and meaningful way some dogmatism is necessary, as well as considerable selection of material from the avalanche of new and specialised knowledge which is tending to engulf even the expert. Our main aim is to present a coherent picture. We believe that this can now be done and that, when the details are subsequently filled in, the outline we draw now will remain substantially the same.

THE STRUCTURE OF PROTEINS

The building blocks from which all proteins are constructed are the amino acids. There are twenty different 'essential' amino acids, all of which the cell must be able either to synthesise for itself or to obtain from an exogenous source in order to grow. These amino acids are listed in Table 15, p. 358. Nearly all the proteins normally found in cells contain only the L-isomers of these amino acids. The majority contain all or most of the twenty amino acids although the proportions in which they are present differ from protein to protein. However in some proteins, such as silk, a few amino acids predominate while many are absent altogether; other examples are the flagellar proteins of the bacteria *Proteus vulgaris* and *Bacillus subtilis* which contain no detectable histidine, tryptophan, proline or cysteine, while that of *Salm. typhimurium* lacks cysteine alone. A few proteins incorporate amino acids which are additional to the twenty 'essential' ones; thus hydroxyproline is found in collagen and in the flagella of some bacteria; the flagellar protein of certain *Salm. typhimurium* strains is unique in containing the amino acid ε-N-methyl lysine, which replaces lysine and is not found in any other biological material so far examined (Kerridge, 1961). These aberrant amino acids, however, are synthesised from their normal precursors, by the hydroxylation of proline and the N-methylation of lysine respectively, and so need not be regarded as 'essential.'

All amino acids share a common structure in the form of a carbon atom, called the α-carbon, to which is attached a carboxyl group (COO^-), an amino group (NH_3^+), a hydrogen atom and a side-chain. It is the nature of the side-chain which determines the differences between the various amino acids. In Fig. 66 (upper diagram) three different amino acids, distinguished by side-chains R′, R″ and R‴, are represented. Amino acids can link together by attachment of the α-carbon amino group of one to the α-carboxyl group of the other, with the elimination of water (Fig. 66, lower diagram). This linkage is known as a peptide bond and results in the formation of a peptide; two amino acid residues joined in this way form a dipeptide, three

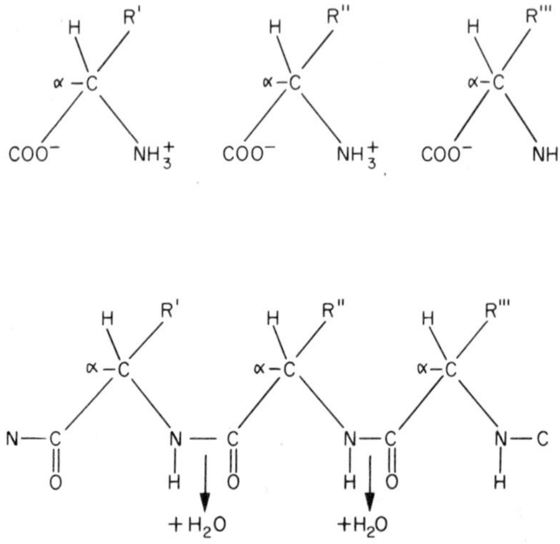

FIG. 66. A representation of the nature of amino acids, and of their linkage by peptide bonds to form peptides.

R′, R″ and R‴ represent different side-chains which distinguish one amino acid from another.

form a tripeptide, and so on. When a large number of amino acid residues become linked, the polymer is called a polypeptide. The effect of acid or enzymatic hydrolysis of polypeptides is to reintroduce water into the peptide bond, so that the linkage is broken and the amino acids are liberated. In fact, proteolytic enzymes are able to operate in the reverse direction and synthesise peptide bonds, provided the peptides are removed from the mixture as they are formed as, for example, by being insoluble.

Polypeptides are therefore long chains of amino acid residues connected together by peptide linkages. Natural proteins may be made up of a single polypeptide, or of two or more polypeptides connected together by cross-linkages between the side-chains of certain of their amino acids. Insulin, one of the smallest of the proteins, consists of two polypeptides, one of which contains thirty amino acid residues and the other twenty-one residues; ribonuclease isolated from bovine pancreas, with a molecular weight of 13,683, is a single polypeptide of 124 amino acid residues; haemoglobin, of molecular weight about 65,000, is built up of two pairs of polypeptide chains ($\alpha_2\beta_2$; see p. 152), each monomer having some 150 amino acid residues. The specificity of proteins is governed by the sequence of amino acid residues in the polypeptide chain or chains of which they are composed. Since there are twenty amino acids, the number of different sequences in which they can be arranged is enormous. Just how enormous is evident when one thinks of all the words that can be made from the restricted use of a twenty-six letter alphabet!

This linear sequence of amino acid residues, which is fixed for each protein, is termed the *primary structure* of the protein. Proteins, however, are far more complicated than this. A new dimension was added to our knowledge of their structure by the method of X-ray diffraction analysis (p. 231) which was initially used to estimate the distances and angles between the atoms of amino acids and of small, synthetic peptides, and then to ascertain how the amino acid residues are oriented to one another in space. It turned out that the dimensions of some synthetic polypeptides revealed by crystallography are incompatible with a fully extended polypeptide chain, and that the chain must be coiled helically around an axis to form what is known as an α-helix, whose geometry is maintained by hydrogen bonding between the CONH linkage of each residue and that of the third residue beyond it. Thus while the distance between the amino acid residues of an *extended* polypeptide chain is about 3·5 Å, in the α-helix they are actually separated by only 1·47 Å along the long axis of the helix, a complete turn of which contains 3·7 residues (Pauling and Corey, 1951a, b; Pauling *et al.*, 1951; Perutz, 1951). This coiling of the polypeptide chain is termed its *secondary structure*. It seems that *fibrous proteins* such as make up hair, silk, keratin and other structural features of cells, and occur naturally as fibres, consist of many strands of α-helices twisted about one another.

However, the biologically functional proteins, such as enzymes or haemoglobin for example, are not fibrous but *globular* in shape. Our present knowledge of the structure of globular proteins has been achieved by correlating their shape and dimensions, obtained by X-ray diffraction analysis, with the number and sequence of their amino acid residues as determined by chemical analysis. The most comprehensive crystallographic study to date is that of myoglobin whose general structure has been defined with a resolution as high as 1·4 Å (see Kendrew, 1960; Plate 11). It will be seen that the polypeptide chain shows long, straight runs of helix but that, at certain regions, it is folded and bent back on itself to give a complicated, three-dimensional architecture. From the dimensions of this model and from the known number of amino acid residues, as well as from high resolution diffraction data, it can be inferred that most of the straight portions, at least, of the polypeptide chain must have the α-helical configuration. This folding of the α-helix in globular proteins therefore adds another dimension to the molecule which is known as its *tertiary structure*. It should be pointed out, however, that many proteins, such as lysozyme, contain very little α-helix. As we have seen in the case of haemoglobin, a *quarternary structure* may be superimposed on the tertiary by the aggregation of a number of such sub-units to form a complete protein of larger size than and different shape from, the sub-units themselves. In the case of some globular proteins the aggregation of individual protein molecules, called 'sub-units', goes much further than this, and results in the formation of a structure of defined morphology which may comprise hundreds of subunits. By altering the pH or the ionic environment, these structures may often be dissociated into their constituent molecules which then re-aggregate to reproduce the structure when the original conditions are restored. Good examples of globular proteins which behave like this are the sub-units of tobacco mosaic virus (p. 229) (Fraenkel–Conrat and Williams, 1955) and the 'flagellin' molecules which make up bacterial flagella (Abram and Koffler, 1964). In addition, the activity of many enzymes depends on the aggregation of small numbers of polypeptide sub-units.

It is almost certain that the biological activity and specificity of proteins is determined partly by the linear disposition of reactive chemical groupings, and partly by the folding of the molecule which fixes the spatial arrangement of these groupings at the molecular surface. It is becoming increasingly clear, however, that although quite small chemical substitutions may alter the antigenic specificity of proteins, considerable liberties may be taken with their structure without interfering appreciably with their activity as enzymes or hormones. Thus the adrenocorticotrophic hormone, ACTH, containing thirty-nine amino acid residues, will tolerate the removal of

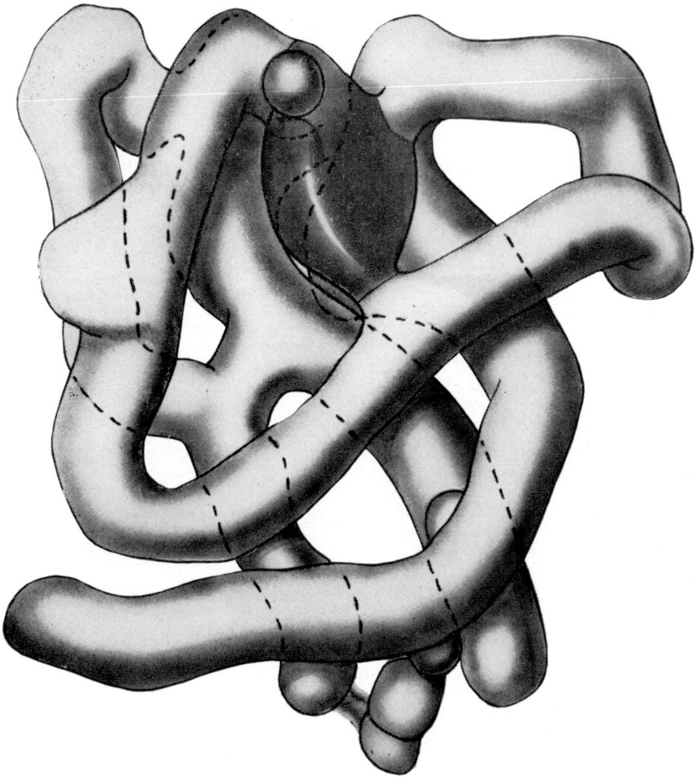

PLATE 11. Drawing of a myoglobin molecule, constructed from X-ray diffraction data.

The dark area (top centre) represents the prosthetic haem group.

(Kindly provided by Dr J. C. Kendrew)

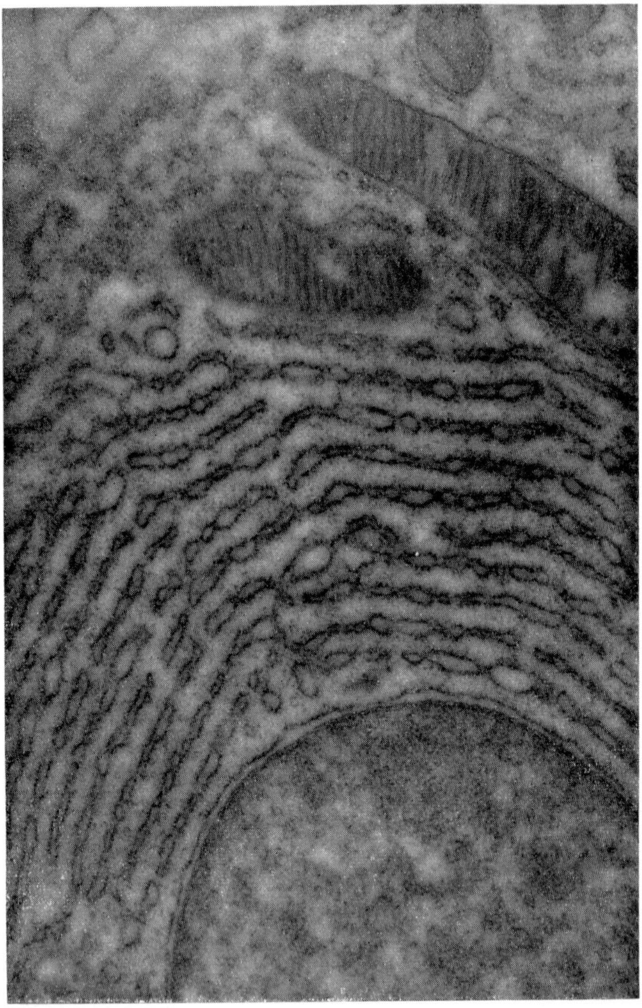

PLATE 12. Electron micrograph of a cross-section of an exocrine cell from guinea-pig pancreas, showing ribosomes.

The ribosomes appear as numerous, minute dots lying between the layers of the endoplasmic reticulum which are seen sandwiched between the nucleus (bottom right) and mitochondria (top right and centre). Magnification about ×24,000.

(Kindly provided by Dr G. E. Palade)

fifteen residues from one end of the chain without loss of hormonal activity. Similarly, the intact ribonuclease molecule has two free 'tails' at either end of its polypeptide chain, one containing twenty-five and the other fourteen amino acid residues (Fig. 67); a number of these

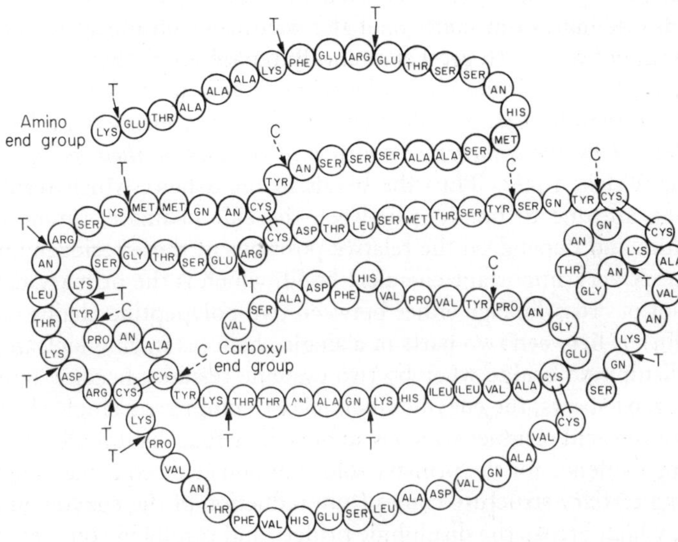

FIG. 67. The structure of bovine pancreatic ribonuclease as determined by sequence analysis.

ALA = alanine(12)
AN = aspartic acid side-chain NH$_2$(11)
ARG = arginine(4)
ASP = aspartic acid(15)
CYS = cysteine(8)
GLU = glutamic acid(12)
GLY = glycine(3)
GN = glutamic acid side-chain NH$_2$(6)
HIS = histidine(4)
ILEU = isoleucine(3)
LEU = leucine(2)
LYS = lysine(10)
MET = methionine(4)
PHE = phenylalanine(3)
PRO = proline(4)
SER = serine(15)
THR = threonine(10)
TYR = tyrosine(6)
VAL = valine(9)

The numbers in brackets after each amino acid indicate the number of residues in each polypeptide chain.

The double black lines indicate disulphide linkages between pairs of 'half'-cystine residues.

$\overrightarrow{\text{T}}$ indicates sites of trypsin cleavage.

$\overrightarrow{\text{C}}$ indicates sites of chymotrypsin cleavage. (Data and general layout from Anfinsen, 1959).

residues can be nibbled away from either end by controlled hydrolysis without affecting the enzymic behaviour of the molecule (see Anfinsen, 1959, p. 126).

From the genetic point of view it is crucial to decide how the tertiary structure of proteins, on which their function depends, is controlled, for on this hangs our concept of the way in which the genetic code specifies function. At present the only tenable hypothesis, which is compatible with all the known facts and is seriously contradicted by none, is that *the tertiary structure of globular proteins is exclusively determined by the sequence of amino acid residues in their polypeptide chains* (Crick, 1958). Thus the bends in the α-helix which result in folding are due to bonding between the side chains of amino acid residues and depend on the relative positions of these residues in the chain. A very important chemical bond, which is the primary determinant of cross-linkage either between two polypeptide chains (as in insulin) or between two parts of a single chain (as in ribonuclease), is the disulphide bridge whereby two cysteine residues become united. As Fig. 67 shows, the enzyme ribonuclease comprises a single chain of 124 amino acid residues which is stabilised by four disulphide bridges. Strong evidence for the primary role of amino acid sequence in determining tertiary structure comes from reduction of the enzyme in $8M$ urea, which breaks the disulphide bridges and results in complete loss of enzyme activity, followed by re-oxidation which yields 100 per cent restoration of activity and a product indistinguishable from the original material. Since a random, pairwise bonding of the eight cysteine residues could occur in 105 different ways, it is clear that some other factor must be instrumental in determining that only specific pairs of residues are brought into apposition so that the disulphide bridges reform in a unique manner (White, 1961).

Apart from cysteine bonds, do we know of any rules, connected with amino acid sequence, which govern the specific configurations of polypeptides? It is probable that globin chains have the same three-dimensional shape in myoglobin and haemoglobin from all vertebrates. Yet, strangely enough, a comparison of amino acid sequences of globins from divergent sources reveals that less than 7 per cent of sites are occupied by the same amino acid in all the globins. However, the most striking feature to emerge is that amino acids with polar (hydrophilic) side-chains are almost totally excluded from the interior of the molecule. At surface sites, on the other hand, many amino acid replacements are possible, including the substitution of polar by non-polar residues and *vice versa*, without altering the tertiary structure (Perutz, Kendrew and Watson, 1965). This general rule seems also to hold in the case of egg-white lysozyme which is the first *enzyme* whose amino acid sequence and structure have been elucidated (Blake *et al.*, 1965) and whose active centre has been identified with a surface cleft which can specifically accommodate the substrate, in a way which suggests a precise model for enzyme action (Johnson and Phillips, 1965; Blake, 1966; see also Blake *et al.*, 1967a, b and accompanying discussion).

If we assume that the effect of a mutation is, in general, to substitute one amino acid for another in the polypeptide chain, the sequence hypothesis allows us to draw certain inferences from what we have learnt about protein structure. One is that we may expect many mutations to be cryptic, producing no detectable alteration of function. For example, it is easy to imagine that many amino acid substitutions along straight runs of α-helix in the molecule will not change the α-helical structure in any way and will interfere with function only if functionally reactive groupings are involved. Again, we have seen that some protein molecules are insensitive to the removal of an appreciable proportion of their residues, so that even mutations which result in the actual deletion of amino acids need not necessarily be harmful (see pp. 351, 352). On the other hand, it is obvious that certain parts of the polypeptide chain, such as the regions of folding and the reactive groupings themselves, are likely to be highly vulnerable to amino acid substitutions, while the removal of some key amino acid such as cysteine may lead to a drastic distortion of structure. In our study of genetic fine structure we have already encountered some correlations between the position of mutational sites in the gene and the degree or type of resulting functional defect (p. 166).

SEQUENCE ANALYSIS OF PROTEINS

Sequence analysis means the determination of the sequence of amino acid residues in a protein by chemical methods. It is outside the scope of this book to attempt to describe these methods in any detail. In view of the importance of the procedure as the complement of genetic analysis, however, we must outline the general principles involved. The first protein to be fully described in terms of its amino acid sequence was insulin, following ten years of sustained and brilliant work by Sanger and his colleagues (review, Sanger, 1956). It was this work that laid the foundation of modern sequence analysis which is now sufficiently advanced to be applied, with reasonable chance of success, to any protein of modest molecular weight.

The first step in the analysis is to obtain pure and homogeneous samples of the protein and to determine the total number and nature of its amino acid residues, as well as the number of repeats of each residue, by chromatographic analysis of the fully hydrolysed protein. The number of polypeptide chains must then be ascertained. The clue to this is given by the presence and number of cysteine residues which, by the formation of disulphide bonds, are the primary means of linking two polypeptide chains together. These bonds can be broken and blocked by oxidation with performic acid, so that if two chains are present they will separate and remain apart, thus reducing the mean molecular weight. If the molecular weight remains the same after this treatment, it is likely that only a single polypeptide chain is present. This can be checked by what is called 'end group' analysis. The fact that the carboxyl group of one amino acid is bound to the amino group of its neighbour means that the terminal residue at one end of a polypeptide chain will have a free amino group, while the residue at the other end will have a free carboxyl group; thus every chain has only a single free amino group. This group can react with reagents such as dinitrofluorobenzene so that the N-terminal residue, with a free amino group, is converted to a pigmented dinitrophenyl (DNP) derivative. Subsequent hydrolysis of the protein, and separation and identification of its individual amino acids by chromatography will reveal the N-terminal amino acid by its pigmentation. The presence of only a single N-terminal amino acid supports the existence of a single polypeptide.

The next step is to find out the amino acid sequence. This is the most difficult and arduous part of the analysis. The basic procedure is first to open out the polypeptide chain by cleaving the disulphide bridges, and then to break it up into small peptides by partial hydrolysis with acids or enzymes. These peptides are separated and defined by two-dimensional paper chromatography, or on an ion exchange column, and are then eluted, degraded further, and their content of individual amino acids determined.

The order of the amino acids in small peptides can be elucidated by means of 'end group' analysis as described above. Suppose, for example, that a sample of peptide, which separates at a particular spot in two-dimensional paper chromatography, is found to contain three amino acids, A, B and C. 'End group' analysis of this tripeptide shows that B is the N-terminal residue; on chromatography of a partial hydrolysate of the tripeptide, the DNP derivative of B is found either alone or in association with amino acid A. It follows that the order of amino acids must be BAC.

When the various peptide fragments of the polypeptide chain have been defined in this way, the final stage is to attempt to fit them together in their proper order by looking for overlapping amino acid sequences. For example, Fig. 68a is a schematic representation of the sequences of ten amino acid residues, A–J, in eight hypothetical peptides, 1–8, which have been defined by chromatographic analysis in the way we have described. From 'end group' analysis of the intact polypeptide it is known that the N-terminal residue is amino acid B, so that the sequence must begin with B. The problem is now like a jig-saw puzzle in which the parts must be fitted together by observing their overlapping patterns. Only the arrangement in Fig. 68b is compatible with

a. The seqences of amino acid residues in eight peptides of a hypothetical polypeptide chain.

1. D D A
2. B F B C
3. A E B
4. B A C D
5. C D DD
6. G G
7. B C G
8. G H I J

b. Order of peptides deduced from overlapping sequences.

```
4. B A C D
5.     C D D D
1.         D D A
3.             A E B
2.                 B F B C
7.                     B C G
6.                         G G
8.                           G H I J
```

c. Sequence of amino acid residues

B A C D D D A E B F B C G G H I J

d. Presumptive structure of polypeptide chain if "C" is a cysteine residue.

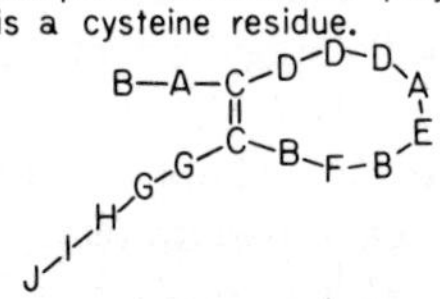

FIG. 68. The principle of sequence analysis (see text).

the sequences of residues in the individual peptides, giving an overall sequence of seventeen residues as shown in Fig. 68c. If we know that the polypeptide chain contains seventeen amino acid residues, while the number of repeats of each residue corresponds with that inferred from the sequence analysis (2A, 3B, 2C, 3D, 1E and so on), then the sequence is confirmed. Finally we may guess that a disulphide bridge will form between the two cysteine (C) residues to yield the kind of structure shown in Fig. 68d.

This, of course, is rather a gross simplification of the situation as it is usually found in practice. However, there exist several chemical tools which can greatly mitigate the formidable task of reconstruction. For example, samples of the protein may be hydrolysed by two or more enzymes which split it at different but specific points so that distinct sets of fragments, which are necessarily overlapping, are obtained; thus trypsin disrupts the chain on either side of a lysine or arginine residue, while chymotrypsin specifically severs peptide bonds involving tyrosine. The results of a complete sequence analysis of bovine pancreatic ribonuclease is represented in Fig. 67 (p. 263). For those readers who wish to learn more about the chemical analysis of protein structure a simple but excellent account, together with recommendations for further reading, is given by Anfinsen (1959, pp. 98–163).

THE INVOLVEMENT OF RIBONUCLEIC ACID (RNA) IN PROTEIN SYNTHESIS

The evidence that DNA carries the genetic information which determines the synthesis of specific proteins poses two questions: where, in the cell, does protein synthesis take place, and how is it effected? One of the earliest experimental approaches to these questions came from the work of Hammerling and of Brachet on a huge alga called *Acetabularia mediterranea*. This organism, which may reach a length of 5 cm, consists of a single cell which is differentiated into a root-like structure containing the nucleus, a stalk, and an umbrella-like 'cap'. If, before the cap is formed, the nucleus is cut off from the growing end of the organism by tying the stalk, specific protein synthesis continues in the anucleate part and a normal cap develops. This implies that the information contained in the nucleus is not interpreted directly, but is transmitted to some cytoplasmic intermediary which directs the synthesis and organisation of cellular structure.

A likely candidate for this intermediary role is ribonucleic acid which is not only a predominant constituent of the cytoplasm of all cells, but is also found in the nuclear nucleolus (p. 9) and is closely related to DNA in its chemical composition (p. 271). Cytochemical studies by Brachet, Caspersson and others (review, Brachet, 1957), as well as quantitative biochemical analysis (see Davidson, 1947; Chantrenne, 1953), strongly supported this view by establishing correlations between the RNA content of cells and their involvement in active protein synthesis. Moreover, Brachet showed that the enzyme ribonuclease (RNase) can be taken up by a variety of living cells and that this is followed by complete cessation of protein synthesis as well as of the incorporation of labelled amino acids into protein; the inhibition is reversible, however, by the addition of fresh RNA to the cells (see Brachet, 1957).

The essential features of these experiments have been confirmed and extended by studies of the metabolism of intact bacteria and other microorganisms. For example, when bacteria are cultivated under different conditions of growth, the amount of RNA and protein per bacterial nucleus rises markedly with the growth rate, while the amount of DNA per nucleus remains constant (Maaløe, 1960); this shows that RNA and protein synthesis are correlated, but dissociated from the synthesis of DNA. Secondly, DNA synthesis is not required for either protein or RNA synthesis, since both the latter continue unabated under conditions of thymine starvation which specifically suppresses the synthesis of DNA. Nevertheless the occurrence of protein synthesis does appear to depend on the *integrity* of the gene, at least in *E. coli* (Riley *et al.*, 1960). Thirdly, RNA synthesis is needed for protein synthesis; the specific suppression of RNA synthesis, either by addition of uracil analogues (Slotnick *et al.*, 1953) or by deprivation of uracil (Pardee, 1954), completely stops the induced formation of β-galactosidase in *E. coli*. Fourthly, RNA synthesis continues, at least temporarily, following the arrest of protein synthesis by chloramphenicol (Wisseman *et al.*, 1954). It seems, however, that both protein and RNA synthesis cease concomitantly in the absence of certain required amino acids (Pardee and Prestidge, 1956); this, however, does not necessarily imply that protein synthesis precedes that of RNA, but may mean either that RNA synthesis requires the manufacture of certain enzymes, or else that amino acids may play a key role in the regulation of RNA synthesis.

This array of circumstantial evidence gave birth to the plausible hypothesis that the DNA of the nucleus impresses its information on

RNA which then migrates to the cytoplasm where it forms a template on which specific proteins are built up. This *template hypothesis*, which implies a one-way transfer of information from nucleic acids to protein and which Crick (1958) has immortalised by the title 'The Central Dogma', can thus be expressed in the form

$$\text{DNA} \rightarrow \text{RNA} \rightarrow \text{Protein.}$$

The only rival to the template hypothesis supposes that proteins are formed under the direction of a multitude of sequentially operating enzymes, each endowed with the knowledge not only to perform its specific linkage but also to coordinate its activity with that of all the other enzymes. Apart from the inherent improbability of this multiple enzyme hypothesis, it imposes the insuperable difficulty of how the enzymes themselves are formed (Spiegelman, 1957).

THE RIBOSOMES

An assembly line for protein synthesis suggests a factory. When cells are disrupted, their various structural constituents, such as nuclei, mitochondria and so on, can be separated and isolated by differential centrifugation. In this way Claude (1943) identified, in animal cells, a new class of numerous and very small cytoplasmic particles called *microsomes* which contain the bulk of the cellular RNA. It was later found that, in intact cells, these microsomal particles adhere to a highly involuted membrane, or series of membranes, known as the *endoplasmic reticulum* (Palade, 1955). Plate 12 (facing p. 263) is an electronmicrophotograph of a cross-section of a guinea-pig pancrease cell, showing the endoplasmic reticulum studded with microsomes. In bacteria, about 80 per cent of the RNA is similarly concentrated into minute cytoplasmic particles, as many as 5000 of which may be present in each cell, but the particles do not appear to be attached to membranes as in animal cells (Schackman *et al.*, 1952). Chemical analysis of these particles revealed that they are made up of about 60 per cent RNA and about 40 per cent protein. They are therefore usually referred to nowadays by the alternative name of *ribosomes*.

It was then discovered, by means of tracer studies with radioactive amino acids, both in living animals as well as in *in vitro* experiments, that the amino acids are rapidly taken up by the ribosomes and are there initially incorporated into proteins (Littlefield *et al.*, 1955;

Simkin and Work, 1957). Due to the high rate at which bacteria metabolise, the uptake and incorporation of amino acids by their ribosomes, and the subsequent release of the soluble protein, may occur so quickly that it can easily be missed altogether. For example, it has been calculated, from tracer 'pulse' experiments in *E. coli*, that saturation of the ribosomes with the label may occupy only five seconds, and that it may be as quickly replaced (McQuillen *et al.*, 1959).

To conclude this general account of the role of RNA in protein synthesis we may mention that autoradiographic studies have demonstrated that, in *Neurospora*, ribosomal RNA is synthesised in the nuclei but that protein synthesis occurs mainly in the cytoplasmic ribosomes (Zalokar, 1959a, b). We may therefore regard it as well established that protein synthesis takes place in the ribosomes which contain most of the cellular RNA. Moreover there is good circumstantial evidence for the hypothesis that protein synthesis is mediated by the assembly and polymerisation of amino acids on an RNA template bearing some kind of copy of the genetic information, concerning the amino acid sequence, which is encoded in the DNA.

Assuming the correctness of this hypothesis, a number of very specific questions may now be asked. How is the DNA code passed on to RNA? Is the template RNA the ribosomal RNA itself? How do the amino acids reach the template? What is the mechanism which enables the amino acids to recognise their own code words, and thus to assemble themselves in proper sequence on the template? Before attempting to answer these questions it is necessary to learn something about the chemistry and structure of RNA, and about the various types of RNA that are found in the cell.

THE CHEMISTRY OF RIBONUCLEIC ACID

In the last chapter we looked rather closely at the chemical composition and mode of synthesis of deoxyribonucleic acid (DNA). What we learnt there will prove very useful now (if you remember it) because the chemistry of ribonucleic acid (RNA) is very similar in every respect. As we shall see, this essential similarity becomes very meaningful when we begin to enquire how the genetic information encoded in the DNA is translated into the synthesis of specific proteins.

RNA differs from DNA in only three important ways.

1. The sugar D-ribose 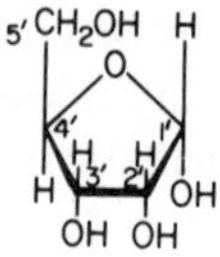 replaces the 2-deoxy-D-ribose, 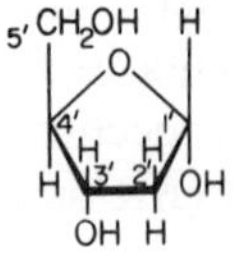 of DNA.

2. The pyrimidine base *uracil* 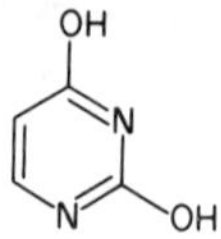 replaces the thymine 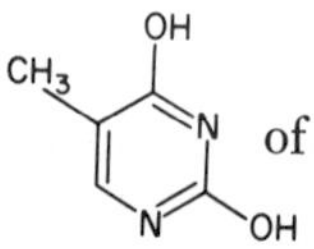 of DNA. As will be apparent from the structure of these two pyrimidines, uracil is an analogue of thymine and, if supplied instead of thymine in the *in vitro* synthesis of DNA by the Kornberg system, pairs with adenine and is incorporated into the DNA double helix. In fact, natural DNAs containing uracil have been reported (Kallen, Simon and Marmur, 1962). The purines adenine and guanine, as well as the pyrimidine cytosine, are common to both RNA and DNA.

3. RNA appears to occur naturally in the form of a single-stranded polymer. However, as we shall see, it is capable of uniting with a complementary strand under certain conditions, to yield a hydrogen-bonded double helix closely similar to that of DNA.

When natural RNA is treated with various diesterases it is broken down into individual ribonucleotides by the severing of phosphodiester bonds, just as in the case of DNA (Fig. 55). With some enzymes, such as the diesterase from snake venom, the products of hydrolysis are ribonucleoside 5′-monophosphates in which the phosphate molecule is attached to the 5′ position of the ribose; with others such as spleen diesterase, as well as on hydrolysis with alkali, riboside 3′-monophosphates are liberated. In general, the four types of nucleotide, riboadenylic, riboguanylic, ribocytidylic and ribouridylic acid, are present in roughly, but by no means precisely, equivalent amounts. RNA is therefore very similar to a single strand of DNA, consisting of a sequence of ribonucleotides strung together by 3′–5′ phospho-ribose linkages but because natural RNA consists of only a single chain, there is no strict relationship between pairs of bases as in DNA. In *E. coli* RNA, for example, the ratio of purines to pyrimidines is not unity but about 1·3.

What kind of structural configuration does RNA adopt? In DNA the presence of two hydrogen-bonded, complementary polynucleotide chains is not only necessary for replication of the genetic material but gives rise to a naturally stable structure. On the other hand one would imagine that the single strand of RNA, with no complementary strand to brace it and to satisfy the hydrogen atoms of its bases, would be an inherently unstable structure. There are two points worth noting here which are likely to have relevance with respect to RNA function. The first is as follows. We shall shortly see (p. 292) that it is now possible to make synthetic RNA polymers of various kinds, including chains incorporating only one type of base such as adenine (called 'poly A') or uracil ('poly U'). If two such synthetic polymers having complementary bases (e.g. 'poly A' and 'poly U') are mixed, the two strands unite by A—U hydrogen bonding to form a double helix which X-ray diffraction analysis shows to be very like that of DNA (Warner, 1956; Rich and Davies, 1956). If more 'poly U', but not 'poly A' or 'poly C' (polycytidylic acid), is added, the 'poly U' wraps itself around the 'poly A–poly U' double helix to form a 3-stranded structure (Felsenfeld and Rich, 1957), thus offering a possible model for the way in which DNA passes on genetic information to a single strand of RNA (see next section; also p. 373). Double-stranded RNA in the sense that there are two separate strands, does not appear to be a normal cellular constituent and is found only following infection with RNA viruses when it comprises the replicative form (p. 442). We will deal with the biological synthesis of RNA on p. 289.

The second point is that analysis of both ribosomal and 'soluble' RNA (see below) by X-ray diffraction and other methods provides strong evidence that parts of the molecule, in some cases over 80 per cent, are folded into a double helical structure (Zubay and Wilkins, 1960; see also Brown and Lee, 1965). Similarly, the study of synthetic polynucleotides, in which the four bases have been randomly incorporated, suggests that about half the bases may specifically hydrogen bond, adenine to uracil and guanine to cytosine. This has prompted a model of single-stranded RNA in which the chain is folded back on itself in such a way that the coincidence of specific base-pairs leads to hydrogen bonding and the formation of short, double-helical regions; those in bases which do not find a partner are 'looped out' of the double helix the manner represented in Fig. 69 (Fresco *et al.*, 1960). It is very likely that the three-dimensional architecture of RNA molecules, like that of proteins, has specificity and is functionally meaningful (see p. 282).

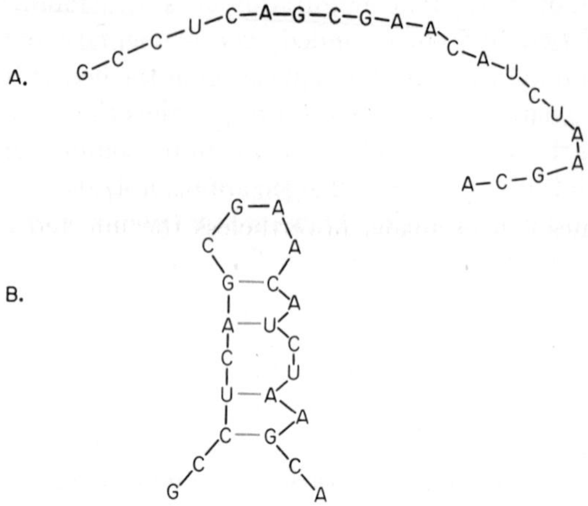

FIG. 69. Demonstrating how a single polynucleotide chain (A) may fold back on itself to form double-helical regions, held together by specific hydrogen bonding (B). Those bases which cannot pair are 'looped out' of the double helix.

The green lines represent hydrogen bonds; the black lines the phosphodiester bonds linking the nucleotides.

TYPES OF RNA AND THEIR RELATION TO PROTEIN SYNTHESIS

So far we have discussed RNA as if it consisted of an undifferentiated family of molecules which are involved in protein synthesis in some undefined way. One of the most exciting and revealing developments in recent years has been the discovery that there exist three distinct types of RNA molecule which have quite different and highly specific roles to play in the cellular protein factory. Before discussing the biosynthesis of RNA we must outline the main characteristics of these RNA types which are distinguished primarily by molecular weight.

MESSENGER RNA

The discovery that proteins are synthesised on ribosomes in the cytoplasm made it clear that the genetic messages inscribed in the base sequence of chromosomal DNA must be transcribed to mobile molecules which then become associated with the ribosomes. The obvious candidate was the ribosomal RNA itself until, for a number of reasons, it became apparent that this could not be so. Thus there is no relationship between the ratios of the various bases in ribosomal RNA and DNA, such as one might expect if the RNA base sequence was a transcript from the DNA; in fact, although widely divergent A+T/G+C DNA ratios may be found in nature (p. 231), the base ratios of ribosomal RNAs are remarkably constant. Again, although infection of *E. coli* with virulent bacteriophages T2 and T4 results in an abrupt and permanent cessation of net RNA synthesis, as well as that of bacterial protein, so that no new ribosomes can be made, nevertheless the infected cells rapidly begin to synthesise *phage* protein, under the direction of the injected phage DNA, at about the same rate as they had previously synthesised their own (p. 427 *et seq.*).

Furthermore, kinetic studies of β-galactosidase production not only showed that enzyme synthesis begins within a minute or so of the introduction of its determinant gene (z^+) into the cell by conjugation (see Fig. 135 and p. 704), and proceeds at maximal rate from the start, but also that the rate of synthesis declines *pari passu* with the ability to yield z^+ recombinants as a result of the decay of ^{32}P incorporated in the DNA, that is, only cells with an intact z^+ gene can synthesis the enzyme (Riley *et al.*, 1960). In view of the known stability of ribosomes, which conserve their RNA for at least three generations (Davern and Meselson, 1960), these experiments meant that the genetic information for enzyme synthesis is carried from gene to ribosome by means of very rapidly produced but very unstable 'messengers' with a high rate of turnover (Jacob and Monod, 1961a).

We can therefore clearly reject the idea that the genetic information for protein synthesis is imprinted on the ribosomal RNA. Strangely enough, positive evidence for 'messenger' RNA molecules (m-RNA) was first obtained many years ago as a result of studies of RNA synthesis following infection of *E. coli* bacteria with the virulent T2 and T4 phages. The earliest observation was that although no further net increase in RNA occurred after infection, a small amount of RNA with a high turnover rate continued to be synthesised (Hershey, Dixon and Chase, 1953). This type of RNA was further investigated by Volkin and Astrachan (1957) who transferred infected bacteria to a ^{32}P-containing medium, so that the isotope became incorporated into the newly synthesised RNA fraction. They then hydrolysed the RNA and analysed the distribution of radioactivity between its four nucleotide bases; they found that the ratio adenine+uracil/guanine+cytosine in this RNA fraction was about 1·7 and approximated closely to that of the phage DNA (A+T/G+C=1·8), but was quite different from that of the bacterial DNA (1·0) or of the bulk of the RNA (0·85).

It was later shown in a novel way that this RNA fraction, which we will now openly call *messenger* RNA, has a very specific affinity for the phage DNA which provoked its appearance. We have mentioned that if DNA is heated, the two strands of the double helix uncoil and separate (p. 244). If the solution is now cooled *very slowly*, the complementary strands come together again to reconstitute specific, double helical DNA. Even more surprising is the fact that single DNA strands derived from different bacterial species can come together in this way to form 'hybrid' double helices, but this only happens if the bacterial species are so closely related as to be capable of genetic recombination. Presumably the specificity required for 'hybrid' duplex formation is determined by a close approximation to complementary base sequences along the two strands (Doty *et al.*, 1960).

It occurred to Hall and Spiegelman (1961) that if single but complementary strands of DNA and RNA could form duplexes in this way, the phenomenon would provide a very specific method of checking the equivalence of base sequence between DNA and messenger RNA. Their technique was to mix ^{32}P-labelled messenger RNA, isolated from bacteria infected with phage T2, with tritium (^{3}H)-labelled phage T2 DNA which had been heated to separate its strands. The mixture was then slowly cooled to allow 'hybrid' DNA–RNA duplexes to form. Now RNA is more dense than DNA and therefore bands at a different level in a caesium chloride density gradient (p. 243). It follows that DNA–RNA duplexes, if formed, should be of intermediate density, yielding a band lying between the DNA and RNA bands in the gradient. Such an intermediate band, containing both the ^{32}P label of the RNA and the ^{3}H label of the DNA, was found. Control experiments, where the messenger RNA was mixed with denatured DNA from uninfected bacteria, yielded no intermediate band. The existence of a similar type of RNA with a high rate of turnover has also been demonstrated in normally growing, uninfected *E. coli* cells (Gros *et al.*, 1961), while a comparable RNA fraction with a base ratio equivalent to that of the DNA has been isolated from yeast (Yčas and Vincent, 1960).

An index of the molecular weight of a particle or large molecule in solution, although shape is also a significant factor, is the *sedimentation coefficient*, s, which is estimated from its rate of sedimentation on ultracentrifugation. Thus m-RNA, which amounts to about 2 per cent of the total cellular RNA and is very heterogenous in size, has a sedimentation coefficient which varies between 8 and 30s, corresponding to molecules about 600 to 5000 nucleotides long (McQuillen, 1965).

Since protein synthesis depends on reading the messages carried by the base sequences of m-RNA, we might expect that the molecules would remain linear and not fold up by hydrogen bonding between the bases. This, indeed, appears to be the case and may also be the reason why m-RNA is most susceptible to attack by nucleases.

Ribosomal RNA

More than 80 per cent of cellular RNA is found in the ribosomes where it is associated with protein in roughly equal parts, although ribosomes from different sources may vary somewhat in composition and properties. Functional ribosomal particles are nearly spherical, with a diameter of about 200Å, and have a sedimentation coefficient of 70s (Tissières, Schlessinger and Gros, 1960; McQuillen, 1961). The electron microscope does not reveal any formal arrangement of protein subunits such as characterises the RNA viruses (Huxley and Zubay, 1960). If the concentration of magnesium ions is reduced to about 10^{-3}M, it is found that the 70s ribosomes dissociate reversibly into two unequal particles, 50s and 30s, both of which are essential for protein synthesis (Tissières *et al.*, 1959). In addition, at high Mg^{++} concentrations, 70s ribosomes may aggregate in pairs to form larger, dissociable 100s particles. When ribosomal RNA(r-RNA) is freed from the protein moiety it is found to comprise two fractions, 23s and 16s, corresponding to molecular weights of $1{\cdot}1 \times 10^6$ and 0.56×10^6 respectively, which differ slightly in base composition; while the 30s particles contain only 16s RNA, and the 50s particles only the 23s RNA, the 70s particles contain a mixture of the two RNA fractions (Kurland, 1960).

We have discussed the evidence that the part played by ribosomes in protein synthesis is a non-specific one, the specificity residing in unstable m-RNA molecules with which they become associated. Proof that this was so came from an elegant experiment employing the heavy isotopes of nitrogen (^{15}N) and of carbon (^{14}C) to label the ribosomes, and pulses of radioactive phosphorus (^{32}P) and of sulphur (^{35}S) to localise m-RNA and newly synthesised protein respectively. *E. coli* bacteria, containing heavy ribosomes following growth in ^{15}N- and ^{14}C-containing media, are infected with phage T4 and immediately transferred to a medium devoid of heavy isotopes so that any newly formed ribosomes incorporate only ^{14}N and ^{12}C and are light. Old, heavy ribosomes can then be distinguished from new, light ones by disrupting the bacteria and centrifuging the ribosome fraction in a caesium chloride density gradient of appropriate mean density. The result of the experiment was that, following infection, messenger RNA and newly synthesised phage protein are found in association with the old ribosomes which previously had been concerned with the synthesis of bacterial protein (Brenner, Jacob and Meselson, 1961). The m-RNA is associated with the 70s particles from which it can be dissociated by lowering the Mg^{++} concentration (Nomura, Hall and Spiegelman, 1960; Brenner *et al.*, 1961). In the apt words of Hurwitz and Furth (1962), we can regard the ribosomes as 'for hire' to messenger RNA molecules of any specificity that wish to occupy them.

What is the nature of the association between m-RNA and ribosomes? This has been studied *in vivo*, using reticulocytes and other cells as well as bacteria, and also in an *in vitro* protein-synthesising system containing amino acids, a cell-free extract of bacterial ribosomes and a synthetic polyribonucleotide as m-RNA (see p. 355). Essentially the same findings have emerged from all these investigations. It turns out that, in the presence of a suitable Mg^{++} concentration (10^{-2}M), m-RNA is taken up by 70s particles, but that the subsequent synthetic activity is exclusively associated with a rapidly sedimenting (140 to 200s) fraction containing aggregates of ribosomes in which pentamers (groups of five) predominate. These aggregates appear to be held together by RNA since treatment with ribonuclease, under conditions which leave the ribosomes undamaged, causes them to disperse (Warner, Rich and Hall, 1962; Warner, Knopf and Rich, 1963; Rich, Warner and Goodman, 1963; Gilbert, 1963a). Moreover, in the *in vitro* system, ribosomal aggregates do not form unless natural or synthetic m-RNA is added to the mixture. All this suggested that protein synthesis is normally carried out by *polysomes* consisting of a number of 70s ribosomes held together by a strand of m-RNA, and electron microphotographs indeed reveal a linear arrangement of ribosomes connected by a fibre of 10 to 15Å diameter (Rich *et al.*, 1963).

The evidence that several ribosomes operate on one m-RNA molecule in the production of protein poses the question of how the system works. One possibility is that the polypeptide chain is passed along from one ribosome to the next as it grows; that is, that all the ribosomes of a polysome cooperate serially in the synthesis of a single protein molecule. This seems a most improbable model since only a fraction of the m-RNA appears to be in contact with the ribosomes which may be separated from one another by as much as 150Å. A much more attractive alternative is that the ribosomes attach in turn to one end of the m-RNA molecule and then move along it, translating the base sequence into the sequential addition of amino acids to the growing polypeptide as they go. When the end of the m-RNA chain is reached the polypeptide is complete. The ribosome then drops off, releases the polypeptide, and again becomes available for attachment to the same or another m-RNA molecule (Rich *et al.*, 1963).

This model is illustrated in Fig. 70 and we will discuss it in more detail shortly when we have completed our review of all the components of the protein-synthesising machinery. For the moment we may note that it allows very efficient utilisation of the relatively small amounts of m-RNA produced and is also compatible with what we know of the molecular dimensions. Thus an average polypeptide chain, such as that of a haemoglobin subunit, contains about 150 amino acids. It is now known that each amino acid is coded for by three nucleotides (p. 350). If the spacing between nucleotides in m-RNA is about 3·5Å and the molecules are really extended linearly as has been postulated, then an m-RNA molecule about 1500Å long is needed to code for each polypeptide chain. This is quite compatible with the arrangement and spacing of ribosomes which have been observed in polysomes. There is a strong presumption that a whole operon, that is, all the contiguously linked genes serving a single biochemical pathway (pp. 716–721), may be transcribed into one long m-RNA molecule, and that correspondingly large polysomes exist for its translation. It has been shown

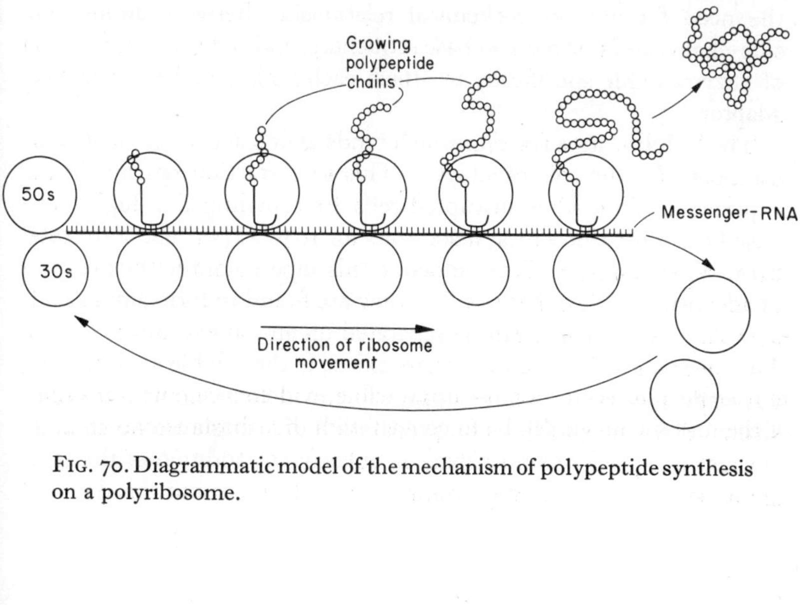

FIG. 70. Diagrammatic model of the mechanism of polypeptide synthesis on a polyribosome.

experimentally that the RNA from the bacterial virus f2, which comprises a single one-stranded molecule, serves as a messenger for at least three proteins, the synthesis of one of which in an *in vitro* system precedes that of the others by several minutes (Ohtaka and Spiegelman, 1963).

Amino acid-transfer RNA

In the cellular factory where proteins are made we have now identified the ribosomes as the work benches where the final products are assembled, and have seen how the templates carrying the specifications for each protein reach them by special messenger RNA molecules. We must now turn our attention to the way in which the raw materials, the amino acids, reach the work benches and are assembled in correct sequence on the m-RNA templates.

The first clue came from the discovery that rat liver cells contain a 'soluble' RNA fraction which has the unique property of binding amino acids (Hoagland, Zamecnik and Stephenson, 1957; Hoagland *et al.*, 1958). An RNA fraction having similar properties was then isolated from *E. coli* (Berg and Ofengand, 1958). At about this time Crick (1958) suggested that the most likely way for amino acids to 'recognise' sequences of nucleotides coding for them was by means of specific 'adaptors', which could well be oligonucleotides carrying base sequences complementary to the amino acid coding sequences which were then thought to reside in the ribosomal RNA. This scheme obviates the need for any stereochemical relationship between amino acid side-chains and nucleic acid base sequences, and requires only a series of enzymes able specifically to attach each amino acid to its correct adaptor.

The 'soluble' RNA fraction which binds amino acids was an obvious candidate for the role of adaptor. This RNA fraction remains in the supernatant fluid when disrupted cells are centrifuged at high speed to sediment the ribosomes; it accounts for 10 to 20 per cent of the total RNA. When amino acids are added to this supernatant in the presence of adenosine triphosphate (ATP), they are found to form amino acid-RNA complexes which can be separated on an ion exchange column. Moreover for each amino acid there exists, in the soluble RNA fraction, a specific species of RNA molecule which will combine with it exclusively. Thus the addition of a preparation of a single amino acid can exhaust its specific RNA so that, on subsequent addition of the same amino acid, no further RNA complexes are formed; but this does not affect the formation of complexes when a preparation of a different amino acid is added. The RNA molecules which unite with amino acids in this way form the greater part of the soluble RNA fraction; when they were first recognised they were given the name of *acceptor* RNA, but this has now been largely changed to *amino acid-transfer* RNA, or simply *transfer* RNA (t-RNA), as their role in protein synthesis has become apparent.

Molecules of transfer RNA have a sedimentation coefficient of about 4S, corresponding to a molecular weight of 25,000 to 30,000, and are about 70 to 80 nucleotides long. A striking characteristic is the uniformity in the ratios of the various bases in t-RNAs from different sources; the purine to pyrimidine ratio is very close to unity, while the ratio A+U/G+C lies between 0·6 and 0·67. This is reflected in the remarkable extent to which various components of the t-RNA-ribosome systems of phylogenetically diverse cells are interchangeable. For example, t-RNA from *E. coli* can cooperate in protein synthesis with ribosomes from rat liver cells and *vice versa* (Nathans and Lipmann, 1961). Assuming that the genetic code is finally translated by the t-RNA, such findings strongly suggest that the code is a universal one throughout Nature (see Chapter 14, p. 364).

Regardless of specificity, all t-RNA molecules terminate at their acceptor end (3′) in the same trinucleotide sequence—cytidylic-cytidylic-adenylic acid (. . .pCpCpA) (Hecht, Stephenson, and Zamecnik, 1959)—while the other (5′) extremity ends in guanylic acid (pG . . .) in the majority of cases. Another characteristic feature is the presence in t-RNA of an unusually high proportion of abnormal bases such as dihydrouridylic acid, inosinic acid and various methylated bases (Dunn, Smith and Spahr, 1960; review, Stacey, 1965b).

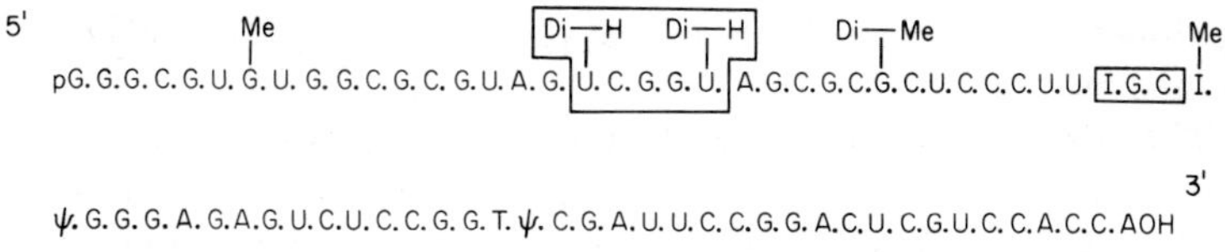

FIG. 71. Nucleotide sequence of an alanine-transfer ribonucleic acid.

A: adenlyic acid	Di—H—U: 5,6-dihydrouridylic acid
C: cytidylic acid	Di—Me-G: Dimethylguanylic acid
G: guanylic acid	Me-G: 1-methylguanylic acid
I: inosinic acid	Me-I: methylinosinic acid
T: thymine ribotide	ψ: pseudouridylic acid
U: uridylic acid	

Such bases can serve as useful 'markers', identifying particular nucleotide fragments, in sequence analysis of t-RNA. A recent landmark in nucleic acid chemistry was the elucidation of the complete nucleotide sequence of yeast t-RNA specific for alanine (Holley *et al.*, 1965). The molecule contains a unique sequence of 77 nucleotides of which nine have abnormal bases (Fig. 71). Since then the nucleotide sequences of serine-specific t-RNA from yeast (84 bases; 13 abnormal) (Zachau, Dütting and Feldmann, 1966) and of yeast tyrosine-specific t-RNA (78 bases; 15 abnormal) (Madison, Everett and Kung, 1966) have been determined.

Examination of the secondary structure of t-RNA by a variety of physical and chemical methods indicates that the greater part of the molecule is looped back on itself to form a double-stranded region or regions. Although the primary structures of the alanine, serine and tyrosine t-RNAs fail to display any long runs of complementary base sequences along different parts of the chains, they can all be folded into a 'clover leaf' type of structure in which the periphery of the leaflets consists of unpaired bases. Such a structure for serine t-RNA is shown in Fig. 72; in fact, two such serine-specific types of molecule have been analysed and found to differ by only three bases indicated by triangles in the figure (Zachau *et al.*, 1966). A postulated structure for alanine-specific t-RNA has one leaflet less (Holley *et al.*, 1965).

We know that t-RNA molecules certainly act as adaptors and that part of the molecule must be able specifically to recognise the messenger RNA nucleotide triplet, or *codon*, which represents the amino acid in the genetic dictionary. Although other possibilities have been suggested (e.g. Woese, 1962), it is now highly probable that t-RNA recognises its codon and becomes bound to messenger RNA because it carries a complementary triplet of nucleotides, or *anti-codon*, which hydrogen bonds to the bases of the codon. Where is the anti-codon located on t-RNA molecules?

The triplets which code for all twenty amino acids are now known with some certainty (Table 15, p. 358). Alanine is represented by GCU, GCC, GCA and GCG, and the first three of these triplets have been shown to be recognised by alanine t-RNA from yeast. When the base sequence of this t-RNA species alone was known, it seemed likely that the sequence CGG, enclosed between two dihydrouracil residues which might isolate the triplet in single-stranded form, was the anticodon (see left-hand box, Fig. 71; Holley *et al.*, 1965; review, Brown and Lee, 1965). However a comparison of the three sequences now

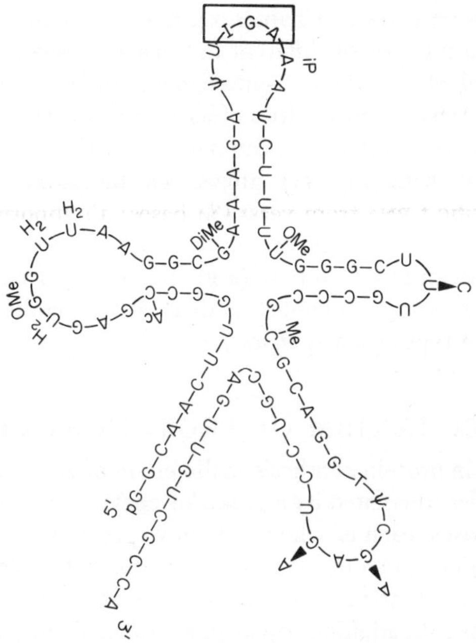

FIG. 72. A secondary structure proposed for serine t-RNA.

Two varieties were isolated and analysed and found to differ in three bases only, indicated by the black triangles. The presumptive anti-codon, IGA, is boxed. According to the 'wobble' hypothesis (Crick, 1966b) this triplet will recognise the serine codons UCU, UCC and UCA when oriented in an anti-parallel way, thus:

5′ 3′5′ 3′ 5′ 3′
AGI .
3′ 5′

From Zachau *et al.*, 1966.

available for study points rather decisively to a triplet in the middle of the chain and at the periphery of one of the leaflets, which in the case of alanine is IGC (right-hand box, Fig. 71; I=inosine, an analogue of guanine), of serine is IGA (box, Fig. 72), and of tyrosine is GψA (ψ=pseudouracil which pairs like uracil) (Madison *et al.*, 1966).

It is also apparent that pairing between codon and anti-codon is 'anti-parallel', as is the case between the two strands of DNA; that is, if the alanine codon is oriented 5′ GpCpC 3′, then the orientation of

| | |

the anti-codon is 3′ CpGpI 5′. We will see when we

study the genetic code in Chapter 14, that each amino acid may be coded for by a number of different but, usually, related triplets and that, in general, these different synonymous triplets differ only in the third base. Because of ambiguities of base pairing at the third position, known as 'wobble', such that U can pair with either A or G, G with U or C, and I with U, C or A (p. 365; Crick, 1966a, b), it is very likely that a single t-RNA species can recognise several synonymous codons. However, some amino acids are represented by triplets which differ in the first or second base, or both (arginine, serine; Table 15, p. 358), and in such cases it is probable that these distinct codons are read by different t-RNA types (p. 364 *et seq.*).

THE MECHANISM OF PROTEIN SYNTHESIS

The first step in protein synthesis is the union of amino acids to their t-RNA molecules, mediated by a group of enzymes called amino-acyl t-RNA synthetases, each of which is highly specific for one of the 20 amino acids. The reaction occurs in two steps, both catalysed by the same enzyme. First, the amino acid is activated by the reaction of its carboxyl group with adenine triphosphate to yield amino-acyl adenylate (AMP) and pyrophosphate, thus:

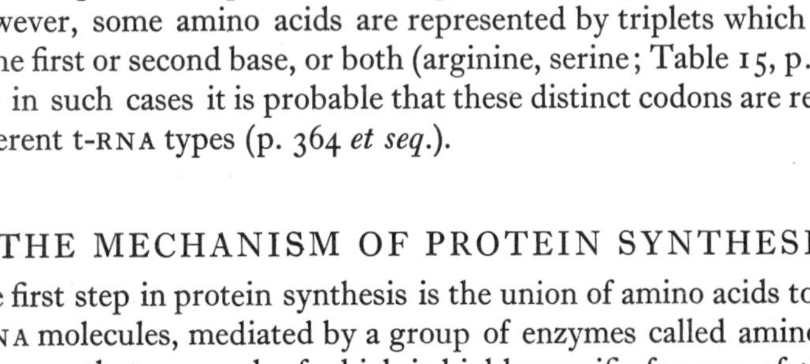

These amino-acyl adenylates remain bound to the enzyme which then recognises and attaches the appropriate t-RNA molecules with great precision. The carboxyl group of the amino acid is finally transferred to the terminal adenylic residue of the C-C-A sequence at the 3′-hydroxyl end of the t-RNA, becoming attached to either the 2′- or the 3′-hydroxyl of the ribose, or possibly alternating between them (McLaughlin and Ingram, 1964). At the same time the AMP and the enzyme are released (reviews: Brown, 1963; Arnstein, 1965; Ingram, 1965; Cold Spring Harbor Symposium, 1963). Although, as we have mentioned, transfer RNA from, say, *E. coli* can cooperate with ribosomes from animal cells, activating enzymes display considerable species specificity for the t-RNA molecules to which they attach their amino acids (Nathans and Lipmann, 1961; see Brown and Lee, 1965).

The next step is the alignment of the aminoacyl-t-RNA complexes in correct sequence on messenger RNA associated with ribosomes. Although a detailed understanding of how this is accomplished will depend on elucidation of the three-dimensional structure of both transfer RNA and ribosomes, about which little is yet known, we are nevertheless now able to describe the process in general terms. We have already mentioned how the 30s and 50s ribosomal components unite to form the functional 70s ribosome which is the basic unit of the polysome. It turns out that inactive 70s ribosomes possess a single site to which a molecule of t-RNA, or an amino-acyl-t-RNA complex, can attach and that this is located on the 50s component; however, ribosomes actively engaged in protein synthesis carry an average of two t-RNA molecules each, one of which firmly binds the growing polypeptide to the 50s component (Cannon, Krug and Gilbert, 1963; Gilbert, 1963a, b; Warner and Rich, 1964).

On the other hand the 30s component has an affinity for the messenger RNA to which it attaches at or near the 5′ end, so that the message is read with a 5′→3′ polarity (see p. 294; Takanami and Okamoto, 1963; Salas *et al.*, 1965). Since any amino-acyl-t-RNA can attach to the 50s ribosomal site, the selection of a particular amino acid for incorporation at any point in the polypeptide chain must be due to specific binding between the anti-codon of the t-RNA and the m-RNA codon. This recognition has nothing to do with the amino acid itself because if cysteine, after attachment to its normal t-RNA, is chemically reduced to alanine and these 'hybrid' complexes are then used in an *in vitro* protein-synthesising system, it is found that alanine is incorporated into the resulting polypeptide in place of cysteine (Chapville *et al.*, 1962).

We can therefore visualise the first amino-acyl-t-RNA as being held in a groove or slot in the ribosome in such a way that the nucleotide triplets of t-RNA anti-codon and m-RNA codon are hydrogen-bonded. The initiation of chain formation now requires the alignment of the next amino acid in sequence, so that we must now imagine a second ribosomal slot adjacent to the first which will accommodate amino-acyl-t-RNA No. 2, specifically bound to codon No. 2 of the m-RNA. This attachment of amino-acyl-t-RNAs to ribosomes requires the prior attachment of m-RNA (Kaji and Kaji, 1963) and appears to depend on the activity of an enzyme (transferase) and of guanosine triphosphate (GTP) as an energy source (Arlinghaus, Shaeffer and Schweet, 1964).

How do amino acids 1 and 2 become linked to initiate chain formation? It is known that the synthesis of haemoglobin polypeptides begins at the amino end and proceeds by the sequential addition of amino acids (see p. 343; Bishop, Leahy and Schweet, 1960; Dintzis, 1961). In conformity with this, we have seen that amino acids are joined to t-RNAs by their carboxyl groups so that, whatever the first amino acid, its amino group is free. Studies of *in vitro* protein synthesis have shown that peptide bond formation occurs between the carboxyl group of amino-acyl-t-RNA No. 1 and the amino group of amino-acyl-t-RNA No. 2. This results in the dipeptide becoming attached through its carboxyl group to t-RNA 2 thus, forming a peptidyl-t-RNA, while t-RNA 1 is released to pick up another amino-acyl-AMP complex (Fig. 73; Gilbert, 1963b). The dipeptide is now connected to the 2nd slot on the ribosome which, of course, is needed for attachment of amino-acyl-t-RNA No. 3, so that extension of the chain requires a return of the peptidyl-t-RNA to the 1st slot; at the same time the ribosome must move forward three nucleotides on the m-RNA so that the codon for amino acid 3 and slot 2 on the ribosome can together select and attach the proper amino-acyl-t-RNA in the same way as before (Fig. 73). Thus the ribosome moves along the m-RNA until it reaches the end when, presumably, it drops off and becomes available for reattachment to the same or another m-RNA molecule. How this movement is effected is quite unknown.

How is the completed polypeptide chain released from the ribosome-m-RNA complex? If every messenger RNA molecule coded for only a single polypeptide chain, as seems to happen in the case of haemoglobin synthesis by reticulocytes, the polypeptidyl-t-RNA would be released when the ribosome came to the end of the message, and the t-RNA could then be removed enzymically. However, it is probable that many m-RNA molecules carry the specifications for several distinct polypeptides so that this hypothesis is not a satisfying one. An alternative is that the genetic code embodies a sequence which is translated as 'end of protein' and this, as we shall see (p. 361), is probably the correct hypothesis. There are two specific triplets of bases, and probably a third, known as 'chain-terminating triplets', which can arise by mutation in normal codons. When such a mutation is present in a gene, N-terminal fragments only of the polypeptide chain are

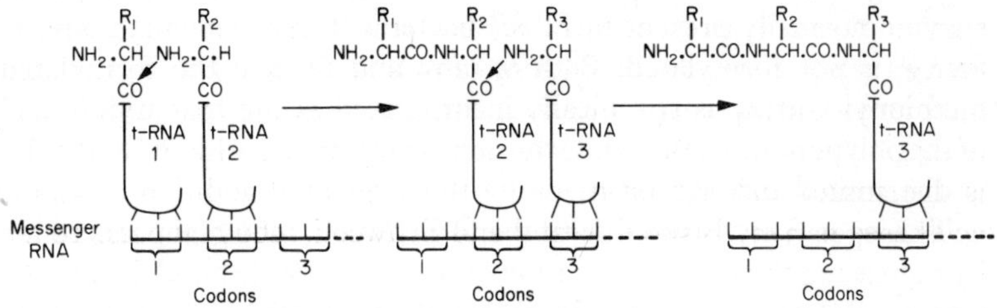

FIG. 73. Diagram illustrating the mechanism of peptide bond formation from amino-acyl-t-RNA during polypeptide synthesis. The three projections from the bottom of each ⌊ represent the three bases of the anticodon on the t-RNA molecules which recognise the appropriate codon on the m-RNA.

From Arnstein, 1965.

synthesised, and these can be shown to be homogeneous in length for any given mutation and to arise from an error in translation and not in transcription of the m-RNA; that is, the m-RNA is not truncated, only the protein (Sarabhai *et al.*, 1964). Assuming that one of these triplets is used for normal chain termination, how they operate is still speculative. A possible mechanism is that they attract specific t-RNAS carrying a special molecule which, either directly or indirectly, releases the chain (see Stretton, 1965).

Unlike chain termination, there did not seem to be any pressing reason why a mechanism for chain *initiation* should exist; the attachment of a ribosome to the 5′end of m-RNA should suffice, while in the case of polycistronic messenger the codon for the N-terminal amino acid of the second or third protein would lie adjacent to the chain-terminating triplet for the preceding protein. Moreover the fact that synthetic m-RNA containing only uracil residues yields polyphenylalanine in an *in vitro* protein-synthesising system, demonstrating that the code for phenylalanine is UUU (p. 356; Nirenberg and Matthaei, 1961), seemed to show directly that specific initiation is unnecessary, even though the rate of synthesis is low; in contrast, in the absence of chain-terminating triplets the polyphenylalanine is not liberated but remains bound to the polysomes.

Despite these arguments, a chain-initiating device has recently been discovered. It turns out that, in *E. coli* at least, there are two kinds of t-RNA which can form amino-acyl-t-RNA complexes with methionine. When methionine is attached to one of these, termed t-RNA$_F$, its amino group can be formylated and blocked (HCO—NH) by an enzyme normally present in *E. coli* bacteria. The other, methionyl-t-RNA$_M$, is not formylated. Both *in vitro* and *in vivo* the formylated methionyl-t-RNA$_F$ is specifically incorporated as the first amino acid of the polypeptide chain, while the non-formylated methionyl-t-RNA$_M$ is distributed throughout it.

This specificity is not due to the formylation, which appears to act by increasing the rate of formation of the first peptide bond, but to the t-RNA$_F$ molecule. This can be shown in two ways. First, in *in vitro* synthesis under conditions where formylation cannot occur, the non-formylated methionyl-t-RNA$_F$ complexes nevertheless continue to initiate. Secondly, using synthetic trinucleotides of defined composition as m-RNA in association with ribosomes (see p. 357), it can be shown that whereas methionyl-t-RNA$_M$, which cannot be formylated, binds only to the trinucleotide AUG, which is the sole methionine codon, methionyl-t-RNA$_F$ binds well to GUG and UUG in addition (see Table 15, p. 358).

The most interesting feature of these two methionyl-t-RNA molecules is that they can be shown, by their different response to the antibiotic puromycin, to attach to different slots on the ribosome. You remember that the polypeptide chain grows by the sequential attachment of each new amino-acyl-t-RNA to slot No. 2 of the ribosome, which is therefore the amino-acyl-t-RNA slot. Once the new amino acid has been connected to the chain by a peptide bond, the resulting peptidyl-t-RNA shifts back to slot No. 1. Puromycin has a close structural resemblance to the terminal amino-acyl adenosine of t-RNA and interferes with protein synthesis, both *in vivo* and *in vitro*, by inserting itself at ribosomal slot No. 2 where it becomes attached by its amino group to the carboxyl end of the polypeptide chain, which is thereupon released from the ribosome with the puromycin still bound to it (review: Collins, 1965). When the two methionyl-t-RNAs were tested for sensitivity to puromycin, that is, whether they are or are not released from m-RNA-ribosome complexes by the drug, it was found that the methionyl-t-RNA$_F$, which can be formylated, is sensitive when bound to ribosomes with AUG trinucleotides, while methionyl-t-RNA$_M$ is relatively resistant. This strongly suggests that methionyl-t-RNA$_F$, unlike all other amino-acyl-t-RNAs, has an initial affinity for the ribosomal slot No. 1 which normally accomodates only peptidly-t-RNAs, and so alone is able to initiate the chain (Clark and Marcker, 1966a, b; Bretscher and Marcker, 1966).

THE BIOSYNTHESIS OF RNA

RNA *polymerases*

We have seen that the cell produces at least three distinct varieties of RNA molecules, clearly distinguishable on the basis of size, composition and function. Moreover each of these varieties is itself heterogeneous, comprising a number of specific molecular types. Thus there are potentially as many types of messenger RNA as there are genes in the genome, while more than 20 specific types of transfer RNA are known to exist. Even ribosomal RNA has at least two molecular types which, because of the different sizes and functions of the complexes they form with protein, may also be assumed to be very specific structures. Two questions may therefore be posed: How are the specificities of all these molecular types of RNA determined, and how are they built up within the cell?

The demonstration by Hall and Spiegelman (1961) that DNA-RNA 'hybrid' duplexes are formed when m-RNA is mixed with homologous, denatured DNA (p. 276) provided good evidence for assuming that m-RNA is synthesised directly on a chromosomal DNA template, from which it transcribes a base sequence complementary to that of one of the DNA strands—presumably in much the same way that a new DNA strand is built up along an old strand during DNA replication. When ribosomal RNA and transfer RNA were similarly tested for hybridisation, by a more sensitive technique making use of the fact that DNA/RNA hybrid molecules are resistant to the action of ribonuclease, it turned out that their base sequences are also determined by regions of chromosomal DNA (Yankofsky and Spiegelman, 1962a, b; Giacomoni and Spiegelman, 1962). Furthermore, it could be shown by the absence of competitive inhibition in hybrid formation, that the 23S and 16S r-RNA components are specified by distinct chromosomal regions (Yankofsky and Spiegelman, 1963). Thus we must now extend our concept of the gene to include the determination of RNA as well as of protein structure. Quantitatively, about 0·3 per cent of the genome of *E. coli* is occupied with the specification of r-RNA, and about 0·02 per cent with that of t-RNA which is enough to code for about 50 distinct types of t-RNA molecule (p. 364; Spiegelman and Hayashi, 1963).

The actual synthesis is accomplished by an enzyme which has now been isolated in a highly pure state and which is variously known as

RNA nucleotidyl-transferase, DNA-dependent RNA polymerase, RNA polymerase and (specifically by Spiegelman) 'transcriptase'. This enzyme has been identified in a wide range of bacteria and animal cells and is very similar to DNA polymerase in its properties (p. 247) (Weiss, 1960; Hurwitz, Bresler and Diringer, 1960; Weiss and Nakamota, 1961; Hurwitz *et al.*, 1961; Hurwitz *et al.*, 1963; Chamberlin and Berg, 1963; Smellie, 1963, 1965).

The synthetic activity of RNA polymerase depends on the presence of all four ribonucleoside 5′-triphosphates (ATP, GTP, CTP and UTP), of Mg^{++} or, preferable, Mn^{++} ions, and of DNA as a template to prime the reaction. Either native or single-stranded (phage ϕX 174) DNA can act as templates but the former appears to promote more intensive synthesis; in fact, it has been reported that RNA can also prime the reaction (Chamberlin and Berg, 1963; Hurwitz *et al.*, 1963). The product is a single-stranded RNA molecule whose base composition varies with that of the template DNA. When a synthetic DNA primer is used in which adenylic and thymidylic residues alternate (poly AT; p. 249), the RNA product incorporates only adenine and uracil; again, the use of single-stranded ϕX174 DNA yields RNA with a complementary base composition. But the most convincing evidence that RNA polymerase mediates transcription of the DNA base sequence into a complementary RNA sequence comes from hybridisation experiments. These have shown unambiguously that the product of *in vitro* RNA synthesis from a double-stranded DNA template consists of two complementary RNA strands which can specifically form hybrid duplexes with both strands of the template DNA, as well as with one another to form RNA double helices.

There is one difference between the synthesis of RNA by RNA polymerase and the replication of DNA in that, in the former reaction, the double-stranded template DNA is conserved intact, the RNA product appearing free in the medium. On the contrary, when the DNA template is single-stranded the first product appears to be a DNA/RNA hybrid duplex which then acts as a template for the synthesis of further single-stranded RNA (Chamberlin and Berg, 1964; Sinsheimer and Lawrence, 1964).

As we shall see when discussing the genetic code, mutational studies have shown that the genetic message is carried by only one of the two strands of the chromosomal DNA, while lack of equivalence between complementary bases in m-RNA strongly implies that it is derived from the transcription of a single DNA strand (p. 368). Direct confirmation of this comes from experiments involving a virulent *B. subtilis* bacteriophage, SP8, which has double-stranded DNA of which one strand, rich in pyrimidines, is more dense than the other, purine-rich strand. Separate preparations of the two strands can thus be made by denaturing the DNA and then centrifuging it in a caesium chloride density gradient (p. 243). When phage m-RNA, isolated from infected cells, is mixed in turn with the preparation of each DNA strand, it is found to form hybrids with one of them only (Marmur *et al.*, 1963).

How can this be reconciled with the *in vitro* finding, mentioned above, that two complementary RNA strands are transcribed from a double-stranded DNA template? A possible clue comes from an *in vitro* study of RNA polymerase activity when various forms of phage ϕX174 DNA are used as primer. In infected host bacteria the single-stranded DNA of this phage acquires a complementary strand and adopts a circular structure, the duplex constituting the replicative form (RF) of the DNA (p. 439). When this circular, double-stranded DNA is used as template, it turns out that the RNA produced is complementary to only one of the two DNA strands—the new one; but if the circular configuration of the template is broken by mechanical shear, then both strands are copied (Hayashi, Hayashi and Spiegelman, 1964). However, the situation may be more subtle and complicated than this suggests since RNA polymerase from *B. megaterium* has been shown to transcribe only one strand of double-stranded DNA from a particular *B. megaterium* phage irrespective of whether the DNA is broken by shear or not, while purified RNA polymerase from *M. lysodeikticus* transcribes both strands of this same DNA (Geiduschek, Tocchini-Valenti and Sarnat, 1964). We will discuss this topic further on p. 368 *et seq.*

In addition to the DNA-dependent RNA polymerases, the synthesis of another group of RNA polymerases, specifically dependent on an RNA template, has been found to be produced by infection, both of bacteria and of animal cells, with RNA viruses (August *et al.*, 1963; Baltimore and Franklin, 1963; Spiegelman and Doi, 1963; Weissman *et al.*, 1963). Since the enzyme does not pre-exist in the infected cell but is determined by the virus genome, the infecting RNA strand must first serve as messenger-RNA for the enzyme to be synthesised. In fact it turns out, in the case of at least one bacterial RNA virus, that the infecting parental strands are completely conserved throughout the lytic cycle, so that instability is not a necessary property of m-RNA (Spiegelman and Doi, 1963). An interesting and important feature of the RNA-dependent polymerases produced by certain unrelated bacterial RNA viruses is that they are not only inactive with host t-RNA or r-RNA templates, but show a high degree of specificity for homologous viral RNA (Haruna and Spiegelman, 1965). How this recognition is effected is not yet understood.

RNA *phosphorylase*

The first isolation of an enzyme capable of catalysing *in vitro* RNA synthesis was by S. Ochoa and his group (Grunberg-Manago, Ortiz and Ochoa, 1956; Ochoa and Heppel, 1957). This enzyme, polynucleotide phosphorylase, is present in a number of bacterial species, including *E. coli*, and acts quite differently from the polymerases we have described above. Ribonucleoside 5′-*di*phosphates (instead of -triphosphates) are required as substrates, the reaction takes place in the complete absence of either DNA or RNA templates, while the presence of all four bases is not obligatory for the synthesis. In fact, if only one kind of 5′-ribonucleotide is present the enzyme will build up a polynucleotide chain containing this base alone. Such a chain containing only riboadenylic acid or ribouridylic acid is briefly designated as 'poly A' or 'poly U' respectively. In the presence of two or more kinds of ribonucleoside 5′-*di*phosphate, polynucleotides are synthesised which contain the bases in about the same proportion as they were present in the mixture. In addition, the results of nearest-neighbour sequence analysis (p. 250) suggest that the bases are coupled together in a random way. A synthetic polynucleotide containing all four bases displays all the physical and chemical properties of natural RNA and is referred to as 'poly AGUC'.

Since the structure of all the main categories of cellular RNA is now known to be coded in the DNA and to be transcribed by RNA polymerase, it is difficult to postulate a function for the phosphorylase. Indeed it has been suggested that it may operate in the breakdown, rather than in the synthesis, of RNA. In any case the enzyme cannot be of great importance to the cell since a viable mutant of *E. coli* which lacks it has recently been reported (Overby *et al.*, 1966). However, the molecular biologist has no cause for complaint on this account for, paradoxically, polynucleotide phosphorylase has proved an invaluable tool for preparing artificial polynucleotides of predetermined base composition which have played a key role in solving the genetic code (p. 356).

THE TOPOGRAPHY OF PROTEIN SYNTHESIS

When we consider the various mechanisms of protein synthesis piecemeal, as we have done so far, each appears adequate for its purpose and seems to form a consistent link in the chain of events. But when we begin to think about the spatial interactions between these mechanisms, difficulties arise. For instance, how is the nascent m-RNA shed by the DNA from which it is transcribed, and how does it find a ribosome to translate its message? If m-RNA and r-RNA are transcribed from the DNA in exactly the same way, what determines the difference in their subsequent history? At what stage in the synthesis of a polypeptide chain does folding commence and functional activity appear?

Kinetic and other studies of *in vitro* RNA synthesis have shown that this differs from natural synthesis in having a much lower rate and in the fact that each polymerase molecule makes only a single RNA molecule which remains bound to the DNA template by a protein component which may be the enzyme itself (Bremer and Conrad, 1964), suggesting that the *in vitro* system lacks some important factor present in the cell.

Stent (1966) has suggested that this factor may be the ribosomes, which play a role in RNA as well as in protein synthesis by attaching to the nascent m-RNA as it emerges from the polymerase, so that polysomes are formed and protein synthesis initiated before formation of the m-RNA molecule is complete. In this way the movement of the ribosomes could provide the energy as well as the mechanism for peeling the RNA from the DNA-polymerase complex. If this is so, *in vivo* complexes containing DNA, RNA, polmyerase and ribosomes should be found. Such complexes have in fact been shown to exist by infecting *E. coli* bacteria with virulent (T-even) bacteriophage, labelling the newly synthesised phage DNA for a time with ^{32}P, and then adding a large excess of non-radioactive phosphate together with ^{3}H-uridine, so that ^{32}P label in the labile, preformed m-RNA would be flushed away and newly formed m-RNA molecules tagged with ^{3}H. The cells were then lysed by freezing and thawing and the lysate, without further treatment, sedimented in a sucrose density gradient. Peaks of ^{3}H activity were found to correspond to sedimentation constants of 70s, 105s and 130s, indicating m-RNA attached to ribosomes and polyribosomes (p. 278), while an appreciable proportion of the DNA ^{32}P label appeared in the ribosome region, with a peak at 130s (Stent, 1966).

Although r-RNA extracted from ribosomes is not an effective stimulant of *in vitro* protein synthesis it has been reported that nascent r-RNA *is* effective (Otaka, Osawa and Sibatani, 1964);. Accordingly, it has been suggested that nascent r-RNA may act as messenger-RNA for ribosomal protein, eventually condensing with its own protein subunits to form a ribosomal component. Thus the newly formed r-RNA would be treated in the same way as m-RNA, being stripped from the DNA-polymerase complex by the movement of ribosomes (Stent, 1966). How transfer-RNA molecules might be dealt with within this scheme is not clear.

An important pre-requisite for this model is that the direction of synthesis of messenger-RNA and that of translation of the message on the ribosome must be the same; that is, the end of the RNA chain to emerge first from the polymerase must be the end to which the ribosomes can attach. We have already seen that the ribosomes attach to, and m-RNA is read from, the 5′ end (p. 285), and since it has also been shown that m-RNA synthesis begins at the same end (Bremer *et al.*, 1965), this necessary condition is fulfilled. We will conclude by mentioning that induced β-galactosidase activity has been demonstrated in association with ribosomes, so that folding of the polypeptide chain probably begins before its synthesis is complete (Kiho and Rich, 1964).

CONCLUSIONS

In this Chapter we have attempted to summarise the present state of our knowledge of the relationship between the nucleic acids and protein synthesis. We hope we have succeeded in fitting together all the major pieces of the jig-saw puzzle so that, although some pieces are still missing, the clear outline of a meaningful picture can be seen. This picture is reproduced in diagrammatic form in Fig. 74. Here is a simple commentary.

The shape and specific activity of a protein molecule is a function of the sequence of amino acids in the polypeptide chain, or chains, of which it is composed. The genetic information specifying this sequence is carried by a region of chromosomal DNA, the gene, in the form of a chemical code. This code is based on the sequence of the four nucleotide bases of the DNA; that is, it is a code of four letters, various combinations of which spell the words indicating each of the twenty amino acids. Thus the sequence of bases along the whole region of DNA lays down the sequence of amino acids in the protein.

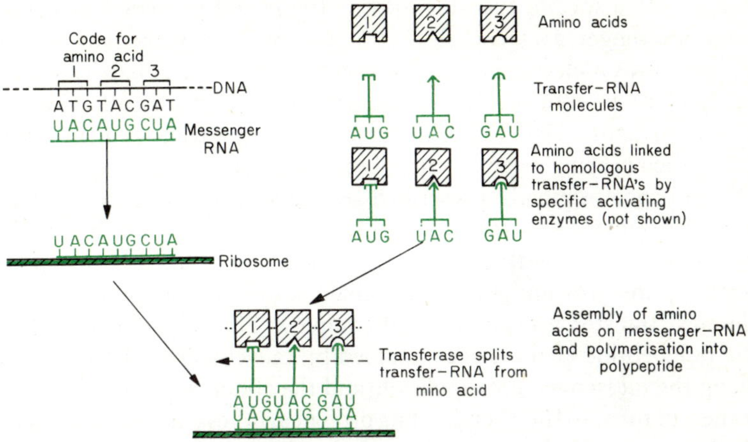

FIG. 74. Schematic representation of the mechanism of protein synthesis.

RNA molecules are shown green.

Bases are indicated by letters: A = adenine, G = guanine, T = thymine (DNA only), C = cytosine, U = uracil (RNA only); hydrogen bonding is restricted to the base pairs A-T, A-U and G-C. Amino acids are indicated by figures (see text).

Proteins are synthesised in association with the ribosomes which are relatively stable, cytoplasmic particles composed of protein and RNA. Any ribosome can serve as a work-bench for the assembly of any protein, and a ribosome which has been the site of synthesis of a particular protein can subsequently participate in the synthesis of a different protein.

The genetic information for protein synthesis is carried from genes to ribosomes by molecules of messenger (template) RNA which, in bacteria at any rate, have a high rate of turnover. Messenger RNA picks up its message in the form of a base sequence complementary to that of the DNA, by being synthesised, from the 5′ to the 3′ end, along one (or possibly both) of the DNA strands in the presence of an enzyme, RNA polymerase. Ribosomes attach to messenger RNA molecules at the 5′ end which is the first to be synthesised, and there is some evidence that this may happen before the synthesis of the RNA molecules is complete.

Amino acids form complexes with adenylic acid, in the presence of ATP and specific activating enzymes. Each activated amino acid then unites with a specific transfer RNA molecule and is carried to a ribosome-messenger RNA complex. It is assumed that some part of each transfer RNA molecule is characterised by a sequence of bases complementary to a sequence on the messenger RNA which codes for the particular amino acid which it carries. Thus the transfer RNA molecules, each bearing its amino acid, become arranged on the template in the sequence originally prescribed by the DNA of the gene. In fact, only one transfer RNA molecule, with its amino acid, is attached to the ribosome at any one time; only when the amino acid has been transferred to the growing polypeptide chain is the ribosome accessible to the next transfer RNA-amino acid complex. This stepwise growth of the polypeptide chain is accompanied by movement of the ribosome along the messenger RNA molecule and the attachment of other ribosomes, in turn, to the 5′ end. Thus protein synthesis is associated with the formation of polyribosomes, with the result that many polypeptide chains are being synthesised at the same time on any given messenger RNA molecule. When a ribosome reaches the end of the RNA chain it falls off and the completed protein attached to it is released. It is almost certain, at least in bacteria, that some messenger RNA molecules carry genetic information specifying several proteins. In such cases it is probable that the end of each protein is indicated by a 'nonsense' codeword in the message, which results in release of the protein from the ribosome. The base sequences of three such 'chain-terminating' codewords are already known. (Reviews: Arnstein, 1965; Attardi, 1967; Hayes, 1967; McQuillen, 1965; Smellie, 1965; Stacey, 1965a.) Rich sources of information on nucleic acids and protein synthesis are: Cold Spring Harbor Symposium, 'Synthesis and structure of macromolecules', Vol. 28, 1963, published by the Cold Spring Harbor Laboratory of Quantitative Biology, L.I., New York; Progress in Nucleic Acid Research, eds. Davidson & Cohn, 1963a,b, 1964, 1965, 1966.

CHAPTER 13

DNA as Genetic Material: The Nature of Mutation

Earlier in this book we reviewed the methods and findings of genetical *analysis*, whereby a series of progressively refined models of the structure and behaviour of the genetic material were inferred from quantitative observations of the inheritance of characters in crosses. The ultimate model to emerge from this approach defined the gene as a heritable unit of function and revealed it as a small, linear segment of chromosome which specifies the structure of a single polypeptide chain. The gene, in turn, is made up of a large number of different mutational sites which are separable by recombination. In the last few chapters on the physico-chemical mechanisms of heredity we have looked at inheritance from the novel viewpoint of what might be called *genetical synthesis*. Starting from a knowledge of how the molecular constituents of the genetic material, as well as those of the proteins it specifies, are put together, we have tried to construct a working model of genetic phenomena at the molecular level.

THE MOLECULAR SCALE OF RECOMBINATION ANALYSIS

How can these two models of the genetic material, the analytical and the synthetic, be equated so that we can translate the abstract results of genetic analysis into concrete molecular terms? In principle the method is simple. Genetic analysis shows that the genetic material of both phage T4 particles and of *E. coli* nuclei constitutes a single chromosome. It is assumed that all the DNA of the particle or cell is present in this chromosome and that it exists in the form of a single double helix; as we shall see, there is convincing evidence that this is so in the case of both phage (p. 535) and bacteria (p. 550). Thus by extracting the DNA from a phage or bacterial suspension, the amount of DNA per chromosome can be estimated by dividing the total amount of

DNA by the number of phage particles, or by the number of bacteria multiplied by the mean number of nuclei per bacterium. In this way the *length* of DNA double helix which comprises the chromosome can be measured in terms either of Ångstrom units or, with more meaning from the genetic point of view, of the number of pairs of bases it contains. It turns out that the chromosome of phage T4 contains about $1\cdot5\times10^5$ base pairs while that of *E. coli* is some 3×10^6 base pairs long. It is worth noting that the actual extended length of DNA double helix within a phage particle or bacterium is of the order 1000 times longer than that of the particle head or cell which contains it.

The next step is to estimate the total length of the chromosome in terms of percentage recombination, or map units (1 map unit = 1 per cent recombination), by summating the map distances between each pair of a large number of genetic loci distributed along it. As has already been explained (pp. 47, 52, 144), the occurrence of double cross-overs, which are not revealed as recombinants, results in total map lengths in excess of 100 per cent (100 map units). In fact, the genetic length of a chromosome is directly proportional to the frequency of recombination per unit of its length. Thus the total length of the chromosome of phage T4 has been estimated as about 800 map units, and of *E. coli* as approaching 2000 map units.

Knowing the length of the chromosome in nucleotide pairs of DNA and in units of recombination, it is easy to relate the one kind of unit to the other. Thus in the case of phage T4

800 per cent recombination (map units) $= 1\cdot5\times10^5$ nucleotide pairs.

Therefore, 1 per cent recombination $= \dfrac{1\cdot5\times10^5}{800} = \text{c.}\,2\times10^2$ nucleotide pairs.

Similarly with *E. coli*,

2000 per cent recombination $= 3\times10^6$ nucleotide pairs.

Therefore, 1 per cent recombination $= 1\cdot5\times10^3$ nucleotide pairs.

From this it should be simple to translate any recombination frequency as, for example, that between the closest recombining sites within a cistron, into molecular distances. In doing this, however, we should bear in mind that we have been forced to make certain assumptions whose validity is uncertain. The two most important of these are that all the DNA in the chromosome is genetically functional, and that negative interference over short chromosomal regions is not significant. The falsity of either of these assumptions would yield an over-estimate of the number of nucleotide pairs which separate the recombining sites.

Benzer (1957) was the first to attempt a correlation between genetical and physical distances, as the ultimate result of his analysis of the fine genetic structure of the r_{II} region of bacteriophage T4. We have seen (p. 171) that the genetic length of the A and B cistrons of this region are 5 per cent and 3 per cent of the total map length and, according to the computation set out above, are therefore equivalent to about 1200 and 700 nucleotide pairs respectively. Again, Benzer found that crosses between independently isolated r_{II} mutants either yield no recombinants or give frequencies of 0·01 per cent or over, although his system is sensitive enough to detect recombination at a very much lower frequency. Thus, in this system, the smallest distance that can be resolved by genetic analysis is about 0·01 per cent recombination. Since 1 per cent recombination is equivalent to about 2×10^2 nucleotide pairs (see above), 0·01 per cent recombination represents the distance between two sites only 2 nucleotide pairs apart. When we consider that this calculation rests on assumptions which tend to set an upper limit to this figure, it seems likely that recombination is ultimately capable of separating adjacent base pairs on the DNA double helix (see below).

There are other ways in which we can juggle with genetical and physico-chemical data in order to establish correlations between them. We have seen from Benzer's studies that each of the two cistrons of the r_{II} region of phage T4 is of the order 1000 nucleotide pairs long. Unfortunately the nature of the products determined by these cistrons remains quite unknown, although they are assumed to be proteins. However, similar estimates may be made for other cistrons, both in phage (see Streisinger *et al.*, 1961) and in bacteria (Garen, 1960), which specify proteins of known composition. Assuming that the sequence of amino acids in such proteins is laid down by the base sequence in the DNA of the gene, the average number of base pairs which code for a single amino acid, known as the *coding ratio*, can be calculated. For example, if a gene is estimated to constitute about 1000 nucleotide pairs of DNA and this gene carries the genetic information for the synthesis of a polypeptide composed of some 300 amino acids, then each amino acid is specified by a sequence of 3 nucleotide pairs. As we shall see in the next Chapter, there is very good reason for believing that the true coding ratio is, in fact, 3. This can be used as the basis for a number of new inferences. For example, on the supposition that the average biological polypeptide contains about 300 amino acids, then the average gene is about 1000 nucleotides long which is the same figure as that derived from rather different data. Assuming that all the DNA is genetically meaningful, this indicates that the DNA of an *E. coli* chromosome, containing 3×10^6 base-pairs, is sufficient to determine the synthesis of about 3000 different polypeptides, and that of phage T4, with 1.5×10^5 base pairs, about 150 polypeptides; similarly the DNA of the small phage, ϕX174 (p. 436), which contains only 1.6×10^3 nucleotides, can code for no more than a few proteins.

By comparing the results of sequence analysis ('fingerprinting') (p. 265) of protein derived from wild type and mutant strains it has been shown, for such divergent material as human haemoglobin (Ingram, 1957), the protein of tobacco mosaic virus (Tsugita and Fraenkel-Conrat, 1960, 1962; Wittman, 1962) and tryptophan synthetase from *E. coli* (Helinski and Yanofsky, 1962), that a single mutation may result in the substitution of a single amino acid by another, in the sequence of amino acids that constitute the protein. In the case of the A component of tryptophan synthetase from *E. coli* (pp. 119, 155), two independent mutations have been reported, one of which results in the replacement of a particular glycine residue by arginine, and the other in the substitution of the *same* glycine residue by glutamic acid. Yet when these two mutant strains are crossed by means of transduction, wild type recombinants arise at a very low frequency (Helinski and Yanofsky, 1962). According to the theory of mutagenesis based on the Watson-Crick model of DNA structure, mutation may result from the substitution of one base-pair by another, so that the code word embodied in the nucleotide sequence is altered (p. 238). The new code word may be meaningless or may stand for another amino acid. If we assume that a sequence of three bases or pairs of bases, which we can designate ABC, codes for glycine, then the arginine substitution must have followed a change in one of these three letters (for example, to BBC) and the glutamic substitution a change in another (to ACC or ABD, for example). Given a coding ratio of 3, therefore, the occurrence of wild type recombinants in crosses between these mutants can only be explained by crossing-over between adjacent base-pairs, thus:

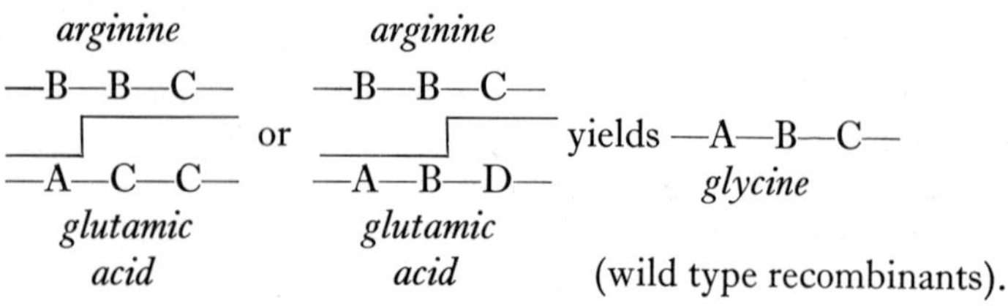

This exemplifies a novel feature of recombination within a coding unit, or *codon*, since the recombinant is not constructed from a re-assortment of the *character* differences of the two parents; it possesses a nucleotide sequence in the coding unit and a corresponding amino acid in its protein which are distinct from those of either parent.

If there are four nucleotides, and the coding unit for an amino acid comprises a sequence of three of these, then there are nine possible ways in which a coding unit can be changed as a result of a single mutation, assuming that a particular base can be replaced by any one of the remaining three. For example, the first base of the wild type triplet ABC could be substituted by B, C or D to give BBC, CBC or DBC, and similarly for the second and third base of the triplet. If we imagine that BBC, CBC and DBC, for example, each codes for a different amino acid, it is evident that three distinguishable mutant types could arise from mutation at the *same mutant site*. Unlike the case mentioned above, however, no wild type recombinants could arise from crosses between these mutants.

Yanofsky *et al.* (1963) have recently examined in great detail the amino acid substitutions, in a single peptide of the A protein of *E. coli* tryptophan synthetase, which result from mutations at the same, or very closely linked, genetic sites. Their work has revealed six different amino acid replacements *at the same position in the protein* as a result of mutations within a single coding unit. They have also shown that three of these mutants stem from alterations at the same genetic site, so that all the possible base substitutions there are accounted for.

An interesting feature of this work is that the mutational reversion of a mutant to the normal, wild phenotype is not necessarily correlated with restoration of the wild type amino acid in the protein. For example, in one mutant, a glycine residue in the wild type protein was replaced by glutamic acid. A revertant of this mutant was isolated which yielded normal enzyme activity, as a result of replacement of glutamic acid by alanine, instead of by the original glycine. A number of partial revertants, producing sub-optimal levels of enzyme activity, were also found. In some of these a third amino acid (such as valine), different from either mutant or wild type, had been substituted at the same position in the peptide.

One partial revertant, however, was found to be due to a second mutation in the same gene, but located some distance from the site of the original mutation. Analysis of the A protein from this revertant showed that the glutamic acid residue which characterised the mutant was retained but that, in another peptide, tyrosine had been changed to cysteine. Thus a second substitution at some distance from the first in the mutant protein had restored some of its functional activity. By crossing this revertant (glutamic acid+cysteine) to wild type (glycine+tyrosine), a recombinant strain was isolated which formed an A protein containing the glycine of the wild type in association with the cysteine of the revertant; this protein was functionally inactive. This not only demonstrates the complexities of protein structure (p. 264), since two different amino acid substitutions which, alone, are detrimental to function can restore it in combination, but also reveals the way in which one type of suppressor mutation may operate (p. 213).

THE MOLECULAR THEORY OF MUTAGENESIS

The concept of DNA as the genetic material and the elucidation of its structure by Watson and Crick offered, for the first time, a firm foundation on which to build a rational theory of mutagenesis. The only irregularity in the DNA polynucleotide which can possibly carry genetic information is the unrestricted sequence of bases along each chain of the double helix. From this it follows that the most likely mechanism of mutation is the substitution of one base by another so that the sequence of bases is altered. You may remember that Watson and Crick (1953a) suggested that spontaneous mutation might result from a tautomeric shift in the position of a hydrogen atom on adenine (A), for example, leading to improper pairing of this base with cytosine (C) instead of with thymine (T); at the next replication the cytosine would pair normally with guanine (G) so that the outcome would be substitution of an A-T base-pair by G-C at a specific point in the DNA chain (Fig. 57c, d, p. 235).

In the following discussion we shall not consider those gross alterations of the genetic material which manifest themselves as deletions, inversions, and so on, but only reverting, single-site mutations which are assumed to involve single base-pairs.

MUTAGENESIS BY BASE ANALOGUES AND OTHER SPECIFIC MUTAGENS

An ideal test of the mutation hypothesis would be to show by direct base sequence analysis that in the DNA from a gene which has suffered a single mutation, just one specific base-pair has been substituted by another. Since this is technically impossible at present one must resort to indirect methods. Among the most promising of these is the use of certain groups of chemicals which are not only highly mutagenic but are also known to interact with the purine and pyrimidine bases of DNA in very specific ways. These chemicals fall into three main categories. These are the base analogues which can replace the normal bases of DNA during replication, substances which chemically alter the bases of resting DNA, and those whose action is to remove DNA bases.

1. *Base analogues*. These are close analogues of normal nucleic acid bases which are, or can reasonably be expected to be, incorporated into DNA without destroying its capacity for replication. However, because the analogue differs from the normal base in the distribution of its hydrogen atoms, it has a greater tendency than normal to improper pairing. The most potent mutagens among the base analogues are 5-bromouracil (BU) and 5-bromo-deoxyuridine which are analogues of thymine and thymidine respectively (Fig. 57, p. 235), in which the methyl group at the 5-position is substituted by a bromine atom, and 2-aminopurine (AP) which is similar to adenine (Fig. 57) except that the amino group is attached to the 2- instead of to the 6-position.

5-Bromouracil is incorporated into the DNA of both bacteria and phage, replacing thymine quantitatively, so that a proportion of A-T base pairs is replaced by A-BU pairs. Its mutagenic behaviour was thought to reflect its greater probability than thymine of accidentally pairing with guanine, due to unstabilisation of its hydrogen bonds by the bromine atom which has a higher electro-negativity than the substituted methyl group (Freese, 1959a). That BU does, in fact, pair with guanine much more frequently than does thymine has recently been shown by means of studies of DNA synthesised on artificial primers in the Kornberg *in vitro* system (p. 245). When a DNA-like primer containing only adenine and thymine residues is used, no incorporation of guanine in the replica DNA is detected; but when BU is substituted for thymine in the primer, the newly synthesised DNA contains guanine residues in the proportion of one per 2000 to 25,000 adenine and thymine nucleotides incorporated (Trautner, Swartz and Kornberg, 1962). However, 'nearest-neighbour sequence' analysis (p. 250) of this DNA reveals that only about 40 per cent of the guanine is incorporated opposite BU, instead of the 100 per cent anticipated by current theory (Fig. 75).

2-Aminopurine (AP), like adenine, pairs normally with thymine by two hydrogen bonds (Fig. 76a). Like adenine, AP may occur in the rare imino-state in which it can form two hydrogen bonds with cytosine (Fig. 76b). Unlike adenine, however, it can also form a single hydrogen bond with cytosine in its normal state (Fig. 76c) so that it may be expected to pair frequently with cytosine (Freese, 1959a, b). Although an efficient mutagen, its incorporation into DNA is very slight.

2. *Substances whose chemical action alters the nucleic acid bases of resting* DNA in such a way that their specific pairing with other DNA bases is changed. The most important of these substances is *nitrous acid* (HNO_2). Unlike base analogues whose mutagenic action depends on their incorporation during DNA replication, nitrous acid is highly

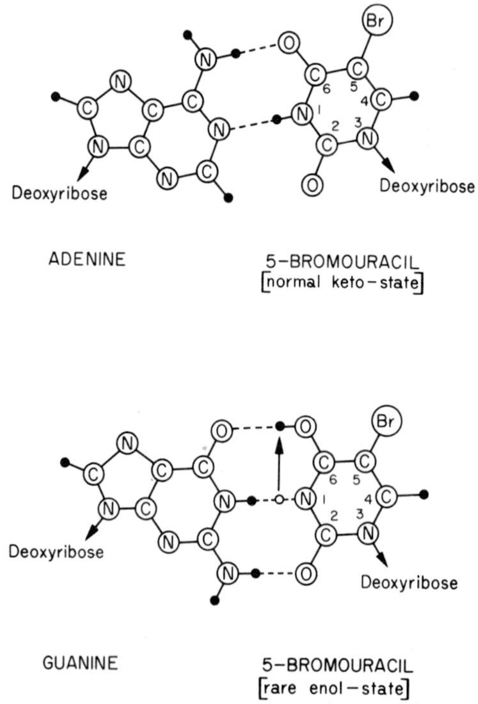

FIG. 75. The base-pairing attributes of 5-bromouracil.
Top: in the normal keto-state, with a hydrogen atom in the N1 position, bromouracil bonds to adenine.
Bottom: in the rare enol-state, a tautomeric shift of this hydrogen atom determines specific pairing with guanine.

The hydrogen atoms are represented as small, black circles, and the hydrogen bonds by interrupted lines.

mutagenic not only for free phage particles (Tessman, 1959; Vielmetter and Schuster, 1960; Bautz-Freese and Freese, 1961) but also for isolated DNA in pneumococcal transformation (Litmann and Ephrussi-Taylor, 1959) and for purified RNA derived from tobacco mosaic virus (Mundry and Gierer, 1958).

Nitrous acid acts directly on nucleic acids by oxidatively deaminating their bases. Thus adenine is deaminated to hypoxanthine which pairs

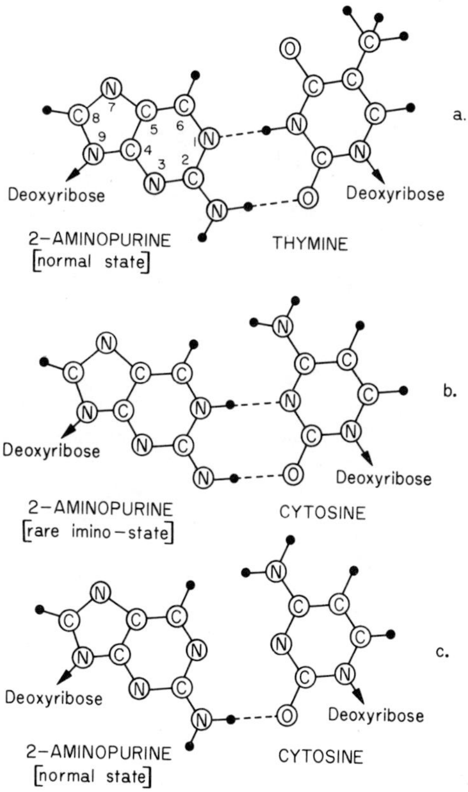

FIG. 76. The base-pairing attributes of 2-amino purine.

a. Normal pairing with thymine by means of two hydrogen bonds.

b. Pairing of the rare imino-form with cytosine, as a result of a tautomeric shift of a hydrogen atom to the N1 position.

c. Pairing of the normal form with cytosine by means of a single hydrogen bond.

The hydrogen atoms are represented as small, black circles, and the hydrogen bonds by interrupted lines.

specifically with cytosine instead of with thymine (Fig. 77a), while cytosine is deaminated to uracil which pairs specifically with adenine (Fig. 77b). Guanine, on the other hand, is changed to xanthine which continues to pair with cytosine, but only by two instead of by three hydrogen bonds (Fig. 77c). Since thymine (as well as its uracil equivalent in RNA) has no amino group, it remains unchanged by treatment. We may thus expect that mutations induced by nitrous acid are predominantly due to its action on adenine and cytosine.

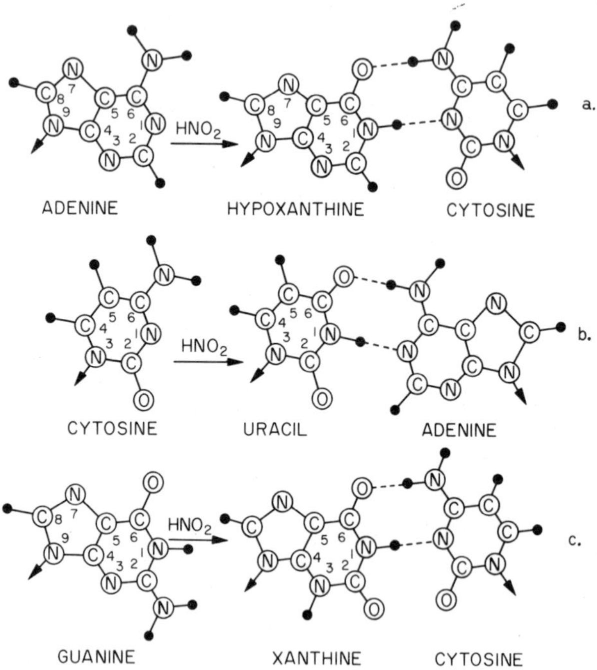

FIG. 77. The oxidative deamination of DNA bases by nitrous acid, and its effects on subsequent base pairing.

a. Adenine is deaminated to hypoxanthine which bonds to cytosine, instead of to thymine.

b. Cytosine is deaminated to uracil which bonds to adenine, instead of to guanine.

c. Guanine is deaminated to xanthine which continues to bond to cytosine, though with only two hydrogen bonds.

Thymine, and the uracil of RNA, do not carry an amino group and so remain unaltered.

The hydrogen atoms are represented as small, black circles, and the hydrogen bonds by interrupted lines.

Another mutagen belonging to this class is *hydroxylamine* (NH_2OH) which is important because it appears to react significantly only with cytosine, which is thought to be altered in such a way that it not only pairs specifically with adenine (Fig. 78), but the N1 hydrogen interferes sterically with guanine pairing. Hydroxylamine has a negligible *in vitro* action on thymine nucleotides and is without effect on the purines (Freese *et al.*, 1961a,b).

3. *Substances or treatments which remove bases from* DNA. The most important of these are the alkylating agents which include the sulphur and nitrogen mustards and ethylene oxides, as well as such more recently developed and much less toxic mutagens as ethyl ethane-sulphonate (EES) and ethyl methanesulphonate (EMS) (Loveless, 1958; Loveless and Howarth, 1959). Alkylating agents appear to act specifically on guanine at the N7 position (Fig. 57, p. 235), labilising the deoxyriboside linkage so that 7-alkylguanine is released from the DNA, even at 37°C. and neutral pH (Brookes and Lawley, 1960); this offers

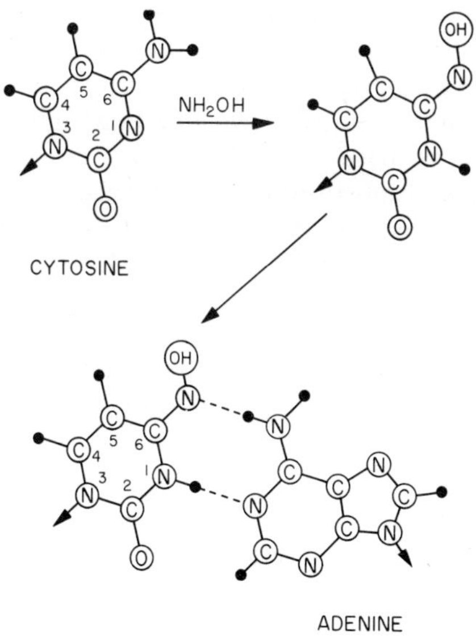

FIG. 78. The postulated action of hydroxylamine on cytosine.

The base formed by the action of hydroxylamine pairs specifically with adenine, instead of with guanine.

The hydrogen atoms are represented as small, black circles, and the hydrogen bonds by interrupted lines.

a plausible explanation of their mutagenic effect since the missing guanines might be replaced by any one of the four bases (Bautz and Freese, 1960).

Mere exposure of DNA to moderately low pH, removes purine bases indiscriminately without loss of pyrimidines, and has been shown to be mutagenic for bacteriophage T4 (Freese, 1959b).

We must now turn to consider the way in which these chemical tools may be used to explore the mechanism of mutation at the molecular level. In particular, we would like to test the molecular theory of mutagenesis by direct experiment, to investigate the different ways in which the genetic code can be changed and, if possible, to pin-point the actual base-pair substitution involved in a particular mutational event. Studies of this kind were pioneered by Freese and his colleagues, working with mutations in the r_{II} region of bacteriophage T4. Before looking at their more significant findings, however, it will be helpful to make some theoretical predictions, on the basis of a model system, with which the experimental results can be compared.

Assuming that a reversible, single-site mutation results from the alteration of a single base-pair, two questions must be considered. First, what types of base-pair substitutions are possible in DNA? Second, which type of substitution may be expected of each mutagenic agent from our knowledge of its chemical interaction with the various nucleotides? For the moment we may ignore the intriguing problem of whether the genetic information is carried by the sequence of the four bases, adenine (A), guanine (G), thymine (T) and cytosine (C), on only one or the other of the two strands of the DNA double helix or, alternatively, by the sequence of the four possible *pairs* of bases, A-T, T-A, G-C and C-G, since either possibility embodies a 4-letter code, while the substitution of a single base by another automatically yields a new base-pair when the DNA replicates (see p. 302). We will therefore regard the base-pair as the unit of alteration in the following discussion.

There are two different ways of changing the base sequence. One is by substituting a purine by another purine, or a pyrimidine by another pyrimidine; this may be written in the general form

$$\text{A-T} \leftrightarrow \text{G-C}$$

since, for example, if A on one chain is replaced by G, at the next replication the G will pair with C, so that the original A-T pair will be substituted by G-C. The same alteration follows the initial replacement of T by C. This type of alteration is called a *transition* (Freese, 1959c). The second type of change, known as a *transversion* (Freese, 1959c), is the substitution of a purine by a pyrimidine or *vice versa*, thus:

$$\text{A-T} \leftrightarrow \begin{matrix}\text{T-A} \\ \text{or} \\ \text{C-G.}\end{matrix}$$

Recognition of the distinction between transitions and transversions is important in the study of chemical mutagenesis because, as we shall see, some mutagens may be expected to induce only transitions while others yield both transitions and transversions; since mutations resulting from a transversion clearly cannot be induced to revert to true wild type by a transition, or *vice versa*, this distinction offers us a way of relating theoretical predictions to experimental results.

Let us now look at the various mutagens at our disposal to see what types of base change they may be expected to impose. The action of base analogues depends not only on their incorporation into DNA in place of a normal base but mainly on their greater probability than normal of making *mistakes in pairing*. There are two ways in which such mistakes can arise. The analogue may initially be incorporated opposite its correct partner but err in the choice of a new partner during a subsequent replication of the DNA, when the strand that contains it serves as a template for the synthesis of a new strand. For example, 5-bromouracil (BU) may be properly incorporated opposite adenine (A), but at some later replication may pair with guanine (G) so that G is substituted for A in the new strand. At the next replication this guanine will pair normally with cytosine (C) so that double-stranded DNA emerges in which an initial A-T base-pair is replaced by a G-C pair (Fig. 79, 1a). The alternative possibility is that the 5-bromouracil mistakenly pairs with guanine *during its incorporation* (G-BU) but at the next replication pairs normally with adenine, so that adenine appears on the complementary strand at a site formerly occupied by guanine. During the following replication this adenine pairs with thymine so that the initial G-C base-pair is replaced by A-T (Fig. 79, 1b). Thus, as Fig. 79 (1) demonstrates, *5-bromouracil should induce mutations by causing transitions in either direction*, depending on whether the pairing error arises during incorporation or during subsequent

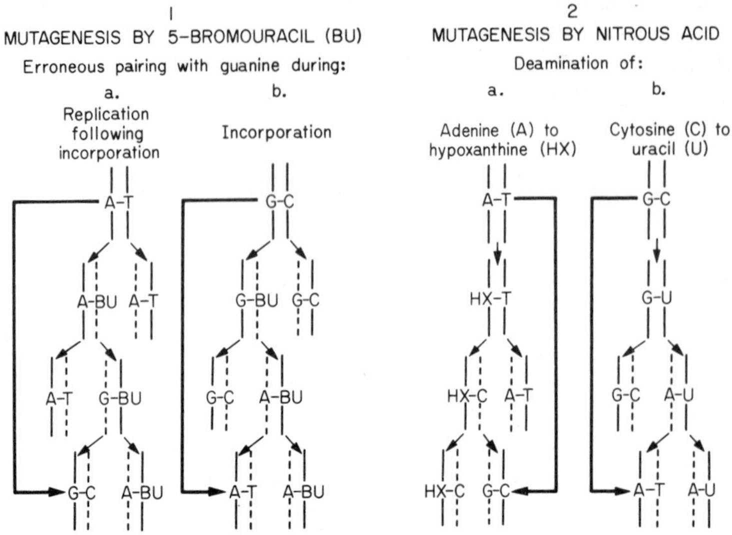

FIG. 79. Showing how base analogues and nitrous acid can lead to base-pairing transitions in either direction, i.e. substitution of A-T by G-C, or of G-C by A-T.

1. In the case of 5-bromouracil, the base pair which will be altered depends on whether the erroneous pairing of BU with guanine (G), due to tautomeric shift, occurs a. during replication following correct incorporation, or b. during its initial incorporation.

2. In the case of nitrous acid, the base pair to be altered depends upon whether the deaminated base is a, adenine or b, cytosine.

replications. The same type of argument may be applied to 2-aminopurine and other base analogues, and there is no apparent mechanism whereby these substances could yield mutations of the transversion type.

Nitrous acid is assumed to be mutagenic because it specifically deaminates adenine and cytosine to yield new bases which have the opposite hydrogen-bonding properties (Fig. 77). Although there is evidence that the deamination of guanine is lethal (Vielmetter and Schuster, 1960), it is thought not to be mutagenic since its product, xanthine, can pair specifically to cytosine by two hydrogen bonds. Nitrous acid acts directly on the bases so that the choice open to base analogues, of making mistakes during incorporation or during subsequent replications, does not arise. Nevertheless the fact that either of the two non-complementary bases, adenine or cytosine, may be deaminated offers another alternative. Deamination of adenine (A) yields hypoxanthine (HX) which, at the first replication, will pair specifically with cytosine (C); at the next replication this cytosine will pair with guanine (G) so that an A-T $\rightarrow$ G-C transition is effected (Fig. 79, 2a). On the other hand, deamination of cytosine to uracil (U), which bonds specifically with adenine, will produce the opposite G-C $\rightarrow$ A-T transition. Thus, like base analogues, nitrous acid appears capable of inducing only transitions, but can do so in both directions. We may therefore anticipate that all mutations induced by any base analogue or by nitrous acid will also be revertible by all these agents. It does not necessarily follow that the *frequency* of reversion will be the same in each case. For instance the rate of deamination of cytosine extracted from phage T2 is about six times faster than that of adenine (Vielmetter and Schuster, 1960). If this same difference held for mutagenic deamination in the intact phage we might expect a great excess of G-C $\rightarrow$ A-T over A-T $\rightarrow$ G-C transitions. On the other hand this could be compensated if, for example, the three hydrogen bonds linking cytosine and guanine rendered the cytosine less accessible than adenine to the action of nitrous acid on intact DNA. It is important to stress here that virtually nothing is definitely known about the interactions that take place when intact phages or bacteria are treated with chemical mutagens.

So far as purely chemical experiments go, hydroxylamine has the interesting property of altering only a single base, cytosine, in such a way that it is thought to pair specifically with adenine instead of with guanine (Fig. 78). A cytosine molecule altered in this way should pair with adenine at the first replication after treatment. At the next replication this adenine should then pair with thymine so that the outcome is a G-C $\rightarrow$ A-T transition. Unlike the base analogues and nitrous acid, therefore, hydroxylamine should induce transitions in one direction only (Freese *et al.*, 1961a, b). If this is really so, two consequences should follow. First, mutations arising from treatment with hydroxylamine should not be revertible by hydroxylamine since a transition can only be reversed to its original form by a second transition in the *opposite* direction. Second, hydroxylamine should be able to revert *some* of the mutations induced by base analogues and nitrous acid (the A-T $\rightarrow$ G-C ones) while others (the G-C $\rightarrow$ A-T ones) should prove stable to its action. In this way it should be possible to specify the particular base-pair alteration responsible for any mutation known to be due to a transition.

When we come to consider depurinating agents, such as ethyl ethanesulphonate (EES) and low pH, we are confronted by a rather different situation. The chemical evidence here suggests that purines, and specifically guanine in the case of EES treatment, are actually stripped from the DNA backbone, leaving a 'purine gap' wherever this has occurred. When the treated DNA replicates, the site on the new strand opposite this gap could, theoretically, be occupied by any one of the four nucleotide bases. If this is so, depurination could lead to any of the four possible base-pair substitutions depicted in Fig. 80. When adenine is

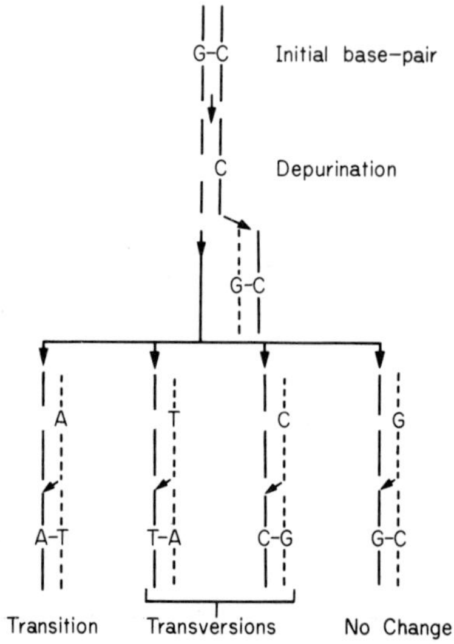

FIG. 80. The types of base-pair substitution which may be expected to follow removal of a guanine base by ethyl ethanesulphonate (EES).

The pairs of vertical lines represent the two strands of the DNA double helix.

Continuous lines represent 'old' strands, and interrupted lines newly synthesised strands.

The 'purine gap' is indicated by a hiatus in the continuous lines.

incorporated opposite the guanine gap, the initial G-C pair is changed to A-T, that is, a transition is effected; when guanine is incorporated, the initial G-C pair is re-established. On the other hand, the incorporation of thymine or cytosine opposite the guanine gap results in replacement of the original G-C base-pair by T-A or C-G respectively, both changes being transversions. If there are no stereo-chemical or other reasons why the site on the new strand, opposite the purine gap, should be filled preferentially by any one type of base rather than another, we should expect about two thirds of mutations induced by EES, being transversions, to be stable to reversion by nitrous acid, hydroxylamine and *all* the base analogues since these can produce transitions only. The postulated ways in which these various chemical mutagens may induce base-pair alterations are summarised in Table 13. We must now see to what extent these predictions are borne out by experiment.

TABLE 13.

The base-pair substitutions assumed to be induced by various chemical mutagens.

Mutagen	Substitutions	Type of substitutions
5-Bromouracil (BU)	A-T ↔ G-C	Two-way transitions
2-Aminopurine (AP)	A-T ↔ G-C	Two-way transitions
Nitrous acid (NA)	A-T ↔ G-C	Two-way transitions
Hydroxylamine (HA)	G-C → A-T	One-way transition
Ethyl ethanesulphonate (EES)	G-C → A-T	One-way transition
	G-C → T-A	Transversion
	G-C ↔ C-G	Transversion

EXPERIMENTAL ANALYSIS OF CHEMICAL MUTAGENESIS

With the exception of studies of the action of nitrous acid on the RNA of tobacco mosaic virus, virtually all experimental work on chemical mutagenesis has utilised the r_{II} system of bacteriophage T4, because of the ease with which r_{II} mutants can be isolated and r^+ reversions of them selected, the very large number of mutants of diverse origin which have already been mapped in the r_{II} region, and the great resolving power of the system in genetic analysis (Benzer, 1957, 1959, 1961; see p. 167 *et seq.*). The use of phage T4 introduces a new factor, namely that cytosine is exclusively replaced by the equivalent pyrimidine base 5-hydroxymethylcytosine (HMC) in the DNA of this phage (p. 426).

Since HMC bonds specifically to guanine in precisely the same way as cytosine does, and since the two pyrimidines appear to interact in a virtually identical manner with the relevant chemical mutagens *in vitro*, we propose to ignore this idiosyncrasy of phage T4 and continue to refer to guanine's partner as cytosine.

In all that has been said so far it has been assumed that revertible mutations arise from the alteration of a single nucleotide of the DNA double helix that constitutes the chromosome. The availability of substances which induce mutations and, at the same time, are known to act chemically on individual bases provides the possibility of checking this assumption experimentally by observing whether or not the production of mutants follows a first-order kinetics. Put another way, if, for a low, relatively non-lethal concentration of mutagen, the frequency of mutation increases linearly with time of treatment, it may be inferred that the mutations arise from single inactivations or 'hits' which means, in effect, the alteration of single nucleotides. Such a first-order kinetics has been confirmed for the action of nitrous acid on the infective RNA of tobacco mosaic virus (Gierer, 1960) and on phage T4 (Bautz-Freese and Freese, 1961), as well as of hydroxylamine on phase T4 (Freese *et al.*, 1961b).

One way of approaching chemical mutagenesis from a genetical standpoint is to map a large number of mutations induced by each agent and to observe whether the distributions of their mutational sites on the map, known as *mutability spectra*, tend to be the same or different. After taking into account the variable and rather small numbers of mutations which have been mapped for each mutagen, as well as the inevitable background of contaminating spontaneous mutations, two significant generalisations have emerged from this approach. The first is that the spectrum of spontaneous mutations is quite different from that of any of the induced series, not only in the distribution of 'hot spots' where many mutations arise at a particular site (p. 325), but also in the virtually complete absence of coincidence of spontaneous and induced sites; crosses between them always yield recombinants. Secondly, although the mutability spectra of the various mutagens show a rather wide overlap in that many of the mutant sites of each series coincide, each does differ significantly from the others, especially in the distribution of hot spots (Benzer, 1961). Such differences are evident, for example, when the spectra for 5-bromouracil, 2-aminopurine and nitrous acid are compared—a surprising finding when one considers the similarity of the base-pair substitutions postulated for these three mutagens (Table 13). Of course the identity of two sites on the map does not imply that they have been altered in the same way, even on the likely assumption that the site corresponds to a single base-pair, for a base-pair such as A-T can be changed in three ways, to G-C, T-A or C-G, although not all of these changes need result in an observable mutation (see p. 301).

TABLE 14.

The revertibility of r_{II} mutations induced by various chemical mutagens. (Collected data of Freese and his colleagues).

Mutations induced by:	Proportion of mutations reverted by:					Base-pair substitutions inferred
	BU	AP	NA	HA	EES	
BU	[2] ±	[2] ++	[6] ++	[3*] —	[4,5] —	G-C ⇌ A-T
AP	[2] ++	[2] ±	[6] ++	[3†] +	[4,5] +	A-T ⇌ G-C
NA	[1] +	[1] +	[6] ++	..	..	A-T ⇌ G-C
HA	[7] +	[7] ++	..	[3‡] —	..	G-C → A-T
EES	[4] Base analogues +		..	..	[4] ±	G-C → A-T G-C ⇌ C-G G-C → T-A

++ = virtually 100 per cent reversion
\+ = majority reverted.
± = minority reverted.
— = virtually no reversion.
BU = 5-bromouracil or 5-bromodeoxyuridine.
AP = 2-aminopurine.
NA = nitrous acid.
HA = hydroxylamine.
EES = ethyl ethanesulphonate.

[1] Freese, 1959b.
[2] Freese, 1959c.
[3] Freese *et al.*, 1961a.
[4] Bautz-Freese, 1961.
[5] Bautz & Freese, 1960.
[6] Bautz-Freese & Freese, 1961.
[7] Freese *et al.*, 1961b.

* Only 7 BU mutants tested.
† 5 out of 9 mutants reverted.
‡ Only 4 HA mutants tested.

A much more specific and promising genetical approach is to induce mutations by means of one mutagen and then to study their revertibility (back-mutation) to wild type by the same and other mutagens. The first thing to look at is the *proportion* of revertible mutations. The principal findings coming from the extensive work of Freese and his colleagues are summarised in Table 14, and may be correlated with the theoretical predictions set forth in Table 13 (p. 313). Let us first consider the three mutagens, 5-bromouracil (BU), 2-aminopurine (AP) and nitrous acid (NA), which were expected to yield transitions in both directions. Table 14 shows that this prediction is fulfilled for NA since virtually all mutations induced by this agent are also revertible by it, as are all the BU- and AP-induced mutations. On the contrary, although all BU-induced mutations are reverted by AP, and *vice versa*, only a minority of these mutations respond to the particular mutagen that induced them. This means that transitions induced by BU are predominantly in one direction, while those due to AP are predominantly in the opposite direction. Can we specify these directions? The key is hydroxylamine (HA) which, as we have seen, interacts chemically with cytosine in a very specific way so that it is presumed to give rise to transitions which are overwhelmingly of G-C→A-T type. This is consistent with the fact that none of the small number of HA-induced mutations tested was reverted by HA. We also find (Table 14) that none of the BU-induced mutations tested was HA-revertible, suggesting that BU also strongly favours G-C→A-T transitions. In this case AP, which produces a predominance of transitions in the opposite direction to BU, should revert all (or virtually all) HA-induced mutations and this, indeed, is what it does do; similarly, about half of a small series of AP mutants were reverted by HA.

When we come to consider ethyl ethanesulphonate (EES) we again find that the predicted patterns of base-pair substitutions are in reasonable accord with the experimental results. You will recollect that EES is thought to remove guanine preferentially and that any base may subsequently be inserted opposite the guanine gap, so that G-C→A-T transitions, as well as G-C→C-G and G-C→T-A transversions, result. The existence of transitions among EES-induced mutations is confirmed by the revertibility of about 70 per cent of them by base analogues, even though this proportion is greatly in excess of the one third anticipated by theory. The revertibility by EES of a majority of AP-induced transitions (mainly A-T→G-C), but of virtually no BU-induced transitions (overwhelmingly G-C→A-T), shows that EES produces transitions by acting on G-C base-pairs. Those EES mutations which are not reverted by base analogues are classified as transversions. Since EES is presumed to act only on guanine, the only kind of EES-induced mutation which could be reverted by EES is the G-C→C-G transversion which should comprise about half the total yield of transversions (Table 13). Experimentally it turns out that those EES-induced mutations which are also EES-revertible cannot be reverted by nitrous acid, showing that they are not transitions; conversely, of five EES-induced mutations which were classified as transversions on the basis of non-revertibility by BU, AP and NA, two were revertible by EES (Bautz-Freese, 1961). Similar results were obtained for those mutations induced by low pH.

It is clear that all these findings add up to a story that is sufficiently self-consistent to serve as a working hypothesis, although the critical reader will note that they do not always accord precisely with the predictions as, for example, in the case of the polarity of BU- and AP-induced transitions, the proportionate excess of transitions induced by EES, and the revertibility of a majority of hydroxylamine-induced mutations by 5-bromouracil (Table 14). We should perhaps make it clear that the experiments themselves were adequately, and in some cases rigorously, controlled to exclude such factors as the selection of certain mutant classes during base analogue induction (e.g. by fluctuation tests and reconstruction experiments) or the confusion of reversions by suppressor mutations with true back-mutation to wild type (e.g. by back-crossing the reversions to wild type; see p. 214), although, of course, only samples of the revertants could be tested in this way. Considering the dearth of exact knowledge about so many biochemical, physiological and environmental aspects of chemical mutagenesis, the appearance of some anomalies is hardly surprising and should not deter us from provisionally accepting the more obvious consistencies at their face value.

In fact we have to accept very little to make progress. We have seen that all transition mutations can be induced to revert either by AP (2-aminopurine) or by BU (5-bromouracil). There is also good evidence that BU and HA (hydroxylamine) induce reversions by acting predominantly on G-C base-pairs. It follows that mutations which are highly revertible by both BU and HA must initially have arisen by an A-T→G-C transition, that is, that the original, wild type base-pair at the mutant site was an A-T one. By the same argument, those mutations which are revertible by AP, but not by BU or HA, must have suffered an initial alteration of a G-C base-pair in the wild type DNA. Thus by testing mutations for reversion by BU, AP and HA it should be possible to define the positions of individual A-T and G-C base-pairs on the DNA of the wild type chromosome. Champe and Benzer (1962b) have used this method to analyse the 339 distinct mutational sites which have been mapped within the r_{II} region of the phage T4 chromosome (Fig. 49, p. 175) and have been able to identify 125 of them as transitions. To 62 of these transitional sites an original base-pair could be unambiguously assigned, 16 being A-T and 46 G-C pairs. The remaining transitional sites either showed weak reversion or, uncommonly, discrepancies between the behaviour of BU and HA, so that the base-pairs could not be definitely specified. The mutant sites not revertible by base analogues are assumed to result from transversions, or from tiny deletions or insertions of a novel type which we will describe in the next section. (For an extension of these studies, see p. 369.)

It should be mentioned here that a comparable study of mutagenesis in phages S13 and ϕX174, both of which have single-stranded DNA, has given results different from those described above. In the case of *in vitro* mutagenesis, the specific action of HA on cytosine, producing a C-T change, was confirmed, but NA was found to mutate all four bases including T-C and G-A transitions (ref. Fig. 77, p. 306; Tessman, Poddar and Kumar, 1964). During *in vivo* mutagenesis the main discrepancy occurred with BU which preferentially induced the T-C transition; this was unequivocally confirmed by the fact that induction was specifically repressed by thymidine which competed with the analogue for incorporation at the thymine site. On the other hand, the results we have described above for phage T4 were confirmed when this phage was tested in the same host as was used for the S13 analysis (Howard and Tessman, 1964a). The reason for these discrepancies is not known, but they serve as a warning against drawing too far-reaching conclusions from the study of one organism.

ACRIDINES AS MUTAGENIC AGENTS

Phase-shift Mutations

Dyes of the acridine series, such as proflavin and 5-amino acridine, are mutagenic for bacteriophage (de Mars, 1953), but not significantly so for bacteria. The reason for this distinction may be because acridines act only on DNA which is undergoing genetic recombination which occurs frequently during the growth of many phages (see below, p. 324; Magni, von Borstel and Sora, 1964). Because of Benzer's brilliant exploitation of the r_{II} system of phage T4 for the study of genetic fine structure (p. 167), work on acridine mutagenesis has been virtually restricted to this system. The facts which have emerged turn out to have profound implications for our concepts about the mechanism of mutagenesis and the nature of the genetic code. These facts can be summarised very briefly. The first is that the mutability spectrum of proflavin-induced r_{II} mutations is unique. There is virtually no coincidence of a proflavin site with any site induced by other chemical mutagens; a small proportion of proflavin sites do coincide with those of spontaneous mutations but, as a whole, they are distributed much more randomly and form few 'hot spots' of any size (Brenner, Benzer and Barnett, 1958; Benzer, 1961).

Secondly, reversion studies have shown that whereas 98 per cent of base-analogue-induced mutations are also base analogue-revertible, 86 per cent of spontaneous and 98 per cent of proflavin-induced mutations, all of which revert spontaneously, cannot be reverted by base analogues (Freese, 1959b) although the great majority respond to acridines (Alice Orgel, personal communication). Moreover, of a small series of spontaneous mutations which failed to respond to base analogues, about one half reverted when treated with ethyl ethanesulphonate (EES), while only one of a comparable series of proflavin-induced mutations gave any response at all (Bautz-Freese, 1961). Acridine-induced mutations therefore clearly form an exclusive class. A possible interpretation is that all those mutations not revertible by base analogues are transversions, those responsive to EES having a G-C pair (Table 13), and those not responding having an A-T pair, at their mutant sites. According to this simple hypothesis to explain the existence of two mutually exclusive classes of mutation (Freese, 1959b), the effect of acridines would be the specific induction of A-T transversions (C-G→A-T, or A-T→T-A).

An alternative view, put forward by Brenner *et al.* (1961), is that acridine mutations and the majority of spontaneous mutations are not caused by base-pair substitutions at all, but *by the deletion or addition of a base-pair*. Obviously such mutations could not be reverted by any agent whose effect was simply to replace one base-pair by another, but only by the restoration of a deleted base or the removal of an extra one. This hypothesis was deduced from two quite different types of observation. The first is that base analogue and acridine mutants differ profoundly in phenotype. The ability of phage particles to infect and lyse bacteria depends on the integrity of a number of proteins such as the tail fibre protein responsible for adsorption to bacterial cell walls, the protein of the phage head which encloses the DNA, and the enzyme lysozyme which enables progeny particles to lyse and escape from the infected host. The structure of each of these proteins is specified by a single gene. These genes can be recognised and mapped because mutations arise in them which alter the adsorption specificity of the tail protein (*host-range*, or *h*, mutants, the permeability of the phage head as shown by altered resistance to osmotic shock (*o* mutants; Brenner and Barnett, 1959), or the conditions necessary for the expression of lysozyme activity (Streisinger *et al.*, 1961) (p. 480). Mutations of this sort not only arise spontaneously but are readily induced by base analogues, and it is obvious that they confer only minor modifications on the structure of the protein which *alters its specificity without destroying its function*; for mutations leading to gross structural changes, or to absence of protein, would be lethal to the phage, that is, the mutation would pass unrecognised. The significant point is that mutations of this modifying kind are not induced by acridines, and the conclusion to be drawn is that acridine-induced mutations wreck the structure of the protein (Brenner *et al.*, 1961). If this is so, why is it that acridine-induced $r^+ \rightarrow r_{II}$ mutations are found? The answer is that complete abolition of the function of the r_{II} region, whatever this may be, is only lethal when the phage is grown on *E. coli* K. We have already seen that extensive deletions in the r_{II} region, which may excise the whole of both cistrons, still permit the development of *r* type plaques on *E. coli* B. Yet even among r_{II} mutants a difference in degree of function may be observed, for while the majority of base analogue-induced mutants tend to be *leaky* in that a few progeny particles may emerge from infection of *E. coli* K, those induced by acridines are *non-leaky* and behave phenotypically like deletion mutants (Crick *et al.*, 1961). The term 'leaky' is an expressive one which is widely applied to mutations which fail to shut off completely the activity of a gene so that some residual expression of its function remains.

What we have said so far should not be construed to mean that base-analogue-induced mutations are always leaky. What is clear is that *many* base analogue-induced, and *some* spontaneous, mutants continue to make functional protein and these are the ones we observe when selection is made for, say, host-range mutants of phage. On the contrary, mutants arising from acridine treatment do not make functional protein and so probably arise by a different mechanism. Although we now know that certain mutations, which terminate the growth of polypeptide chains so that no functional protein is made, can arise by base substitution, (pp. 361–364), the possibility of deletion or addition of *single* base pairs or, at most, of a very small number by acridines was first considered, substantial deletions being ruled out on the ground of revertibility. Can we expect such a mechanism to account for a complete disruption of protein synthesis? An elementary consideration of how the genetic information, carried by the sequence of bases in the DNA, is translated into terms of protein structure suggests that we can. We will consider this matter in detail in the next chapter, but for the moment let us assume that each amino acid is coded for by a specific sequence of three bases which are read off in one direction in groups of three. Thus if A, B and C are three bases, and the sequence ABC codes for leucine, then the extended sequence

$$\overrightarrow{\underbrace{ABC}\underbrace{ABC}\underbrace{ABC}\underbrace{\ldots}}\ldots$$

specifies three adjacent leucine residues in the polypeptide chain. If one of the letters in the first triplet is substituted by another (base-pair substitution), the code is changed to, say,

$$\overrightarrow{\underbrace{ABB}\underbrace{ABC}\underbrace{ABC}\underbrace{\ldots}}\ldots$$

The result is that the last two leucine residues remain unaffected, and only the first may be substituted by some other amino acid. If, however, one letter of the first triplet is deleted thus,

$$\overrightarrow{\underbrace{ABA}\underbrace{BCA}\underbrace{BC\ldots}\underbrace{\ldots}}\ldots,$$

or if an additional letter is added to give

$$\overrightarrow{\underbrace{ABB}\underbrace{CAB}\underbrace{CAB}\underbrace{C\ldots}}\ldots,$$

then *every triplet to the right of the deletion or insertion will be misread* so that no protein can be made. This misreading results from a shift of the reading frame so that the reading device is now out of phase, so that mutations due to such a mechanism have been termed *phase-shift mutations*.

Recent investigations of the interactions of acridines with DNA, by means of X-ray diffraction analysis and other physical methods, have suggested a way in which such deletions or insertions may arise. At high concentrations, acridines are bound to the phosphate groups of the DNA backbone on the outside of the double helix. At relatively low concentrations, however, the evidence strongly suggests that the acridine molecules become inserted, or intercalated, between adjacent base-pairs in the DNA, with some extension and unwinding of the phosphodiester backbone. This has the effect of stretching the distance between neighbouring bases to 6·8 Å, or precisely double the distance between them in normal DNA (Lerman, 1961, 1963; Luzzatti, Masson and Lerman, 1961) (see Fig. 59, p. 237).

If this latter type of interaction is the mutagenic one, we can explain the insertion or deletion of a base-pair in the following way (Fig. 81).

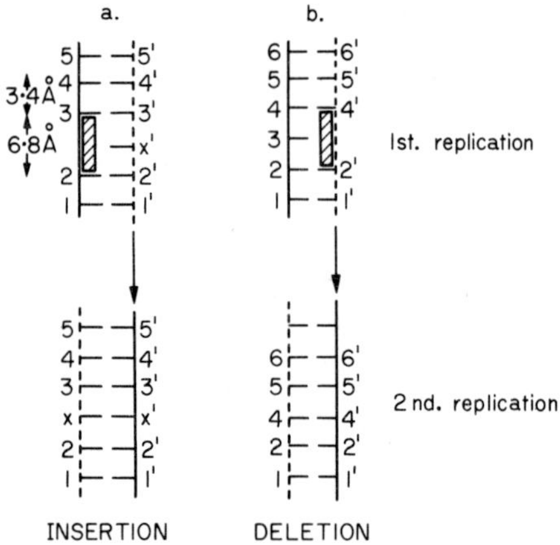

FIG. 81. Showing how acridines may interact with DNA to (a) insert or (b) delete a base-pair.

The acridine molecule is represented by a shaded rectangle; its intercalation between adjacent bases forces them 6·8 Å apart, instead of the normal 3·4 Å.

The continuous vertical lines indicate old template strands, and the interrupted lines the synthesis of new strands.

The short horizontal lines are the bases, designated as 1, 2, 3 etc., the complementary bases on the other chain being 1′, 2′, 3′ etc. respectively.

Suppose that, during DNA replication, an acridine molecule inserts itself between two adjacent bases of the *template strand* while a new strand is being formed along it by base bonding (Fig. 81a). As a result of the intercalation, a gap will be left between the two bases which can be exactly filled by the insertion of an extra base at this position on the new strand. Thus the new strand acquires an extra base which will be perpetuated in the double-stranded DNA when this strand replicates in turn. Alternatively, a base-pair deletion may result if an acridine molecule interposes itself between the bases of the *new strand* as this is being built up, separating them by 6·8 Å. This means that the acridine will lie opposite one of the bases on the template strand and mask it, so that its complementary partner cannot pair with it and so will be missing from the new strand (Fig. 81b). This new strand, on subsequent replication after shedding the acridine, will perpetuate its deficiency as a base-pair deletion in the double-stranded DNA. The fact that the majority of spontaneous mutations are revertible by acridines, and *vice versa*, but not by base analogues, suggests that these, too, arise from the deletion or addition of a base-pair, although we do not know why mistakes of this particular type should be the main ones to occur during normal replication.

How do mutations of this kind revert to wild type? If the mechanism we have postulated for their origin is correct, reversion must be due to restitution of the proper number of base-pairs by the insertion of an extra base where one has been deleted, or by the subtraction of a base where one too many is present, so that the reading device is in phase again. This restitution could occur at the precise site of the initial error or at *some other site located extremely close to it*. Let us take the previous example of the sequence ABCABCABC . . . where the deletion of base C from the first triplet

ABABCABC . . .

had disrupted the correct reading of the entire sequence. If, instead of restoring the original sequence, we introduce an extra base, X, at a close but different position thus,

ABABCXABC . . . ,

it is clear that although the sequence will continue to be misread in the interval between the initial deletion and the new insertion, the third triplet, and every succeeding triplet thereafter, will now be read correctly. Thus if the nature of either the code or of the protein is such that aberrations over very small coding regions can be tolerated without gross loss of function, some kind of function could be retrieved by such a mechanism. Note that this sort of reversion is technically of the *suppressor* type since it is due to the occurrence of a second (suppressor) mutation at a site different from that of the primary event (p. 213).

There are two ways of testing this suppressor hypothesis, both of which correspond with the criteria used to identify suppressor mutations in general. The first is that the phenotype of the revertant strain cannot be expected to be identical with that of the wild type since it still has a defect in its protein; it will tend to display a *pseudo-wild* phenotype. Secondly, genetic crosses between the revertant and the wild type should yield some mutant progeny as a result of crossing-over in the region between the site of the original mutation and that of its suppressor (see Fig. 54, p. 214). It turns out that both these criteria are satisfied in the case of virtually all acridine-induced reversions of r_{II} mutations in phage T4. Although the revertant strains are wild type in so far as they produce plaques on *E. coli* K, not only are the appearances of these plaques on *E. coli* B different from the wild type, but back-crosses to the wild type (r^+) strain yield r_{II} recombinants identical with the original mutants, thus proving retention of the initial mutation by the revertants and, therefore, the existence of suppressors (Crick *et al.*, 1961). Since the operation of these suppressors provides crucial evidence about the nature of the genetic code we will look at them again in more detail in the next chapter (p. 346).

With regard to the apparent mutagenic specificity of acridines for bacteriophage, it has recently been shown that the spontaneous reversion rates of biochemical mutants, as well as the rates of forward mutations, in the yeast *Saccharomyces cerevisiae*, may increase by as much as 20-fold when the cells are undergoing meiosis, and that these mutations are associated with genetic exchanges in the regions of the specific loci involved. Many of the prototrophic revertants are pseudo-wild and have a slower growth rate than wild type in minimal medium. Moreover, mutations due to base substitutions are not reverted in this way. These striking findings were equated with the phenomenon of phase-shift by supposing that 'unequal crossing-over' during meiosis may lead to deletion of bases (Magni and von Borstel, 1962; Magni, 1964). Two types of evidence support this idea. First, the spontaneous reversion of two r_{II} mutants, which were proflavin-revertable, was shown to be associated with recombination for outside markers, while that of an AP-induced and BU-revertable mutant was not (Strigini,

1964; also quoted by Magni, 1964). Secondly, acridines induce both forward and backward mutations in yeast during meiosis, but repress mutation during mitosis (Magni *et al.*, 1964).

It is worth noting that acridines are fluorescent dyes and render a variety of cells, including bacteria and phages, sensitive to the photodynamic action of visible light which is not only lethal but actively mutagenic (Ritchie, 1964, 1965; Nakai and Saeki, 1964).

THE HIGHER ORDER SPECIFICITY OF MUTATIONS

This rather pompous title simply denotes the susceptibility of certain genetic sites to specific mutational change, as exemplified by the distinctive mutability spectra induced by different mutagenic agents and, in particular, by the occurrence of 'hot spots' (Benzer, 1961, see p. 174). This phenomenon has only been demonstrated in phage T4 so that we do not know whether it is a general one, although we may suspect that it is. It means, in effect, that mutations are not distributed randomly along the chromosome as was formerly supposed. We might, indeed, expect some lack of randomness from the fact that, at the molecular level, there are four possible types of site, corresponding to the four base-pairs, and three ways in which these may be altered (transitions, transversions, and deletions or insertions). Moreover, we know that whether or not a genetic alteration is expressed as a mutation depends, among other things, on the particular amino acid substitution involved and its specific position in the polypeptide chain. For example, virtually all mutational changes within coding units specifying amino acids at critical positions in the protein, might be *expressed* as mutations while, at other sites, only a minority of alterations might result in the substitution of an unacceptable amino acid or none at all. Again, a transition at a particular site might be innocuous but a transversion mutagenic, or *vice versa*. Thus we might anticipate that some sites would prove more sensitive to mutation than others, and that mutagens promoting different types of alterations in the DNA would yield different mutability spectra.

However, the matter is far more complicated than this, from both the qualitative and quantitiative points of view. In the first place, mutagens belonging to the same class induce different spectra although some of the sites may overlap. Secondly, in the case of spontaneous mutations, the enormous preponderance of mutations at two sites, and the absence of 'hot spots' at these sites following induction by any class of artificial mutagenic agent, seems inexplicable on our present knowledge. Thirdly, individual mutants vary greatly in their spontaneous reversion frequencies, as well as in their response to induced reversion. For example, when a series of mutants revertible by acridines is compared, the frequency of reversion is found to vary from mutant to mutant. Far from falling into one or the other of a small number of classes, each mutant appears to possess a unique sensitivity (Alice Orgel, 1961, personal communication). This also applies to other classes of mutations and other mutagens.

Let us examine this question of randomness in another way. In an organism such as phage T4, with an A+T/G+C base ratio of 1·8, every wild type gene about 1000 nucleotide pairs long may reasonably be expected to contain about 600–700 A+T combinations, about half of which are A-T and half T-A pairs. If alterations occur randomly among, say, A-T sites, *whether or not these are expressed as mutations*, there is a strict limit to the preponderance of *observed* mutations which can occur at any one site. This is because there are only three ways in which a base-pair can be changed (for example, A-T→T-A, G-C or C-G). Thus even if every alteration at a particular site results in an overt mutation, while only 1 in 3 is mutagenic at some other sites, the number of observed mutations at the most sensitive site cannot exceed those at other sites by a factor greater than 3. Such an excess would hardly qualify as a 'hot spot'. Since the same argument applies to the other base-pairs, we can infer that true 'hot spots' have a highly enhanced, inherent susceptibility to mutation which cannot be ascribed to the type of base-pair involved, the way in which it is altered or the resulting change in the protein.

There is no information at all concerning the nature of this susceptibility, but it must be imposed by some environmental factor whose influence is restricted to the mutable site. One possible factor is the nature of the adjoining bases; for instance a sequence of A-T base-pairs, being less strongly bonded than G-C pairs, may introduce local weaknesses into the double helix (see Pontecorvo, 1958, p. 45). Another is the occurrence of bends at specific points in the DNA chain when this is folded into a tertiary or quaternary structure (see Benzer, 1961).

SUMMARY

We will now summarise briefly the inferences we think may safely be drawn from studies of chemical mutagenesis, concerning the mechanism of single-site mutations at the molecular level. The genetic information specifying the sequence of amino acids in a protein, and therefore the structure of that protein, is carried by the sequence of base-pairs in the DNA that constitutes the chromosome. The code for each amino acid resides in the sequence of a small number of base-pairs which we assume to be 3. Mutation results from an alteration involving one of the base-pairs of a triplet so that the triplet now codes either for a different amino acid or for none at all.

There are two main types of alteration. The first is the *substitution* of one base-pair by another. There are two varieties of substitution, *transitions* and *transversions*. In transitions a purine base on one of the DNA strands is replaced by another purine, or a pyrimidine by another pyrimidine; in transversions, a purine is replaced by a pyrimidine, or *vice versa*. Transitions are typically induced by base analogues and may arise spontaneously as a result of tautomeric shifts of hydrogen atoms on the bases of replicating DNA, leading to mistakes in base pairing. Transversions are typically induced by agents, such as alkylating chemicals and low pH, which remove bases from the DNA backbone, leaving a gap opposite which another base of different type may be inserted at replication, but there is no reason why transversions should not also occur spontaneously from accidental mis-pairing of two purines or two pyrimidines, since this would only lead to a temporary, local distortion of the DNA double helix. It seems likely that the reversion to wild type of mutations arising from base substitutions is mainly due to restoration of the original base-pair, although transitions can only be reversed by transitions in the opposite direction, and transversions by transversions. A considerable proportion of transitions, at least, yield 'leaky' mutants in which the function of the affected protein is merely altered or reduced, but not destroyed.

The second type of mutagenic change is the *deletion or insertion of a base pair*. Mutations of this type, known as *phase-shift* mutations, are induced by acridines but also comprise the majority of spontaneous mutations, at least in bacteriophage. On the assumption that, in translating the genetic code into terms of protein synthesis, the bases specifying each amino acid are read off without punctuation, in groups of three and in sequence from a fixed starting point, the removal or addition of a base will change the transcription of every triplet from the site of the alteration onwards with the result that no protein, or only grossly deranged protein, is formed. In accord with this, mutations involving characters, such as host-range in phage, which require functional protein for their expression are not evoked by acridines, while acridine-induced mutations affecting dispensable characters are 'non-leaky'. The reversion of acridine-induced mutations results from restoration of the proper number of base-pairs to the sequence, by the insertion of an extra base-pair in the case of an initial deletion, or *vice versa*. This secondary insertion or deletion may not be at the site of the original alteration but at a very closely linked site; that is, reversion is due to a type of suppressor mutation.

In the case of phage T4, the great majority of spontaneous mutations are not revertible by base analogues and are therefore not due to transitions. Some of these can be reverted by alkylating agents and are presumably transversions. The remainder, comprising the majority of all spontaneous mutations, are revertible by acridines and are caused by the deletion or addition of a base-pair.

THE PHOTOCHEMICAL ACTION OF ULTRAVIOLET LIGHT ON DNA

The damage inflicted on cells by radiations has proved singularly resistant to analysis, partly because of the difficulty of disentangling general physiological effects from localised injury to specific cellular components, and partly because little precise information exists about the physico-chemical changes induced by radiations in biological macromolecules. This is particularly so for the rather low dose ranges producing biological damage. Recent developments in bacterial and phage genetics, and especially in molecular biology, have given us some new and beautifully refined tools for exploring the biological effects of radiation at the molecular level. One of the most valuable of these is genetic transformation in bacteria which permits the results of physico-chemical analysis of naked DNA to be correlated directly with its biological activity. Conversely, radiations have been employed, though usually rather empirically, to elucidate the mechanism or kinetics of various genetic phenomena, particularly in the study of bacteriophage as we shall see (p. 522). For the moment we propose only to discuss briefly some current studies of the action of ultraviolet light (UV) on DNA.

Nucleic acids display a specific absorption spectrum for ultraviolet light of wavelength about 254 mμ. Analysis of the susceptibility of the various bases of DNA to chemical alteration by UV light showed that while the purines are resistant, the pyrimidines become hydrated at their 4:5 bonds (Fig. 57, p. 235), cytosine nucleotides being more susceptible to this photolytic action than thymine nucleotides (Shugar, 1960). Changes of this sort might well be expected to weaken the stability of the double-stranded DNA molecule which is highly dependent on the integrity and specific hydrogen-bonding properties of the individual base residues. However no cytosine hydrate has been detected in irradiated, natural DNA. A more significant clue to the biological effect of UV light was the discovery that, in the DNA polymer, it is the thymines that appear to be more reactive since the photolytic action induces the thymine rings to coalesce in pairs to form what are known as *thymine dimers*. This dimerisation occurs when isolated thymine residues are irradiated in the frozen state, with the result that the normal absorption spectrum of the solution is shifted (Beukers and Berends, 1960; Wang, 1961), while thymine dimers can also be extracted from irradiated DNA (Wacker *et al.*, 1961) as well as from irradiated bacteria (Wacker, Dellweg and Jacherts, 1962).

Thymine dimers are very stable to acid hydrolysis and so can be removed intact from irradiated DNA and then separated from thymine by paper chromatography. However, more recent work has revealed that other pyrimidines also yield dimers which are less stable and more difficult to recognise, but which probably have biological effects similar to those of thymine dimers. Thus cytosine-cytosine, cytosine-thymine, thymine-uracil and uracil-uracil dimers have now been found in irradiated DNA, the uracil being formed by deamination of dimerised cytosine (Setlow, Carrier and Bollum, 1965; Setlow, 1966).

The question arises whether, in the natural DNA polymer, dimers are formed between adjacent thymines on the same strand, or between thymines on *opposite* strands of the DNA double helix leading to cross-linkages between the strands which would prevent their separation and thus interfere with replication. We have already seen that one of the most striking effects of UV irradiation of intact cells is the specific inhibition of DNA synthesis (Kelner, 1953). However, cross-linkage between thymines would clearly require considerable prior distortion of the double helix since the specificity of base pairing means that thymines on the two strands are never normally opposite one another.

Investigations of the alterations in the physical properties of DNA which accompany UV irradiation suggest that a weakening of the structure of the double helix and cross-linkages between the two strands occur concomitantly (Marmur *et al.*, 1961). For example, we have seen that when DNA is heated the two strands separate (p. 244) and, at the same time, become much more reactive with formaldehyde than is the undenatured DNA. For any double-stranded DNA, the temperature at which this thermal denaturation occurs is very critical and is a function of the tenacity of the hydrogen bonding between the two strands. The effect of UV irradiation on both natural and synthetic DNA (p. 245) is to lower the denaturation temperature and to increase greatly the reactivity with formaldehyde, indicating a weakening of base-bonding and some opening up of the double helix. Moreover this effect is most marked with those DNA species which have a high adenine+thymine content, suggesting that the thymine residues are primarily involved.

The reverse kind of experiment can be performed by making use of the fact that treatment with formamide efficiently separates the two strands of DNA. Since the separated strands have a higher buoyant density than when they are united in the double helix, the degree of denaturation can be estimated from the intensity of the bands at the two density levels when the DNA is centrifuged in a caesium chloride density gradient (p. 243). It turns out that, following high doses of UV irradiation, a proportion of the DNA molecules becomes resistant to strand separation by formamide treatment, this effect being most marked in the case of those DNAs with the highest content of adenine+thymine. This cross-linking of the two strands of the DNA molecule has been confirmed by the use of an enzyme (phosphodiesterase), isolated from *E. coli*, which has the peculiar property of attacking only denatured or single-stranded DNA. It was found that UV-irradiated DNA became decreasingly susceptible to attack by this enzyme with increasing dosage (Marmur *et al.*, 1961).

It is thus undoubtedly the case that one of the effects of ultraviolet light on DNA is to produce some opening-up of the duplex as well as the cross-linking of the two strands, but the extent to which the latter is due to the formation of dimers between pyrimidines on opposite strands is a controversial matter. In natural DNA, dimers are formed predominantly between thymines on the same chain (Setlow and Carrier, 1963). Moreover, cross-links can be produced under conditions of irradiation which yield few dimers and, unlike dimers, are not eliminated by short-wavelength irradiation (see Setlow, 1966).

We can therefore make a plausible, even though circumstantial, case that the most important effect of UV light on DNA is the photochemical formation of thymine dimers. It seems likely that the primary change is the photolysis of pyrimidines, with dimer formation between adjacent thymines, along the two strands, leading to weakening of the bonds between them and distortion of the architecture of the double helix. Once this has happened, dimerisation of thymines on *opposite* strands can occur as a result of the distortion so that replication is no longer possible.

THE CELLULAR REPAIR OF IRRADIATION DAMAGE

Photoreactivation

It has been known for some time that the killing and mutagenic action of ultraviolet light may be reversed by exposure of the irradiated organisms to visible light (Kelner, 1949). However, not all species of organism are photoreactivable in this way. It has recently been discovered that several photoreactivable species that have been examined (*E. coli*, *Saccharomyces cerevisiae*, *Bacillus*) possess an enzyme which, when added to UV-inactivated transforming DNA in the presence of visible light of wavelength 300–400 mμ, can restore about ten per cent of its transforming activity (Rupert, 1961). No reactivation takes place in the absence of light, and the enzyme is devoid of species specificity. Thus enzyme extracted from baker's yeast or from *E. coli* can equally repair UV damage induced in transforming DNA from *Pneumococcus*, *Haemophilus* or *B. subtilis*, all of which lack the enzyme, in DNA from *E. coli*, phage T2 and calf thymus, and even in single-stranded DNA from phage ϕX 174 (p. 436) as well as DNA artificially synthesised in the Kornberg system from a natural primer (p. 245).

The photoreactivating enzyme combines with UV-irradiated DNA in the dark and is thereby stabilised against inactivation by heat or heavy metals. On exposure to light, however, the complex dissociates again with release of active enyzme, and of DNA whose lesions have now been repaired. In contrast, unirradiated DNA neither combines with the enzyme nor stabilises it (Rupert, 1961).

The nature of the enzyme substrate in irradiated DNA has been the subject of considerable controversy. The fact that single nucleotides, completely digested DNA or synthetic adenine-thymine (A-T) deoxyribose co-polymers completely fail, after irradiation, to bind the enzyme, at first suggested that the enzyme does not act on thymine dimers. However, efficient dimer formation requires that the thymines are adjacent and oriented in a particular way with respect to one another, and does not occur between individual thymine molecules in solution. Moreover, the enzyme may require a minimum chain length of DNA for binding.

An indirect method for studying possible substrates is to test various irradiated, synthetic polynucleotides for *inhibition* of biological photoreactivation, and to show that this inhibitory capacity is itself destroyed by illumination in the presence of the photoreactivating enzyme. The method has revealed that G-C polymers, devoid of thymine, are acted on by the enzyme (Rupert, 1961, 1964). Furthermore, other experiments of the same kind have shown that only synthetic polymers which contain *adjacent* pyrimidines, including thymine, are photoreactivable, while their ability to compete for enzyme is correlated kinetically with dimer formation.

Finally it has been demonstrated that, following treatment with purified photoreactivating enzyme, a variety of pyrimidine dimers, and especially thymine dimers, are in fact eliminated from DNA and probably monomerised (Setlow, Carrier and Bollum, 1965). There is thus little doubt that the main biological effect of UV irradiation is due to formation of pyrimidine dimers and that the action of the photoreactivating enzyme is to uncouple them. In this respect it is interesting to note that the enzyme also restores to normal the cross-linking of DNA strands induced by UV light (see above: Marmur *et al.*, 1961).

Dark Repair

Both bacteria and certain bacteriophages possess a mechanism for the repair of UV-induced damage which can operate in the dark. This mechanism was discovered by the isolation of mutant strains of *E. coli* which not only were themselves much more radiation-sensitive than the wild type, but also had lost the capacity to 'reactivate' certain irradiated bacteriophages (dependent virulent and temperate, such as T1, T3 and λ; see p. 527) which lack repair mechanisms of their own; the ability of the irradiated phage suspension to form plaques on the mutant as compared with wild type bacteria is greatly reduced (Hill, 1958; Howard-Flanders, Boyce and Theriot, 1962). The wild type strains which can exert 'host cell reactivation' of phage are termed Hcr^+, and the sensitive mutants Hcr^-; the presumptive genetic locus controlling UV-resistance is designated *uvr*$^+$ and its sensitive mutant alleles as *uvr*. A semi-selective method of isolating Hcr^- mutants of *E. coli* is to mutagenise the bacteria and then to plate them with an excess of UV-inactivated phage T1; the Hcr^+ cells reactivate the phage and are killed, while the Hcr^- cells (and, of course, phage-resistant mutants) have a high probability of surviving (Howard-Flanders and Theriot, 1962).

When wild type strains of *E. coli* are irradiated with UV light and then incubated in a nutrient medium, the thymine dimers begin to be lost from the acid-precipitable, native DNA and to appear in association with short, single-stranded DNA fragments in the acid-soluble fraction; the time at which all the dimers have been excised in this way coincides with the time of resumption of DNA synthesis following its UV-induced delay (Setlow and Carrier, 1964; Boyce and Howard-Flanders, 1964a).

Clearly the excision of single-stranded DNA fragments containing dimers must be followed by repair synthesis in which the gap is 'patched' by copying the residual, intact strand. This was looked for, and found, by following the incorporation of 5-bromouracil (BU), which is heavier than thymine and can replace it in DNA, after UV irradiation (Pettijohn and Hanawalt, 1964). Both degradation and synthesis of DNA commence immediately after irradiation and proceed simultaneously; when the newly synthesised DNA was analysed in a caesium chloride density gradient it was found to remain normal in density at first, and later to shift into a density region intermediate between that of normal and hybrid DNA, and to consist of a population of molecules heterogeneous with respect to density. You will remember that in normal, semiconservative replication (pp. 239–244) newly synthesised DNA is of hybrid density from the beginning and is homogeneous, and this, in fact, is what is found when DNA synthesised after photoreactivation of UV damage is examined by the same method. The molecular heterogeneity associated with dark repair indicates that DNA synthesis is proceeding at many random sites along the chromosome (p. 554), while the failure of the DNA which has taken up BU to reach hybrid density implies that, in the majority of molecules, the BU was incorporated into a fraction of one strand only; that is, the repaired regions are very short.

Density-distribution analysis of these BU-containing molecules after fragmentation by sonication and separation of the strands by heat (p. 244) confirmed that the BU had been incorporated into very short segments along single strands. A rough computation suggested that the replaced segment might include about 20 nucleotides in addition to the thymine dimer (Pettijohn and Hanawalt, 1964).

Sensitivity to irradiation could arise from interference with either excision of dimers or repair synthesis. It turns out that all the sensitive mutants so far analysed have had a defect in their excision mechanism (Howard-Flanders, Boyce and Theriot, 1966); it may be that the presumptive polymerase mediating repair synthesis performs some essential normal function so that mutations affecting it are lethal. Unexpectedly, *uvr* mutations have been found to be located at three widely separated loci on the chromosome of *E. coli* K12, designated *uvrA*, *uvrB* and *uvrC* (Howard-Flanders, 1964; van de Putte *et al.*, 1965; Mattern, van Wisiden and Rörsch, 1965), which map in proximity to the *met*, *gal-bio* and *his* loci respectively and are co-transducible with them Fig. 127, p. 666; Howard-Flanders *et al.*, 1966).

A, B and C mutants appear to have identical phenotypes while double mutants, involving any pair of the three loci, are scarcely more sensitive than single ones. However the mutants complement one another and can readily be distinguished by the development of UV resistance in the transient zygotes during conjugal crosses with Hfr strains carrying complementary *uvr* mutations. This means that the wild type allele of all three genes is dominant and suggests that the function of the genes is to determine enzymes such as nucleases. It is possible, of course, to imagine the participation of three enzymes in the excision process, but experiment has failed to reveal any sequential order of action for them (Howard-Flanders *et al.*, 1966).

Although mutants affecting the excision of pyrimidine dimers are highly sensitive to UV light but retain most of their resistance to X-rays and other ionising radiations, they are not entirely specific for UV-induced lesions. For instance they are also much more sensitive than wild type to mitomycin C which is supposed to cross-link purines in DNA (Iyer and Szybalski, 1964) and produces extensive DNA breakdown (Boyce and Howard-Flanders, 1964b).

Another, and very different, type of *E. coli* K12 mutant has recently been described in which high sensitivity to both UV and ionising radiations is associated with inability to undergo genetic recombination (Rec^- phenotype) (Clark and Margulies, 1965; Howard-Flanders and Theriot, 1966). While *uvr* mutants are specifically sensitive to UV light and show normal recombination, it appears that all mutants which are sensitive to X-rays are also UV-sensitive and that the great majority of them, at least, are defective in recombination. Similarly, the few mutants initially identified on the basis of inability to form conjugal recombinants with certain Hfr strains have usually turned out to be sensitive to X-rays and UV light, although some mutants resistant to UV light have been reported (A. J. Clark, personal communication, 1965; Erskine, 1966). At the moment these UV-resistant mutants have not been well enough characterised to classify them as recombination-deficient. Mutations to the Rec$^-$, radiation sensitive phenotype may be located at one of at least three well separated *rec* loci which are quite distinct from the *uvr* loci, and are recessive to wild type in zygotes (van de Putte, Zwenk and Rörsch, 1966; Howard-Flanders and Theriot, 1966).

The mechanism of repair which is disrupted in these mutants is not yet understood, although it is clearly one which is utilised in the process of recombination. As we shall see in Chapter 15 (p. 394), evidence is accumulating that recombination in bacteria and phages is mediated by breakage and reunion of the parental DNA strands; part of the reunion process, while possibly requiring the completion of a DNA strand of one of the parents by copying the complementary strand of the other, will certainly demand the formation of covalent bonds to connect the phospho-diester backbones of the recombinant strands. Since the greater part of the damage inflicted on DNA by ionising radiations results, unlike UV damage, from double-strand breaks, one could postulate that the X-ray- and UV-sensitive *rec* mutations block the activity of one of several enzymes rejoining the DNA backbone—a step necessary for the completion of all types of repair or reunion. One might further speculate that the strand breakage that initiates recombination is the function of a recombination-specific enzyme whose inactivation by mutation would yield a recombination-defective strain of normal resistance to both UV and ionising radiation. Unfortunately for these speculations, the outstanding metabolic peculiarity of those radiation-sensitive, Rec$^-$ mutants so far investigated is a spontaneous breakdown of DNA, and an exaggerated breakdown and repair delay following exposure to UV light—a feature which has earned these mutants the punning epithet 'reckless'! (Howard-Flanders and Theriot, 1966). There are also 'cautious' Rec$^-$ mutants which break down their DNA at a lower rate than wild type after irradiation (Howard-Flanders and Boyce, 1966).

We could hardly end this review of the repair of radiation damage without mentioning the remarkable resistance to both UV and ionising radiations of *Micrococcus radiodurans*, a red-pigmented coccus originally isolated from food which had been 'sterilised' by radiation. *M. radiodurans* remains virtually unaffected by a dose of UV light sufficient to reduce the survivors in a population of non-lysogenic *Salm. typhimurium* by a factor 10^{-4} to 10^{-5}, and can suffer a one per cent. conversion of its DNA thymine residues to dimers without death; again, the dose of X-radiation which allows 10 per cent. survival of

Salm. typhimurium and *M. radiodurans* is about 24 krad and 1000 krad respectively (Setlow and Duggan, 1964; Moseley and Laser, 1965).

The survival curve of irradiated *M. radiodurans* is characterised by a very striking 'shoulder', while the exponential part of the curve is much less steep than usual (p. 525 and Fig. 106). This shoulder, which is found to follow both UV and X-irradiation, is probably due to an extremely efficient repair mechanism or mechanisms which can cope with a very high level of DNA lesions. Non-pigmented mutants of *M. radiodurans* have been isolated which retain high radiation resistance; the X-ray survival curve of one of these was found to have lost its shoulder and to be exponential throughout. However, it did not fall off at a constant rate but suddenly increased its slope to that of the exponential part of the curve given by the wild type strain. This may indicate two distinct types of damage to the genetic apparatus, each with a specific repair system (Moseley and Laser, 1965). In this connection it is interesting that *M. radiodurans*, unlike normal bacteria, is approximately as sensitive to irradiation at 2800Å, which is an absorption peak for protein, as at 2650Å which is strongly absorbed by nucleic acid (Setlow and Bolling, 1965). Despite these developments in our knowledge, however, the mechanism of the extraordinary resistance of *M. radiodurans* remains mysterious.

ULTRAVIOLET-INDUCED MUTATION

The recent and rapid growth of information about the molecular and photochemical results of UV irradiation appear, at present, merely to have increased the field and scope of controversy concerning the mechanism of UV-induced mutation. Instead of there being no hypothesis, there are now three main hypotheses and conflicting evidence both for and against them all. One incriminates the formation of pyrimidine dimers, which appear to be the main cause of death, as the primary premutational event. The photoreactivability and dark repair of many UV-induced premutations favours this view. Dimer formation, and consequent intra-molecular cross-linking (p. 329), might be thought to cause mutation by distorting the DNA template so that copying errors of various sorts would be introduced at replication. Although DNA replication is inhibited following exposure to UV light, there is no reason to think that replication is necessarily delayed until *all* the lesions have been eliminated. Moreover, an Hcr^- strain of *E. coli* has been reported in which inability to repair damage to phage DNA, due to defective excision of thymine dimers (Setlow, 1966), is correlated with failure to repair premutational lesions (Hill, 1965).

Against all this is the fact that although dimer formation is induced by irradiation of isolated DNA, transforming DNA is not mutagenised by UV treatment (Setlow, 1966) although treatment with nitrous acid is effective (Litman and Ephrussi-Taylor, 1959); moreover, UV light is an ineffective mutagen for free phage particles unless the host bacteria are themselves irradiated (phage S13; Howard and Tessman, 1964b) or are multiply infected (T4: Krieg, 1959).

A second hypothesis is that dimer formation is only indirectly the cause of mutation, which results from errors in replication during the repair that follows dimer excision. It is possible that the DNA polymerase responsible for this 'patching' process lacks the precision of the enzyme which duplicates the chromosome. We have already noted (Chapter 10, p. 208) that protein synthesis is required for the expression of UV-induced mutations. While this mechanism seems to account for most of the facts, there is no direct evidence to support it.

The third hypothesis is based on the fact that one of the photochemical effects of UV irradiation of DNA is to change cytosine to uracil (p. 306). Such an alteration, if it occurred naturally, would not of itself constitute a lesion in the DNA and might be found in template strand regions exposed by excision of dimers. Under such conditions it would introduce a mutation during repair synthesis because uracil pairs with adenine instead of guanine; at the next replication this adenine would pair with thymine so that an initial CG base pair would be altered to TA. This mechanism of UV mutagenesis should therefore behave like hydroxylamine, yielding only GC→AT transitions, and there is some evidence that this may be so. A reversion analysis by various mutagens (as in Table 14, p. 315) of 16 UV-induced mutants of phage S13, which possesses single-stranded DNA, showed that in eleven the mutation had resulted from an initial C→T change; in the remaining five the initial change was either a C→T transition, or not a transition at all (Howard and Tessman, 1964b). A similar investigation in phage T4 revealed that 47 of 99 UV-induced mutations were GC→AT transitions, only six were AT→GC transitions, while the remainder were non-transitions (Drake, 1963). The main objection to this hypothesis, which is otherwise an attractive one, is that the production of uracil is usually secondary to the formation of dimers containing cytosine. The formation of cytosine hydrate offers an alternative mechanism to that of uracil. The irradiation of polydeoxycytidine (poly C) has been shown to alter its *in vitro* template properties in such a way that RNA polymerase incorporates adenine and ceases to incorporate guanine into the RNA product. This has been ascribed to the hydration of cytosine which alters it pairing affinity (Ono, Wilson and Grossman, 1965) and so could be a cause of mutation, although the amount of cytosine hydrate formed by irradiation of native DNA is too small to be revealed by spectroscopic analysis (Setlow and Carrier, 1963).

Reviews on the topic of radiation damage and its repair are: Rupert and Harm, 1966; Adler, 1966; Joset, Moustacchi and Marcovich, 1966; Setlow, 1966; Radiation Research Supplement 6 (1966) on 'Structural defects in DNA and their repair in microorganisms'.

CHAPTER 14

DNA as Genetic Material: The Nature of the Genetic Code

In previous chapters we have introduced the general proposition, arising from our knowledge of the physico-chemical structure of nucleic acids and of proteins, that the sequence of four nucleotides along the polynucleotide chain specifies the sequence of amino acid residues in the derivative polypeptides. This is the famous 'sequence hypothesis' first propounded by F.H.C. Crick (1958). We have also seen that correlations between genetic and physico-chemical measurements indicate that each amino acid is coded for by the sequence of a small number of nucleic acid bases or base-pairs and have assumed that this number, termed the *coding ratio*, is 3 (p. 299). We must now study the evidence for this assumption and then go on to consider the subtle means that have recently been evolved to 'break the genetic code' by assigning particular sets of bases as code words for each amino acid. The problem of the genetic code can be treated as an exercise in the theory of coding or, more directly, as a matter susceptible to direct genetical or biochemical analysis. We will consider each of these approaches in turn.

THEORETICAL ASPECTS OF CODING

The sequence hypothesis states that a linear array of four nucleotides codes for a linear sequence of twenty amino acids. This can be translated into more general terms by representing the amino acids as words and the nucleotides as the letters from which these words are constructed. Thus we have an alphabet of four letters, A, B, C, D, and a language of twenty words.

The first question is, how many letters are there in each word, assuming that all the words are made up of the same number of letters? If the words were single-letter ones it is obvious that only four words,

A, B, C and D, are possible. If each word has two letters, a maximum of 16 words can be made as follows:

AA	AB	AC	AD
BA	BB	BC	BD
CA	CB	CC	CD
DA	DB	DC	DD.

On the other hand, as many as sixty-four distinct 3-letter words can be constructed from an alphabet of four letters, so that a 3-letter, or triplet, code is the minimum that will satisfy our need for 20 words. Of course this does not preclude the use of longer words, but since a triplet code is the most economical way of using the alphabet we will assume it to be the most likely.

The next question is, what kind of a code are we dealing with? Let us consider first what is called an *overlapping code*, that is, a code in which the letters of consecutive words overlap one another. For example, in the sequence of triplets

..ABCABC..,

the first word would be ABC, the second BCA, the third CAB, and so on. At first, the idea of an overlapping code seemed a good one because the distance separating amino acid residues in an extended polypeptide (but not in an α-helix; see p. 261) is very similar to that between bases in the DNA (3.4Å) (Gamow, 1954). However, it is evident that a code of this sort must impose certain restrictions on the sequence of words since the word ABC in the example would always have to be followed by another word beginning BC., while the third word in sequence must begin with C. If such restrictions are inherent in the genetic code, the fact may be verified experimentally in two ways. First, if we examine the known sequences of amino acid residues in a number of biological proteins such as insulin, thyroxin, ribonuclease and so on (p. 265 *et seq.*), we should find significant associations of particular amino acids lying next to one another in the polypeptide chains. It turns out that no such associations occur. Thus all overlapping codes can be excluded, provided that the amino acid sequences of proteins from different species reflect the same genetic code so that it is valid to compare them (Brenner, 1957a). Second, if the code is an overlapping one, the alteration of a single letter will change more than one word. For instance, in the example given above, substitution of the first C of the sequence by, say, D changes the first *three* words to ABD, BDA and DAB. But we have already noted, in the case of haemoglobin (Ingram, 1957), tobacco mosaic virus protein (Tsugita and Fraenkel-Conrat, 1960, 1962; Wittman, 1962) and *E. coli* tryptophan synthetase (Helinski and Yanofsky, 1962) for example, that a single mutation usually results in the substitution of a single amino acid only. This is emphasised by the finding that on the few occasions when *two* amino acids are altered in tobacco mosaic virus protein following treatment of the virus RNA with nitrous acid, their positions are not adjacent to one another in the polypeptide chain; presumably in such cases the RNA molecule must have suffered two separate deaminations. It therefore seems certain that the genetic code is not an overlapping one.

If we are dealing with a *non-overlapping triplet code*, how is it read? For example, in a sequence ABCABCABC, indicating a repetition of the same amino acid coded for by the letters ABC, there must be some way of distinguishing the meaningful triplets from the others. One kind of code which would permit unambiguous interpretation would be a *punctuated code* such as ABC,ABC,ABC, in which every forth base represented a comma (see p. 356). Alternatively, an *unpunctuated triplet code* could be deciphered if two-thirds of the possible combinations of three letters were meaningless. For example if the sequence ABCABC-ABC is read from left to right and the triplets BCA and CAB make nonsense, the only sense that emerges is ABC ABC ABC. Similarly, in a code of this sort, all triplets of the same letter, AAA,BBB,CCC and DDD, must be regarded as meaningless since adjacent triplets such as AAAAAA would give overlapping sense and would be read as four identical words instead of two (A A A A A A). From this it follows

that, in order to read correctly an unpunctuated triplet code employing four letters, of the sixty-four possible triplets we must first eliminate four, and then two-thirds of the remaining sixty, as being 'nonsense' triplets. This variety of code therefore allows precisely twenty words to be written unambiguously, and this is just the number of amino acids which the four nucleotide bases are required to specify (Crick, Griffith and Orgel, 1957). Despite the elegance of the reasoning and the beauty of the fit, however, there are two good reasons why this theoretical solution to the coding problem proved unacceptable. In the first place, evidence is accumulating that the code is *degenerate*, meaning that more than one sequence of bases can specify a single amino acid, so that more than twenty words are required. Second, if two-thirds of all triplets are 'nonsense' and do not specify an amino acid, the substitution of any letter by another has a 2/3 probability of changing a meaningful triplet into a 'nonsense' one; but there is good evidence, especially from the study of lysozyme mutants of phage T4 (Streisinger *et al.*, 1961; Terzaghi *et al.*, 1966; see p. 360) as well as of tryptophan synthetase mutants in *E. coli* (Yanofsky *et al.*, 1963; see p. 301), that the great majority of base analogue-induced mutations result in the substitution of one amino acid by another, that is, that most alterations of a single code letter lead to 'mis-sense' rather than to 'nonsense'.

There is, however, another way of translating an unpunctuated, non-overlapping triplet code, by starting at a prescribed point and then reading off the letters *in groups of three* in one direction. Thus if X indicates the starting point for translating the sequence XABCABCABC, we get

$$\text{X}\,\underline{\text{A B C}}\,\underline{\text{A B C}}\,\underline{\text{A B C}}\;\longrightarrow$$

so that the code is read ABC ABC ABC. This method of reading the code is singularly attractive because it bears a close formal resemblance to the experimental model of how the information carried by messenger RNA is translated into protein synthesis (Figs. 73, 74, pp. 287, 295). You will remember that, according to this model, the sequence of DNA bases, transcribed into a complementary base sequence on the messenger RNA, is actually read by molecules of amino acid-transfer RNA. Each of these transfer RNA molecules is specifically joined to a particular amino acid and also carries a base sequence complementary to that which codes for the same amino acid on the messenger RNA. Thus the transfer RNA molecules seek and bond to the appropriate, complementary base sequences along the messenger RNA and, in doing so, arrange their attached amino acids in the proper order for polymerisation into a polypeptide chain. In the case of a triplet code, therefore, transfer RNA molecules behave as a kind of automatic translating machine which reads off the appropriate sequences of letters in groups of three, as represented in Fig. 82. But is there any evidence that they are read off *in sequence* as theory demands?

We have already seen how an oriented, sequential reading of the code, in groups of bases equivalent to the coding ratio, accounts neatly for the disruption of protein synthesis which is found in acridine-induced mutants (p. 318). In addition there is the direct evidence, which we will not review again here, that not only are the transcription of messenger RNA, and the translation of its message on the ribosomes, polarised, but both have the same $5'-3'$ polarity (pp. 285, 294). In confirmation of this it has also been shown directly that polypeptide synthesis is a polarised process. If, during synthesis of a protein, a 'pulse' of radioactive amino acid is injected over a small fraction of the time required for synthesis of a complete polypeptide chain, the labelled amino acid will be incorporated into that part of the chain which is being built up during the period of exposure to the pulse.

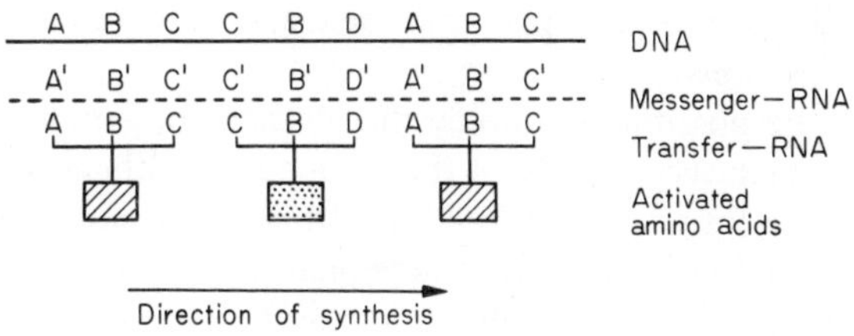

FIG. 82. Illustrating the hypothesis that transfer RNA translates the genetic code by reading the sequence of letters, transcribed into complementary sequence on messenger RNA sequentially and in groups of three.

Imagine a family of polypeptide chains which are half completed at the time the label is injected. If synthesis always begins at, say, the amino end of the chain and proceeds towards the carboxyl end, then the label will be incorporated only into the distal, carboxyl half of the finished molecules. On the other hand, if synthesis starts at many foci and spreads randomly in either direction, the label will be distributed throughout the completed chains. In practice, of course, the synthesis of protein molecules is not synchronised in this way; at the time the label is injected molecules in all stages of completion will be present. Nevertheless, oriented, sequential synthesis should reveal itself by a *gradient* of labelling from one end of the chains to the other when the completed protein molecules are extracted and the distribution of the label determined in a sequence or end-group analysis (p. 265). The synthesis of mammalian haemoglobin has been studied in this way, both in an *in vitro* ribosome system (Bishop, Leahy and Schweet,

1960) and in intact, anucleate reticulocytes (Dintzis, 1961). The results agree in showing that the peptide chains of haemoglobin are built up by the steady sequential addition of amino acids to growing chains and that the process terminates at or near the free carboxyl end.

GENETIC ANALYSIS OF THE CODE

In Chapter 13 (p. 299) we saw how estimates of the coding ratio can be derived from genetic analyses, by first equating the genetic length of a gene with the number of base-pairs of DNA which constitute it, and then dividing this number of base-pairs by the number of amino acids in the protein which the gene determines. Such estimates lie between 3 and 6 base-pairs per amino acid and tally well with the theoretical expectations.

It should be possible to employ the methods of genetic analysis to prove or refute the correctness of the sequence hypothesis itself which, after all, is the foundation for all current concepts about the genetic code. This hypothesis states that the sequence of bases, or of base-pairs, along the DNA double helix which constitute a gene, determines the sequence of amino acids along the derivative polypeptide chain. It follows that if we map the relative positions of a series of mutational sites within the gene and, at the same time, establish the position within the polypeptide of the amino acid substitution effected by each mutation, the sequence of the mutational sites should correspond with that of the equivalent amino acid changes (Fig. 83). This has now been proven for the tryptophan synthetase A protein of *E. coli*, by C. Yanofsky and his colleagues (pp. 118, 300). Mutants producing enzymatically defective protein were isolated and the amino acid alteration in the protein ascertained, by chemical analysis of the particular peptide in which the mutant and wild type proteins differed. In some cases different mutations were found to involve substitutions in the same peptide, but usually the relative positions of the different peptides had to be mapped by a sequence analysis (p. 265) which finally covered a large region of the protein, containing 75 amino acids out of a total of about 300. The genetic sites of nine mutations were then mapped. Not only were the sequence of the mutations and that of the corresponding amino acid substitutions the same, but the recombination frequencies turned out to be surprisingly proportional to the linear distances separating the relevant amino acids on the polypeptide chain (Yanofsky *et al.*, 1964).

At about the same time, co-linearity between gene and protein was demonstrated for the head protein of phage T4 by a quite different and technically simpler approach, which does not involve determining the amino acid sequence (Sarabhai *et al.*, 1964). There is a type of mutation known as 'amber', which we will discuss later in this Chapter, which alters the code word for certain amino acids into a new code

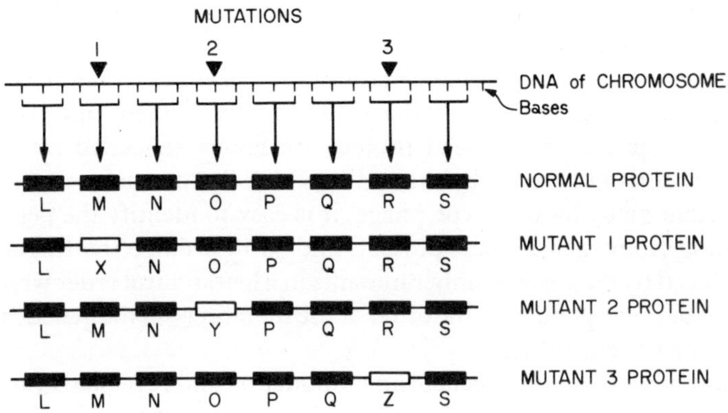

FIG. 83. The relationship between the sequence of mutational sites within a gene and the sequence of amino acid substitutions in the derivative protein, as postulated by the sequence hypothesis.

The positions of mutational sites 1, 2 and 3 are indicated thus, ▼

The black rectangles represent the amino acid residues, L-S, in the wild type protein determined by the gene; the white rectangles, X, Y and Z, are substituted residues resulting from mutation.

The sequence of the mutational sites on the chromosomes coincides with that of the equivalent amino acid substitutions in the protein.

word meaning 'end of protein' (see p. 361); in consequence, synthesis of the polypeptide proceeds from its N-terminal end as far as the amino acid specified by the altered codon and then stops (Stretton and Brenner, 1965). It follows that if there is colinearity between gene and protein, different amber mutations should produce homogeneous populations of polypeptide chain fragments whose lengths are proportional to the distances of the mutations from one end of the gene. Thus in Fig. 83, assuming the N-terminal amino acid to be J (off the Fig. on the left), an amber mutation at site 1 will yield JKL fragments, at site 2 JKLMN fragments, and at site 3 JKLMNOPQ fragments.

It is possible to isolate such phage mutants because of their ability to grow in certain 'permissive' bacterial hosts which carry a suppressor mutation for the amber character. This suppression is probably due to production of a transfer RNA species which mis-reads the chain-terminating, amber codon as the codon for an amino acid (pp. 215, 361).

Experimentally, the non-permissive host bacteria are infected with the amber mutants and are then fed a radioactive amino acid late in the lytic cycle, when most of the protein synthesised by the cells is phage head protein. The total protein in the crude cell lysate is then separated from nucleic acids and digested with a proteolytic enzyme; the resulting peptides of the head protein are finally separated by paper ionophoresis and located by autoradiography. By comparing the results with that given by wild type phage, it is easy to identify the peptides missing from each mutant protein, and so to arrange the fragments produced by the various amber mutants in a heirarchical order which is found to correspond with the order of the mutations on the genetic map (Sarabhai *et al.*, 1964).

Recent work on the mechanism of induction and reversion of mutations by acridine dyes, which we have already reviewed in part (p. 318), has greatly illuminated the nature of the genetic code and the way in which it is read. We must now examine in more detail this brilliantly imaginative study by Crick, Barnett, Brenner and Watts-Tobin (1961) which ranks as a masterpiece of genetical analysis.

THE RELATIONSHIP OF ACRIDINE-INDUCED MUTATIONS AND THEIR SUPPRESSORS TO THE CODE

We saw in Chapter 13 (p. 318) that acridine-induced mutations in phage T4, as well as the majority of spontaneous mutations in this organism, probably arise from the insertion or deletion of a single base-pair (or a small number of base-pairs; see Okada *et al.*, 1966) which results in a disruption of protein synthesis. This disruption is best explained by supposing that the sequence of letters comprising the genetic code is read in one direction, from a specific starting point, in groups of three (or whatever the coding ratio may be); thus the removal or addition of a letter would displace the reading device by one letter from the site of alteration onwards so that every triplet thereafter would be misread. We have also seen that transfer RNA is a device which appears well adapted to reading the code in this way (Fig. 82). Reversion of acridine-induced mutations is nearly always by suppressor mutations at a different but closely linked site, which are assumed to work by adding a base where one has been deleted, and *vice versa*, so that the proper wild type number of bases is restored.

Crick *et al.* (1961) started their investigation by isolating independent, spontaneous reversions of a proflavin-induced r_{II} mutation of phage T4, located in the B1 segment of the B cistron (Fig. 49, p. 175). This was done by plating about 10^8 particles of the mutant phage on *E. coli* K bacteria on which only wild type, r^+, particles grow. The small number of revertant plaques which developed were picked and purified by plating on *E. coli* B on which, you will remember, both wild type

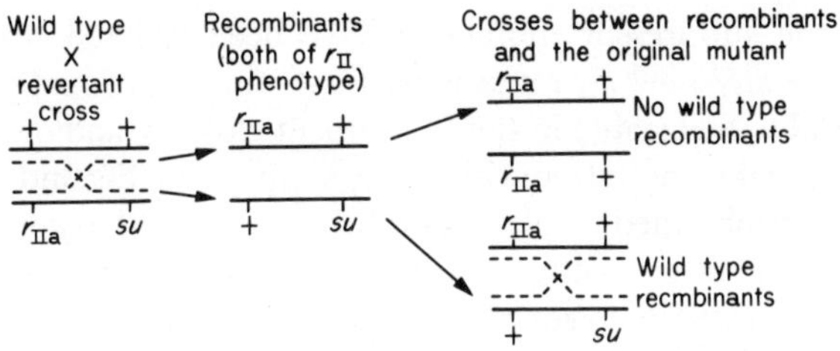

FIG. 84. Illustrating the isolation and characterisation of suppressors of acridine-induced r_{II} mutations.

The interrupted lines indicate the formation of recombinant chromosomes.

su indicates the suppressor site (see text).

and r_{II} mutants grow but form distinctive plaque types (p. 169). Of 20 revertants tested, 18 were found to differ from true wild type in the appearance of their plaques on *E. coli* B, that is, they were *pseudo-wilds*. Each of these pseudo-wild revertants was then crossed with the wild type phage in order to demonstrate the presence of recombinants of the original r_{II} mutant type whose recovery was predicted by the suppressor hypothesis. Mutant recombinants were in fact recovered but were found to be of *two genetically distinct r_{II} types*. One type yielded no wild type recombinants when crossed with the original r_{II} mutant strain, with which it was therefore identical; the other type did produce wild type recombinants in crosses with the original mutant, showing that its r_{II} phenotype resulted from mutation at a new site (Fig. 84) (see p. 214). In this way, all eighteen pseudo-wild revertants were shown to be due to suppressor mutations, and when these suppressors were isolated in the manner we have described, each was found to correspond to a new, non-leaky r_{II} mutation. We thus find that two distinct r_{II} mutations operate together in restoring wild type function, but this situation is quite different from complementation which only occurs when the two mutations are in different cistrons and in the *trans* position on opposite chromosomes. When *E. coli* K is mixedly infected with two phage particles, one carrying an r_{II} mutation and the other its suppressor, nothing happens; restoration of function requires an association of the two mutations *on the same chromosome*. This is exactly the result expected on the hypothesis that the suppressors compensate the original mutation by inserting or deleting a base-pair.

Once the suppressor mutations have been isolated, they can be treated like any new r_{II} mutations. First, they were mapped and all were found to be located in the B_1 segment close to, and on either side of, the original r_{II} mutation which they suppressed. Second, reversions of them were obtained by plating on *E. coli* K and the revertant strains were analysed in the same way as before. Once again the reversions were found to be due to suppressor mutations, that is, to *suppressors of suppressors*. When these 'second generation' suppressors were isolated they proved, like the first generation suppressors, to be typical non-leaky r_{II} mutations mapping within the B_1 segment. They, in turn, were reverted and *their* suppressors—*suppressors of suppressors of suppressors*, or 'third generation' suppressors—were isolated. In this way some eighty new r_{II} mutations of the non-leaky, but spontaneously reverting, acridine type were obtained.

On the assumption that the original acridine-induced mutation resulted from, say, the insertion of an extra base-pair, and that the reversions were due to a compensating deletion of a base-pair at other sites close to it, we can assign a '+' sign to the original mutation and a '—' sign to its suppressors. Similarly all the suppressors of these suppressors would be '+' mutations, while the third generation suppressors would be '—' mutations. It is immaterial whether the original mutation is really an insertion or a deletion; the important point is that a mutation and its suppressor have opposite signs. According to this convention, of the eighty new suppressor r_{II} mutations, some were + and some —. What happens if new combinations of these mutations are synthesised by recombination? We would expect that combinations of two + (++) or two — (——) mutations would always yield the mutant, r_{II} phenotype because the effect would be to add or subtract *two* base pairs and this could not shift into its proper reading sequence a device which reads in groups of three. This was confirmed experimentally. Fourteen pairs of $++$ or $--$ mutations were constructed and all turned out to be of r_{II} phenotype.

On the other hand, if the coding ratio is really three, or a multiple of three, the addition or subtraction of *three* bases should not throw the reading mechanism out of alignment with the code. For example if we imagine, for simplicity, that the wild type sequence is

A B C A B C A B C A B C A B C

then the addition of three bases (X, Y, Z) changes the sequence to

Insertions

A B C X A B Y C A B Z C A B C A B C;

similarly the deletion of three bases gives the sequence

Deletions

A B C B C A C B C A B C · · · .

In both cases the triplets are read correctly beyond the distal insertion or deletion. The code is, of course, misread in the region between the outermost insertions or deletions but the situation here is no worse than when a $+$ and a $-$ mutation, which together restore the wild type, are separated by the same interval, thus,

Deletion Insertion

A B C B C A B C A B C X A B C · · ·

In all, six triple mutants, five of $+++$ type and one of $---$ type, were constructed by Crick *et al.* (1961); all formed plaques on *E. coli* K and displayed a pseudo-wild phenotype on *E. coli* B. The dramatic nature of this result is stressed by the fact that each of the individual mutations of one $+++$ combination gave wild type in association with a particular — mutation, but in isolation and in all pairwise combinations with one another only the mutant phenotype was produced. Thus the coding ratio is probably 3 or, if more than one base-pair is added or removed at a time, some multiple of 3. There is, as we have seen, some indirect evidence which favours the usual involvement of a single base-pair, such as the physical estimate that intercalation of an acridine molecule exactly doubles the distance between adjacent bases (p. 322) and the fact that both acridines and spontaneous events can result in the same alterations. Moreover, as we shall shortly see, biochemical experiments also support a coding ratio of 3.

Latitudes and restrictions of the code

There are several features of the work of Crick *et al.* (1961) which we must now peruse more closely. The hypothesis explaining the reversion of r_{II} mutants to pseudo-wild phenotype by means of suppressor mutations implies that, in the revertant, the genetic code continues to be deranged in the segment of chromosome contained between the sites of the two mutations. Although the distance between the two sites may be very short, it can also be as long as the entire B_1 segment which constitutes about 1/10 of the B cistron. If we assume that the B cistron determines the structure of a protein, and that the coding ratio is 3, it is easy to calculate that this distance carries the coding sequence for some twenty amino acids. It is hardly likely that these amino acids can simply be omitted from the presumptive protein specified by the B cistron and still leave it functional. This suggests that although the sequence is completely deranged in the interval between the sites of the original mutation and its suppressor, nevertheless it does code for amino acids even though these may be different from those in the wild type protein. Since the reading of each of some 20 triplets may be affected, this means that the code is probably *degenerate* in the sense that some amino acids, at least, can be specified by more than one triplet sequence.

Another implication of the pseudo-wild revertants is that that part of the protein which is specified by the B_1 segment cannot be of much functional importance if such liberties can be taken with its structure. There is a deletion (No. 1589, Benzer, 1961; Crick *et al.*, 1961) which extends over part of the A cistron as well as the whole of the adjacent B_1 segment of the B cistron into which all the studied r_{II} suppressor mutations fall. By means of complementation tests it can be shown that although this deletion prevents the A cistron from expressing its function, it fails to suppress B cistron function (Fig. 85, a and b; Champe and Benzer, 1962a). Now normally the A and B cistrons are functionally independent so that a mutation or a deletion in one has

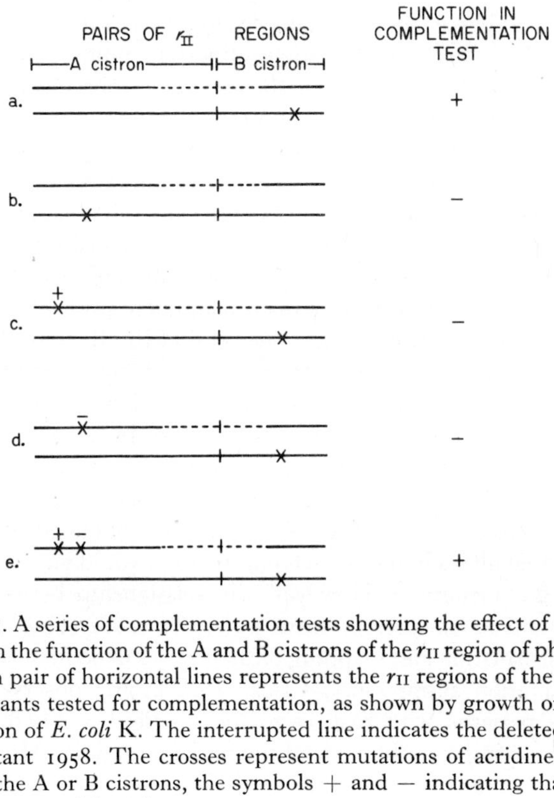

FIG. 85. A series of complementation tests showing the effect of deletion 5819 on the function of the A and B cistrons of the r_{II} region of phage T4.
Each pair of horizontal lines represents the r_{II} regions of the pairs of r_{II} mutants tested for complementation, as shown by growth on mixed infection of *E. coli* K. The interrupted line indicates the deleted region in mutant 1958. The crosses represent mutations of acridine type in either the A or B cistrons, the symbols + and − indicating that one is a suppressor of the other.

no effect on the function of the other. In the deletion mutant 1589, however, the introduction of an r_{II} mutation of acridine type into the A cistron abolishes the function of the B cistron (Fig. 85, c and d), but this is restored if a + and a − mutation are present together in the A cistron (Fig. 85, e). Everything behaves as if deletion 1589 has cut out some kind of punctuation between the two cistrons which normally ensures that the A and B codes are read individually. The result is that the two genes become joined so that the two codes are read off from left to right as a continuous sequence. If the wild type genes really specify protein structures, and were it possible to examine the product, it is likely that only a single protein would be found (Crick *et al.*, 1961). This kind of mechanism could well explain the development of such differences as exist between the tryptophan synthetase produced by *Neurospora* and by *E. coli*, for example (Fig. 43, p. 156). It also strongly reinforces the hypothesis that an acridine-type mutation disrupts the correct, sequential translation of the code over long distances. However, the fact that deletion 1589 eliminates the entire B_1 segment and yet permits the B cistron to function in the new conjoint protein, does *not* mean that this segment is dispensible under *normal* conditions when the product of the B cistron exists as an individual entity. There is no doubt, however, that the segment is an unusually malleable one and, therefore, well adapted to the demonstration of reversion by suppressors. Crick *et al.* (1961) have made a tentative study of an acridine-induced mutation located at the other end of the B cistron, in the B_9 segment, and find that this is also reverted by suppressor mutations, but these appear much more closely linked than those in the B_1 segment.

All the selected pseudo-wild reversions arose, by definition, from the association of two mutations of opposite sign (+ —). If these mutations are isolated, and + and — mutations then reassociated in combinations other than those in which they were originally obtained, it might be supposed that all such new + — combinations would yield the wild or pseudo-wild phenotype. However, the relationship between an r_{II} mutation and its suppressor turns out to be more subtle than this. During the isolation and mapping of their series of 80 first, second and third generation suppressors, Crick *et al.* (1961) observed certain restrictive patterns with respect to the distribution on the map of suppressors lying on either side of the initial mutations. Moreover the restrictions were correlated with the *polarity* of the + and — signs. This rather complicated matter is reduced to its essentials in Fig. 86 where the results of mapping a hypothetical series of first and second generation suppressors is represented. You will see that no + suppressors are found beyond a certain 'barrier' region when the pairs of mutations lie to the right and have a + — polarity. On the other hand, the 'barrier' allows suppressors to cross it if the polarity is reversed; that is, if the — mutation is to the left of the barrier, then + suppressors can lie to the right of it (Fig. 86, bottom line). Recognition of certain barrier regions in this way allowed Crick *et al.* (1961) to predict which new combinations of + and — mutations would, or would not, produce wild type phage, and all these predictions were confirmed by experiment. For example, the findings of Fig. 86 would predict that the + mutation on the extreme left of the bottom line would fail to act as a suppressor for any of the — mutations to the right of the 'barrier', whereas all other +— and —+ combinations would be compensatory.

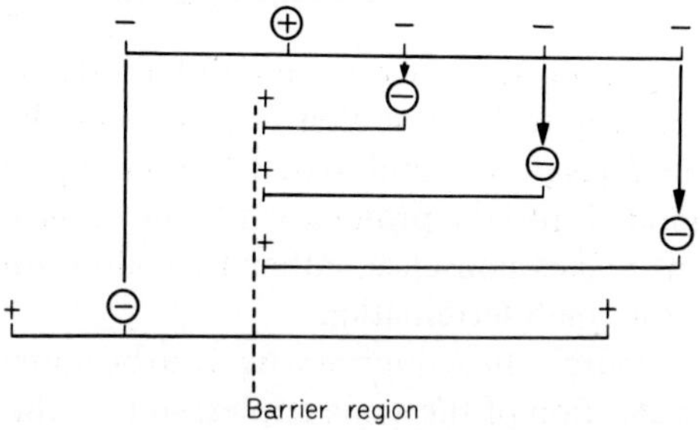

FIG. 86. Diagram illustrating the recognition of a barrier region across which insertions (+) and deletions (—) of a given polarity (in this case + —) are no longer compensatory.

Each line indicates the extreme distances of suppressor mutations from an initial mutation indicated by a ring. Note that no + suppressors, *lying to the left of an initial mutation*, cross a certain region designated the barrier region. On the other hand, the region offers no barrier to + suppressors when these lie to the *right* of an initial — mutation. See text for explanation.

The maps are entirely arbitrary and hypothetical.

What is the nature of these barriers which discriminate between the polarity of mutations? If we look at a coding sequence such as

A B C A B C A B C A B C A B C ,

it is evident that although it can be read correctly in only one way, as shown, there are *two* ways in which it can be read incorrectly, and these two ways depend on whether the reading device is shifted one letter to the left by removal of a letter, or one letter to the right by addition of a letter. For example, if we delete a letter on the left and compensate for this by adding a letter to the right, thus,

Deletion (—) Insertion (+)

A B A B C A B C A B X C A B C ,

the sequence between the deletion and insertion is misread as BCA BCA. Alternatively, if the polarity of the alterations is reversed we get

Insertion (+)   Deletion (−)

A B X C A B C A B C A B A B C ,

so that the intervening sequence is now differently misread as CAB CAB CAB. It is easy to imagine that one of these altered sequences (such as BCA BCA) is meaningful so that a series of permissible amino acids is incorporated into the protein, while the other sequence (CAB CAB and so on) makes nonsense, either by coding for unacceptable amino acids or for chain termination.

In practice, of course, the sequence is likely to be much more complex than a simple repetition of the same triplet, so that the probability of encountering a *single* forbidden triplet as a result of a change of polarity is increased. If, therefore, we assume that a particular barrier is due to such a triplet, we can investigate it experimentally in two reciprocal ways. Firstly, it should be possible to change the unacceptable triplet into an acceptable one by altering one of its bases by means of a chemical mutagen. This has been accomplished by Crick, Brenner, Barnett and Shulman (personal communication, 1962), by constructing a phage strain having a + and a − mutation on either side of a 'barrier' region, so that it continued to display the mutant, r_{II} phenotype. When this strain was treated with 5-bromodeoxyuridine and plated on *E. coli* K, wild type plaques were recovered. Subsequent genetic analysis revealed that the mutation to wild type was due to an alteration which mapped within the barrier region. Secondly, base analogue-induced r_{II} mutations in the B_1 segment may be presumed to have changed a normal triplet into an unacceptable one by the alteration of a single base. If an acridine-induced r_{II} mutation and its suppressor are introduced into such a strain, one on either side of the base analogue-induced mutation, the reading device should be shifted in such a way that the previously unacceptable triplet might now be read as acceptable, with the result that the mutant reverts to wild type. This prediction has also been confirmed experimentally (Crick *et al.*, personal communication, 1962).

All the important conclusions which emerge from these beautiful and elegant studies (reviewed in detail by Barnett *et al.*, 1967) have now been confirmed directly, by the sequence analysis of lysozyme synthesised in infected cells by a pseudo-wild double mutant (—+) of phage T4. This double mutant was constructed by recombination between two proflavine-induced mutants having closely linked mutations of opposite sign. A comparison of the primary structures of the pseudo-wild and wild type proteins showed conclusively that they differ by five consecutive amino acids (Terzaghi *et al.*, 1966). We shall see later that a consideration of the precise amino acid substitutions resulting from the frame-shift, in terms of specific nucleotide triplets proposed for these amino acids from bio-chemical studies, has unambiguously defined the codons actually used by phage T4 (p. 361). In addition it has indicated which mutation is the deletion (—) and which the addition (+), has shown that, in this case, only a single base was initially deleted, and has confirmed that the 5′ to 3′ polarity of the nucleotide triplets must parallel the N-terminal to C-terminal polarity of the amino acids (Terzaghi *et al.*, 1966). Frame-shift mutations are specifically induced by certain acridine mustards (Ames and Whitfield, 1966) but, unlike all other types of mutations, are not reverted by nitrosoguanidine (Whitfield, Martin and Ames, 1966).

BIOCHEMICAL ANALYSIS OF THE CODE

Our discussion so far, however informative it may have been about the general nature of the genetic code, has scarcely touched upon the ultimate problem of the actual nucleotides which code for each amino acid. Spectacular and unexpected advances in this aspect of coding have come from the study of *in vitro*, cell-free systems which can synthesise protein. If disrupted *E. coli* bacteria are centrifuged at high speed and the ribosomes (isolated from the deposit) are mixed with the supernatant (containing transfer RNA and enzymes) in the presence of ATP and an ATP-generating system, it is found that ^{14}C-labelled amino acids are incorporated into protein at a high rate. This incorporation is severely depressed by ribonuclease or chloramphenicol, and is stimulated by addition not only of an amino acid mixture but also of relatively high molecular weight RNA. RNA isolated from tobacco mosaic virus is much more efficient than ribosomal RNA in this respect. The inference is that the added RNA serves, like messenger RNA, as a template for protein synthesis, tobacco mosaic virus RNA being more effective because it carries the full specifications for a protein. When precautions are taken to exclude pre-existing messenger RNA from the system, the synthesis of protein becomes dependent on the addition of such extraneous RNA (Nirenberg and Matthaei, 1961; review, Nisman and Pelmont, 1964).

We have already described how artificial polyribonucleotides can be synthesised *in vitro* from ribonucleoside diphosphates by means of the enzyme polynucleotide phosphorylase (Ochoa and Heppel, 1957; see p. 292). This enzyme can couple together any pair of bases so that precursor nucleotides are incorporated randomly into the emerging polynucleotide in proportion to their initial concentrations. Thus in the presence of ribouridylic acid (U) alone, a chain consisting solely of this nucleotide (poly-U) is built up; but if a second base such as adenylic acid (A) is added, then the resulting poly-AU chain will contain AA, UU and AU sequences with a frequency proportional to the initial concentration of uridylic and adenylic precursors. The exciting discovery was made, and first reported by M. W. Nirenberg at the International Congress of Biochemistry at Moscow in 1961, that the addition of synthetic poly-U as sole template RNA in the *in vitro* protein-synthesising system, leads to the specific uptake of ^{14}C-labelled L-phenylalanine to form a highly insoluble polypeptide which, on analysis, yields only phenylalanine residues (Nirenberg and Matthaei, 1961; Lengyel, Speyer and Ochoa, 1961). It was further shown that the poly-U does not affect the activation of phenylalanine or its union with transfer RNA which takes place normally prior to incorporation (Nirenberg, Matthaei and Jones, 1962; see p. 284), that synthetic RNA copolymers containing other bases in addition to U are less effective templates, and that one phenylalanine residue is incorporated into the polypeptide for approximately every 3·5 bases in the poly-U chain (Lengyel *et al.*, 1961). Moreover, if the preparation of poly-U is mixed with a complementary poly-A preparation so that double- or triple-stranded RNA molecules are formed by hydrogen bonding (Warner, 1956; Rich and Davies, 1956; see p. 273), no template activity is found, suggesting that messenger RNA must be single-stranded (Nirenberg *et al.*, 1962).

This discovery obviously meant that, assuming a triplet code, a succession of three uracils (UUU) in messenger RNA, equivalent to three adenines (AAA) in the chromosomal DNA, codes for phenylalanine. The homogeneous nature of the RNA template also ruled out the possibility of a punctuated code. The next question was whether other RNA base sequences could be constructed which would reveal the code for other amino acids. This investigation was undertaken by two groups of workers and, within months of the initial observation, bases corresponding to 19 of the 20 essential amino acids were deciphered (Martin *et al.*, 1962; Speyer *et al.*, 1962a, b, 1963; Nirenberg *et al.*, 1963; Wahba *et al.*, 1963). This remarkable progress was achieved in the following way. As we have said, the evidence suggests that, in the *in vitro* synthesis of polyribonucleotides, precursor bases are randomly incorporated into the chain according to their proportion in the mixture. Thus if uracil and cytosine are present in the proportion of, say, U:C=3:1, the most frequent triplet in the derived polynucleotide will be UUU; UUC, UCU and CUU associations will be less frequent, and those of 2C+1U, and 3C, decreasingly rare. When such a polynucleotide is used as messenger RNA in a protein-synthesising system, if one of the 2U+1C triplets codes for an amino acid, we will expect this amino acid to be incorporated into the polypeptide at about one third the frequency of phenylalanine. Experiment shows that in the presence of poly-UC having a presumptive U:C ratio of 3:1, serine appears in the polypeptide, the ratio phenylalanine:serine being 3 (Speyer *et al.*, 1962a). It follows that the code for serine is probably 2U+1C, but whether the triplet sequence is UUC, UCU or CUU cannot be inferred.

Further work showed that poly-UC having uracil in excess also promotes the incorporation of leucine, together with traces of proline, suggesting 2U+1C as the code for leucine and 2C+1U or 3C for proline. Conversely, a UC-copolymer containing an excess of cytosine stimulates uptake of proline and threonine which can thus be provisionally allotted the formula 2C+1U since poly-C, poly-G and poly-A are without significant effect. However, various UA- and UG-copolymers lead to the incorporation of further amino acids, while some are incorporated only in the presence of copolymers such as poly-UAG, poly-UGC or poly-UAC which contain three RNA bases.

Remarkable as these advances were, they revealed only the base composition and not the base sequences of the codons, while low levels of incorporation, from which certain codons were inferred, were found to depend on such environmental factors as ionic concentration and temperature and were clearly unreliable.

We owe our present very precise knowledge of the genetic code to two further developments. The first of these was the finding that the addition of particular trinucleotides to an *in vitro* protein-synthesising system causes specific binding to the ribosomes of the amino acid-transfer RNA complex for which the trinucleotide codes (Nirenberg and Leder, 1964). For example, the trinucleoside diphosphate

TABLE 15. The genetic dictionary.

Here are the 64 possible triplets of the messenger RNA bases, uracil (U), cytosine (C), adenine (A) and guanine (G), together with the amino acids they code for.

Ala: alanine	Gly: glycine	Pro: proline
Arg: arginine	His: histidine	Ser: serine
Asp: aspartic acid	Ileu: isoleucine	Thr: threonine
AspN: asparagine	Leu: leucine	Trp: tryptophan
Cys: cysteine	Lys: lysine	Tyr: tyrosine
Glu: glutamic acid	Met: methionine	Val: valine
GluN: glutamine	Phe: phenylalanine	

The triplets 'ochre' and 'amber' code for suppressible nonsense (see text).

From Crick, 1966a.

First base	Second base								Third base
	U		C		A		G		
U	UUU	Phe	UCU	Ser	UAU	Tyr	UGU	Cys	U
	UUC		UCC		UAC		UGC		C
	UUA	Leu	UCA		UAA	Ochre	UGA	?	A
	UUG		UCG		UAG	Amber	UGG	Trp	G
C	CUU	Leu	CCU	Pro	CAU	His	CGU	Arg	U
	CUC		CCC		CAC		CGC		C
	CUA		CCA		CAA	GluN	CGA		A
	CUG		CCG		CAG		CGG		G
A	AUU	Ileu	ACU	Thr	AAU	AspN	AGU	Ser	U
	AUC		ACC		AAC		AGC		C
	AUA		ACA		AAA	Lys	AGA	Arg	A
	AUG	Met	ACG		AAG		AGG		G
G	GUU	Val	GCU	Ala	GAU	Asp	GGU	Gly	U
	GUC		GCC		GAC		GGC		C
	GUA		GCA		GAA	Glu	GGA		A
	GUG		GCG		GAG		GGG		G

GpUpU attaches only valine to ribosomes; the trinucleotides
5′ 3′
UpUpG and UpGpU do not attach valine, but leucine and cysteine
5′ 3′ 5′ 3′
respectively. Apart from preparing the trinucleotides, the method is a very simple one which does not involve peptide analysis. To 20 tubes containing the cell-free system and a particular trinucleotide, is added a mixture of all 20 amino acids of which a single one, different for each tube, is radioactive. The ribosomes are then separated from the unbound amino-acyl-t-RNA complexes by filtration on a nitrocellulose membrane which retains the ribosomes, and the particular amino acid bound is identified by its radioactivity. Although some trinucleotides give doubtful results, the method has permitted the amino acids specified by nearly 50 of the 64 possible triplets to be unambiguously defined (Nirenberg *et al.*, 1965; Söll *et al.*, 1965).

The code words arrived at in this way have been confirmed and extended by several other kinds of method. The first was developed by Khorana and his colleagues who succeeded in synthesising long RNA molecules having precisely repeating sequences of bases, and then observed the amino acid sequence of the derived polypeptide when such polynucleotides were used as messenger RNA in a cell-free system. For instance, the repeating dinucleotide sequence UGUGUGUGUG . . . should give the message UGU-GUG-UGU . . . so that the polypeptide should comprise an alternation of the two amino acids coded for by these triplets; it turned out that the product was cysteine-valine-cysteine-valine . . . so that UGU is confirmed as the code for cysteine and GUG for valine (Nishimura, Jones and Khorana, 1965). A repeating *tri*nucleotide sequence, on the other hand, should yield homopolypeptides containing only one amino acid but, because of ambiguity about the starting point, three different homopolymers may be produced. For example, the nucleotide sequence AUCAUCAUC could be read as AUC-AUC-AUC, UCA-UCA-UCA and CAU-CAU-CAU; experimentally, this sequence yields polyisoleucine, polyserine and polyhistidine, thus confirming the previously postulated coding sequences of AUC for isoleucine, UCA for serine and CAU for histidine (Nishimura *et al.*, 1965; see also Morgan, Wells and Khorana, 1966). Table 16 shows the 'genetic dictionary' which has come from all these studies, as well as from mutational studies (see below).

Two other methods which we will mention briefly are genetic, and are of particular value because they check the code elucidated by *in vitro* experiments with artificial polynucleotides, against the results of events which have occurred in the living cell. The first sort of comparison examines the amino acid substitutions which result from mutations to see if they can be accounted for, as theory demands, by a single base change in the postulated coding triplets. Extensive data are available for this purpose, particularly from analysis of abnormal human haemoglobins by many workers, of the protein of tobacco mosaic virus (Tsugita and Fraenkel-Conrat, 1960, 1962; Wittman, 1962; Wittman and Wittman-Liebold, 1963) and of the A protein of *E. coli* tryptophan synthetase (Yanofsky, 1963). Despite the diverse organisms from which these proteins were derived, it turned out that only two cases were found where the observed amino acid substitution could not have resulted from a single base substitution in the proposed coding triplet. The exceptions themselves provided reinforcing evidence for the code. One involved an abnormal haemoglobin in which the amino acid substitution Lys→Asp was reported. Reference to Table 15 shows that the codes for these amino acids differ by two of the three bases. However, because of this discrepancy, the substitute amino acid was re-examined and found, in fact, to be Glu whose code differs from that of Lys by only a single base (Beale and Lehmann, 1965; quoted by Stretton, 1965). The second case was a Glu (GAA or GAG)→Met (AUG) change in *E. coli* tryptophan synthetase, but the mutant turned out to be the only one of all those studied which does not revert to wild type and may therefore be regarded as a rare case of double mutation (Yanofsky, 1966; Yanofsky, Ito and Horn, 1966).

The most recent genetic check on the code comes from an analysis of the altered amino acid sequence of lysozyme produced by a pseudo-wild, frame-shift (—+) type double mutant of the bacterial virus T4, which we have already recounted (p. 348). The double mutation was found to have changed the amino acid sequence from

Lys-Ser-Pro-Ser-Leu-AspN-Ala-Ala

to Lys-Val-His-His-Leu-Met-Ala-Ala.

Assuming that the upper sequence has been changed to the lower by the deletion and addition of a base, if we now turn to Table 15 and try to see what codons would fit the observed change of sequence, we find that there is a unique set for both wild type and double mutant which alone can account for the facts, thus:

```
        −
Lys   ↑ Ser   Pro   Ser   Leu   AspN  Ala
A A ?(A)G U  C C A  U C A  C U U  A A U  G C ?

A A ? G U C  C A U  C A C  U U A  A U G (G)C ?
Lys   Val   His   His   Leu   Met   ↑Ala
                                     +
```

As mentioned earlier, this unique solution requires that the base deletion and addition must have occurred as shown, so that it is now possible to attribute the correct sign (+ or −) to any phase-shift mutation by reference to the parents of this double mutant, and also that the N-terminal end of the protein corresponds to the 5′ end of messenger RNA (Terzaghi *et al.*, 1966).

Chain-terminating Triplets

We have seen how the so-called 'amber' mutations involving the head protein of phage T4 were used to prove the co-linearity of the polynucleotide and polypeptide chains (p. 345; Sarabhai *et al.*, 1964; for the derivation of the name 'amber', see p. 484). These phage mutations have three unusual features. Firstly, they may arise at any region of the chromosome and involve any character indiscriminately. Secondly, they are suppressible so that mutants carrying them can grow in certain permissive (su^+) hosts but not in non-permissive (su^-) hosts, the Su character being determined at a single locus; accordingly they have been called 'conditional lethals' (p. 483). Thirdly, they are nonsense mutations which interrupt the genetic message, as was first shown by their ability, when in the $r_{II}A$ cistron, to abolish *B* cistron function in the r_{II} mutant 1589 in which the two cistrons are joined by a deletion (p. 351; Fig. 85; Benzer and Champe, 1962). Finally, their role in terminating polypeptide synthesis was demonstrated (Sarabhai *et al.*, 1964; Stretton and Brenner, 1965).

How do we know whether amber mutations act at the level of transcription or of translation? Put another way, are only N-terminal fragments of protein synthesised because only equivalent fragments of messenger RNA are presented to the ribosomes, or because the mutation causes the growing polypeptide chain to be released from an otherwise normal messenger? We have seen how a pair of closely linked phase-shift mutations of opposite sign alters every codon located between them. If two such mutations were introduced on either side of an amber mutation, the chain-terminating triplet would be expected almost certainly to be wiped out by the phase shift and to be substituted by an amino acid codon. In several cases where this has been done the amber mutation has disappeared, suggesting strongly that a normal m-RNA molecule is transcribed and that chain termination results from the reading of a special triplet which means 'end of protein'. Since transfer RNA molecules are the only ones we know that specifically recognise codons in the m-RNA, it is logical to think that the amber triplet attracts a species of t-RNA which does not carry an amino acid but some special component which releases the growing polypeptide chain when this is transferred to it (see p. 286; Brenner, 1966). Amber mutations are by no means restricted to phage T4, but have also been demonstrated in other phages such as λ (Campbell, 1961) as well as in *E. coli* (alkaline phosphatase mutants; Garen and Siddiqi, 1962). There is no reason to suppose that they will not be found wherever they can be looked for. In the case of bacteria, the nature of the mutation is recognised by its suppressibility by specific amber suppressors when these are introduced into the mutant by recombination.

More recently, another class of suppressible nonsense mutations has been discovered in phage T4 and has been given the name of *ochre*. Ochre mutations have a range of suppressors of their own and are not suppressed by amber suppressors, although amber mutations may be suppressed by some ochre suppressors. Ochre mutations have been shown to disrupt the reading of the genetic message, by observing their effect on $r_{II}B$ cistron function in deletion mutant 1589 (see above), but have not yet been proved to be chain terminating (Brenner and Beckwith, 1965).

The nucleotide triplets which code for the amber and ochre characters have been elucidated by a most ingenious line of reasoning and experiment which, however, is too intricate to describe here, beyond saying that it is based mainly on the specific mutagenic action of hydroxylamine (p. 307) and on a study of the substitutive relationships between amino acids of the phage T4 head protein and amber mutations (see Stretton, 1965; Brenner, 1966). The amber triplet turns out to be UAG and the ochre triplet UAA (Table 15). Since the two triplets differ by only a single base they should be inter-convertible by a single mutational step and, indeed, the ochre to amber mutation has been demonstrated (Brenner, Stretton and Kaplan, 1965). The presence of the trinucleotides UAG and UAA does not lead to the attachment of any amino acid to ribosomes while, in the case of amber mutations, the amino acids which the mutation replaces or reverts to are fully compatible with the triplet UAG. For example, in the head protein of phage T4, transition mutations connecting amber to tryptophan (UGG) and glutamine (CAG), and a transversion to tyrosine (UAU, UAC) have been observed (Brenner, 1966), while in the case of *E. coli* alkaline phosphatase substitutions between amber and glutamine and tryptophan (transitions), and tyrosine, leucine (UUG), serine (UCG) and glutamic acid (GAG) (transversions) have been shown (Weigert and Garen, 1965).

It has recently come to light that the third unassigned triplet UGA (Table 15) is also a 'nonsense' triplet and does not code for an amino acid, at least so far as *E. coli* and phage T4 are concerned. From a comparison of the behaviour of UGA mutations with ochre mutations located at the same sites it seems highly probable that the UGA is also a chain-terminating triplet. UGA is not suppressed by amber or ochre suppressors but a mutant strain of *E. coli* which does suppress it specifically and strongly has been isolated (Brenner *et al.*, 1967; Sambrook, Fan and Brenner, 1967).

A question which naturally arises is, does any of the amber, ochre or UGA triplets act as a normal chain terminator in protein synthesis? We have already mentioned that UGA can be efficiently suppressed. Amber suppressors also happen to be rather efficient so that, in su^+ strains, between 30 and 65 per cent. of initiated polypeptide chains proceed to completion. This virtually rules out the UGA and amber triplets as natural chain terminators since, in an efficient su^+ strain, a majority of protein molecules translated from a polycistronic message would not be terminated but would remain joined to others, so that the cell would hardly be viable. On the other hand, the ochre triplet could serve this role since ochre suppressors are weak (Kaplan, Stretton and Brenner, 1965). Alternatively, natural chain termination might be due to some quite different mechanism. From the evolutionary point of view these nonsense triplets must be assumed to possess some positive survival value, for otherwise they would have become amino acid codons. Like phase-shift mutations, chain-terminating mutations have polar effects, but for a different reason which turns out to have important implications. We will discuss this polarity on p. 718.

THE NATURE OF THE CODE

The most interesting and important features of the genetic code are its degeneracy and its universality. The word 'degeneracy' denotes the fact that the code is imprecise in so far as all the amino acids except methionine and tryptophan are represented by more than one triplet and, in some cases (arginine and leucine) by as many as six (Table 15). However, there is a considerable measure of order in this degeneracy; the assignment of triplets is far from random. For all the amino acids, the first two bases are fixed. In the case of the eight amino acids which have a 'box' to themselves in Table 15, the first two bases of the triplet uniquely indicate the amino acid while the third may be any one of the four bases. In all cases there is no discrimination between U and C in the third position, and in some (leucine, lysine, glutamic acid and arginine) A and G are also interchangeable but, again, only when in the third position. However, although the first two bases of the various codons are far from being specific for particular amino acids, and although it is not known for certain that the same triplet may not sometimes represent more than one amino acid, it is clear that the code as a whole is highly specific in the sense that it leaves little room for errors in protein synthesis.

In practice, the messenger RNA code is read and interpreted by the molecules of transfer RNA. If you will refer back to the sequence of five amino acids in phage T4 lysozyme which are altered by the two phase-shift mutations (p. 361), you will see the striking fact that this organism may use more than one codon for the same amino acid. Thus serine is represented by AGU and UCA, leucine by CUU and UUA, and histidine by CAU and CAC. This raises two possibilities; either there is more than one type of t-RNA which can bind the same amino acid but recognise different synonymous codons, or else a single t-RNA molecule can recognise synonymous codons as meaning the same amino acid. It is almost certain that both these mechanisms operate, for not only have t-RNA molecules specific for certain amino acids, such as leucine, been separated into types with different physical properties, but the single type of alanine-specific t-RNA, whose nucleotide sequence is known (p. 281), has been shown to recognise at least three of the alanine codons which differ by their third base only (review: Brown and Lee, 1965; see also Crick, 1966a). It is therefore probable that the use of two codons differing in the first or second base, or in both, for the same amino acid, as in the case of serine and leucine in lysozyme (above), indicates the operation of separate types of t-RNA molecules. On the other hand, where only the third base is involved, and especially if the difference is between U and C as in the two histidine codons (above), then it is likely that both codons are recognised by the same t-RNA type. This ambiguity has been ascribed by Crick (1966a, b) to a 'wobble' in the pairing between the third bases of codon and anti-codon due to different possibilities for hydrogen bonding which exist between the various bases. Thus it is suggested that, at the third position, U on the anti-codon may pair with A or G, G with U or C, and I (inosine, an analogue of guanine found in t-RNA) with U, C or A. These contraventions of the strict specificity of base pairing found in DNA arise from the likelihood that the physical interactions between t-RNA and m-RNA in codon recognition will be very different from the rigid structural requirements imposed by the DNA double helix (see p. 284).

The genetic code is remarkable in combining accuracy in directing protein synthesis with a high degree of flexibility in its response to mutational change. Thus the degeneracy of the code minimises the likelihood of mutation leading to amino acid substitution. We have already observed a parallel and even more striking degeneracy in the amino acid composition of proteins, in the form of the haemoglobin molecule, in which extensive amino acid substitutions may occur without affecting structure or function (p. 264).

The evidence that the genetic code is a universal one is two-fold. First of all the translating machinery (essentially t-RNA molecules and associated enzymes) reads the message UUUUUU . . ., for example, as polyphenylalanine, irrespective of whether it is derived from bacterial or animal or plant cells. The same point is made by the fact that all these cells can be infected by, and can synthesise the proteins of a wide variety of DNA and RNA viruses of base composition different from that of their own DNA. Secondly, extensive data on amino acid substitutions following mutation in such diverse organisms as the bacterial virus T4, the RNA-containing tobacco mosaic virus, the bacterium *E. coli* and man, are all compatible with the code set out in Table 15.

At first sight it might seem paradoxical that a universal code could be compatible with the wide variation in DNA base composition which exists between different species and, even among the bacteria, can range from about 25 per cent to 80 per cent G+C (Sueoka, 1961b; Hill, 1966). In practice, of course, this flexibility is a natural consequence of the degeneracy of both the code and the amino acid constitution of proteins which permits innocuous base substitutions which may favour either AT or GC DNA base-pairs. For example, two synonymous leucine codons are UUA and CUG, corresponding to the DNA triplets

AAT and GAC respectively; two mutations can change this DNA coding triplet from a predominantly AT to a predominantly GC one without altering the amino acid. Presumably those organisms with a low G+C content utilise maximally those DNA triplet synonyms which contain A and T, reflected by U and A in the m-RNA codons, and *vice versa*. Of course, whether or not a particular codon can actually be used depends on the presence of a type of t-RNA which can read it, and there is evidence that some cells may be unable to translate part of the information coded by DNA of very different base ratio because they lack the proper t-RNA type or types. For example, animal cells having DNA of 42 per cent G+C content have been found to produce new t-RNA types when infected with herpes simplex virus (68 per cent G+C content). Presumably these RNA molecules are coded for by the virus genome and there is evidence that they include t-RNAs for arginine and serine, but not for lysine (Subak-Sharpe, Shepherd and Hay, 1966). Reference to Table 15 will show that arginine and serine are two of the three amino acids whose codons are of two distinct types. In the case of arginine these codon types not only differ in the first base, but whereas one type (AGA, AGG) corresponds to a DNA 50 per cent G+C content, the other (CGU, CGC, CGA, CGG) is equivalent to 83 per cent GC; the two types of serine codon have no bias with respect to GC content but they do differ in both the first and second bases so that it is very unlikely that a single t-RNA type could read them both.

The adaptor hypothesis implies that there is no relationship between amino acids and their codons, and that the original assignments were determined in a random manner by historical accidents. In other words, if life were to originate all over again, we might expect a quite different code to arise. But whatever may have been the origins of the code, the fact is that, far from being random, it now shows a high degree of order, in the sense that similar triplets generally code for amino acids that are structurally or functionally related (Nirenberg *et al.*, 1965). This is particularly striking when we look at the amino acids whose codons have the second base in common. For example, in Table 15 the hydrophobic amino acids are found in columns U and C, those with branched methyl groups being in the U column and those with hydroxyl groups in the C column; hydrophilic acid and basic amino acids are found in the A and G columns. Moreover, amino acids which have a common derivation, such as ala, ser, phe, tyr, cys, trp and his from propionic acid, glu and glun from glutaric acid, and asp and aspn from succinic acid, are interconvertible by a single base substitution (Pelc, 1965). Again, Woese (1965; Woese *et al.*, 1966) has found a correlation between related codons and amino acids which show similar chromatographic behaviour (R_F values) which he takes to reflect a 'composite of all the interactions of which an amino acid is capable'.

The code, then, appears not to be random but to display a 'logic' which must have evolutionary significance. What is this logic? Two possibilities, which are by no means exclusive, have been suggested. One is that natural selection has operated on the code in such a way as to minimise the likelihood of lethal mutations. Such selection would eliminate all 'nonsense' triplets except those needed for punctuation and would maximise for base substitutions which result either in no amino acid change, or in substitution by a structurally or functionally related amino acid (Sonneborn, 1965). We have seen that the relationships of hydrophobic and hydrophilic (polar) amino acids is probably the most important factor determining the specific folding of polypeptide chains (p. 264; Perutz *et al.*, 1965). On the assumption that mutation leads to entirely random base substitutions, it has been calculated from the data of Table 15 that both one-step and two-step mutations predominantly substitute hydrophobic for hydrophobic, and hydrophilic for hydrophilic, amino acids, rather than hydrophobic for hydrophilic and *vice versa* (Epstein, 1966). Thus the pressure of natural selection on an initially random code at a very early stage of evolution could have produced the ordered and universal genetic dictionary we now observe.

The second possibility is that there exists a structural relationship between amino acids and the trinucleotides which code for them; that is, that the code possesses an intrinsic logic which determined its nature from the beginning. Some molecular models constructed by Pelc and Welton (1966; Welton and Pelc, 1966) suggested that amino acids might be fitted stereochemically to their trinucleotide codons, provided that the $H_2N.CH.COOH$ group, which is shared by all amino acids (Fig. 66, p. 260), is first bonded to the relatively non-specific third base of the codon; it therefore seemed possible that t-RNA molecules might directly recognise amino acids by carrying the appropriate triplet of codon bases in addition to the anti-codon. However, a recent reappraisal of some of their stereochemical relationships has refuted their validity (Crick, 1967a). A final judgment as to whether amino acids and t-RNA molecules have any structural affinity probably depends on elucidation of the three-dimensional configuration of the latter.

For details of every aspect of the genetic code, see Cold Spring Harbour Symposium on Quantitative Biology, 1966, Vol. 31, which is entirely devoted to this subject. Other recent reviews are by Stretton, (1965), Brenner (1966), and Crick (1967b).

THE TRANSCRIPTION OF THE CODE

Before concluding this Chapter on the nature of the genetic code we must consider two more relevant and closely related questions. What is the precise form in which the genetic information is carried by the DNA, and how is this information transcribed into messenger RNA? We have seen that the mechanism of reading the code in terms of protein synthesis is best interpreted on the basis of a single strand of messenger RNA, and that this is supported by a considerable weight of experimental evidence. The DNA, however, comprises *two* strands which carry complementary base sequences. It follows that the information can be inscribed in the DNA in one or the other of two ways. Either the sequences of bases along only one of the two strands is meaningful, in which case the sequence along the other strand must be nonsense since it is quite different, or else it is the sequence of the four *pairs* of bases, A-T, T-A, G-C and C-G, along the intact double helix which is important. Let us examine these two possibilities in more detail, assuming that the code is ultimately read from a single-stranded RNA transcript.

The principal clues came from two studies of messenger RNA. The first of these stemmed from the discovery that the *phenotypes* of certain r_{II} mutants of phage T4 could be temporarily reverted to wild type if the mutants were grown on *E. coli* K in the presence of 5-fluorouracil (5FU); however, the progeny phages which emerged from the infection retained their r_{II} *genotype* since they failed to grow on *E. coli* K in the absence of 5FU (Benzer and Champe, 1961). 5-Fluorouracil is an analogue of uracil which is readily incorporated into RNA, usually in place of uracil, but not into DNA; unlike uracil, however, it is capable of hydrogen-bonding to guanine. Since infection with phage T4 leads to immediate and almost total arrest of the synthesis of both ribosomal and transfer RNA while, at the same time, the synthesis of phage-directed messenger RNA is initiated (Nomura,

Hall and Spiegelman, 1960), it is probable that the 5FU is incorporated only into messenger RNA. Accordingly, Champe and Benzer (1962b) proposed a mechanism to explain how the reversal of mutant phenotype is effected. Fig. 87 illustrates how the transition of a cytosine-guanine (C-G) base-pair in the wild type DNA to a thymine-adenine (T-A) pair in the mutant may alter the corresponding base in the messenger RNA from cytosine (C), which pairs with guanine (G) in transfer RNA, to uracil (U) which pairs with adenine (A), thus leading to incorporation of amino acid Y instead of amino acid X into the protein. If, however, the messenger RNA is synthesised in the presence of 5-fluorouracil, U may be replaced by 5FU which, by its capacity to pair with G, will restore the wild type protein.

According to this model, only those mutant strains having an A-T base-pair at their mutant sites should show phenotypic reversion in the presence of 5-fluorouracil since it is only in such cases that the wrong message is represented by the substitution of uracil for some other wild type base in the messenger RNA. Moreover, only those mutations which have resulted from G-C→A-T *transitions* (p. 309) should respond because, had a *transversion* to A-T occurred, the equivalent to U in the wild type messenger RNA would then have been G, pairing with C in the transfer RNA; since FU cannot pair with C, phenotypic reversion due to this analogue would not be possible.

We have already recounted how Champe and Benzer (1962b) succeeded in assigning unambiguously either an A-T or a G-C base-pair to 62 transitional mutant sites on the chromosome of phage T4 on the basis of their specific revertibility by various chemical mutagens (pp. 317–318). Of the 20 mutants which gave a wild phenotypic response in the presence of 5-fluorouracil, 17 had an A-T pair at their mutant sites. The remaining three, which showed only a weak response, were A-T→G-C transitions, but these may easily be explained by the occasional incorporation of 5FU in place of cytosine rather than of uracil in the messenger-RNA (cf. Fig. 79, 1, p. 310). Thus the main predictions of the model are satisfied, in that only transition mutants are responsive to 5FU, while all those which respond well, comprising the great majority, result from G-C→A-T transitions. However, 29 of the total of 46 G-C→A-T transition mutants did not respond to 5FU. If every A-T pair was represented by U in the messenger RNA, irrespective of the orientation of the two bases, then all should respond similarly to 5FU. The fact that rather more than half do not, suggests that responsiveness is determined by whether A or T resides on a

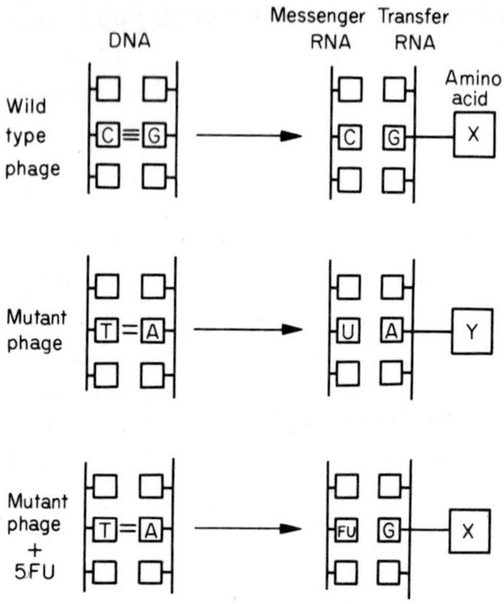

FIG. 87. Proposed mechanism for reversal of mutant phenotypes in the presence of 5-fluorouracil (5FU).

The G-C base-pair in the wild type DNA is transcribed as C in the messenger RNA; this C, as one of a coding triplet of bases, bonds specifically to G in the corresponding transfer RNA which carries amino acid X. In the mutant phage a G-C →A-T transition has occurred in the DNA; the new A-T base-pair is now transcribed as U in the messenger RNA so that the wrong transfer RNA, carrying amino acid Y, is attached. In the presence of 5-fluorouracil, U may be substituted by 5FU in the messenger RNA; since 5FU can bond to G, this messenger is indistinguishable from the wild type and so specifies incorporation of amino acid X. The DNA, however, remains mutant. (Modified from Champe and Benzer, 1962b.)

particular DNA strand; it is only when A is present on this strand that U is specified at the equivalent position on the messenger. In other words, the results approximate to what we might expect if *only one strand of the* DNA *is used as a template for physiologically active messenger* RNA.

This work of Champe and Benzer (1962b) tells us that the genetic information is probably carried by only a single strand of DNA, but it does not exclude the possibility that RNA copies of both strands are made, only one of which can be used as an effective messenger. Bautz and Hall (1962) succeeded in isolating and purifying the messenger-type RNA produced after infection of *E. coli* with wild type phage T4, and in analysing its base content. The method of initial isolation was based on the formation of specific DNA-RNA duplexes when the phage DNA is denatured to the single-strand state by heat and then allowed to cool in the presence of the RNA (see p. 276). If both strands of the DNA are represented in the messenger RNA fraction then, among the bases in this fraction, equivalent amounts of adenine and uracil, and of guanine and cytosine, should always be found. On the contrary, it turns out that the base composition of the messenger RNA fraction is different from that of the DNA; for example, 21 per cent guanine and 15 per cent cytosine were found (Bautz and Hall, 1962). This clearly indicates that only one strand of the DNA is copied. The ratio A + U/G + C, however, was 1·8 and thus identical with the equivalent ratio for the double-stranded DNA of the phage (p. 275; see also p. 374).

It is interesting to note that, to a first approximation, the quantitative results of these two quite different kinds of analysis are very similar. Thus the proportion U:A in the mutant messenger RNA studied by Champe and Benzer (1962b) corresponds to the proportion C:G in the wild type messenger (Fig. 87). As we have seen, the proportion U:A is given by the proportion of A-T mutant sites which responds to 5FU, that is 17:29, as compared to a C:G ratio of 15:21 as estimated directly by Bautz and Hall (1962). Incidentally, we may mention that the action of 5-fluorouracil in altering phenotype is not restricted to r_{II} mutants of phage T4 but also occurs in other systems. For instance, *E. coli* grown in the presence of 5-fluorouracil produces modified β-galactosidase (Bussard *et al.*, 1960) while Champe and Benzer (1962b) have demonstrated the phenotypic reversion of a proportion of *E. coli* phosphatase mutants by the analogue.

The copying of only one DNA strand by messenger RNA under natural conditions has been elegantly confirmed by the use of a *B. subtilis* phage, SP8, possessing DNA in which the density of the two strands is different because one (the denser) is rich in pyrimidines and the other in purines. When this DNA is denatured by heat the two strands can be isolated separately by centrifugation in a caesium chloride density gradient. When the separated strands are independently tested, only one is found capable of forming hybrid duplexes, with messenger RNA extracted from phage-infected bacteria (Marmur *et al.*, 1963).

If the base sequence of only a single DNA strand carries the genetic information, and if this strand alone is copied by messenger RNA, how is the transcription accomplished? We have already discussed some of the contradictory results which have been obtained when *in vivo* and *in vitro* RNA synthesis by DNA-dependent RNA polymerase are compared, and have seen that single-strand copying may sometimes depend

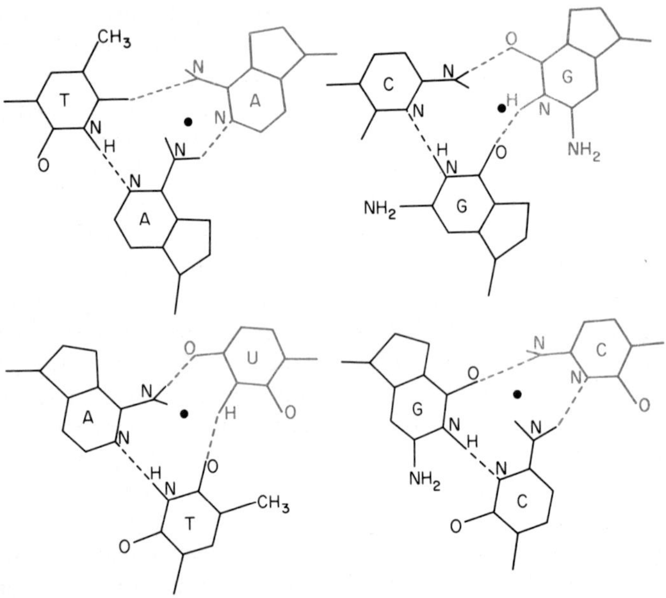

FIG. 88. Showing the hydrogen-bonding arrangements of DNA base-pairs with RNA bases, during synthesis of a 3-stranded, helical intermediate in messenger RNA formation, as postulated for bacteria (Zubay, 1962).

The black structures indicate DNA, and the green structures RNA, nucleotides.

T = thymine; A = adenine; C = cytosine; G = guanine; U = uracil.

The dotted lines represent hydrogen bond.

The black dots indicate the position of the helix axis.

on the preservation of an intact, circular DNA molecule or, in some cases, on the nature of the polymerase itself (p. 289). In all such cases, however, there has been an implicit assumption that the synthesis must be preceded by the separation and partial unwinding of the two DNA strands, even though this may be a local and transient change.

An alternative possibility is that the DNA does not unwind, but that the more or less intact *duplex* acts as a template for the synthesis of a third RNA strand, to yield an unstable, triple-helical structure composed of two DNA strands and one strand of messenger RNA. We have already seen that 3-stranded, hydrogen-bonded, helical structures are indeed possible since they are known to be formed by interaction of the appropriate, synthetic polyribonucleotides (p. 273). We must admit that there is as yet no direct evidence for such a mechanism, but it embodies so many plausible and attractive features as to merit further discussion.

It is possible to construct molecular models which satisfy reasonably well the atomic distances and bond angles required for the specific base bonding, as well as for the structural integrity, of triple helices of the proposed type. Two such theoretical models have been described by Zubay (1962); these appear to be equally stable but differ in the RNA bases which pair with the adenine-thymine and guanine-cytosine base-pairs of the DNA. The nucleotide triplets which characterise one of these models, postulated for bacteria, are shown in Fig. 88. It will be seen that one of the hydrogen bonds joining the complementary base of the two DNA strands is broken, each of the atoms thus released participating in new hydrogen bonds with a specific RNA base.

Let us look at the model represented in Fig. 88. Its base-pairing arrangements can be summarised thus:

DNA base-pairs		Messenger RNA base
A-T	—	A
T-A	—	U
G-C	—	G
C-G	—	C

Observe that every base in the RNA is the same as (or equivalent to) the corresponding base *on one particular* DNA *strand* and complementary to that on the other strand, so that the RNA has, in effect, copied only one DNA strand. Secondly, the A + U/G + C ratio of the single-stranded RNA must be the same as the A + T/G + C ratio of the double-stranded DNA, because every appearance of A and T (or of G and C) in the DNA is transcribed as either U or A (or as C or G) in the RNA. These are precisely the requirements which the *E. coli*-phage T4 data appear to demand (see p. 371).

Although the concept of a 3-stranded DNA-RNA intermediate of some kind or other in the synthesis of messenger RNA is, at present, a purely speculative one, there is a good deal of additional circumstantial evidence of a general nature which favours it. Most of the nuclear DNA appears to exist in the double-stranded form under natural conditions, and it is difficult to account for the rapidity of the cellular response to specific enzyme induction, for example, if enzyme synthesis is dependent on prior separation of DNA strands. Moreover, the RNA strand of a 3-stranded structure would be relatively loosely bound to the DNA, from which it could readily dissociate after synthesis (Zubay, 1962).

Before leaving the problems of genetic transcription we must mention a paradox which has recently come to light through the discovery that inversion of segments of bacterial chromosome can arise without affecting the expression of the genes involved. These inversions have so far been restricted to the lactose (*lac*) region of the *E. coli* chromosome and have usually been obtained by such genetical tricks as infecting bacteria having a deletion of the *lac* region with a sex factor carrying a functional *lac* region; Hfr strains are then isolated in which the sex factor has become inserted into the chromosome the 'wrong' way round from the point of view of the lactose region, as judged from the polarity of chromosome transfer (pp. 218, 726; Cuzin and Jacob, 1964; Berg, 1966; Berg and Curtis, 1967; Signer, Beckwith and Epstein, 1966). Under these conditions the lactose region, which is not only inverted but usually transposed as well, remains functional.

The paradox arises from the fact that the two strands of the DNA double helix run in opposite directions. Suppose that strand A has $5' \rightarrow 3'$, and strand B $3' \rightarrow 5'$, polarity. If we cut out a segment of DNA and invert it, the polarity of strand A will now be $3' \rightarrow 5'$, and of strand B $5' \rightarrow 3'$. It follows that the only way this inverted segment can be put back into the chromosome is to insert strand A of the segment ($3' \rightarrow 5'$) into the continuity of strand B of the chromosome ($3' \rightarrow 4'$), and *vice versa*. But if only strand A of the chromosome, with the $5' \rightarrow 3'$ polarity, is the 'sense' strand and is transcribed into messenger RNA, this means that strand B of the inverted segment, which is the 'anti-sense' strand but now carries the $5' \rightarrow 3'$ polarity, will be transcribed, so that no sense should emerge. But it does. Why? The probable explanation is that transcription does not proceed around the chromosome in a continuous, polarised way but starts independently at numerous 'operators' where its polarity is determined. It is known that, in *Salmonella*, different groups of linked genes which are transcribed as units (that is, different 'operons') may in fact be transcribed with opposite polarities (see Sanderson, 1967).

CHAPTER 15

DNA as Genetic Material: The Nature of Recombination

MODELS OF RECOMBINATION

Recombination is the central phenomenon of genetics and the principal tool of genetic analysis. It is therefore paradoxical that its nature remains one of the few dark corners of cellular behaviour which the recent developments in molecular biology have, so far, done little to illuminate. In a sense, our concepts about the mechanism of recombination are more confused now than they were a quarter of a century ago, when Darlington (1935) developed a theory of genetic exchange which was based on the observed behaviour of the chromosomes at meiosis and seemed in good accord with the facts of inheritance as they were then known.

According to this theory, when homologous chromosomes from the two parents pair and begin to contract and thicken, at the beginning of the first meiotic division (Fig. 3A, B, p. 15), the coiling of the paired chromosomes around one another creates tension in the strands. This tension is relieved by the breakage of two homologous strands (chromatids) at precisely the same point, in much the same way as the strands of a twisted rope will tend to fracture. The forces that promoted coiling then subside and the broken ends of the fractured chromatids unite crosswise, so that reciprocally recombinant chromatids, as well as two parental chromatids, result (Fig. 89A). This mechanism, known as *breakage and reunion*, as well as agreeing with the facts in a general way, also explained the occurrence of chiasma interference (p. 50), because chromatid breakage at one point would relieve the tension for some distance along the twisted bivalents, thus reducing the probability of a neighbouring break. Note that, according to this model, the time of recombination must coincide with that at which chromosomal coiling is observed, and probably occurs after chromosome duplication is complete, that is, at the 4-strand stage.

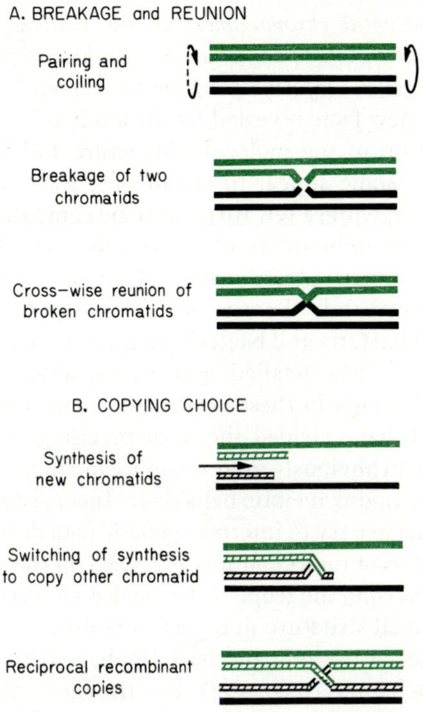

FIG. 89. Diagrammatic representation of two possible mechanisms of genetic recombination.

An alternative mechanism, suggested earlier by Belling (1931), related recombination to the duplication process itself. A modern modification of Belling's mechanism supposes that, during replication of the paired chromosomes, the new structure being formed along, say, the paternal chromosome switches to the maternal one which it thereafter proceeds to copy; when the replica of the maternal chromosome reaches the switch point it has no option but to follow suit so that, once again, two reciprocally recombinant strands are formed (Fig. 89B). This mechanism has been aptly called *copying choice* (Lederberg, 1955), since a growing replica strand has a choice of copying either parental strand. The model restricts the recombination process to the time of chromosome duplication but does not necessarily tie it to any morphologically observable stage of meiosis. However, as we shall see. the really vital distinction between the two models is that, *following breakage and reunion, the recombinant chromosomes inherit physical material from the two parental chromosomes;* on the contrary, *copying choice yields recombinant chromosomes which are synthesised from new material and inherit only genetic information from the parental structures.*

Of the many new facts revealed by the study of microbial genetics, and by elucidation of the molecular structure and behaviour of the genetic material, some appear to favour one, and some the other, of these two theories, while a few turn out to be compatible with neither. One of the major ambiguities which confuses our thinking is that recombination concerns *chromosomes* and we do not really know what chromosomes are. While there is now convincing evidence that the chromosomes of bacteria and bacteriophages consist of a single duplex of DNA, we have little detailed information about the outcome of recombinational events in these organisms. Conversely, those higher organisms which have yielded the most precise genetic data possess chromosomes of an obviously more complex and highly organised type, which display orthodox meiotic behaviour. Inconsistencies are mainly encountered when we try to interpret genetic data derived from higher organisms in terms of molecular models based on the study of bacteria and phage. It is beyond the scope of this book to describe what is known about chromosomal structure in higher organisms. Various aspects of the chromosomes of bacteria and bacteriophage will be considered in more detail in Chapter 19 (p. 534). For the moment we will merely assume that all chromosomes are structures which carry genetic information in the form of a single double-helix of DNA, and analyse the phenomenon of recombination in the most general way from this point of view.

THE MECHANISM OF CHROMOSOME PAIRING

The production of recombinants is the outcome of two distinct kinds of interaction between the genetic material of the two parents. The first of these, an obviously essential preliminary to recombination, is *pairing* which brings homologous regions of the two chromosomes into intimate contact. The second is *genetic exchange* which leads directly to recombinant formation. Theoretically speaking, the frequency of recombination may be considered as a function of the product of the efficiency of these two steps. Thus the same frequency of recombination (10 per cent) will be expected between two loci whether pairing is always complete while genetic exchange along the intervening segment

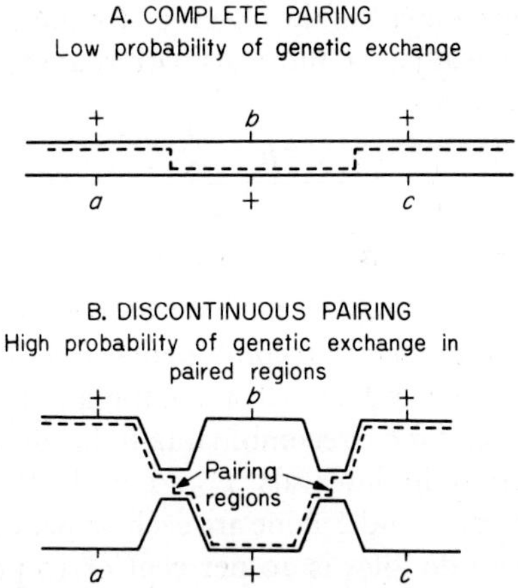

FIG. 90. Diagram illustrating how the frequency of recombination in two intervals may be judged the same, whether

A. pairing is complete but the frequency of genetic exchange low, or

B. pairing is discontinuous, but the frequency of genetic exchange is high with the small paired regions.

The interrupted lines trace the constitution of recombinants inheriting the + + + alleles of the two parents; the vertical components indicate the positions of recombination events.

occurs with a random but low probability (0·1) (Fig. 90A), or, alternately, pairing is discontinuous and only arises with a random, low probability (0·1) in the interval between the loci, while genetic exchange occurs in the paired region with a probability of 1·0 (Fig. 90B).

The traditional view of what happens during meiosis, based largely on cytological evidence, has always assumed that recombinants arise from crossing-over between chromosomes which are paired, point for point, along their entire length and, until recently, this view has dominated our concepts about recombination. The pioneer work of Pritchard (1955; reviews, 1960a, b) on the phenomenon of negative interference in *Aspergillus*, which has since been shown also to occur in bacteriophage, *Neurospora*, yeast, bacteria and other organisms, has now shifted the balance of evidence in favour of discontinuous pairing as the main factor determining recombination frequency.

NEGATIVE INTERFERENCE IN RELATION TO PAIRING

Interference means the effect which a recombination event has on the probability of another recombination event occurring close to it, and can be measured by comparing the observed frequency of concomitant recombination in two adjacent intervals with that expected by chance. For example, in the cross

$$\begin{array}{ccc} + & b & + \\ \hline & 1 \quad\quad 2 & \\ \hline a & + & c \end{array}$$

the frequency of recombination in interval 1 alone is given by the percentage of + + *c* and *a b* + recombinants, and that in interval 2 by the percentage of + *b c* and *a* + + recombinants; similarly the percentage of + + + and *a b c* recombinants indicates the frequency of double recombination in intervals 1 and 2. If the recombination frequencies in intervals 1 and 2 alone are each 10 per cent, the expected frequency of random doubles is 10 per cent of 10 per cent, or 1 per cent. If the observed number of doubles is less than 1 per cent, it means that the occurrence of recombination in one interval tends to suppress that in the other and this is called *positive interference.* Conversely, if the observed frequency of doubles exceeds one per cent, then *negative interference* exists, implying that recombination in one interval increases the probability of a simultaneous event in the other.

In *Aspergillus* no inference of any sort is found for loci more than about 0·5 map units apart. Over distances shorter than this, however, a high degree of negative interference comes into play so that the observed frequency of double events exceeds that expected by as much as 100-fold. This negative interference is thus localised to very small regions of chromosome equivalent to not more than a few cistrons, beyond which it cuts off sharply. Two possible explanations of this phenomenon have been suggested. One is that it is the heterozygosity itself which stimulates recombination (see Pontecorvo, 1958, p. 89), but this is not in good accord with the genetic data (but see p. 403). The other is that *effective* pairing between chromosomes, that is, the kind of pairing which actually mediates recombination, is normally restricted to very small regions within which the probability of recombination is high. On this theory, the occurrence of simultaneous recombination in two intervals requires an effectively paired region in both of them if the genes are far apart, and these will arise randomly (Fig. 90B). On the other hand, if the genes are very close together, then *the two recombination events can arise within a single paired region*, so that the probability of their simultaneous occurrence is no longer random. A hypothetical example will clarify the situation.

Suppose that, in the cross represented above, *a*, *b* and *c* and their alleles are sites in the same or adjacent genes and that, among a total of 10,000 progeny only 1000 have emerged from zygotes in which effective pairing had occurred in the *a*—*c* region. Let us assume that we can recognise the 1000 progeny from this special class of zygote and find that 190 are recombinants, of which 90 arise from recombination events in interval 1, 90 from events in interval 2, and 10 from recombination in both intervals. Thus the frequency of recombination in each interval is $90+10(=100)/1000$, or 10 per cent, while the expected frequency of double events (1 per cent) is the same as that observed (10/1000) so that no interference is evident. But in practice, of course, we do not usually exclude from consideration the 9000 progeny emerging from zygotes with pairing in other regions, but calculate recombination frequencies on the basis of the *total* progeny of the cross. Thus the recombination frequency in intervals 1 and 2 would normally be estimated as 100/10,000, or 1 per cent. From this it follows that the expected frequency of double events is 1 per cent of 1 per cent, or 1/10,000, so that only a single double recombinant should be found as against the 10 that are actually observed. From this we infer that negative interference is present although, in fact, there has been no interference but only the normal, high probability of recombination within effectively paired regions (Pritchard, 1955, 1960b).

THE EVIDENCE FOR DISCONTINUOUS PAIRING

Apart from its neat and elegant disposal of the awkward problem of negative interference as such, Pritchard's hypothesis to explain negative interference is supported by some direct evidence, and also plausibly interprets several unrelated genetic anomalies. The direct evidence concerns what happens when *three* homologous chromosomes are involved in meiosis, as in triploids or trisomics (p. 41). If it is assumed that only two chromosomes can pair together *at any point*, then, if pairing is complete, its efficiency cannot be increased by adding a third homologue so that the frequency of recombination in a triploid should be no higher than in a diploid. On the other hand if pairing between chromosomes only occurs at a few points of random, homologous contact along their length, the probability of such points of contact being made is increased 3-fold if three homologues are present. The frequency of recombination, which the hypothesis relates to the frequency of effective contact formation, should therefore increase. This has been confirmed experimentally in the case of a trisomic in

Aspergillus, and accords with the more general observation in other organisms that chiasma frequencies are usually higher in triploids and trisomics than in diploids (Pritchard, 1960b). Moreover, phage T4 displays the same type of high, localised negative interference as is found in *Aspergillus* (Chase and Doermann, 1958; Barricelli and Doermann, 1960); there is also evidence for *group mating* in bacterial and phage crosses, that is, that more than two genomes can participate in a single mating event (Hershey, 1958; Hausmann and Bresch, 1960; Fischer-Fantuzzi and di Girolamo, 1961), which implies incomplete pairing.

The genetic length of a linkage map in map units is directly proportional to the frequency of recombination per unit length of the chromosome (pp. 52, 144). If the process of recombination is the same in different organisms and is a function of DNA structure and behaviour, one might think that the map unit of each organism should correspond to roughly the same length of DNA double helix. But this is not the case. We have seen that the number of nucleotide pairs per map unit is 2×10^2 for phage T4 and $1{\cdot}75 \times 10^3$ for *E. coli* (p. 695); the comparable figure for *Aspergillus* is 4×10^4 (Pritchard, 1960b), for *Drosophila* 3×10^5 and for the mouse 3×10^6 (Pontecorvo, 1958) (see Table 16).

TABLE 16.

A correlation between genetic and physical estimates of chromosome distance in various organisms. (Data from Pontecorvo, 1958.)
The datum for *Aspergillus* in columns 1 and 3 is that amended by Pritchard, 1960b.

Organism	Total linkage map (in map units)	Total chromosomal DNA (in nucleotide pairs)	Nucleotide pairs per map unit
Phage T4	800	$1{\cdot}5 \times 10^5$	$2{\cdot}0 \times 10^2$
E. coli	1,800	3×10^6	$1{\cdot}75 \times 10^3$
Aspergillus	1,000	4×10^7	$4{\cdot}0 \times 10^4$
Drosophila	280	8×10^7	$3{\cdot}0 \times 10^5$
Mouse	1,954	5×10^9	$3{\cdot}0 \times 10^6$

If we accept the traditional view that recombination is mediated by a low probability of crossing-over along completely paired homologues, and graft on to this view the reasonable idea that pairing primarily involves two DNA duplexes, then we must conclude that, as between one organism and another, there is no correlation between the paired length of genetic material and the probability of genetic exchange.

One could, of course, account for this lack of correlation by supposing that, as organisms increase in complexity, an increasing proportion of their chromosomal DNA becomes non-genetic (see Table 16). Alternatively, if we assume the correctness of Pritchard's hypothesis of negative interference, the recombination frequencies used to estimate map length in fact indicate the frequency of discontinuous pairing, and in no way reflect the frequency of exchange within effectively paired regions, so that the logic of a correlation between map units and nucleotide pairs disappears. It is not surprising that various organisms should differ in the frequency with which their homologous chromosomes establish random contact, which may be expected to vary with such factors as chromosome length and structure. But within the effectively paired regions themselves a true correlation between recombination frequency and DNA content should be found. This may be studied in microorganisms where selective methods permit high resolution genetic analysis within those intervals over which effective pairing is complete, as judged by the occurrence of negative interference along them. In the case of *Aspergillus* and phage, map units calculated on this basis are smaller, by factors of 150 and 10 respectively, than those obtained by considering the population as a whole, so that the estimated amounts of DNA per adjusted map units correspond much more closely (Pritchard, 1960b). The similarity of these values reinforces the view that large differences between recombination frequencies in different organisms is mainly due to disparities in the efficiency of pairing. This is strikingly exemplified by comparing the results of meiotic and mitotic recombination in *Aspergillus* (p. 79), which are very similar except that the frequency of recombination at mitosis is lower than that at meiosis by a factor 10^4. However, when recombinants for different sites *within the same cistron* are selected and the frequency of recombination in an adjacent interval is scored among these, the difference between mitotic and meiotic recombination disappears (Pritchard, 1955, 1960a, b).

Despite the simplicity of this hypothesis, and its aptness in explaining certain anomalies of bacterial crosses (see below), it fails to explain some evidence concerning the frequency with which different strands are involved in multiple exchanges within small regions (Perkins) 1962). Moreover a fundamentally different model, based on a study of the fungus *Ascobolus*, could also account for negative interference. This model supposes that chromosomes are divided into small regions by 'linkers' (p. 536) and that recombination occurs by copying choice during polarised chromosome replication; a recombination event occurring between two linkers tends to be corrected by a second event when the replication process reaches the next linker (Rizet, Lissouba and Mousseau, 1960). Thus a high probability of double exchanges in the intervals between linkers becomes compatible with uninterrupted pairing.

ANOMALIES OF RECOMBINATION FOLLOWING CONJUGATION AND TRANSDUCTION

The most striking distinction between the various systems mediating recombination in bacteria lies in the length of donor chromosome transferred to the recipient cell. For example, in transduction and transformation the donor bacterium contributes only about 1/100 of its genome to the zygote, while in conjugation the mean contribution of the donor is about 10 to 50 times greater. In *E. coli* strain K-12 it is possible to transfer the same, closely linked, donor genes to the same recipient bacteria by either conjugation or transduction by phage P_I, and to compare the results of recombination between them in the two systems. In these circumstances we would imagine the same process of genetic exchange to operate in both cases, and that no differences would be found. In practice, however, closely linked donor genes appear to be more often separated by recombination in transduction than in conjugation. For example, in *E. coli* crosses mediated by phage P_I transduction, when recombinants inheriting the closely linked donor loci, thr^+ and leu^+ (ability to synthesise threonine and leucine) are selected, intermediate donor loci such as *ara* (ability to ferment arabinose) are excluded from the thr^+leu^+ recombinants with significant frequency (Lennox, 1955; Gross and Englesberg, 1959). On the other hand, when the same two genes are selected in crosses mediated by conjugation between the same strains, virtually all the thr^+leu^+ recombinants inherit the donor *ara* locus (Cavalli, Lederberg and Lederberg, 1953; Maccacaro and Hayes, 1961).

These findings are explicable only on the basis of incomplete pairing, if it is assumed that the two crosses are identical except for the disparate lengths of the two donor chromosome fragments. In all merozygotic systems, two recombination events are required for incorporation of a donor fragment into a recombinant chromosome (p. 57; Fig. 14). For thr^+leu^+ recombinants to arise, recombination must occur to the left of *thr* and to the right of *leu* (Fig. 91A); in conjugation, the length of chromosome available in these regions is much greater than in transduction so that we may expect a higher proportion of zygotes which have received a *thr—leu* fragment to yield recombinants (Fig. 91A). For exclusion of the donor *ara*$^+$ locus from *thr*$^+$*leu*$^+$ recombinants, two further recombination events are required, one in the interval *thr—ara* and the other in the interval *ara—leu* (Fig. 91). If pairing is complete, and *provided the probability of recombination per unit length of the paired region is the same* in both cases, then the expectation that *ara*$^+$ will be excluded is the same for transduction and conjugation, irrespective of the paired lengths of chromosome outside the *thr—leu* region. Nor does

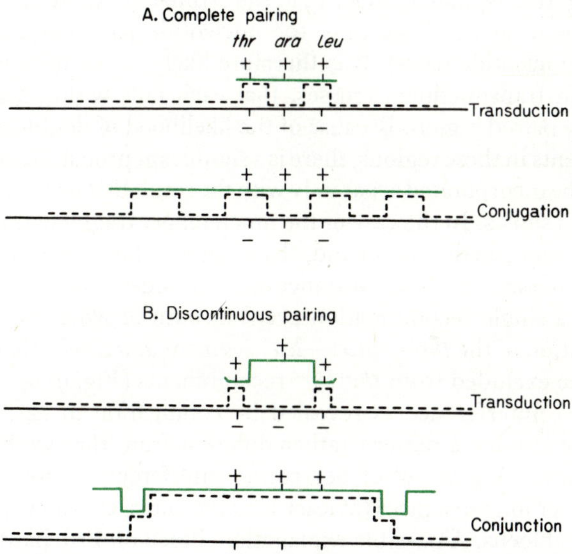

FIG. 91. Demonstrating the theoretical outcome of crosses mediated by transduction and conjugation, when pairing of the donor fragment of chromosome is (A) complete, and (B) discontinuous.

The donor chromosome fragment is shown green, and the recipient chromosome black.

The interrupted lines trace the inheritance of donor and recipient alleles among recombinants; the vertical components indicate the positions of recombination events, the distance between them being roughly proportional to frequency.

thr (threonine synthesis), *ara* (arabinose fermentation) and *leu* (leucine synthesis) are closely linked loci on the *E. coli* chromosome, the distance *thr—leu* being close to the maximum length of a transducible fragment (see text).

the possibility of negative interference in the *thr—leu* segment, due to heterozygosity alone (p. 380), affect the issue since, in complete pairing, there is no distinction between the crosses in this respect.

Let us now look at these crosses from the point of view of discontinuous, small regions of effective pairing within which recombination events occur with high probability (Fig. 91B). It happens that the distance between the selected genes, thr^+ and leu^+, approaches the maximum length of a transducible fragment, so that these genes must be very close to the extremities of the transduced piece of donor chromosome. This fragment, however, is of the order 10^5 nucleotide pairs long, that is, much longer than the estimated extent of regions of effective pairing in phage (4×10^3 nucleotide pairs) or *Aspergillus* (2×10^4 nucleotide pairs). It is therefore likely, when both genes are selected in transductional crosses, that each falls within a separate, effectively paired region. Because of the likelihood of double recombination events in these regions, there is a significant probability that each gene will be incorporated separately with the result that ara^+ is excluded as Fig. 91B shows. In the case of the much longer fragment transferred by conjugation, on the other hand, chance greatly favours the formation of regions of pairing at some distance on either side of the short *thr—leu* segment, a single recombination event in both of which will lead to incorporation of the thr^+—ara^+—leu^+ segment *as a whole*, that is, ara^+ will not be excluded from thr^+leu^+ recombinants (Fig. 91B).

Negative interference as revealed by conjugation in *E. coli* K-12 appears to assume a pattern rather different from that we have described above. A *gradient* of decreasing interference is found which extends over long distances (at least 1/10 the entire chromosome) from the selected locus. A possible explanation, based on Pritchard's hypothesis, is that 'as soon as (random) pairing has occurred at any region, the probability of pairing at any other region is no longer random but is increased in inverse proportion to its distance from the first region of pairing' (Maccacaro and Hayes, 1961). It seems reasonable to think that the homologous apposition of two chromosomes at one point would so align them as to facilitate homologous contact in neighbouring regions.

WHAT IS THE PHYSICAL BASIS OF PAIRING?

This is a key question, for the answer to it clearly determines the nature of recombination at the molecular level. We now have two clues of a rather decisive kind. The first is the fact that the genetic information, which is the important thing to be exchanged in recombination, is carried by DNA; the second is that, to account for the results of high-resolution genetic analysis of phage and bacteria, the precision of pairing must be such as to discriminate accurately between adjacent nucleotide pairs (pp. 299, 300). Theoretically, we are driven to the conclusion that the specificity of chromosome pairing is determined by hydrogen bonding between complementary bases.

From this point of view let us re-examine the studies of Crick *et al.* (1961) on the nature of suppressors of acridine-induced mutations in phage T4. You will recall that acridine-type mutations result from deletion (—) or insertion (+) of a base-pair in the DNA, and that pseudo-wild revertants arise from the insertion or deletion, respectively, of a base-pair at a different but closely linked site, so that the proper number of base-pairs is restored (pp. 318, 346). When such revertants are crossed with wild type phage, both containing equal numbers of base-pairs, two recombinant types of r_{II} mutant can be isolated; one corresponds to the original mutant in having, say, a base-pair too few (—), while the other has an additional base-pair (+) at the suppressor site (Fig. 84, p. 347). The problem is, how can this disparity in the nucleotide content of the recombinants arise by recombination? If we suppose that pairing is precise outside the minute region between the deletion and the insertion in the revertant, thus,

```
—A—B—C—A—B—C—A—B—C—A—B—C—A—B—C—   Wild-
..........................  .....................................
                           ..                                     type
..........................  .....................................
—A—B—C—B—C—A—B—C—A—X—B—C—A—B—C—   Re-
       ↓                  ↑                                        vertant
       —                  +
```

it is obvious that only recombinants having the normal number of base-pairs can be produced. Moreover, legitimate base-pairing cannot arise in the region between insertion and deletion, so that we would not expect recombination there. There are, however, two other possible pairing schemes, both of which would yield + and — recombinants of the required type. One is attained by shifting the revertant chromosome one space to the right, so that base-specific pairing is achieved only in the minute region between deletion and insertion, thus:

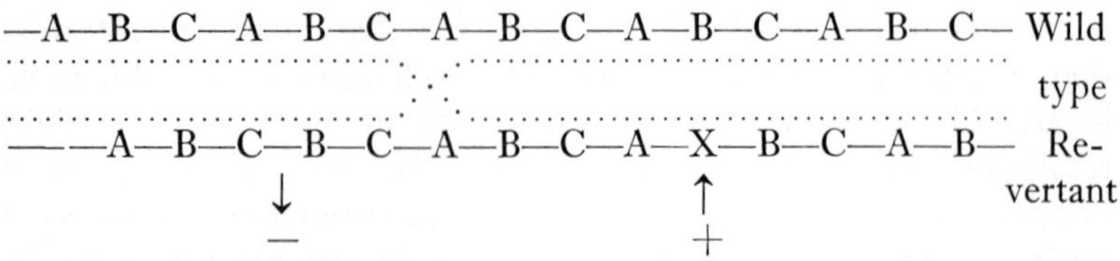

If, as seems likely, this minute region is usually very small compared with that normally involved in effective pairing, this mechanism is not an appealing one.

The third scheme allows complete specificity of pairing over the whole region, by 'looping out' one, or a few, bases (or base-pairs) from the DNA of each chromosome, as follows (p. 260):

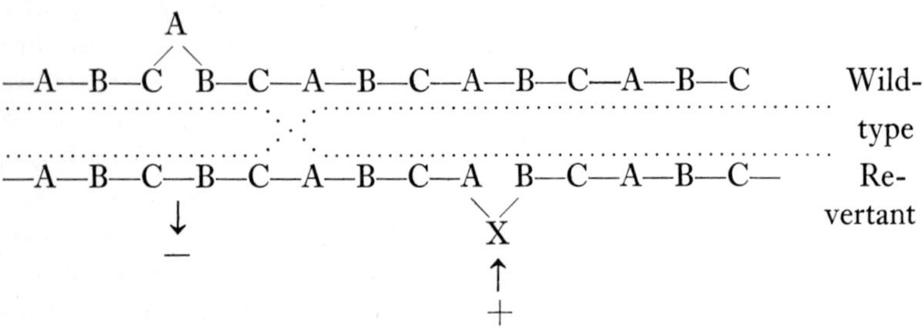

If this third model is the correct one, it implies that pairing possesses not only a specificity, but also a strength of bonding, such as exists between two complementary single strands of DNA or RNA. This suggests a possibility that the strands of the parental double helices may separate, perhaps only locally, and that the 'Watson' strand from one parent pairs, by hydrogen bonding, with the 'Crick' strand from the other parent, and *vice versa* (p. 398).

THE MECHANISM OF GENETIC EXCHANGE

Of the two general models which have been proposed to account for genetic exchange, that based on the breakage, and reunion in new combinations, of pre-existing chromosomal structures (Fig. 89A) dominated genetical thinking for many years. However, with the growth of phage and bacterial genetics, new aspects of the recombination process were revealed which seemed more in keeping with a modification of Belling's hypothesis, whereby newly synthesised chromosomal material, which begins by copying one chromosome, may switch to copy the other (Fig. 89B) (Levinthal, 1954; Lederberg, 1955). In particular, the new concept of a chromosome in molecular terms, as a two-stranded DNA fibre which replicates semi-conservatively, appeared singularly well adapted to the copying-choice model, since recombination could be regarded as a simple variation of normal, mitotic chromosome replication, consequent upon meiotic pairing.

Let us continue in the assumption that, in all organisms, the fundamental attributes of a chromosome reside in a single DNA duplex, and that all genetic phenomena can be interpreted in terms of DNA structure and behaviour. From this point of view, let us examine the two models of genetic exchange more closely, to see to what extent each is compatible with what we know about the recombination process.

THE KINETICS OF RECOMBINATION

In higher organisms, the only way of looking at the sequence of events during recombination is by microscopic observation of chromosomal behaviour during meiosis. What is observed is that pairing occurs early in the first meiotic prophase and is followed by the appearance of chiasmata, each of which clearly involves crossing-over between two of the four chromosome strands which are now visible (Fig. 3A, B, C; p. 15). The high degree of correlation between the results of genetical and cytological studies leaves no doubt that chiasmata reflect recombination events (pp. 35–39). If recombination is mediated by copying-choice, it must occur during chromosome duplication which has been shown experimentally to coincide with the time of DNA synthesis in *Lilium* (Mitra, 1958). By means of autoradiographic studies of the uptake of radioactive isotopes (such as ^{32}P or tritiated thymidine which are incorporated into DNA) by cells undergoing mitosis or meiosis, it is possible to correlate DNA synthesis with the cytological behaviour of the chromosomes. In the two organisms investigated in this way (*Lilium* and *Tradescantia*), it turns out that DNA synthesis is restricted to a fraction of the division cycle and is complete before cytologically observable pairing commences (Howard and Pelc, 1951; Taylor, 1957). Similarly, in mitosis, both DNA synthesis and chromosome duplication are over before the characteristic features of prophase become visible (Fig. 1, p. 11).

At first sight this might seem to rule out recombination by copying-choice. But if we accept the view that *effective* pairing involves only a few small regions of homologous chromosomes, there is no reason why such contact should not occur before or during replication, when the chromosomes are in an extended and hydrated state, and thus not be seen. On this interpretation, pairing and genetic exchange are already completed before the elaborate, observable machinery of meiosis begins to operate. This machinery would therefore play no role in the recombination process itself, but would operate to ensure the precise *segregation* of the products of recombination and their distribution to the daughter cells. Chiasmata would be the natural consequence of prior genetic exchanges, irrespective of the time at which these had occurred. While there is no evidence that this reconstruction of the classical picture is the correct one, we wish to make the point that it is an equally valid interpretation of the facts (see p. 435). Mitosis, in which chromosome duplication occurs equally early, presents a closely analogous sequence of observable stages, except that there is usually no pairing and, perhaps because of this, the chromosomes are *seen* to be already divided in early prophase; they are not held together by crossover strands (review; Pritchard, 1960a, b).

ANALYSIS OF THE PRODUCTS OF RECOMBINATION

Apart from cytological studies, the only direct information we possess about individual recombination events comes from the analysis of tetrads in which the products of each event are isolated in a single ascus (pp. 68–77). The important features of genetic exchange which have been revealed by tetrad analysis and which concern us here are, first, that the products of meiosis are pure; they do not segregate different genotypes on further division. Secondly, either of the two strands of each parental chromosome can participate in exchange, as is shown by the fact that two recombination events can yield four recombinant, and no parental, products (4-strand doubles; see Fig. 4, p. 19 and Fig. 20, p. 75). Thirdly, although genetic exchange invariably leads to the formation of recombinants which are reciprocal with respect to well separated loci, this reciprocity is often not found when recombination between very closely linked loci is studied.

There is evidence that when chromosomes divide mitotically, and even meiotically, the distribution of isotopic labelling between daughter *chromosomes* is similar to that which follows the replication of DNA *molecules* in *E. coli* and other organisms (p. 245; Taylor *et al.*, 1957; Taylor, 1965). This suggests not only that chromosomes consist of two sub-units which replicate semi-conservatively, but that these sub-units correspond to the two strands of a DNA duplex. But when we attempt to interpret recombination between chromosomes of this kind on the copying-choice model, and compare the predicted outcome with the facts of tetrad analysis, we run into serious difficulties. As Fig. 92A shows, the reciprocal switching of the newly synthesised replica strands to copy the old strand of the other parental chromosome automatically

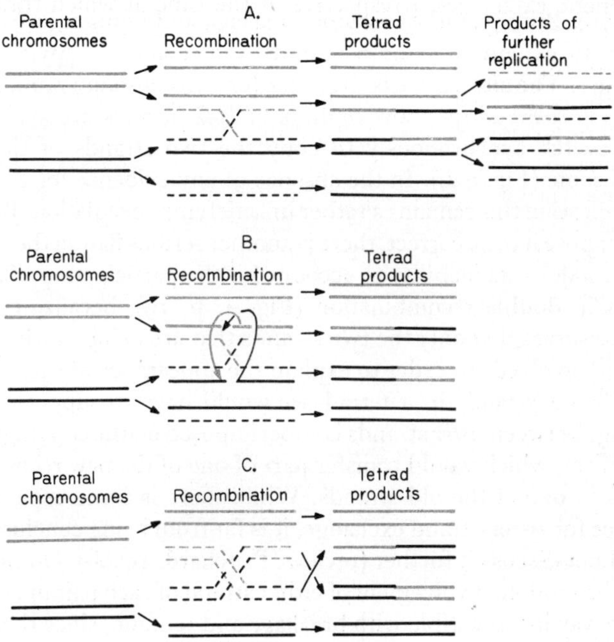

FIG. 92. Illustrating the genetic constitution of the four products of meiosis produced by various mechanisms of genetic exchange based on the copying-choice model.

The chromosomes are assumed to consist of two identical sub-units or strands (which could be the two strands of a DNA duplex).

The green lines represent the sub-units of one parent, and the black lines those of the other parent.

Continuous lines indicate template (old) strands, and interrupted lines newly synthesised strands.

A. Uncomplicated copying-choice during semi-conservative replication.

B. Copying-choice during semi-conservative replication, associated with breakage and crosswise reunion of the old, template strands.

C. Copying-choice during *conservative* replication, in which both original strands serve as templates for the synthesis of a completely new chromosome (see text).

yields two *hybrid* chromosomes, each with one parental and one recombinant strand; at the next replication each will give rise to a parental and a recombinant chromosome, that is, they are genetically impure.

There are two ways out of this predicament. One is to suppose that semi-conservative replication and copying-choice is accompanied by breakage and crosswise reunion of the old strands at the switch point (Fig. 92B). But if we must assume breakage and reunion, it would be simpler to extend it to both strands and abandon copying-choice altogether. The alternative is that DNA (or chromosome) replication at meiosis is *conservative*; new replicas of *both* strands are synthesised which switch synchronously to copy the two strands of the other chromosome (Fig. 92C). In the absence of any evidence for conservative replication this remains a rather unsatisfying speculation. But even if we suppose it to be correct, there is another serious flaw in the copying choice model—its inability to account for the participation of all four strands in double recombination (Fig. 4, p. 19), because copying-choice assumes that only the two strands that are being newly synthesised are involved. In order to explain the appearance of three or four recombinant strands in a tetrad, we would have to suppose that an exchange between *sister* strands is superimposed on the copying-choice mechanism, which would transfer part of one of the new recombinant strands to one of the old strands. While there is some experimental evidence for sister-strand exchange, it is far from being conclusive and we will not discuss it further (review; Pritchard, 1960b). On the other hand, the random involvement of either strand of each pair in exchange is in no way incompatible with breakage and reunion, since there is no *a priori* reason why any one strand should break rather than another.

Non-reciprocal recombination

Despite the poor performance of the copying-choice model in these critical tests, it does account very well for one phenomenon revealed by tetrad analysis—the non-reciprocity of recombinants for very closely linked (intra-cistronic) sites. This phenomenon was originally described as occurring in recombination between pyridoxin-less mutants of *Neurospora* (Mitchell, 1955) and had been tentatively called *gene conversion* because of its non-conformist nature. It has since been found to occur during recombination within many loci of *Neurospora*, *Aspergillus*, yeast and even *Drosophila*. However, its occurence is variable, depending on the organism, the sites involved and the experimental conditions. In yeast, for example, both reciprocal and non-reciprocal recombination have been found between the same two sites (Roman, 1958) while, in *Aspergillus*, intra-cistronic recombination is predominantly reciprocal (review; Pritchard, 1960a, b).

The usual characteristic of non-reciprocal recombination is that the tetrad contains three parental spores and only one recombinant spore with respect to the intra-cistronic sites (that is, a 3:1 ratio), while the usual reciprocity is found for less closely linked, outside markers (1:1 ratio), thus:

Cross	*Tetrad products*	*Type*
	$a\ b_1\ +\ c$	Parental
$a\ b_1\ +\ c$	$a\ b_1\ +\ +$	Recombinant for a and c; Parental for b_1 and b_2.
$+\ +\ b_2\ +$	$+\ +\ +\ c$	Recombinant for both a and c, and b_1 and b_1.
	$+\ +\ b_2\ +$	Parental.

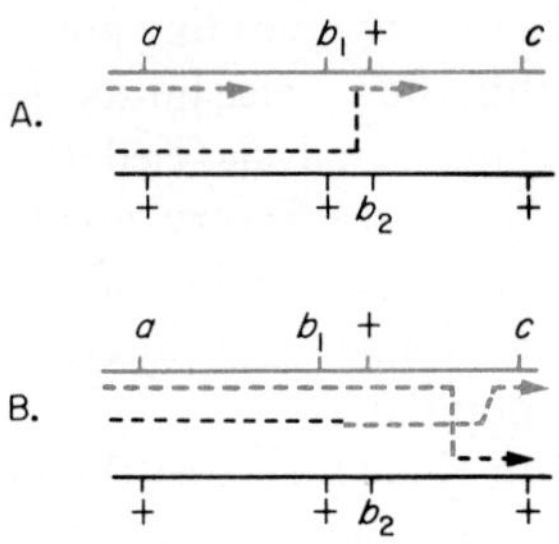

FIG. 93. Non-reciprocal recombination explained by asynchronous replication during copying-choice.

A. The replication of the lower strand (or chromosome, in conservative replication) is in advance, and has switched, to copy the upper, between the intra-cistronic sites b_1 and b_2.

B. The leading replica has separated from its template by the time the second replica reaches the switch point. The second replica therefore continues to copy the upper strand (or chromosome) and only switches to the lower at some other point between b_2 and c. Thus part of the upper strand (or chromosome) is copied twice.

For simplicity, the separation of the upper replica from its template is not shown (see text).

While there is no obvious interpretation of this kind of result on the bases of breakage and reunion, it is readily explained by the copying-choice model if asynchrony in the replication of the paired strands is assumed. For example, in Fig. 93, the replication of the lower strand is more advanced than that of the upper, and switches to copy the upper strand between the sites b_1 and b_2. By the time the upper strand replica reaches the switch point, the more advanced copy has already separated from it, so that there is no obligation for it to switch down; it therefore proceeds along the upper strand, part of which is copied twice. Eventually the lagging, upper strand replica does switch down, either spontaneously or because it catches up with the first strand and so finds its template occupied. However, an alternative and preferable molecular model has recently been developed, and this we will discuss on p. 398 *et seq*.

MOLECULAR MODELS FOR RECOMBINATION

The original reasons for resurrecting copying-choice as a preferred mechanism of recombination were, firstly, that it appeared at the time to offer a better explanation for the formation of phage heterozygotes (see p. 496; Levinthal, 1954) and, secondly, it seemed difficult to understand how breaks in corresponding positions on the two chromosomes could occur within the precise limits required to account for the facts of fine structure genetic analysis (Lederberg, 1955). In retrospect, these reasons do not now seem very good ones, but nevertheless because of them, as well as of a number of experiments which seemed to support it but are no longer very interesting, the copying-choice model became widely popular.

The turning point in this story was the clear demonstration that recombination in bacteriophage λ involved breakage and reassembly of the pre-existing parental chromosomes (Meselson and Weigle, 1961; Kellenberger, Zichichi and Weigle, 1961a). Phage particles of greater than normal density can be prepared, by propagation in bacteria growing in medium containing the heavy isotopes of carbon (^{13}C) and of nitrogen (^{15}N) instead of the normal isotopes, ^{12}C and ^{14}N. A mixture of such particles with normal particles may be centrifuged in a caesium chloride density gradient without loss of viability; the intact particles will separate into discrete bands according to their density. The particles within each band can then be recovered, and their number assayed, by perforating the bottom of the centrifuge tube, collecting the drops in sequence as they emerge, plating them and counting the number of plaques. The heavier particles will band in the more dense regions of the gradient towards the bottom of the tube and will be the first to appear. As the sequence of drops is assayed the titre of phage particles begins to rise, increases rapidly to a peak and then falls off again as the band containing them is passed; this is followed by a second, similar peak corresponding to the emergence of drops containing the band of lighter density particles (see Fig. 94).

It is clear that if, in crosses between heavy and light phages, recombinants arise by the breakage and reunion of the heavy and light chromosomes, these recombinants will have densities different from those of the parental types and will reveal themselves in new, intermediate peaks in a caesium chloride gradient. This, in fact, is what was found, recombinants of reciprocal genotype being isolated at density levels corresponding to those expected from genetic exchanges within the appropriate regions of the linkage map.

A further important point to emerge from these density studies is that when light bacteria are infected *with a high multiplicity* of heavy phage, up to one per cent of the progeny phages may contain heavy DNA. At low multiplicities of infection only progeny containing DNA of hybrid and light densities are found. This means that, at high multiplicities, a proportion of the infecting phage genomes do not replicate but are incorporated unchanged into protein coats. When crosses are mediated by a high multiplicity of infection, recombinant classes are found whose density can only be explained by the inheritance of a large fragment of intact DNA duplex from the heavy parent, so that 'recombination by chromosome breakage may occur without separation of the two subunits of the parental chromosome', that is, that genetic exchange may be divorced from replication (Meselson and Weigle, 1961).

This kind of result is illustrated in Fig. 94 which shows the density distributions of total phage progeny (upper curve) and of wild type recombinants (lower curve) from a two-factor cross between heavy and light parental phages. The genetic markers were in a gene located about 20 per cent. of the length of the linkage map from one extremity, and were arranged so that the formation of wild type recombinants entailed the inheritance of 80 per cent. of the map from the heavy parent, thus:

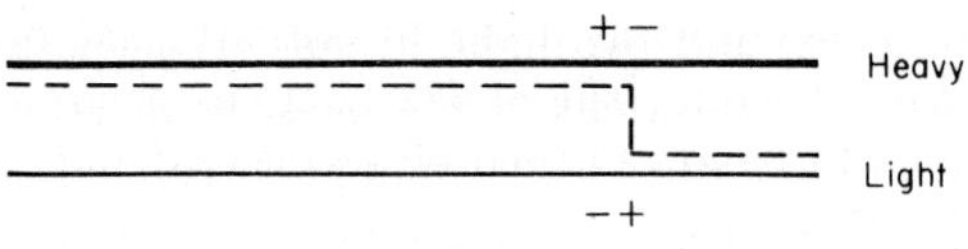

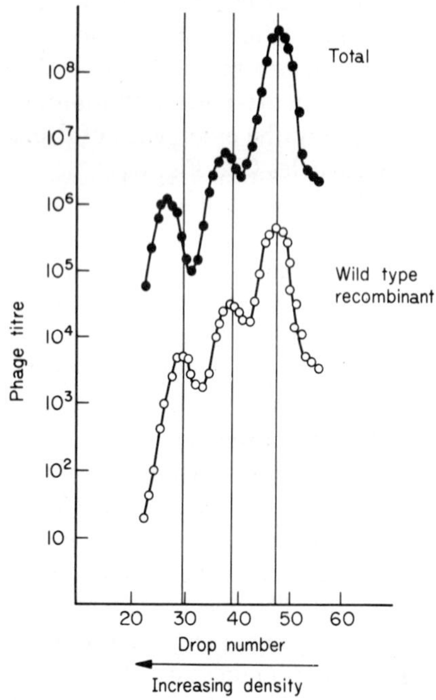

FIG. 94. Illustrating the analysis of recombination in bacteriophage λ by means of density gradient centrifugation.

The curves show the density distributions of total phage progeny (o—o) and of wild type recombinant particles (o—o) from a two-factor cross between heavy ($^{13}C^{15}N$) and light mutant, parental phages, having different mutations in the same gene located about 20 per cent. of the length of the linkage map from one extremity.

The peaks of the upper curve correspond, from left (dense) to right (light), to particles in which both DNA strands are heavy, one is heavy and one light (hybrid), and both strands are light, respectively.

For interpretation, see text.

From Meselson, 1967, by kind permission.

Both curves reveal three peaks corresponding to phages whose DNA contains two heavy strands (left peak), one heavy and one light strand (hybrid; middle peak) and two light strands (right peak). However, when we compare the positions of the peaks with reference to the density axis, we find that whereas the light peaks coincide as we should expect, the hybrid and heavy peaks do not; the recombinant phages which inherit one or two heavy strands are less dense than the equivalent parental phages. In fact the unreplicated recombinants with two heavy DNA strands turn out to have about 80 per cent., and the hybrid recombinants about 40 per cent., heavy DNA (see Meselson, 1967).

While these experiments prove that the recombinant chromosome inherits pre-existing genetic material from at least one of the parents, the possibility remains that a fragmented chromosome might be completed by copying the missing region from the homologous portion of the other parental chromosome ('break and copy' mechanism; Meselson and Weigle, 1961). This was excluded by showing that when *both* parental phages were labelled with heavy isotopes and 2-factor crosses made in light media, recombinants were formed which were entirely, or almost entirely, heavy. This means that these recombinant chromosomes were reconstituted from the DNA of both parents and thus confirms the occurrence of the breakage and reunion of double-stranded DNA molecules (Meselson, 1964). However, in these experiments the peak representing the heavy recombinant phages was not quite so sharp as that containing unlabelled recombinant particles, but showed a slight 'shoulder' which suggested that recombination involved a small amount of DNA breakdown and resynthesis. In the case of bacterial recombination, the incorporation of donor fragments of DNA into the recipient chromosome has been demonstrated in transformation (Fox and Allen, 1964), even though these fragments may be single-stranded (pp. 600–603), while there is also evidence for incorporation following conjugal transfer (Siddiqi, 1963; Oppenheim and Riley, 1966; Bresler and Lanzov, 1967).

In chapter 18 we shall discuss the occurrence of heterozygous phage particles which have been studied particularly in phages T2, T4 and λ (p. 496). When the particles emerging from a phage cross are plated, a small proportion of the resulting plaques are neither pure parental nor recombinant in phenotype, but contain a mixture of the two parental types with respect to any particular marker that may be looked at; that is, the infecting particle was heterozygous for this marker, the two parental alleles segregating on replication of the chromosome. When two markers are very closely linked, heterozygotes for one usually involve both; from data of this kind it is found that the length of a heterozygous region is not more than a few cistrons, that is, about the distance over which localised negative interference operates. Since there is a high correlation between heterozygosis for a particular marker and recombination for outside markers, while heterozygous regions are frequently and exclusively found close to the sites of selected recombination, it is very probable that heterozygous chromosomes are original recombinant structures which were trapped in phage head protein before replication occurred. It is therefore of great interest to know the physical nature of such chromosomes. There is now convincing direct evidence that they comprise uninterrupted DNA duplexes with a small region of heterozygosity where homologous strands from the two parental DNA molecules overlap and pair by hydrogen bonding (Tomizawa and Anraku, 1964; Meselson, 1967), thus:

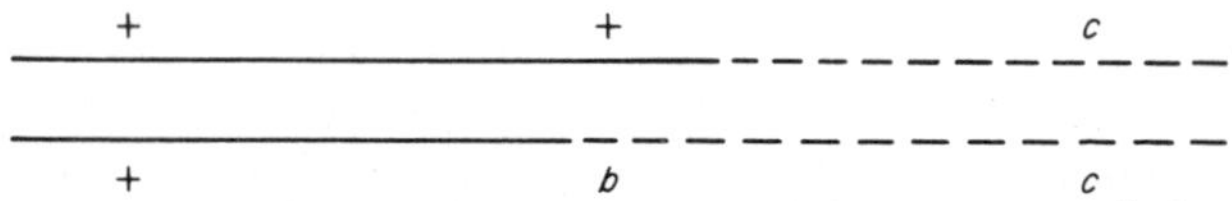

Surprisingly, a phenomenon very similar to heterozygosis in phage, called *post-meiotic segregation*, has been described in ascomycetes, although its occurrence is rare. In tetrad analysis, one of the four haploid products of meiosis is found to be heterozygous for a given marker which only segregates at the subsequent mitotic division, although the outside markers are homozygous and usually recombinant. Thus instead of the eight final products showing a 4:4 segregation ratio for the marker, the ratio is 5:3 (see Lissouba *et al.*, 1962).

When discussing non-reciprocal recombination, or gene conversion (p.392), we saw that this resulted in a 3:1 segregation ratio (or a 6:2 ratio if we look at all eight products of the mitotic division following meiosis), and seemed best explained by a copying-choice model in which one allele in a heterozygous region is copied twice. However it could equally well arise from a breakage and exchange event if the complementary strands of the two parental DNA duplexes in effect break at different positions and then pair crosswise by hydrogen bonding so that the Watson strand from one parent pairs with the Crick strand from the other, and *vice versa*. It is then supposed that, before reunion occurs, a terminal part of one or more of the unjoined strands are excised by nucleases, the excised portion being subsequently made good by copying the complementary strand and, finally, given continuity by covalent bonding. It is clear that if an excised portion of strand happened to be part of a heterozygous region, this would become homozygous as a result of the excision and repair. This kind of situation is illustrated in Fig. 95, in which the pair of lines represents two parental (1 & 4) and two recombinant (2 & 3) DNA duplexes which result from meiosis, the DNA strands derived from

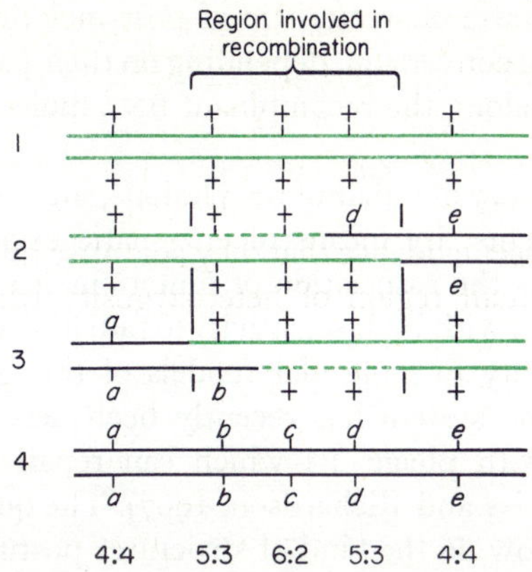

FIG. 95. Illustrating the possible structures of DNA duplexes, comprising the four chromatid products of meiosis, following a single recombination event in which the two middle duplexes (2 & 3) have been involved.

Each pair of lines represents the two strands of a DNA double helix. The DNA polynucleotide strands of the two parents are distinguished by the continuous green and black lines. The interrupted lines indicate strand fragments newly synthesised on parental template strands of the same colour (in this case, green).

The letters *a*, *b*, *c*, *d*, *e* represent mutational markers, and + their wild type alleles.

The two vertical lines delimit the extent of the chromosome region involved in the recombination event.

For explanation, see text.

After Meselson, 1966, by kind permission.

the two parents being distinguished by green and black. As a result of this particular recombination event it is supposed that two of the black strands which have paired with their green counterparts (the upper in duplex 2, and the lower in duplex 3) have undergone excision from opposite ends, the newly synthesised repair regions, which have copied the green strand, being indicated by the interrupted green lines. Five sites of heterozygosity (*a*, *b*, *c*, *d*, *e*, and their wild type alleles, +) are also shown, of which *a* and *e* are outside the region involved in recombination, while *b*, *c* and *d* lie within it. With respect to these markers, each strand defines the genotype of one of the eight products of mitosis, so that by counting the alleles in a vertical direction, the segregation ratio for each marker may be assessed. Observe that while the outside markers, *a* and *e*, show normal reciprocal recombination, markers within the region involved in recombination can show the 5:3 ratio characteristic of post-meiotic segregation or the 6:2 ratio of gene conversion, depending on their particular relation to repair synthesis along the recombinant DNA molecules (Meselson, 1967).

The recent discovery of repair of DNA lesions, induced by ultraviolet light and other agents, by means of enzymatic excision and repair synthesis, as well as the association of mutations leading to inability to recombine (*rec*—) with high sensitivity to radiation damage (p. 334), lend great plausibility to molecular models of this general type. In addition, an enzyme system has recently been described in *E. coli* bacteria infected with phage T4 which can repair single-stranded breaks in DNA (Weiss and Richardson, 1967). The question we must now consider is, How do the kind of structures postulated in Fig. 95 actually arise? Although we do not yet have any precise knowledge of how this is accomplished, there is no doubt that, during the last three years, we have progressed from a state of virtual ignorance to one where we can construct models which represent at least a first approximation to what probably happens (Meselson, 1963; Whitehouse, 1963; Holliday, 1964; Howard-Flanders and Boyce, 1966). All these models are based on the breakage of parental DNA strands, the formation of regions of hybrid DNA between broken complementary strands of the two parents, and some mechanism of 'repair' type for the correction of heterozygosity in order to explain gene conversion. One model of this general type, which embodies breakage of parental strands at different points, as well as two steps of DNA synthesis within the region of recombination, is shown in Fig. 96.

A subsequent complication was the discovery that conversion in the fungus *Ascobolus* (Lissouba *et al.*, 1962) and recombination within small regions of chromosome in *Neurospora* (Murray, 1963) and *Aspergillus* (Siddiqi and Putrament, 1963) are polarised processes; that is, within a given small region, these events occur with an increasing probability in a fixed direction on the genetic map, although the direction may differ in different regions. Such regions which show polarity of conversion or recombination have been called *polarons* and appear to be about one or a few genes long. To account for this polarity it was supposed that strand breakage occurs at the *same* points on the two DNA molecules, unlike the situation shown in Fig. 96, and that these points are located at the beginnings of genes. The synthesis of new strand fragments would then proceed in the same direction from corresponding points on the two parental molecules, thus introducing

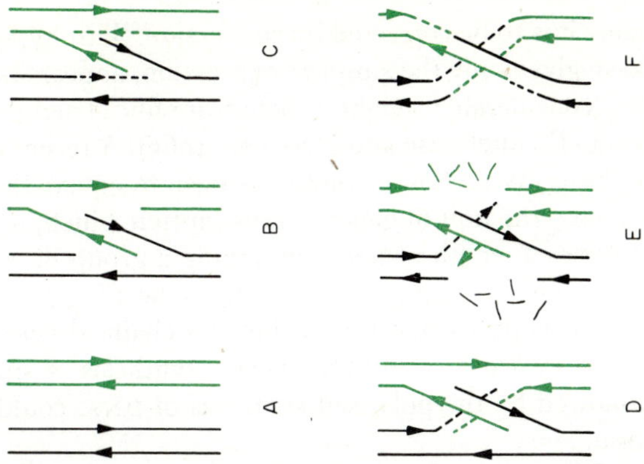

FIG. 96. A model of how recombination might occur between two DNA molecules.

The polynucleotide strands of the two parental molecules are distinguished as black and green.

The arrow-heads indicate the 5′-3′ polarity of the strands as well as the direction of synthesis of the new DNA.

The interrupted lines represent newly-synthesised DNA of complementary base sequence to the paired strand of opposite polarity.

The fragments in E. indicate DNA breakdown of the outer, unpaired parental strands.

For simplicity, the base pairs which unite the strands of the double helices by hydrogen bonding are not shown.

A. Homologous regions of the parental DNA molecules.
B. Breakage at different sites, of two parental strands of opposite polarity, and their crosswise pairing to form a hybrid duplex region.
C. New synthesis of DNA to copy unpaired regions of strands.
D. Separation and cross-wise pairing of newly synthesised regions, reiterating B.
E. Excision of redundant, unpaired regions of parental strands.
F. Repair synthesis and covalent bonding of strands of the same polarity.

After the first model of Whitehouse (1963), by kind permission.

polarity into the system (Hastings and Whitehouse, 1964). For example, we have seen that a new strand fragment which begins by copying the homologous complementary strand (Fig. 96C) may complete its synthesis on a complementary strand from the other parent (Fig. 96E, F) and that it is this second stage of synthesis that leads to conversion (Fig. 95); therefore, if we assume that the amount of synthesis on the homologous template varies widely, it follows that the more distant a mutational site from the point of breakage and initiation of synthesis, the more likely it is to be corrected by conversion. This hypothesis, of which we have given only the simplest expression, appears to be compatible with a considerable weight of data from fine structure analysis of various fungi (Whitehouse and Hastings, 1965). A recent extension of this hypothesis (Whitehouse, 1966) has been to equate the polaron with the operon—the unit of genetic transcription which, in bacteria, usually comprises a length of DNA containing a group of contiguous genes, and is transcribed into a single molecule of messenger RNA (pp. 166, 716). It is postulated that if the DNA chains dissociate at the operator region in order to initiate m-RNA synthesis, a similar dissociation, followed by the polarised synthesis of DNA, could mediate recombination.

A feature of operons is that the synthesis of m-RNA is controlled by a repressor, determined by a regulatory gene (R) which may be unlinked to the operon. In inducible enzyme systems the repressor attaches to the operator and prevents m-RNA transcription until it is inactivated by the inducer; mutations in the R gene which lead to production of defective repressor are recessive and permit unrestricted activity of the operator so that enzyme is synthesised continuously. Recently, mutations have been found in *Neurospora* which *increase* intragenic recombination by a factor $\times 10$ or more (unlike the *rec* mutations in *E. coli*), are recessive to wild type and appear to be specific for certain chromosomal regions (Catcheside, Jessop and Smith, 1964; Jessop and Catcheside, 1965; Smith, 1966). Whitehouse (1966) has analysed these data in terms of his operator model of crossing-over and claims that they provide direct evidence for it, on the supposition that the *rec* mutation has resulted in depression of the gene (*his*-1) within which the recombination frequency is increased. These models of Whitehouse are much more complicated and sophisticated than we have been able to present them here, so that interested readers should refer to the original papers.

The recent and fundamental change which has arisen in our conception of recombination at the molecular level is that we no longer believe it to be an operation which occurs at some precise point on the paired parental chromosomes. Instead, there is now good reason for thinking that recombination comprises a complicated series of enzyme-mediated events, including the breakage, synthesis and covalent bonding of parental DNA strands, which are spread over a region of chromosome at least as long as a single gene, that is, of the order 1000 or more nucleotide pairs. The details of what goes on within this region are still mysterious. This general hypothesis and the experimental facts which gave it birth suggest that the overall features of recombination may be universal but, on the other hand, that two distinct mechanisms of recombination may exist. Thus recombinants for outside markers result from the sum total of the events within the recombination region, while those for sites which happen to lie within the region itself may be differently determined as, for example, by DNA excision or synthesis which lead to correction of heterozygosity.

Holliday (1964) first postulated a mechanism for gene conversion in fungi, which might or might not be associated with recombination for outside markers, in which the pair of bases at heterozygous sites of hybrid DNA, being unable to form proper hydrogen bonds, might rotate out from the duplex and so be recognised by a repair enzyme which would substitute the correct base. Since such recognition would not discriminate between wild type and mutant bases, one would expect the hybrid DNA to become homozygous mutant or wild with equal probability. Recently, an experiment designed to demonstrate which of the two strands of the DNA of phage λ is the 'sense' strand from which m-RNA is transcribed, has yielded strong evidence that this kind of correction of heterozygosity actually occurs (Hogness *et al.*, 1966).

The two strands of phage λ DNA differ slightly in density and so can be separated and isolated by denaturation and centrifugation in a caesium chloride density gradient (pH 12) (p. 372). In this way preparations of heavy (H) and light (L) strands were obtained from wild type (+) phage and from a mutant (−) which is unable to grow in a particular host due to a defective 'early' function which is necessary for m-RNA synthesis. It happens that, under certain special conditions (p. 544), the isolated DNA of phage λ is infective for sensitive bacteria so that the biological activity of the DNA can be measured by means of plaque counts. The four single-stranded DNA preparations, H^+, L^+, H^- and L^- were mixed and reannealed to give the four possible types of duplex, the parental homoduplexes H^+/L^+ (wild) and H^-/L^- (mutant) and the two heteroduplexes H^+/L^- and H^-/L^+. Infection with the homoduplex H^+/L^+ yielded plaques (activity $=1{\cdot}0$), while H^-/L^- gave no plaques (zero activity).

It was anticipated that activity would be expressed by whichever of the two heteroduplexes carried the wild type allele in the 'sense' strand, while the other heteroduplex would be inactive; thus if the H strand carried the genetic message, a normal message would be given by the duplex H^+/L^-, but no message could be transcribed from the H^-/L^+

duplex. Experimentally, it turned out that both heteroduplexes showed about half normal activity. Guessing that this unexpected result might be due to a repair mechanism which was correcting the heterozygosity by changing the mis-matched base on either strand with equal probability, the experiment was repeated using host cells whose repair mechanisms had been inactivated by a heavy dose of ultraviolet light. Under these conditions the heteroduplex H^+/L^- was active while H^-/L^+ was not, implying that the heavy strand, H, is the one which carries the genetic message (p. 372) (Hogness *et al.*, 1966).

Part VII
The Physiology and Genetics of Bacteriophage

CHAPTER 16

The Physiology of Bacteriophage Infection

Bacteriophages were discovered independently, by Twort in England and d'Herelle in France, nearly a half century ago and for many years afterwards there was great controversy as to whether they were inanimate, autocatalytic agents, or, as d'Herelle strenuously maintained, were self-reproducing organisms like viruses (see Stent, 1963). The latter hypothesis prevailed and, with its wide acceptance, interest in bacteriophages waned. Although sporadic research into their behaviour continued, and unsuccessful attempts were made to apply them to the treatment of some infectious bacterial diseases—a role in which they were dramatised by Sinclair Lewis in his novel, 'Martin Arrowsmith'—they were for many years generally regarded merely as a laboratory curiosity. The revival of interest in bacteriophages, culminating in their present unprecedented status both as a model virus and as a fundamental tool in biological research, is largely due to the stimulus provided by Max Delbrück in introducing and exploiting new methods of experimental analysis some 25 years ago.

Since then an immense amount of detailed study has been devoted to the mechanisms of phage infection and reproduction. The successful outcome of these researches is a consequence of several features of phage-bacterium systems which give them a unique advantage over other host-virus systems as experimental material. The most important of these are, firstly, that, unlike all other viruses, every phage particle has a probability approaching 1·0 of causing infection of a particular host and, secondly, that the infectious particles can be counted by simple and accurate means so that results can be expressed quantitatively. In addition, the infected host population is very amenable to quantitative chemical analysis during the course of the infection so that the chemical kinetics of the process can be studied, while the genetic constitutions of both phage and bacteria, and the interactions between them, are accessible to genetic analysis.

We have referred to bacteriophages as bacterial viruses and we should, perhaps, begin this account of them by justifying this assumption. What are viruses? In the early days of virology, when little was known about their true nature, viruses were primarily defined by the two properties by which they could be recognised—filterability and pathogenicity. It turns out, however, that neither of these properties is definitive. Filterability is mainly an index of size and does not help us to distinguish what we now regard as a virus from, say, minute organisms such as *Rickettsia*; pathogenicity can readily be lost by mutation without changing the fundamental nature of the virus in any way. Three further properties of viruses in general have now become pre-eminent and serve, when taken together, to distinguish them clearly from either cells or cell organelles. These are infectivity, the absence from them of a built-in metabolic system so that they are dependent for reproduction on the biochemical machinery of the host, and their possession of only one type of nucleic acid. Thus, in the words of Lwoff (1959), 'a virus is a virus'. In all these respects, as we shall see, bacteriophages are viruses.

CLASSIFICATION AND GENERAL PROPERTIES OF BACTERIOPHAGE

Viruses in general vary widely in their size and shape, in the location within the cell (nucleus or cytoplasm) where they start to multiply, and in the type of nucleic acid (DNA or RNA) which they contain, and are classified according to these criteria. Until very recently none of these distinctions were observed among phages, all of which were of comparable size and comprised a tadpole-shaped protein sheath enclosing DNA within its head. A number of very small phages have now been discovered, however, which appear to lack a tail and closely resemble the small animal viruses in size and morphology; moreover one group of these phages contains RNA as genetic material (pp. 436, 440). In addition, a number of filamentous phages, endowed with single-stranded DNA and probably related, have recently come to light (p. 444). Because of the previous, apparent absence of morphological or chemical criteria for classification, categories of phage have been primarily distinguished, on the basis of their interaction with host bacteria, as *virulent* or *temperate*.

The character of virulent phages is invariably to lyse the bacteria they infect; they have no alternative but to follow what is called the *lytic cycle*. Temperate phages, on the other hand, do have an alternative; each infecting particle has a choice, following infection, either of entering the lytic cycle like a virulent phage, or of establishing a kind of symbiotic relation with the host cell which results in its perpetuation thereafter in all the descendants of the bacterium. Bacteria which carry temperate phage in this way are said to be *lysogenic*, while the form of the carried phage is known as *prophage*. The decision as to whether the lytic or the lysogenic response supervenes, on infection by a temperate phage, is largely determined by the physiological state of the bacteria and can be profoundly modified by altering the environmental conditions (see Lieb, 1953a, b; Bertani, 1953).

The distinction between virulent and temperate phages is a convenient one, and the study of its basis has yielded novel and important concepts; but it is probably of little taxonomic significance since a virulent phage can arise from a temperate phage as the result of a single mutational step. This raises the interesting problem of the taxonomy of viruses in general. Nothing certain is known about their origin and evolution. However, as we shall see, the genetic homology which has been found to exist between temperate phages and their hosts, as well as the increasing difficulty of drawing a demarcation line between some prophage mutants and fragments of bacterial chromosome which have attained an autonomous existence within the cell (Jacob, Schaeffer and Wollman, 1960), strengthens the hypothesis that phages may have evolved by a series of mutational steps from normal bacterial DNA (see Chapter 24, p. 806). By analogy, it is plausible to think that the different characteristics of DNA and RNA viruses in general may have been conditioned by the ancestral cellular constituents from which they arose; the direct operation of tobacco mosaic virus RNA as messenger RNA in cell-free, protein-synthesising systems (p. 355) and also, presumably, in the infected plant, is a striking case in point.

If this hypothesis is correct it implies that many strains of virus have independent origins and, therefore, have no phylogenetic relationships with one another, even though they resemble one another closely. According to this view, such similarities between viruses could be ascribed to their having arisen from comparable nucleic acid components within the cell, and to their independent adaptation to similar environments, rather than to a common line of descent. This, of course, is not to say that speciation in viruses cannot occur by mutational variation from a common ancestor.

METHODS OF STUDYING PHAGE BEHAVIOUR

We do not intend here to describe any of the techniques used for studying phage, except where this is necessary for the understanding of particular experiments. Nor do we wish to discuss the innumerable details which are of interest to the specialist, but only to provide such a general account of phage structure and behaviour as will give the reader a sound insight into the conceptual basis of phage research, and enable him to read original papers with understanding. Many excellent and detailed reviews are now available on every aspect of the subject, many of them bound within a single volume (Hershey, 1957b; Adams, 1959; Brenner, 1959; Garen and Kozloff, 1959; Jacob and Wollman, 1959; Levinthal, 1959; Stahl, 1959; Stent, 1959; Sinshieimer, 1960; Kellenberger, 1961; Mahler and Fraser, 1961; Zinder, 1965). In particular, a comprehensive account of phage methodology and techniqueis given by Adams (1959), while Stent (1963) has produced a superbly written and extremely readable book covering the whole field for university students and non-specialists in this subject. There are, however, some general and elementary procedures which not only form the basis of nearly all phage experiments but also reveal the principal manifestations of phage infection, and these we must introduce before progressing to an analysis of phage behaviour.

THE RECOGNITION AND ENUMERATION OF PHAGE PARTICLES

Phage particles can be identified and counted either by means of electron microscopy, or by observing their effects on populations of sensitive bacteria. If a suspension of virulent phage is added to a growing culture of sensitive bacteria in broth, so that the number of phage particles exceeds the number of bacteria (*multiplicity of infection* $= > 1$), after an interval which varies with the phage and the experimental conditions, the culture becomes clear, as a result of the infection and lysis of virtually all the bacteria. If incubation of the culture is continued, however, turbidity may reappear after many hours, due to growth of rare, pre-existing bacterial mutants which are *resistant* to infection (p. 482). Infection with temperate phage under similar conditions may lead to a transient reduction of turbidity, but this is rapidly restored, and normal bacterial growth is resumed by that proportion of the population which becomes lysogenised by the phage and thereby acquires *immunity* to subsequent lytic infection (p. 463).

Viable phage particles may be counted by mixing a small volume of a suitable dilution of the suspension with a few drops of a concentrated broth culture of sensitive bacteria, in about 2 ml of molten, 'soft' agar, at 46°C. Since the bacteria are present in gross excess, each of the small number of phage particles will infect a single bacterium. The mixture is poured over the surface of a nutrient agar plate to form a thin 'top layer' which is allowed to harden so that the bacteria are immobilised. The plate is incubated at 37°C. Uninfected bacteria multiply to form a confluent film of growth over the surface of the plate. Each infected bacterium, however, bursts after a short time and liberates several hundred progeny phage particles which immediately infect adjacent bacteria which, in turn, are lysed. This 'chain reaction' spreads locally through the bacterial population until brought to a halt by a decline in bacterial metabolism on which phage multiplication depends. The result is a visible, circumscribed area of clearing in the confluent bacterial growth, known as a *plaque*. Whereas *plaques produced by virulent phages are nearly always clear* (although otherwise they may vary widely in size and form), since all infected bacteria are lysed, *infection by temperate phages yields turbid plaques* from the growth of immune, lysogenic bacteria on the plaque floor.

By counting the number of plaques produced, the number of *plaque-forming units* in the original phage suspension can be calculated. This is termed the *plaque titre* of the suspension. By comparing the plaque titre with the actual number of particles present as estimated, for example, by electron microscopy, the proportion of total phage particles able to produce infection, called the *efficiency of plating* (EOP), can be estimated. In general, the efficiency of plating of phages is found to lie between 0·5 and 1·0 under optimal conditions (Luria, Williams and Backus, 1951; Kellenberger and Arber, 1957). It might be thought that the efficiency of plating of temperate phage would be markedly low since every *initial* infection which resulted in lysogeny would fail to produce a plaque. This is not usually found, however, when plaque counts are made under optimal conditions for bacterial growth, when the probability of lytic response is highest.

The revival of interest in bacteriophage, and the great output of research that accompanied it, was mainly due to the development of two types of experimental method which survive today as standard procedures for studying the kinetics of phage multiplication in infected bacteria. Before we describe them, however, we must comment on the phage strains used initially in these experiments and, subsequently, as model phages throughout most of the fundamental work discussed in this chapter.

At the commencement of this period of quickening research, various workers began to isolate independent strains of phage, virulent for *E. coli*, in order to carry out their own investigations. Delbrück foresaw that knowledge would advance more quickly if research was concentrated on a small number of strains, and prevailed upon the others to restrict their attention to the limited number of strains then in use. For this reason, most research into the nature of phage and the mechanism of phage infection has employed a particular set of seven virulent phages, known as the T series (T1–T7), which infect *E. coli*.

These seven phages fall into three categories. In one category are the 'T-even' phages (T2, T4 and T6) which are closely similar in their serological properties and genetic constitutions, and are characterised by two important features. Firstly, their intra-cellular developments is quite independent of the integrity of the bacterial nucleus which, in fact, is rapidly destroyed following infection so that specific bacterial metabolism ceases (Luria and Human, 1950); phages of this type have therefore been called *autonomous virulent* (Whitfield, 1962). Secondly, the pyrimidine, 5-hydroxymethylcytosine (HMC), uniquely replaces ordinary cytosine in the DNA of the T-even phages. Phage T5 resembles the T-even phages in being autonomously virulent, but retains ordinary cytosine in its DNA. Phages T1, T3 and T7, on the other hand, are serologically and genetically unrelated, both to one another and to the T-even phages, and have cytosine in their DNA; they also differ fundamentally from the T-even phages in that not only does bacterial metabolism continue during most of the lytic cycle, but phage development appears to demand an intact bacterial nucleus. These strains are therefore termed *dependent virulent* (Whitfield, 1962). Phages T2 and T4 have been mainly employed in the following studies.

THE ONE-STEP GROWTH EXPERIMENT

This experiment was devised by Ellis and Delbrück (1939) to study the kinetics of phage multiplication in bacterial populations. A dilution of the phage suspension is added to a young broth culture of sensitive bacteria. After a few minutes, during which the phage particles become adsorbed to, and infect, the bacteria, residual free phage particles are neutralised by the addition of antiserum; the mixture is then highly diluted into warm broth and maintained at 37°C. Samples are withdrawn at intervals thereafter and assayed for the number of plaque-forming units they contain, in the manner described above.

In the case of phage T2 and its host *E. coli* B, the plaque count remains constant for about 25 minutes after infection, known as the *latent period*, and then rises sharply to reach a plateau about 10 to 20 minutes later. The number of plaques at the plateau represents a 100- to 200-fold increase over that found during the latent period (Fig. 97).

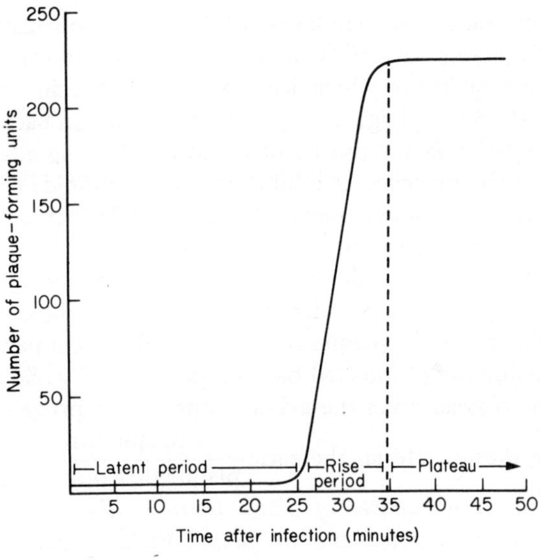

FIG. 97. A theoretical graph illustrating the one-step growth experiment. (See text for explanation.)

What do these results mean? Throughout the latent period free phage particles are absent from the samples, due to the previous addition of antiserum as well as to the high dilution, so that the plaques which are formed arise solely from infected bacteria which lyse *after plating* and liberate progeny phage particles. The particles which emerge from these individual bacteria are localised on the plate at the site of the burst and therefore give rise to only a single plaque; that is, *each plaque represents an infected bacterium* (*infectious centre*) and not a single phage particle. At about 25 minutes after infection, however, infected bacteria begin to lyse in the diluted mixture so that the liberated phage particles are no longer localised but become distributed throughout the broth. Thus, samples plated at this time contain *free phage particles, each of which generates a plaque*. When the plateau is achieved, all the infected bacteria have burst and released their content of phage; these free particles are prevented from adsorbing to uninfected bacteria by the low population density of the latter in the diluted mixture.

If all the infected bacteria burst at the same time, the increase in the plaque titre to plateau level would be instantaneous. The rise period (Fig. 97) therefore represents the scatter in the burst times of individual phage-bacterium complexes; while some of this scatter may be ascribed to inherent variability, it is also partly due to variation in the initial adsorption times which leads to asynchrony in the onset of phage growth. Since phage multiplication depends on bacterial metabolism, this latter factor can be obviated by allowing adsorption to take place in the presence of inhibitors such as cyanide (Benzer and Jacob, 1953) or chloramphenicol (Hershey and Melechen, 1957), the agent being later diluted out.

We see, then, that the duration of the latent period indicates the time which elapses between infection and the beginning of lysis of the infected cells. Since the plaque count during the latent period reveals the average number of infected bacteria per sample, while the plaque count at the plateau gives the average number of progeny particles released by these bacteria, the ratio $\frac{\text{plaque titre at plateau}}{\text{plaque titre during latent period}}$ indicates the *mean number of phage particles liberated per infected bacterium*. This is known as the *burst size*. Both latent period and burst size vary widely with the strain of phage and bacterium used, as with the physiological state of the bacteria. We will give two examples of such variation which are important in the study of phage genetics, one demonstrating the effect of a genetic difference in the phage, and the other the effect of altering the host's metabolism. It so happens that, in both these examples, an extension of the latent period is associated with an increase in the burst size, but such a direct correlation is not always found (see Delbrück, 1940; Héden, 1951).

We have referred many times to the series of exciting genetic studies, initiated by Benzer, on the structure of the r_{II} region of phage T4, but have so far refrained from discussing the nature of the r^+ and r phenotypes (p. 167 and Table 5). We shall do so now. When one-step growth experiments are performed on a young broth culture of *E.*

coli B infected *at low multiplicity* with the wild type (r^+) phages T2, T4 or T6 and then highly diluted, normal latent periods of about 25 minutes are found. If the infected culture is not diluted, however, the latent period is greatly prolonged. This is due to an unexplained phenomenon called *lysis inhibition* whereby the adsorption of the early liberated phage particles to previously infected bacteria greatly extends the time these bacteria require to burst; when lysis does occur however, the burst size is considerably greater than normal (Doermann, 1948). The occurrence of lysis inhibition is reflected in the morphology of plaques of r^+ strains, which are small and have a turbid halo. Mutants of these wild type phages arise which produce larger, clearly defined plaques which are easily distinguishable from the r^+ type (Hershey, 1946). Phage isolated from such mutant plaques is not subject to lysis inhibition and shows a normal latent period at all multiplicities of infection. These mutants are designated *r* because of their property of causing *rapid lysis*. Plaques of r^+ and *r* type are compared in Plate 22 (p. 489).

Our second example of a factor influencing the duration of the latent period concerns substances, such as chloramphenicaol and 5-methyltryptophan, which specifically inhibit protein synthesis. If one of these substances is added to a culture of infected bacteria at about the middle of the latent period, the synthesis of those enzymes which promote cellular lysis is arrested, whereas the synthesis of phage DNA continues. The latent period may thus be extended to an hour or more. Shortly after the drug is diluted out, however, the cells begin to lyse, producing many times the normal yield of phage (Hershey and Melechen, 1957).

THE SINGLE BURST EXPERIMENT

The one-step growth experiment provides information about the minimum latent period and average burst size occurring in large *populations* of infected bacteria, but tells us nothing about what is happening in *individual* phage-bacterium complexes. For this latter purpose the *single burst experiment*, originally devised by Burnet (1929) and later developed by Ellis and Delbrück (1939), must be used.

The principle of the experiment is, first, to infect a bacterial culture with phage, residual free particles being neutralised with antiserum as in the one-step growth experiment. The mixture is then highly diluted and small samples transferred to each of a large series of broth tubes, in such a way that each tube has rather a low probability of receiving an infected bacterium. The tubes are incubated until it is judged that every infected bacterium has certainly burst. The entire contents of each tube are then plated separately for plaques. By means of Poisson's formula, the proportion of those tubes yielding plaques which received only a single infected bacterium may be calculated from the proportion of all the tubes which yield no plaques at all and, therefore, did not receive a phage-bacterium complex. For example, if one-third of the tubes yield no plaques, then each tube contained, *on the average*, one infected bacterium (compare the similar basis for estimating mutation rates on p. 195). From the range of variation in the plaque counts emanating from these tubes, the range of burst sizes of individual complexes can be assessed. It turns out that burst sizes vary markedly and are not correlated with the sizes of the bacteria involved (Delbrück, 1945).

The single burst experiment is an important tool in phage genetics since it permits analysis of the recombinant progeny emerging from individual doubly-infected cells, so that such problems as whether a single recombination event yields reciprocal recombinants can be analysed much more rigorously than in the population as a whole (pp. 63, 494).

THE ANATOMY OF BACTERIOPHAGE

By far the most intensive studies of the structural components of bacteriophage, and of the way in which these are fitted together to form a co-ordinated infective apparatus, have been made on the T-even phages, and on phages T2 and T4 in particular. We will therefore concentrate on a description of this phage, merely noting such points of difference from other phages as our scanty knowledge of them allows.

MORPHOLOGICAL COMPONENTS

The overall picture of phage T2, revealed by electron microscopy, resembles that of a rather angular tadpole. The head has the form of an elongated bipyramidal, hexagonal prism about 1000 Å long and 650 Å wide which, strictly, is probably a prolate icosahedron (see Kellenberger, 1966b); to the head is attached a straight tail about 1000 Å long and 250 Å wide (Fig. 98 and Plates 13, 14). The volume of the phage is thus about one thousandth that of its bacterial host. Phages

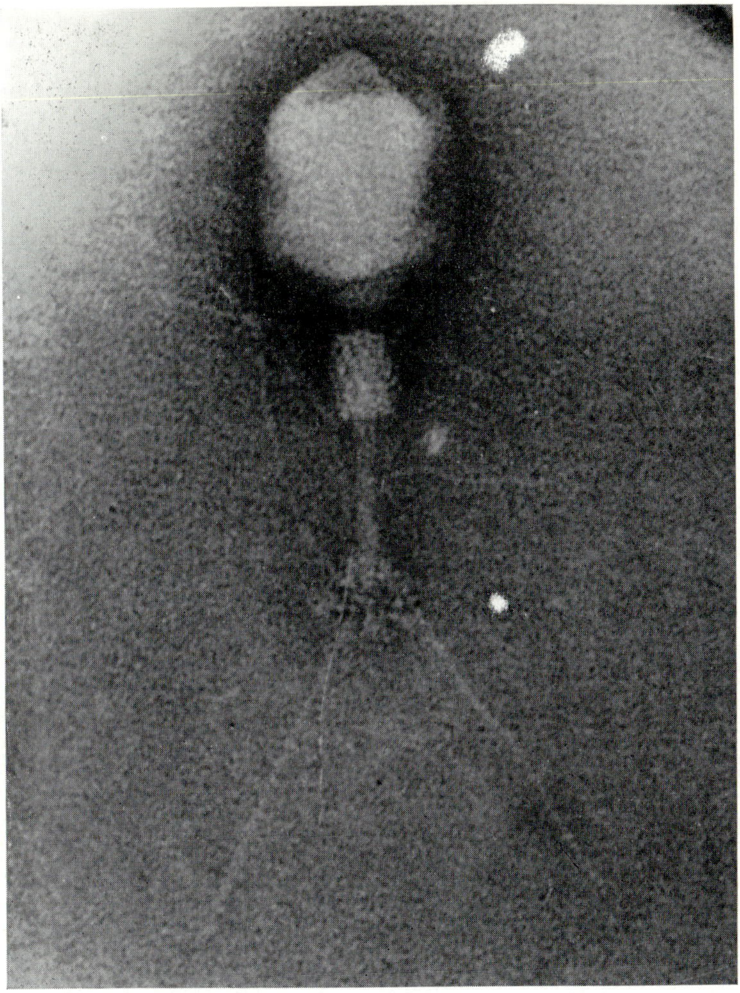

PLATE 13. Electron micrograph of an isolated particle of phange T2, showing the relations of the filled head, contracted sheath, core and tail fibres. The phage has been treated with hydrogen peroxide and embedded in PTA. Magnification about $\times$ 399,000.

(Reproduced from Brenner *et al.*, 1959, by kind permission of the authors, and the editors of the *Journal of Molecular Biology*. Original photograph kindly provided by Dr R. W. Horne)

facing p. 416

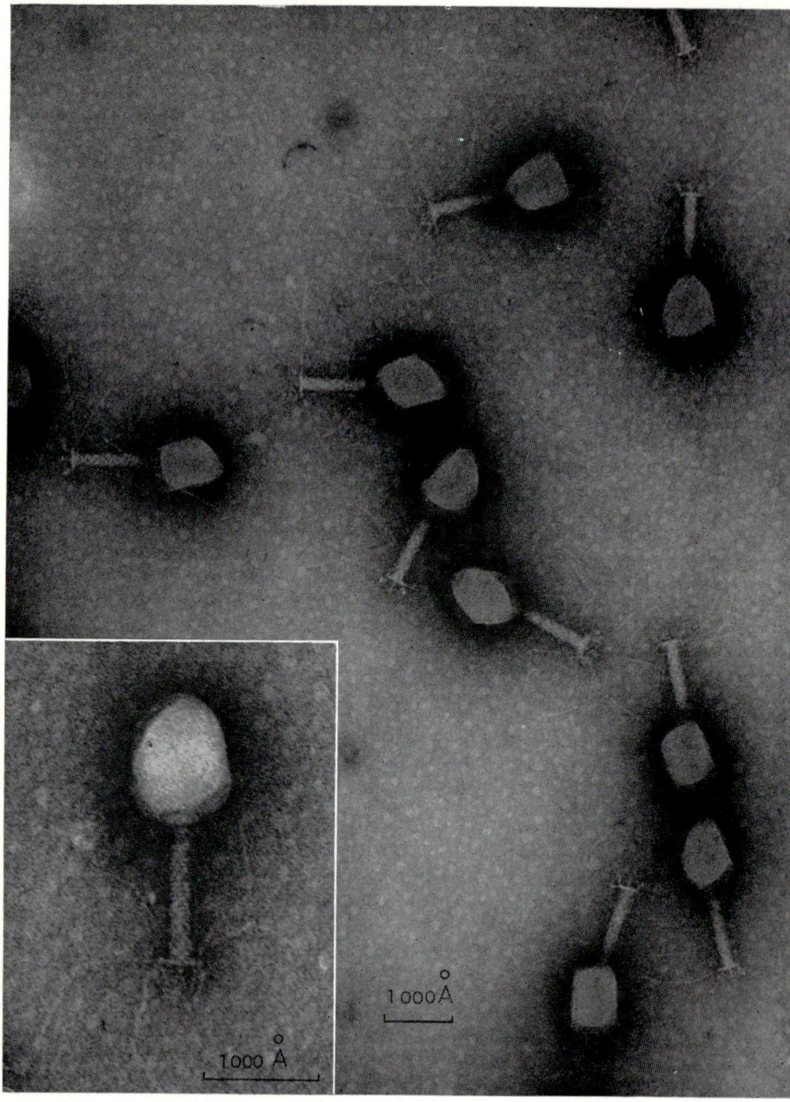

PLATE 14. Electron micrographs of particles of phage T4, showing filled heads, uncontracted sheaths, base plates, spikes and tail fibres. Embedded in PTA. Magnification of larger picture about × 130,000, and of the inset about × 156,000.

(Kindly provided by Dr E. Kellenberger)

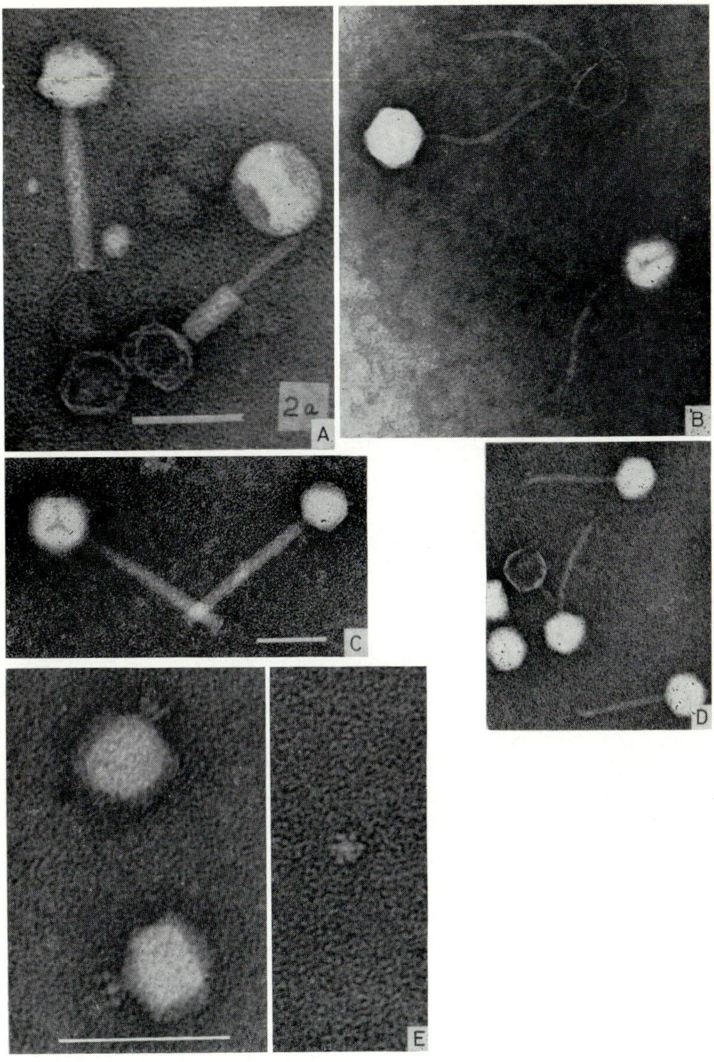

PLATE 15. Electron micrographs of particles of various types of bacteriophage.

A. Phage P2: a normal particle and one with empty head and contracted sheath are shown. The tail terminates in a base plate to which a group of very thin fibres (not seen) is attached. The scale at the bottom is 1000 Å long.

B. Phage T5: note the thin and flexible tails which appear to terminate in a single sharp point. No contractile sheath, base plate or tail fibres have been observed and it is not known how the phage DNA penetrates infected bacteria. Magnification about ×68,250.

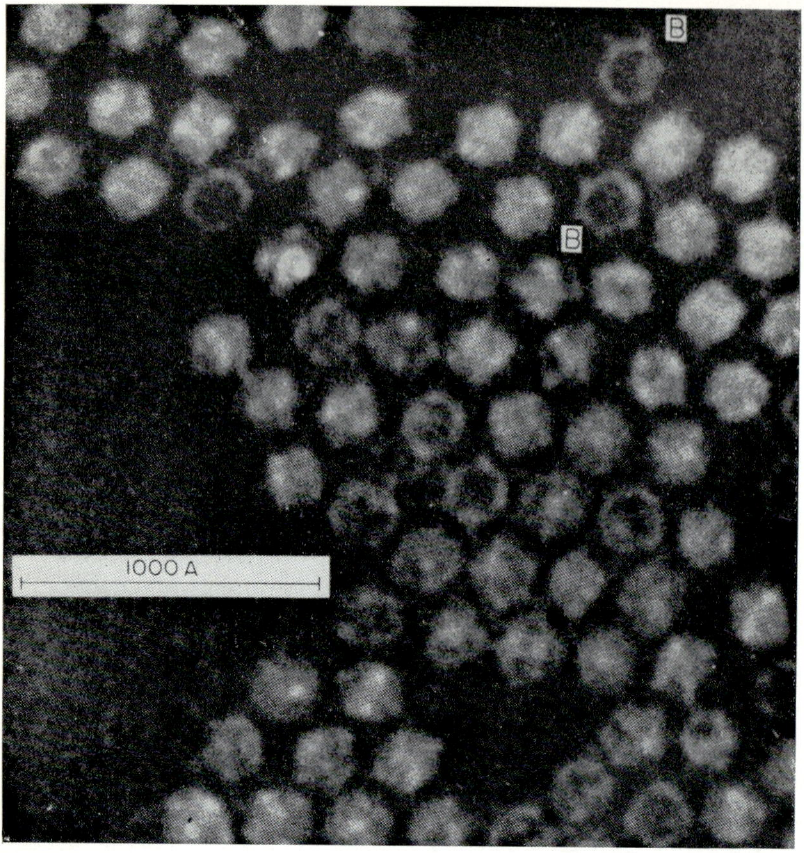

PLATE 16. Electron micrograph showing the morphology of phage $\phi X174$. The external shape of the virus appears polygonal, with a maximum linear dimension of 225 Å. Some particles, as at B, are oriented so as to present a clearly hexagonal outline. The ring-like particles, with a dense central core, are thought to have lost their nucleic acid so that the phosphotungstate has penetrated to their interior. The particles marked B are also seen to have rod-shaped projections, but there is no evidence of a tail structure. Embedded in PTA.

(Reproduced from Tromans & Horne, 1961, by kind permission of the authors, and Academic Press, Inc.)

C. Phage P1: this is the largest phage so far examined. The sheath is contractile and a flat base plate with attached fibres has been observed. A peculiarity of this phage, shown in the photograph, is that particles with two different head sizes are found, even in

T4 and T6 are identical in appearance with phage T2. Phages T1 and T5 have longer and more slender tails (Plate 15B), while those of phages T3 and T7 are very short and rudimentary (cp. Plate 15E) (Anderson, 1953, 1960; Williams, 1953; Williams and Fraser, 1953).

Phage particles become attached to bacteria by means of their tails (Anderson, 1953), but are also able to adsorb to cell wall material isolated from sensitive hosts. This adsorption can be seen to be accompanied by a striking change in the appearance of the phage tail, the upper part of which becomes contracted and thickened to reveal a protruding *core* about 70 Å in diameter; that part of the tail which undergoes contraction is known as the *sheath* (Kellenberger and Arber, 1955). The phage tail thus turns out to be a more complicated structure than was at first thought. Further details of the anatomy of the tail were resolved as a result of two technical advances. The first of these was the dissociation of various tail components by treatment with certain chemical and physical agents (Kozloff and Henderson, 1955; Kellenberger and Arber, 1955; Williams and Fraser, 1956; Brenner *et al.*, 1959); the second lay in the development of a method of negative staining suitable for electron microscopy, by 'embedding' the preparation in phosphotungstate, which gives far more refined visualisation of ultra-structure than metal shadowing (Brenner and Horne, 1959).

By the use of these methods, the earlier finding that the tail comprises a core surrounded by a contractile sheath has been confirmed. The sheath itself is a hollow cylinder, built up from a number of helically arranged sub-units, which conserves a constant volume on

---

fresh preparations grown from a single plaque; the volume of the smaller head is about half that of the larger. The scale at the bottom is 1000 Å long.

D. Phage λ: the particles resemble those of phage T5, shown at B. The tails terminate in a point, and no contractile sheath, base plate or tail fibres have been observed. Magnification about × 59,500.

E. Phage P22: the particles possess a well defined, hexagonal base plate with spikes, which is joined to the head by a short neck about 70 Å long by 70 A wide. No contractile sheath has been observed. The particles shown have been treated with hydrogen peroxide and alcohol. On the right, an isolated base plate is seen end-on. The scale at the bottom is 1000 Å long.

All preparations were photographed after embedding in PTA

(The electron micrographs were kindly provided by Dr T. F. Anderson. Photographs A, C and E appeared in Anderson, 1960)

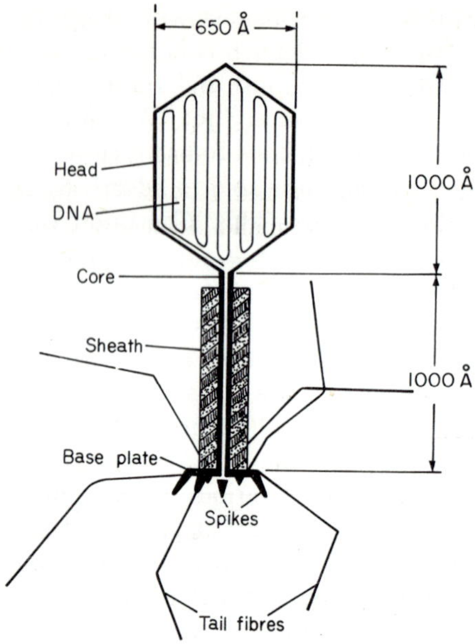

FIG. 98. Diagram showing the morphological components of phage T2, and their arrangement in the intact structure.
Drawn roughly to scale.
For comment see text, and compare with Plates 13, 14 and 15.

contraction and appears not to be connected to the head. The core is a hollow cylinder extending the length of the tail, and having an axial hole 25 Å in diameter. The proximal end of the core is attached to the head, while to its distal end is fitted a hexagonal *base plate* from which project six short *spikes* and, distinct from the spikes, six long *tail fibres*, each 1300 Å in length and 20 Å wide and with a kink in the middle (Brenner *et al.*, 1959; see also Kellenberger, 1961, and Bradley, 1965a, b.). These features are represented diagrammatically in Fig. 98, and most of them can be seen in Plates 13, 14 and 15. The temperate phages P1, P2 and P22 also possess a contractile sheath. On the contrary, phage T5 and the temperate phage λ have thinner and more flexible tails with only a single spike and no tail fibres at their tips and, in them, a sheath has not been demonstrated (Anderson, 1960). The appearance of some of these phages is shown in Plate 15.

In addition to these relatively large phages, which are all of the same order of size, some minute bacterial viruses of quite different appearance have recently been described. Among these are the DNA-containing phage ϕX174, notorious for carrying its DNA in single-stranded form (Sinsheimer, 1959a; see p. 436), and a group of RNA phages, first described by Loeb and Zinder (1961) which specifically attack male strains of *E. coli* K-12 (p. 440). The morphology of phage ϕX174, infective for *Shigella* and *E. coli* C, has been shown to resemble closely that of many small animal and plant viruses. Similar in size to poliomyelitis virus, with a mean diameter of 200 Å, it has a polygonal architecture, being constructed of twelve morphological sub-units located at the vertices of an icosahedron, and lacks a tail (Tromans and Horne, 1961; see Plate 16). The appearance of the RNA-containing phages is broadly similar to that of phage ϕX174 (see Plate 34B, p. 773).

A number of filamentous bacterial viruses have also recently been isolated, most of which, like the spherical RNA viruses, infect only male strains of *E. coli*, or other *Enterobacteriaceae* carrying the E. coli sex factor (pp. 103, 670), and are probably closely related (Zinder *et al.*, 1963; Marvin and Hoffmann-Berling, 1963; Hofschneider, 1963); an exception is a rod-shaped *Pseudomonas* phage which appears not to be male-specific (Takeya and Amako, 1966). These phages are about 800 Å long and 60 Å in diameter, and contain single-stranded DNA (Plate 20). We will discuss the physiological aspects of these various unorthodox phages later on in this Chapter (p. 440). A broad review of phage morphology is by Bradley (1965b).

CHEMICAL COMPONENTS

When highly purified preparations of classical bacteriophages are analysed chemically they are found to be composed of only two substances—protein and DNA. Only negligible amounts of lipoid and RNA are detectable and these can be wholly accounted for by contamination of the preparations with bacterial debris. This has been demonstrated for all of the T-series of phages, as well as for the temperate phages P22 and λ (see Stent, 1958). A particle of phage T2 weighs 5×10^{-13} grams, of which about 60 per cent is protein and 40 per cent DNA; a particle of λ phage weighs less than half as much and consists of equal parts protein and DNA (see Kellenberger, 1961).

When phage particles are subjected to osmotic shock, by rapid transfer from high salt concentrations to distilled water (p. 485), the DNA dissociates from the protein and, at the same time, becomes sensitive to the action of deoxyribonuclease to which it was formerly resistant. Examination of osmotically shocked particles by electron microscopy reveals that they retain their basic anatomical features, except that the heads now appear collapsed and empty, presenting a 'ghost-like' appearance (Anderson, Rappaport and Muscatine, 1953); these 'ghost' particles are enmeshed in a tangled mass of DNA fibres (Williams, 1953). This appearance is admirably displayed by means of a beautiful technique whereby the discharging DNA thread is gently diffused on a spreading, molecular film of protein on water (see Plate 24, p. 544; Kleinschmidt *et al.*, 1962).

The phage 'ghosts' produced by osmotic shock retain all the protein of the original particles and can still adsorb to sensitive bacteria although, like all ghosts, they are 'dead' and can no longer produce infection (Hershey, 1955). It is therefore evident that the DNA is contained within the protective protein coat which constitutes the phage head (Fig. 98). The amount of DNA accommodated within the head of phage T2, for example, if extended as a single, continuous double helix, is about 52μ long (p. 537) or more than 500 times the length of the head itself, and so presents a formidable packing problem which we will shortly consider. In addition to DNA, osmotically shocked T2 particles release small amounts of protein (*internal protein*) into the environment. This protein is presumably associated with the DNA inside the head membrane and includes the polyamines, spermidine and putrescine (Ames *et al.*, 1958), an acid-soluble polypeptide containing only aspartic acid, glutamic acid and lysine, and an acid-insoluble protein (Hershey, 1955, 1957a). The function of these substances is probably to neutralise the acidity of the DNA and to provide some sort of a matrix which promotes its condensation (p. 430) (see Kellenberger, 1961).

The ability to separate the protein components of phage from one another, and to obtain highly purified preparations of each, permits the various types of protein structure that make up the phage coat to be analysed. A number of different proteins have been identified. The head membrane is homogeneous and composed of a large number of repeated sub-units of molecular weight 80,000; the sheath comprises about 200 repeating sub-units, each of about 50,000 molecular weight; the tail fibres have a protein sub-unit of molecular weight not less than 100,000 (Brenner *et al.*, 1959). The head protein and two tail components have also been distinguished serologically; only antisera which interact with phage tails are able to neutralise infectivity (Lanni and Lanni, 1953; Kozloff, Lute and Henderson, 1957).

THE INFECTIVE PROCESS

The process of phage infection may arbitrarily be divided into a number of stages which can be analysed experimentally. These stages are adsorption, penetration, intracellular development, maturation and lysis, and we will consider each of them in turn.

THE ADSORPTION OF PHAGE PARTICLES

The first step in infection is a random collision between phage and bacterium. Since phage infectivity tends to be highly host-specific and to depend on the antigenic structure of the bacterial surface, collisions must be followed by a more precise and durable attachment of the phage tail. What components of the cell surface and of the phage tail are involved? The bacterial contents are contained by two membranes, an outer, rigid *wall* and an inner, *semipermeable membrane*. Phage is found to adsorb to the purified cell walls of sensitive bacteria as actively as to the bacteria themselves, but not to bacterial protoplasts from which the cell walls have been stripped; protoplasts, however, can support phage growth if infection occurs before removal of the cell walls (Weibull, 1953; Zinder and Arndt, 1956).

Chemical analysis of *E. coli* cell walls, especially by Weidel and his colleagues, has revealed that they are made up of two layers; the outer is composed of lipoprotein containing a wide range of amino acids but no diaminopimelic acid. The inner layer, which gives the cell wall its rigidity, is mainly lipopolysaccharide in nature; it comprises a network of large units built from phospholipid, glucose, glucosamine and L-gala-D-mannoheptose, linked together by smaller units containing glucosamine, muramic acid, alanine, glutamic acid and diaminopimelic acid (review: Rogers, 1965). Cell walls from which the outer lipoprotein layer has been extracted by phenol no longer adsorb phages T2, T5 and T6, although this treatment actually enhances the adsorption of phages T3, T4 and T7. The receptors for T2, T5 and T6 must therefore reside in the outer layer and those for T3, T4 and T7 in the inner one. Mutants of *E. coli* B (designated B/3, 4, 7) are found which have acquired simultaneous resistance to the three latter phages and, in these, L-gala-D-mannoheptose is completely absent from the inner cell wall layer. This sugar is therefore a common and essential receptor for all these phages, but it is not the only one because other mutants, resistant to each phage individually and retaining the sugar, are also found (Weidel, Koch and Lohss, 1954; Weidel and Primosigh, 1958).

The tail fibres appear to be the components responsible for specific attachment of the T-even phages. Thus amber mutants of phage T4 which lack tail fibres fail to adsorb, while complete tails and isolated tail fibres do; moreover, when cores are tested, only those which are seen to have retained their tail fibres adsorb. Finally, isolated tail fibres agglutinate bacteria which are sensitive to the phage (Williams and Fraser, 1956; Kellenberger and Séchaud, 1957; Wildy and Anderson, 1964; Kellenberger *et al.*, 1965) When intact particles of phages T2 or T4 are examined microscopically, tail fibres are not seen and are thought to be wrapped around the base of the tail. It is only when the particles interact with cell wall material, or are treated in rather specific ways, that the fibres unwind and make their appearance. What it is that triggers this unwinding is not clear, although there is evidence that zinc complexes in the cell wall may be responsible (Kozloff and Lute, 1957). It should, of course, be remembered that many phages do not normally possess tail fibres; their mechanism of attachment is quite unknown. The role played by the base plate and its attached spikes also remains unknown.

The environment is an important factor in adsorption. In particular, inorganic salts must be present and probably act by neutralising the net negative charges on bacterium and phage so that initial contact is facilitated. Some phages have specific cationic requirements, such as that of phages T4 and λ for Ca^{++} and Mg^{++} respectively. L-Tryptophan is needed as an *adsorption co-factor* in synthetic media by certain strains of phages T4 and T6. This requirement is remarkably specific since anthranilic acid, indole and D-tryptophan are ineffective. The role of tryptophan appears to be to release the tail fibres from association with the sheath, so that they are free to interact with the bacteria surface. In phage T4 the fibres can be seen to retract to form a 'jacket' around the sheath in the absence of tryptophan, and to extend in its presence (Brenner *et al.*, 1962; Kellenberger *et al.*, 1965). Mutations to tryptophan-independence occur, so that the requirement for tryptophan is genetically determined (Anderson, 1948; Delbrück, 1948; see Brenner, 1959; see p. 513 for genetic analysis).

PENETRATION BY THE PHAGE

The key to understanding the true nature of the infective process, as well as of the genetic system of phage, was forged by the famous experiment of Hershey and Chase (1952), which showed that only the phage DNA penetrates the infected bacterium, the protein coat remaining outside.

Two types of preparation of phage T2 were made, one labelled with radioactive phosphorus (^{32}P) which is incorporated only into the DNA, the other with radioactive sulphur (^{35}S) which is present exclusively in the protein coat. A culture of *E. coli* B was infected with one or the other of these two preparations, unadsorbed phage being removed by centrifugation. The bacteria were then agitated violently in a Waring blendor to shear off adsorbed phage particles. This treatment was found to have no effect on the plaque-forming ability of the infected cells. Finally, the distribution of the radioactivity between the bacteria and the extraneous medium was assayed.

It was found that the blendor treatment removed about 80 per cent of the sulphur but only about 20 per cent of the phosphorus of the infecting particles. Moreover, nearly all of the 20 per cent of the phage sulphur which remained bound to the bacteria could be ascribed to failure of the treatment to remove all the attached phage coats (Hershey, 1953), while most of the 20 per cent of phage DNA found in the medium was contained in intact particles which had not been adsorbed (Hershey and Burgi, 1956). In confirmation of this, about 50 per cent of the ^{32}P label (that is, of the DNA) of infecting particles is found to reappear in the progeny phages which emerge from the infection, while less than 1 per cent of the protein label is so transferred.

The experiment of Hershey and Chase proved what had previously been suspected from analysis of the transforming principle of *Pneumococcus* (p. 575), that the genetic material of phage is its DNA and that phage infection is, in essence, a genetic phenomenon. But it is not exclusively so, for it turns out that about 3 per cent of the total phage protein is injected along with the DNA. This protein, however, is not part of the phage envelope but comprises those polyamines and other constituents which are contained within the phage head and are liberated by osmotic shock (internal proteins, p. 420).

What is the mechanism of infection? The morphological evidence that the core, through which the DNA must pass, is surrounded by a contractile sheath and actually appears to penetrate the cell wall following attachment, supports the view that the phage coat behaves like a 'micro-syringe', as originally suggested by Hershey. An important clue as to how penetration occurs was the discovery that the tails of the T-even phages (Barrington and Kozloff, 1954; Koch and

Weidel, 1956; Weidel and Primosigh, 1958), as well as of phage λ (Fisher, 1959), contain a lysozyme-like enzyme, which can be liberated by freezing and thawing and which acts on the deeper, lipo-polysaccharide layer of the cell wall, releasing large amounts of alanine, glutamic acid and diaminopimelic acid which, you will remember, constitute the chemical links holding together the network of large molecules to which phages T3, T4 and T7 attach (p. 421). While the specific phage receptors are unaffected by the enzyme, there is evidence that some of the substances released by the enzyme can trigger the discharge of phage DNA in the absence of the cell walls themselves (see Garen and Kozloff, 1959). The current model of phage infection may therefore be summarised as follows. The phage tail makes contact with the cell wall and so is brought into contact with zinc complexes, thus activating the unwinding of the tail fibres which then adhere specifically to certain molecular groupings. At the same time, the phage enzyme is released and bores a hole through the rigid layer of the wall. The products of this lytic activity finally trigger contraction of the sheath, leading to penetration of the cell wall by the core and to discharge of the DNA into the cell. The nature of the forces responsible for propelling a viscous DNA thread, $5{\cdot}5 \times 10^5$ Å long and 20 Å wide, through a tube 1000 Å long and only 25 Å in diameter in a matter of a minute or so, remains a mystery.

THE INTRACELLULAR DEVELOPMENT OF PHAGE

The first direct study of the intracellular events which follow phage infection was by Doermann (1952). During the course of a one-step growth experiment with phage T4, he broke open the infected cells at intervals during the latent period, by treating them with cyanide together with an excess of phage T6 (see *lysis from without*, p. 435), and then assayed these *premature lysates* for the presence of infective phage T4, by plating for plaques on indicator bacteria which were sensitive to phage T4 but resistant to phage T6. Other methods of disrupting the infected bacteria are by sonic oscillation (Anderson and Doermann, 1952), chloroform (Séchaud and Kellenberger, 1956), streptomycin (Symonds, 1957), and by forming protoplasts from previously infected cells and then lysing them by osmotic shock (Brenner and Stent, 1955). The result of this kind of experiment is that infectious particles do not begin to appear until about half-way through the latent period, at 12 minutes after infection in the case of

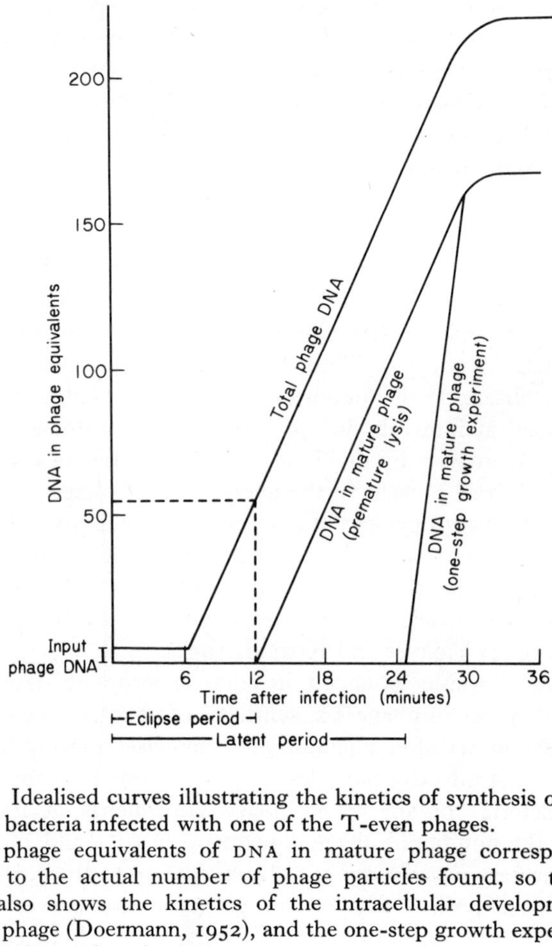

FIG 99. Idealised curves illustrating the kinetics of synthesis of phage DNA in bacteria infected with one of the T-even phages.

The phage equivalents of DNA in mature phage correspond, of course, to the actual number of phage particles found, so that the figure also shows the kinetics of the intracellular development of mature phage (Doermann, 1952), and the one-step growth experiment. (See text for explanation.)

phage T4, but thereafter their number increases linearly to reach plateau level at the normal burst time (Fig. 99).

The experiment reveals two important features of intracellular phage growth. The first is that during the early *eclipse period*, when no infective phage is found, the phage must exist in a state different from that of the infecting particle, as, indeed, we would expect from the fact that only the DNA is injected; this non-infective form of phage is known as *vegetative phage*. Secondly, the fact that the subsequent rate of increase in the number of intracellular infective particles is *linear*, and not exponential, suggests that these particles are not produced by successive replications of a complete structure but, rather, by the assembly of component parts which are independently synthesised (see below). The next step therefore lay in a chemical analysis of how the various components of phage are synthesised.

The chemical analysis of phage growth

A way of studying the synthesis of phage DNA in infected cells was provided by the discovery that the DNA of the T-even phages contains 5-hydroxymethylcytosine (HMC) in place of the cytosine of *E. coli* DNA (Wyatt and Cohen, 1953). Thus the growth of phage DNA can be followed by observing the amount of HMC in infected bacteria, prematurely lysed at intervals during the course of a one-step growth experiment. Moreover, by dividing the total amount of HMC by the amount of the base present in the average *intact* phage particle, the amount of phage DNA present at any time may be expressed in terms of *phage equivalents*, that is, the amount corresponding to a complete phage genome, so that the kinetics of DNA synthesis and of the appearance of infective particles can be directly correlated (Hershey, Dixon and Chase, 1953; Vidaver and Kozloff, 1957).

The kind of results obtained in such experiments is shown in Fig. 99. In the case of phage T2, synthesis of phage DNA commences at about six minutes after infection and then rises sharply so that by the time the first infective particles begin to appear at 12 minutes, the infected bacteria contain 50–80 phage equivalents of HMC–DNA. Thereafter the number of phage equivalents of DNA and, of infective particles, increase *linearly and at the same rate* up to the normal burst time. If normal lysis is delayed (p. 415), the two curves continue to rise at the same rate as before. The reason why both curves rise linearly and maintain a steady state equilibrium, although the DNA itself presumably replicates exponentially (p. 239), is that when the DNA *pool* reaches a certain size, completed phage genomes begin to be irreversibly withdrawn from it and incorporated into protein coats at a rate proportional to the synthesis; since the withdrawn genomes cease to replicate, the growing DNA pool suffers constant depletion. This interpretation is supported by the fact that a 'pulse' of ^{32}P label introduced into the DNA pool *early* in the latent period is transferred very efficiently to the mature, intracellular phage particles and is not returned again to the pool (Hershey, 1953).

Tracer experiments using radioactive isotopes (^{32}P and tritiated thymidine) have shown that while about two-thirds of the DNA present in a population of the T-even phages is built from precursors in the medium (Cohen, 1948), about one-third comes from breakdown products of the host DNA which are mainly incorporated into those particles formed early in the latent period (Stent and Maaløe, 1953; Hershey *et al.*, 1954).

Let us now turn to the synthesis of phage protein, which can be identified and separated from bacterial protein by its specific interactions with phage antisera and its ability to adsorb to the walls of sensitive bacteria. Moreover, virtually all the protein of phage is built from precursors in the medium and is not derived from constituents of the host; since the synthesis of bacterial protein ceases abruptly after infection with the T-even phages, as judged by the cessation of inducible enzyme formation (Benzer, 1953), that of phage protein can be followed quantitatively by the uptake from the medium of radioactive sulphur (^{35}S) which is not incorporated into nucleic acid. As in the case of DNA, the amount of phage protein in infected cells can be expressed in phage equivalents, by dividing the total amount detected by the average protein content of an intact particle.

In the case of infection with phages T2 and T4, serologically specific phage protein begins to be detected in the infected bacteria shortly after the commencement of DNA synthesis, at about nine minutes from the beginning of the latent period, and rises to 30–40 phage equivalents by the time infectious particles start to appear. Thereafter it increases in amount at the same rate as does the DNA. By injecting short 'pulses' of radioactive sulphur (^{35}S) into the medium during the latent period, it can be shown that the rate of uptake of the sulphur into intracellular protein is constant from the beginning and that, from about the tenth minute onwards, half of it is transferred to phage particles, so that the greater part of the protein, at least, consists of phage coat precursor material (Maaløe and Symonds, 1953; Hershey *et al.*, 1955.). If, however, the fate of 'pulses' of ^{35}S added *early* in the latent period, before the emergence of serologically indentifiable phage protein, is followed, although it is incorporated into non-bacterial protein, very little of it is ultimately transferred to phage particles. Thus new protein is synthesised very early after injection of the phage DNA, *but this is not phage precursor protein* (Hershey *et al.*, 1954). What is its nature and function?

The nature of non-precursor (early) phage protein

We have mentioned that the addition of specific inhibitors of protein synthesis, such as chloramphenicol or 5-methyltryptophan, to infected bacteria at about the middle of the latent period represses the synthesis of phage protein without affecting that of DNA (p. 415). If, on the other hand, these inhibitors are added either before, or up to five minutes after infection, neither protein *nor* DNA synthesis occurs (Burton, 1955; Tomizawa and Sunakawa, 1956; Hershey and Melechen, 1957). Initiation of the synthesis of phage DNA is therefore dependent on early protein synthesis, and it is plausible to infer that the early proteins comprise a series of enzymes, specified by the infecting phage genome, which are required for building up new HMC-containing replicas of it. In fact many such proteins have now been identified which are not included in the envelopes of the mature particles but are necessary for DNA synthesis. Many, if not all, of these are synthesised *de novo* in the infected cell (reviews: Kellenberger, 1961; Mahler and Fraser, 1961; Champe, 1963). In the T-even phages the most important are as follows: (*a*) an enzyme which hydroxymethylates deoxycytidine 5′-phosphate as an essential step in HMC synthesis (Flaks and Cohen, 1959; Flaks, Lichtenstein and Cohen, 1959); (*b*) an enzyme which phosphorylates hydroxymethyl-deoxycytidine 5′-phosphate, leading to synthesis of the triphosphate; (*c*) an enzyme which breaks down deoxycytidine triphosphate, thus preventing the incorporation of bacterial cytosine into the phage DNA; (*d*) an enzyme involved in the direct glucosylation of the HMC–DNA, the incorporation of glucose being a characteristic feature of the DNA of the T-even phages (see p. 520; Kornberg *et al.*, 1959) and (*f*) a DNA polymerase which is distinct from that normally found in *E. coli* (Aphosian and Kornberg, 1962) (p. 247). Initial attempts to demonstrate phage polymerase *in vitro* were unsuccessful due to breakdown of added deoxycytidine triphosphate (dCTP) by the phage enzyme mentioned under (*c*). The polymerase was revealed only when this enzyme was inactivated by fluoride, or the dCTP in the system substituted by dHMCTP. In addition, presumptive early enzymes include those responsible for dissolution of the host DNA (Kellenberger, 1961), for manifestation of the phenotype of r_{II} mutants (Steinberg, quoted in Stahl, 1959) and for the synthesis of at least one of the internal proteins which are injected along with the DNA (Koch and Hershey, 1959).

In the case of the T-even phages the formation of early enzymes, as a prelude to DNA synthesis, is obviously necessitated by the foreignness of the new DNA which the biochemical machinery of the infected bacterium is expected to synthesise. But is this so in the case of other phages, and especially of temperate phages such as λ, whose DNA is indistinguishable from that of the host bacterium with respect both to its constituent bases and the ratios between them? To find the answer, a more subtle technical approach is needed, not only on account of the impossibility of differentiating phage from bacterial DNA by chemical means, but also because bacterial DNA may continue to be synthesised after infection.

One approach, devised for phage λ by R. Thomas (1959, 1962b), depends on the fact that if an infected bacterial population in a one-step growth experiment is later superinfected, at the same multiplicity, with a distinguishable (host-range) mutant of the same phage, the ratio of the progeny of the two types of phage in the lysate is proportional to the ratio of their DNA pools at the time of superinfection (see also p. 453). For example, if superinfection is instituted before the first phage has started to multiply, the first phage will have no advantage over the second so that the two will be liberated in equal numbers at lysis; but if, at the time of superinfection, the pool size of the first phage has increased 20-fold, this disparity will be reflected in the proportion of the two phages in the output. By adding chloramphenicol to the culture at various times after the initial infection, and then super-infecting later with the second phage, the effects of the drug on the kinetics of development of the DNA pool of the first phage may be studied. In this way it has been shown that, as with the T-even phages, the replication of phage λ DNA also depends on the formation of early protein. Moreover, the protein made by the first phage suffices for the replication of the superinfecting mutant, that is, the early protein is not mutant-specific (Thomas, 1959, 1962b).

As regards how phage protein synthesis is mediated, we have already discussed, in considerable detail, the formation of messenger RNA which follows phage infection (p. 275 *et seq.*; Hershey *et al.*, 1953; Volkin and Astrachan, 1957). It is only necessary to add here that the first synthesis to be discerned in infected complexes is that of phage-specific messenger RNA, commencing one to two minutes after phage penetration, while there is some evidence that the site of synthesis is associated with the bacterial cell walls (Mahler and Fraser, 1961).

PHAGE MATURATION

We have seen that phage develops intracellularly in the form of precursor pools of DNA and protein which increase linearly with time; when these pools attain a certain size, infective phage particles begin to appear and increase in number at the same linear rate as the total phage DNA and protein, so that the pool sizes remain constant. All this suggests very strongly that the various components of phage are synthesised separately and are then withdrawn from the pool and assembled into mature particles. All the evidence supports this view. For example, in the phenomenon known as *phenotypic mixing*, mixed infection with two related phages of different host specificity may generate progeny particles which carry the genome of one parent and the host specificity (tail protein) of the other. Quantitative aspects of the phenomenon indicate that the DNA and tail proteins of the two parents become randomly associated in the formation of the mature particles. We will consider phenotypic mixing later from the standpoint of genetic analysis (p. 511).

Again, *defective phages* may arise by mutation which can promote synthesis of all the recognisable constituents of the phage, yet cannot yield phage particles (see Jacob and Wollman, 1959). In the case of some phages this behaviour can be mimicked by the addition of various acridine dyes to infected cells during the latent period; the synthesis of phage DNA and protein appear unaffected by the drug and the bacteria lyse normally, but no infective, mature particles are released (Foster, 1948; De Mars, 1955). Kinetic studies of the effect of 9-aminoacridine on the maturation of phage T4 have shown that sensitivity to the drug extends over a continuous period commencing shortly after infection up to the time when the first mature particle is produced. The action of the drug is reversible, and its removal during the sensitive period delays maturation for a time corresponding to the duration of treatment (Susman, Piechowski and Ritchie, 1965).

This process of phage morphogenesis (*maturation*) confronts us with a very complicated physico-chemical problem about which virtually nothing is known. One of the most fascinating aspects of maturation is how the DNA becomes compressed into the phage head. Plate 17 shows a series of electron microphotographs of ultra-thin sections of *E. coli*, infected with phage T2 and sampled at intervals during the latent period. The first two pictures (A and B) show the dissolution of the bacterial nuclei which are represented by the white regions of low electron density. Shortly after the time at which synthesis of new phage DNA begins (Plate 17C), the developing DNA pool can be seen as vacuolated areas filled with finely fibrillated material of low density, very similar in appearance to the bacterial nuclei. This is the picture presented by replicating DNA which is in a highly hydrated state. A little later, however, when mature phage particles begin to be formed, very dense, small bodies are found in the DNA pool which increase in number thereafter (Plate 17, D and E). These bodies are composed of DNA of low water content which has *condensed* to about one-fifteenth of its former volume, and represent individual phage genomes (Kellenberger, 1961).

When chloramphenicol is added to the infected cells after DNA synthesis is under way, the DNA pool is observed to grow until it occupies almost the entire volume of the cell, but it does not condense to form compact bodies. About 15 minutes after removal of the chloramphenicol, however, numerous compact bodies appear. But if the bacteria are now artificially ruptured, no compact bodies, and only very few phage coats, can be identified in the released material. The inference is that the compact bodies are not completed phage particles, nor even heads, but represent condensed DNA moieties which are devoid of a stable protein coat (Kellenberger, Séchaud and Ryter, 1959). This unstable condition persists for about five minutes and is then transformed, in about one minute, to the mature, infective state in the absence of any major addition of protein (Koch and Hershey, 1959).

The condensation of DNA is a high energy process and it is hard to imagine how the aggregation of identical sub-units of head protein, and the kinetics of their assembly, could account for it. Nevertheless, as the inhibitory action of chloramphenicol shows, it is dependent on protein synthesis. One suggestion is that condensation results from the cross-linking of DNA into closely packed layers on a kind of protein 'scaffolding' termed the *condensation principle*, which may comprise one or more of the internal proteins (p. 420); it is this process which determines the architecture of the phage head, around which the sub-units of the head membrane then begin to aggregate automatically, in a manner analogous to the *in vitro* reconstitution of tobacco mosaic virus particles from purified protein and RNA fractions (p. 229). (For a detailed discussion of this complicated problem, see Koch and Hershey, 1959, and Kellenberger, 1961.) Normal lysates of the temperate phage λ contain many inviable particles comprising tailless heads, smaller than those of λ and devoid of any kind of nucleic acid. Nevertheless these heads retain the usual polyhedric shape, have a coat protein of normal antigenic structure and clearly contain 'something'. It has been suggested that this 'something' may be the condensing principle wrapped up in normal phage head protein (Karamata *et al.*, 1962).

Recent studies of amber(*am*) and temperature-sensitive(*ts*) conditional lethal mutants of phage T4 have revealed the remarkable fact that more than 40 genes are concerned with phage morphogenesis (see Fig. 104, p. 491). When these mutants are grown under restrictive conditions (that is, at high temperature for *ts* mutants and in a suppressing, su^{+} host in the case of *am* mutants), no mature phage particles are produced; however, the lysate contains an accumulation of recognisable phage components from which a particular one is usually missing or clearly abnormal. Thus lysates from certain mutants contain filled heads but no tails, other heads and tails but no tail fibres, yet others complete tails but no heads (Epstein *et al.*, 1963). It turns out that lysates of different morphologically defective

---

PLATE 17. Electron micrographs of ultra-thin sections of *E. coli* bacteria at various times after infection with phage T2, showing the stages of phage development.

A. *Before infection.* The white regions of low electron density are the bacterial nuclei.

B. *4 minutes after infection.* Dissolution of the bacterial nuclei is under way.

C. *10 minutes after infection.* The bacterial nuclei have disappeared. The finely fibrillated, low density material, scattered throughout the cell, represents the developing DNA pool of the phage. An empty phage coat is seen attached to the outside of the cell wall (above, right).

D. *14 minutes after infection.* The phage DNA has begun to condense, as indicated by the appearance of very dense bodies about the size of the phage head. Mature phage particles begin to be found about this time when the bacteria are prematurely disrupted.

E. *30 minutes after infection.* A large number of condensed phage genomes have been produced, many of which are actually mature phage particles. Lysis of the bacterium and liberation of the mature particles is imminent. Magnification about ×21,600. See text.

(Reproduced from Jacob & Wollman, 1961b, by kind permission of *Scientific American*. Original photographs kindly provided by Dr E. Kellenberger)

facing p. 432

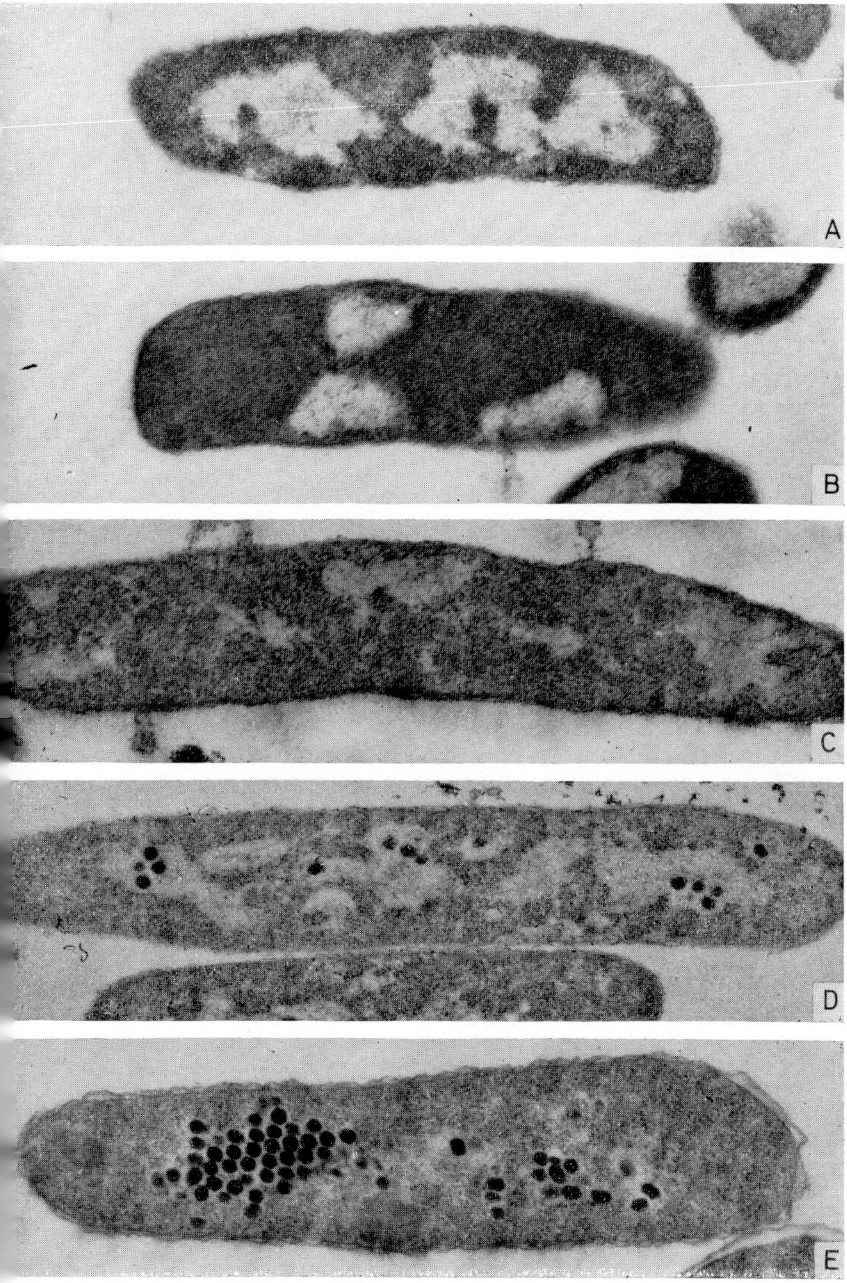

Plate 17—*facing p.* 432

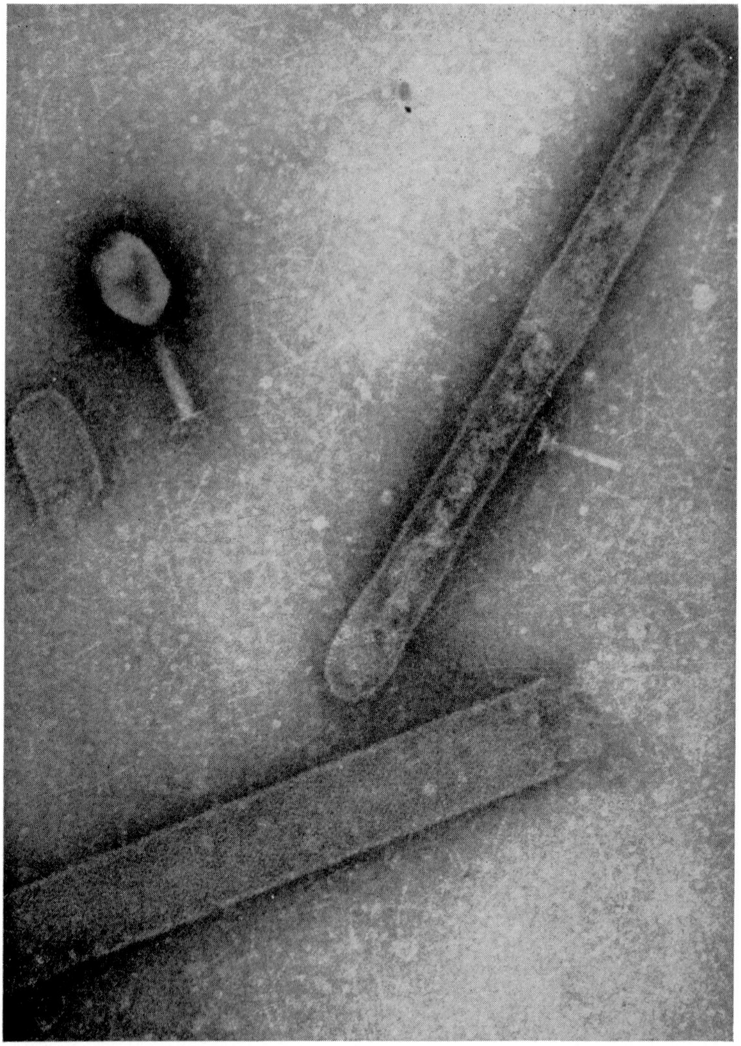

PLATE 18. Electron microphotograph of polyheads of phage T4, in a lysate of a restricting strain of *E. coli* infected with an amber mutant. Normal phage T4 particles, one of which is seen here, have been added to the lysate for comparison. Cores, base plates and tail fibres are also present in the lysate.
Negative staining. Magnification × 190,000. See text for explanation.
(Reproduced from Favre, R., Boy de la Tour, E., Segre, N. and Kellenberger, E. (1965). *J. Ultrastructure Research,* **3**, 318, by permission. The original photograph was kindly provided by Dr. E. Kellenberger).

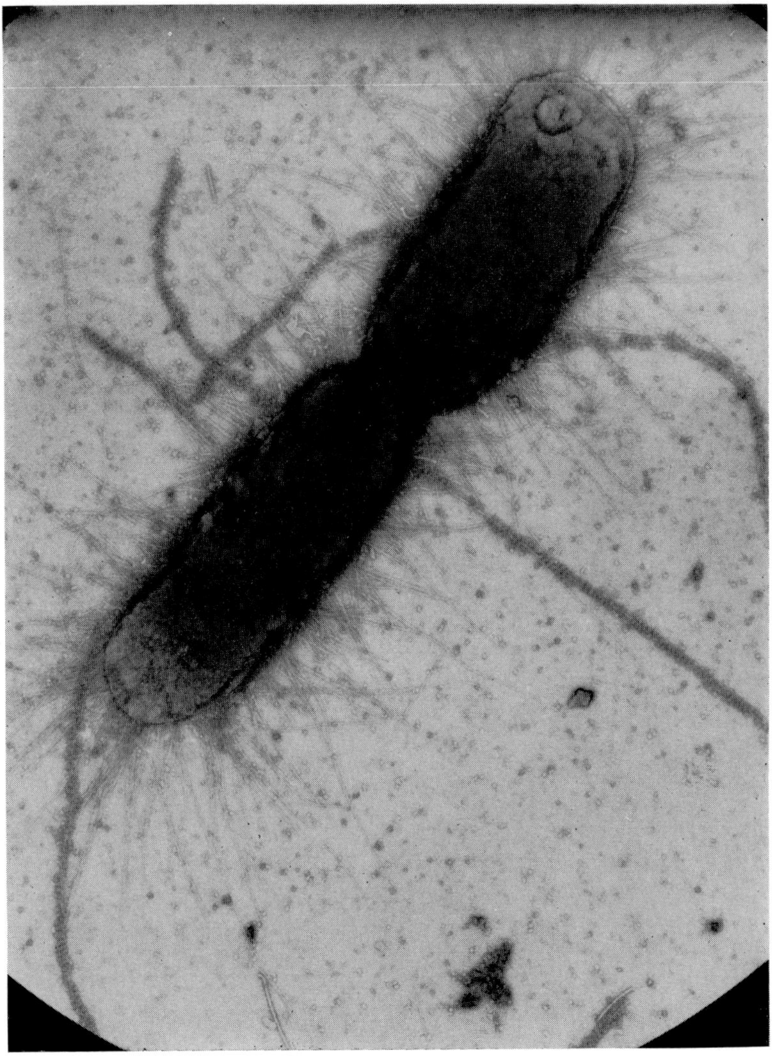

PLATE 19. Electron microphotograph of particles of male-specific RNA phage adsorbed to sex fimbriae (pili) of an Hfr strain of *E. coli*.
The predominant, hairlike appendages which cover the organism, and are devoid of adsorbed phage, are common fimbriae.
Negative staining. Magnification × 34,750. The photograph was taken and kindly provided by Dr. Lucien G. Caro.

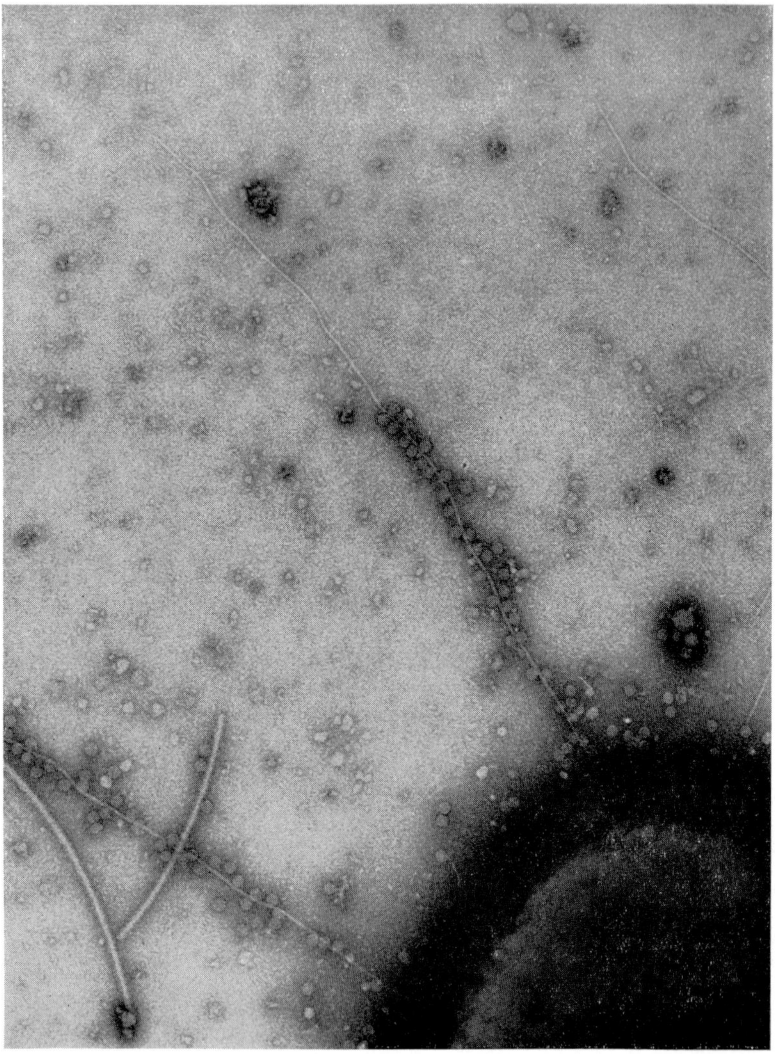

PLATE 20. Electron microphotograph of attachment of both RNA phage particles (MS2), and of a filamentous, DNA-containing male-specific phage (f1), to sex fimbriae of an Hfr strain of *E. coli.*
Two sex fimbriae are revealed by their coating of spherical, RNA phages. The filamentous extension of the upper fimbria, devoid of particles, is the filamentous phage. The two, thicker, curved rods at lower left are fragments of flagella.
Negative staining. Magnification × 87,750. The photograph was taken and kindly provided by Drs. Lucien G. Caro and D. P. Allison.

groups are often able to complement one another when mixed *in vitro*, yielding active phage with high efficiency, while mixtures of lysates showing the same defect are inactive (Edgar and Wood, 1966; Wood and Edgar, 1967). This *in vitro* complementation has also been demonstrated for phage λ (Weigle, 1966). It is therefore apparent that some of the structural protein components of the phage coat are capable of self-assembly.

However, the overall situation is more complicated than this. For example, the structure of the head protein subunit of phage T4 is determined by a single gene (No. 23; see Fig. 104, p. 491), but the actual shape of the structure formed by the subunits can be profoundly modified by mutations in other genes. Thus a mutation in gene No. 20, associated with wild type gene No. 23, leads to the formation of long tubular, head protein structures called polyheads (Plate 18); some factor required for closing the ends of the growing tube to form an icosahedron is missing. Again, a mutation in gene No. 66 leads to production of abnormally short heads, in gene No. 21 to heads that contain no DNA, and in gene No. 31 to head protein subunits which aggregate into clumps and therefore produce no head at all. Evidently the head protein itself carries little shape-specifying information, the actual architecture of the head depending on the presence of a number of secondary components. In the same way, enormously elongated sheaths, called polysheaths, result from mutation in a different gene from that determining the sheath protein itself (see p. 417). (Epstein *et al.*, 1963; Kellenberger, 1966a, b).

An interesting and novel clue of a quite different kind concerning assembly of the phage head comes from genetic and physical studies of the chromosome of T2 and T4 phages. We will discuss this topic in some detail later (pp. 537–542). For the moment suffice it to say dogmatically that the heads of these phages appear to contain a fraction more than a complete genome—each chromosome begins and ends with the same nucleotide sequences, this region of *terminal redundancy*, as it is called, comprising about one to three per cent. of the whole genome (Streisinger, Edgar and Denhardt, 1964; MacHattie *et al.*, 1967). Moreover, the chromosomes of a phage population are not all the same but are *circular permutations* of one another; that is, the redundant extremities are not fixed regions but may be derived from any region of the chromosome at random. This means that, although each chromosome is a linear structure in the phage head, any pair of markers which could be postulated as extremities will be found linked on the majority of chromosomes so that the linkage map will be circular (p. 489) (Streisinger *et al.*, 1964; Thomas and MacHattie, 1964). These facts suggest that the individual phage chromosomes are formed by chopping off fixed lengths of DNA, rather longer than the complete genome, from a long chain (*concatenate*) of genomes linked together, thus;

Genomes

123456789012345678901234567890123456789012345678 90

Chromosomes

In fact, there is some evidence that phage T4 DNA replicates as a long polymer (Frankel, 1966a, b). This naturally leads to the hypothesis that the chopper is the protein head which, on completion, in some way snips off any DNA in excess of a 'headful' (Streisinger, Emrich and Stahl, 1967). That the length of DNA to be included in the head is measured by an agency which is not the DNA itself is shown by the fact that chromosome deletions in phage T4 increase the length of the regions of redundancy (Nomura and Benzer, 1961), thus:

Genomes

123478901234789012347890

Chromosomes

(12 digits long, as before)

Finally, several head sizes have been found among normal populations of those phage species that have been looked at from this point of view. In the case of phage T4, a smallheaded type containing only 70 per cent. of the normal DNA content has been isolated. This is inviable on single infection but, in multiple infections, several of these short genomes can complement one another to yield normal, infective phage particles (Mosig, 1963). In this case, circularly permuted DNA fragments shorter than the complete genome are being chopped off, the implication being that this may be the outcome of a physiological error in the packing of the head protein subunits. An even more striking example is found in the case of the generalised transducing phage P1, whose chromosome is also circularly permuted. About one quarter of the particles of normal populations are of small size, having only 40 per cent of the normal DNA content. A proportion of these small particles are transducing, but carry a diminished length of bacterial DNA (see p. 640). This reduction of DNA content irrespective of its source, clearly indicates a primary defect in the head structure (Ikeda and Tomizawa, 1965c).

In addition to the anatomical abnormalities outlined above, many others have been described including such freaks as two-tailed 'monsters' (Kellenberger, 1966a, b).

From the genetic standpoint, one of the interesting features of maturation is that the condensation of DNA observed in phage, and the thickening and contraction of chromosomes which occurs in the mitotic and meiotic prophase, may be analogous phenomena (Kellenberger, 1961). If this is so, it is highly relevant to the mechanism of recombination, since replication and recombination are certainly functions of vegetative phage and condensed DNA plays no part in them (Chapter 15; see p. 389).

THE LYSIS OF INFECTED BACTERIA

We have seen from the one-step growth experiment (Fig. 97, p. 413) that the latent period is terminated, at a particular time under standard conditions, by the bursting of the infected bacteria with liberation of, usually, several hundred progeny phages. This bursting is not the result of mere distension because it is quite uncorrelated with burst size. To take two extreme examples, the burst size may be greatly increased if the latent period is extended by treatment with chloramphenicol which is later diluted out (p. 415); conversely, mutants of phage T4 are known which multiply and mature normally within infected bacteria, but the bacteria fail to burst (see below).

The discovery that phage tails contain an enzyme which can break down the bacterial cell wall (p. 424), and the earlier observation that attachment of a high multiplicity of T-even phages produces almost immediate *lysis from without* (Delbrück, 1940), suggested that normal lysis might result from synthesis, directed by the phage genome, of a lytic enzyme within the infected cell. Two types of investigation have now offered direct evidence that this is so. If *E. coli* K-12 bacteria, lysogenic for phage λ, are treated with small doses of ultraviolet light, the prophage is *induced* to enter the vegetative state and multiply, so that the bacteria are lysed and progeny phages liberated (p. 451). Two types of defective prophage mutants have been described, neither of which can yield progeny phage following UV-induction of the bacteria harbouring them; in one case, induction is followed by lysis at the end of the usual latent period and the lysate is found to contain a factor which dissolves chloroform-treated *E. coli* cells. In the other case induction is not followed by lysis; the bacteria die but no lytic factor can be extracted from them (Jacob and Fuerst, 1958).

By an ingenious technique, Streisinger *et al.* (1961) have demonstrated the existence of a class of mutants of phage T4 which, as a result of mutations within a single cistron determining the structure of a lysozyme-like enzyme, are unable to escape from the bacteria in which they multiply. A bacterial population is infected with a suspension of phage T4; lysis is allowed to occur normally at 37°C. and the liberated phage particles are inactivated by antiserum. The lysate is then centrifuged to concentrate any residual bacteria, some of which are assumed to have escaped lysis because they were infected by lysozyme mutants. These bacteria are then plated with sensitive, indicator bacteria in the presence of extraneous, egg-white lysozyme to break open the killed but unlysed cells, and incubated at 25°C. It is found that plaques arise, the majority of which consist of phage which can produce only a heat-sensitive lysozyme, so that they can lyse their hosts and produce plaques at 25° but not at 37°C. We have already noted how analysis of the altered amino acid sequence of the mutant protein from one of these 'lysozyme mutants', following a phase-shift mutation, has revealed for the first time some of the codons of the genetic dictionary which an organism actually uses to specify the amino acids of its proteins (p. 361).

PHYSIOLOGICAL ASPECTS OF THE MINUTE BACTERIOPHAGES

We have already mentioned the existence of a morphological group of bacteriophages which are characterised by small size and the apparent absence of a tail, although they differ decisively in other respects such as having single-stranded DNA or RNA as genetic material (p. 419). Nothing is known about how these phages penetrate host cells. It seems certain that phage ϕX174 is devoid of a tail and therefore, presumably, does not inject its single-stranded DNA in the way we have described for larger, tailed phages. Two alternative mechanisms of penetration are possible; either the intact phage particle is taken up by the bacterium to which is has attached or else, after adsorption, the protein sheath is enzymatically digested, the released nucleic acid then entering the cell in the same way as the transforming principle of bacteria (see below).

BACTERIOPHAGE ϕX174

The single-stranded nature of the DNA of this phage, whose discovery Sinsheimer (1962) likened to finding 'a unicorn in the ruminant section of the zoo', was originally inferred from two kinds of evidence. Firstly, when the viability of one of the tailed phages such as phage T2, or of bacteria, is scored as a function of the decay of radioactive phosphorus (^{32}P) incorporated into their DNA, only about one in ten of the $^{32}P \rightarrow ^{31}P$ disintegrations is found to be lethal. The probable explanation is that a 'lesion' at the same site in *both* strands of the DNA double helix is needed to destroy its ability to replicate (Stent and Fuerst, 1955; see p. 523). However, in the case of phage ϕX174, as well as of the closely related phage S13, every disintegration of the incorporated ^{32}P is lethal. Secondly, we have seen that in double-stranded DNA, the rigorous pairing of adenine with thymine, and of guanine with cytosine, means that the ratio of purine to pyrimidine bases is always unity. On the contrary the DNA of phage ϕX174 contains adenine, thymine, guanine and cytosine in the ratio 1:1·33:0·98:0·75 respectively (Sinsheimer, 1959a, b). The single-stranded nature of this DNA has since been confirmed by other physical and enzymatic criteria.

The single-stranded DNA of each particle has a molecular weight of about $1{\cdot}6 \times 10^6$ and is only about 4500 nucleotides long; since this is equivalent to only a few cistrons, the synthetic potentialities of the phage must be strictly limited and, in fact, the phage appears to possess only a single species of structural protein (Carusi and Sinsheimer, 1961). As in the case of some animal and plant viruses, the DNA of phage ϕX174 is, by itself, infectious for the protoplasts of its host, though at a much lower efficiency than the intact particles. The single-stranded DNA occurs naturally in the form of a continuous loop and this continuity must be preserved during extraction for the DNA to remain infectious (Sinsheimer, 1962; Sinsheimer *et al.*, 1962; Denhardt and Sinsheimer, 1965a). During the greater part of the latent period following infection, the host cells (*E. coli* C) continue to multiply and to synthesise their DNA, RNA and protein at the normal rate. The synthesis of virus protein is initiated at about 8 minutes after infection and reaches a maximum at about 15 minutes (Rueckert and Zillig, 1962). The first mature particles begin to appear at about 8 minutes after infection, but neither particles nor single strands of DNA can be found in the infected cells if chloramphenicol is present. The cells lyse at about 20 minutes after infection to liberate several hundred new particles (Sinsheimer *et al.*, 1962).

The mode of synthesis of the phage DNA is particularly interesting. From what we have learnt about DNA replication we might predict that the single DNA strand would first serve as a template for the synthesis of a complementary strand and that the resulting double-stranded structure would then duplicate in the normal way (p. 239), and in fact this is just what happens. The circular infecting strand is converted to a double-stranded *replicative form* (RF) by a host enzyme, and this is then replicated semi-conservatively by a host polymerase to

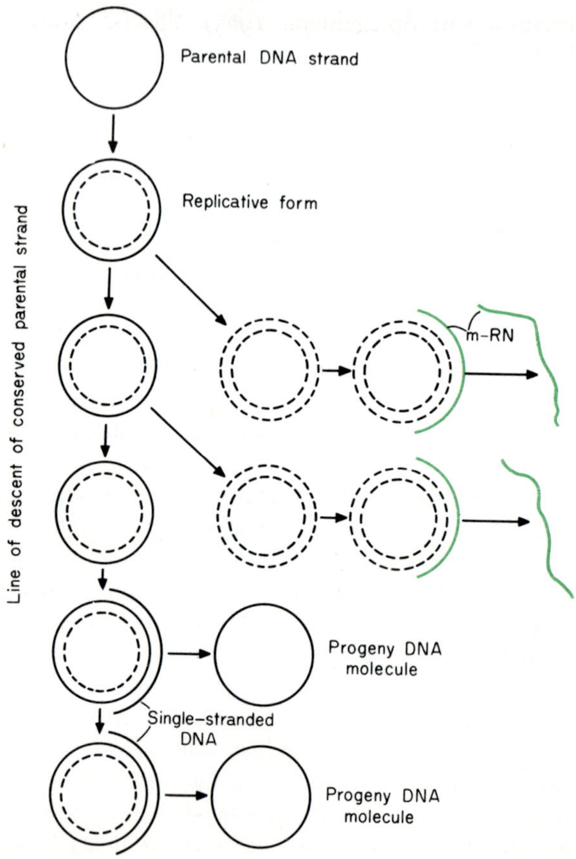

FIG. 100. The postulated reproductive mechanism of phage $\phi X174$. The single and double circles represent circular molecules of single- and double-stranded DNA respectively. The series of continuous, circular lines on the left shows the line of descent of the conserved strand of parental DNA; the interrupted, circular lines are newly synthesised strands.
The green lines indicate messenger RNA molecules.
The segmental lines represent the synthesis of single-stranded DNA (black) and messenger RNA(green) on double-stranded DNA templates.

yield two progeny duplexes (Sinsheimer *et al.*, 1962; Denhardt and Sinsheimer, 1965a). But from this stage onwards, this conformist behaviour changes. First of all, there is evidence that the daughter duplex not containing the parental strand may not replicate again but, instead, may act as a template for the synthesis of messenger RNA having the same base sequence as the parental, single-stranded DNA (Hayashi, Hayashi and Spiegelman, 1963). Secondly, not only is radiophosphorus (^{32}P) incorporated into the parental, single-stranded DNA not transmitted to the DNA of the progeny phages (Kozinski, 1961), but the infective complex remains sensitive to the decay of ^{32}P in the infecting DNA up to the end of the eclipse period (Denhardt and Sinsheimer, 1965b). This strongly suggests that the parental replicative form is the sole template for the synthesis of all the single-stranded DNA molecules, identical with the infecting strand, of the progeny particles (see Fig. 100). Since a system of this kind resembles a 'stamping machine' and does not involve a replicating pool of phage genomes, mutations would not usually be replicated and so should show a random rather than a clonal distribution, and this has indeed been demonstrated for reversion of an amber mutant of phage ϕX174 after a single growth cycle in a permissive host (Denhardt and Silver, 1966) (cp. p. 502).

Although a stamping machine model might, at first sight, appear to be less efficient than an exponentially replicating one, it may in fact be more economical in producing single-stranded copies of a single-stranded parental molecule. Moreover it has been shown to be a feasible model by a calculation which demonstrates that it requires the same rate of synthesis (5 nucleotide links per millisecond) to yield the usual phage output in the normal latent period as is required by *E. coli* to replicate its chromosome semi-conservatively every 20 minutes (see Denhardt and Silver, 1966).*

We have already noted that the conversion of the single-stranded DNA to the double-stranded replicative form is mediated by a host enzyme and occurs even in the presence of chloramphenicol. On the other hand, chloramphenicol blocks the formation of progeny single strands if added early, and under these conditions the synthesis of RF continues; but once single-strand synthesis is initiated it can continue in the presence of the drug. It appears that, in normal infection, semi-conservative replication of RF comes to a halt with the

* For mechanism of *in vitro* of infectious phage ϕX174 DNA, see footnote to p. 257.

commencement of single-strand synthesis. However, if chloramphenicol is added at this stage the amount of RF may begin to rise again. It is likely that this is due, not to a resumption of the original semi-conservative replication, but to conversion of the single-stranded progeny molecules into RF in the same way as the parental strand, under conditions where protein subunits, which would normally invest them, are not being made (Hutchinson and Sinsheimer, 1967).

It is interesting that, as in the case of the circular, double-stranded DNA molecules of mouse polyoma virus and of vegetative phage λ, a predominant fraction of the replicative DNA of phage φX174 has been found to exist in a 'supercoiled', tightly twisted form, due to the presence of an abnormal number of turns per unit length of DNA (Vinograd *et al.*, 1965; Roth and Hayashi, 1966). A single scission of one strand by treatment with deoxyribonuclease releases the torsion within the molecule so that, although the uncut strand preserves its circularity, it is no longer twisted on itself. The significance of super-coiling is not yet known.

We will discuss the genetic properties of the single-stranded phages φX174 and S13, in Chapter 18 (p. 510). Although these viruses have, only recently come into prominence because of the nature of their DNA, it is of some historic interest that phage S13 was first isolated by Burnet 40 years ago, while its diameter was estimated to be about 20 mμ by ultrafiltration in 1932, before the invention of the electron microscope (review; Stent, 1963). The main importance of these small phages, including the RNA-containing phages which we will discuss in the next section, is, firstly, that they closely resemble many small animal and plant viruses so that knowledge of their behaviour is likely to be widely applicable. Secondly, their low content of nucleic acid implies an ultimate simplicity of function and holds out the promise that there is at least one form of life which, shortly, we will fully understand.

RNA BACTERIOPHAGES

A necessary condition for sexual conjugation in *E. coli*, and other organisms which harbour sex factors, is possession by the donor or male bacteria of special filamentous appendages called *sex fimbriae* (or *pili*) whose synthesis is determined by the sex factor DNA (p. 670).* Minute bacteriophages which specifically infect male bac-

* The terminology of these appendages whether normally present (see Plate 19) or determined by the sex factor, is controversial. The word 'pilus' has now been widely adopted. However, we intend to use the word 'fimbria'

teria carrying the *E. coli* K12 sex factor were first isolated from sewage in New York (Loeb, 1960; Loeb and Zinder, 1961) and, shortly afterwards, in Milan (Dettori, Maccacaro and Piccinin, 1961). Since then, similar viruses have been isolated wherever they have been looked for, and have been designated by a variety of symbols such as f2 to f7, μ2, MS-2, fr, R-17, M-12, Fh5 and β (review: Zinder, 1965). The infectivity of all these viruses for male bacteria alone, is due to their specific adsorption to sex fimbriae (Crawford and Gesteland, 1964; Brinton, Gemski and Carnahan, 1964). All these viruses show considerable antigenic overlap and therefore can be said to be related (Scott, 1965). On the other hand, a quite distinct RNA coliphage, Qβ, has recently been isolated (see Overby *et al.*, 1966) while six RNA coliphages isolated by Bishop and Bradley (1965) fell into two serological groups. In addition to coliphages, RNA viruses have been reported which infect *Pseudomonas aeruginosa* (Feary, Fisher and Fisher, 1963; Bradley, 1966) and the stalked bacteria, *Caulobacter* (Schmidt and Stanier, 1965); in the case of the *Pseudomonas* phage, at least, adsorption is to fimbriae but, although this organism displays sexuality (p. 753), these are probably not sex fimbriae since the virus lacks sex specificity.

Although the RNA phages which attack *E. coli* are characterised by their exclusive infectivity for male strains, they show a remarkable lack of species specificity and can readily infect bacteria of other species and even genera, such as *Salmonella*, *Shigella* and *Proteus*, to which the *E. coli* sex factor has been transferred. Moreover, the purified virus RNA is also infective for protoplasts (sphereoplasts) of thse organisms, as well as of female *E. coli*, even though they lack the sex factor (Davis, Strauss and Sinsheimer, 1961). Since the specificity of these phages is due only to adsorption, Zinder (1965) has suggested that the contrast between their obiquity and the rarity of male strains of *E. coli* under natural conditions may arise from the fact that *E. coli* is not their natural host. On the other hand, recent studies show that sex factors (transfer factors) are much more widespread in Nature than was formerly believed (Anderson, 1965a), and that many of these permit infection by RNA coliphages (p. 773) (Meynell and Datta, 1966a).

Like phage ϕX174, the RNA coliphages are polyhedral in shape and

---

throughout this book since it has indisputable priority and is derivationally appropriate and correct. The case for 'pilus' is presented by Brinton (1965), and for 'fimbria' by Duguid, Anderson and Campbell (1966) and Duguid and Anderson (1967).

about 20 mμ in diameter. Their protein coats are constructed from about 150 subunits of a single protein species (molecular weight = 15,000) characterised by the absence of histidine in the case of phage f2, and enclose single RNA strands of about 10^6 molecular weight and a chain length of about 3,500 nucleotides. On infection of male bacteria at high multiplicity, the sex fimbriae can be seen to be thickly coated with virus particles along their length (Plates 19, 20, 34, pp. 432, 433, 773). It is probable that the protein coat remains attached to the fimbriae and that only the RNA enters the cell, since the phage becomes sensitive to the action of ribonuclease in the presence of sensitive bacteria while the fimbriae and attached virus coats can be sheared off the cells by blending without preventing infection (Edgell and Ginosa, 1965). It has been reported that female bacteria can become infected if the phage is added to a conjugating mixture (Sironi, Galucci and Maccacaro, 1964).

Following infection there is an eclipse period of about 15 minutes and phage particles begin to be released by lysis at about 25 minutes, the number continuing to rise for some time thereafter. Lysis occurs more efficiently if the bacteria are incubated at 45° than at 37°C. When bacteria are infected at high population densities, lysis inhibition occurs (p. 415) with the result that as many as 20,000 or more particles may subsequently be released by each bacterium; under these conditions crystals of phage particles may be seen packing the infected cells (Schwartz and Zinder, 1963). This enormous burst size, dramatic as it may seem at first sight, is actually not much different from the outcome of infection by a T-even phage in terms of viral mass synthesised. An appreciable proportion of the progeny RNA virus particles turn out to be inviable, although the reason for this is not apparent (Zinder, 1965).

One of the principal problems raised by the discovery of RNA phages is the mechanism of replication of RNA viruses in general. One question is, does synthesis proceed directly, using parental RNA as a template, or is a DNA intermediate required, possibly provided by a sequence on the host DNA corresponding to that of the viral genome? The intervention of DNA was ruled out since inhibition of DNA synthesis has no effect on phage production (Cooper and Zinder, 1963), no hybridisation occurs between phage RNA and the DNA of either normal or infected bacteria (Doi and Spiegelman, 1962), while actinomycin, which prevents transcription of DNA by RNA polymerase, fails to interfere with development of RNA phage (Haywood and Sinsheimer, 1963). Then three groups of workers, at about the same time, isolated and purified an enzyme system, synthesised only by phage-infected bacteria, which acted as an RNA-dependent RNA polymerase, catalysing the incorporation of nucleotide triphosphates into RNA (August *et al.*, 1963; Weissman *et al.*, 1963; Spiegelman and Doi, 1963). A novel feature of the polymerase isolated by Spiegelman and Doi is its high specificity for homologous RNA as template (p. 291)—a feature which must be very advantageous to the virus since cellular RNA is ignored (Haruna and Spiegelman, 1965). Using this enzyme system, Spiegelman and his colleagues have achieved the *in vitro* synthesis of self-propagating viral RNA which is infectious for protoplasts (Spiegelman *et al.*, 1965).

Initially there was some controversy as to whether replication occurs through the intermediary of a double-stranded form of RNA which then serves as a template for the synthesis of a single-stranded copy of the parental strand, much as in the case of phage ϕX174 (p. 437). Spiegelman *et al.*, (1965) at first suggested, because of the lack of contrary evidence from their own data, that only a single enzyme operates in their system and that double-stranded RNA may not be involved. However there now seems little doubt that double-stranded RNA *is* produced in phage-infected cells (Weissman *et al.*, 1964a; Amman, Delius and Hofschneider, 1964; Kaerner and Hoffmann-Berling, 1964) while Weissman *et al.* (1964b) reported that part of the product of the enzyme isolated by them is double-stranded. In fact it seems very likely that two enzymes are involved in the replication of phage RNA; one converts the parental (+) RNA strand to double-stranded form ($\pm$) by the synthesis of a complementary (—) strand while the other uses this duplex as a template for the synthesis of single-stranded progeny (+) molecules identical with the parental strand, in the same way that DNA-dependent RNA polymerase transcribes m-RNA from one strand of a DNA duplex. A proportion of these progeny strands are then converted in turn to the double-stranded form by the first enzyme, thus increasing the number of templates available for the synthesis of more progeny strands (Lodish and Zinder, 1966a). In agreement with this general sequence of events, Spiegelman's group have now discovered in their *in vitro* system a non-infectious complex containing parental and newly synthesised RNA. The appearance of this complex accompanies an eclipse period of about six minutes during which the parental RNA becomes non-infective, and after which infectious progeny strands begin to appear. Furthermore this complex, as we might expect of double-stranded RNA, releases infective RNA on heat denaturation (Mills, Pace and Spiegelman, 1966).

From all this it is apparent that the parental, infecting RNA strand plays two roles; it serves as its own messenger RNA for the synthesis of early enzymes (polymerases) and then, after conversion to the double-stranded form, as the template for progeny strands. If this is so, we would not expect any parental RNA molecules to appear among the progeny particles and this is indeed what is found (Spiegelman and Doi, 1963). It should also follow that synthesis of the protein coat should mainly be directed by progeny strand messengers, since the parental strands are otherwise engaged, and should thus occur later. Again there is evidence that this sequence of events occurs; it is significant that histidine, which is absent from the coat protein of phage f2, is essential during the early stages of infection but becomes dispensible later (Cooper and Zinder, 1963; Lodish, Cooper and Zinder, 1964; Lodish and Zinder, 1966b).

Using the spreading technique of Kleinschmidt (p. 537) and highly purified preparations of RNA from phage and infected cells, it has proved possible to obtain electron microphotographs of the RNA molecules. Not only may single- and double-stranded molecules of the correct length (c. 1μ) be found, but also double-stranded molecules showing randomly distributed branches which, unlike the double-stranded molecules themselves, can be removed by treatment with RNase and may be presumed to be single strands in the process of synthesis on, or displacement from, the double-stranded template (Granboulan and Franklin, 1966).

Although no genetic recombination has been found among the RNA phages, a considerable amount of genetic and functional analysis has been possible by the isolation of conditional lethal (amber and temperature-sensitive) mutants and their use in complementation tests. We will briefly discuss some of these results in Chapter 18 (p. 509).

FILAMENTOUS BACTERIOPHAGES

A rod-shaped coliphage, fd, was first reported by Marvin and Hoffmann-Berling in 1963; a previous isolation by Loeb (1960), called f1, was not described until later (Zinder *et al.*, 1963). Since then there have been many and widespread isolations (M13 by Hofschneider, 1963; Bradley, 1964; Dettori and Neri, 1965). All these phages appear to be very similar, although serologically distinguishable (Salivar, Tzagoloff and Pratt, 1964); they contain single-stranded DNA which is probably arranged as a continuous molecule and, like the spherical RNA phages, specifically infect male strains of *E. coli*. However, a rod-shaped *Pseudomonas* phage has recently been found which probably contains single-stranded DNA but is not male-specific (Takeya and Amako, 1966).

The filamentous coliphages are about 8000 Å long and 60 Å wide (Zinder *et al.*, 1963; Salivar *et al.*, 1964). In preparations negatively stained with phosphotungstic acid, a thin dark line can often be seen running centrally along the long axis of the phage, suggesting an axial hole (Bradley, 1964). The DNA contained by these phages has a molecular weight of about 1·5 to 2×10^6. Caro and Schnös (1966) have shown that the phages adsorb only to the terminal end of sex fimbriae, thus accounting for their exclusive infectivity for male bacteria, and do not compete for attachment sites with the spherical RNA phages which are arranged all along the long axis of such fimbriae (see Plate 20, p. 433). Occasionally two, and very rarely three, filamentous phages may attach to the extremity of the same pilus, but since such multiple attachments are found with a much lower than chance probability, it is likely that the adsorption of one phage interferes with that of another.

As Plate 20 shows, the virus is very similar to the fimbriae in appearance but can be distinguished from them by its uniform length and greater flexibility; the length of sex fimbriae may vary enormously, ranging from about one to over 20μ under certain conditions. If sex fimbriae are fragmented by violent agitation in a mixer and then mixed with a phage suspension, it turns out that the phages attach to only one of the two free ends of the pieces, and since attachment can still occur to the broken ends of fimbriae still connected to bacteria, it follows that the phage recognises a repeating molecular configuration characterising the distal ends of fimbrial fragments. In addition, the viral filaments appear to have terminal affinities for one another since they may join up end to end to form long polymers with no apparent junction between them, but clearly distinguishable from sex fimbriae by their inability to adsorb RNA phage (Bradley, 1964; Caro and Schnös, 1966).

During infection, most of the protein of filamentous phage remains outside the cell; the time required for DNA entry is very variable, ranging from as little as 20 seconds to several minutes (Tzagaloff and

Pratt, 1964). This scatter in injection time could be due to variation in the length of the sex fimbriae if it is assumed that these, like common fimbriae, possess an axial hole through which the single strand of DNA passes to initiate infection (Brinton, 1965; Caro and Schnös, 1966).

A striking and unique feature of infection by these phages is that the infected bacteria do not lyse and, indeed, die only infrequently. Instead, they continue to grow and divide while the filamentous progeny particles are extruded longitudinally through the cell wall which gives the appearance, superficially at least, of being heavily fimbriated. Bacteria releasing phage in this way aggregate to form clumps. If infected bacteria are maintained under conditions where reinfection is prevented they tend to cure spontaneously, losing their capacity to produce phage (Hofschneider and Preuss, 1963; Hoffman-Berling and Maze, 1964.)

The existence of six genes has been demonstrated in phage M13 by analysing 99 conditional lethal (amber and temperature-sensitive) mutants for their ability to complement one another. In the case of the temperature-sensitive mutants, which were a majority (66) of those studied, shifting the infected cells from the low to the high temperature at various times after infection revealed that one of these genes served an early function while four performed late ones (pp. 474, 510; Pratt, Tzagaloff and Erdahl, 1966).

In this chapter we have aimed at presenting the more important facts concerning bacteriophages and their development, with special reference to virulent phages. We must point out, however, that although the general features we have described are applicable to a wide range of phages, nearly all strains of phage differ from one another in the particulars of their morphology or development. It is beyond the scope of this book to consider such details, except in so far as they may reveal an interesting feature of a model system or throw light on some wider genetic phenomenon. The reader who is interested in a particular phage-bacterium relationship must consult one of the comprehensive reviews which are available (p. 410), and from there proceed to original papers.

CHAPTER 17

Temperate Bacteriophages and Lysogeny

The virulent phages have proved an admirable model for investigating and demonstrating the mechanisms involved in bacteriophage infection, if only because of the homogeneity they impose on populations of infected cells. Since virulent phages invariably destroy the bacteria they infect, the kind of environment best suited to their survival is one which is continually replenished with new, sensitive bacteria. Thus phages virulent for intestinal bacteria are prevalent in nature, and the prime source from which they may be isolated is sewage.

Within a few years of the discovery of bacteriophages, however, it became apparent that some bacteria were themselves an initial source of phage which they seemed to carry in some intracellular form. Cultures of these bacteria always contained free phage particles, although the bacteria seemed otherwise normal. Moreover, it proved impossible to rid them of their carried phage by such measures as the serial propagation of single-cell lines, treatment of the bacteria with specific anti-phage sera or, in the case of sporing bacilli, by heating the spores at temperatures known to destroy the phage. Bacteria which behaved in this way were called *lysogenic*.

The two outstanding properties of lysogenic bacteria are, firstly, that they carry the potentiality to produce and release phage as a stable, heritable trait and, secondly, that they are immune to lytic infection by the same or closely related phages (Bertani, 1953). Lysogeny is now known to be so prevalent and widespread among the strains of different bacterial species that it must be regarded as the normal, rather than the exceptional, state. In addition, lysogeny is not limited to the carriage of a single type of phage by each bacterium; many cases of double or triple lysogeny have been described, while one strain of *Staphylococcus* has been reported to carry as many as five different phage types (Rountree, 1949).

Because of the stable, hereditary nature of lysogeny, and of the immunity which it confers, its recognition usually depends on the chance isolation of a sensitive, indicator strain on which the phage will form plaques. For instance, the K-12 strain of *E. coli* had been used in laboratory experiments for many years before its lysogenisation by phage λ was realised as a result of the accidental discovery that cultures of the strain produced plaques when plated with a radiation-induced, auxotrophic K-12 mutant which had been concomitantly 'cured' of the phage and thereby became sensitive to it (E. M. Lederberg, 1951).

Once a sensitive bacterial strain is available, the existence of lysogeny can be detected, and the phage assayed, in various ways. One is to separate the free phage particles from the parent bacteria by filtration, or by treating a centrifuged supernatant of the culture of lysogenic bacteria with chloroform which, in many cases, kills the residual bacteria but not the phage; the phage is then assayed by plating with the indicator bacteria in the same manner as for virulent phage (p. 410). Alternatively, if the lysogenic bacteria are sensitive to streptomycin and the indicator bacteria are resistant, the two can simply be plated directly together on streptomycin-agar; the presence of the lysogenic bacteria can be ignored since their ability to grow, as well as to produce phage, is suppressed by the drug, while that of the phage particles to yield plaques is unaffected (Bertani, 1951). Finally, a high dilution of the culture of lysogenic bacteria may be plated on an excess of indicator bacteria so that isolated colonies of the former develop on a confluent 'lawn' of sensitive cells. Under these conditions free phage particles give rise to normal, turbid plaques, while the colonies of the lysogenic strain are surrounded by a halo of lysis caused by new phage particles liberated during growth of the lysogenic clone. Such *colony-centred plaques* are usually most strikingly seen when the growth rate of the indicator is slower than that of the lysogenic bacteria.

PSEUDO-LYSOGENY AND THE CARRIER STATE

In practice the existence of lysogeny should be judged by rather rigorous criteria of its stability, since the interaction of some virulent phages with their hosts may superficially simulate the condition. For example, if a population of sensitive bacteria, exposed to virulent phage at a low multiplicity of infection, is able to acquire *phenotypic resistance* at an early stage in its growth, that is, can readily produce physiologically (but not genetically) resistant *phenocopies*, the cells first infected will liberate phage normally, but the ability of these progeny phages to reinfect new cells will become increasingly difficult with time. An equilibrium is thus established in the population between uninfected bacteria and free phage particles. On subculture, however, the phenotypically resistant bacteria become sensitive again and the process is repeated, so that bacteria and phage can be propagated together indefinitely. This type of relationship is found when *Shigella dysenteriae* is infected with phage T7; it appears that lysis of the initially infected bacteria releases an enzyme which removes the phage receptors from the cell walls of the remaining, uninfected bacteria, so that they are rendered resistant until the enzyme is diluted out by subculture (Koibong, Barksdale and Garmise, 1961). A comparable state of affairs arises when cultures of male strains of *E. coli* K-12 are infected with male-specific RNA phage (p. 440), but in this case the development of phenotypic resistance is a natural property of male cells which tend temporarily to lose the male antigen (fimbriae or pili) to which the phage adsorbs (*F^- phenocopies*, p. 658) (see Dettori *et al.*, 1961; Marilyn Monk, personal communication, 1962).

In the type of pseudo-lysogeny we have just described, the coexistence of virulent phage with genotypically sensitive bacteria is associated with temporary inability of the bacteria to *adsorb* the phage; the phage is present in the population as free particles, but is not carried *inside* individual bacteria. A rather different phenomenon is the ability of certain phages, classified as virulent because they cannot establish lysogeny, to be propagated for many generations within some of the cells they infect; these cells, on division, give rise to a proportion of uninfected daughters as well as to daughter cells which lyse and liberate phage. Since stable, lysogenic clones cannot arise, every line of descent which propagates the phage must ultimately end in lysis, but the process can be perpetuated in the population as a whole by the reinfection of sensitive segregants. A likely mechanism for this type of response to infection is that the phage, although unable to form a stable, inheritable relationship with the cell, is nevertheless capable of provoking some degree of immunity in the infected bacteria so that its free, vegetative replication, which would normally result in lysis, is partially repressed in a proportion of them (see p. 463 *et seq.*). In pseudo-lysogeny of this kind, which has been reported in the case both of phage T3 and *E. coli* B (Fraser, 1957) and of a virulent mutant of phage P22 and *Salmonella* (Zinder, 1958; see also Luria *et al.*, 1958), the bacteria responsible for persistence of the phage are said to be in the *carrier state* which must be regarded as a kind of abortive lysogeny. In fact, the carrier state may sometimes be superimposed on the first type of pseudo-lysogeny (Koibong *et al.*, 1961).

A temperate transducing phage of *B. subtilis* has also been shown to exist in the carrier state since growth of the lysogenic bacteria in phage antiserum leads to loss of phage and phage immunity (Bott and Strauss, 1965; see p. 478; see also phage λ*b2* mutant, p. 470).

The persistence of both types of pseudo-lysogeny in bacterial populations depends on the presence of free phage, so that phage-free, sensitive clones can usually be isolated by growth in the presence of phage antiserum, or by the serial subculture of single colonies. On the other hand, the immunity of truly lysogenic bacteria, as well as their potentiality to produce phage, is quite unaffected by such procedures.

THE NATURE OF PHAGE RELEASE

The stability of the lysogenic state, and the immunity which it confers against lytic infection by the carried phage, raises the question of why free phage particles are always present in lysogenic cultures. In their absence, of course, lysogeny could not be recognised. There are two possibilities; either infectious particles are present in every parasitised bacterium whence they continually leak into the surrounding medium, or else the balanced phage-bacterium relationship breaks down in a small proportion of the population at each generation so that the lytic cycle supervenes.

That lysogenic bacteria do not harbour infectious phage particles was first demonstrated by Burnet and McKie (1929) who showed that no phage is released when the bacteria are disrupted. At about the same time it was mooted that *lysogeny might represent an hereditary potentiality of the bacteria to generate phage* (E. Wollman, Snr., 1928). The correctness of this concept was established beyond doubt by Lwoff and Gutman (1950) as a result of a pedigree analysis of individual bacteria from a lysogenic strain of *B. megaterium*. By means of a micromanipulator, single cells were seeded in micro-drops of broth and their successive progeny distributed to fresh micro-drops over many generations; the micro-drops were kept under continual observation, samples of bacterium-free fluid being periodically removed and assayed for free phage. The essential findings were: (1) that as many as 19 generations can pass without the release of a single phage particle, although each bacterium subsequently yields a lysogenic clone; (2) that the appearance of phage particles in a micro-drop is correlated with the rapid disappearance of a bacterium; (3) that the mean number of phage particles liberated by the lysis of each bacterium is high.

It is therefore clear that any bacterium of a lysogenic culture inherits the capacity to liberate phage, but this capacity is expressed with only a low probability by any particular bacterium. It is now known that this probability varies for different strains of temperate phage, ranging from about 10^{-2} to 10^{-5} per cell per generation. However, for any given phage the probability of lysis remains rather constant under standard conditions so that the ratio of free phage particles to bacteria tends to remain the same throughout the growth cycle of a broth culture. Nevertheless, alteration of the environmental conditions may profoundly influence the probability of lysis, as we shall now see.

THE INDUCTION OF LYSIS

During their pedigree studies, Lwoff and Gutman (1950) observed that the proportion of bacteria which liberate phage tends to vary from experiment to experiment, and they suggested that phage production might be *inducible* by external factors. It was then discovered that exposure of cultures of lysogenic *B. megaterium* to doses of ultraviolet light too small to affect the growth of non-lysogenic bacteria (*c.* 500 ergs per mm^2 of wavelength 2537 Å) was followed, after a latent period of one or two divisions, by lysis of virtually the entire population and the liberation of 70 to 150 phage particles per bacterium (Lwoff, Siminovitch and Kjelgaard, 1950). It turned out, however, that the capability of UV light for induction depended very much on the nutritional environment; maximal lysis occurred only when the bacteria were grown in a rich medium both before and after exposure, and none was found if minimal medium was used, even though this could support bacterial growth. Many other effective inducing agents have since been found, including X-rays, γ-rays, nitrogen mustards, hydrogen peroxide and organic peroxides (see Jacob and Wollman, 1959). In addition, induction follows treatment with the base analogue 6-azauracil (Zgaga and Miletić, 1965) or any agency which leads to inhibition of DNA synthesis, such as fluorodeoxyuridine, Mitomycin C or deprivation of thymine (Korn and Weissbach, 1962).

The latent, non-infectious form in which temperate phage is carried by lysogenic bacteria is called *prophage* (Lwoff and Gutman, 1950). By assuming that the same sequence of events precedes spontaneous and induced lysis, the phenomenon of UV-induction enables these events to be synchronised throughout the population and the kinetics of prophage development to be studied. It turns out that induction is followed by a short lag but, thereafter, the kinetics of phage development, both in premature lysates and as a result of natural lysis, are very much the same as when sensitive bacteria are infected from without by free phage particles (Jacob and Wollman, 1953). This implies that the lysis of lysogenic bacteria results from the transition of prophage to the vegetative state which ushers in the normal lytic cycle. We will see later that although we cannot distinguish between prophage and vegetative phage from the structural point of view, both being identical with the phage genome, there is reason to believe that, in the prophage state, the phage genome is prevented from expressing its function by a cytoplasmic immunity substance or repressor.

Not all lysogenic bacteria are inducible. The property of inducibility appears to be a function of the genetic constitution of the phage rather than of the bacterium since the same bacterial strain may be inducible when lysogenised by one type of phage, and non-inducible when lysogenised by another. Alternatively, an inducible phage which is able to lysogenise a number of different hosts is inducible in all of them, and *vice versa.* As examples, we may mention phage λ which is one of the most intensely investigated of inducible phages; the transducing phages P1 and P22 are also inducible but the proportion of lysogenised cells which lyse following UV-irradiation is usually smaller than in the case of phage λ. On the other hand, the *Shigella* phage P2, which can also lysogenise some strains of *E. coli*, is non-inducible. Among a series of phages lysogenic for *E. coli* K-12, isolated at random from 500 strains of *E. coli* derived from human fæces, some were inducible and some were not (Jacob and Wollman, 1956a, 1957). There is evidence that the type of immunity set up in bacteria lysogenised by non-inducible phages is different from that imposed by inducible strains.

THE PROPHAGE-BACTERIUM RELATIONSHIP

The fact that every bacterium of a lysogenic culture contains prophage, restricts the number of possible models of the virus-host relationship to the few which can adequately account for the great stability of the association. We can take the line that prophage multiplies independently of the bacterial genome and is distributed randomly among daughter cells when the bacteria divide. In this case the number of prophage units per bacterium must be large in order to avoid the chance segregation of non-lysogenic cells. The alternative is that the number of prophage units per bacterium is small. This implies that both the replication of prophage and its distribution to daughter bacteria must be tied in some way to the replication and distribution of the bacterial genome. There are three variations of this second model: in one the prophage, which we may assume to be composed of DNA, is so closely associated with the bacterial chromosome that it behaves in all relevant respects as if it were a part of it; secondly, the behaviour of prophage might be regulated by a bacterial gene in much the same way as *kappa* particles are controlled in *Paramecium* (Sonneborn, 1950); thirdly, chromosome and prophage, although independent, could both be subject to the same cellular mechanism of regulation and segregation (pp. 735, 568).

There is indirect evidence that the number of prophage units per bacterium is small and that no bacterial gene is concerned in their maintenance. When sensitive bacteria are simultaneously infected with two distinguishable mutants of the same temperate phage, but at different multiplicities, the ratio of the two types of particle among the progeny after lysis is the same as the input ratio. If, now, lysogenic bacteria carrying the prophage of one of the mutants is induced and, at the same time, is superinfected with different multiplicities of the other mutant, it is found that the carried phage and the superinfecting phage are liberated in equal numbers when the multiplicity of the superinfecting phage is about three. Since, in this experiment, the mean number of nuclei per bacterium was three, the result suggests that the lysogenic bacteria carry only one prophage per nucleus (Jacob and Wollman, 1953). Again, rare non-lysogenic 'mutants' can be isolated from lysogenic bacteria following heavy irradiation or the decay of ^{32}P incorporated into their DNA. If prophage maintenance is the function of a bacterial gene, we would expect a proportion of the non-lysogenic derivatives to have arisen from mutation in this gene and, thus, to be incapable of re-lysogenisation by the same phage. In fact, all such 'cured' bacterial strains are readily re-lysogenised.

Definitive information about the presumptive association of prophage with the bacterial chromosome first came from genetic crosses between strains of *E. coli* K-12, one lysogenic for phage λ and the other non-lysogenic, in which the inheritance of lysogeny among recombinants could be scored like that of any other genetic determinant. It was found that lysogeny and non-lysogeny *did* segregate among recombinants, and that their inheritance was highly correlated with that of the parental alleles of a particular locus, *gal*4, determining galactose fermentation. Thus lysogeny appeared to be determined at a genetic locus very closely linked to the *gal* locus on the bacterial chromosome (E. M. Lederberg and Lederberg, 1953; Wollman, 1953). This locus could represent either the prophage itself, or the site of a gene necessary to maintain the prophage; in the latter case, if one or more prophage units were usually distributed *cytoplasmically* to the recombinant bacteria, lysogeny would be preserved only in those recombinants which had also inherited a functional maintenance gene. Discrimination between these two possibilities was achieved by further crosses in which the two parents, instead of being distinguished by lysogeny and non-lysogeny, were each lysogenised with a different mutant of λ phage. The two types of prophage segregated among recombinants and were found to be closely linked to their parental *gal* alleles, in a manner similar to lysogeny and non-lysogeny in the previous crosses (Appleyard, 1954). Thus the determinant of lysogeny is the prophage itself and not some bacterial gene which controls it; the prophage occupies a specific site on the bacterial chromosome and is inherited as if it were a bacterial gene.

This remarkable conclusion was reached as a result of crosses employing F^+ male strains of *E. coli* which not only yield a very low proportion of recombinants but also show many anomalies of recombination (see Chapter 22, pp. 653, 659). With the discovery of male strains termed *Hfr* (high frequency of recombinants), every bacterium of which can transfer its chromosome to females in such a way that the genes enter the zygote in sequence from one particular extremity, genetic analysis of *E. coli* became a precise and unequivocal procedure. Moreover, by interrupting mating cultures at intervals after mixing, and then scoring for the presence of recombinants inheriting particular male genes as a function of time of mating, the relationship between genes and their distance apart can be measured rather accurately in terms of absolute chromosome length (p. 65 and Fig. 15; also Chapter 22, p. 689).

Using these newer methods, Jacob and Wollman (1957) mapped the prophage locations of 14 different temperate phages (including λ) isolated from *E. coli* strains of fæcal origin. Seven of these phages were inducible and seven non-inducible and none of them showed cross-immunity, that is, lysogenisation with one strain did not confer immunity to lytic infection by any of the others. Each prophage was found to have a different location on the chromosome, the close linkage of phage λ to the *gal* region being confirmed. The locations of some of these prophages are shown on the linkage map of *E. coli* in Fig. 127 (p. 666).

We have mentioned that individual bacteria can carry more than one type of prophage. In addition, double and even triple lysogenisation with mutants of the same prophage is possible. In double lysogenisation of *E. coli* K-12 with mutants of phage λ, both prophages appear to be located at the same site on the bacterial chromosome. In the case of the non-inducible phage P2, however, independent genetic studies have revealed the existence of at least three alternative prophage locations which show a preferential order of occupancy (Bertani, 1955).

PHAGE MUTATIONS AFFECTING LYSOGENY

We have said that prophage is inherited in bacterial crosses *as if it were* a gene occupying a specific locus on the bacterial chromosome. The first thing to understand is that prophage is not a gene but appears to be identical in its physical structure to the vegetative phage genome which comprises many genes. Thus if radioactive phosphorus (^{32}P) of high specific activity is incorporated into the DNA of lysogenic *Hfr* bacteria, which includes the λ prophage DNA, and the prophage is then transferred to non-radioactive female bacteria, the rate at which it is subsequently inactivated as a function of ^{32}P decay can be measured. It turns out that the sensitivity of prophage to the disintegration of ^{32}P atoms is identical with that of infectious phage particles. This means that prophage and infectious phage contain the same number of phosphorus atoms and, therefore, equivalent amounts of DNA (Stent, Fuerst and Jacob, 1957). The prophage thus consists of a DNA molecule approximately 50,000 nucleotide pairs long which could accommodate at least 50 genes (p. 543). When a lysogenic bacterium becomes non-lysogenic, following irradiation for example, the situation is quite different from the mutational alteration of a bacterial gene to a non-functional allele; *the entire prophage genome is eliminated* from the cell which can only become lysogenic again through reinfection with the phage. On the other hand, mutational defects can arise in prophage genes which interfere with the development of the phage after induction, but have no effect on carriage of the defective prophage or on the immunity if confers.

We owe most of our basic knowledge about the nature of the link between prophage and bacterial chromosome to the outstanding early genetic studies of F. Jacob and E. L. Wollman, and their associates, on the *E. coli*-phage λ system (reviews: Jacob and Wollman, 1957, 1959, 1961a). These studies, some of which we shall shortly review, employed phage mutations involving such characters as host-range, plaque size and morphology, and various reproductive defects. It was soon established that phage λ possesses a single, linear linkage group (Jacob and Wollman, 1954; Kaiser, 1955). More recently two additional factors have further refined our concepts of the structure and behaviour of the λ prophage. The first of these was the isolation by Campbell (1961) of a series of conditional lethal mutants, originally called *sus* mutants, which grow only in certain 'suppressor' host strains and are now known to be amber mutants; it is the genes revealed by such amber mutations which are mainly shown in the map of the phage λ chromosome in Fig. 101. The second factor was the development of methods and techniques for exploring the physical behaviour of the phage chromosome, and this we will discuss principally in Chapter 19 (p. 535).

Among the more readily isolated and recognised mutants of phage λ are those which interfere with the ability of the phage to lysogenise sensitive cells; these *C* mutants, as they are called, are virulent, producing clear instead of turbid plaques on sensitive bacteria, but they are usually unable to infect lysogenic cells, that is, they are still susceptible to prophage immunity. Genetic analysis of the sites of these *C* mutations reveals that they are all located close together within a very small region of the phage chromosome called the *C* region. The ability of *C* mutants to complement one another can be investigated by seeing whether mixed infection by various combinations of pairs of them can co-operate in establishing lysogeny. The result is that *C* mutants fall into three functional groups, *c*1, *c*2 and *c*3, the mutational sites of each group mapping together in different parts of the *C* region; that is to say, the *C* region comprises three cistrons (Fig. 101). Infection with pairs of mutants from different groups, but not from the same group, can lead to the lysogenisation of sensitive cells at the normal, wild type rate. This shows that each of the three mutational groups is blocked with respect to a different function which is mediated by some diffusible product; two mutant strains which are unable to fulfil different functions can, between them, provide all the products necessary for lysogenisation (Kaiser, 1957). Analysis of clear-plaque mutants of the *Salmonella* phage, P22, have yielded similar results (Levine, 1957).

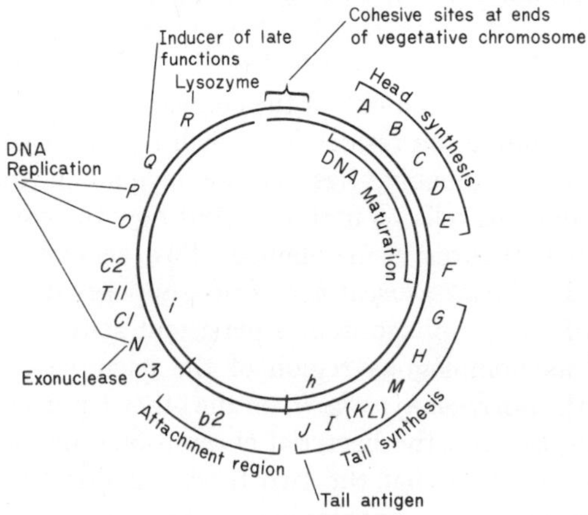

FIG. 101. A genetic and functional map of the phage λ chromosome.

The capital letters represent genes based on analysis of Campbell's (1961) *sus* mutants; c_1, c_2 and c_3 are genes concerned with establishment and maintenance of lysogeny, while *i* and *h* indicate regions determining immunity and host range respectively, as established by earlier work.

The extremities of the linear vegetative map are *A* and *R*, and of the prophage map *c3* and *J*.

From the data of Dove, 1966; Franklin, Dove and Yanofsky, 1965; Joyner *et al.*, 1966.

You will recollect the seven inducible phages, mentioned above, which were found to have different prophage locations and to show no cross-immunity. These phages nevertheless possess considerable antigenic and genetical similarity and, in mixed infections, can produce progeny which are recombinant for a wide range of characters. With one exception, all the known characters of phage λ can be transferred by recombination to any of the other phages. The exception is the *C* region. In the case of some phages the c_2 and c_3 cistrons are transferable, but the c_1 region appears to be strictly specific and to lack genetic homology (Kaiser and Jacob, 1957). When sensitive bacteria are lysogenised with a recombinant phage which carries the *C* region of, say, phage 434 but all the other markers of phage λ, the location of the prophage on the bacterial chromosome, as well as the prophage immunity, are found to be those of 434 and not λ. It is therefore clear that the location of the prophage is determined by only a small fraction of the phage genome which was at first thought to be the c_1 region itself (Kaiser and Jacob, 1957; see p. 466).

THE NATURE OF THE PROPHAGE CONNECTION

We have seen that the specificity of the connection between prophage and bacterial chromosome resides in a small region of the prophage genome which, presumably, has its homologue on the bacterial chromosome, but this tells us nothing about how the connection between the two structures is maintained. Two general models have been suggested. The *attachment hypothesis* supposes that the attachment region of the prophage is in a persistent state of synapsis, or pairing, with its homologous region of the bacterial chromosome. According to the *insertion hypothesis*, on the other hand, the prophage becomes *integrated* into the bacterial chromosome by some kind of recombination event so that the two form a continuous structure. Although attachment was considerably in vogue a few years ago, the insertion hypothesis has now won hands down, so let us not waste time by recounting a dead controversy but proceed directly to the insertion mechanism and the evidence which has led to its general acceptance.

If, in lysogeny, the phage and bacterial chromosomes form one continuous structure, it follows that recombination between two prophage markers should be associated, more often than by chance, with recombination for markers on the bacterial chromosome which lie on either side of, and close to, the prophage, and *vice versa* (*X* and *Y*, Fig. 102). This can be tested by crossing *Hfr* male and female bacteria which carry not only genetically marked strains of the same prophage but also different alleles of outside bacterial genes. When this experiment was performed using prophage λ and the bacterial markers *gal* and *try* (Fig. 127, p. 666), the results suggested that recombinants for the *gal* and *try* alleles may indeed be correlated with reassortment of the prophage markers so that all the genes can be arranged in a definite order (*X*-*a*-*b*-*c*-*Y*, Fig. 102), although the *try* marker was too distant from the prophage for this to be altogether convincing (Calef and Licciardello, 1960). Recent studies with more closely linked bacterial markers have confirmed this (Rothman, 1965).

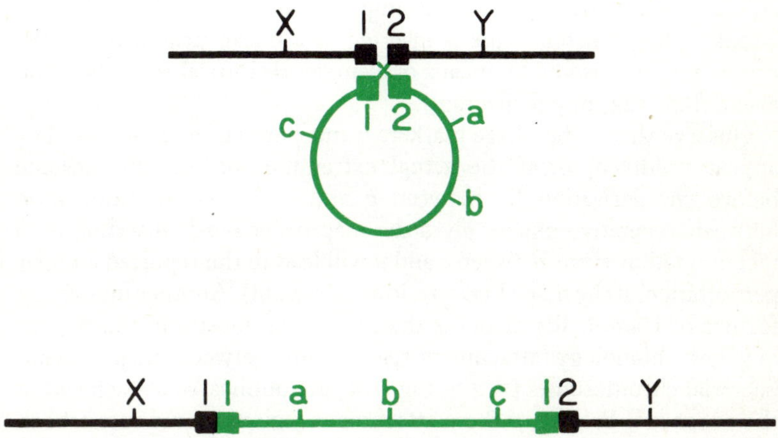

FIG. 102. Illustrating how a circular phage chromosome could, theoretically, become inserted as a linear structure into the bacterial chromosome.

The bacterial chromosome is shown black, and carries the genes X and Y on either side of the prophage attachment region 1, 2.

The phage chromosome is shown green, and carries the three markers a, b and c; the green figures 1, 2 represent the attachment region of the prophage and indicate homology with the corresponding bacterial region 1, 2 (black).

In the *upper diagram*, the attachment region of the phage has paired with its location on the bacterial chromosome and reciprocal recombination is occurring in the 1–2 interval.

The *lower diagram* shows the outcome of the recombination, and demonstrates the relationships of the prophage to the bacterial chromosome.

See text for further explanation. (After Campbell, 1962.)

However, the interesting point is that a different order of the same markers on the phage chromosome is found (*c-a-b-*, Fig. 102) in crosses between the same phage mutants during *vegetative* growth following mixed infection (Calef and Licciardello, 1960; see also Rothman, 1965); the two orders are *circular permutations* of one another. This, together with the fact that the linkage maps of both *E. coli* and phage T4 were already known to be circular, prompted A. Campbell (1962) to propose a simple, theoretical model in which a circular phage chromosome is inserted as a linear structure into the bacterial chromosome by means of a single, reciprocal recombination event (Fig. 102, upper diagram).

Observe that if the phage markers *b* and *c* are far apart so that they appear unlinked, or if the actual extremities of the chromosome before circularisation lie between *b* and *c* so that recombination between vegetative phages gives the map order *c-a-b*, insertion by a recombination event between *c* and *a* will lead to the reported circular permutation of the map (Fig. 102, lower diagram). Another interesting feature of Campbell's model is that, following insertion, the regions of genetic homology (attachment regions: *att*) between the phage and bacterial chromosomes (*1*, *2* in Fig. 102) are duplicated at each end of the prophage. Pairing of these attachment regions would re-establish the situation shown in the upper diagram of Fig. 102, where another recombination event would lead to release of the phage genome, thus offering a mechanism for induction. As a bonus, the model neatly explains how transducing phage particles acquire fragments of bacterial chromosome in localised transduction (Fig. 124, p. 634).

A prediction which follows from the insertion hypothesis is that the presence of prophage should decrease the apparent linkage between bacterial markers which straddle it. This can be assessed by comparing linkage data from conjugal crosses in which both parents are lysogenic on the one hand, and both non-lysogenic on the other. Alternatively, a more sensitive reflexion of linkage is the co-transduction frequency, comparing lysogenic and non-lysogenic strains as donors, of two markers on either side of the prophage attachment site. It happens that λ prophage is closely linked to the galactose (*gal*) region on one side and to a gene determining biotin synthesis (*bio*) on the other (Fig. 127, p. 666). Rothman (1965) has shown that the frequency of joint *gal-bio* transduction by phage P1 is reduced from 0·47 to 0·03 following lysogenisation with λ phage. An equally unambiguous effect has been demonstrated, by both conjugation and phage P1 transduction, in the case of phage $\phi80$ whose attachment site (att_{80}) lies in a very well marked region close to the tryptophan gene cluster (*try*) (Fig. 127, p. 666; Matsushiro, 1963; Signer, Beckwith and Brenner, 1965); while lysogenisation reduces the joint transduction frequency of markers which span att_{80} from as high as 0·5 to the limit of detectability (about 0·01), irrespective of which marker is selected, the co-transducibility of pairs of markers on the *same* side of att_{80} remains unaffected (Signer, 1966).

An elegant proof of the continuity of bacterial and prophage chromosomes in lysogeny would be the demonstration that deletions can arise which intrude simultaneously into both structures. Mutations to resistance to phage T1 in *E. coli* K-12 (and especially in *E. coli* strains B and C) are often associated with a requirement for tryptophan and are known to be due to deletions extending into the adjacent *try* region. The attachment site for phage $\phi80$ is located on the opposite side of the *T1* locus to *try* and is very closely linked to it. It was initially observed, in phage $\phi80$ lysogens, that occasional *T1-r —try* deletions were associated with a prophage defect so that, although phage immunity was retained, the ability to yield infective phage particles was greatly reduced or abolished (Franklin, Dove and Yanofsky, 1965). Now, although phage $\phi80$ and phage λ show no cross-immunity, they are able to undergo recombination, so that hybrid prophages can be obtained which insert themselves at the att_{80} site and possess $\phi80$ immunity but carry many λ phage markers (Signer, 1964). *E. coli* K-12 was lysogenised with such a hybrid prophage which probably carried about half the phage λ genome since it was shown to possess the λ genes *A*, *B*, *C*, *E*, *F* and *h* (Fig. 101). A high proportion of *T1-r* mutants of this strain, some of which were also *try* mutants, carried defective prophages. When these were examined for their ability to contribute their λ genes to phage recombinants, they were found to have resulted from a series of deletions extending for various distances into the prophage from one extremity. When these deletions were used for mapping the prophage genes, in a manner similar to the deletion mapping developed by Benzer (see p. 170) they revealed the same permuted order of the vegetative linkage map of phage λ already described by others (Franklin *et al.*, 1965; see Campbell, 1963; Rothman, 1965). Recently λ prophage itself has been similarly mapped from the other end by a series of deletions involving the galactose region (Fig. 127, p. 666; Shapiro, 1967).

The insertion of a circular phage chromosome into the continuity of the bacterial chromosome by recombination at once suggests that double lysogeny should arise by the same mechanism, and it has been shown that double lysogens are in fact formed by the linear integration of a superinfecting phage genome into the resident prophage (Calef, Marchelli and Guerrini, 1965). It might be thought that insertion of the second prophage would occur more readily than the first, since the resident prophage offers a complete region of homology with which any part of it can pair; yet double lysogeny is actually much more difficult to achieve (Campbell, 1962). As we shall see, this paradox is probably due to the production of a repressor by the resident prophage which shuts off a synthetic activity of the superinfecting phage which is needed for recombination.

The key feature of the Campbell model is, of course, the circularity of the phage chromosome and there is now good evidence that this exists not just as a circularly permuted linkage map but as a real physical structure. Hershey, Burgi and Ingraham (1963) first showed that the linear DNA molecule present in particles of phage λ has *cohesive ends* which can adhere to form a circle. Alternatively, the ends of different molecules can link together into large circles consisting of two, three or more chromosomes which can be clearly seen and measured by electron microscopy using Kleinschmidt's technique (p. 537; Ris and Chandler, 1963). Such chromosome polymers have been called *concatenates* (Latin: con = together; catena = a chain). Half molecules of phage λ DNA can be obtained by shear breakage and, since the 'right' and 'left' halves differ in density, they can be separated. It turns out that 'right' and 'left' half molecules stick to one another but not to themselves, so that the adhesion sites at the ends of complete molecules must be complementary in structure (Hershey and Burgi, 1965). The circular molecules become linear again if they are rapidly cooled after heating to 75°C. and reform at about 60°. It now seems highly probable that the DNA molecule terminates in single-stranded regions which have complementary base sequences so that adhesion is due to hydrogen bonding between the bases (Fig. 101; Strack and Kaiser, 1965). Cohesive sites appear to be a general feature of temperate coliphages at least, and it is interesting that all those 'lambdoid' phages (424, 434, 21, Fig. 127 p. 666) which can recombine with phage λ as well as among themselves have very similar, if not identical, cohesive regions so that they can form mixed concatenates, even though they show no cross-immunity (Baldwin *et al.*, 1966).

It is obviously unlikely that the inserted λ prophage, on which the circularity of the bacterial chromosome depends after lysogenisation, is held together by hydrogen bonding alone; we may anticipate the formation of covalent bonds uniting the backbones of the DNA strands. Young and Sinsheimer (1964) first succeeded in isolating two DNA components, characterised by very different sedimentation rates, from bacteria infected with ^{32}P-labelled λ phage and, by analogy with phage ϕ174, suggested that the fast-sedimenting component comprised covalently bonded circular molecules. It was then found that phage λ DNA, which is itself infective in the presence of particles of defective 'helper' phage (p. 587; Kaiser and Hogness, 1960), lost its infectivity shortly after injection into sensitive bacteria, suggesting that it undergoes some alteration in molecular structure (Dove and Weigle, 1965). Finally, three types of phage λ DNA, based on sedimentation properties, have been isolated following superinfection of λ lysogens. Analysis of these types revealed linear molecules, predominant only shortly after infection, circular molecules with only one strand covalently bonded, and circular molecules with both strands covalently bonded. Since this last type of molecule also follows infection with virulent λ phage, its formation is not specifically related to the immunity of the lysogenic host (Ogawa and Tomizawa, 1967).

THE NATURE OF PROPHAGE IMMUNITY

ZYGOTIC INDUCTION

The principal clue to the nature of the specific immunity which accompanies lysogenisation came from the observation that, in conjugal crosses between lysogenic and non-lysogenic strains of *E. coli*, the genetic outcome is profoundly influenced by whether the prophage is carried by the *Hfr* male (donor) or by the female (recipient) parent. When the male strain is lysogenic for an inducible phage such as λ, as soon as the prophage, at its location on the male chromosome, penetrates a sensitive female bacterium it enters the vegetative state and multiplies so that, after a normal latent period, the zygote lyses and liberates infective phage particles. Such zygotes constitute infectious centres which can be scored by the number of plaques they produce when plated on sensitive indicator bacteria. Since all the zygotes which inherit the prophage are destroyed, those bacterial genes to which the prophage is closely linked on the male chromosome are lost, and fail to appear among recombinants. This phenomenon is known as *zygotic induction* (Jacob and Wollman, 1956a; Wollman and Jacob, 1957). On the other hand, when the male strain is non-lysogenic and the female is lysogenic, or when both parents are lysogenic, no zygotic induction occurs; lysogeny and non-lysogeny, or the two types of prophage, segregate normally among recombinants (p. 454). These findings are summarised in Table 17.

TABLE 17

Illustrating the dependence of zygotic induction on the sexual polarity of *E. coli* crosses.

Cross	Genetic constitution of zygotes	Zygotic induction
♂ly^+ × ♀ly^-	ly^+/ly^-	Present
♂ly^- × ♀ly^+	ly^+/ly^-	Absent
♂$ly_a{}^+$ × ♀$ly_b{}^+$	ly^+/ly^+	Absent

The symbol ly^+ indicates lysogeny by an inducible prophage, and ly^- the absence of lysogeny (sensitivity): $ly_a{}^+$ and $ly_b{}^+$ symbolise genetically marked strains of the same prophage.

Observe that the occurrence of zygotic induction is not related to the genetic constitution of the zygotes with respect to lysogeny; the zygotes are genetically ly^+/ly^- whether the male or the female parent is the lysogenic one. Zygotic induction must therefore be determined by the *cytoplasm* of the female strain; *it is only when the prophage is introduced into the cytoplasm of a non-immune bacterium that the transition from prophage to vegetative phage occurs.* It follows that, in lysogenic bacteria, a cytoplasmic *immunity substance* is synthesised under prophage direction, which not only interferes with the vegetative function of superinfecting phage, but is also responsible for maintenance of the prophage itself. In its absence, prophage automatically becomes vegetative phage within a very short time. Zygotic induction is also found when an inducible prophage is transferred from lysogenic to non-lysogenic bacteria by means of transduction (Jacob, 1955; see p. 626).

These simple yet profound observations and deductions of Jacob and Wollman formed the basis of a unifying hypothesis whereby the phenomena of lytic infection, lysogeny and immunity may be regarded as different expressions of a single process—the regulation of the functional activity of the phage genome. We shall see in Chapter 23 (p. 713; see also p. 615) that this hypothesis is a very general one which has been applied with great success to the control of bacterial

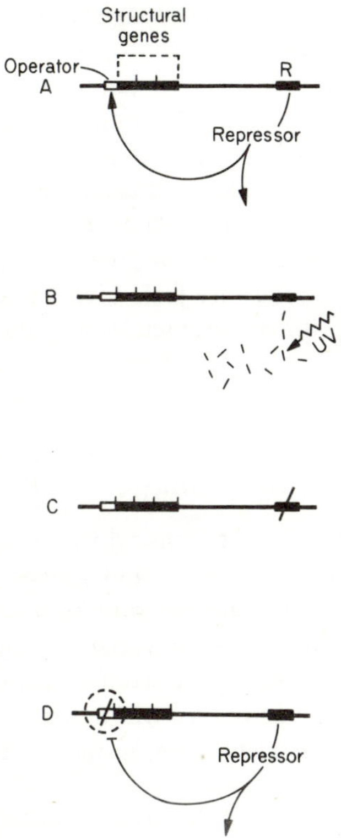

Normal Prophage: repressor acts on sensitive operator.
Inactive structural genes.
Cell immune to superinfection.

UV-*Induction:* repressor inactivated.
Operator switches on activity of structural genes.
Phage synthesis leads to lysis and liberation of normal, temperate phage.

Mutation in Repressor Gene: no repressor produced.
Structural genes permanently active in absence of repressor.
Phage virulent for non-lysogenic cells, but not for lysogenic cells which contain cytoplasmic repressor.

Mutation in Operator: repressor produced but operator insensitive.
Structural genes permanently active.
Phage virulent for both non-lysogenic and lysogenic cells.

FIG. 103. Illustrating the hypothesis of the genetic control of phage function by means of a cytoplasmic repressor. (See text for explanation).

metabolism, where many facts have been found to support it. The essence of the hypothesis, represented diagrammatically in Fig. 103A, is that the activity of certain genes, or co-ordinated groups of genes, is controlled by cytoplasmic *repressors*, whose synthesis is determined by *regulator* (*R*) *loci* which may or may not be closely linked to the genes which the repressors control. Each repressor acts specifically on a sequence of *structural genes* which mediates a single biochemical pathway, shutting off their functional activity. The site of action of repressor is not the structural genes themselves, but a small adjacent region of chromosome called the *operator*, whose function is to switch the activity of the structural genes off or on depending on whether the repressor is present or absent. At the molecular level, it seems that the operator is the site which controls transcription by messenger RNA of the information carried by the gene sequence. If the operator is blocked by attachment of repressor, no messenger RNA can be made so that synthesis of the proteins determined by the structural genes ceases. It is possible to identify and titrate phage-specific messenger RNA by estimating the amount of RNA, incorporating a radioactive label after infection or induction, which will form RNase-resistant hybrid molecules when mixed with denatured phage DNA(p. 276). Using this method, phage λ m-RNA has been shown to increase from a very low to a very high level in lysogenic bacteria following induction (see Sly, Echols and Adler, 1965).

PHAGE MUTATIONS INVOLVING THE ESTABLISHMENT AND CONTROL OF LYSOGENY

We have already referred in general terms to mutations which arise in the small *C* region of the chromosome of phage λ and prevent transition from the vegetative to the prophage state, so that the mutants carrying them form clear plaques. These mutations map in one or another of three functional units, the c_1, c_2 and c_3 cistrons, which can complement one another to promote lysogeny. We will now look more closely at the functions which these cistrons serve, in the light of the repressor hypothesis outlined above.

The function of the c_1 cistron can be shown to differ from those of the other two in several respects. In the first place, although mutation in the c_2 and c_3 cistrons interferes with lysogenisation, it does not prevent it entirely; the proportion of lysogenic bacteria which result from infection is merely reduced to a small fraction of normal. On the contrary, the ability to lysogenise is abolished by c_1 mutations. Secondly, activity of the c_2 and c_3 cistrons is not necessary for the maintenance of the prophage state once this has been achieved; mixed infection by c_2 and c_3 mutants can yield cells which are singly lysogenic for either type. In the case of mixed infection by c_1 and c_1^+ mutants, however, lysogeny by the c_1 mutant alone is never found; doubly lysogenised cells may be isolated which subsequently lose one prophage, but it is always the c_1 type which is lost (Kaiser, 1957). This suggests that c_1 mutants have lost the ability to synthesise immunity substance (repressor) although they remain sensitive to that produced by other prophages. The c_2 and c_3 cistrons do not determine repressor synthesis but serve fringe functions in the transition of vegetative phage to prophage, whose nature remains unknown (see Bode and Kaiser, 1965). We can thus equate the c_1 cistron with the repressor (*R*) locus of the model shown in Fig. 103C.

Mutations of another kind have been found in phage λ which also map within the c_1 region but, paradoxically, have exactly the opposite effect to the c_1 mutations described above. Mutant phages carrying this type of mutation lysogenise sensitive bacteria in the normal way and are characterised by the fact that *they are insensitive to induction by* UV *light*, so that they are symbolised as ind^-. Despite this, ind^- mutants remain normally susceptible to zygotic induction so that they clearly continue to synthesise repressor, but this repressor appears insensitive to the effect of UV light. When bacteria carrying a normal (ind^+) prophage are induced by UV light and then superinfected with an ind^- mutant, the induction is arrested and the normal prophage state restored; similarly, bacteria doubly lysogenised with a wild type and an ind^- prophage are insusceptible to UV induction, so that ind^- is a character which is dominant to ind^+, but the immunity which it confers remains strictly specific for phage λ. For example, if bacteria carrying some other prophage, such as 434 or 21, are induced and then superinfected with phage λ ind^-, the induction proceeds normally. On the other hand, if the ind^- mutation is introduced by recombination into a clear-plaque c_1 mutant, the resulting double mutant, c_1ind^-, is unable to lysogenise due, presumably, to failure to produce repressor (Jacob and Campbell, 1959). These results may be understood if we assume that c_1 and ind^- mutations both involve the same gene which specifies repressor structure, but whereas the former leads to the formation of inactive repressor, the latter yields a modified repressor of the same specificity but with the new property of resistance to the inducing effect of UV light. In addition, a temperature-sensitive mutant of phage λ has been isolated which maintains normal lysogeny at 32° but is induced at temperatures above 38°C. This mutation also maps in the c_1 region and presumably leads to the synthesis of a heat-sensitive repressor (Sussman and Jacob, 1962).

The last type of phage mutation we will consider here is one which leads to virulence for *lysogenic* bacteria; that is, the virulence overrides the prophage immunity. According to the repressor hypothesis such a mutation would be sited in the operator so that the mutant would become insensitive to the controlling action of repressor (Fig. 103D, p. 465). Such a virulent mutant of phage λ has been isolated, but it appears to have resulted from the summation of several mutations, one in the c_1 region but the other(s) outside the *C* region altogether (Jacob and Wollman, 1954; 1961a, p. 303).

NATURE OF THE REPRESSOR

When sensitive bacteria are infected with temperate phage, the proportion of the bacterial population which becomes lysogenic varies markedly with the cultural conditions (see Lieb, 1953a, b; Bertani, 1953). Under conditions conducive to optimal bacterial growth the great majority of the cells may be lysed, but when growth is retarded, as, for example, by lowering the temperature or interfering with metabolism in other ways (review: Jacob and Wollman, 1959), lysogenisation predominates. The introduction into the medium of agents such as chloramphenicol, which specifically inhibits protein synthesis, leads to virtually 100 per cent lysogenisation (L. E. Bertani, 1957). The efficiency of zygotic induction is similarly affected by the environment in which the cross is made; if prophage is transferred to non-immune, female bacteria in the presence of chloramphenicol, the transition from prophage to vegetative phage does not occur and the prophage segregates among the bacterial recombinants. All this suggests that whereas chloramphenicol blocks vegetative phage function by preventing the formation of early protein, it does not interfere with the synthesis of repressor. Comparable observations have been made with respect to the synthesis of a repressor of β-galactosidase formation in *E. coli* (Chapter 23, p. 714). From this one might think repressor is not a protein. On the other hand, permissive strains of *E. coli*, carrying suppressors which allow the growth of amber mutants and so intervene at the level of protein synthesis (pp. 345, 361), also permit lysogenisation by a small proportion of *c* mutants of phage λ which, in non-permissive strains, cannot make repressor (Jacob, Sussman and Monod, 1962). There is now little doubt as to the nature of the phage λ repressor, since it has recently been isolated in pure form and been shown to be an acidic protein of molecular weight about 30,000. The evidence that it is really the product of the c_1 gene is that it is not produced at all by phage having a *c*I amber mutation, while a temperature-sensitive *c*I mutant synthesises a modified substance (Ptashne, 1967a). Moreover, this protein has been shown to bind specifically to phage λ DNA; it does not bind to DNA from a recombinant phage which carries the c_{I} region of phage 434, although most of the rest of its genome comes from phage λ (Ptashne, 1967b). The repressor of the lactose region of *E. coli* has also recently been isolated and shown to be a protein (Gilbert and Müller-Hill, 1966).

A clever method of using the conjugation system of *E. coli* to separate repressor from the lysogenic cells in which it is made, so that some of its properties can be studied in isolation, was devised by Fisher (1962). F^+ male strains of *E. coli* K-12 conjugate with female cells and transfer various non-chromosomal (cytoplasmic) genetic determinants to them with high efficiency; unlike *Hfr* males, however, they do not transfer their chromosomes (p. 672). F^+ bacteria are lysogenised with a defective mutant of phage λ which confers immunity against infection by free, wild type λ particles without liberating infective phage into the medium. These immune, male cells are then mated with non-lysogenic, non-immune females. Note that, under these conditions, only cytoplasmic elements can be transferred to the females, prophage transfer being excluded. At intervals after mixing, samples are removed and the male cells destroyed by treatment with streptomycin and one of the virulent, T-even phages to which they are sensitive but the females are resistant. The female cells, which are left intact, are finally challenged with wild type phage λ and their susceptibility to lytic infection assayed. The result is that the initially sensitive female population acquires immunity as a function of the duration of mating, until a high proportion is immune. This passive immunisation is not correlated with inheritance of any particular one of several other non-chromosomal determinants which are also transferred. The immunity decays spontaneously and rather rapidly in a temperature-dependent way, but is *immediately destroyed by a dose of* UV *light* equivalent to about one-third the optimal inducing dose for lysogenic bacteria. This is not due to a direct susceptibility of the transferred repressor to UV light, however, since the irradiation of non-immune females *before* mating prevents their acquisition of immunity but not of other cytoplasmic determinants. The result of repeating these experiments with an F^+ male strain lysogenised by an *ind*$^-$ mutant of phage λ was most revealing. The female bacteria become passively immunised, and the immunity subsequently decays spontaneously, with approximately the same kinetics as before, but in this case the immunity is not significantly affected by UV irradiation.

THE MECHANISM OF PROPHAGE INSERTION AND RELEASE

In the early days of mapping the genetic fine structure of the *C* region of phage λ, it was thought that the *c*I gene not only was responsible for immunity but was also the region of attachment to the bacterial chromosome since, in crosses between lambdoid phages, the determinants of immunity and of chromosomal location appeared to be inseparable. The first exception was a recombinant from a cross between phages λ and 21 which occupied the 21 attachment site and conferred immunity only against phage λ (Jacob and Wollman, 1961 a p. 292). The determinants of these two functions were then unequivocally separated and mapped by the isolation of a mutant strain of phage λ, called λ*b2*, whose particles contained 18 per cent. less DNA than wild type and displayed a low frequency of apparent lysogenisation, but nevertheless retained immunity (Kellenberger, Zichichi and Weigle, 1961b).

It then turned out that phage *b2* did not lysogenise infected bacteria in the proper sense of becoming inserted into the chromosome, but showed abortive lysogeny. When sensitive bacteria are infected with this phage at multiplicities of 5 to 10, the prophages persist but do not replicate because they produce repressor. If the infected cells are diluted and allowed to multiply in the presence of λ antiserum, the number of infectious centres which are found after induction rises to a plateau which is equivalent to, but never exceeds, the phage input; thereafter sensitive bacteria begin to appear as the prophages become diluted out. Under the same conditions wild type prophages, after a lag, are inserted into the host chromosome and replicate with it so that the number of infectious centres rises at the same rate as the number of bacteria. Thus prophage *b2* is inherited unilinearly, just as in the case of abortive transduction (pp. 96, 641). Since it is not inserted into the chromosome, prophage *b2* shows no zygotic induction, nor is it capable of producing transducing particles (Kellenberger *et al.*, 1961b).

Because the DNA deficiency of phage λ*b2*, which yields particles of lower density, is an inheritable defect which is associated with abortive lysogeny, it is reasonable to guess that the extra DNA in the wild type contains the attachment region, and there are two experimental facts which confirm this. First, the location of the deletion has been mapped and found to lie in the region between the markers *h* and *c3* (Fig. 101, p. 457); in fact the deletion results in a reduction of the recombination frequency between these markers as compared with wild type (Jordan, 1965). If you look at the linkage map (Fig. 101) and imagine insertion occurring by a recombination event in this region, as proposed in Fig. 102 (p. 459), you will observe that the order of markers on the inserted prophage will be *h—F—A—R—O—c3*. On the other hand, the extremities of the linear, vegetative chromosome (cohesive sites) lie between *A* and *R*, giving an order *A—F—h—c3—O—R*. This is precisely the circular permutation of the λ chromosome which was previously observed (Jordan, 1965).

Secondly, mixed infection with *b2* and wild type ($b2^+$) phages permits insertion of *b2* into the chromosome, but no lysogenic segregants are found which carry prophage *b2* alone (Zichichi and Kellenberger, 1963). This, of course, could be due to the fact that prior lysogenisation with prophage $b2^+$ provides a region of phage homology for insertion of prophage *b2*; alternatively, it could follow enzymatic complementation as in the case of the similar behaviour of *c*I mutants (p. 456). However, the chance isolation of a strain of *E. coli* K-12 which is diploid for the λ attachment region (*att*λ/*att*λ) allowed a decision to be made. Following mixed infection with two genetically marked, ordinary λ mutants, the two prophages are generally found to be inserted at different attachment sites, as one would expect. But prophages *b2* and $b2^+$ are always found together at the same site after mixed infection; phage $b2^+$ fails to mediate insertion of *b2* in the *trans* position, showing that a structural interaction rather than an enzymatic complementation is involved (Campbell, 1965).

If prophage is inserted into the chromosome by recombination, its release on induction is presumably due to a realignment of the regions of homology flanking the prophage (*1*, *2* in Fig. 102, p. 459) and a second recombination event. It has been shown that deletions involving one of the prophage termini (h^+, Fig. 101, p. 457) continue to permit the preferential synthesis of phage DNA after induction, but that the replicating molecules cannot be matured into defective phage particles. This suggests that the prophage may be unable to escape from the chromosome because one of its two regions of homology is missing (Dove, 1967).

If release is due to recombination, progeny particles should be found which contain either one or both of the DNA strands of the parental prophage. This was investigated in the following way. It happens that superinfection of a λ-lysogen with a recombinant of phages λ and 434 (434*hy*), which carries the immunity region of phage 434 in a predominantly λ genome, induces the production of a very small proportion of phages of λ-immunity type in addition to the normal 434*hy* particles. The low yield of the resident phage suggests that it does not replicate much on release from the chromosome, so that the chance of finding particles containing the original prophage DNA is relatively good. Lysogenic bacteria were therefore grown in medium containing heavy isotopes ($^{13}C + ^{15}N$), superinfected with phage 434*hy*, and the lysate then centrifuged to equilibrium in a cæsium chloride density gradient. Finally, the density distribution of particles of λ-immunity type was assessed. Three classes of phage λ particles, containing heavy, hybrid and light DNA, were found. To demonstrate that the density-labelled DNA had actually been a part of the bacterial chromosome, the heavy lysogenic bacteria were allowed to replicate in light medium for one generation before superinfection. This should convert all chromosomally replicated prophages from heavy to hybrid density, but have no effect on repressed, cytoplasmic prophage DNA. The outcome was the virtual elimination of the heavy class of phage λ progeny (Ptashne, 1965a; see also Prell, 1965).

Granted that prophage insertion and release are mediated by recombination, a number of further questions arise. Why, for instance, does spontaneous release of prophage not occur with the same high frequency as insertion, thus leading to unstable lysogeny? Again, if the prophage state is maintained by the production of a protein repressor directed by the c_1 gene, why do factors whose only common denominator is inhibition of DNA synthesis all lead to induction? With regard to the first question it seems that phage φ80, at least, synthesises a product which specifically promotes its insertion at the att_{80} site located close to the *try* region (Fig. 127, p. 666). The evidence for this comes from exploiting an ingenious but complicated system in which transposition of the *lac* region (*lac*$^+$) to a location close to att_{80} (see p. 461) permits the formation of defective strains of phage φ80 which carry the *lac*$^+$ region (φ80*dlac*) in a manner analogous to *gal*-transducing particles of phage λ(λ*dg*) (pp. 100, 633). Thus, potentially,

ϕ80*dlac* carries affinities for both of the widely separated *lac* and att_{80} sites at either of which it could become inserted. Experimentally, in single infections ϕ80*dlac* is preferentially inserted at the *lac* site, but switches to insertion at the att_{80} site in mixed infections with wild type phage ϕ80. Clearly the wild type phage possesses a function promoting insertion at the att_{80} site which is lacking from the defective transducing phage. Moreover, wild type phage ϕ80 can lysogenise rec^- strains of *E. coli*, which are unable to undergo recombination (p. 334); on the contrary, ϕ80*dlac* cannot by itself become inserted at either the *lac* or the att_{80} site of rec^- bacteria, but acquires this ability in mixed infection with wild type phage (Signer and Beckwith, 1967). In normal lysogeny the repressor presumably shuts off the synthesis of this recombination-promoting substance so that prophage release is prevented.

The mechanism of induction is controversial at the time of writing. The effect of ultraviolet light was at first assumed to be due to direct destruction of repressor, but this hardly explains the similar effect produced by such a negative procedure as deprivation of thymine (Korn and Weissbach, 1962), or by treatment with Mitomycin C or naladixic acid, all of which stop DNA synthesis.

An interesting variation on this theme was provided by the isolation of a mutant of *E. coli* in which induction of inducible prophages (λ, 434, 424, 21) occurs on raising the temperature from 30° to 40°C (Goldthwait and Jacob, 1964). The mutation is in the bacterial genome and not in the prophage; the liberated phage is not induced at 40°C following lysogenisation of wild type bacteria nor, for that matter, do the mutant bacteria lyse at 40°C if they have been cured of lysogeny. Lysis at 40°C does not occur if the mutant bacteria are lysogenised by non-inducible phages, including λind^-. Since non-lysogenic mutant bacteria grow normally at 40°C in minimal medium supplemented only with the nutritional needs of the wild type, the inducing effect is not due to a temperature-sensitive requirement for thymine, nor to any overt defect in DNA synthesis. It turns out, however, that the addition of guanosine, cytidine or uridine to a culture of the lysogenic mutant bacteria at the time of shifting to the inducing temperature markedly protects against the induction; on the contrary, addition of adenine, adenosine or deoxyadenosine accentuates induction and antagonises the protective action of the other nucelosides. Since one of the effects of inhibition of DNA synthesis, at least by thymine deprivation, is the specific accumulation of adenine derivatives, notably ATP and deoxy-ATP, Goldthwait and Jacob (1964) postulated that induction results from inactivation of repressor by adenine derivatives which accumulate when the regulation of DNA synthesis is disturbed (but see Pritchard, 1966).

Another aspect of what is probably the same phenomenon is the failure of certain λ-lysogenic *rec*⁻ bacterial strains to be induced by ultraviolet light. However, when these strains are lysogenised by a mutant phage which is inducible at 40°C, normal induction follows the temperature shift, so that the failure of UV induction is not due to inability of the prophage to 'recombine out'. Moreover, it is also found that superinfection fails to be productive after irradiation of these strains, as it normally is, showing that the repressor is still intact. These *rec*⁻ strains are characterised by excessive breakdown of their DNA (Howard-Flanders and Thériot, 1966) which would presumably produce accumulation of deoxyribonucleotides which might protect the repressor (Hertman and Luria, 1967; Brooks and Clark, 1967; Ben-Gurion, 1967). In this connection it appears that certain mutations in gene *N* (Fig. 101, p. 457), which determines synthesis of a phage λ exonuclease, prevent heat-induction of thermo-inducible phage mutants involving the c_1 gene, suggesting that the products of these two genes may interact in some way to produce repression (Lieb, 1966). The complexity of the induction mechanism is further stressed by the finding that induction fails to occur in the absence of protein synthesis, even though repressor has been destroyed as judged by the initiation of phage m-RNA synthesis (Green, 1966), possibly due to failure to produce a specific recombinase necessary for prophage release (see Weisberg and Gallant, 1967).

THE PHYSIOLOGICAL GENETICS OF PHAGE LAMBDA

This means the study of the functions exercised by the various phage genes, and how these functions are regulated during the course of phage development following infection or prophage induction. This kind of analysis was initiated by Jacob, Fuerst and Wollman (1957) who observed the steps blocked in a series of defective λ prophages which were unable to produce infective particles following induction, and first distinguished between 'early' functions needed for chromosome replication, and 'late' functions not affecting replication. A major step forward was the isolation by Campbell (1961) of a large number of '*sus*' mutants, so called because they grow normally in certain 'suppressor' strains of *E. coli*, which are now known to be of amber type (p. 361). These 'conditional lethal' mutants carry the advantage that essential functions can be identified by observing what stage of development fails to be effected following infection of a non-permissive host. For example, DNA and m-RNA synthesis can be followed chemically or by polynucleotide hybridisation methods, the maturation of DNA by assaying its infectivity in the presence of 'helper' phage (p. 587), the synthesis of phage antigens by serological methods and of lysozyme by the effect of lysates on the optical density of sensitised bacteria, while failure to form structural components can be observed by electron microscopy.

Campbell's '*sus*' mutants were found, by complementation tests, to be distributed among 18 genes, *A* to *R*, which are arranged in sequence on the phage chromosome as shown in Fig. 101 (p. 457), which also indicates what is known of their functions. It will be seen that genes *A* to *D*, located at one end of the vegetative chromosome, are concerned with formation of the phage head and *G* to *J* with tail structure, while gene *J* determines the synthesis of a tail antigen; the proper functioning of genes *A* to *F* are needed for maturation of the phage DNA. On the other side of the attachment region lie genes *N*, *O* and *P* which determine DNA replication and are located close to the *c* genes; gene *N* also controls exonuclease production while a mutation, *T11*, which maps between *c1* and *c2* and likewise produces a block in DNA synthesis, concurrently yields a great increase in exonuclease level and so may involve some regulatory function. Gene *R*, at the opposite end of the vegetative chromosome to *A*, determines lysozyme synthesis so that cells infected with *R* mutants fail to lyse. It is thus apparent that there is a considerable degree of clustering of functionally related genes on the phage λ chromosome (cp. pp. 163, 490) (Brooks, 1965; Dove, 1966; Harris *et al.*, 1966; Joyner *et al.*, 1966).

It turns out that mutations in genes *N*, *O*, *P* and *Q* are pleiotropic in effect, since they switch off not only DNA synthesis but also all the late functions associated with maturation and lysis (genes *R*, and *A* to *J*). This effect appears to be due to the production of a very low level of m-RNA when any of these early genes are blocked. On the other hand, mutations in gene *Q* do not interfere with DNA replication so that this gene appears to act directly in inducing the formation of m-RNA serving late functions. It may possibly determine the synthesis of a new RNA polymerase and be itself activated by a certain stage in the process of DNA synthesis in a way not yet understood (Dove, 1966; Joyner *et al.*, 1966; see also Eisen *et al.*, 1967).*

We can therefore distinguish six types of function performed by the phage λ chromosome; (1) early functions, mainly or wholly connected with DNA replication (*N*, *O*, *P*); (2) late functions, serving the synthesis of phage protein components, maturation and lysis (*A* to *J* and *R*); (3) regulation of the sequence and level of these functions (*Q*, *t11*); (4) repression, necessary for the establishment and maintenance of lysogeny (*c1*, *c2*, *c3*); (5) homology for the attachment site on the bacterial chromosome, required for insertion of prophage by recombination, carried by the *b2* region; (6) circularisation, necessary for insertion and also, probably, for other purposes, due to the presence of complementary single DNA strands (cohesive sites) at the extremities of the linear, vegetative chromosome. We shall encounter many of these features again when we come to review the physiological genetics of phage T4 in the next chapter (p. 490).

Since so many vital processes, such as DNA-directed protein synthesis and the replication of nucleic acids, can progress efficiently in the test tube when all the necessary ingredients are provided, one may ask to what extent a more complicated system such as phage development is dependent on the integrity of the infected cell. Mackal, Werninghaus and Evans (1964), reported that the addition of phage λ DNA to mechanically disrupted cells of an *E. coli* strain, normally resistant to infection by either λ particles or DNA, resulted in the formation of infective particles, but with rather a low efficiency. An appreciable yield of phage in the presence of protoplasts of this strain was found only if the protoplast suspension was first diluted with distilled water or mechanically disrupted.

*This is supported by the finding of Shalka (1966) that, during the early stages of infection, phage λ-specific m-RNA, extracted from the host bacteria, forms hybrids only with denatured DNA derived from that half (the 'right' half) of the λ DNA molecule which codes for early functions (see p. 544). Later in infection, λ-specific m-RNA of different base composition, which hybridises with both half molecules of λ DNA, is produced. A corollary to this is that *in vitro* transcription of phage λ DNA by *E. coli* RNA polymerase is mainly confined to the 'right' half molecule (Cohen, Maitra and Hurwitz, 1967). These experiments suggest that phage λ uses its host's RNA polymerase to initiate early functions, but makes its own for the transcription of genes serving late functions.

More recently, Zgaga (1967) has succeeded in obtaining high titres of phage λ following addition of λ DNA to a cell-free extract of *E. coli* K-12S prepared from osmotically ruptured protoplasts. Unlike the infection by DNA of the protoplasts themselves, this system is reported to give about a hundred times greater yield of phage, is sensitive throughout to the action of DN-ase, and is also susceptible to RN-ase activity. After addition of the DNA there is a latent period of about 30 minutes before infective particles begin to appear; the titre then rises to reach a plateau about 90 minutes later (Zgaga, 1967).

LYSOGENY IN SOME OTHER SYSTEMS

Virtually all the work we have described concerning the mechanism of prophage attachment and immunity has been based on phage λ and a few other inducible phages related to it. We have seen that this work has given birth to an elegant and unified hypothesis which explains the important aspects of phage behaviour in terms of the regulation of gene activity by means of cytoplasmic repressors. Unfortunately there exist a number of temperate phages which refuse to conform to this pattern of behaviour. These rebellious phages fall into two categories, one of which displays a different immunity mechanism, and the other a different relationship to the bacterial chromosome, from phage λ. Very little is yet known about these phages so that you must be content with a very brief summary of their salient features.

NON-INDUCIBLE PHAGES

Among these are the *Shigella* phage, P2, which can also lysogenise some *E. coli* strains, and the seven non-inducible phages, lysogenic for *E. coli* K-12, to which we have already referred (p. 454). The failure of UV light to induce bacteria carrying these phages would not, by itself, be enough to place them in a different category since, as we have seen, ind^- mutants of phage λ are not induced by this means. The real test is that these phages are not amenable to zygotic induction, which suggests that the prophage state is not controlled by a cytoplasmic repressor. Despite this, bacteria lysogenised by the non-inducible phages are specifically immune to superinfection; moreover, as in the case of inducible phages, two types of virulent (clear-plaque) phage mutant can be isolated, one of which remains susceptible to prophage immunity while the other overrides it.

The most thoroughly investigated of the non-inducible phages is P2. Unfortunately this phage shows such a low frequency of recombination in mixed infection that no genetic analysis of the phage chromosome has been made, but it can infect and lysogenise strains of *E. coli* C, in which conjugal crosses are possible, so that information about its relationship with the bacterial chromosome is available (review; Bertani, 1958). It turns out that the P2 prophage can attach itself to any one of at least three distinct locations, but does so with different frequencies. In double lysogeny the two prophages occupy different sites and the condition is more stable than is the case with phage λ. When single lysogeny is established at location 2 or 3 the bacteria are fully immune even though the preferred location 1 remains vacant. Moreover, by recombination between phage P2 and a defective prophage found to be present in *E. coli* B, a 'hybrid' phage was obtained which occupies the same chromosomal locations as P2, but confers a different specific immunity (Six, 1961). Immunity is therefore a function separable from attachment, and is not due to steric blockage of a bacterial locus which must be contacted by superinfecting phage as a prelude to multiplication.

PHAGES MEDIATING UNRESTRICTED TRANSDUCTION

The transducing ability of phage λ is restricted to a cluster of *gal* loci closely linked to its location on the bacterial chromosome (pp. 99, 454, 627). Some other temperate phages, however, can act as vectors for the transfer, to recipient bacteria, of virtually any region of the host chromosome, that is, they are capable of *unrestricted or generalised transduction* (Chapter 21, pp. 627, 633). Among the most important of these are the *Salmonella* phage P22, and the indistinguishable *E. coli* phages P1 and 363, all of which are inducible. While phage λ can acquire the property of transducing the *gal* region only when it is in the prophage state, transducing particles of phages capable of unrestricted transduction arise as a result of lytic infection. We should therefore expect to find some difference in the relationship of these two types of transducing phage to the bacterial chromosome, which would reflect their distinctive behaviour in transduction. It turns out that, of all the inducible and non-inducible phages which lysogenise *E. coli* K-12, only in the case of phage 363 can no prophage location be mapped. In reciprocal crosses between various *Hfr* male and female strains, lysogenic and non-lysogenic for this phage, neither the ly^+ not the ly^- character of the male parent is transmitted to recombinants, irrespective of the male marker selected (Jacob and Wollman, 1961a, p. 176).

By a rather complicated procedure it is possible to obtain particles of phage P1 which have incorporated the lac^+ region of the *E. coli* chromosome, in much the same way that transducing phage λ particles carry the *gal* region. These P1-lac^+ particles are able to lysogenise F^+ male cells of *E. coli* K-12 so that, following conjugation with non-lysogenic, lac^- female bacteria, inheritance of the lac^+ marker serves to trace those females to which the P1 prophage may have been transferred. It is found that the P1-lac^+ prophage is, in fact, transferred at a low frequency which would be difficult to detect in the absence of the lac^+ marker, and that this transfer is not associated with that of the bacterial chromosome (Boice and Luria, 1961). This suggests that the P1-*lac* prophage may exist as a cytoplasmic entity in lysogenic cells, but whether normal P1 prophage can do so is another matter.

We have already mentioned the existence of a generalised transducing phage for *B. subtilis* which exists in its host in the carrier state (p. 450), rather like phage λb_2 (p. 47), so that it is lost if the lysogenic bacteria are grown in the presence of phage antiserum. The phage DNA can be distinguished from that of the host and is detectable in the lysogenic bacteria, but no homology exists between phage m-RNA and the bacterial DNA (Bott and Strauss, 1965).

What we have said here should not be taken to imply that generalised transducing phages as a class lack a chromosomal location, for prophage P22 has been shown to occupy a site on the *Salm. typhimurium* chromosome close to a cluster of proline genes (Smith and Stocker, 1966).

CHAPTER 18

Bacteriophage as a Genetic System

In previous Chapters we have often described, in considerable detail, the results of genetic analyses in bacteriophage wherever these seemed to embody important conclusions relevant to general genetic mechanisms and principles as exemplified, for instance, by Benzer's analysis of the fine genetic structure of the r_{II} region of phage T4 or by the fundamental studies of chemical mutagenesis in this organism. Our present aim is to give a much more introspective account of bacteriophage genetics. We will look at recombination in bacteriophage as if it were a self-contained system which has many novel features of interest for their own sake. This does not mean, of course, that what we have to say may be ignored by fundamentalists, for nature is well known to drop her most vital clues in the most unexpected places.

The starting point of all genetic studies is the recognition of inheritable character differences, which arise by mutation and serve as genetic markers in recombination analysis. We will therefore begin with a brief account of the main types of mutation that are found in phage. Throughout most of this Chapter we will take the virulent phages, T2 and T4, as model systems demonstrating the main features of recombination which, indeed, appears to be basically the same in all phages, whether virulent or temperate.

MUTATIONS IN PHAGE

The classical characters employed in most of the early, fundamental studies of phage genetics, are those involving differences in *plaque-type* and *host-range* (*h*). More recently, mutants of phage T4 resistant to the inhibitory effect of acridine dyes on maturation (*ac*) (p. 430), as well as an extensive series of temperature-sensitive (*ts*) and amber (*am*) mutants, have been described. We have already mentioned those mutations, peculiar to temperate phages, which interfere with lysogenisation (*c*) or development (*sus*) (Fig. 101, p. 457) and will not discuss them further here.

PLAQUE-TYPE MUTANTS

Phage plaques are much more than mere circumscribed zones of clearing in a 'lawn' of bacterial growth. They show reproducible individualities of structure which are used to identify phage strains as well as different mutants of the same strain. Firstly, plaques may vary in *size*; for example, plaques of phages T1, T3 and T7 are strikingly larger than those produced by the T-even phages. Mutants of phages T2 and T4 are found which form minute plaques (*m*) (Hershey and Rotman, 1949) and this type of mutation has also been used as a marker for phage λ (Jacob and Wollman, 1954).

Secondly, plaques may differ in *morphology*; their edges may be clearcut or fuzzy, abrupt or shelving, and may be circumscribed by halos of partial clearing or turbidity. These individualities reflect physiological differences in the phage-bacterium relationship which are usually determined by the phage genome and are phage-specific, but it is important to realise that plaque size and morphology can change drastically when a particular phage is grown on different susceptible hosts, and even on the same host under different cultural conditions.

The most famous types of mutant involving plaque morphology are the *r* mutants of the T-even phages. When wild type (r^+) strains of these phages are plated on *E. coli* B they produce small plaques with a turbid halo since, due to the occurrence of lysis inhibition, only a fraction of the infected bacteria burst before the phase of active bacterial metabolism is over (Doermann, 1948; see p. 415). Among populations of r^+ particles mutants are found, with a frequency of about one in 10^4, which are not restrained by lysis inhibition and so are called *r* (rapid lysis) (Hershey, 1946); these *r* mutants produce larger, clear plaques which are easily distinguished from those of r^+ particles, as Plate 22 (p. 489) shows. We have already noted that *r* mutations fall into three groups, r_{I}, r_{II} and r_{III}, which map in different regions of the chromosome and are distinguished phenotypically by their ability to be expressed in different host strains of *E. coli* (Table 5, p. 167).

Other distinct, genetically determined differences from wild type plaque morphology are exemplified by the turbid (*tu*) mutants of phage T4 which yield plaques which are circumscribed by a turbid ring (Doermann and Hill, 1953), and the 'star' mutants described by Symonds (1958) which produce sectored plaques. Comparable varieties of plaque-type mutants arise in phage λ (Jacob and Wollman, 1954). Mutants of phage T1 have been found which introduce an attractive range of colour differences into plaque morphology when a rather complicated mixture of dyes is incorporated into the medium; presumably these mutants affect the nature and distribution of components of the lysate which interact with the dyes (Bresch, 1953).

HOST-RANGE MUTANTS

If a sensitive culture of *E. coli* B is plated with a gross excess of wild type (h^+) phage T2, confluent lysis occurs but a minute fraction of the bacteria escape infection and grow up to form colonies. These colonies are found to consist of mutant bacteria, designated B/2 (B 'bar' two), which possess the inheritable inability to adsorb the phage as a result of alteration in the constitution of their cell walls, so that they are *resistant* to infection. If, now, an excess of wild type T2 particles is plated with a culture of B/2 bacteria, no confluent lysis is found, but a few plaques arise from infection by mutant phage particles, designated T2*h*, which have the ability to infect and lyse both wild type B and B/2 bacteria. Again, by plating B/2 bacteria with T2*h* phage particles in the same way as before, a new set of bacterial mutants (B/2/*h*) can be isolated which are resistant to both h^+ and *h* phage types. By a further selective step, another series of phage mutants, designated h', can be obtained from phage T2*h*; these mutants plaque normally on B, B/2 and B/2/*h* bacteria (Baylor *et al.*, 1957). The infectivity of these *host-range mutants* of phage T2 for wild type *E. coli* B and its resistant mutant strains is summarised in Table 18. By alternate, serial selections of this kind, similar series of host-range mutants can be obtained

TABLE 18

The infectivity of host-range mutants of phage T2 for wild type, and various resistant mutant strains of *E. coli* B.

Mutant types of: Phage	Bacteria B	B/2	B/2/*h*
h^+	+	−	−
h	+	+	−
h'	+	+	+

'+' indicates ability, and '−' inability, to form plaques.
For the derivation of the various mutant types, see text.

from phage T3, while host-range mutants have also been isolated from other phages, including temperate ones such as phage λ.

If the host-range character is to serve as a genetic marker in phage crosses, it is necessary that both the h^+ and *h* alleles be recognisable among the progeny particles by producing distinctive plaques. But the difference between h^+ and an *h* mutant is an all-or-none one —the phage either plaques or does not plaque on a particular host. Thus we can see from Table 18 that if the progeny are plated on *E. coli* B, the *h* and h^+ alleles are indistinguishable, while if they are plated on *E. coli* B/2, only the *h* particles form plaques so that those carrying the h^+ allele are lost. This quandary may be overcome by the simple expedient of plating on a *mixture of* B *and* B/2 *bacteria.* Since *h* particles can lyse both bacterial strains indiscriminately they will yield *clear* plaques; h^+ particles can infect and lyse the B bacteria in the mixture and so can form plaques, but these will be *turbid* from the background of resistant B/2 cells (see Plate 21). In the same way, mutational reversions from *h* to h^+ may be isolated by plating large numbers of *h* particles on a mixture of B + B/2 bacteria and picking the rare, turbid plaques which arise.

Genetic analysis of phage T2 has shown that host range is controlled by a single gene (Streisinger, 1956b; Streisinger and Franklin, 1956) and, therefore, probably depends on the configuration of a single protein sub-unit of the tail fibres (p. 418). An interesting point brought out by this analysis is our arbitrary and artificial use of the word 'wild type'. The starting point of the investigation was the naturally occurring strain of phage and a laboratory strain of *E. coli* B which was sensitive to it, and it was only on this basis that the strains were called wild type. It turns out, however, that crosses between *h* mutants of independent origin never yield h^+ recombinants, implying that the DNA base sequence and its derivative tail protein are the same in all *h* mutants. On the other hand, when independently isolated h^+ revertants of *h* mutants are crossed, *h* recombinants, forming plaques on B/2 bacteria, may be readily recovered. We will expand this discussion of the host-range character in the section dealing with phenotypic mixing (p. 511).

CONDITIONAL LETHAL MUTANTS

These are mutants which behave as lethals under one set of conditions termed *restrictive*, and as wild type under other, *permissive* conditions. They comprise mainly temperature-sensitive (*ts*) and amber (*am*) mutants. Temperature-sensitive mutants, of which a large number have been isolated and mapped in the case of phage T4 by R. S. Edgar and his colleagues, are unable to grow and form plaques at 42°C although they do so normally at 25°C. The efficiency of plating of wild type phage T4 is the same at both temperatures. Since the mutant strains are not inactivated by exposure to the higher temperature, it is only their functional potential and not their structural integrity which is susceptible to heat. Temperature-sensitive mutants may be isolated by exposing wild type phage to the action of a chemical mutagen and, after growth at low temperature to allow segregation of mutant particles, plating with sensitive bacteria at 25°C. At this temperature only small plaques develop. The plates are then transferred to the higher temperature at which wild type plaques increase greatly in size. Those plaques which fail to develop further are presumed to have arisen from *ts* mutants. Of course mutant stocks must be propagated, and crosses involving them performed, at 25°C (Edgar and Lielausis, 1964).

Amber mutants of phage T4 were first isolated and studied by R. H. Epstein and his colleagues and were named after the mother of one of these colleagues, H. Bernstein, whose name is the German equivalent of 'amber', in fulfilment of a promise to do so in return for help in a search for this kind of mutant, should the search be successful (see Edgar, 1966). As Edgar suggests, no doubt the sheer irrelevancy of the name of amber mutants, and of their ochre cousins (p 362), will ensure their immortality, while more scientifically based designations change with the fickleness of hypotheses!

Amber mutants were initially characterised by ability to grow and form plaques on a particular strain of *E. coli* K-12 (CR63) but not on *E. coli* B. As we now know, they result from 'nonsense' mutations which alter a nucleic acid base triplet coding for an amino acid into one which terminates the growing polypeptide chain. However, the permissive host carries a species of transfer RNA molecule which translates the chain-terminating triplet as a compatible amino acid so that functional polypeptide is produced (pp. 361-364). These T4 amber mutants and the phage λ *sus* mutants of Campbell (1961), discussed in the previous chapter, belong to the same mutant class since they share the same permissive hosts.

Conditional lethal mutants are becoming increasingly important in the genetic analysis of phage, and especially in the study of physiological genetics. The reasons for this are three-fold. (1) The mutations are not restricted to the limited number of genes which determine modifiable characters such as plaque-type or host-range, but extend to many, if not all, genes irrespective of whether their function is dispensible or not. Thus, in theory, these mutations permit the complete genetic mapping of an organism. (2) Complementation tests can be carried out by mixed infections, as in the case of r_{II} mutants which are really conditional lethal mutants of a third type (p. 169), so that different genes can be identified at the functional level. This allows the study of physiological genetics in the RNA phages, for example, where genetic recombination has not been found. Amber mutants are particularly suitable for the identification of separate genes by complementation tests, since their failure to produce complete polypeptides prevents the occurrence of interallelic complementation (p. 149). (3) The ability to isolate and maintain mutants involving essential functions allows these functions to be analysed, following infection of the restrictive host, by observing what synthetic or morphogenetic step is blocked (pp. 432, 492–494).

As might be expected, the sites of both *ts* and *am* mutations appear to be widely and more or less randomly distributed in the phage T4 genome, while both types of mutation may be found in many genes. At the same time, amino acid substitutions leading to temperature-sensitive proteins, and base changes generating a specific, chain-terminating triplet, might be thought of as rare events, and in fact the sites of conditional lethal mutations are found to constitute only a very small proportion of total mutable sites (Epstein *et al.*, 1963; Edgar, Denhardt and Epstein, 1964; see also Alikanian *et al.*, 1966a).

Due to the isolation of *ts* and *am* mutants, some 70 genes of phage T4 have now been identified and mapped. A simple calculation shows that this number begins to approach about half the possible genes on the T4 chromosome. Thus direct measurements show that the T4 chromosome consists of a DNA double helix about 55μ or $5{\cdot}5 \times 10^5$Å long. Since the distance between base pairs is 3·4 Å, the total number of base pairs is $1{\cdot}6 \times 10^5$. Assuming that an average gene is about 1000 base pairs long, coding for a protein containing some 300 to 350 amino acids, the phage DNA can accommodate 160 such genes.

HEAD PROTEIN MUTANTS

The last type of mutant we will describe here, shows an altered resistance of the protein head membrane to osmotic shock, and is designated *o*. The head membrane of the wild type T-even phages is only slowly permeable to NaCl so that when the particles are equilibrated with a high salt concentration and then rapidly diluted into distilled water, the water is osmotically absorbed and the heads burst (p. 419). In the case of *o* mutants the permeability of the head membrane is increased, to the extent that absorbed NaCl can readily escape so that rupture does not occur on transfer to distilled water. This resistance to osmotic shock enables *o* mutants to be selected from wild type populations. On the other hand, the increased permeability allows penetration of the head by substances of higher molecular weight to which wild type phage is impervious. Thus *o* mutants are much more susceptible than wild type to the lethal action of nitrogen mustard, by the use of which o^+ recombinants can be selected and counted in crosses between parents carrying different *o* mutations (Brenner and Barnett, 1959). As one would expect from the homogeneity of the head protein (p. 420), *o* mutants do not complement one another and the mutations map together within a single, small region of the chromosome.

MATING AND GENETIC RECOMBINATION IN PHAGE

The discovery that recombinant particles arise when bacteria are mixedly infected with phages distinguished by two or more mutational differences, was made independently by Delbrück and Bailey (1946) and by Hershey (1946) and first reported at the same meeting at which genetic recombination in *E. coli* was also disclosed. The first definitive study of genetic recombination in phage was undertaken by Hershey and Rotman (1949), using phage T2, and revealed most of the essential features of phage crosses as we know them today.

MAKING THE CROSS

The first requirement for setting up a phage cross is a suitably marked pair of parental phages. In order to reveal recombinants, the parents must differ with respect to two characters. The classical mutations employed to study the genetics of the T-even phages are those involving host range (*h*) and plaque-type (*r*). These are particularly useful on two counts. First of all, the inheritance of any combination of the two character differences can be observed directly from the appearance of individual plaques (see below); secondly, since *r* mutations fall into three well separated loci (r_{I}, r_{II}, r_{III}), their judicious use can provide a wide range of information about linkage relationships.

The two types of mutation may be arranged in the parents in two ways. Either each parent carries a single mutation and the wild type allele of the other ($h^+r \times hr^+$), yielding recombinants which are wild type (h^+r^+) or doubly mutant (hr) for both characters, or else one parent is wild type and the other doubly mutant so that the recombinants are of h^+r or hr^+ type. The outcome of the cross is the same in both cases (Hershey and Rotman, 1949). Doubly mutant strains of phage can be isolated *de novo* in two mutational steps, or obtained as recombinants from the $h^+r \times hr^+$ cross.

The cross is set up by adding a mixture of the two parental phages to a liquid culture of a bacterial host which is equally sensitive to both. The conditions of infection should be such that nearly all the bacteria are simultaneously infected with both types of phage. To ensure that each bacterium is mixedly infected, an average multiplicity of five particles of each parental phage is usually used. The desirability of synchronous infection is imposed by a phenomenon known as *superinfection breakdown* which appears to be restricted to phages T2, T4, T6 and T5. If bacteria infected with phage are superinfected, a few minutes later, with a genetically marked strain of the *same* phage, not only do the markers of the superinfecting particles fail to appear among recombinants, but their DNA is broken down to low molecular weight molecules, probably as a result of increased deoxyribonuclease activity stimulated by the initial infection (reviews: Adams, 1959, p.212; Stent, 1959). The operation of this phenomenon in phage crosses, as a result of asynchronous adsorption, could obviously perturb the genetic outcome by suppressing recombination in a fraction of the infected cells. Fortunately there is a simple remedy. If the metabolism of the host bacteria is arrested during the infective porcess, by allowing infection to occur in the presence of cyanide-broth, not only does superinfection breakdown not occur, but the initiation of phage development is prevented until the cyanide is diluted out, so that synchronous infection is achieved (Benzer and Jacob, 1953).

When infection is thought to be complete, residual unabsorbed phage is removed by centrifugation or by the use of antiserum. The infected bacteria are then diluted into fresh broth and allowed to proceed to lysis. Suitable dilutions of the lysate are finally plated for plaques, which are scored for the markers characterising the parental phages. This gives the mean output of parental and recombinant particles for the population of infected bacteria as a whole. In an alternative type of investigation, the progeny emerging from *single* bacteria can be analysed in a single burst experiment, by transferring small samples of a high dilution of the infected culture, *before lysis*, to each of a series of broth tubes, in such a way that each tube has rather a low probability of receiving an infected bacterium (p. 415). When lysis is complete, the contents of each tube are plated separately for plaques which are then scored for phenotype as before.

THE OUTCOME OF THE CROSS

When *E. coli* B bacteria are mixedly infected with two mutant strains of phage T2, of genotype h^+r and hr^+, the progeny phage particles in the lysate are found to give rise to four distinct types of plaque, corresponding to the two parental and two reciprocal recombinant classes, when plated with a mixture of *E. coli* B and B/2 indicator bacteria which are together required to distinguish the h^+ from the h phenotype (p. 483 and Table 18). The appearance of each of these four types of plaques is shown in Plate 21. The h^+r parental particles yield large, turbid plaques—large because of the rapid lysis imposed by the r mutation, and turbid because the h^+ allele determines inability to lyse the B/2 bacteria in the indicator mixture. The hr^+ parental particles give small, clear plaques since both types of indicator bacteria are lysed (h phenotype) but lysis inhibition operates (r^+ phenotype). In the same way, the two recombinant classes produce large, clear plaques (hr phenotype) and small, turbid plaques (h^+r^+ phenotype).

It is found that the two reciprocal recombinant classes appear in approximately equal numbers among the progeny. The proportion of the number of recombinants to that of the total progeny of the cross, expressed as a percentage, indicates the *recombination frequency*, and whenever the same cross is performed under strictly standardised conditions, this frequency is constant. However, when different genetic markers are used as, for example, by substituting an r_{III} mutation for an r_{II} in one of the parental phages, a quite different recombination frequency is found. Over a wide range of different combinations of genetic markers, the recombination frequency may vary from about 0·01 per cent for adjacent sites in the same cistron (p. 299) to about 40 per cent for distant markers. Thus by applying orthodox genetic principles, according to which the distance between two genes is proportional to the frequency of recombination between

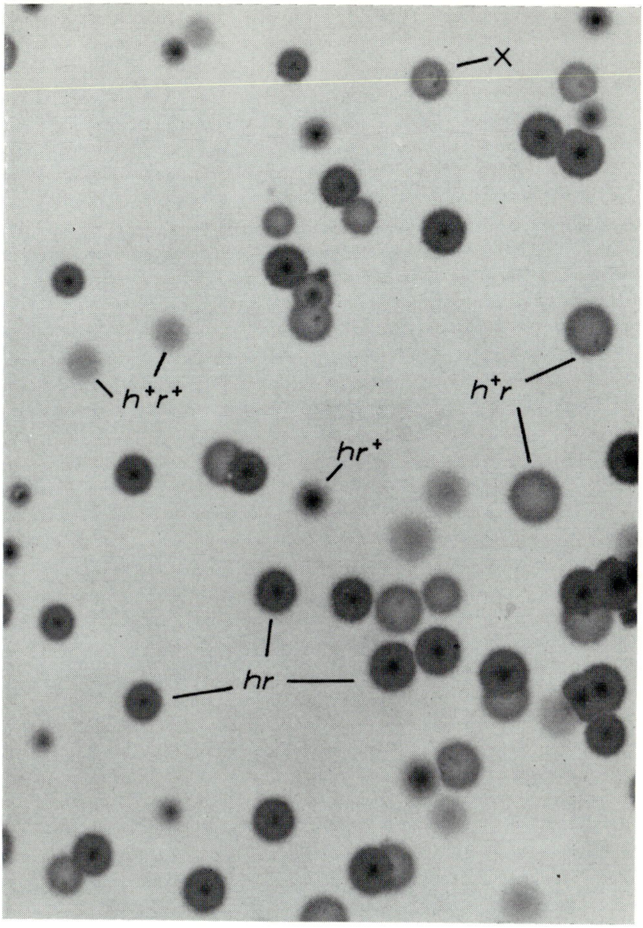

PLATE 21. Photograph showing the types of plaque produced by progeny particles of a cross between h^+r and hr^+ parental strains of phage T2.

The cross is performed by mixed infection of *E. coli* B with the two parental types of phage. The progeny particles are plated on a mixture of B and B/2 bacteria. The appearance of the plaques produced by the parental (h^+r and hr^+) and the two reciprocal recombinant (h^+r^+ and hr) classes of particles are indicated on the photograph. The small clearings in the h^+r plaque marked 'X' originate from h mutants which arise during development of the plaque.

Magnification about ×2. For further description, see text.

(Kindly provided by Mrs Maureen de Saxe & Miss Janet Mitchell)

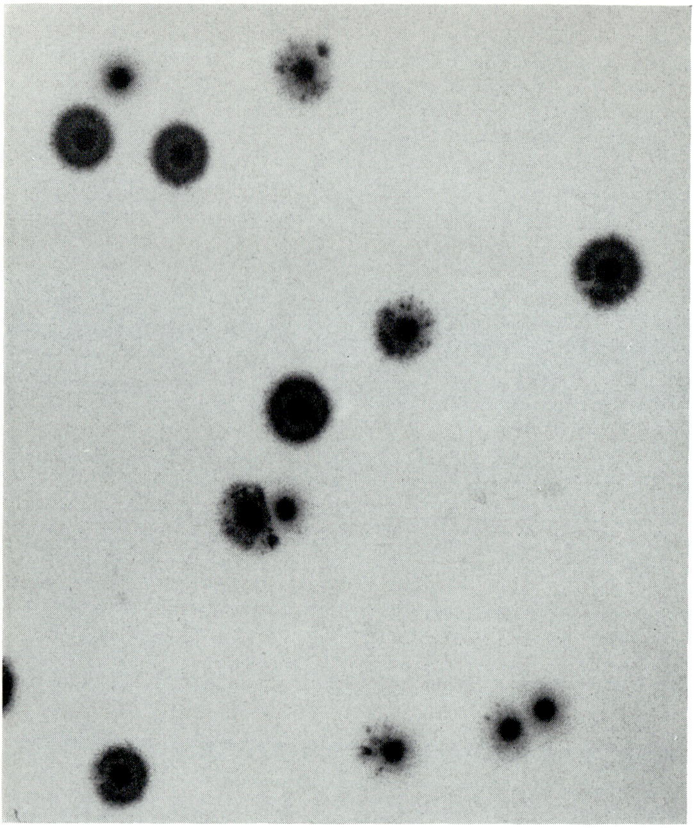

PLATE 22. Photograph of mottled plaques arising from mixed infection of *E. coli* B with r^+ and r particles of phage T2.

These are identical with plaques produced by particles heterozygous at an r locus. Pure r^+ and r plaques, arising from singly infected bacteria, are also clearly distinguished. Magnification about $\times 3\frac{1}{2}$.
(Kindly provided by Mrs Maureen de Saxe & Miss Janet Mitchell)

them, the locations of the various genes can be mapped relative to one another on the phage chromosome.

The phage linkage group: a circular chromosome

The initial analysis of phage T2 by Hershey and Rotman (1949) disclosed the presence of three distinct linkage groups, within each of which the loci appeared to be arranged in a linear order as judged by approximate additivity. By using more sensitive tests of linkage (see next paragraph), markers on these groups were found to be themselves linked, so that the separate linkage groups merged into a single linear linkage group. In other words, the results of genetic analysis are compatible only with the presence of a single chromosome. A similar conclusion has been reached in the case of phage T4 (Streisinger and Bruce, 1960).

The most widely separated loci on the chromosome map, defined by arranging the various linked loci in sequence according to their recombination frequencies in pairwise crosses, appeared to show no linkage; that is, the chromosome could best be represented as a linear structure with two extremities. A more sensitive test of linkage is provided by the use of 3-factor crosses (p. 139) of the type

$$\frac{a \quad b \qquad c}{+ \quad + \qquad +}$$

in which *a* and *b* are definitely linked, while linkage between *b* and *c* is doubtful when judged by recombination frequency alone. Recombinants for the *a* and *b* loci (*a*+ and +*b*) are selected and scored for inheritance of the two alleles of the *c* locus. If *b* and *c* are linked, the *c* allele should appear with a significantly higher frequency among +*b* than among *a*+ recombinants, and *vice versa*. By applying this test to loci situated at the extremities of the phage T4 linkage map, Streisinger Edgar and Harrar (1961) detected unambiguous linkage betweeen them and thus established that the linkage map is a circular one. This circularity has now been confirmed by the mapping of over 400 *ts* and *am* mutations whose relative locations are shown in Fig. 104 (Streisinger, Edgar and Denhardt, 1964). The question arises whether such circular maps necessarily reflect circular physical structures. The answer is that they do not although, of course, the most straightforward interpretation is that the chromosome *is* circular, at least at the time of mating. Alternatively, a circular map would emerge from a population of linear chromosomes which were circular permutations of one another. We have already indicated that the linear chromosomes in the heads of T2 and T4 phage particles are indeed circularly permuted, but there is a repetition of structure at the two ends—the so-called terminal redundancy; the evidence suggests that both the circular permutation and the terminal redundancy stem from a mechanism which chops off, from a chromosome polymer or concatenate, segments which are longer than a single genome for incorporation into the phage head (see p. 433). We will discuss the structure of the phage T4 chromosome further and in more detail in the next chapter (p. 535). We may note that, formally, a requirement for an even number of genetic exchanges in recombination would also yield a circular linkage map (see Stahl and Steinberg, 1964).

THE CHROMOSOME AS A FUNCTIONAL STRUCTURE

We have already seen in the case of phage λ how the genetic and physiological study of conditional lethal mutants can illuminate chromosome function and control during phage development. However, investigation of the physiological genetics of phage T4 by means of *ts* and *am* mutants not only preceded that of phage λ but has been much more extensive. Some 70 genes have been defined by complementation tests, involving over 400 mutants, and mapped, and their functions determined by observing the occurrence or absence of DNA synthesis, production of tail fibre antigens, lysis, and the appearance of various structural phage components during growth of the mutants in the restricting host. A summary of the main results are presented in Fig. 104. As in the case of phage λ, there are two major functional classes of gene, one concerned with early functions which are primarily related to DNA synthesis, and the other with the production of structural components, maturation and lysis. There is rather a marked tendency for genes determining related functions to map in clusters, while those responsible for early and late functions as a whole are separated on different segments of the map. For instance, a great majority of maturation genes (black rectangles, Fig. 104) are located together on the right hand side of the map, while within this class all the genes determining head (capsid) morphology (except 31), as well as groups of tail and sheath genes, are clustered. Similarly, most of the genes concerned with early functions (DNA synthesis) (open rectangles, Fig. 104), lie together on the left of the map. However, this functional distribution is far from perfect. Thus gene 31, which provides a factor which solubilises the head protein subunits, and a group of genes (34–38) determining tail fibres, lie between DNA genes, as do genes 49 and *e*, concerned with sheath and lysozyme (endolysin) production respectively. It is noteworthy that while *ts* mutations in structure-determining genes, such as gene 23 which

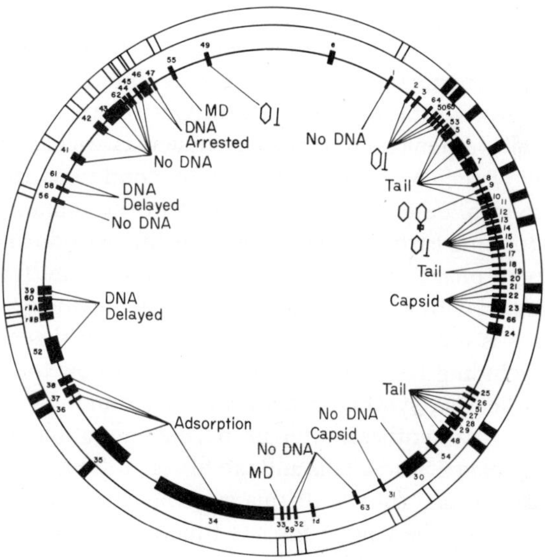

Fig. 104. The circular linkage map of phage T4, showing the functions served by the various genes, as determined by recombination, complementation and physiological analysis of conditional lethal mutants.

The inner circle shows the arrangement of the genes on the linkage map, the circumferential length of the black rectangles representing the length of the genes as estimated by recombination frequencies: the reason for the relatively large apparent length of some genes, such as No 34, is not known.

The functions of the various genes are indicated: CAPSID = head structure; MD = maturation delayed; *e* = endolysin (lysozyme); the diagramatic drawings of structural components show the products found in lysates of the restrictive host, in cases where the actual function of the gene is unknown; the $r_{II}A$ and *B* genes are located between genes 52 and 60.

On the outer circle, the white rectangles indicate genes determining early functions, and the black rectangles late functions: the criterion of early and late functions is the time of major m–RNA synthesis as judged by reversal of phenotype by 5-fluorouracil (5FU) (see Fig. 87, p. 370).
Kindly provided by Dr. R. H. Epstein.

specifies head protein, may permit the formation of morphologically recognisable components, these are rarely found in the case of *am* mutations in the same genes; while *ts* mutations may yield merely faulty protein, *am* mutations, being chain-terminating, produce no protein at all (Epstein *et al.*, 1963).

We have already described some of the morphogenetic studies based on these *am* and *ts* mutations of phage T4, and especially those involving formation of the phage head (p. 433; Epstein *et al.*, 1963; Kellenberger, 1965, 1966) and will not reiterate them here, but will go on to discuss the regulation of chromosome activity. The principal feature of this regulation is that, in cells infected with mutants defective in DNA synthesis, the genes responsible for the synthesis of the structural components of mature particles, as well as of lysozyme, fail to be activated. Moreover under these conditions the synthesis of early enzymes, other than that blocked in the mutant, continues instead of stopping early in development, as happens during infection with wild type phage or with mutants blocked in late functions. In bacteria infected with a particular *am* mutant, involving gene 39, there is a lag which is followed by a net DNA increase; in this case the late functions, including lysis, are not inhibited but show the same delay in onset as the DNA synthesis; again, an *am* mutation in gene 46, which arrests DNA synthesis after its normal initiation, permits the formation of normal heads, tails and tail fibres, although lysis does not occur (Wiberg *et al.*, 1962). All this suggests that some step in the synthesis of DNA is the trigger for late functions, but the situation is probably more complicated than this since there are at least two genes (33 and 55) in which mutations allow normal DNA synthesis but stop the manifestation of all late functions. Thus DNA synthesis is not by itself a sufficient trigger (Epstein *et al.*, 1963; Levinthal, Hosoda and Shub, 1967).

So far we have talked about early and late functions as if all the proteins mediating each class behaved in an identical fashion. The rate and duration of synthesis of various phage proteins can be followed by exposing infected bacteria to short pulses of radioactive aminoacids (or ^{35}S) at frequent intervals following infection; extracts of the bacteria are immediately made, the proteins separated by high resolution gel electrophoresis and those that have taken up the radioactivity revealed by autoradiography, the degree of darkening of the film reflecting the rate of synthesis. The particular proteins corresponding to the various bands after electrophoresis can be identified by seeing which band is missing when different *am* or *ts* mutants are tested in a restrictive host (Cohen and Levine, 1966). Using this novel method, at least four different patterns of protein synthesis have been revealed in phage T4. One of these is that of the late proteins, the synthesis of all of which begins at the same time and continues until lysis. Three classes of early protein can be distinguished; synthesis of two of these begins immediately upon infection, stopping shortly afterwards in one class but, in the other, being prolonged until about half way through the latent period; in the third class, initiation of synthesis is delayed for a few minutes and stops shortly after the production of late proteins begins (Levinthal *et al.*, 1967).

If, after adding the radioactive pulse in this method, the culture is flushed with a 'chaser' of cold amino acid so that no more radioactivity is incorporated, and incubation is then continued for some time before the extract is made for electrophoresis, the radioactive label may be found to have remained in its band or, alternatively, to have disappeared. The latter case could be due to instability of the protein which had incorporated the label, or to its having become part of a larger structure with a different mobility in the gel electrophoresis. These two possibilities may be distinguished by the use of amber mutants involving, for example, the head protein of phage T4, the subunit of which is determined by gene 23. It turns out that the radioactivity can be 'chased' from the hand corresponding to this protein when mutants making normal heads are used, but not when the phage carries mutations in genes necessary for the formation of intact head structures. Since genes 22 and, to some extent, 24 make products which can be chased from their bands in the same way as the head protein subunit, it may be inferred that these products are minor structural components of the head (see p. 433; Levinthal *et al.*, 1967).

The use of this chasing method also enables functional relationships between genes to be traced. For instance, there is a chaseable band which disappears when a mutation in either gene 10 or 11 is present, so that these two genes cooperate in forming a structural component; but in the presence of a mutation in genes 7 or 8, the band persists after chasing. Presumably the band consists of an intermediate component destined to become part of the tail structure (Levinthal *et al.*, 1967).

Two hypotheses have been suggested to explain the switch from the synthesis of early to late proteins. One is that, in the absence of

DNA synthesis, there are not enough gene copies to support late protein synthesis but, as we have seen, certain mutations can block late functions even though DNA synthesis is normal; moreover it was shown many years ago that the rate of protein synthesis remains the same throughout the lytic cycle (Koch and Hershey, 1959). The second hypothesis is that late protein synthesis is effected by means of relatively stable m–RNA molecules which can monopolise the ribosomes when the labile m–RNA molecules serving early functions have decayed. Against this hypothesis it has been shown experimentally, in the case of a virulent *B. subtilis* phage which also produces early and late proteins, that both are translated from m–RNA molecules of equally short life (Levinthal *et al*, 1967).

We mentioned earlier that one of the first manifestations of infection by the T-even phages is break-down of the host DNA and the arrest of host protein synthesis (p. 412), and it might be thought that the second of these effects was the result of the first. This is not so, for mutations in genes 46 and 47 prevent host DNA degradation (Wiberg, 1966) but not the shutting down of protein synthesis which occurs after infection by all amber mutants of phage T4 so far examined (Levinthal *et al.*, 1967).

Reciprocal recombinants

We have seen, from tetrad analysis in higher organisms, that a single recombination event always results in the formation of two reciprocally recombinant chromosomes except, sometimes, when recombinants for very closely linked sites are selected (p. 392). We have also noted that reciprocal recombinants appear in approximately equal numbers among the progeny of phage crosses. This, of course, means nothing in terms of individual acts of recombination. There is simply a statistical equivalence between the numbers of the two recombinant types as, indeed, is the case for all organisms in which tetrad analysis is not possible (p. 68). This equivalence could arise because both types always stem from a single event, that is, true reciprocality, or because each type has an equal probability of emerging from independent events. The distinction between these two possibilities has an obvious bearing on the mechanism of recombination in phage.

The outcome of a phage cross is normally based on the progeny liberated from a large number of infected bacteria. If true reciprocality exists, equal numbers of reciprocal recombinants should also be found among the particles issuing from *single*, mixedly infected bacteria, but this is usually not the case. In single burst experiments involving markers which are not closely linked, so that the number of recombinants is large, the two recombinant types are generally present but in unequal proportions. The significance of this is not easy to assess because, as we shall see, phage recombination occurs in a pool of replicating genomes, each of which may participate in more than one recombination event. Thus reciprocal recombinants formed early in the latent period are likely to have their initial genotypes obliterated by subsequent rounds of recombination.

On the other hand, when recombinants for closely linked markers are scored in single burst experiments, many of the bursts are found to contain only one of the two reciprocal types (Hershey and Rotman, 1949). In this case the close linkage precludes the likelihood of more than one round of recombination involving the region, so that we are usually looking at the outcome of a single event. But another factor begins to operate here. You will remember that, during the latent period, the growing DNA pool is constantly depleted by the random withdrawal of genomes for incorporation into mature particles and that, at the time of bursting, half of the total DNA remains in the pool and never appears in the progeny phage (Fig. 99, p. 425). This means that each phage genome has a probability of 0·5 of representation among the progeny. Thus when the number of recombinants for a given region is very small, we may expect that only one type will often be found among the progeny particles, even though the two reciprocal types of genome were present in the pool.

The assessment of reciprocality is therefore statistical. From this point of view the evidence is rather convincing, from the study of closely linked markers in phage T2 (Hershey and Rotman, 1949) as well as in the case of phage T1 where the overall frequency of recombination is inherently low (Bresch, 1955), that reciprocal recombinant types do not issue from a single recombination event. For example, the correlation between reciprocal types is found to be no greater than that observed between pairs of other recombinants which bear no relation to one another so far as single events are concerned (Bresch, 1955). On the other hand, strong reciprocity is found in recombination between prophages in partially diploid bacteria where the causes of distortion we have described for vegetative recombination do not operate (Meselson, 1967).

Phage heterozygotes

Wild type (r^+) T-even phages produce small plaques with a turbid halo while r mutant particles yield larger, clearer plaques. If bacteria mixedly infected with r^+ and r particles are plated with indicator bacteria *before* they lyse, the plaques which arise are formed by a mixture of the two parental types of particle which emerge from the burst. These plaques have a *mottled* appearance, being composed of zones of lysis inhibition interspersed with zones of clearing. Hershey and Chase (1951) discovered that about 2 per cent of the plaques which arise when the *individual* progeny particles from an $r^+ \times r$ cross are plated, have this mottled appearance (Plate 22). When the particles contained in these mottled plaques are plated in turn, they give rise to about equal numbers of pure r^+ and r plaques and, again, about 2 per cent of mottled plaques. Since the mottled plaques are initiated by infection with single phage particles, these particles must carry both the r^+ and r alleles and are therefore *heterozygotes* for the r region. Heterozygotes are unstable since, on replication in the cells they infect, the alleles segregate to yield progeny of the two parental types and, again, a small proportion of newly formed heterozygotes.

The essential features of heterozygotes, most of which were described for phage T2 by Hershey and Chase (1951), are as follows:

1. Heterozygotes are not restricted to any particular pair of alleles, but are found to arise with the same frequency with respect to every locus examined.

2. When crosses are made between phages which differ by *two* markers, as in the general case of $++$ (wild type) $\times$ ab (double mutant) crosses where heterozygotes for the b locus yield mottled plaques, the constitution of the initial heterozygote can be inferred by analysis of the progeny particles in the plaque. Two types of heterozygote are possible. One is heterozygous for both loci, $\dfrac{+\,+}{a\;\,b}$, and will segregate the two parental types on replication and so is called a *non-recombinant heterozygote*. However, during development of the mottled plaque the two parental segregants go through many growth cycles together so that the situation is equivalent to a mixed infection with the parental types. The result is that the mottled plaque will contain the two parental types and, in addition, some recombinant progeny ($+b$ and $a+$) whose frequency will depend on the distance between the

markers. The alternative possibility is that the initial heterozygote is homozygous at the *a* locus $\left(\frac{++}{+b} \text{ or } \frac{a+}{ab}\right)$; it is a *recombinant heterozygote.* Observe that in this case the segregants of the heterozygote (++ and +*b*, or *a*+ and *ab*) differ only at the single locus selected for heterozygosity (*b*), so that the population of particles in the mottled plaque can comprise only two phage types, one parental and one recombinant with respect to the original cross; the other parental and recombinant classes are excluded. Non-recombinant and recombinant heterozygotes can thus be clearly distinguished on the basis of whether the particles in the mottled plaque comprise a majority of both parental types, or of only one parental and one recombinant type, but it must be emphasised that their structure can never be examined directly but only deduced from the data emerging from a long chain of subsequent events (see Steinberg and Edgar, 1962). It turns out that the type of heterozygote produced depends very much on the distance between the two markers. When the markers are very closely linked, non- recombinant heterozygotes $\left(\frac{++}{a\ b}\right)$ predominate or, to put it another way, particles which are heterozygous at a particular locus are also heterozygous with respect to very closely linked loci, that is, they are *double heterozygotes.* On the other hand, if the markers are located at some distance from each other, recombinant heterozygotes are the rule (Hershey and Chase, 1951).

3. By scoring the frequency or occurrence of double heterozygotes as a function of the distance between two markers, the average length of the heterozygous region can be estimated. This appears to extend over a few cistrons, which is the same order of distance over which high negative interference is found (p. 380; see Steinberg and Edgar, 1962; Chase and Doermann, 1958; Barricelli and Doermann, 1960).

4. When crosses are made between phages which differ by *three* markers (3-factor crosses of the type $+++ \times abc$) which are not very closely linked, particles heterozygous for the middle marker $\left(\frac{+}{b}\right)$ appear to be usually recombinant for the outside markers; that is, they behave as if their structure was $\frac{++c}{+b\ c}$ or, alternatively, $\frac{a++}{a\ b\ +}$. This has been found in the case both of phage T2 (Levinthal, 1954) and of phage T1 (Trautner, 1958) and strongly

suggests that recombinants and heterozygotes are produced by the same mechanism.

5. We will see in the next section that the frequency of recombination between two markers, especially when these are closely linked, has been shown to increase with the total amount of DNA and proportionately to the burst size. Thus the frequency is lowest in premature lysates and highest when lysis is delayed. In contrast, the frequency with which heterozygotes are found remains constant under these different conditions. The inference is that heterozygote structures are continually being formed in the mating pool but are lost at the same rate as a result of segregation. Assuming that these heterozygotes segregate, by replication, at the same rate as the normal population of genomes replicate, it has been calculated that they can account for most, if not all, of the recombinants that emerge from the cross (Levinthal, 1954). This is another way of saying that heterozygotes may represent a normal stage in the production of recombinants, and that those we observe among the progeny particles are a proportion which are trapped before they can segregate, as a result of random withdrawal from the pool and incorporation into protein coats.

Until recently it was assumed that all phage heterozygotes, whatever their genetical characteristics, possess a similar physical structure. The most popular of the proposed structures was Levinthal's (1954) *heteroduplex* model, in which the heterozygous region comprises a normal DNA duplex of which the two strands (or the information carried by them) are derived from different parents. We have already considered such a heteroduplex structure as a likely intermediate in a region involved in recombination in general, and have seen how it may be recombinant for outside markers and segregates into two homozygous structures at the first semi-conservative replication (Fig. 95; pp. 397–400). An alternative model embodied a truly diploid heterozygous region due to the presence in the phage head of double helices from both parents. This was suggested by Steinberg and Edgar (1962) because they found, in crosses between multiple r_{II} mutants of the type $\frac{+\quad +r+\quad +}{a\quad r+r\quad c}$, where *double* recombination is necessary in the r_{II} region to yield $+++$ and *rrr* progeny and mottled plaques, that the heterozygotes were frequently not recombinant for outside markers, just as one would expect in classical meiosis.

Theoretically, these models should be distinguishable by looking at the density distribution of heterozygous particles in a phage population; the density of particles containing heteroduplex DNA and normal DNA should be the same, whereas heterozygous particles containing duplications of the genetic material should be denser than normal. Experiments yielded conflicting results. Although phage λ displays a very low frequency of recombination and a negligable proportion of heterozygotes (10^{-4}), this proportion can be raised to about 2 per cent by UV-irradiation of the host bacteria, the region of heterozygosity extending over about one tenth of the genome. Under these conditions the density distribution of heterozygotes was found to be the same as that of normal phage particles after centrifugation in a caesium chloride density gradient (Kellenberger, Zichichi and Epstein, 1962). By contrast, not only have heterozygous particles of phage T4 been shown to possess higher than average density, but the greater the density the longer is the region of heterozygosity (Doermann and Boehner, 1964).

As often happens when evidence appears to support both sides on controversial issues, both sides turn out to be right. There are two basically different types of heterozygote so far as phages T2 and T4 are concerned. One corresponds to the heteroduplex model and is called an *internal* heterozygote. It represents an intermediate stage in recombination which would normally disappear at the first replication, but is trapped by maturation of the phage head. Internal heterozygotes can be recognised by two criteria. First, if DNA synthesis is prevented, by adding fluorodeoxy-uridine (FUDR) to the infected cells for example, the loss of heterozygotes by replication is prevented so that they accumulate and increase in frequency (Séchaud *et al.*, 1965). Second, the formation of this type of heterozygote is prevented by the presence of a deletion in the region, probably due to interference with the strand complementarity of the heteroduplex (Nomura and Benzer, 1961). In fact, the magnitude of 'frame-shift' mutations, involving the addition or deletion of small numbers of bases, can be measured by the extent of their effect on the frequency of internal heterozygotes (Drake, 1966).

The second type of heterozygote derives from terminal redundancy (pp. 433, 539) when the duplicated regions at the two ends of the chromosome come from different parents. They are the outcome of completed recombination events and so are recombinant for neighbouring markers lying outside the diploid region. Thus heterozygotes from a cross between *abc* and $a^+b^+c^+$, where the *b* region is redundant, will be expected to be either bc——a^+b^+ or b^+c^+——ab, both of which are recombinant for the markers *a* and *c* adjacent to the two ends. Since heterozygotes of this type arise and are lost only through recombination, their frequency remains unaffected by inhibition of DNA synthesis; moreover, the presence of deletions in the region does not prevent their formation.

The distinction between the two types of heterozygote is very apparent if the effect of treatment with FUDR on the percentage of heterozygotes is compared for crosses between point mutants on the one hand and, on the other, between mutants having deletions in the same region, as shown in Table 19. In the deletion-mutant cross only

TABLE 19

The distinction between heteroduplex and redundancy heterozygotes in phage T4.

Type of cross	FUDR treatment	Percentage heterozygotes among wild type recombinants	Type of heterozygotes formed
Deletion X deletion	−	5·8	Redundancy only
	+	5·8	
Point mutant X point mutant	−	16	Redundancy + heteroduplex
	+	63	

Data from Shalitin and Stahl, 1965.

redundancy heterozygotes arise and these are unaffected by FUDR treatment. In the cross between the point mutants, both types of heterozygote are formed but the increased frequency resulting from FUDR treatment involves only the heteroduplex variety and therefore is considerably greater than the figures would appear to show (Séchaud *et al.*, 1965; Shalitin and Stahl, 1965; Stahl *et al.*, 1965). Since the chromosome of phage λ does not have a terminal redundancy, presumably only heteroduplex heterozygotes are formed, the large increase in their number following UV irradiation being a result both of the stimulation of recombination (Roman and Jacob, 1958) as well as of inhibition of DNA synthesis (p. 329).

THE KINETICS OF MATING AND RECOMBINATION

So far we have treated phage crosses as if they were no different in principle from crosses in other organisms, even though we have run into difficulties concerning reciprocal recombinants and heterozygotes. Nevertheless, such an approach is sufficient to establish the existence of a continuous, linear linkage group as well as the approximate, relative distances between genes. Among the earliest analyses of phage crosses, however, a number of quantitative anomalies were observed which were at variance with orthodox genetical theory. We will give a brief account of some of these anomalies and show that they can be explained by supposing that mating occurs randomly in a replicating pool of phage genomes wherein recombinants, as well as parental types, can multiply and participate in successive acts of recombination. Genomes are withdrawn from this pool, randomly and at a constant rate, for maturation and thereafter become genetically and physiologically inert. Observe that this formulation of the genetic behaviour of phage is consistent with the physiological picture of phage reproduction which we have discussed on pp. 424–435.

Recombination in phage thus becomes a complicated problem in *population genetics* which is best approached through algebraic analysis, especially in the case of the more informative 3-factor crosses which can yield eight different genotypes among the progeny. Such an analysis has been made by Visconti and Delbrück (1953; Visconti, 1966), whose paper is well worth reading for its clear definition of the problem alone, and although some of the assumptions on which the analysis is based are not necessarily true, the general findings retain a high measure of validity (see p. 62). We will not deal at all with mathematical considerations here, but will try to give an account of the main conclusions in general terms.

Before discussing some of the specific anomalies of phage crosses, it will be well to consider the genetic pool in broad terms, from the point of view of population growth. We have seen that when the DNA pool starts to grow and reaches about 50 phage equivalents in size, mature phage particles begin to be formed and thereafter increase linearly at the same rate as the total DNA (p. 424 and Fig. 99). Moreover, DNA withdrawn from the pool by maturation is not returned to it. From the chemical point of view, the failure of the DNA to increase exponentially was assumed to be due to the withdrawal of phage genomes from it at the same rate as they were formed by replication. This has been confirmed at the genetic level by a statistical analysis of the distribution of phage mutants in single burst experiments, equivalent to the fluctuation test in the case of bacterial mutants (p. 181). The mutants showed a clonal distribution which was consistent only with their production, with a uniform probability, and subsequent multiplication in an exponentially growing population of genomes (Luria, 1951; see also Levinthal, 1959).

An elegant demonstration of the clonal distribution of mutants in the temperate phage λ came from study of the back-mutation to wild type of defective prophages in lysogenised bacteria. In some defective strains the initial mutation prevents DNA synthesis altogether. In the case of such strains the only bacteria which can liberate phage following UV induction are those in which *the prophage itself* has reverted to wild type, so that all the bursts should be of the same size, yielding a Poisson distribution of particles from single bursts. With another type of defective phage, normal replication occurs after induction but mature phage cannot be formed because the defect involves maturation (p. 430); in this case reversions to wild type should be possible at any time during vegetative growth so that clones of variable size should be found among single bursts. The experimental findings accord perfectly with these predictions (Jacob, Fuerst and Wollman, 1957). Thus the phage genetic material is identified with the replicating DNA pool.

On the other hand, the distribution of *recombinants* among single bursts of mixedly infected bacteria is partly random (Poisson distribution) and partly clonal, and varies with the degree of linkage between the markers; for very closely linked markers the recombinants are almost randomly distributed among the progeny from individual bacteria so that each recombinant appears to have arisen from an independent event (Hershey and Rotman, 1949). This apparent disparity between mutation and recombination, both random events which directly involve the genetic material of the phage, can be reconciled with the kinetics of DNA synthesis and maturation in the following way. Mutations, although rare, arise as a function of the size of a population; since each genome has an equal probability of mutation per generation, the larger the population the greater the chance that a mutation will arise in it. From about the middle of the latent period onwards the size of the DNA pool remains constant, due to maturation occurring at the same rate as replication. Any mutation arising during this period has, therefore, an equal probability of withdrawal from the pool or of replicating, so that clone size will be severely restricted and the distribution of mutants will appear part uniform and part clonal. On the other hand, during the period *before* infective phage particles begin to appear, the DNA pool has been increasing exponentially, by as much as 50-fold, so that mutants arising during this period can multiply freely to form large clones, with the result that the overall distribution is clonal.

A rather different situation is found when we look at recombination in terms of population growth, due to the fact that recombination depends on the *interaction of two genomes*. Of several factors which may operate here, probably the most important is the need for the two pools to make physical contact before recombination can occur; in an extreme case, where the parental genomes start to replicate at opposite poles of the infected bacterium, the initiation of recombination may be greatly delayed (see *Topographical Aspects of Recombination*, p. 507). This delayed onset of recombination means that the opportunity for early recombinants to multiply prior to maturation is much less than in the case of mutation, so that a clonal distribution is less evident. However, the fact that some clonal distribution does exist implies that some recombinant genomes *do* multiply prior to incorporation into mature particles. If recombination were a terminal event preceding maturation, each recombinant would be isolated as soon as it was formed so that an exclusively Poisson distribution would result. A similar distribution would be found if recombinant formation always preceded the onset of replication. The smaller clonal distribution of recombinants for closely linked markers is a consequence of their rarity—they are less likely to arise early in the genetic pool. Moreover, as we shall see, the clonal distribution of recombinants for more distant loci tends to be ironed out by multiple rounds of mating.

There is more direct evidence of a general nature that recombinants may not only multiply but may themselves engage in further matings. For example, in 2-factor crosses where one parent is in great excess of the other, it is found that, among the progeny, the number of recombinants may exceed that of the minority parental type (Doermann and Hill, 1953; Visconti and Delbrück, 1953). In such a case some of the alleles of the minority parent must have increased in number after they had been transferred to recombinants. Again, *triparental crosses* can be set up in which the bacteria are infected with three parental types of phage, $a++$, $+b+$ and $++c$, each of which carries a different marker. From crosses of this kind recombinants of type *abc* may be isolated (Hershey and Rotman, 1949). While such recombinants could arise from a single mating event involving three particles (*group mating*; see Hershey, 1958; Hausmann and Bresch, 1960), the result strongly suggests the alternative possibility that they are formed by successive matings. Thus a primary mating of $a++$ and $+b+$ genomes would yield an $ab+$ recombinant which, in a subsequent mating with a $++c$ genome would generate the observed *abc* type. We will now turn to two anomalies of phage crosses which support this contention.

Negative interference

We have previously defined negative interference as an excess of recombination in two adjacent intervals, that is, of double recombination, over that expected on the basis of random exchanges. So far, however, we have dealt only with a particular variety of this phenomenon termed *high, localised negative interference* which is restricted to very small chromosomal regions and can be explained by the hypothesis that pairing is discontinuous and confined to these regions, within which the probability of recombination is high (p. 379 *et seq.*). This kind of negative interference is also found in phage crosses (Chase and Doermann, 1958; Barricelli and Doermann, 1960) and can be interpreted in the same way. We are here concerned with a fundamentally different phenomenon since the excess of doubles is not very high, while much longer regions of the chromosome are involved.

There are two reasons for the excess of double events in this latter sort of negative interference. The first is due to the random nature of mating. If a bacterium is infected by two parental types of phage at equal multiplicity, each genome has the same probability, in any mating event, of pairing with one of its own kind instead of with a genome of the other parent. As a result only half the population can beget recombinants and we should properly restrict our estimate of the expected number of double recombinants to this half. Suppose, on this basis, that the true probability of a single recombination event in each of two intervals is p_1 and p_2, then the actual expected probability with which doubles will arise is $p_1 \times p_2$. But *among the population as a whole* the *observed* proportion of the single events is reduced by a factor 2, so that the probabilities of their occurrence are estimated as $\frac{1}{2}p_1$ and

$\frac{1}{2}p_2$, and the probability of doubles as $\frac{1}{4}(p_1 \times p_2)$. This means that when recombination frequencies are assessed in terms of the entire progeny of the cross, the expected number of double events is automatically calculated as only one-half of the number which actually arises from heterozygous matings; in other words, the random nature of mating in phage carries a built-in, but spurious, negative interference which can easily be compensated for.

Another factor leading to an apparent excess of double recombination in phage crosses is the participation of recombinants in further mating events. Two successive single exchanges, arising from different matings, will appear as a double exchange in the recombinant particle. For instance, in a 3-factor cross of the type $\frac{+\,+\,+}{a\;\;b\;\;c}$, a single recombination event in the region *a—b* will produce a recombinant of genotype, say, $+bc$. If this recombinant now mates again with another $+++$ parent in the pool $\left(\frac{+\,+\,+}{\times\;\;b\;\;c}\right)$, a single exchange in the *b—c* interval yields a $+b+$ recombinant which will be scored as a double among the progeny, or as parental if the outside markers only are scored. Moreover, each double recombinant produced in this way arises at the expense of a single recombinant. If we were to assume that such recombinants arose from single mating events, it is obvious that the number of doubles would exceed the expected number and we would say that negative interference was present. Note that this sort of negative interference, unlike the localised variety, should become greater with distance since the further two markers are apart the greater is the probability that the interval will be involved in two or more successive acts of recombination.

Because the effect of negative interference is to reduce the recombination frequency and, therefore, the apparent distance between two loci, since double exchanges in the interval regenerate the parental genotype, linkage values diverge from the additivity expected of a linear arrangement. In their mathematical analysis of the kinetics of the mating process in growing populations of vegetative phage, Visconti and Delbrück (1953) assumed that intact phage genomes mated in pairs, and randomly and repeatedly with respect to partners, in the DNA pool. When the effect of repeated matings was thus compensated for in the experimental results, negative interference disappeared and the recombination values per mating became strictly additive (see also Doermann and Hill, 1953).

Drift to genetic equilibrium

Another consequence of repeated rounds of random mating in the replicating pool of vegetative phage, is that the percentage of recombinants, with respect to two markers, should increase throughout the latent period. If the latent period could be extended indefinitely, and assuming no selective pressures, an equilibrium would be reached at which half the population would be recombinant and half parental. We have already studied a good example of this *drift to genetic equilibrium* in the case of mutant increase in exponentially growing bacterial cultures (p. 199 *et seq.*) and the phenomenon is a well known one in the population genetics of higher organisms.

Let us look more closely at the genetic pool at the time when the genomes of the two parental types begin to mix. If the two are present in equal proportion, each genome has the same chance of mating with another genome of the same (homozygous) or of different (heterozygous) genotype. Only in the latter event can a recombinant arise, so that the maximum probability of recombination between distant (unlinked) loci *at this first round of mating* is limited to 25 per cent instead of the usual 50 per cent. Both the recombinant and the parental genomes replicate and enter another round of mating, so that those genomes which were previously homozygously mated now have another chance of enjoying a heterozygous mating. It is evident that, after a few rounds of mating, the great majority of genomes will have found a partner of the opposite type in at least one mating event, so that the frequency of recombination for unlinked markers will approach the maximum of 50 per cent. However, due to the fact that, at each generation, every genome has an even chance of being removed from the mating pool by the process of maturation, equilibrium sets in at about 40 per cent and the theoretical maximum is never reached.

This drift to equilibrium can be studied experimentally by estimating the recombination frequency among populations of particles obtained from premature lysates at various times throughout the latent period (p. 424). In addition, the latent period of phages T2 and T4 can be extended from 25 minutes to about 2 hours, during most of which phage replication continues, by means of lysis inhibition (p. 415). The recombination frequency for unlinked markers is found to rise from 32 per cent at the time when infective particles first appear to 42 per cent at the time of normal lysis (Doermann, 1953). The initial figure of 32 per cent suggests that more than one round of recombination precedes the onset of maturation.

In the case of linked markers we should anticipate a proportionately greater increase for two reasons. Firstly, due to the lower probability of recombination, equilibrium will not be reached so soon; secondly, as the interval between two loci becomes smaller, the lower is the probability that it will be involved in another recombination event that obliterates the result of the first. The expectation is realised in practice. Using a pair of closely linked markers, and lysis inhibition to prolong the latent period, Levinthal and Visconti (1953) reported a recombination frequency of 2 per cent at 20 minutes and of 9 per cent at 80 minutes after infection, an increase of more than 4-fold which exactly paralleled the increase in average burst size over the same period. In the case of extremely closely linked markers, such as adjacent mutant sites in the same r_{II} cistron, as much as a 20-fold increase in recombination frequency (0·02 to 0·4 per cent) may be found under conditions of lysis inhibition (Symonds, 1962a). It is obvious that, in phage crosses, an isolated statement of recombination frequencies has little meaning.

Topographical aspects of recombination

We said earlier (p. 503) that the difference between the distributions of mutant and recombinant phages in single burst experiments was mainly due to delay in the initiation of recombination imposed by the need for the two parental pools to interact. It is reasonable to think that if two parental particles inject their DNA into *contiguous* regions of the bacterium, the expanding DNA pools will make contact sooner, and give rise to more recombinants, than if the sites of injection are far apart. Two kinds of evidence favour such a topographical influence. Firstly, increase in the multiplicity of infection should enhance the likelihood of contiguous injection and it is indeed found, in the case of phage T1, that the recombination frequency rises with increasing multiplicity (Trautner, 1960).

More subtle evidence comes from the study of heterozygotes. You will recollect that when the two parents of a cross are characterised by an r^+ and an r marker, as well as by alleles of a closely linked gene, a small proportion of the progeny yield mottled plaques which arise from infection by single, heterozygous particles which segregate the two parental types of genome (p. 496). The mottled plaques thus contain particles which are recombinant for the two markers in addition to parental particles. Infection by a non-recombinant heterozygote differs from ordinary mixed infection in that the two parental genomes (or their immediate precursors) are physically adjacent when they are injected. It turns out that the recombination frequency obtained from infection by such heterozygotes is markedly higher than that resulting from mixed infection by the parental phages (see Steinberg and Edgar, 1962).

A HETERODOX MODEL OF RECOMBINATION

In this section we have looked at recombination in phage in terms of a rather formal system of population genetics based on classical principles, and have explained various peculiarities and anomalies which have been found as resulting from the special situation in which these principles operate. Although, of course, it is very probable that orthodox genetic exchange is the principal mechanism of recombination in phage, as the existence of heteroduplex heterozygotes suggests, nevertheless the formation of terminal redundancy heterozygotes in the T-even phages hints at the possibility of a quite different and unique form of recombination. We have seen that the most likely current hypothesis to explain terminal redundancy is that lengths of chromosome longer than one genome are cut off from a linear concatenate of chromosomes in the DNA pool during incorporation into the phage head (pp. 433–434). Following mixed infection, it is likely that concatenates would comprise chromosomes of both parental types, built up by successive recombination events between permutations of a circular chromosome, thus:

(1) A B C D E F G H I J A B

g h i j a b c d e f g h

(2) A B C D E F G H I j a b c d e f g h

E F G H I J A B C D E F

(3) A B C D E F G H I j a b|c d e f g H I J A B C D|E F

Note that recombination in this system will rarely produce viable, reciprocal recombinants. When such concatenates as are shown in the bottom row (3) are chopped up into chromosomes of greater length than one genome (as indicated by the vertical lines), redundancy heterozygotes will be formed which are also recombinant for outside markers.

If recombination of this kind takes place it should, of course, be restricted to those phages whose chromosomes show terminal redundancies and which yield circular linkage maps, and may account, at least in part, for the exceptionally high recombination frequencies which characterise the T-even phages. Thus according to the analysis of Visconti and Delbrück (1953), phage T2 undergoes some five rounds of replication, the mean generation time being about two minutes, and these are accompanied by about five rounds of mating if recombination is by breakage and reunion (see Levinthal, 1959). In contrast, the frequency of recombination in phage T1 corresponds to about one round of mating (Bresch and Trautner, 1955), and in phage λ to 0·5 rounds (Kaiser, 1955), although the burst size of both is comparable to that of phage T2.

GENETIC ANALYSIS OF OTHER PHAGES

We have recounted in some detail the genetic and functional analyses of phages T4 and λ as representatives of virulent and temperate viruses. Until comparatively recently, little information was available about the genetics of other phages, due partly to the concentration of research on T4 and λ, and partly to the fact that some other phages proved hard to study because of the difficulty of obtaining clearcut and stable host-range and plaque-type mutants, or of the low frequency of recombination.

However, during the last few years the ease of isolation and wide chromosomal distribution of temperature-sensitive and amber mutants has revolutionized the study of recombination and physiological genetics in many phages which had previously proved difficult to investigate. Moreover, the use of these mutants, especially of the 'non-leaky' amber variety, in complementation and physiological tests alone, permitted genetic analysis of RNA phages, for example, in which genetic recombination has not been demonstrated. Thus conditional lethal mutants of the small spherical, *E. coli* male-specific, RNA phage, f2, (p. 440), whose genome can code for no more than a few protein species, have been shown to fall into a number of different complementing and functional groups. The mutations of one group block RNA synthesis and, as in the case of early mutants of DNA phages (pp. 474, 492), this is associated with absence of expression of late functions so that neither phage RNA nor coat antigen are produced in restrictive hosts. The mutations of a second group prevent synthesis of coat protein with the result that, although no particles are found, phage

RNA which is infective for protoplasts can be shown to be synthesised. A third group of mutations involves a late function other than coat protein synthesis and yields antigenic but defective particles (Zinder and Cooper, 1964; Valentine, Englehardt and Zinder, 1964; Englehardt and Zinder, 1964; Notani *et al.*, 1965; Horiuchi, Lodish and Zinder, 1966).

Similarly, although recombination has not yet been reported in the case of the filamentous, male-specific, single-stranded DNA phage, M13 (p. 444), six genes have been identified by complementation tests involving 63 *ts* and 33 *am* mutants. The time at which a particular gene begins to act can be ascertained by transferring a temperature-sensitive mutant involving it from the low to the high temperature at intervals during the latent period, and looking to see when infectious progeny are produced. By means of such temperature-shift experiment, four of five genes in which *ts* mutations were available were found to be concerned with late functions and one with an early function (Pratt *et al.*, 1966).

The occurrence of recombination has been known for a long time in the spherical, single-stranded DNA phage ϕX174 (p. 436; Pfeiffer, 1961) as well as in its close relative S13 (Tessman and Tessman, 1959; see also I. Tessman *et al.*, 1964; Howard and I. Tessman, 1964 a, b). It is only recently, however, that the isolation of *ts* and *am* conditional lethal mutants of these phages has initiated study of their physiological genetics. Four complementation groups concerned with host-range, adsorbability and ability to lyse have been indentified in phage S13 and confirmed by recombination analysis (E. S. Tessman, 1965). In the case of phage ϕX174 it has been shown that lysis is due to a chloramphenicol-sensitive synthesis occurring later in the latent period than synthesis of the coat protein; however, it is dependent on DNA synthesis, but only during the early stage of infection when the replicative form (RF) is being made (p. 439). Non-complementing *ts* mutants affecting the synthesis of coat protein, but which do not prevent lysis, have also been isolated (Markert and Zillig, 1965).

We have confined our brief remarks on the genetic analysis of phages other than T4 and λ to the small RNA and single-stranded DNA viruses, not only to complement the intensive biochemical and biophysical studies of the replication of their genetic material which we described in Chapter 16 (pp. 436–446), but because the limited information which their genomes can carry encourages the hope that a full understanding of their development and its control may soon be possible. With regard to other, more classical phages such as T1, T5, P2, P22, suffice it to say that although genetical studies utilising conditional lethal mutants are in progress, no new findings of importance have been published at the time of writing, apart from an unusual feature of the chromosome structure of phage T5 which we will comment on in the next chapter (p. 542).

PHENOTYPIC MIXING

This phenomenon, illustrated diagrammatically in Fig. 105, is seen among the progeny of a phage cross when the parental phages are distinguished by the different specificity of some component of their protein coats. The two parental components are synthesised together in the protein pool but, at maturation, are unable to recognise their own parental type of DNA so that genomes and protein coats become indiscriminately associated in the progeny particles. Thus, statistically, one-half of the particles will acquire corresponding genomes and protein coats, while in the second half the protein coat of one type encloses a genome of the other. When a single particle of such a phenotypically mixed phage infects a fresh bacterium, the injected DNA alone determines the synthesis of new phage protein, so that only homologous protein will be made and the progeny particles will be pure. Phenotypic mixing is a complicating factor which must be taken into account in genetic analyses involving genes determining coat structure and behaviour, since the progeny phenotype which is scored may not reflect the genotype, which is what we want to know. The progeny of the cross must therefore be passed through another cycle of single infection to ensure that the phenotype and genotype correspond (Fig. 105). However, we do not mean to convey the impression that phenotypic mixing has only a nuisance value in genetical research. On the contrary it is a most valuable tool in functional analysis as some examples will demonstrate.

The phenomenon was first observed in the study of mixed infection of *E. coli* B with phages T2 and T4. These phages are closely related since they can mate and generate recombinants, but they possess different tail structures. Thus although wild type *E. coli* B is sensitive to both phages and can yield mutant strains which are resistant to one or the other as a result of specific cell wall changes, these bacterial mutants show no cross-resistance; bacteria resistant to phage T2(B/2) remain sensitive to phage T4, while those resistant to phage T4 (designated B/4) adsorb phage T2 normally. When wild type bacteria are mixedly infected with both phages, about half the total progeny particles display one or the other parental phenotype, forming plaques only on B/2 (T4 phenotype) or only on B/4 (T2 phenotype).

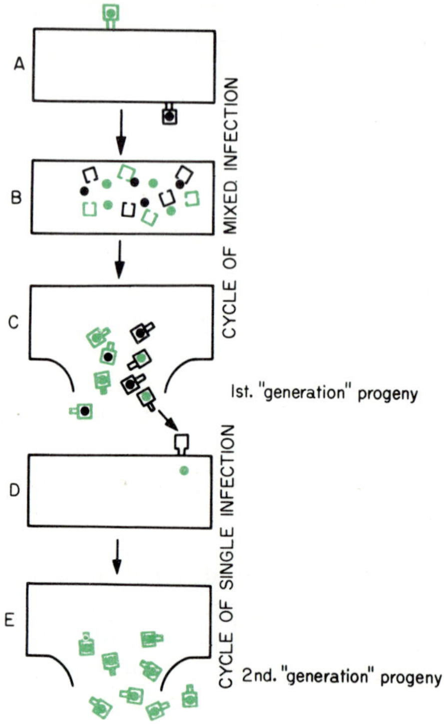

FIG. 105. Illustrating phenotypic mixing.

The large black rectangles indicate sensitive bacteria. The small squares, complete or incomplete, represent phage coats; green and black indicate specific differences in a component of the coat whereby the phenotypes of the parents are recognised. The green and black dots represent the genetic material of the two parents, determining the synthesis of green and black protein components respectively.

During the cycle of mixed infection (A, B, C) the two types of genetic material become randomly associated, during maturation, with the two types of coat so that, in one half of the progeny particles the specificity of the protein coat is different from that of the genome it carries. When such a particle singly infects a new bacterium (D), the synthesis of phage protein is determined only by the injected genetic material so that the progeny particles (E) are pure. Their phenotype indicates the genotype of the infecting particle. (See text).

However, about half the B/2 bacteria liberate 2nd 'generation' progeny which no longer plate on B/2 but only on B/4, and must therefore have been infected with particles having T4 tails, permitting adsorption on B/2, but T2 genomes. Similarly, about half the 1st 'generation' particles which plated only on B/4 (T2 phenotype) yield 2nd 'generation' progeny of pure T4 phenotype (Fig. 105; Delbrück and Bailey, 1946; Novick and Szilard, 1951).

So far we have discussed only half the total progeny of the mixed infection, of which one-half shows phenotypic mixing. The other half is even more interesting for it does not resemble either parent in phenotype but can infect *both* B/2 *and* B/4 bacteria, but the 2nd 'generation' particles which emerge from these infections are always either pure T2 or pure T4 with respect to both phenotype and genotype. This can only be explained if the phage tail possesses *two anatomical sites* determining host range, each of which may independently acquire either T2- or T4-specific components in a random manner when the tail is being assembled. Thus the probability is that half the completed protein structures in mixedly infected bacteria will have protein components of the same specificity, and the other half components of different specificity, at the two sites; these structures, in turn, are randomly associated with the two parental types of genome (Streisinger, 1956b; Brenner, 1957b).

We have mentioned that some strains of phage T4 require tryptophan as a co-factor for adsorption (p. 422). Mutations to tryptophan-independence occur, showing that the requirement is genetically determined. When independently isolated, trytophan-requiring (*c*) strains of phage T4 are crossed, tryptophan-independent (c^+) recombinants arise so that the mutational sites determining the co-factor requirement can be mapped. They lie together in a small region of chromosome distinct from that which specifies the host range-protein, closely linked to gene 34 (Fig. 104). Moreover, different *c* mutants do not complement one another since most of the progeny of a mixed infection are of dependent (C) phenotype, so that the co-factor requirement appears to be determined by a single gene. However, when bacteria are mixedly infected with c^+ and *c* strains of phage, the co-factor requirement shows phenotypic mixing; about three-quarters of the progeny particles are of independent (C^+) phenotype but, among these, the c^+ and *c* genotypes are equally distributed as shown by the behaviour of 2nd 'generation' progeny (Brenner, 1959). This suggests that the co-factor activates some protein component of the phage tail which is different from that determining host range. Moreover, the fact that three-quarters, instead of one-half, of the progeny of mixed infection with c^+ and c genotypes are phenotypically C^+ implies that the protein is duplicated in each phage particle and that the C^+ protein is dominant. As in the case of the host-range protein, half of the tail structures formed in the mixedly infected bacteria will have one C^+ and one C component, the single C^+ component sufficing to confer tryptophan-independence; half of the remaining tail structures, or one-quarter of the total, will have two C^+ components so that the observed three-quarters of the progeny are phenotypically independent. Only one-quarter will possess two C components and require trytophan for adsorption. The c^+ and *c genotypes*, on the other hand, will be randomly fitted with these three types of tail structure.

Phenotypic mixing with respect to the structure of the head protein of phages is also found when bacteria are mixedly infected with wild type (o^+) and *o* mutants of phages T2 and T4 (p. 485; Brenner and Barnett, 1959). We have already seen that the phenomenon of phenotypic mixing can also form the basis of complementation tests in the case of two phage mutants of different host range (p. 105). If the two mutant genomes can complement one another in mixed infection, wild type tail structures will be synthesised and fitted to both parental and wild type recombinant genomes. In the absence of complementation, most of the tail structures synthesised will be of mutant type so that the small number of wild type genomes emerging from recombination will mostly be fitted with them; the result is that the wild phenotype, from which the occurrence of recombination is deduced, can be recognized only after a second cycle of infection.

Phenotypic mixing has also been demonstrated in the *in vitro* reconstitution of 'hybrid' strains of tobacco mosaic virus, providing early evidence that RNA can act as a carrier of genetic information (p. 229). Within its limited context, one of the advantages of the phenomenon as an analytical tool is that it provides useful information from mixed infections with phage strains which differ with respect to a single character only, so that recombinational analysis is precluded.

HOST-CONTROLLED MODIFICATION

In this phenomenon, variously called *phenotypic modification* or *host-induced modification*, a single cycle of phage growth in a particular bacterial host may alter the host range of virtually all the progeny.

This type of alteration differs fundamentally from mutational modifications of host range (p. 482) since it is imposed by the host cell in which the virus grows but is not inheritable; as soon as the phage is grown in some other host, the modification is lost (Luria and Human, 1952). What happens may best be illustrated by a general example. A strain of phage plates with an efficiency of 1·0 on a particular host, A, but with an efficiency of only 10^{-5} on host B; the host range of the phage is *restricted* to A. If one of the rare plaques which form on host B is picked, it is found to consist of particles which now plate with an efficiency of 1·0 on both A and B bacteria; these particles have acquired an extended host range. So far there is nothing to show that the particles did not stem from rare mutants in the original population. After a single cycle of growth on host A, however, about 99 per cent of the progeny have reverted to the restricted host range of the original phage. Only about one in 100 retains the ability to form plaques on B bacteria.

Host-controlled variation is a common phenomenon and many temperate and virulent phages have been found to be subject to it (reviews; Luria, 1953; S. Lederberg, 1957; Arber, 1965a). In several cases the difference between the two host strains, with respect to acceptance and modification of the phage, is determined exclusively by the lysogenisation of one of them by an *unrelated* prophage. For example, phage λ plates with high efficiency on *E. coli* K-12, but with an efficiency of only about 10^{-4} on the same strain lysogenised by phage P1 (referred to as K(P1)); the plaques which do arise on K(P1) consist of particles which plate with equal efficiency on K and K(P1), but after growth on K the ability to plate efficiently on K(P1) disappears. On the other hand, there are some cases where neither of a pair of hosts appears to be lysogenic. A few examples of systems which show host-controlled variation are given in Table 20.

There are two important features of the phenomenon which must be understood before we proceed to study its mechanism. The first is that it has nothing to do with the ability of phage to adsorb to host bacteria. The particles of restricted host range, which initially produce plaques with low efficiency on host B, adsorb normally to the B bacteria and inject their DNA, but this DNA is unable to replicate in the great majority of infected cells. Secondly, the ability of a small fraction of the infecting particles to grow normally in host B, and to become modified so that their progeny can initiate a spreading infection which results in plaque formation, results from the *infection of exceptional bacteria which do not reject the phage* DNA and not from infection by rare particles which can overcome the immunity of the host. For instance those particles, emerging from single bursts of infected A bacteria, which can form plaques on host B show a random distribution instead of the clonal one to be expected of exceptional (mutant) particles (Luria, 1953). Moreover, the number of plaques which arise

TABLE 20

Some examples of phage-host systems which show host-controlled variation.

	Hosts	
Phage	A	B
λ	*E. coli* C	*E. coli* K-12
λ	*E. coli* K-12	*E. coli* K-12 (P1)
P2	*Shigella*	*E. coli* B
P22	*Salm. typhimurium*	*Salm. gallinarum*
P22	*Salmonella*	*E. coli-Salmonella* 'hybrids'
T1, T3, T7, P2	*Shigella*	*Shigella* (P1)
Salm. typhi. Vi II typing phage	*Salm. typhi.* Type A	Some other *Salm. typhi.* types

The suffix '(P1)' indicates lysogenisation by phage P1. References to these and other systems will be found in the text, or in Luria (1953) and Lederberg (1957).

on host B is profoundly affected by the physiological state of the bacteria and by the conditions under which they are grown (Luria, 1953; Lederberg, 1957); in particular, UV irradiation of host bacteria prior to infection greatly increases the rate of acceptance and modification of the restricted phage (Bertani and Weigle, 1953), as also does heating the host cells to 49°C or over (Schell and Glover, 1966a). Finally, mutant strains of bacteria may be isolated (by selecting, strangely enough, for resistance to streptomycin) which have an increased probability, to as high as five per cent, of propagating restricted phage (Lederberg, 1957; Dussoix and Arber, 1962).

Recent experiments by Arber and Dussoix (1962) and Dussoix and Arber (1962), employing phage λ, and *E. coli* K-12 and *E. coli* K-12 (P1) as host bacteria, have thrown considerable light on the molecular basis of this interesting and important phenomenon. There are two aspects of host-controlled variation—the nature of acceptance or rejection of the infecting genomes, and the mechanism of modification of those that are accepted. It turns out that when phage λ particles of restricted host range infect the P1-lysogenic host, the injected phage DNA is rapidly broken down, a fraction of the degraded material diffusing into the medium. In mixed infections with restricted and unrestricted (modified) phage some of the markers of the restricted phage, including pairs of closely linked markers, may be 'rescued' by incorporation into recombinant genomes which are basically of the modified type (Dussoix and Arber, 1962).

The most interesting feature, however, concerns the modification itself. We will adopt the abbreviated nomenclature of Arber and Dussoix (1962), referring to the two host strains as K and K(P1), and to phage propagated on each as λ.K and λ.K(P1) respectively. Phage λ.K has an efficiency of plating of 1·0 on K and of 2×10^{-5} on K(P1); phage λ.K(P1) plates on both strains with an efficiency of 1·0. However, when λ.K(P1) is passed through a single cycle of growth on K, at a multiplicity of infection of about 1·0, the progeny particles plate with an efficiency of about 10^{-2} on K(P1); that is, not all the particles have been modified, but about one in every burst of 100 retains its λ.K(P1) phenotype. If the multiplicity of infection is increased, the number of particles which can plaque on K(P1) rises proportionately. This suggests that the infecting genomes preserve their unrestricted host range, and that it is only their vegetative replicas that lose it. To test this idea, λ.K(P1) particles, density-labelled with the heavy isotope of hydrogen, deuterium, were used to infect K bacteria in normal media. By centrifuging the progeny in a caesium chloride density gradient, the minority of particles in which one of the heavy parental strands of DNA had been conserved ('hybrid' DNA) could be separated from the majority in which both DNA strands had been newly synthesised from light material (pp. 243, 396). It was found that only the 'hybrid' particles retained their ability to plaque with high efficiency on K(P1). It thus transpires that modification of host range results from an alteration of the phage DNA which renders it acceptable to the host bacteria. The preservation of a single modified strand in the DNA duplex suffices for acceptability (Arber and Dussoix, 1962).

The modification can be impressed on non-replicating DNA. If K bacteria are infected with unmodified, deuterium-labelled phage λ.K *at high multiplicity*, a proportion of the injected genomes reappears unchanged among the progeny particles; these particles retain the heavy density of the parental particles so that their DNA has not replicated (p. 395). If, during the latent period, the infected K bacteria are *superinfected with phage* P1, *all* the progeny, including those carrying unreplicated parental genomes, are found to have been modified to the λ.K(P1) type. Phage P1 can therefore exert its modifying action very quickly. On the other hand, K bacteria will continue to accept phage λ.K for a considerable time after they have been infected with phage P1, so that the functions of breakdown and of modification of DNA which are induced by the phage appear to be separable (Arber and Dussoix, 1962).

This duality of function is borne out by genetic analysis of the phenomenon. One cannot isolate and study those K(P1) bacteria, for example, which accept and propagate unmodified phage since they are destroyed by the lysis which liberates the phage. However it turns out that not only phage but also sex factors such as F, and colicinogenic factors such as *col* I (p. 755), are similarly restricted when transferred by conjugation from K to K(P1) bacteria, although usually only by a factor 10^{-2} instead of the usual 10^{-4} to 10^{-5} found in the case of phage λ(Glover *et al.*, 1963). Thus if K(P1) bacteria which cannot ferment lactose (*lac*⁻) are conjugally infected with an F-prime factor carrying a wild type lactose region of chromosome, those cells which have accepted and harbour the F-*lac*⁺ factor can be recognised by their acquisition of the *lac*⁺ character. Analysis of such K(P1)*lac*⁺ clones shows that about one third are altered in their ability to restrict, and that these belong to two types. One has lost the ability to restrict but not to modify (r^-m^+) so that λ.K plates on it with an efficiency of 1·0, but the emergent phage is λ.K(P1); the other type can neither restrict nor modify (r^-m^-) so that the progeny of λ.K infection retain K specificity. These defects can be shown to be inherent in the P1 prophage and not in the bacterial genome by isolating P1 phage from the altered cells and using it to lysogenise K bacteria (Glover *et al.*, 1963).

In the case of host-controlled modification not due to prophage, a controlling locus on the bacterial chromosome was first demonstrated by Zinder (1960a) in crosses betweeen *E. coli* and *Salmonella* (p. 751), and has since been confirmed for *E. coli* K-12 and B and found to be located close to the threonine (*thr*) region of the bacterial chromosome in both strains (Fig. 127, p. 666) (Boyer, 1964; Colson *et al.*, 1965; Hoekstra and de Haan, 1965; Wood, 1965, 1966). As in the case of K(P1) bacteria, non-restricting variants of *E. coli* K-12 strains may be obtained by isolating clones which have accepted an F-prime factor from *E. coli* C (see Table 20). It turns out that, as in the case of phage P1, these non-restricting strains not only fall into modifying (r^-m^+) and non-modifying (r^-m^-) types but these two types arise with about the same frequency; one explanation is that both result from single mutational events within a single, bifunctional locus. In support of this, both types of mutation map in the same small region. Moreover, restriction and modification appear to be allelic characters in *E. coli* K-12 and B since recombinants carrying dual specificity have not been isolated from crosses between these strains (Colson *et al.*, 1965; Wood, 1965). Nevertheless it has been shown that the K and B modifications *can* coexist on the same DNA molecules, for if phage λ, whose DNA carries K specificity and is physically labelled with heavy isotopes, is permitted a single growth cycle on a non-restricting but modifying (r^-m^+) strain of *E. coli* B, emergent particles of parental density, which thus contained un-unreplicated parental DNA (p. 395), were found to carry K as well as B specificity (Kellenberger, Symonds and Arber, 1966). The fact that no variants which can restrict but not modify (r^+m^-) have been isolated, may be due to the fact that mutants of this type, being unable to impress a protective modification against endogenous nuclease upon their own DNA, would be expected to be lethal. On the other hand, no selection method for isolating m^- phenotypes has yet been employed.

The most interesting theoretical aspect of host-controlled modification is the nature of the alteration imposed on the DNA which prevents its recognition by specific nucleases. Although, as yet, nothing is known for certain about the chemical basis of modification, there is very suggestive evidence that, in some cases, it is due to the methylation of DNA bases. For example, phage λ grown in *E. coli* K-12 auxotrophs normally plates with an efficiency of 1·0 on both *E. coli* K-12 and *E. coli* C which is not restrictive; but if the K-12 host requires methionine, which is the source of the methyl group in bacterial DNA, and is deprived of methionine for most of the latent period, early maturing particles of phage λ, released by premature lysis soon after restoration of the amino acid, plate normally on *E. coli* C but fail to plate on the host strain, K—they lack the K modification. Similar results are obtained in the case of methionine auxotrophs of *E. coli* K(P1) and B. On the other hand, deprivation of amino acids other than methionine is ineffective in inhibiting modification (Arber, 1965b).

Another type of experiment provides indirect evidence that methylation is the basis of modification of phage T1, by *E. coli* lysogenised by P1 prophage. As we have said, methionine is the source of the methyl group in the methylation of bacterial DNA, which occurs by a transmethylation reaction involving S-adenosylmethionine as the final methyl donor. It happens that infection with phage T3 uniquely yields an 'early' enzyme which rapidly breaks down S–adenosylmethionine with the result that no methylation of either DNA or RNA can be detected in phage T3-infected bacteria (Gefter *et al.*, 1966). When *E. coli* B(P1) bacteria are mixedly infected with phages T1 and T3, the progeny T1 particles are found to be modified in only a small proportion of the cells which produce them. On the other hand, the ability of another strain of *E. coli* to impose a different specific modification on phage T1 remains unaffected by concomitant T3 infection. Similar results have also been obtained with phage λ(Klein and Sauerbier, 1965; Sauerbier and Klein, quoted by Stacey, 1965b).

A type of host-controlled modification which does not depend on methylation concerns the T-even phages which are unusual not only in carrying the base hydroxymethyl cytosine (HMC) instead of cytosine but also because some (or all, in the case of *T*4) of these hydroxymethyl groups have glucose residues attached to them. This glucosylation requires the presence in the cell of uridine diphosphoglucose (UDPG), which is an intermediate in galactose metabolism (p.110) as well as being involved in normal cell wall synthesis. It happens that mutants of *E. coli* B, selected for concurrent resistance to phages T3, T4 and T7(B/3,4,7), lack the enzyme UDGP pyrophosphorylase and can neither ferment galactose nor glucosylate phage DNA. When infected with normal phage T2 or T6, these mutants can support phage growth and give a normal yield of phage, but these progeny particles are unable to undergo another growth cycle in the B/3,4,7 bacteria so that no plaques arise from the primary infection. This is because the DNA of the first generation progeny is not glucosylated and is broken down following infection of either the initial host or wild type *E. coli.* B. However, these non-glucosylated particles can be revealed by plating on *Shigella* which does not restrict them, so that they plaque normally (Luria and Human, 1952; Erikson and Szybalski, 1964; Hattman and Fukasawa, 1963; Shedlovsky and Brenner, 1963; Symonds *et al.*, 1963; Stacey *et al.*, 1964). Mutants of the T–even phages, called T*, have been isolated which fail to produce the 'early' phage enzyme, glucose transferase, which is also required for glucosylation. Like phage grown in *E. coli* B/3,4,7 strains, these mutant phages grow normally in *Shigella* but are restricted, and their DNA degraded, on infection of wild type *E. coli* B (Fukasawa, 1964; Hattman, 1964: reviews; Arber, 1965a; Stacey, 1965b).

Several kinds of experiment support the likelihood that the nucleases responsible for restriction are associated with the cell surface and act on DNA only as it penetrates the cell and not subsequently. For example, treatment with the detergent EDTA, followed by washing several times at 4°C, which is known to release surface-localised enzymes, can increase the efficiency of plating of phage λ.C on *E. coli* K(P1) by a factor greater than 10^4 (Schell and Glover, 1966b). Again, in the same system, if the cells are infected at 50°C., at which temperature restriction is in abeyance, and then restored to conditions at which restriction is again active, as much as a 10^5-fold increase in the efficiency of plating may be found (Schell and Glover, 1966a).

Many resistance-transfer factors (RTFS; p. 760) are restrictive for a wide range of phages, including λ, P1, P22, T1 and T7, and some are also able to modify the DNA of these phages (Watanabe, 1966a). It has recently been shown, by the use of an *E. coli* mutant which lacks the ability to produce endouclease I which is a major cellular component responsible for *non-specific* DNA degradation, that extracts of cells infected with a restricting resistance-transfer factor can specifically break down unmodified phage λ DNA *in vitro* (Takano, Watanabe and Fukasawa, 1966). An extensive survey of the susceptibility of 28 coli-phages to four systems of host modification is by Eskridge, Weinfeld and Paigen (1967).

Host-induced modification has practical as well as theoretical implications. In the *Salm. typhi* phage-typing system, which is an indispensable tool for tracing the epidemiology of outbreaks of typhoid fever, the bacterial type is determined by its susceptibility to one, or a few, of a range of virulent typing phages. Each type of phage is obtained from an ancestral, wild type phage by selective propagation of the minority of plaques which develop on each bacterial type. It turns out that the host specificity of some of the typing phages results from host-range mutation, while that of others is due to phenotypic modification. The distinction between these two types of variation has been beautifully revealed by the clonal distribution of host-range mutants among single bursts of the wild type phage, as opposed to the random distribution obtained when the phage acquires a new specificity as a result of host-controlled modification (Anderson and Fraser, 1956).

The *Salm. typhi* typing phages adsorb specifically to the Vi (virulence) antigen of this organism, and each can attach to, and inject its genetic material into, all the bacterial types which are usually killed thereby. The various bacterial types, however, possess a range of specific immunity patterns, each of which blocks the replication of all but the homologous typing phage, which has acquired the capacity to override the immunity. The specificity of host-range mutants, therefore, has nothing to do with their ability to adsorb, as is the case with host-range mutants of the T phages (p. 482). It turns out that the kind of immunity which the majority of host-range mutants overcome is conferred on the bacteria by lysogenisation with a specific prophage. These prophages are quite unrelated to the typing phages themselves (Anderson and Felix, 1953; review: Anderson, 1962). The situation is thus akin to the immunity which lysogenisation of *E. coli* K-12 with phage λ confers against lytic infection by r_{II} mutants of phage T4, and which is overcome by reversions of the phage to the r^+ type (Table 5, p. 167). On the contrary, phenotypic modification of the *Salm. typhi* typing phage does not appear to be induced by a carried prophage though this, of course, is almost impossible to exclude with certainty.

The extent to which host-induced modification may play a role in determining variation in viruses other than phage is difficult to assess (see Luria, 1953). It is certain, however, that the phenomenon is not restricted to viruses. The existence of DNAs of different host specificity turns out to be a decisive factor in the capacity of some bacterial crosses to yield recombinants. For example, large numbers of recombinants are produced in crosses between *Hfr* male and female strains of *E. coli* K-12, but if the same females are lysogenised with phage P1 the number of recombinants is dramatically reduced although there is no doubt that the chromosomes of the two parents remain genetically homologous. In fact, the DNA of the male chromosomes is broken down on transfer to the K-12 (P1) females in just the same way as is the DNA of phage λ.K (Dussoix and Arber, 1962; Arber, 1962, 1964, 1965a).

RADIOBIOLOGICAL ASPECTS OF PHAGE GENETICS

Radiobiology is a complicated and highly specialised subject and, despite intensive research in recent years, little of a precise nature is yet known about the effects which radiations may have on cells. So far as the action of radiations on the genetic material is concerned, we know of only two cases where a precisely defined alteration at the molecular level has been shown. One is the formation of pyrimidine dimers induced by ultraviolet light (p. 329). The other is the structural change inflicted on polynucleotide chains as a result of the decay of incorporated radiophosphorus; but, as we shall shortly see, this latter effect is probably not so much due to radiation *per se* as to the atomic disintegration which substitutes a sulphur for a phosphorus atom.

One of the difficulties inherent in radiobiological studies is that of analysing the effects of radiation on specific cell components, such as the genetic material, against a background of general damage. It is largely for this reason that microbial systems have become such popular tools in radiobiology since they provide scope for studying the genetic effects of radiation in comparative isolation. For example, in the *E. coli* conjugation system, irradiated *Hfr* males continue to transfer their chromosomes to females so that the effect on recombination of irradiating only one of the chromosomal homologues of a zygote can be analysed; transformation permits the biological effects of irradiating pure DNA to be assessed.

Bacteriophage is a particularly favourable tool in radiation studies because of the genetic nature of the infection, the detailed knowledge available about the mechanism and kinetics of phage growth, and the ease with which accurate, quantitative data can be acquired. Although intensive research has done little to elucidate the nature of radiation damage, the results of these studies form an important, if not very decisive, part of phage literature and have provoked several stimulating hypotheses. In this section only those experimental findings which have led to meaningful conclusions about the nature of phage infection are summarised. We will not attempt to analyse the action of different types of radiation but will concentrate on the effects of ^{32}P decay and of UV light (reviews: Stent, 1958; Adams, 1959; Stahl, 1959; Stent and Fuerst, 1960).

THE DISINTEGRATION OF RADIOACTIVE PHOSPHORUS

Hershey *et al.* (1951) observed that when bacteriophages incorporate radioactive phosphorus into their DNA, as a result of growth in a medium containing a high specific activity of ^{32}P, the infectivity of the population declines exponentially with time, at a rate proportional to the concentration of radioactive atoms in the medium. Since the mean number of ^{32}P atoms in the DNA of each phage particle can be estimated, and the rate at which the radioactive atoms decay is known (half-life about two weeks), it is possible to calculate the probability with which a phage particle is inactivated by the disintegration of a single, incorporated ^{32}P atom. The result is that about one in ten disintegrations is lethal. The same result is obtained for a range of different phage strains (T1, T3, T5, T7, λ and P22: Stent and Fuerst, 1955; Garen and Zinder, 1955) as well as for bacteria (*E. coli*: Fuerst and Stent, 1956), despite a wide disparity in the amount of DNA involved.

The high efficiency of inactivation, as well as reconstruction experiments in which particles were subjected to the same amount of external radiation as that produced by the incorporated radiophosphorus, show that the lethal effect is not due to the ionisations which follow the emission of β electrons during disintegration. Moreover, the efficiency of inactivation depends on the temperature at which the disintegration occurs, rising from 1 in 25 at −196°C., through 1 in 10 at 4°C., to about 1 in 3 at 65°C. (Stent and Fuerst, 1955). The outcome of radioactive decay on atomic structure is to transmute the radioactive phosphorus to sulphur (^{32}S), so that its obvious effect is to break the phospho-diester bond of the polynucleotide strand at the site where the ^{32}P atom was incorporated (Fig. 55, p. 230). The fact that only one in every ten disintegrations is lethal at 4°C. was therefore ascribed to the 2-stranded nature of DNA; the DNA retains its physical continuity and functional activity unless *both* strands are disrupted at the same point (Stent and Fuerst, 1955). The inactivating effect of every ^{32}P disintegration in the single-stranded DNA of phage ϕX 174 supports this view (p. 437).

There are two hypotheses to explain how a ^{32}P disintegration in one strand can result in breakage of the opposite strand but we do not know which, if either, is correct. One ascribes the effect to the recoil energy liberated by emission of the β particle; the other suggests that interruptions in the continuity of each strand may be a normal feature of double-stranded DNA, and that it is only when a ^{32}P disintegration in one strand coincides with such an interruption in the other that the duplex as a whole is fractured.

In discussing experiments involving the effects of ^{32}P decay on genetic structure and function it is customary, and certainly economical, to refer to phage particles or bacteria which have incorporated ^{32}P (or any other radioactive element) as 'hot', and to those which contain only normal phosphorous (^{31}P) as 'cold'. In all such experiments, the population of 'hot' organisms must be stored over a period of 10 to 14 days during which about half the phosphorous atoms may be expected to decay; samples are removed at intervals, and the proportion of survivors assayed and plotted as a function of the time of storage which, of course, is equivalent to the rate of ^{32}P disintegration. In the case of free phage particles, storage at 4°C. in a special medium is adequate (Stent and Fuerst, 1955). Many experiments, however, are designed to show the effects of ^{32}P decay at various stages during the growth of phage in infected bacteria, or during the development of bacterial zygotes. In such cases bacterial metabolism must be completely arrested throughout the long period required to measure the effects of decay, so that storage at 4°C. is inadequate. It transpires that thawing from temperatures of the order —40°C. is highly lethal to bacteria. In experiments involving bacteria or phage-bacterium complexes, therefore, samples are frozen in liquid nitrogen (—196°C.), thawing from which does not significantly lower bacterial survival.

TARGET THEORY

The biological effect of radiations is usually inferred from the slope and shape of curves constructed by plotting the number of surviving organisms as a function of dosage, which is usually expressed in terms of the duration of exposure to the irradiation or ^{32}P decay. When free phage particles are irradiated, or subjected to the decay of ^{32}P atoms, the number of survivors (infective particles) decreases exponentially with dosage over the entire dose range, that is, if the *logarithm* of the number of survivors is plotted against dosage a straight line is obtained. This is termed a *one hit* type of response and signifies that a quantum of radiation, or the decay of a single ^{32}P atom, has a constant probability of inactivating a phage particle which thus behaves as if it were a single, susceptible *target*. If the target is less susceptible or, to put it another way, if the particle is more *resistant* to irradiation, the probability of its receiving a lethal hit is reduced; any given quantum of radiation has a lower probability of inactivating the particle but this probability remains constant at the lower level. The result is that the inactivation curve is still an exponential one, but it falls at a lower rate; it is less steep (Fig. 106; compare curves A, B and E).

In the case of ^{32}P decay we have seen that a single disintegration has the same probability of causing inactivation in both bacteria and phage, implying that a single hit is lethal *irrespective of the amount of* DNA *involved*. Since the bacterial chromosome contains about 100 times more DNA than that of the phage it follows that, for a given concentration of ^{32}P atoms per unit length, the bacterium will be inactivated at a higher rate than the phage and will show a steeper survival curve. The bacterium is less resistant than the phage. This is another way of saying that the bacterium presents a bigger 'target'. The same kind of relationship holds in the case of radiations. The quanta streaming from the source may be likened to a hail of bullets from a machine-gun. The greater the size of the target the greater is the probability of a lethal hit. We can thus relate the slopes of exponential survival curves to the size of the target which, in the case of

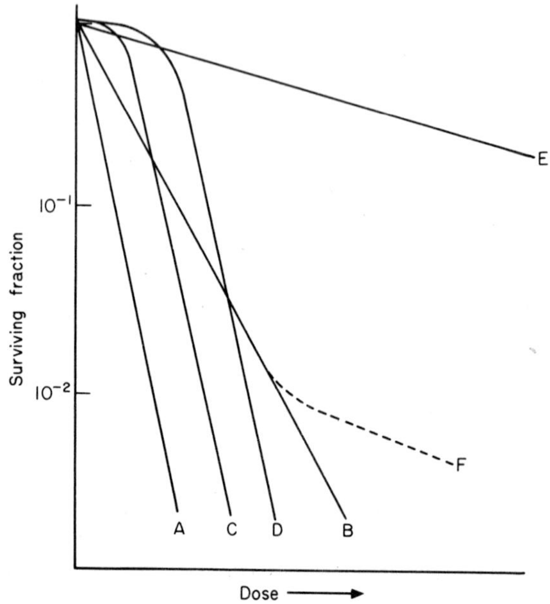

FIG. 106. A series of theoretical curves illustrating the relationships between radiation dosage and the number of surviving organisms.

The logarithms of the numbers of survivors are plotted.

Curves A, B and E decline exponentially throughout (single-hit curves), indicating single, susceptible targets of increasing order of resistance.

Curves C and D show 'shoulders' before they begin to decline exponentially at the same rate as curve A (multiple-hit curves), indicating multiplicity of targets, each of the same sensitivity as the single target of curve A. The extent of the shoulder increases with multiplicity.

Curve B, merging into curve F, indicates a mixture of two components of different sensitivity. The exponential part of the slope of curve F reveals a minority component of the same resistance as the organism yielding curve E, since the two curves are parallel.

See text for explanation.

phage, may be construed as its DNA content. While it is generally true that the sensitivity of a phage, especially to ^{32}P decay and to ionising radiations, tends to be proportional to its DNA content, and that the sizes of many phages have been quite accurately assessed by this means, there are also many complicating factors.

Among the more interesting and important of these factors is the existence of genes which determine the repair of irradiation damage. For example, phage T2 is more than twice as sensitive to the action of UV light as is phage T4, although both phages contain the same amount of DNA. Genetic crosses show that this disparity is due to an allelic difference between the two phages (Streisinger, 1956a; Harm, 1959). Phage T4 possesses a gene (u^+), lacking in phage T2 (u^-), which mediates the partial reversal of UV damage. However, if the phages are grown in the presence of 5-bromouracil, which replaces thymine in their DNA, their sensitivity becomes the same. This suggests that the repair process may be due to an enzyme, determined by the u^+ gene, which specifically uncouples thymine dimers (pp. 329, 331; Stahl *et al.*, 1961). We have already discussed similar gene-controlled differences in sensitivity to UV light between wild type and various *uvr* mutant strains of *E. coli* (p. 333). These differences are revealed not only when the sensitivities of the bacteria to irradiation are directly compared, but also by the capacity of the *unirradiated* bacteria to support the growth of irradiated bacteriophage (see pp. 332, 532). When the autonomous virulent phages T2, T4, T5 and T6 are irradiated they display the same plating efficiency on both wild type (resistant) and mutant (sensitive) bacteria. On the other hand, the capacity of the mutant bacteria to 'reactivate' irradiated, dependant virulent or temperate phages is much less than that of wild type (Hill, 1958; Howard-Flanders, Boyce and Theriot, 1962).

Let us now see what happens if the viable unit consists of two or more targets instead of the single one we have considered so far. Such a situation is found in bacteria, which usually possess more than one nucleus per cell (p. 546), and in phage infection when the infected cell contains a number of phage genomes, the survival of any one of which may permit a normal burst. At low doses of irradiation, the probability that all the targets will be simultaneously inactivated by hits is negligible so that the survival curve remains flat instead of falling. As the dosage becomes larger an increasing proportion of the targets receives a hit until a stage is reached when there is a high probability that only a single target remains unscathed in each viable unit. Thereafter the number of viable units declines exponentially. The resulting *multiple-hit curve* is therefore different from the one-hit type. At low dosage it has a 'shoulder' which becomes steeper as the dosage increases until it merges into an exponential curve of the same slope as the single-hit curve. The greater the multiplicity of targets, the more the shoulder is extended and the higher the dosage at which the exponential part of the curve starts (Fig. 106; compare curves A, C and D). The actual number of targets present can be estimated from the point at which an extension of the exponential part of the curve intercepts the co-ordinate.

Before we leave this theoretical consideration of survival curves, there is one further type which is sometimes found. The curve initially falls exponentially at a certain rate and then, over a small dosage range, becomes less steep although still exponential. Such a *discontinuous type of curve* is represented in Fig. 106 by the alteration from curve B to curve F and signifies that the population contains a minority component which is much more resistant than the majority one. In this example, the small number (about one per cent) of more resistant particles is not observed until it has become a significant fraction of the survivors; with increasing dosage, the sensitive particles rapidly disappear, leaving only the resistant particles which now decline exponentially at their own rate.

MULTIPLICITY REACTIVATION

E. coli B, infected with phage T2 or T4 preparations which have been inactivated by high doses of UV light, yields virtually no bursts when the multiplicity of infection is much less than unity, as one would expect. When the multiplicity is increased so that each bacterium is infected by more than one inactivated particle, however, phage is liberated with high probability (Luria, 1947b). Luria explained this *multiplicity reactivation* by supposing that each particle is built of a number of different sub-units of equal sensitivity, the integrity of all of which is needed for reproduction. The inactivation of a single sub-unit would thus prevent growth when each infected bacterium receives only a single, inactivated particle. In the case of multiple infection, however, it is likely that a different sub-unit is inactivated in each infecting particle, so that the great majority of bacteria receive *at least one complete set of intact sub-units* which could reassemble to yield active phage. This hypothesis was formulated before the discovery of genetic recombination and linkage in phage, and Luria (1947b) supposed that the postulated sub-units were independently reproduced. Now, of course, we can interpret the sub-units as genes, and recombination as the mechanism whereby active vegetative phage is regenerated.

Assuming that each gene is equally susceptible to a hit, target theory anticipates that as the dosage of UV light increases so will the likelihood of inactivation of the *same* genes in any pair of particles. Thus if the mean multiplicity of infection is 2, a stage should be reached at which multiplicity reactivation cannot occur. The survival curve should then decline at the same rate whether single or double infections are scored; that is, with increasing multiplicity of infection a family of curves similar to curves A, C and D of Fig. 106 should be obtained. This theoretical prediction is, in fact, fulfilled in the case of the virulent *Salm. typhi* typing phage, for example (Bernstein, 1957). However, the T-even phages, and phage T5, yield curves of decreasing steepness as the multiplicity of infection increases. How can this be explained?

One answer is that the functional activity of different phage genes may not be equally sensitive to UV inactivation. Two types of gene were therefore postulated, one of which ('reactivable sub-unit') is reactivable with high efficiency in multiple infection while the other ('vulnerable centre') is not (Barricelli, 1956). In single infection, where no reactivation is possible, the phage is inactivated if either type of gene receives a hit. In multiple infection, on the other hand, hits in the same 'reactivable sub-unit' tend to be efficiently compensated by recombination, and only those in the same 'vulnerable centres' lead to inactivation. Thus in multiple infection only a *proportion* of the genome is susceptible to effective inactivation at high dosage, so that less steep curves are obtained. The curves for phage T4 were found to fit the data if about 40 per cent of the genome (target) comprises 'vulnerable centres'.

This concept fits well into what we know about the functional prerequisites for phage development. We have seen that, in the case of the T-even phages, a large number of 'early' enzymes must be synthesised before DNA replication can begin (p. 428). In the absence of DNA replication, recombination and multiplicity reactivation are impossible. The genes determining the structure of these early enzymes are therefore well qualified for the role of 'vulnerable centres'. Provided DNA synthesis can occur, however, damage sustained at different sites within other homologous pairs of genes is reparable by recombination, although in *single* infection such damage would be lethal in the same way that single mutations are lethal. A comparison of multiplicity reactivation of irradiated phage T4 in *E. coli* B and *E. coli* K supports this hypothesis. You will recollect that phage growth in the latter host requires the functional activity of the A and B cistrons of the r_{II} region (p. 169) so that the wild type r_{II} region should constitute an additional 'vulnerable centre' when *E. coli* K is host, but not in *E. coli* B. The experimental results are consistent with this behaviour (D. Kreig, quoted by Stahl, 1959).

The estimate that 'vulnerable centres' constitute about 40 per cent of the phage genome may seem excessive until we examine their nature as a target. In the case of a pair of homologous genes determining an early enzyme, damage to each *at any site* will interfere with function and stop DNA synthesis, so that the target size is the total extent of the gene. Reactivable genes, on the other hand, can yield functional recombinants so long as the damaged sites are not identical in the two genes, so that the effective target size is very much smaller.

Unlike T-even phages and phage T5, phages T1, T3 and T7 show little or no multiplicity reactivation, probably because they have a much lower frequency of recombination.

CROSS-REACTIVATION

This phenomenon, more precisely termed *marker rescue*, is very similar to multiplicity reactivation and is found when bacteria are mixedly infected with two genetically marked phages, of which only one type is irradiated. The progeny phages are, of course, predominantly of the unirradiated type but, among these, recombinants are found which have incorporated genes from the irradiated parent. The genes are 'rescued' by recombination. In addition to demonstrating the recombinational nature of reactivation, the phenomenon provides scope for quantitative studies of the genetic damage inflicted by UV light. For instance, when the phage strains are marked by a number of well mapped allelic differences, it can be shown that unlinked genes are knocked out independently, and with about the same probability, by the radiation, while closely linked genes are knocked out together. Thus the majority of photon hits produce discreet and strictly localised damage to the phage genome (Doermann, Chase and Stahl, 1955).

THE LURIA-LATARJET EXPERIMENT

This experiment was devised by Luria and Latarjet (1947) as a means of investigating the intracellular growth of phage. Instead of irradiating free phage particles, the bacteria are first infected at low multiplicity. Samples are then withdrawn at intervals during the latent period and exposed to varying doses of UV light. The surviving complexes are scored as infectious centres; that is, each bacterium which contains a single, surviving phage genome will yield a burst which is scored as a single plaque when the bacterium is plated with sensitive indicator bacteria (p. 413). The effect of the radiation on the bacterial host need not be considered since irradiated bacteria which can no longer form colonies nevertheless support the growth of autonomous virulent phage (see p. 532). It was thought that, as the phage multiplied, the increasing number of genomes would be reflected in a family of multiple-hit curves from which the number of genomes present at the time of irradiation could be calculated (Fig. 106, curves A, C and D; p. 526). This is the result actually given by phages T1, T3 and T7 (see Benzer, 1952).

In the case of phages T2, T4, T5 and T6, however, the pattern is quite different. Up to six minutes after infection the sensitivity of the complexes remains more or less constant, but resistance then begins to rise rapidly so that, at about nine minutes, it has increased 20-fold. Instead of a family of multiple-hit curves, a series of straight lines of decreasing slope emerge (Fig. 101, curves A, B and E; Luria and Latarjet, 1947; Benzer, 1952). These unexpected results can be resolved in terms of what we now know about phage growth and multiplicity reactivation. The constant slope of the inactivation curve up to six minutes after infection is to be expected, since DNA replication does not commence until then. Once replication begins, two factors come into play. The first is that damage to the 'vulnerable centres' no longer restrains growth since their function—the synthesis of early enzymes—is already completed. Secondly, the rising probability of recombination with growing pool size leads to the regeneration of normal genomes from damaged ones with increasing efficiency. The family of multiple-hit curves which is found in the case of phages T1, T3 and T7 is satisfactorily accounted for by the small number of rounds of mating which these phages display, so that their capacity for multiplicity reactivation is insignificant.

The effects of ^{32}P decay on complexes singly infected with radioactive phage T2 are very striking. For the first five or six minutes after infection the ability of the bacteria to release phage shows virtually no change in resistance, while at nine minutes almost complete resistance to ^{32}P decay is attained (Stent, 1955; Symonds and McCloy, 1958).

The obvious explanation, that at nine minutes nearly every infected bacterium contains at least one ^{32}P-free replica of the phage genome, cannot be sustained since the same degree of stabilisation is found if both the bacteria and the growth medium, as well as the phage, contain a high specific activity of ^{32}P. At one time it seemed that multiplicity reactivation could also be excluded as the cause of stabilisation because this phenomenon is not displayed by ^{32}P-inactivated phage, even when such pre-infective factors as interference with injection are ruled out by following the effects of phosphorus decay *after* infection. However, it has now been shown that multiplicity reactivation of ^{32}P-inactivated phage does occur if the host bacteria are also made radioactive (Symonds and Ritchie, 1961).

THE EFFECT OF RADIATION ON THE BACTERIAL HOST

The entire series of T phages plaque with equal efficiency on *E. coli* B. In contrast, bacteria which have been sterilised by irradiation with UV light or X-rays, or by the decay of incorporated ^{32}P, so that they can no longer multiply to produce colonies, show great variation in their *capacity* to support the growth of these different phages. In general, the capacity of *E. coli* B for the T-even phages is highly resistant to irradiation while, in the case of phages T1, T3, T7, P22 and λ, it is relatively sensitive. An example will make the situation clear. When only about one in 10^{13} bacteria of an irradiated culture survive as colony formers, one-third can still be productively infected by the T-even phages. On the contrary, a dose of UV light which scarcely affects the capacity of the bacteria for the T-even phages may reduce that for phages T1 and T3 10- to 100-fold. Because of this, the interpretation of Luria–Latarjet curves, with respect to the latter phages, becomes rather meaningless.

We have seen that one of the main differences between these two groups of phages is that the T-even phages and phage T5 rapidly break down the nuclei of the bacteria they infect, while phages T1, T3 and T7 do not (Luria and Human, 1950). It therefore seems likely, in the case of those phages which are sensitive to host inactivation, that phage development depends, in some way not yet fully understood, on the production and activity of host enzymes which, in turn, depend on the integrity of the bacterial genome. This is particularly apparent from the action of ^{32}P disintegration in reducing the host capacity, since the lethal effect of the radioactive decay is due almost exclusively to disruption of the chromosome. It further transpires that those phages whose development is little affected by irradiation of the host bacteria are themselves highly sensitive to irradiation, and *vice versa.* This probably means that phage genomes, in general, do not specify very effective mechanisms for the repair of their own radiation damage but that, while damage to the genomes of temperate and dependent virulent phages is susceptible to host repair enzymes, that to the genomes of independent virulent phages is not.

CHAPTER 19

Chromosome Structure in Bacteriophage and Bacteria

There is now an incontrovertible weight of evidence that the material basis of inheritance, as well as the fundamental properties of genetic material, reside in molecules of DNA. The results of recombination analysis and of mutation studies imply that the genetic material *per se* consists of only single DNA duplexes which are coextensive with the chromosomes, but tell us nothing about chromosomal structure and organisation.

Morphological and chemical investigations of the chromosomes of higher organisms reveal that they are composed of more than DNA. They are, in fact, complicated nucleoprotein structures made up of bundles of fibrils, each about 200 Å thick. Moreover, the DNA of chromosomes is intimately associated with certain basic proteins (protamines and histones), some of which are wound helically around the duplex in the lesser of its two grooves (Fig. 59A, p. 237; Feughelman *et al.*, 1955) and double in amount at the same time as DNA replication occurs. Additional proteins also appear to combine periodically with these basic proteins, and with the phosphate groups of the DNA backbone, and are probably in part responsible for alterations in the staining properties of the chromosomes during the division cycle (reviews: Alfert, 1957; Ris, 1957, 1966; Ris and Chandler, 1963; Taylor, 1963; Callan, 1963; Pelling, 1966: recent chromosome models: De, 1964; DuPraw, 1965, 1966).

It seems to us unlikely that these complexities are associated with the way in which genetic information is carried, or with the basic mechanisms of replication and recombination which are inherent properties of DNA, but rather that they signify the elaboration of mechanisms to restrict recombination to the meiotic cycle and to ensure the precise distribution of the products of both meiosis and mitosis to progeny cells. In the case of bacteria (and of phage) the chromosome appears to consist of virtually naked DNA with the result that the introduction into the same cell of two allelic segments leads to recombination between them with high probability (e.g. p. 686). Under these conditions the preservation and evolution of stable, heterozygous diploid cells is clearly impossible and the main advantages of sexuality cannot be realised. On the other hand we have seen that even in such a simple organism as the fungus *Aspergillus*, which displays orthodox mitotic and meiotic behaviour, the occurrence of recombination in diploid, vegetative nuclei is a very rare event (p. 383). We may therefore hope that anything we may learn about the chromosomes of bacteria and phages may help us to distinguish those features of chromosomal behaviour in more sophisticated creatures which are of basic genetic significance, from those which represent mere refinements of cellular organisation.

Because of their essential simplicity of structure, it has been suggested that the genomes of bacteria and bacteriophages should be clearly distinguished from the equivalent but more complicated structures of plant and animal cells which are confined by a nuclear membrane, divide by mitosis and whose DNA is combined with basic protein, and for which the name 'chromosome' should be reserved. The word *genophore* has been proposed for the bacterial and phage genome, and as a general term to indicate the physical counterpart of the linkage group (Ris and Chandler, 1963). It is questionable whether this semantic distinction is a useful one since, as we shall see, the bacterial genophore is itself organised within the cell into a highly compressed, fibrillar, tertiary structure which, on replication, is transmitted to daughter cells with great precision, and which stains in a manner traditionally regarded as characteristic of chromosomes.

In this Chapter we will summarise the more important features of the bacteriophage and the bacterial chromosome as we now see them. The simpler case is that of the phage chromosome and we will examine it first.

THE PHYSICAL NATURE OF THE BACTERIOPHAGE CHROMOSOME

PHAGES OF THE T SERIES

The fact that the DNA within phage heads, but very little else, is injected into bacteria to which the phage attaches, while phage crosses show the existence of a single linkage group, provide *a priori* evidence for a phage chromosome consisting of a single, long molecule of DNA. Early attempts to characterise the state of the DNA in the heads of phage T2 were made by releasing the DNA by osmotic shock, purifying it and then analysing it with respect to homogeneity and molecular weight. In general, the DNA turned out to be very homogeneous and to have a molecular weight of about 10^6. Since the DNA contained by a T2 phage head (c.2×10^{-16} gram) would have a molecular weight of about 1.2×10^8 if it existed as a single molecule, this suggested that the phage chromosome might be constructed of a hundred or so small DNA molecules of uniform size, strung together by non-DNA 'linkers' which are broken by the extraction procedure. This possibility seemed to be reinforced by the fact that the extracted DNA of *E. coli* bacteria, which also possess a single linkage group, likewise proved to consist of homogeneous molecules of about the same molecular weight as those obtained from phage.

However, it was shown that large DNA molecules happen to be highly sensitive to shearing stresses (Davison, 1959). Even the forces imposed by the osmotic explosion of the phage head, or by the use of a syringe to transfer the extracted DNA, can markedly reduce its mean molecular weight. DNA can be extracted from phage in a pure and homogeneous state by treatment with phenol. When special precautions are subsequently taken to protect the DNA from shear, it is found that all the DNA from each phage particle is present in a single structural entity of molecular weight about $1{\cdot}3 \times 10^8$ (Davison *et al.*, 1961; Rubenstein, Thomas and Hershey, 1961). If such high molecular weight preparations are examined after stirring at various speeds, the DNA molecules are found to break successively into halves and then into quarters, and so on, at critical rates of shear until they resist the maximum shearing force applied (Hershey and Burgi, 1960). This behaviour appears to be incompatible with a continuous loop of DNA so that, in spite of the circularity of the phage T4 linkage group revealed by genetic analysis (p. 489), it seems unlikely that the DNA, of phage T2 at any rate, exists as a continuous structure in the phage head (Rubenstein *et al.*, 1961). On the other hand, the manner in which the molecules fracture, by successive mid-point breaks until an equilibrium is reached for a given shearing force, is adequate to account for the homogeneity of the low molecular weight preparations formerly encountered. The need to postulate 'linkers' disappears.

We are thus left an uncomplicated picture of the phage chromosome as a single, enormously long DNA duplex of molecular weight about $1{\cdot}2 \times 10^8$. If we assume that the average molecular weight of each base is 357 and that base pairs are separated by 3·4 Å (Fig. 59B, p. 237), it is easy to calculate that the length of the extended molecule is about 56–58 μ. An elegant direct measurement has recently been made by the autoradiography of phage T2 DNA, heavily labelled with radioactive, tritiated [^{3}H] thymine extracted with phenol in the presence of excess 'cold' phage under conditions of minimum shear, and spread with the photographic emulsion by a special technique (Cairns, 1961). The resulting pictures reveal extended molecules of DNA whose mean length can be estimated, after various minor adjustments, to be about 52 μ, corresponding to a molecular weight of $1{\cdot}1 \times 10^8$ (Plate 23).

Another beautiful procedure has permitted platinum-shadowed electron-microphotographs to be made of the entire chromosome of phage T2. The phage particles are ruptured osmotically in a spreading monolayer of protein at a water–air interface, so that the DNA thread is gently dispersed over a confined region in the vicinity of the phage 'ghost'. A photograph of one of these regions is reproduced in Plate 24. It can be seen that the DNA presents the appearance of a single thread with two free ends, whose length has been estimated as 49 ± 4 μ (Kleinschmidt *et al.*, 1962).

There is as yet little precise information as to how this DNA is organised within the phage head. We have seen, however, that during phage maturation the DNA condenses to about one fifteenth of its volume in the replicating pool, and it seems likely that this process results from the orderly cross-linking of the double helix on some kind of protein scaffolding (condensation principle) which is represented in intact phage by part of the internal head protein (p. 431). Studies of the birefringence properties of intact phage suggest that the DNA fibre is predominantly arranged in bundles parallel to the long axis of the head, as represented diagramatically in Fig. 98, p. 418 (Bendet, Goldstein and Lauffer, 1960).

We have already mentioned, in a rather dogmatic way, that the chromosomes of phages T2 and T4 are not only longer than complete genomes through having the sequence of DNA bases at one extremity repeated at the other (*terminal redundancy*), but are also circularly permuted, as if each particle in the population received a standard length of DNA randomly excised from a long chain of chromosomes attached end to end (*concatenate*) (p. 462). We have seen how this model of the chromosome fits current concepts of morphogenesis (p. 434) and of heterozygote behaviour (p. 508). We want now to discuss some of the elegant physical experiments which have been devised to show directly the existence of terminal redundancy and circular permutation, as well as other peculiarities of chromosomal anatomy.

All these experiments depend on the fact that fragments of a single polynucleotide chain can be 'annealed' to regions of another chain which have complementary base sequences, to yield a double-stranded structure (p. 244). Suppose a chromosome population which is not circularly permuted, so that all the chromosomes possess similar base sequences at their extremities. If these chromosomes are labelled with ^{32}P and their middle parts removed so that only the ends remain, and these end fragments are then denatured to single strands and annealed with the single strands of a denatured preparation of unfragmented chromosomal DNA, it is clear that the terminal fragments can adhere only to the extremities of the intact strands, so that the amount of ^{32}P label bound will be limited. On the other hand, denatured fragments of comparable size, labelled with ^{14}C and derived randomly from every part of the chromosome, will adhere to every part of the intact, denatured DNA molecules so that the amount of ^{14}C-DNA bound will greatly exceed that of the ^{32}P-DNA. In the case of a circularly permuted population of chromosomes, however, the findings will be quite different since the base sequences at the end of some chromosomes will be present at the middle of others, so that fragments derived from the extremities can now anneal with any region of the intact, denatured molecules and so will no longer be at a disadvantage in competition with random fragments—equivalent bonding of terminal ^{32}P-labelled fragments and of random ^{14}C-labelled fragments should be found. Terminal regions of chromosome were obtained by breaking the chromosomal DNA molecules into two by shearing and separating out the smallest fragments by column fractionation. These fragments were then broken into short, single chains by sonication and denaturation, and their ability to bind to whole denatured molecules, fixed in an agar gel (Bolton and McCarthy, 1962), compared with that of similar short chains derived by sonication and denaturation from the entire DNA. The results showed clearly that the terminal and random fragments of phage T2 DNA molecules bound equally well, indicating circular permutation, whereas in the case of DNA from phage T5, used as a control, the random DNA fragments were preferentially bound (Thomas, 1963). It is interesting that genetic analysis of phage T5, by means of temperature-sensitive mutants, has revealed a single, noncircular linkage map which is in keeping with the absence of circular permutation (Fattig and Lanni, 1965).

Another consequence of circular permutation is that if a population of entire DNA molecules is denatured to single strands which are then allowed to re-anneal, the specific bonding of the terminal regions of some strands to the more central regions of others will produce molecules longer than the mean chromosome length which are double-stranded in the middle but have complementary, single-stranded extremities. As Fig. 107 shows, these terminal strands should unite by hydrogen bonding to yield a double-stranded, circular structure, each strand of which is interrupted at a different location. On the contrary, a population of identical DNA molecules should not form circles in this way. Experimentally, phage T2 DNA molecules were denatured by alkali, re-annealed at neutral pH and, finally, examined in the electron microscope by the Kleinschmidt technique (p. 537). Numerous circles were found, in contrast to their complete absence from control, undenatured preparations. As expected, the contour lengths of the circles were the same as those of the entire, linear control molecules (c. 55 μ), while the linear molecules in the circularised preparation showed a variety of lengths as a result of variation in overlap of those annealed structures whose ends had not yet joined (Fig. 107, B). No circles were found when either T5 DNA, or T2 DNA molecules which had been broken by shear, were treated in the same way (Thomas and MacHattie, 1964). More recently, the chromosomes of phages T3 and T7 have been shown to be non-permuted by the same method (Ritchie *et al.*, 1967).

An elegant demonstration of the existence of terminal redundancy in phage T2 involves the use of the enzyme exonuclease III which, as we have previously noted (p. 256), attacks double-stranded DNA from the 3′-hydroxyl ends, so that nucleotides are progressively removed from opposite strands at the extremities of the DNA molecule. As Fig. 108 shows, this treatment exposes single strands at the two ends which carry the complementary base sequences of the redundant regions. Hydrogen bonding between these complementary regions should therefore produce circular molecules, and this has indeed been shown to occur in the case of phages T2, T3 and T7 DNA (MacHattie *et al.*, 1967; Ritchie *et al.*, 1967).

Circular phage T7 DNA molecules formed in this way are illustrated in Plate 25. It will be seen from Fig. 108 that if the exonuclease III degradation extends beyond the region of redundancy on the two chains, the resulting circle will have two single-stranded regions enclosing a double-stranded region between them, those length corresponds to that of the redundant segments. In some of the circular molecules a pair of small regions of apparent interruption, on either

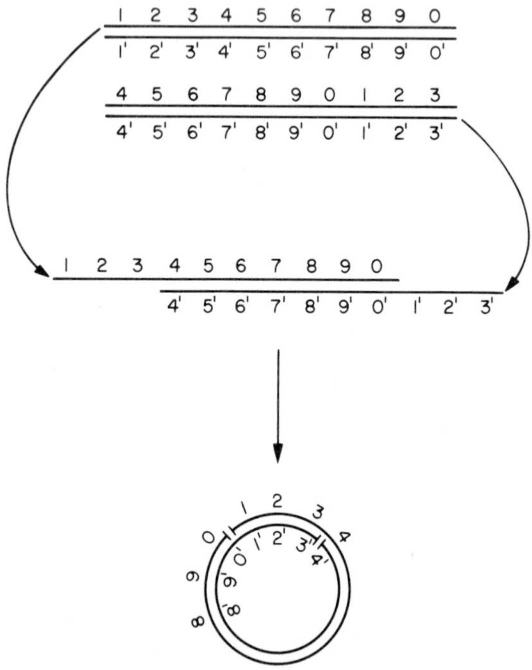

FIG. 107. Illustrating the demonstration of circularly permuted DNA molecules by denaturation and annealing. The paired, parallel lines represent the two strands of a pair of circularly permuted DNA molecules (top), the figures 1 and 1′, 2 and 2′, and so on, indicating complementary pairs of bases. When the DNA molecules are denatured, so that the two strands separate completely, and are then annealed, complementary regions of circularly permuted strands can pair as shown (middle), to yield a duplex having complementary single strands at its extremities. These single strands will pair by hydrogen bonding to form circles of double-stranded DNA (bottom).

side of a region of normal duplex, can be observed which probably represent such single-stranded regions. Measurements of the intervening duplex, based on this assumption, give the length of the

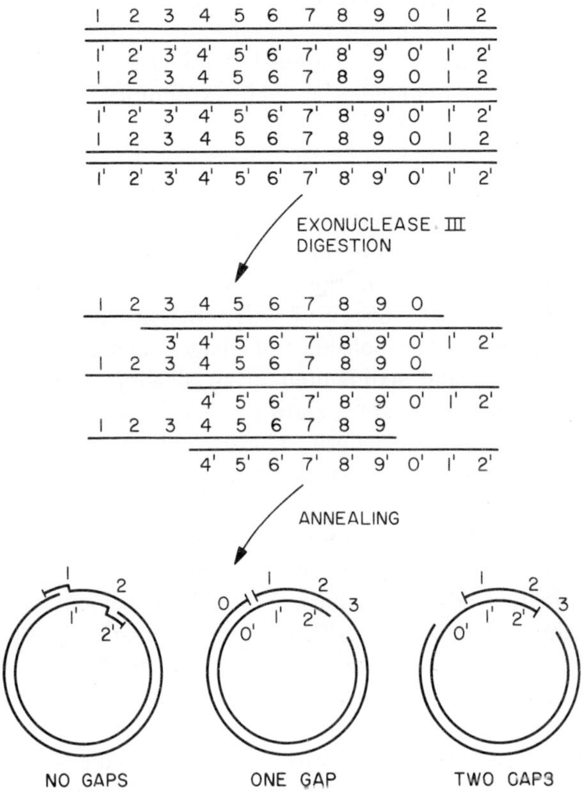

FIG. 108. Illustrating the demonstration of terminally redundant DNA molecules by the action of the enzyme exonuclease III.

The paired, parallel lines represent the two strands of a number of identical, terminally redundant, DNA molecules (top), the figures 1 and 1′, 2 and 2′, and so on, indicating complementary pairs of bases. The action of exonuclease III, which breaks down DNA strands from the 3′ ends, exposes the complementary 5′-ended strands at the extremities (middle). Thus, on annealing, these strands pair by hydrogen bonding to form circles (bottom). If the degradation proceeds beyond the limits of the terminal repetition, then two single-strand gaps will bracket a duplex segment, the length of which is the length of the terminal redundancy. (Reproduced from Ritchie *et al.*, 1967, by kind permission of the authors and the *Journal of Molecular Biology*).

terminally redundant region of the molecule as about 0·7 per cent of the entire molecule (12·5 μ), or about 260 nucleotide pairs, in the case of phage T7 (Ritchie *et al.*, 1967), as compared with one to three per cent for equivalent measurements in the case of phage T2 (MacHattie *et al.*, 1967). The DNA molecules of phages like T3 and T7, which are terminally redundant but not circularly permuted, all have identical base sequences at their extremities. It follows that, after treatment with exonuclease III, in addition to bonding between the ends of the same molecules to yield circles, the ends of different molecules may join up so that chains of molecules, or *concatenates*, may form. This, too, has been demonstrated. Whereas only circles are found at low molecular concentrations, dimers, trimers and higher 'concatemers' appear when the concentration is increased during annealing (Ritchie *et al.*, 1967; review, Thomas and MacHattie, 1967).

When the entire DNA molecules of phages such as T2, P1, λ and T7 are denatured and sedimented in an alkaline sucrose gradient, they sediment in a single zone, showing that the single strands of the DNA are homogeneous in length. The DNA of phage T5 is peculiar, however, since on denaturation its sediments in at least four different zones. This indicates the existence of breaks, or gaps, in one or the other, or both of the DNA strands, and is in keeping with an earlier finding that phage T5 DNA shows preferred breakage points when subjected to hydrodynamic shear. From the behaviour of the fractions obtained in sucrose gradients, various possible models of the distribution of the gaps in the strands of the phage T5 DNA molecule can be inferred. All the models feature four gaps, one on one strand and three on the other, and have an uninterrupted 4 μ length of duplex at one end. Since the entire T5 DNA molecule is 38·8 μ long, this terminal fragment of duplex represents ten per cent of the genome (Abelson and Thomas, 1966). This information about the physical structure of phage T5 DNA is, no doubt, relevant to certain unusual aspects of infection by this virus. If infected cells are treated early in a high speed mixer (blendor) so as to shear off the attached phage particles soon after adsorption, no phage DNA synthesis occurs and no plaques subsequently arise, but there is massive and rapid breakdown of the host's DNA (Lanni and McCorquodale, 1963). By the use of radioactive labelling it was shown that the effect of agitation in a mixer is to isolate in the infected cell an exclusive, double-stranded fragment of DNA, amounting to about ten per cent of the total phage DNA, which carries a gene or genes determining degradation of the bacterial DNA. In the absence of agitation, or if agitation is applied late in infection, the entire T5 DNA molecule is transferred to the host (Lanni, McCorquodale and Wilson, 1964). Presumably the ten per cent fragment is terminal and is separated from the rest of the DNA molecule by a point of weakness which is easily severed in the mixer; it is tempting to equate this fragment with the terminal ten per cent fragment defined by pure physical methods.

PHAGE LAMBDA

We have already dealt in some detail with the chromosome of phage λ both as a genetical and a physical structure (p. 457). We have seen that the linear DNA molecules in the phage heads possess *cohesive ends* which are almost certainly complementary, single-stranded regions so that, on heating and slow cooling for example, the ends of individual molecules anneal to form circles; alternatively, the complementary ends of different molecules may anneal so that concatenates are produced (Hershey, Burgi and Ingraham, 1963; Ris and Chandler, 1963; MacHattie and Thomas, 1964; Hershey and Burgi, 1965). In fact, phage λ DNA molecules are similar in structure to those of phages T3 and T7 which have been treated with exonuclease III, as described in the last section. Since all the molecules of a population have the same cohesive ends, which in fact are shared by other, related lambdoid phages so that mixed concatenates can form (Baldwin *et al.*, 1966), the phage λ chromosome is not circularly permuted, except in the sense that the vegetative and the prophage genetic maps, both linear, are circular permutations of one another (see p. 460). In addition, various closed forms of λ DNA molecules, including circles with both strands unbroken (Young and Sinsheimer, 1964; Ogawa and Tomizawa, 1967) as well as concatenates (Smith and Skalka, 1966), have been found in infected cells.

By measuring the lengths of unbroken and untangled DNA molecules, as revealed in electron microphotographs by the Kleinschmidt or other techniques, it is possible to make a direct estimate of their mean molecular weight. The mean length of the molecules is about 17·3 μ; if the highly hydrated 'B' configuration is assumed for the DNA, this length contains about 50,000 base pairs and corresponds to a molecular weight of 33 million, which is the same as that derived by other methods. Using this technique, the DNA molecules of deletion mutants of phage λ have also been examined (pp. 218, 470). In the case of one such mutant which had been shown by density measurements to have lost 23 per cent of the wild type DNA content, the molecules were shown to be just over 13 μ long, that is, about 23 per cent shorter than normal (MacHattie and Thomas, 1964; Caro, 1965).

One of the principal aims of molecular genetics is to describe genetic phenomena in purely physico-chemical terms. A notable contribution to this end has been the direct demonstration that various functional regions on the DNA molecules of phage λ are arranged in the same order as their equivalent loci on the vegetative linkage map. If a preparation of wild type phage λ DNA is added to sensitive *E. coli* bacteria which have previously been infected with defective, mutant λ particles which by themselves are unable to produce plaques, the wild type molecules are taken up by the cells, giving rise to active infection and plaque formation. The DNA alone is not infective and requires the addition of the mutant, 'helper' phage to penetrate the cell (Kaiser and Hogness, 1960). This procedure was later developed as a method for the biological assay of *fragments* of λ DNA molecules; if a particular fragment carries the gene for which the 'helper' phage is defective, it can restore normal phage function and plaque formation. Thus the genes carried by DNA fragments can be assessed by using 'helper' phages which are mutant with respect to a range of different genes.

In the first experiments, the linear DNA molecules were sheared into half molecules which, because of differences in base composition, could be separated into two types by chromatography on methylated serum albumin columns. One type, the 'left half,' was found to carry the *m*6 marker which is now known to be located in the *A-B* region of the vegetative linkage map, while the 'right half' fragments carry the immunity (*c*I) and *R* regions (Fig. 101, p. 457) (Kaiser, 1962; Hogness and Simmons, 1964; historical review, Kaiser, 1966). Recently the resolution of this type of analysis has been increased to the extent that the genes can be ordered on the molecule, by analysing a wide range of fragments of different but overlapping sizes, separated in a sucrose gradient. Each fraction is then assayed for the presence of different pairs of genes on the *N-R* half of the vegetative linkage map (Fig. 101, p. 457). Since the presence of a cohesive end is necessary for DNA uptake in the presence of 'helper' phage (Strack and Kaiser, 1965; Kaiser and Inman, 1965), all the active fragments must carry the gene *R*. It follows that a correlation should exist

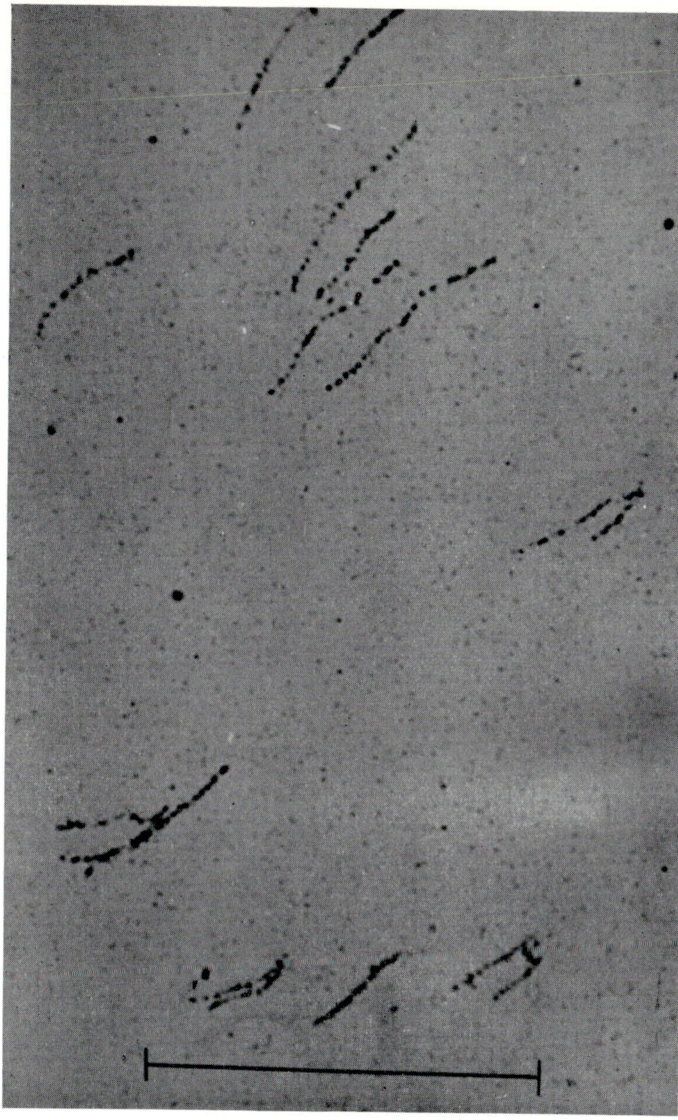

PLATE 23. Autoradiograph of DNA from phage T2 which has incorporated radioactive (^{3}H) thymidine. The scale is 100 μ long. See text.

(Reproduced from Cairns, 1961, by kind permission of the author, and the editors of the *Journal of Molecular Biology*)

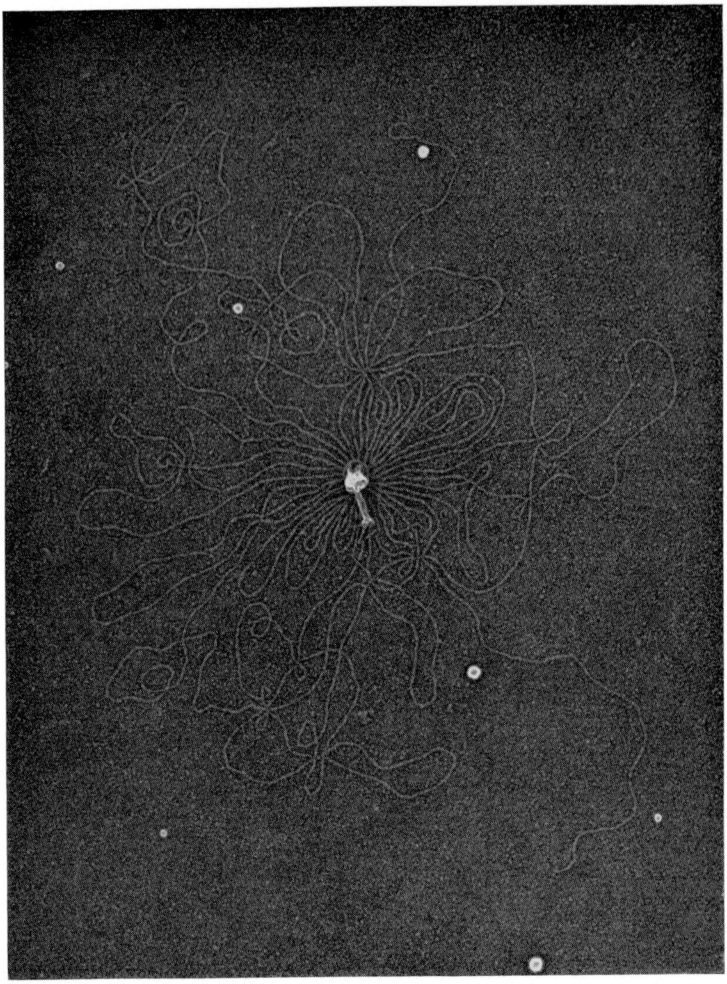

PLATE 24. Electron micrograph (shadowed) of phage T2 DNA, released by osmotic shock. Magnification about ×42,000. See text for account of method used.

(Reproduced from Kleinschmidt, Lang, Jacherts & Zahn (1962) *Biochim. biophys. Acta*, **61**, 857, by kind permission. Original photograph kindly provided by Dr A. K. Kleinschmidt)

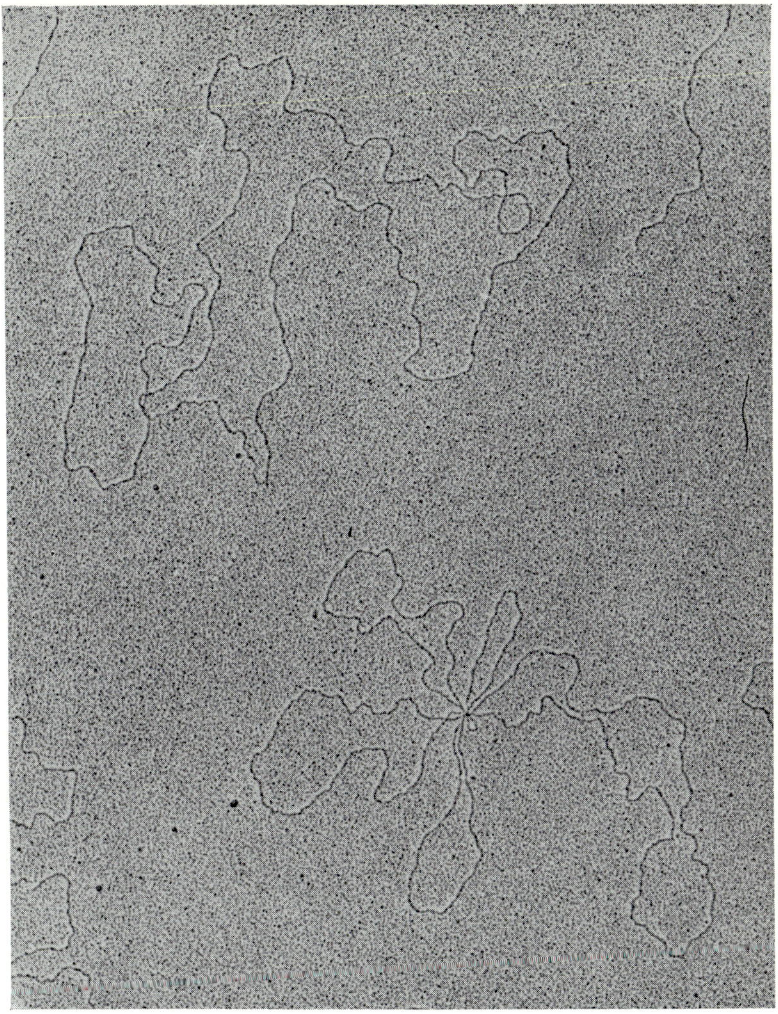

PLATE 25. Electron microphotograph of a circular, phage T7 DNA molecule prepared by the Kleinschmidt method. The DNA molecule extracted from a phage T7 particle is a linear duplex 12·5 μ long (Mol. wt. 24×10^6 daltons). Conversion to circular molecules occurs upon annealing, following partial hydrolysis of the 3′ ends with exonuclease III. See Ritchie *et al.*, 1967.

(Original photograph kindly provided by Dr D. A. Ritchie).

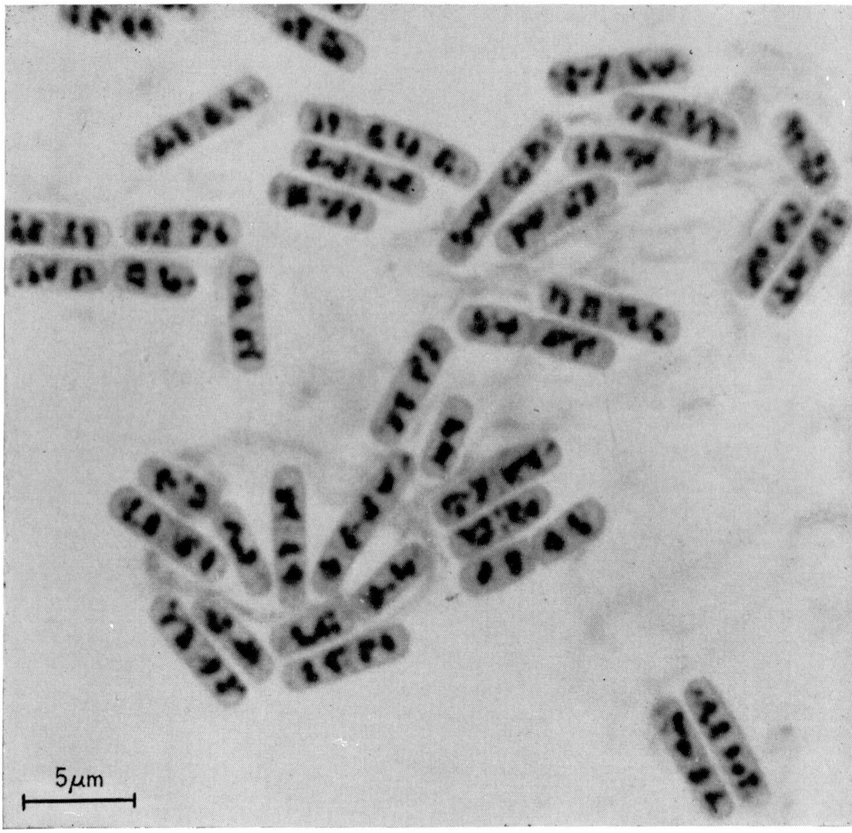

PLATE 26. Microphotograph showing the morphology of the chromatin bodies of *Bacillus cereus*.

Preparation made from 2½ hours growth on Difco heart infusion agar at 37°C., fixed in OsO_4 vapour and treated with HCl before staining with Giemsa's stain. There has been some shrinkage of the chromatin bodies which are seen in all stages of division. Magnification about ×2,880.

(Reproduced from Robinow, 1962. Original photograph kindly provided by Dr C. F. Robinow and the British Medical Bulletin)

PLATE 27. Autoradiograph of a replicating chromosome of *E. coli* strain K 12 (Hfr ♂). The thymine-requiring bacteria were fed tritiated thymidine, which is specifically incorporated into DNA, for two generations before the DNA was extracted and subjected to autoradiography (2 months exposure). The autoradiograph is interpreted by the diagram above it, inferred from measurements of grain density; the continuous lines represent tritium-labelled strands, and the interrupted lines unlabelled strands, of the DNA double helix. During the first generation in the presence of tritiated thymidine only one strand (the newly synthesised one) of each daughter chromosome becomes labelled; during the next generation, the daughter chromosome built on the labelled strand as a template will have both strands labelled, while its sister based on the unlabelled strand will have only one strand labelled. Thus in the diagram

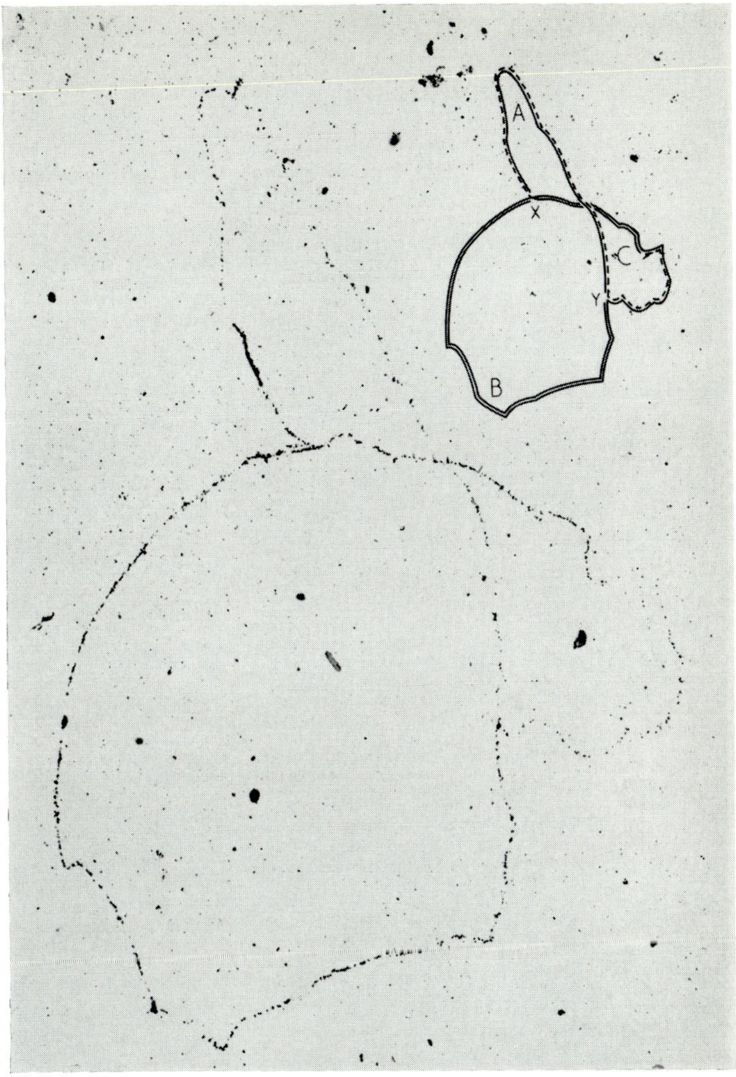

the chromosome is about $\frac{2}{3}$ replicated, XBY (doubly labelled) and XAY (singly labelled) being the daughter replicas, and XCY being the parental double helix. Observe that part of XCY, XC is double labelled. This is because the label was first introduced when chromosome replication had reached point C, towards the end of a cycle; when this point is again reached on one of the daughter chromosomes, double labelling of the region CX will begin before the 2nd round of replication has started *from the initiation point*. The inference is that X must be the initiation point and Y the growing point of replication. The length of the chromosome is 1100 microns. Reproduced from Cairns (1963*b*) by kind permission of the author and the Cold Spring Harbor Laboratory of Quantitative Biology.

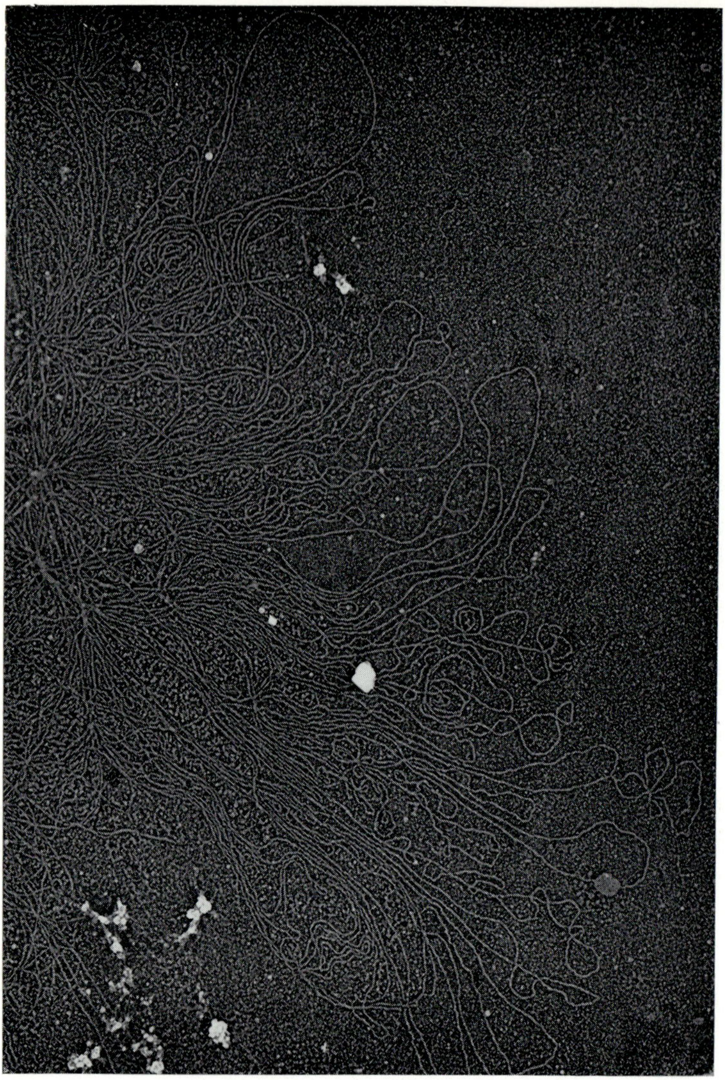

PLATE 28. Electron micrograph (shadowed) of DNA released from *Micrococcus lysodeikticus*. Magnification about ×31,200. See text for account of method used.

(From Kleinschmidt, Lang & Zahn (1961) *Z. Naturforsch.* **16b**, 730, by kind permission. Original photograph kindly provided by Dr A. K. Kleinschmidt)

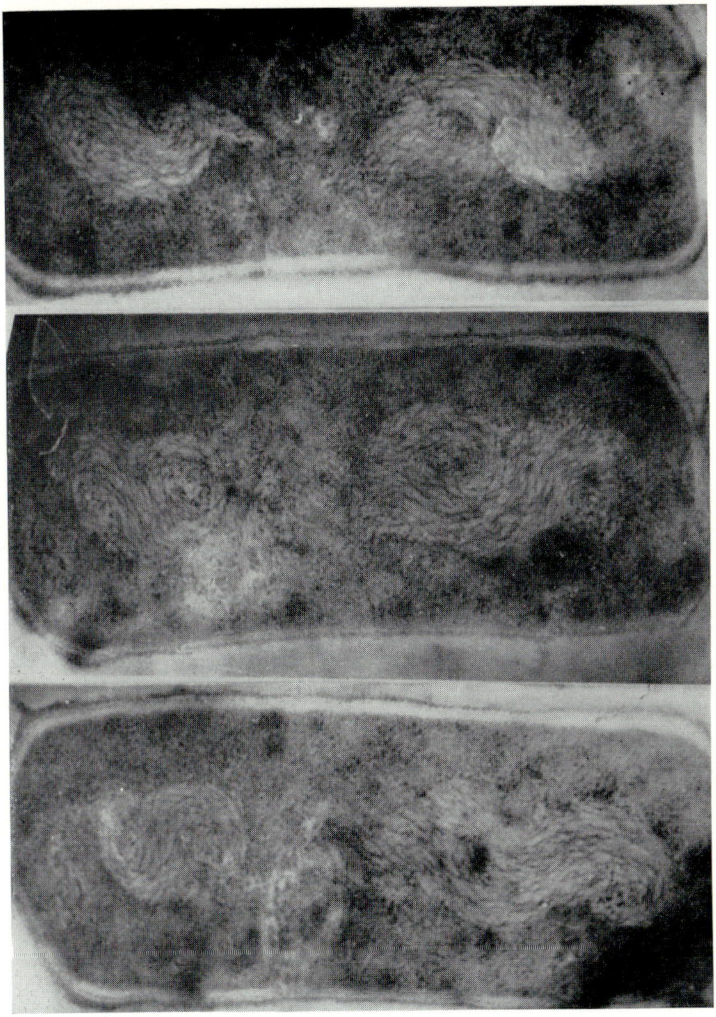

PLATE 29. Electron micrographs of ultra-thin, serial sections of chromatin bodies of *Bacillus subtilis*.

(From Fuhs, 1962. The original negative was kindly provided by Dr G. W. Fuhs)

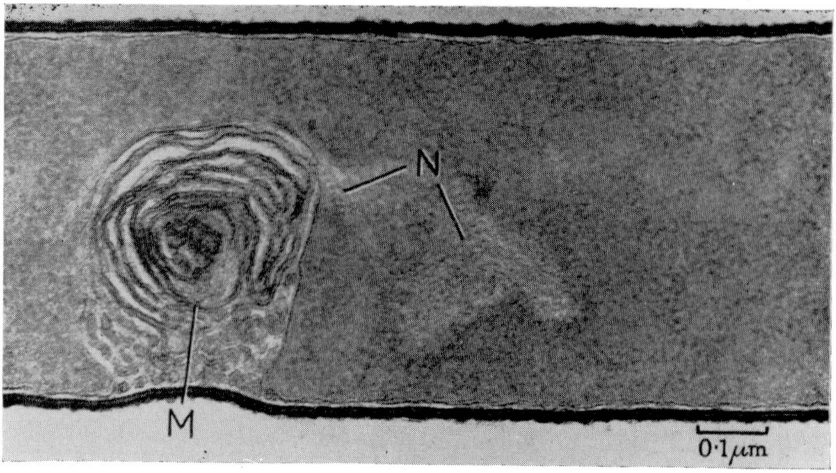

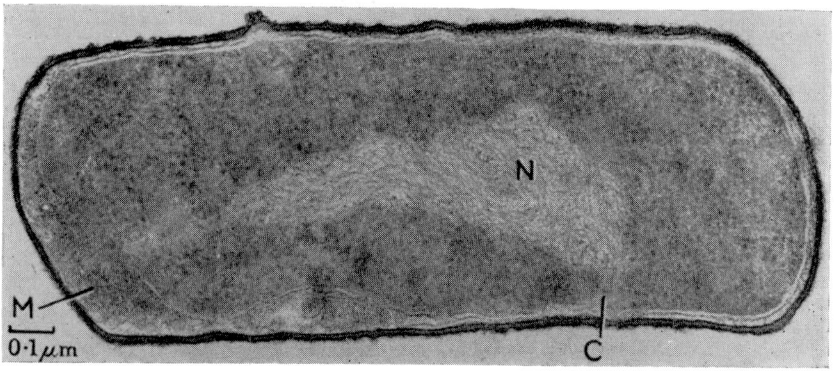

PLATE 30. Electron microphotographs of ultra-thin sections of growing *Bacillus subtilis*, showing attachment of the chromosome (N = nucleus) to the cell membrane through mesosomes (M).

The upper photograph shows the connection between chromosome and mesosome.

The lower photograph is of a section taken after exposure of the bacteria to 0·5 M sucrose for 30 minutes. The mesosomes have become evaginated and disentangled by the osmotic pressure, so as to be indistinguishable from the rest of the cell membrane, to which the chromosome can now be seen to be attached at C.

(Reproduced from Ryter and Jacob, 1964, by kind permission)

between the map distance between two markers and the size of the molecular fragments which carry them; for example, the smallest fragments might carry *R* alone, larger fragments *R* and *Q*, still larger fragments *R* and *P* or *R* and *O*, and only the largest fragments *R* and *N*. This correlation was demonstrated experimentally (Hogness *et al.*, 1966).

REPLICATION OF THE PHAGE CHROMOSOME

Studies of the density of DNA molecules, obtained from bacteria labelled with heavy nitrogen (^{15}N) and multiplying in a medium containing only normal nitrogen, have shown very clearly that molecules of bacterial DNA replicate semi-conservatively (Meselson and Stahl, 1958a, b); all the heavy molecules, labelled in both strands, disappear by the end of the first generation and are replaced by 'hybrid' molecules having one heavy parental strand and one newly synthesised, light strand (p. 241). Attempts to apply this method of investigation to the replication of phage DNA in infected bacteria yielded results which, at first sight, seemed to exclude semi-conservative replication.

In the case of phage T4, the DNA extracted from infected cells at intervals after infection, and having a mean molecular weight about 2×10^7, failed to yield a 'hybrid' band on centrifugation in a caesium chloride density gradient. However, if the molecular weight of the extracted DNA was first reduced to about 10^6 by sonication, then 'hybrid' molecules appeared (Kozinski, 1961). There is little doubt that the existence of 'hybrid' molecules indicates that semiconservative replication occurs. How is it that these molecules are not observed unless the DNA is broken into small fragments? The probable answer is that the 'hybrid' molecules, instead of remaining together in the chromosome to which they initially belong, become dispersed among chromosomes which are not 'hybrid'. When these chromosomes are broken down into relatively large fragments, their 'hybrid' content is only a fraction of the whole, insufficient to affect their density significantly. But when the fragments are further reduced to a size comparable to that of the 'hybrid' components, the latter are released and band at their own buoyant density.

Since the genetic material of phage T4 undergoes at least five rounds of mating in the DNA pool during the growth cycle, it is likely that this dispersion of 'hybrid' molecules can be fully accounted for by genetic recombination. Remember that only a small number of 'hybrid' *chromosomes* will be produced (two from infection by a single particle), and this number will remain constant in a pool which contains about 50 phage equivalents of unlabelled DNA by the time mature phage begins to be formed, so that the probability of mating with unlabelled chromosomes is very high. The temperate phage, λ, undergoes only about 0·5 rounds of mating during the growth cycle, and here the development of 'hybrid' DNA molecules during replication can readily be demonstrated without sonication (pp. 244, 509; Meselson and Weigle, 1961).

THE MORPHOLOGY OF THE GENETIC APPARATUS OF BACTERIA

Attempts to define the genetic apparatus of bacteria in terms of an observable, morphological structure have had a turbulent and controversial history which began many years before the birth of bacterial genetics. In fact, it is only quite recently that studies of this sort have acquired the status of a respectable and important corollary of genetic analysis. Bacteria tend to stain uniformly with basic dyes so that, for a long time, it was generally believed either that they constituted naked nuclei or else that they did not possess any form of discrete, organised nucleus, in spite of many isolated reports to the contrary (see Dubos, 1945). The realisation that the uniform staining of bacteria is due to their rich cytoplasmic content of ribonucleic acid (RNA), which possesses the same basophilic staining properties as deoxyribonucleic acid (DNA), together with the development of cytochemical methods for the selective staining of DNA, eventually led to the unambiguous demonstration, in all the bacterial species examined, of discrete chromatin bodies which could plausibly be interpreted as nuclear analogues (Piekarski, 1937; Robinow, 1942, 1944, 1945). We do not intend to discuss here any very controversial matters (reviews: Robinow, 1956a, b; 1960; 1962; Kellenberger, 1960; Fuhs, 1962, 1965), but only to present such facts as now seem established concerning the nature of the bacterial chromosome as an observable structure.

THE DEMONSTRATION OF CHROMATIN BODIES

The chromatin bodies of bacteria are most convincingly revealed in the case of bacilli, such as *E. coli*, *Proteus vulgaris* and aerobic sporing bacilli. Bacteria from young cultures are fixed with osmium tetroxide, treated with N/1 hydrochloric acid to hydrolyse selectively the cytoplasmic RNA, and then stained with Giemsa's solution or some other basic dye (Piekarski, 1937; Robinow, 1942). The chromatin bodies appear as darkly staining rodlets, arranged more or less at right angles to the long axis of the bacilli. Each bacillus normally contains two to four or more rodlets, but the number may vary widely with the age of the culture and the conditions of growth. The rodlets often appear bifid, as if dividing, and tend to be arranged in pairs; the formation of a cellular transverse septum, which precedes bacterial division, can often be seen to separate two pairs of rodlets. The rodlets themselves may adopt widely different forms, even within a single bacterium. Apart from the bifid V-shaped structure, dumb-bell-, club-, pistol-, U- and T-shaped forms are common (Robinow, 1960). Most of these features are clearly shown in Plate 26 (between pp. 544 and 545).

These nuclear bodies can be observed in living, unstained bacteria by means of phase-contrast microscopy which allows their behaviour to be followed during bacterial growth and division (Mason and Powelson, 1956). They appear more diffuse than in the acid-treated, stained preparations and are less dense than the surrounding cytoplasm. They divide regularly during each division cycle, prior to division of the cell, so that they behave in the way one would expect of nuclei. Similar rather diffuse chromatin bodies become visible on staining when fixed bacteria are treated with ribonuclease instead of with acid (Tulasne and Vendrely, 1947), and it is obvious that the bodies revealed by all these methods are the same.

We should point out that although the appearances we have described are rather typical and are widely found, the morphology of chromatin bodies may be drastically altered by environmental influences. When *E. coli* is grown in broth containing 2 per cent sodium chloride they may assume a globular shape, while in the presence of 4 μg/ml aureomycin each bacterium may contain only a single, spherical chromatin body; yet under both conditions the chromatin bodies may still be able to reproduce. Again, following small doses of ultraviolet light, the chromatin bodies appear to coalesce into a single, irregular bar of chromatinic material extending almost to the extremities of the cell (see Kellenberger, 1960).

CHROMATIN BODIES AS GENETIC STRUCTURES

Two general lines of evidence show, beyond doubt, that the chromatin bodies carry the hereditary determinants of bacteria. The first of these is that chromatin bodies not only contain DNA, to which they owe their staining properties, but that all (or nearly all) the DNA of the cell is concentrated within them. For example, chromatin bodies give a positive Feulgen reaction (see Robinow, 1960), and fluoresce with the yellow-green colour typical of DNA when stained with acridine orange (Anderson, 1957; Anderson, Armstrong and Niven, 1959), while their ability to be stained by any dye is abolished by treatment with deoxyribonuclease (Munson and Maclean, 1961).

The mean number of chromatin bodies per bacterium can be varied by changing the nutritional properties of the culture medium or the temperature of incubation. It is found in the case of *Salm. typhimurium*, under different conditions of 'balanced growth' where the generation time ranges from 0·6 to 2·8 generations per hour and the number of chromatin bodies per bacterium from one to three, that the amount of DNA per chromatin body remains constant despite wide fluctuations in mass and in RNA content (Schaecter, Maaløe and Kjeldgaard, 1958; see also Maaløe and Kjeldgaard, 1966). Again, the spores of aerobic sporing bacilli are known to contain a single chromatin body only, while the amount of DNA per spore is remarkably constant for a given species. Vegetative bacilli resulting from spore germination, in which the mean number of chromatin bodies is found to be two, contain just twice the amount of DNA per cell (Young and Fitz-James, 1959). Finally, autoradiographs of sections of *E. coli* bacteria which have incorporated tritium(^{3}H)-labelled thymidine into their DNA show that the tritium is concentrated into the same locations of the cell as the chromatin bodies revealed by electron microscopy (Caro, van Tubergen and Forro, 1958).

The only direct demonstration of the genetic role of the chromatin bodies was made by Witkin (1951). In old (resting) cultures of *E. coli* the mean number of chromatin bodies per bacterium is one. On subculture to a fresh medium, the number rises to two during the lag phase, and then to four when the growth becomes exponential. Assuming that the chromatin bodies are haploid, if a mutation arises in one of them, the purity of the emergent line of descent with respect to the mutant character will be a function of the number of chromatin bodies in the bacterium at the time the mutation occurred. Thus if the bacterium possesses only a single chromatin body which suffers a mutation, all the progeny of the cell will inherit the mutation so that, on solid medium, a colony of pure mutant phenotype will emerge (Fig. 109A). If the bacterium has two chromatin bodies, only one of which becomes mutant (Fig. 109B), then the mutant and non-mutant bodies will segregate into separate cells at the first division so that the final population contains equal numbers of mutant and non-mutant bacteria. If the mutant character is of a kind directly observable by colony inspection, a half-sectored colony will be seen. Similarly, a mutation in a bacterium with four chromatin bodies will yield one mutant and three non-mutant bacteria at the end of the second division, resulting in a quarter-sectored colony (Fig. 109C).

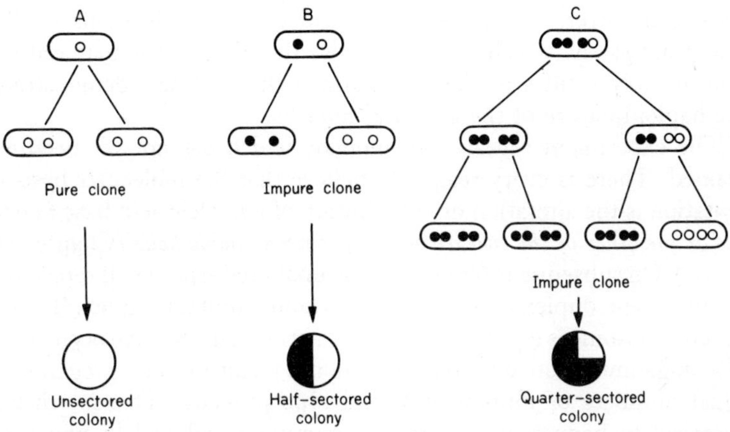

FIG. 109. Shows how the proportion of mutant bacteria in colonies derived from a single bacterium in which a mutation has occurred, reflects the number of chromatin bodies present in the bacterium at the time of mutation.

The black circles indicate wild type chromatin bodies; the white circles are chromatin bodies which have experienced, or inherited, a mutation. See text. (From Hayes, 1960.)

Witkin irradiated populations of lactose-fermenting (lac^+) *E. coli*, having different mean numbers of chromatin bodies per bacterium, with large doses of ultraviolet light to induce mutations to inability to ferment lactose (lac^-). The bacteria were then plated on eosin-methylene blue (EMB) medium containing lactose, on which colonies of lac^+ and lac^- bacteria are readily distinguishable (p. 101; Plate 7). She found the following correlations.

1. The ratio of unsectored to sectored colonies is proportional to the ratio of cells containing only one chromatin body to those containing more than one.

2. Populations showing mainly two chromatin bodies per cell yield a predominance of half-sectored colonies; in those with mainly four chromatin bodies per cell, quarter-sectored colonies predominate.

3. The greater the number of chromatin bodies per cell of the population, the higher is the total yield of mutants, as would be expected from the greater number of mutable sites in multinucleate populations. Control experiments showed that, in a small proportion of cases, sectored colonies may arise from causes other than simple segregation of the chromatin bodies. The frequency of sectoring from such causes, however, is too low to invalidate the implications of this important study, which not only shows that the genetic material unquestionably resides in the chromatin bodies but also demonstrates the haploid nature of the bacterial nucleus.

The experiment reveals one anomaly which has not yet been explained. There is every reason to believe that the molecular basis of mutation is the alteration or substitution of a nucleic acid base *in only one or the other of the two strands of the* DNA *double helix* (Chapter 13, p. 302). On subsequent replication the unaltered strand will reproduce the wild type duplex and the altered strand a mutant duplex. Thus if the chromosome is equivalent to double-stranded DNA, the occurrence of a single mutation at a given site should result in the segregation of equal numbers of mutant and wild type progeny. This is what is observed to happen in the case of mutations induced in phage by nitrous acid. It follows that mutations in bacteria containing only a single chromatin body should yield half-sectored colonies and not the pure mutant clones observed by Witkin.

THE STRUCTURE AND ORGANISATION OF THE BACTERIAL CHROMOSOME

Genetic analysis of *E. coli* has shown that all of the large number of genes now mapped are located on a single, circular linkage group (Fig. 127, p. 666). By analogy with phage we might anticipate that the bacterial chromosome is, basically, a single molecule of DNA of immense length. The methods which we have already described for phage, whereby DNA can be gently liberated and dispersed, and then visualised by autoradiography or by the electron microscope (p. 537) have also been applied to bacteria with striking results. In the autoradiographic method *E. coli* bacteria, which have incorporated tritiated [^{3}H] thymidine of high specific activity into their DNA, are gently lysed in the presence of excess, unlabelled DNA from phage T2 which acts as a protective, supporting 'carrier' for the labelled molecules. After treating to remove protein, the bacterial DNA is allowed to disentangle slowly on the wall of a dialysis membrane, influenced only by the gentle stresses of Brownian movement, convection currents and the drag of the meniscus as the dialysis chamber is slowly drained. The membrane is then overlaid with photographic emulsion. When the resulting autoradiographs are examined, the *E. coli* DNA, revealed by the silver grains provoked by the tritium disintegrations, is found mainly in the form of tangled masses. However, some continuous DNA threads about 1,100 μ or more long are also found (see Plate 27, between pp. 544 and 545) (Cairns, 1962, 1963a, b). Calculations based on chemical estimations of total DNA, and allowing for the fact that each bacterium contains several such bodies and that DNA synthesis continues throughout the division cycle (see next section), show that, if each *E. coli* chromatin body contains a single continuous DNA duplex, this should have a maximum extended length of not more than 1,400 μ. The fact that the size of the silver grains (c. 1 μ) is about 500 times greater than the diameter of the DNA molecule which they reflect, makes it difficult to exclude small tangled regions, so that Cairn's pictures provide indisputable direct evidence that the chromosome of *E. coli* is indeed a continuous, circular thread of double-stranded DNA.

The electron microscopic method of visualising the chromosome has the advantage that the diameter of the chromosomal thread can be assessed. In the case of the bacterium *Micrococcus lysodeikticus*, protoplasts were lysed in a spreading monomolecular layer of basic protein (cytochrome *c*), which combines with the DNA, on a water surface. The emerging DNA is thus very gently dispersed and is finally photographed after metal-shadowing. As Plate 28 (between pp. 544 and 545) shows, the chromosome is seen as an enormously long, tangled thread, of a diameter compatible with that of double-stranded DNA, in which no free ends can be seen (Kleinschmidt, Lang and Zahn, 1961). These experiments, of course, give no indication of the extent to which protein or RNA may be associated with the DNA as structural or functional (e.g. messenger RNA) components of the chromatin body (see p. 293).

Assuming that the chromosome consists of a single molecule of DNA, the next question to ask is how this molecule is organised in the chromatin body. Since the dimensions of chromatin bodies are close to the limits of resolution of the light microscope, little information about their fine structure and organisation can be obtained by this means. Electron microscopy of intact cells is not much more revealing, since this involves visualisation of the whole thickness of a body which is less electron-dense than the surrounding material, while it is difficult to assess to what extent artefacts are introduced by the need for fixation and dehydration.

Most of the information we possess about the organisation of the chromatin bodies comes from electron microscopy of *ultra-thin sections* of bacteria. The cells, after fixation and careful dehydration, are permeated by a monomeric solution of a synthetic resin which is then allowed to polymerise. Sections of the order 0·05 μ thick are cut, and examined without removal of the resin. The adoption of a fixation procedure which avoids artefact formation is obviously crucial and raises the question as to how we may recognise what is normal or abnormal. An ingenious solution is to use phage-infected bacteria as a control, and to take as standard a procedure which reveals the finely fibrillar state of the phage DNA pool (p. 431 and Plate 17) (Kellenberger, Ryter and Séchaud, 1958; Ryter and Kellenberger, 1958a, b; review: Kellenberger, 1960).

These methods, developed by E. Kellenberger and his colleagues, have since been used by several groups of workers in the study of a number of bacterial species, especially *E. coli* and aerobic sporing bacilli, with the result that some important features of the ultrastructure of chromatin bodies have been brought to light.

1. The bodies are not confined by a nuclear membrane. This suggests that their shape may be determined, at least in part, by their internal organisation.

2. There is no evidence that any one body contains the analogue of more than one chromosome which, as we shall see in Chapter 22 (p. 666), is in full agreement with the results of genetic analysis.

3. The chromatin body itself has the appearance of a finely fibrillar 'nucleoplasm' which closely resembles that of the phage DNA pool (Plate 17, p. 432) (Kellenberger *et al.*, 1958; Ryter and Kellenberger, 1958a; van Iterson and Robinow, 1961). The fibrils composing it have a diameter of the order 20 to 60 Å, but the impregnation surrounding every fibril precludes accurate measurement (Kellenberger, 1960). However, it may be assumed from the evidence we have already reviewed that these fibrils are constructed mainly, if not exclusively, of DNA.

4. Although the fibrils often appear to be arranged randomly, giving 'the impression of being part of a felt or sponge' (Robinow, 1962), recent studies of chromatin bodies of *E. coli* and *B. subtilis* show clearly that they are organised into bundles which may contain up to 500 parallel DNA fibres (Kellenberger, 1960; Fuhs, 1962, 1965). Serial sections, particularly of *B. subtilis* protoplasts, have revealed that the fibres are not twisted about one another in the long axis of the bundle to form a rope-like structure, as was at first supposed, but that the twisted appearance results from the wrapping of two bundles around each other (Plate 29). Similarly, the existence of a multi-stranded ring structure has been excluded. The most likely type of organisation is that of a simple bundle composed of a DNA fibre folded back and forth along itself (Fuhs, 1965).

5. Unlike the DNA contained in phage heads, that of the chromatin bodies is not condensed but resembles the hydrated DNA of the replicating phage DNA pool (Plate 17, facing p. 432); appearances suggesting condensation are artefacts due to dehydration. In fact, the fibrillary structure of chromatin bodies retains the same fundamental texture throughout the division cycle, as can also be seen by phase contrast microscopy. There is no basic change in chromosome structure corresponding to that observed during mitosis in the cells of higher organisms (Kellenberger *et al.*, 1958).

The relative compactness of chromatin bodies, despite the absence of a restricting nuclear membrane, and the way in which the DNA seems to be folded, imply some kind of organisation into a tertiary structure, although this must be rather flexible to permit not only the many variations in shape which chromatin bodies can adopt, but also the continuous semi-conservative replication of the DNA which we will discuss in the next section. Nothing is yet known about the nature of this organisation but, on general principles, we would suspect that basic protein is involved in it.

REPLICATION OF THE CHROMOSOME AND ITS CONTROL

We have seen that autoradiographic analysis of dividing plant cells shows that DNA synthesis occurs during late interphase and early prophase and occupies only a fraction of the division time (Chapter 15, p. 389). Initial studies of DNA synthesis in synchronously dividing bacteria suggested that in them, too, synthesis is restricted to only part of the division cycle (review: Campbell, 1957b). This, however, was the result of interference with metabolic control imposed by the synchronisation procedure (temperature shifts), and is not the case with normal bacterial populations except under conditions where growth is very slow (Maaløe and Kjeldgaard, 1966). On the contrary, it turns out that DNA synthesis continues throughout the division cycle. When a growing, *unsynchronised* culture of a strain of *E. coli* which requires thymine for growth is exposed to a pulse of [^{3}H]-thymidine during only a small fraction of the generation time, only those bacteria engaged in DNA synthesis at the time of exposure will take up the radioactive base and incorporate it into their DNA. Yet autoradiography shows that the chromatin bodies of the great majority of individual bacteria become radioactive under these conditions. If it is assumed that the cells of an unsynchronised culture are distributed randomly with respect to DNA synthesis, this result means that DNA replication must continue during most, if not all, of the division cycle (Schaecter, Bentzon and Maaløe, 1959; but see p. 567). Similar results have been obtained by following the rate at which bacterial populations are rendered inviable as a function of radioactive decay in their DNA, following exposure to short 'pulses' of ^{32}P during growth (p. 523; McFall and Stent, 1959). Thus the kinetics of DNA synthesis in bacteria accord well with the uniform microscopic texture of their chromatin bodies throughout the division cycle, and indicate quite clearly that the bacterial chromosome does not divide by mitosis.

We have already seen how the famous experiment of Meselson and Stahl (1958a, b) showed that molecules of *E. coli* DNA reproduce themselves in the semi-conservative manner predicted by Watson and Crick (1953b) from the structural features of their model; that is, the two complementary polynucleotide strands unwind and separate, each then acting as a template for the synthesis of a new complementary strand so that two identical progeny duplexes arise, in each of which one of the parental strands is conserved. However in this experiment the mean molecular weight of the DNA molecules was 7×10^6, whereas that of the intact chromosomal DNA is about 500 times greater. It follows that the experiment only tells us how minute fragments of the chromosome replicate and provides no direct information about replication of the chromosome as a whole. For example, we would like to know whether replication begins at only one or at many points on the chromosome, and whether it proceeds in the same or in opposite directions along the two strands. In fact the Meselson-Stahl experiment provides more information than is at first apparent—for example, the delayed appearance of light ($^{14}N—^{14}N$) DNA until all the original heavy ($^{15}N—^{15}N$) DNA has been replaced by hybrid ($^{15}N—^{14}N$) DNA implies that a new cycle of replication is not initiated until the previous cycle is completed and, therefore, that replication is sequential (see Meselson and Stahl, 1958a). However we will not discuss this experiment further here but proceed to some recent and more direct demonstrations of the mechanisms of chromosome replication.

Of these, the most direct is an extension of the autoradiographic method of Cairns. Actively multiplying, thymine-requiring *E. coli* cells are allowed to incorporate radioactive [^{3}H]-thymidine into their DNA during a period of about two division cycles, before being gently lysed and the DNA embedded in photographic emulsion as described in the previous section (p. 550). Let us imagine a bacterium whose chromosome is commencing to replicate semiconservatively at the time of addition of the [^{3}H]-thymidine. At the end of the first replication cycle there will be two daughter DNA duplexes each having one newly synthesised, radioactive (hot) polynucleotide strand and one non-radioactive (cold) conserved strand. During the next replication cycle the pre-existing hot strand will acquire a hot complementary strand so that this grand-daughter duplex will be labelled in both strands; however, the other grand-daughter duplex will inherit one cold strand and so will be only half labelled. If the presumption of semi-conservative replication is correct, therefore, it should be possible to distinguish these two grand-daughter duplexes by autoradiography since one, being twice as radioactive, should yield double the grain density of the other. A beautiful example of the kind of result obtained is shown in Plate 27. Not only is it clear that the single, circular DNA molecule which constitutes the chromosome is indeed replicated by a semi-conservative and sequential process from a single starting point, but the existence of only one growing point shows that synthesis proceeds along both parental strands in the same direction (Cairns, 1963a, b). As we have previously mentioned, this suggests that the Kornberg DNA polymerase may not be the enzyme responsible for chromosome replication since *in vitro* studies show that it can only copy the template strand from its 3′ end (p. 253). However, Mitra *et al.* (1967) suggest an escape from this dilemma by supposing that once replication has been initiated by a molecule of Kornberg polymerase and is proceeding along the 3′–5′ strand, other molecules of the same enzyme could then copy the exposed strand of opposite (5′–3′) polarity in the opposite (i.e. 3′–5′) direction by a kind of patching process such as occurs in the repair of damage from ultra-violet irradiation (p. 333). A final interesting feature of these auto-radiographs is that the daughter chromosomes appear to remain joined at the starting point of replication, at least throughout the greater part of the replication cycle. As is demonstrated in Plate 27, it is sometimes possible, from the autoradiographic pattern, to distinguish which of the two forks of the replicating chromosome is the starting point (Cairns, 1963b).

At this point we must digress for a moment to consider the relation between protein synthesis and chromosome replication. If thymine-requiring mutants of *E. coli* are deprived of thymine in an otherwise complete medium, they continue to synthesise protein and RNA but, after a short lag, begin to die exponentially—the so-called 'thymine-less death' (Barner and Cohen, 1957). This phenomenon has been demonstrated in many strains of *E. coli* and in *B. subtilis* and is probably general among bacteria. If a population of thymine-requiring *E. coli* bacteria is provided with thymine but prevented from synthesising protein by, for example, withholding a required amino acid, and samples are then removed at intervals and tested for susceptibility to thymine-less death by withdrawing the thymine, it turns out that an increasing proportion becomes resistant with time until the entire population is resistant; that is, under conditions when DNA synthesis can continue, inability to synthesise protein leads to progressive protection against thymine-less death. Throughout the period needed for the development of resistance by the entire population, the total DNA increases by 40 to 50 per cent. When the resistant population is allowed to resume protein synthesis in a complete medium, susceptibility of the bacteria to thymine-less death gradually reappears (Maaløe and Hanawalt, 1961).

Although the reason for thymine-less death remains a mystery, it could be ascribed, rather semantically, to 'some irreparable mistake in *attempted* DNA synthesis' and since, normally, one round of replication succeeds another with little or no pause in individual, multiplying bacteria, at any one time a negligible proportion would be in a 'resting' phase when DNA synthesis would not be attempted, so that virtually the whole population would be susceptible. The development of resistance to thymine-less death when protein synthesis is stopped therefore suggested that DNA synthesis is dependent on protein synthesis. Moreover, the 40 to 50 per cent increase in total DNA which accumulates during the acquisition of resistance by all the cells of the population, is just the amount of increase that would be expected in an unsynchronised population if previously initiated replication proceeded to completion, but the initiation of another round of replication could not occur. It was therefore proposed that *protein synthesis is necessary to initiate a new round of* DNA *replication, but not to sustain it once it has started* (Maaløe and Hanawalt, 1961; Maaløe, 1961).

Various predictions emerging from this hypothesis have been confirmed by experiment. For example, autoradiographic analysis of individual bacteria shows that they continue DNA synthesis for very different periods following the arrest of protein synthesis. Again, whatever proportion of a culture is resistant to thymine-less death, the same proportion is unable to synthesise DNA when protein synthesis is stopped. Finally, the usual amount of DNA is synthesised in the absence of protein synthesis if 5-bromouracil is substituted for thymine (see p. 248), but analysis shows that none of this DNA has 5-bromouracil in *both* strands; a proportion of the DNA is 'hybrid' with respect to thymine-containing and 5-bromouracil-containing strands, and this proportion is almost exactly that expected if DNA synthesis is completed in every cell of an unsynchronised population and then stops (Hanawalt *et al.*, 1961).

An ingenious way of testing the idea that replication of the *E. coli* (strain $15T^-$) chromosome proceeds sequentially from a unique starting point and requires protein synthesis for its initiation was developed by Lark, Repko and Hoffman (1963). An unsynchronised, exponentially growing culture is given a 'pulse' of [^{3}H]-thymidine for about one tenth of a generation time, so that the particular region of chromosome being replicated at the time can be identified by its radioactive label, and is then immediately transferred to medium containing 5-bromouracil (BU) instead of thymine (T). At intervals thereafter samples are removed, the DNA extracted and the newly synthesised 'hybrid' (BU-T) DNA, which is denser than the pre-existing T-T DNA, separated in a caesium chloride density gradient. This hybrid DNA, which began to be synthesised immediately after incorporation of the [^{3}H] label, is then examined for radioactivity. It is found that the label only begins to appear in the hybrid DNA towards the end of a generation time following transfer to the BU medium, as shown by curve A of Fig. 110. This means that replication must proceed right round the chromosome before the label can be picked up by the newly synthesised DNA and, therefore, that replication is sequential (see Fig. 111, a).

In a variation of this basic procedure the bacteria, labelled as before, are deprived of a required amino acid for a period, to allow all the

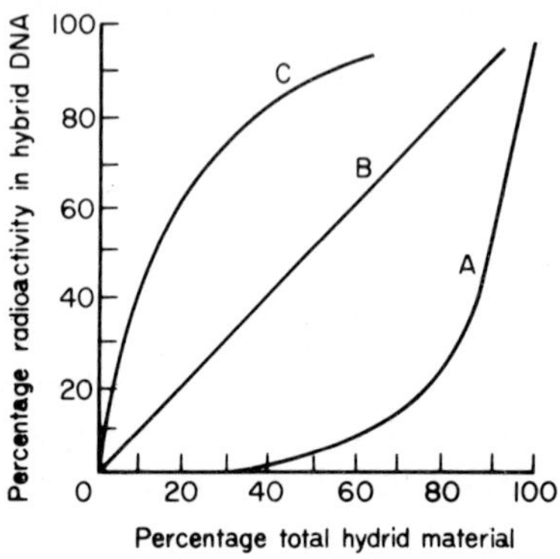

FIG. 110. Curves showing the kinetics of incorporation of a pulse of tritiated thymidine into DNA subsequently synthesised by a thymine-requiring strain of *E. coli* ($15T^-$) under various conditions.

The newly synthesised DNA is labelled by uptake of bromouracil (BU) instead of thymine, and is separated from the preformed DNA by centrifuging in a density gradient. The 'percentage total hybrid material' indicates the percentage of chromosome synthesised.

A. An exponentially growing, unsynchronised culture is exposed to ^{3}H-thymidine for a short fraction of a generation time and is then immediately transferred to BU medium. The curve indicates the amount of radioactivity appearing in the DNA synthesised at intervals thereafter. The bulk of the label appears late.

B. After incorporation of the pulse label, the bacteria are deprived of an amino acid so that all the chromosomes complete replication. The amino acid is then restored and the bacteria transferred to BU medium. The label appears in direct proportion to the amount of new DNA synthesised.

C. The bacteria are first starved of an amino acid so that all the chromosomes complete replication. The pulse label is administered immediately after restoration of the amino acid so that it should be incorporated at the starting point. After several generations, amino acid starvation is again imposed. Replication is then allowed to resume in BU medium. The bulk of the label appears early.

See text for explanation.

(Data from Pritchard and Lark, 1964)

chromosomes to complete replication, before being allowed to grow in BU medium. It follows that, in this case, the synthesis of the hybrid DNA should begin at the same starting point in all the cells. But since the label was incorporated randomly around the chromosomes of the unsynchronised population, it should now appear in the hybrid DNA from the beginning and in proportion to the amount of new DNA synthesised (Fig. 111, b). As curve B of Fig. 110 shows, this is what is found. In a third type of experiment the bacteria are first subjected to a period of amino acid starvation, and the [^{3}H] pulse introduced immediately after protein synthesis is restored, so that the label should be adjacent to the starting point in every chromosome. The bacteria are then allowed to multiply normally for several generations before again being deprived of the amino acid to line up the chromosomes at the starting point once again. The amino acid is then restored and BU added to the medium. In this case the label should be picked up by the hybrid DNA synthesised early (Fig. 111, c), and this is what is found experimentally (Curve C, Fig. 110). Note that in this last experiment a number of generations elapse between the two periods of amino acid withdrawal. Since the experiment shows that chromosome replication was initiated at the same starting point on both occasions, it follows that the starting point is fixed and genetically inherited (Lark *et al.*, 1963; Lark, 1966).

By using this method it has been shown that a period of thymine deprivation has the effect of re-initiating chromosome replication at the starting point, as well as its continuation from the growing point, when the thymine is restored (Pritchard and Lark, 1964). Thus in the first type of experiment (Fig. 111, a), if a period of thymine deprivation is interspersed between incorporation of the [^{3}H] label and transfer to BU medium, instead of the label appearing only in the hybrid DNA synthesised late (Fig. 110, curve A), it now appears in the hybrid DNA from the beginning, just as if the chromosomes had been lined up at the starting point by amino acid withdrawal (Fig. 110, curve B). In fact the curve obtained following thymine starvation resembles a composite of curves of curves A and B of Fig. 110. Quantitative aspects of this phenomenon make it very likely that the reinitiation of synthesis involves only one of the two daughter chromosomes (Lark and Bird, 1965a). This effect of thymine starvation explains an earlier and paradoxical finding that thymine-requiring bacteria which had first been permitted to synthesise protein for some time in the absence of thymine, were able to more than double their DNA, instead of limiting

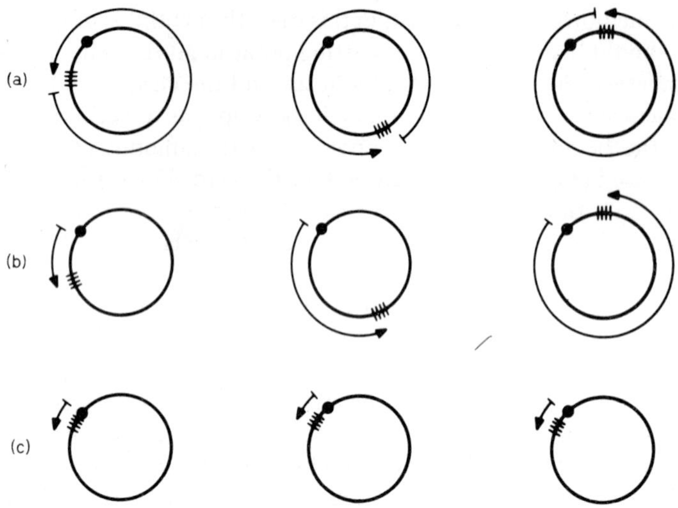

FIG. 111. Diagrammatic representation of the events yielding the curves of Fig. 110, which shows the kinetics of incorporation of an [^{3}H]-thymidine pulse into DNA subsequently synthesised by a thymine-requiring strain of *E. coli*, under various conditions.

Each row of bold inner circles represents the chromosomes of an unsynchronised bacterial population, each of which is presumed to have the same starting (initiation) point, indicated by the black circle. The cross-lines represent the incorporated pulses of tritium, and the outer light lines the commencing points, direction and extent of new DNA synthesis which picks up the label in different types of experiment.

(a) The tritium pulse is incorporated at random points around the chromosomes of the unsynchronised population. The newly synthesised DNA is examined for pick-up of the label as a function of time after its incorporation. A replication cycle should elapse before the bulk of the label is picked up (see Fig. 110, curve A).

(b) After incorporation of the label, chromosome replication is allowed to run to completion by amino acid deprivation. When new DNA synthesis starts at the starting point, the label should now be picked up randomly (see Fig. 110, curve B).

(c) The tritium pulse is incorporated immediately after a period of amino acid deprivation, so that it is located adjacent to the starting point in all the chromosomes. After a number of replication cycles, replication is allowed to run to completion by stopping protein synthesis again. When new DNA synthesis starts at the starting point, the label is picked up early (see Fig. 110, curve C).

the increase to 50 per cent, when thymine was restored and protein synthesis stopped (Nakada, 1960).

Assuming that replication of the chromosomal DNA, of *E. coli* at least, proceeds semi-conservatively and sequentially from a fixed starting or *initiation point*, as summarised in Fig. 112, we must next consider whether it is possible to locate the initiation point on the

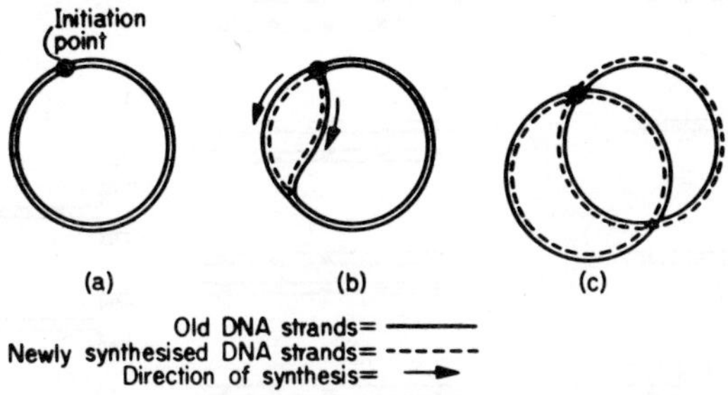

FIG. 112. Diagrammatic representation of how the circular chromosome of *Escherichia coli* replicates. The chromosome consists of a continuous loop of double-stranded DNA. In any particular bacterium, chromosome replication begins at a single starting, or initiation, point and continues in the same direction along both strands of the parental DNA until the entire chromosome has been duplicated. Throughout the replication cycle the two daughter chromosomes remain attached at the starting point.

(a) Parental chromosome; (b) partially replicated chromosome; (c) completely replicated chromosome; —— = old DNA strand; ---- = newly synthesised DNA strand; → = direction of synthesis. Compare this diagram with the autoradiograph shown in Plate 27.

linkage map, and whether the direction of replication is always the same in relation to the genetic markers, that is, whether replication is polarised. Two ingenious approaches to these problems have been developed by N. Sueoka and his colleagues, using the *B. subtilis* transformation system (Yoshikawa and Sueoka, 1963; Sueoka and Yoshikawa, 1963; Oishi, Yoshikawa and Sueoka, 1964). The first method rests on the assumption that if replication begins at one end of the chromosome, or at a fixed point on a circular chromosome, and runs at uniform speed towards the other, and is continuous, in the great majority of bacteria of a randomly growing, unsynchronised population, the growing point at any given moment will lie somewhere between the two extremities, as is shown in Fig. 113a where four replicating chromosomes are represented, the black circles, *A*, *B*, *C*, *D*, *E*, indicating genes. Because of the continuous nature of DNA synthesis, a negligible proportion will have just completed a replication cycle and not yet have started another. It follows that, in the population,

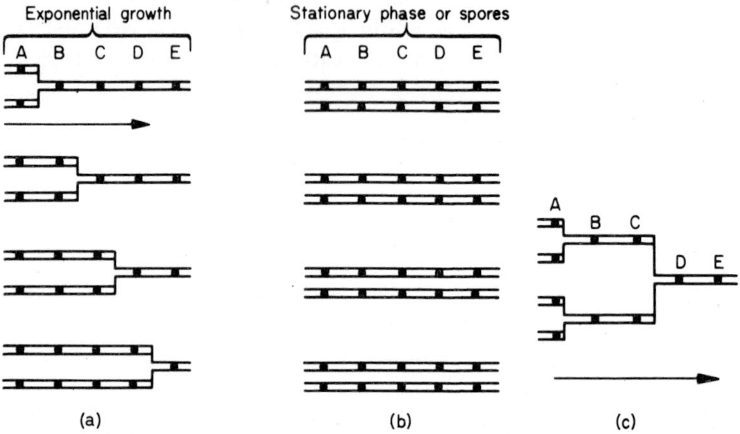

FIG. 113. Diagrammatic comparison of the state of chromosome replication among individual *Bacillus subtilis* bacteria in unsynchronised exponentially growing, and stationary phase, populations. The paired lines represent the two polynucleotide strands of the chromosomal DNA. The black circles, indicated *A*, *B*, *C*, *D*, *E*, represent various genes distributed along the chromosome. The arrows indicate the direction of chromosome replication, and the fork the position of the growing point of replication. It is assumed that replication begins at a specific point, is polarised and is continuous.

(a) In a randomly growing population during exponential growth, the growing point of replication lies somewhere along the chromosome; in very few bacilli will the chromosome have completed one cycle of replication and not have started another. For every copy of gene *E* there will be two copies of gene *A*, while the number of copies of intermediate genes will lie on a 2:1 gradient in proportion to their distance from *A*.

(b) In a stationary phase (or spore) population, when growth ceases the current cycle of chromosome replication is completed; a new cycle is initiated only on transfer to a fresh medium.
The figure shows that the ratio of the number of copies of a gene in an exponentially growing population to the number in a stationary phase population varies from 1·0 close to the starting point, to 0·5 near the completion point.

(c) This diagram explains the relationship between an observed 4:2:1 gene ratio, and initiation of a new replication cycle on the two daughter chromosomes before completion of the initial cycle.

(From Hayes, 1966).

there will be twice as many copies of gene *A*, located close to the starting point, than there are of gene *E* which is close to the finishing point. Similarly, the number of intermediate genes should lie on a 2 to 1 gradient in proportion to the distance of each from the starting point. Clearly the existence of such a gradient could be assessed from the number of transformants produced by the extracted DNA, on the reasonable assumption that the number of transformants with respect to any particular gene is proportional to the number of copies of that gene per unit volume of the transforming preparation. Since, however, the frequency of transformation of different characters may vary widely for other reasons, a true relative index of gene frequencies requires a comparison, not of the *absolute* numbers of transformants for different genes, but of the *ratios* of the number of transformants given by DNA from an exponentially growing culture, to that from a culture in which replication of all the chromosomes is complete so that all the genes are present in equal numbers, as shown in Fig. 113b. It happens that this latter condition is fulfilled by stationary phase cultures of *B. subtilis*, so that the ratio of the transforming efficiency of exponential to stationary phase cultures for each gene is determined. Experimentally, the ratios for eleven genes were in fact found to lie on a gradient between 1·0 and 0·5 and to be reproducible (Yoshikawa and Sueoka, 1963; Sueoka and Yoshikawa, 1963). This not only provides evidence for polarised chromosome replication from a fixed starting point in *B. subtilis*, but also indicates the relative positions of the genes on the chromosome—an advantage which the fragmentary nature of chromosome transfer normally denies to transformation systems.

These ratios were estimated for bacteria grown in synthetic medium in which the generation time is about 40 minutes. Surprisingly, it turned out that when the bacteria were grown in nutrient broth, in which the generation time is halved, a ratio of 4:2:1 was found, implying that there were now four copies of genes near the starting point for every one near the finishing point. As Fig. 113c shows, this can be interpreted as indicating that chromosome replication keeps pace with the increased growth rate by initiating a new cycle of replication at the starting point on each of the two daughter chromosomes, at a time when the first cycle is only about half completed, rather than by increasing the speed of chromosome replication (Oishi, Yoshikawa and Sueoka, 1964). This phenomenon has been called *dichotomous replication*, and we shall have more to say about it soon.

The implications of these experiments were confirmed by a more direct approach. *B. subtilis* bacteria are allowed to attain the stationary phase of growth, or to spore, in a medium rich in the heavy isotopes deuterium and ^{15}N, so that both strands of the chromosomal DNA are denser than normal. They are then transferred to a medium containing only light isotopes. Throughout the first generation time after transfer, samples of the culture are removed at intervals, the DNA extracted from the bacteria, and the newly synthesised hybrid (heavy-light) molecules separated from the old (heavy-heavy) molecules in a caesium chloride density gradient. The transforming ability of this newly synthesised DNA fraction for a series of genetic markers is then analysed. As Fig. 114 demonstrates, if transfer to the new, light medium initiates synchronous, polarised replication from a fixed starting point, the various genes should appear in the newly synthesised DNA in a strict and reproducible sequence, indicated by *A*, *B*, *C*, *D*, *E* in the diagram; at the beginning of the replication cycle the DNA should transform with respect only to genes near the starting point, the ability to transform for all the genes being delayed until the cycle is complete. This, in fact, is what was found experimentally, the synchronisation of replication being especially clearcut when heavy spores were transferred to the light medium (Sueoka and Yoshikawa, 1963; Oishi *et al.*, 1964).

When heavy spores were transferred to broth instead of to synthetic medium in the latter type of experiment, the induction of dichotomous replication (that is, the initiation of a new cycle of replication before completion of the old one) was demonstrated by the presence together of heavy, hybrid and light molecules in the same DNA preparation (Oishi *et al.*, 1964). You will recollect that the co-existence of heavy, hybrid and light DNA molecules is not normally found in the Meselson-Stahl experiment—a fact from which the sequential replication of chromosomal DNA was first deduced (Plate 10, p. 233; p. 555).

Is there any evidence for a fixed starting point, or for polarity, in replication of the *E. coli* chromosome? Nagata (1963a, b) devised an ingenious experiment to determine, in a synchronously dividing culture of *E. coli* lysogenised by two prophages (λ and 424, Fig. 127, p. 666), located some distance apart on the chromosome, whether one prophage replicates before the other. The method was to treat samples, removed at intervals during the division cycle, with ultraviolet light in order to induce the prophages to enter the vegetative stage, and then to count the number of particles of each type liberated; it was assumed that doubling of the prophage would be reflected in a doubling of the number of particles liberated on induction (see p. 453). The experiment appeared to show, in the case of two Hfr male strains, in which the sex factor in inserted at different chromosomal sites and which transfer the chromosome with opposite polarity at conjugation, that replication of the chromosome is initiated at or near the sex factor

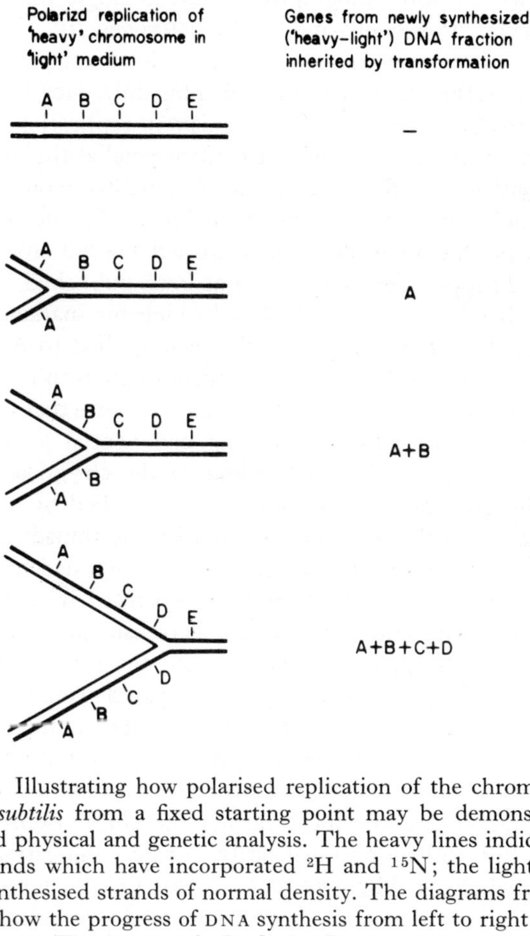

FIG. 114. Illustrating how polarised replication of the chromosome of *Bacillus subtilis* from a fixed starting point may be demonstrated by combined physical and genetic analysis. The heavy lines indicate dense DNA strands which have incorporated ^{2}H and ^{15}N; the light lines are newly synthesised strands of normal density. The diagrams from top to bottom show the progress of DNA synthesis from left to right along the chromosome. The letters, *A*, *B*, *C*, *D*, *E*, represent a series of genes distributed along the chromosome, whose presence in extracted DNA can be recognised by the transformation of recipient bacilli carrying mutant alleles of these genes. For further explanation, see text.
(From Hayes, 1966)

itself and proceeds in a reverse direction to that of transfer. In contrast, when a female strain of *E. coli* was similarly examined, no specific time for prophage doubling could be found—the ratio of plaque counts of the two phages after induction remained constant throughout the synchronised division cycle (Nagata, 1963a, b). In view of the fact that the chromosomes of female as well as of male *E. coli* bacteria can be shown by autoradiography to have a single initiation point (Cairns, 1963a), this could indicate a heterogeneity of the female population with respect to the location of the starting point on the chromosome, rather than any fundamental difference between Hfr male and female bacteria in the mechanism of replication.

These interesting conclusions are controversial at the moment and have been criticised on the grounds that they are based on the assumption that synchrony of cell division and of DNA coincide; so far as we know, the experiment has not been confirmed, nor has any convincing experimental support for its conclusions been published.

On the other hand, the method of looking for marker gradients, developed for *B. subtilis*, has recently been applied to *E. coli* using transduction by phage P1 instead of transformation to measure gene frequency (Berg and Caro, 1967). In order to ensure that the markers being sampled were replicated during exponential growth, *before* phage infection, the replicating regions of the chromosomes of an exponentially growing culture were density-labelled by a prolonged pulse of bromouracil prior to infection with the transducing phage. After lysis, the more dense phage particles, which had taken up the bromouracil-containing DNA, were then separated from the normal particles in a density gradient and used as transducing phage to estimate the marker frequency. This rules out the possibility that the markers whose frequencies are being compared were incorporated into the transducing phage particles only after completion of the replication cycle. The results of these BU-pulse experiments showed no differences in marker frequency between three different, isogenic, Hfr male strains, two of which transfer the chromosome from the same point but with opposite polarity during conjugation. On the other hand, when a culture was pulsed with BU during amino acid starvation; or during recovery from amino acid starvation, opposing gradients of marker frequency were revealed. This pointed to polarised replication from a fixed starting point distinct from the site of sex factor insertion. An interesting additional finding to emerge from these studies was that a distantly related substrain of *E. coli* K-12 appeared to replicate its chromosome from a different point and in the opposite direction (Berg and Caro, 1967; Caro and Berg, 1968).

K. G. and C. Lark, and their colleagues, have studied chromosome behaviour in a particular strain of *E. coli* (15T⁻) growing in media in which the generation time could be varied from 40 to 270 minutes by changing the energy source. In general, the number of chromosomes per bacterium was estimated from DNA content, the chromosome replication time from the generation time and the percentage of the population not synthesising DNA (as measured by autoradiography of pulse-labelled cells; see p. 554), and the number of chromosomes being replicated at any one time by the uptake of [^{3}H]-thymidine following withdrawal of a required amino acid (p. 557). The results are summarised in Table 21. In aspartate and acetate media, the

TABLE 21

Replication of the chromosome of *E. coli* growing in various media (Data from Lark, 1966).

Medium	Generation time (mins)	Chromosome replication time (mins)	Number of chromosomes per cell	Number of chromosomes replicating at one time
Glucose	40	40	2	2
Succinate	70–75	70	2	1
Aspartate	120	110	1	1
Acetate	270	160	1	1

bacteria contain a single chromosome which is replicated once every division cycle, but in the latter case during only a fraction of the generation time. On the other hand, bacteria grown in either glucose or succinate media possess two chromosomes and DNA synthesis proceeds throughout the generation time. The interesting point is that, when protein synthesis is stopped in the succinate culture, only half as much residual DNA is synthesised, and at half the rate, as in the glucose culture. This suggested that whereas both chromosomes of the glucose-grown bacteria are replicated at the same time, in the succinate medium replication of the two chromosomes is alternate—the replication time for each is the same as in glucose, but the replication of one does not commence until that of the other is completed. This inference was confirmed by comparing the cellular distribution of pulse-labelled, radioactive DNA in bacteria subsequently grown for several generations in glucose and succinate medium (Lark and Lark, 1965).

SEGREGATION OF THE BACTERIAL CHROMOSOME

We have mentioned that bacteria do not possess a mitotic apparatus. Nevertheless, not only daughter chromosomes but also other, independent, replicating units of DNA, such as the *E. coli* autonomous sex factor, are distributed with great precision to daughter cells. How is this segregation accomplished? As we shall see when we discuss the mechanism of conjugation in *E. coli* (p. 799), the virtually 100 per cent efficiency with which transfer of the autonomous sex factor follows conjugal union suggested that the sex factor must be attached to the cell membrane, where it produces a local surface change which determines that cellular contact, and the subsequent connection with a female bacterium, occurs at that point. Thus the sex factor is already waiting at the door when the door is opened. Furthermore, the evidence that the actual process of transfer is mediated by replication of the sex factor (p. 799), suggested that the initiation of normal chromosome replication might also be a membrane function.

A generalisation of this idea led to a search for a physical association between the chromosome and the cell membrane in bacteria. Electron microscopic study of serial, ultra-thin sections of *B. subtilis* revealed that every chromatin body is connected to one or two structures called *mesosomes*, which are convoluted invaginations of the cell membrane. This association persists throughout the growth cycle. If the bacteria are placed in 0·5 M sucrose, the mesosomes are unravelled by the osmotic pressure, pulling the chromatin bodies with them so that they are now seen to attach directly to the cell membrane (see Plate 30, facing p. 545) (Ryter and Jacob, 1964; Jacob, Ryter and Cuzin, 1966).

By superimposing transparent drawings of serial sections through multiplying bacteria, a three-dimensional model of the bacteria can be reconstructed in which both the size of the chromatin bodies as well as their mesosomal connections are apparent. A series of such chromatin bodies, graded according to size, can be taken to represent what happens to a single body throughout a division cycle. Small chromatin bodies are found to be associated with a single mesosome, but with increasing size they appear to become connected to two adjacent mesosomes which then move farther apart until the chromatin body divides into two, each again being associated with a single mesosome. This strongly suggests that the segregation of daughter chromosomes is mediated by the synthesis and growth of new membrane from an equatorial region between their attachment sites, as represented in Fig. 115. Evidence for such equatorial growth comes from exposing growing bacteria for a short time to potassium tellurite which is deposited as needles of metallic tellurium on the membrane. Electron microscopic examination of samples of the growing bacteria taken at intervals thereafter, shows that the needles are not separated uniformly by growth but, instead, are progressively distributed, with little loss of density, towards the poles of the cells, as the hypothesis requires (Jacob *et al.*, 1966).

In addition, various fractionation procedures involving pulse-labelled DNA support the hypothesis that the growing point of chromosome replication, which can be equated with the polymerase (replicase), is bound to the membrane as first suggested by Jacob, Brenner and Cuzin (1963) (p. 256). In the case of thymine-requiring *E. coli*, for example, newly replicated DNA fragments, containing the growing point, can be isolated from a randomly growing population exposed to bromouracil instead of thymine, together with a ^{32}P pulse, since they possess two arms of hybrid density as well as an unreplicated fraction of normal density, and so will separate in a radioactive fraction intermediate in density between hybrid and light fragments in a caesium chloride density gradient. It was found that the isolation of this fraction is greatly enhanced if the lysate is first treated with lipase and proteolytic enzymes which, it is inferred, releases the growing point from the membrane (Hanawalt and Ray, 1964; Smith and Hanawalt, 1967). Similarly, in the case of growing *B. subtilis* cells, DNA fragments labelled with very short pulses of ^{3}H-thymidine are found predominantly in the pellet at the bottom of a sucrose gradient, which mainly contains membranes and broken cell wall fragments, but the label is quickly 'chased' from this fraction by exposure of the cells to 'cold' thymidine. On the other hand, if care is taken to prevent shearing stresses following lysis, all the DNA of the bacteria separates in the pellet (Ganesan and Lederberg, 1965).

We know that bacteria may carry small, independent, self-replicating chromosomes (*plasmids* and *episomes*), such as colicin factors and the *E. coli* sex factor, in addition to their own chromosome. Since these factors are usually very stably inherited, and confer distinctive characters on their bacterial hosts, we might regard them as small, supernumerary chromosomes and ask whether they segregate randomly or

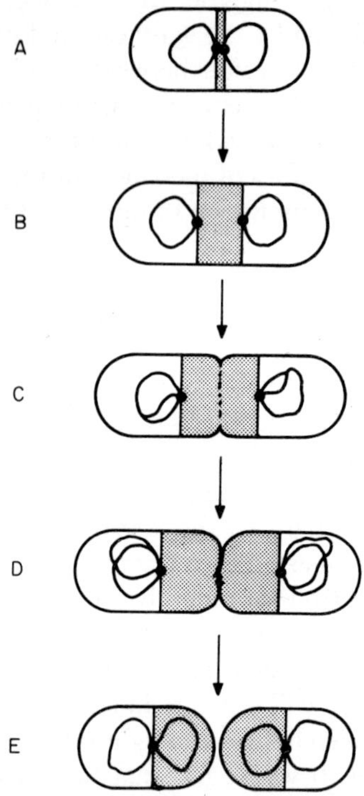

FIG. 115. A model of the mechanism suggested by Jacob and Brenner (1963) whereby both replication and segregation of the bacterial chromosome are regulated by its attachment to the cell membrane. The inner continuous lines represent bacterial chromosomes attached to the cell membrane at a specific chromosomal region. The sequence (A) to (E) shows a series of events over a complete division cycle. The shaded regions represent newly synthesised cell membrane.

(A) The two products of chromosome replication have separated at their point of attachment.

(B) The cell membrane, growing outwards from the equatorial region, is dragging the daughter chromosomes apart towards opposite poles of the growing cell.

(C) A feed-back mechanism, emanating from completion of a particular step in the division cycle, has provoked initiation of chromosome replication.

(D) Division of daughter cells is almost complete.

(E) Chromosome replication and separation of daughter cells is complete.

(Based on a Figure in Jacob *et al.*, 1963).

by the same mechanism as the bacterial chromosome. This has been elegantly studied in the following way. *E. coli* bacteria, carrying a temperature-sensitive and readily identifiable sex factor ($F^{-}lac^{+}$) which cannot replicate and so is eliminated at 42°C, are allowed to incorporate a high specific activity of radioactive phosphorus (^{32}P) into their DNA at low temperature. The cells are then transferred to a non-radioactive medium at 42°C so that the bacterial chromosomes, but not the sex factor DNA, can replicate; at intervals, sets of samples are removed and frozen, each set being subsequently assayed for bacterial survival as a function of ^{32}P decay (see p. 523). At the end of the first generation in the 'cold' medium each chromosome will possess one 'hot' and one 'cold' DNA strand instead of two hot ones, so that the bacteria should have twice the probability of survival than at the time of transfer. This, in fact, is found. At each subsequent generation an increasing number of chromosomes will arise which are devoid of radioactivity; only the diminishing proportion which inherit one of the conserved radioactive strands will be subject to damage by ^{32}P decay. Due to the relatively very small amount of the sex factor DNA, it is assumed to have only a low probability of destruction from ^{32}P decay as compared with the chromosome, but since the sex factor is unable to multiply at 42°C it follows that it can be inherited along with only one of the two daughter chromosomes at each division, so that bacteria which lack it will appear after a few generations. In the experiment, when the susceptibility to ^{32}P decay of bacteria carrying the sex factor (lactose-fermenting) and of those lacking it (non-lactose-fermenting) were compared, it turned out that virtually only the former were sensitive, dying at the same rate as after one generation. This result means that the sex factor did not segregate randomly but remained associated with one of the original 'hot' chromosomal DNA strands (Jacob *et al.*, 1966; Cuzin and Jacob, 1967c).

All these findings strongly suggest the operation of a single, precise mechanism in bacteria for the segregation of both chromosomes and episomes, and there is good evidence that this mechanism is a membrane function. In addition, it seems probable that the regulation of chromosome replication is also a membrane function. The fact that the duration of replication is by no means proportional to the division time (Table 21, p. 567), and that when the division time is short a new replication cycle may be initiated before the existing cycle is complete, certainly indicate that the chromosome is not the pace maker in the growth cycle. Moreover, temperature-sensitive mutant strains of *E. coli* have been isolated which cannot support the replication of the sex factor at high temperature, although replication of the bacterial chromosome is normal, so that some specific bacterial component, which could be a membrane site, plays an essential role in sex factor replication (Jacob *et al.*, 1963).

From a molecular point of view, we can ask whether there is any polarity with respect to the segregation of individual DNA strands; whether, for example, during DNA synthesis, only the strand being used as template is attached to the membrane, subsequent attachment of the newly formed strand being always oriented in the direction of the future septum which will divide the cell, as illustrated in Fig. 116a.

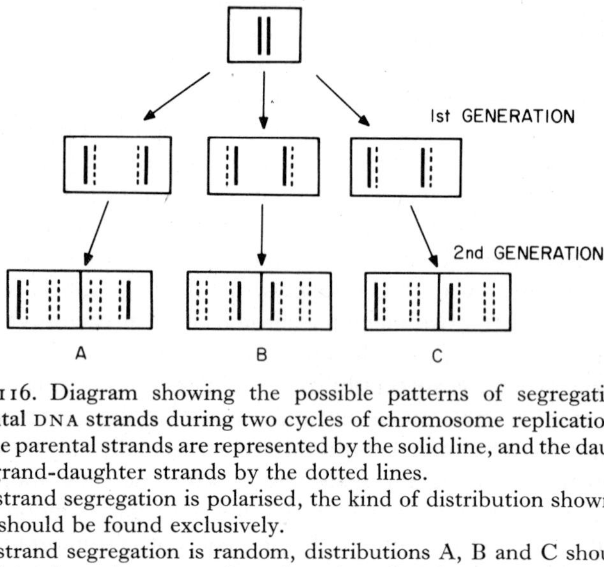

FIG. 116. Diagram showing the possible patterns of segregation of parental DNA strands during two cycles of chromosome replication.

The parental strands are represented by the solid line, and the daughter and grand-daughter strands by the dotted lines.

If strand segregation is polarised, the kind of distribution shown at A or B should be found exclusively.

If strand segregation is random, distributions A, B and C should be found in the ratio 1:1:2, since there is a fourth pattern, the mirror image of C (not shown), which is indistinguishable from it.

Compare Plate 31.

Such oriented segregation can be tested in a bacillus such as *B. subtilis* in which the daughter cells tend to remain attached for several generations, since the pattern of segregation of parental DNA strands, labelled with radioactive thymidine, can be followed by autoradiography and electron microscopy. In one such experiment, starting with *B. subtilis* spores, all three patterns of segregation represented in Fig. 116 were found, in the proportion 45A:46B:109C which is

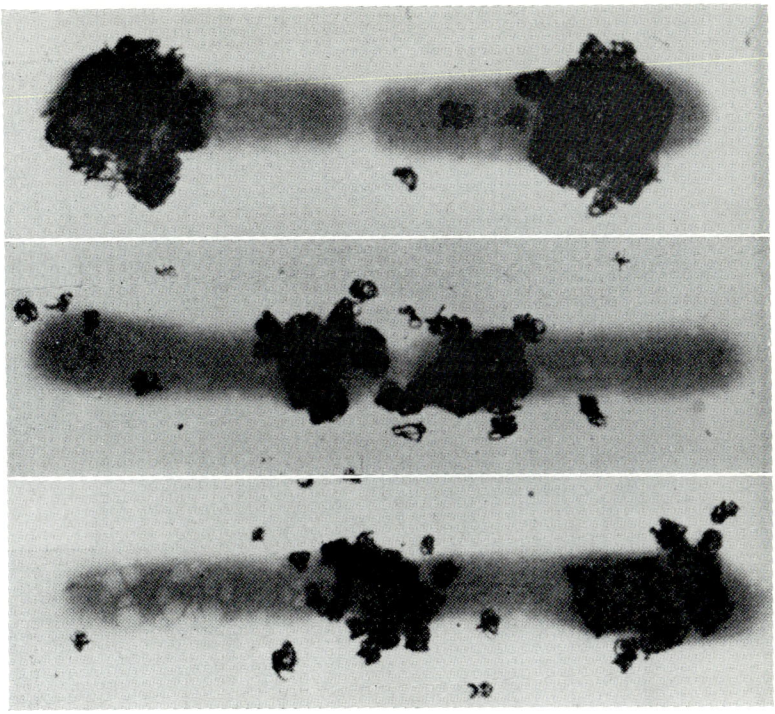

PLATE 31. Autoradiographs of *Bacillus subtilis* bacteria, which have emerged from spores labelled with tritiated thymidine, after growing for two generations in non-radioactive medium (5 months exposure). The radioactivity, shown by the black, heavily exposed regions, indicates the positions in the daughter bacteria of the labelled, parental DNA strands. The three pictures illustrate the three patterns of segregation observed, in conformity with the schema of Fig. 116. The finding of all three patterns, and especially the ratio between them, indicates random segregation of the DNA strands.

(Reproduced from Ryter and Jacob, 1964, by kind permission)

strikingly close to the 1:1:2 ratio expected from random strand segregation (Plate 31; Ryter and Jacob, 1966). However, a somewhat similar experiment by Eberle and Lark (1966), but using a different strain of *B. subtilis*, yielded quite different proportions which they interpret in terms of a model of non-random strand segregation developed from the results of an autoradiograph experiment with *E. coli* (Lark and Bird, 1965b; Lark, 1966). While we consider that the result of the experiment of Ryter and Jacob (1966) points unambiguously to random strand segregation, we must stress the fact that different strains, and especially different species, may utilise different mechanisms and that differences in method and technique may exert considerable influence on the behaviour of cells.

We will discuss the regulation of chromosome replication further in Chapter 23, in the context of general models for the genetic control of biosynthesis. Some useful original papers and reviews about DNA synthesis and its control are: isolation of thymine-requiring mutants, Okada, Yanagisawa and Ryan, 1960, 1961; Stacey and Simson, 1965: mapping of the thymine synthetase and other loci concerned with DNA synthesis, Ishibashi, Sugino and Hirota, 1964; Alikhanian *et al.*, 1965, 1966b: thymineless death, and the regulation of DNA synthesis, Jacob, Brenner and Cuzin, 1963; Jacob, Ryter and Cuzin, 1966; Lark, 1966; Maaløe and Kjeldgaard, 1966; Pritchard, 1966.

CHAPTER 20
Transformation

The discovery of transformation by Griffith in 1928 is a landmark in the history of genetics, for not only did it show clearly and for the first time that hereditary determinants could be transferred from one bacterium to another, but also laid the foundation for the subsequent recognition of the hereditary material as DNA. (Historical reviews: Hayes, 1966; Hotchkiss, 1966.)

The virulence of pneumococci for mice or men depends on the production of a polysaccharide capsule which encloses the organism and protects it against phagocytosis in the animal body. Capsulated pneumococci yield colonies which are smooth and glistening in appearance and are designated *S*. There are a large number of distinct types of *Pneumococcus* which differ from one another in the chemical consisttution of their capsular polysaccharide and can be distinguished serologically. Mutant strains of pneumococci frequently arise which have lost the capacity to synthesise capsular polysaccharide, and these strains are avirulent when injected into mice; they produce colonies which have a matt, rough surface, quite distinct from those of capsulated *S* bacteria, and are designated *R*. Griffith (1928) found that when mice were inoculated either with cultures of living, non-capsulated type II pneumococci (II-*R*), or with a heat-killed suspension of capsulated organisms of some other type such as type I (I–*S*), none of the animals succumbed to infection. However, when a *mixture* of the living II–*R* and the killed I–*S* bacteria was injected, a proportion of the mice died from septicaemia, and from their blood he isolated pure cultures of capsulated pneumococci of either type II or type I. It was clear that some substance present in the suspension of dead I-*S* bacteria was capable either of repairing the hereditary defect of type II capsule synthesis, or of bestowing a new hereditary ability to synthesise type I capsule, on type II bacteria which had lost their capacity to make capsule. Griffith speculated that this *transforming principle* might be the polysaccharide itself which is needed as a primer in autocatalytic polysaccharide synthesis, or, possibly, a protein which activates the reaction (Griffith, 1928). Strangely enough, the fundamental notion that the transforming principle mediates the transfer of an *hereditary* property seems to have eluded Griffith, as well as other workers for several years afterwards. It was later shown that the transformation of *R* pneumococci of one type to *S* pneumococci of another can be effected in the test tube (Dawson and Sia, 1931), and that the transforming principle is present in cell-free extracts of the capsulated donor strain, is precipitated by alcohol and acetone and withstands heating at 80°, but is destroyed by boiling and by enzymes liberated by autolysis of pneumococci—properties which clearly distinguish it from the capsular polysaccharide (Alloway, 1933).

A systematic analysis of the chemical nature of transforming principle, by O. T. Avery and his colleagues during the next ten years, culminated in a formidable weight of evidence that it possessed all the properties of deoxyribonucleic acid (DNA) (Avery, MacLeod and McCarty, 1944). Not only this, but its biological activity was unaffected by treatment with either proteolytic enzymes or ribonuclease (RNase). Conversely, treatment with the enzyme deoxyribonuclease (DNase) was shown to destroy transforming activity (McCarty and Avery, 1946). Despite this evidence, the possibility remained that the transforming principle might be some minor impurity in the DNA preparations, or a nucleoprotein whose protein moiety was resistant to proteolytic enzymes.

The application to transforming preparations of more exacting methods for the purification of DNA revealed that the transforming activity increased *pari passu* with the purity of the DNA, and that progressive removal of proteins and other serologically active substances, and of RNA, did not reduce its potency at all. The degree of purity attained was such that less than one molecule of a contaminating substance of molecular weight 10^6 would have to be assumed to be capable of effecting a transformation, so that protein activity can virtually be excluded on analytical grounds alone (Hotchkiss, 1949, 1952; Zamenhof, Alexander and Leidy, 1953). Finally it has more recently been shown that when the DNA of a transforming preparation is labelled with radioactive phosphorus (^{32}P), not only is the label incorporated irreversibly into the DNA of recipient bacteria, but the extent of the incorporation is directly proportional to the number of transformants (Goodal and Herriot, 1957a; Lerman and Tolmach,

1957). The transforming principle, if not DNA, must therefore be a substance so bizarre that its nature is difficult to imagine.

At the same time as these observations were being made on the chemical nature of the transforming principle, analysis of transformation as a genetic phenomenon, initiated by Harriet Ephrussi-Taylor (1951a, b) who was then working in Avery's laboratory, demonstrated unambiguously that the transforming molecules behaved in every way like fragments of genetic material, generating recombinants by pairing and undergoing genetic exchange with homologous regions of the recipient genome. For example; 1. $S \rightarrow R$ transformations can occur in addition to the usual $R \rightarrow S$, showing that transformation does not consist in the simple addition of a character determinant, but in the substitution of one allele by another; 2. a single transformation system may yield more than one type of transformant, which is compatible with the segregation of different recombinant types as a result of genetic exchanges in different regions; 3. certain pairs of characters are found to be inherited together at characteristic frequencies in transformation, which is explicable only if the determinants of these characters are physically linked so that they are located, with a fixed probability, on the same transforming fragment of genetic material; 4. transformation between two *R* strains can yield capsulated, *S* transformants, as would be expected from a genetic exchange between different, but closely linked, mutational sites involving the same biosynthetic pathway. Thus the evidence accumulated that the genetic material was DNA, and the stage was set for intensive research into the physico-chemical structure of this nucleic acid.

THE RANGE AND IMPORTANCE OF TRANSFORMATION

The occurrence of transformation has now been confirmed in a considerable number of bacterial genera and species, including *Pneumococcus*, *Streptococcus*, *Haemophilus*, *Neisseria*, *Bacillus* and *Rhizobium*. A list of these may be found in recent reviews (Ravin, 1961; Schaeffer, 1964; Spizizen, Reilly and Evans, 1966). Since most of the features of transformation appear to be common to all those organisms which display it, we do not intend to describe the details of any particular system in this chapter, but will utilise information derived from several systems to build up a general picture of the phenomenon.

Most of our basic knowledge has come from studies of transformation in *Pneumococcus* and *Haemophilus influenzae* (Alexander and Leidy, 1951), although the *Bacillus subtilis* system (Spizizen, 1958) is becoming increasingly exploited because of the ease with which this organism can be cultivated and genetic markers obtained in it. *H. influenzae* is a small Gram negative bacillus which, like *Pneumococcus*, is divisible into a number of serological types on the basis of specific capsule formation. As in the case of pneumococci, transformation of capsular type was the first to be recognised in this species which, as an experimental tool, has the advantage over pneumococci in being less exacting in the nutritional and environmental conditions which it requires both for growth and transformation. However, it was soon found that transformation is by no means restricted to characters involving synthesis of polysaccharides, but can be demonstrated for *any* character whose inheritance by the recipient bacteria is readily detectable.

For reasons which we will shortly discuss, the proportion of cells of a recipient population which is transformed, with respect to any character, by DNA extracted from donor bacteria, tends to be rather low. Thus, precise *quantitative* studies of the number of transformed bacteria as a function of DNA concentration and other variables, on which progress depended, required donor characters which, unlike specific capsule formation, could be easily selected for and scored among the recipient population. Resistance to various antibiotics and drugs such as streptomycin, cathomycin, erythromycin, sulphonamides and aminopterin fulfilled this role admirably in *Pneumococcus* and *Haemophilus*, neither of which displays the nutritional independence which makes organisms like *E. coli* and *Salmonella* such excellent genetic tools. On the other hand, *B. subtilis*, several transformable strains of which have now been found, is a species of Gram positive, aerobic sporing bacillus which can grow in a simple, chemically defined medium, while auxotrophic mutants, requiring the addition of various amino acids for growth, can easily be isolated from it. Thus the way is open to exploration of the genetics of defined biochemical pathways in this organism, similar to those studied in *Salmonella* and *E. coli* by means of transduction (Chapter 8, p. 158 *et seq.*).

Transformation, however, is something more than just another way of performing genetic analysis in bacteria, and of extending it to a few species which are not amenable to transduction or conjugation. It is the only method we possess of correlating the effects of physical or chemical alterations in the structure of DNA with its biological activity and, therefore, of investigating directly the relation between DNA structure and function. We have already seen how transformation may be used to elucidate the action of ultraviolet and visible light on the behaviour of genetic material (Chapter 13, p. 328 *et seq.*).

In the following sections we will give a rather general account of the mechanism of transformation. For details and discussions of the physiological and genetic aspects of specific systems the reader is referred to various reviews (Austrian, 1952; Ephrussi–Taylor, 1955, 1960a, b; Hotchkiss, 1955; Ravin, 1961; Thomas, 1962a, Schaeffer, 1964) and to the original papers cited in the text.

THE INTERACTION OF BACTERIA WITH TRANSFORMING DNA

In all systems which mediate genetic recombination in bacteria we are concerned with two aspects of the parasexual process, either of which may profoundly affect the outcome of the cross. The first of these is the mechanism of *transfer* of genetic material from donor to recipient bacteria; the second is the nature of the recombination event itself whereby the transferred donor genes are *integrated*, or *incorporated*, into the chromosomes of the recombinant progeny. We will first analyse the various interactions which occur between transforming DNA and recipient bacteria, and which lead to entry of the transforming molecules into the cells.

THE SIZE OF TRANSFORMING MOLECULES OF DNA

In transduction and conjugation the mechanism of transfer is pre-eminent in determining the size and nature of the fragments of donor chromosome which enter the merozygotes. In transformation, the size of the transforming fragments is predetermined by the degree to which the chromosomal DNA is broken down during extraction from the donor bacteria and subsequent purification. We have seen that the shearing stresses to which DNA is normally subjected during its extraction from either bacteria or bacteriophages result in homogeneous preparations of rather uniform molecular weight (p. 536). The mean molecular weight of transforming preparations of pneumococcal DNA has been estimated to be about 5×10^6 on the basis of inactivation by ionising radiations (Fluke, Drew and Pollard, 1952); using other methods, the rather higher value of 15×10^6 has been obtained for *H. influenzae* DNA (Goodgal and Herriot, 1957a;

but see p. 592). When fragments of this size are assessed as a fraction of the total amount of DNA per bacterium, on the assumption that each bacterium has two chromatin bodies it turns out that each molecule of the transforming preparation is equivalent to about 1/200 to 1/500 part of the entire genome. Since each genome almost certainly contains only one representative of each gene, we might predict that each bacterium making effective contact with a single molecule of DNA would have a less than 1 in 200 chance of being transformed with respect to any given donor marker or, alternatively, that a bacterium that had taken up more than 200 molecules might have a high probability of yielding transformed progeny. In practice it is found that the proportion of recipient bacteria transformed is never greater than 10 per cent. in the case of *Pneumococcus* and, in the case of *Haemophilus* and *B. subtilis*, is usually less than 1 per cent under optimal conditions.

This raises a number of pertinent questions. For instance, does effective contact with a single transforming DNA molecule suffice for transformation, or must a number of molecules be taken up? Is there a limit to the number of molecules which recipient bacteria can absorb, so that non-transforming molecules compete with those carrying the gene whose inheritance is scored? Are all the bacteria of a recipient population, or only a fraction, receptive?

TRANSFORMATION AS A FUNCTION OF DNA CONCENTRATION

The introduction of selective markers into transformation experiments enables the number of transformants inheriting a particular marker, such as resistance to streptomycin, to be accurately related to the concentration of DNA molecules in the transforming preparation carrying the marker. A population of streptomycin-sensitive recipient bacteria is exposed to a range of low concentrations of DNA (1 to 1000 mμg/ml) extracted from streptomycin-resistant donor bacteria. After allowing time for the expression of resistance in the transformed cells, samples of the mixture are plated on medium containing streptomycin. The number of transformant colonies which arise is finally counted. It turns out that, for concentrations of DNA below about 100 mμg/ml, the number of transformants is directly proportional to concentration as Fig. 117 shows. This type of curve indicates that each transformant arises from the interaction of a recipient bacterium with a single unit, or molecule, of the transforming DNA (Hotchkiss, 1957; Goodgal and Herriot, 1957a; Schaeffer, 1957).

Another experimental approach, which we will study in more detail shortly, is to label the DNA of the donor bacteria with radioactive phosphorus (^{32}P) and then to see what proportion of the label is taken up from the transforming DNA preparation by the recipient cells. In the case both of *Pneumococcus* (Lerman and Tolmach, 1957) and of *Haemophilus* (Goodgal and Herriot, 1957a; Schaeffer, 1958) it is

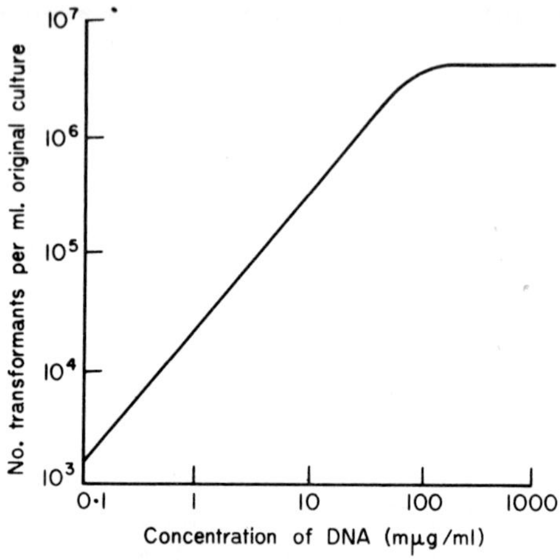

FIG. 117. Illustrating the relation between the concentration of transforming DNA and the number of transformants obtained.

found that, over a wide range of different concentrations of DNA and populations of bacteria, the amount of ^{32}P irreversibly bound to the bacteria is strictly proportional to the number of transformants obtained. Assuming a mean molecular weight of 15×10^6 for the DNA, it can be estimated that between 100 and 200 molecules are taken up *by the population as a whole* for every bacterium transformed to streptomycin resistance. These experimental results tally very well with the notion that each molecule of the DNA preparation corresponds to a 1/200 part of the genome of a donor bacterium.

COMPETITIVE INHIBITION

Let us now return to Fig. 117. We have seen that over a range of low concentrations of DNA, the number of transformants is related linearly to the concentration. However, above about 100 mμg/ml DNA, equivalent to some ten molecules per bacterium under the experimental conditions employed, the curve flattens to a plateau so that increasing concentrations of DNA fail to yield more transformants. So far as ability to be transformed is concerned, the bacteria have become saturated with DNA. Nevertheless, at the plateau concentration, no more than about ten per cent. of the recipient population, and usually much less, become transformants. Assuming that all the bacteria are capable of being transformed, it looks as though the number of molecules that each bacterium can adsorb is strictly limited. Since the bacteria can be transformed with respect to a wide variety of characters there is no *a priori* reason to suppose that, so far as DNA uptake is concerned, the cell should distinguish between a molecule carrying, say, a streptomycin-resistance marker and any other molecule. Thus, *once the saturation point is reached*, transforming molecules must compete with a 200-fold excess of non-transforming molecules for the available receptors so that, in practice, no further transformants are found with increasing DNA concentration.

If this explanation of the plateau is correct, it should be possible to lower the plateau by the addition of homologous but non-transforming DNA, extracted from the *recipient* bacteria, to the transforming preparation; that is, the non-transforming DNA should behave as a *competitive inhibitor of transformation, but only at concentrations of total* DNA *which saturate the available receptor sites*. This has been demonstrated experimentally in both *Pneumococcus and Haemophilus* (Hotchkiss, 1954, 1957; Alexander, Leidy and Hahn 1954; Schaeffer, 1957). The results of a typical experiment are illustrated in Table 22 which shows, for each concentration of transforming DNA, the excess of non-transforming DNA which must be added to obtain a significant inhibition of the number of transformants. You may remember that under the experimental conditions shown in Fig. 117 to which these results are related, the plateau was reached at a DNA concentration of about 100 mμg./ml. Table 22 reveals that no inhibition occurs until the total DNA reaches this concentration. For example, when the concentration of transforming DNA is 0·01 mμg/ml, a 10^4 times excess of non-transforming DNA (i.e. 100 mμg/ml) is needed to inhibit (line 1); similarly, the transforming activity of 0·1 mμg/ml DNA is first inhibited by a 10^3 times excess (line 2), and of 1 mμg/ml by a 10^2 times excess (line 3) of non-transforming DNA; that is, by 100 mμg/ml non-transforming DNA in every case. But when the concentration of *transforming* DNA approaches the plateau level of 100 mμg/ml, then the addition of an equal concentration of non-transforming DNA is inhibitory (line 5) (Schaeffer, 1957).

TABLE 22

Illustrating competitive inhibition of the uptake of transforming DNA by non-transforming DNA isolated from the recipient bacteria.

Concentration of transforming DNA (mμg/ml)	Presence or absence of inhibition					
	Relative conc. of non-transforming DNA added:					
	× 1	× 10	× 10^2	× 10^3	× 10^4	× 10^5
0·01	−	−	−	−	+	++
0·1	−	−	−	+	++	
1	−	−	+	++		
10	−	+	++			
100	+	++	++			
1000	+	++				

Inhibition is judged by the percentage reduction in the number of transformants yielded by each concentration of transforming DNA alone: − = no significant inhibition; + = approx. 30 to 50 per cent inhibition; ++ = 80 per cent inhibition or over.
The minimum saturating concentration of transforming DNA alone = 100 mμg/ml (see Fig. 117).
The data are those of Schaeffer (1957), employing *H. influenzae* strain *Rd* and streptomycin-resistance as a genetic marker for selection of transformants. Bacteria (4 × 10^8/ml) at the peak of competence were exposed to mixtures of transforming and non-transforming DNAs in the given relative concentrations.

We have seen that the uptake of a single molecule of DNA carrying the selective marker can result in transformation, that transformation becomes a very probable event when the DNA equivalent of approximately a single genome has been taken up by the population as a whole, and that there is a limit to the amount of DNA that can be adsorbed by a single receptive bacterium. If all the bacteria in the population are transformable, the saturating concentration of DNA found by experiment suggests that each bacterium can take up no more than about ten molecules. This explains why never more than 10 per cent, and usually much less, of recipient bacteria become transformants in practice, for if each bacterium were able to absorb the equivalent of one or two genomes of DNA (say, 200 molecules), there is no obvious reason why the frequency of transformation should not approach 100 per cent. On the other hand, we could equally well account for the facts if only 10 per cent of the recipient population were capable of being transformed, while each receptive, or competent, bacterium could take up 100 or more molecules. We must now examine this question of competence.

THE COMPETENCE OF RECIPIENT POPULATIONS

Very early in the study of transformation it was discovered that the ability of recipient bacteria to be transformed depends on their physiological state (McCarty, Taylor and Avery, 1946). A cell which is able to take up a molecule of transforming DNA, and to be transformed by it, is said to be *competent*. The competence of a recipient population can be investigated in the following way. At intervals during the growth of a culture of recipient bacteria in a suitable liquid medium, samples are diluted into an excess of transforming DNA. After allowing time for the expression of the selective marker, the mixtures are plated and the number of transformants scored. The kind of results obtained in the transformation of *Pneumococcus* to streptomycin-resistance is shown in Fig. 118, curve *b*; towards the end of the period of exponential growth of the culture, the ability of the bacteria to be transformed falls rapidly. On the other hand, if an excess of transforming DNA is added to the growing culture from the beginning and, at intervals, samples are treated with deoxyribonuclease (DNase) to destroy the transforming potential of the DNA, the type of response shown in Fig. 118, curve *c*, is found; when the DNA is removed by the DNase at any time during the first eighty minutes of growth, no transformants arise, but after this time their number rises rapidly to a plateau. These two curves, taken together, show that the bacteria of a growing population are at first insusceptible to transformation. Towards the end of the exponential phase of growth, however, susceptibility rises rapidly to a maximum and then falls off again equally rapidly (Thomas, 1955). In other words, *competence is a transitory state of the recipient population, and its duration is restricted to a small fraction of the growth cycle*. This wave of competence can be demonstrated in a single experiment by diluting samples of a growing culture into an excess of transforming DNA and then adding DNase a few minutes later (Thomas, 1955). When the numbers of transformants emerging from each sample are related to the age of the recipient population, the type of curve shown in Fig. 119 is obtained; this curve is really a composite of curves *b* and *c* of Fig. 118. Results of a comparable nature are found for the kinetics of the development of competence in *Haemophilus* (Schaeffer, 1956) and in *B. subtilis* (Schaeffer and Ionesco, 1959; Spizizen, 1959), although developing spores of the latter organism are also competent.

So far we have seen that the population of recipient bacteria as a whole is subject to a wave of competence of rather short duration. To

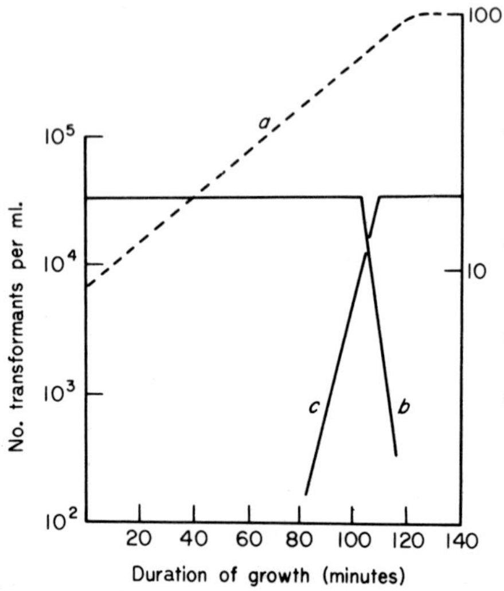

FIG. 118. Illustrating how the ability of bacteria to be transformed (competence) is related to a particular phase of growth of the culture.

Curve *a* (interrupted line) represents the growth of the recipient population, expressed as a percentage of maximum.

Curve *b* shows the number of transformants obtained when samples of the population are transferred at various times to an excess of transforming DNA, in which they are left for about ninety minutes to allow time for expression of the donor character.

Curve *c* shows the number of transformants arising from samples exposed to DNA from the beginning, and then treated with DNase at various times thereafter. (After Thomas, 1955.) See text.

what extent do the kinetics of this wave reflect the competence of individual organisms? For example, if the peak of the wave (Fig. 119) indicates an overlap of the duration of competence in *all* the potentially competent cells, then the time over which an individual bacterium remains competent can be assessed from the distance between those points on the rising and falling components of the curve at which half the peak value of transformants is found; in the case of *Pneumococcus* this turns out to be about fifteen minutes, or about half the generation time. The alternative is that the duration of competence is much shorter than this, so that the peak of the curve only represents the instant when the majority of potentially receptive cells happens to be competent. In this latter case, the number of transformants found at

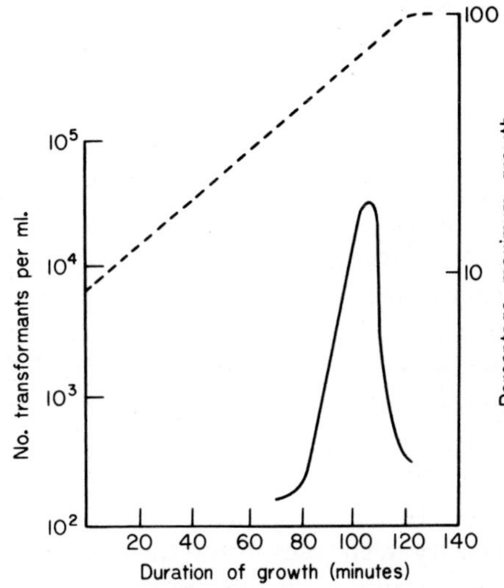

FIG. 119. Illustrating the rise and fall of competence during the growth cycle of a culture of recipient bacteria.

The interrupted curve represents the growth of the recipient population, expressed as a percentage of maximum.

The continuous curve shows the number of transformants obtained when samples of the population are removed at intervals during the growth cycle, exposed for a few minutes to an excess of transforming DNA and then treated with DNase.

The competence curve has, in fact, been artificially constructed to conform to the data of Fig. 118, curves *b* and *c*, with which it should be compared. (see text.)

the peak will be less than the total, cumulative number obtainable by permitting a prolonged interaction of the cells of the growing culture with the DNA, as in the experiments portrayed in Fig. 118, curves *b* and *c*. In practice, the number of transformants found at the peak of the non-cumulative curve (Fig. 119) and at the plateaux of the cumulative curves (Fig. 118) are closely similar (Thomas, 1955; reviewed, 1962a).

Hotchkiss (1954) found that if synchronous division is induced in the cells of a competent culture of *Pneumococcus*, the culture also shows a cyclical variation in competence which is commensurate with the division time and which may be marked. Comparable periodic fluctuations in competence have also been observed by Thomas (1955) in the absence of deliberate synchronisation. It therefore seems that, in *Pneumococcus* at least, competence is a property of the bacteria which develops only in populations which are approaching the end of exponential growth, and may be associated with some particular phase of the division cycle. However, a recent study of pneumococci growing in minimal medium with varying degrees of enrichment, suggests that the sharp onset of 'synchronous' competence is a function of population density rather than of growth phase or generation time (Tomasz, 1965).

In the case of *H. influenzae* bacilli growing exponentially in a rich medium, competence begins to develop about 40 minutes after transfer to a chemically defined, 'non-growth' medium containing various amino acids and sodium fumarate, reaching a plateau at 150 minutes. However, the appearance of competence requires an energy source and protein synthesis, and is lost almost completely after one generation following return to the rich medium (Spencer and Herriot, 1965). In addition, the presence of divalent cations appears to be necessary in all cases. From the metabolic point of view, therefore, competence appears to be associated with some kind of 'resting state' in the growth cycle. This is borne out by a report that, in *Pneumococcus*, the development of competence is associated with arrest of DNA synthesis and a pronounced lag in the incorporation of lysine, a cell wall constituent, although protein synthesis as a whole continues at the normal rate (Ephrussi–Taylor and Freed, 1964). In the case of thymine-requiring *B. subtilis* bacteria there is also good evidence that DNA synthesis is arrested in competent cells, at least after uptake of donor DNA; unlike the population as a whole, the number of transformants is not reduced by thymine-less death nor by 5-bromouracil incorporation (Bodmer, 1965). Again, Nester (1964) has claimed that competent *B. subtilis* cells are more resistant to the action of penicillin, which interferes with growth of the cell wall, than the rest of the population, suggesting that cell wall synthesis may be in abeyance.

At the peak of the competence curve, are all the bacteria competent, or only a proportion of them? In the absence of autoradiographic studies of the ability of individual bacteria to take up radioactive DNA, we have no direct information on this point; but it is an important point for, as we have seen (p. 582), a computation of the number of DNA molecules that can be adsorbed by each competent bacterium hinges upon it. In the case of *Haemophilus*, Schaeffer (1957) has approached the problem indirectly by studying the relative concentrations of non-transforming and transforming DNA at which inhibition of transformation occurs. Provided that the number of transforming molecules per competent bacterium is low, the *relative* concentration of non-transforming DNA required to inhibit transformation if all the bacteria are competent, should be 100 times greater than if only 1 per cent of the population is competent, for in the former case 100 times as many bacteria must be saturated before the probability of effective contact between a transforming molecule and a competent bacterium is reduced. The results, shown in Table 22, agree well with those predicted for a culture of *Haemophilus* at the peak of competence, in which all the cells are competent. Using another method (see p. 609), this high degree of competence has been confirmed for both *Haemophilus* (Goodgal and Herriot, 1961) and *Pneumococcus* (Fox and Hotchkiss, 1957), where the frequency of single transformants at saturating DNA levels may be as high as 5 per cent or more. In the case of *B. subtilis*, on the other hand, the frequency of transformants for single markers is only 0·1 to 0·5 per cent at best, and that of competent cells less than 10 per cent (Nester and Stocker, 1963).

THE NATURE OF COMPETENCE

There are two current hypotheses concerning the physiological basis of competence which are far from being mutually exclusive. According to one, competence results from alterations in the structure of the cell wall which render it permeable to large molecules, and there is now considerable evidence that various sorts of wall change do accompany the development of competence (review: Spizizen *et al.*, 1966). Also favouring this view is the fact that although DNA extracted from transducing particles of phage λ*dg*, which carry the bacterial galactose (*gal*$^+$) region on their chromosomes, is unable by itself to transform *gal*$^-$ *E. coli* K12 to galactose fermentation, nevertheless *gal*$^+$ transformants do arise if the *gal*$^-$ bacteria are infected with non-transducing phage λ particles at the time of exposure to the DNA (p. 544; Kaiser and Hogness, 1960). The infecting phage acts as a 'helper' in facilitating entry of DNA into the cells, possibly by making a 'hole' in the wall through which the DNA can pass. However, the matter is more complicated than this explanation suggests, since the helper phage does not assist the entry into the cell of DNA extracted from *gal*+ *bacteria*; this specificity of uptake, which requires that the *gal*+ region of bacterial chromosome be incorporated into the phage DNA, seems to depend on the presence of at least one extremity of the phage chromosome which carries a 'cohesive site' (see p. 462; Strack and Kaiser, 1965). In addition, there are other kinds of specificity inherent in DNA uptake which rule out a simple permeability hypothesis. For example, such a hypothesis would predict that the smaller the size of DNA molecules the more readily they would be taken up but, as we shall see, the opposite is the case (Litt *et al.*, 1958). Moreover, competent cells reject molecules of single-stranded DNA and of RNA (Lerman and Tolmach, 1957, 1959), while strains of *Haemophilus* show considerable ability to discriminate, during uptake, between homologous DNA and that from other species (Schaeffer, 1958).

The second hypothesis is that competence is determined by the synthesis of specific receptor sites at the bacterial surface. This view originated from the fact that protein synthesis is required for the development of competence, as well as from the results of a detailed study of the kinetics of interaction of competent pneumococci with transforming DNA, which appeared to conform to the known reaction kinetics of enzymes and their substrates (Fox and Hotchkiss, 1957). This hypothesis has the advantage of accounting for the specificity of uptake of double-stranded DNA molecules of a certain minimum size, and is not incompatible with the finding that the actual uptake of DNA by *Haemophilus* cells no longer depends on protein synthesis once competence has been achieved, although the requirement for energy remains (Stuy, 1962).

An important, though still unresolved clue to the nature of competence was the discovery that competent cultures may produce an extracellular factor which is able to induce competence in non-competent cells. This *competence factor* was first demonstrated in the case of Group H haemolytic streptococci, which become competent only when grown in a serum-supplemented medium. The addition of supernatant fluid from a competent culture to a serum-free culture provokes optimal competence in about 30 minutes. The competence factor is effective not only with cells of the same strain, and of some other related strains which have the potential to achieve competence in its absence, but also with certain specific streptococcal strains which are not naturally transformable. The competence factor can be precipitated by ammonium sulphate (80 to 90 per cent saturated), is slowly destroyed at temperatures above 100°C, is non-dialysable, and appears to be sensitive to proteolytic enzymes; a low molecular weight protein, it is inactive at temperatures of 20°C or under, as well as in the presence of chloramphenicol (Pakula and Walczak, 1963; Pakula, 1965; see also Dobrzanski and Osowiecki, 1967).

Competence factors have since been found to be produced by *Pneumococcus* (Tomasz and Hotchkiss, 1964; Tomasz, 1965; Kohoutová and Málek, 1966), *Bacillus cereus* (Felkner and Wyss, 1964) and *Bacillus subtilis* (Charpak and Dedonder, 1965). In the case of *Pneumococcus* the initial observation was that, in a mixed culture of two strains carrying different genetic markers, in which the inoculum of one strain greatly exceeded that of the other, both strains developed competence at the same time; on the other hand, in control, unmixed cultures, competence developed in that with the larger inoculum more than 30 minutes before the other. The provoking substance appeared to be unstable and mainly associated with the cells since, in general, cell-free filtrates proved relatively ineffective. However, when a competent culture was separated from an incompetent one by a Millipore filter, which holds back the bacteria but allows free passage of the medium, the incompetent bacteria were rendered competent, even at 30°C at which competence does not normally develop. The factor is of high molecular weight and non-dialysable, is destroyed by proteolytic enzymes, and acts specifically on pneumococci and the closely related Group D streptococci. During activation by the factor, recipient cells acquire a new antigen and develop the capacity to bind DNA. An interesting corollary to these findings is that cells which have passed through the phase of competence produce a heat-stable, non-dialysable *inhibitor*, which blocks the action of the competence factor and may account for its apparent instability (Tomasz and Hotchkiss, 1964; Tomasz, 1965).

Somewhat different results have been obtained with *Pneumococcus* by Kohoutová (1965; Kohoutová and Málek, 1966) who recognised two distinct kinds of effect which culture filtrates of transformable strains may have on the frequency of transformation. Firstly there is a factor which appears in culture supernatants *before* the development of competence which markedly stimulates the frequency of transformants if mixed with the transforming DNA for a very short time (3 to 10 minutes), prior to adding recipient cells which have not yet reached the peak of competence. Exposure of the DNA to the filtrate for a longer period than is required for the optimal response, results in a rapid and progressive fall in the number of transformants, while the effect is not found if the recipient culture is fully competent. It is suggested that the effect may be due to DNase which accumulates in the culture during the competence cycle, and that competence of the recipient is somehow connected with the interaction of this nuclease with the donor DNA (Kohoutová and Málek, 1966). We shall shortly see that the uptake of transforming DNA by pneumococci is, in fact, followed by its breakdown into single strands and nucleotide fragments, and that only a single strand of the donor DNA is probably incorporated into the recipient chromosome.

The second type of filtrate effect reported by Kohoutová is the actual induction of competence in young, non-competent pneumococcal cultures by sterile filtrates of highly competent cultures. As much as a thousand-fold increase in the number of transformants was shown to follow ten minutes exposure to undiluted filtrate. An interesting feature of this induction is that filtrates from cultures which have passed through the phase of competence, so that their efficiency as recipients has dropped to one per cent. of peak level, nevertheless continue to provoke maximal competence in non-competent cultures. Unlike the competence factor described by Tomasz and Hotchkiss (1964), this factor is thermostable, withstanding 100°C. for at least 30 minutes. It is not known to what extent these two effects are due to a single factor or multiple factors, or are interdependent phenomena under natural conditions.

The idea was first put forward by Thomas (1955) that competence might be due to the formation of partial protoplasts; we have already seen that protoplasts are often able to take up virus DNA and RNA to which the parental bacteria are normally impermeable (pp. 437, 441), and there is considerable evidence that competence is associated with cell wall alterations (see Spizizen *et al.*, 1966). However in the case of *B. subtilis* it has been clearly shown that protoplasts of highly competent bacteria are completely refractory to transformation. Moreover, if competent cells which have taken up ^{32}P-labelled transforming DNA, so that the transformation cannot be reversed by DNase, are washed and protoplasted, all the transforming DNA is lost, as is the ability to yield transformants, although there is no leakage of ^{3}H-thymidine-labelled recipient DNA and the recipient population retains its viability. This implies that transforming DNA is held, for a considerable time after irreversible uptake, at a location where it is protected from the action of DNase but which, nevertheless, is outside the cell membrane, and suggested that mesosomes (p. 568) might fulfill these requirements since, of course, they are expelled from the cytoplasm during protoplasting (Miller and Landman, 1966). More recently, autoradiography of thin sections of *B. subtilis* exposed to ^{3}H-thymidine-labelled DNA has demonstrated that, during the phase of maximal competence, the DNA molecules are indeed associated with, but on the outside of, the cytoplasmic mesosomes; in addition, during this phase, the total volume of membranous invagination, as well as the frequency with which it can be seen to be connected with the bacterial chromosome, are markedly increased. On the basis of these observations it has been suggested that the mesosomes may be the sites of production of enzymes necessary for the breakdown and incorporation of transforming DNA (Wolstenholme, Vermeulen and Venema, 1966).

Finally, competence is a character which is under genetic as well as physiological control, since mutants have been isolated from competent strains which show enhanced or reduced inheritable levels of competence, or its complete loss with inability to bind DNA at any phase of growth (Young and Spizizen, 1961; see Spizizen *et al.*, 1966).

ANALYSIS OF DNA UPTAKE

Let us now examine in more detail the interaction of competent cells with transforming preparations and, in particular, trace the sequence of events that ensues between the instant of primary contact of the bacteria with the DNA molecules and the time when the transforming molecules begin to be integrated into the chromosomes of the recipient cells.

THE SPECIFICITY OF UPTAKE

When the donor bacteria are grown in the presence of radioactive phosphorus (^{32}P), the ^{32}P atoms are incorporated into their RNA and DNA with equal efficiency. Yet the uptake of ^{32}P per transformant is found to be the same from crude extracts, containing protein as well as about four times more RNA than DNA, as from highly purified DNA from the same preparations (Lerman and Tolmach, 1957). Thus competent bacteria appear to have an exclusive affinity for DNA. To what extent is this affinity dependent on molecular weight and structure? With regard to molecular weight, it turns out that a certain minimum size is necessary. For example, when the mean molecular weight of DNA preparations is reduced to about 5×10^5 by treatment with ultrasonic sound, or by spraying through an atomiser, uptake ceases and no more transformants are found (Litt *et al.*, 1958; Rosenberg, Sirotnak and Cavalieri, 1959). Similarly, uptake is rapidly abolished by treatment with DNase. In the case of *B. subtilis*, controlled hydrodynamic shearing of transforming DNA indicated a much greater minimum size for transforming fragments, since DNA of mean molecular weight of less than ten million turned out to be virtually devoid of transforming activity. The disparity between this result and those previously found for *Pneumococcus* may be partly due to much greater heterogeneity of fragment size in the pneumococcal DNA preparations, due to incomplete shearing and the lack of precise fractionation procedures (Syzbalski and Opara-Kubinska, 1965). The molecular weight of the *B. subtilis* chromosome is about 10^9, adjusting for the fact that most chromosomes are in the process of replicating in young cultures. When the usual procedure for extracting the DNA, and removing the protein and RNA, is employed, the mean molecular weight of the DNA is reduced to about 25×10^6, so that the chromosome gets broken into about 40 fragments on the average (Szybalski and Opera-Kubinska, 1965).

On the other hand, there seems to be no restricting upper limit to the size of transforming DNA molecules. As we shall see, when special steps are taken to prevent fragmentation of the DNA, markers which are unlinked in normal preparations can be shown to be linked in transformation (p. 611; Kelly and Pritchard, 1965). Again, DNA extracted from certain *B. subtilis* phages is able to infect competent cells of this organism and produce normal bursts of progeny phage—a process known as *transfection*. It must be presumed, in successful infection, that the whole phage genome, consisting of a DNA molecule of molecular weight 120×10^6 in the case of one phage (SP8), enters the cell.

The effect of shearing is the transverse severence of the double-helical molecules. Another way of altering the molecules is to separate the two strands of the double helix by heating (p. 244). If the heated, denatured DNA is cooled rapidly the strands remain separate so that a single-stranded preparation of half the initial molecular weight is obtained. Such single-stranded DNA is virtually inactive in transformation, and this is correlated with loss of its ability to be taken up by competent bacteria (Lerman and Tolmach, 1959; Marmur and Lane, 1960). On the other hand if the denatured DNA is cooled *slowly*, complementary single strands reunite to yield the normal, double-helical structure and, *pari passu*, the ability of the DNA to be adsorbed and to yield recombinants is restored. Furthermore, if denatured, transforming DNA, carrying a genetic marker such as streptomycin-resistance (*str-r*), is slowly cooled in the presence of an *excess* of similarly denatured, non-transforming, streptomycin-sensitive (*str-s*) DNA from the same species, the single strands carrying the *str-r* marker will tend to encounter and re-unite with complementary *str-s* strands from the non-transforming DNA. The result is that those double-stranded molecules which now carry the *str-r* marker are *hybrid* with respect to it, that is, they have the mutant *str-r* base sequence on one strand and the wild-type (*str-s*) base sequence on the other. Nevertheless such molecules are not only taken up by competent *str-s* bacteria but are able to transform them to streptomycin-resistance (Marmur and Lane, 1960). These fascinating experiments clearly show that although the uptake of DNA depends on the integrity of the double-helical structure, only one of the two strands is required for the transformation event itself provided it can penetrate the recipient cell. We shall see later on that only one of the two immigrant strands appears, in fact, to be involved in the recombination process (see pp. 600–604).

Another question we can ask about DNA uptake is whether or not it is specific for homologous DNA. In the case of *Pneumococcus* the answer is that any double-stranded DNA of proper molecular weight is irreversibly bound by, and presumably enters, competent cells. For instance one can hardly conceive a DNA more foreign to pneumococci than that from calf thymus, yet this DNA competitively inhibits the uptake of pneumococcal transforming DNA (Hotchkiss, 1954). Similarly, ^{32}P-labelled *E. coli* and pneumococcal DNAs are irreversibly adsorbed by competent pneumococci with almost equal efficiency although the former, of course, is not transforming (Lerman and Tolmach, 1957). A comparable lack of specificity has been found for *Streptococcus* (Pakula, Hulanicka and Walczak, 1960) and *B. subtilis* (see Schaeffer, 1964). On the contrary, in the case of *Haemophilus*, uptake and competitive inhibition appear to be largely restricted to DNAs from rather closely related species (Shaeffer, Edgar and Rolfe, 1960).

THE KINETICS OF UPTAKE

Competent bacteria are exposed for some minutes, at 37°C., to DNA which is labelled not only with a genetic marker but also with radioactive phosphorus (^{32}P). The bacteria are then washed in the cold until the level of residual radioactivity, indicating the amount of bound DNA, remains constant. On re-warming to 37°C. it is found that a proportion of the bound DNA is not only released from the cells spontaneously but can very rapidly be removed by treatment with DNase. Thus the DNA taken up by the bacteria is present in two states—one reversibly attached and the other permanently bound and insusceptible to DNase. Comparison of the amount of permanently bound DNA with the frequency of transformants shows that the two are strictly proportional for a wide range of different experimental conditions (Lerman and Tolmach, 1957). Moreover, as we have seen, the probability that a molecule of DNA carrying a given genetic marker will yield a transformant is very high once it has been permanently bound, since one transformant is found for the equivalent of every one to two genomes of DNA taken up irreversibly (Fox and Hotchkiss, 1957; Lerman and Tolmach, 1957). However one should be cautious in interpreting the results of experiments involving the use of ^{32}P, since transforming preparations may be contaminated with phosphorus-containing polymers other than DNA, such as teichoic acid from the cell walls which is not taken up irreversably by competent bacteria but bands at roughly the same buoyant density as DNA in a density gradient (see Spizizen *et al.*, 1966).

We are primarily concerned here with the irreversably bound DNA whose fate is followed, in most kinetic experiments, by exposing competent bacteria to the transforming preparation and then treating them with DNase, as a prelude to examining their various potentialities as a function of time. The irreversible uptake of transforming DNA by competent bacteria is very quickly accomplished. In the case of *Pneumococcus*, for example, if DNase is added to the DNA at the same time as the bacteria, no transformants arise; but if the addition of the bacteria precedes that of the DNase by as short a time as ten seconds, transformants can subsequently be isolated (Hotchkiss, 1954). Similar kinetic experiments with *Haemophilus* have shown that about four to five seconds is required for the irreversible uptake of DNA molecules having a mean molecular weight of 8 to 9×10^6, and it has been calculated that each bacterium has about two uptake sites on the average. If the DNA is added even one second *after* addition of DNase, no transformants arise. Moreover, the uptake is a metabolically active process since it is strongly temperature-dependent and is reversibly inhibited by 2,4-dinitrophenol which prevents energy production (Stuy and Stern, 1964).

In the case of *B. subtilis* the much longer mean time of 2·5 *minutes* appears to be necessary for the irreversible uptake of DNA molecules carrying any single marker, but whether or not this is due to the use of a different mechanism of uptake is unknown. However, the time required for uptake in this organism appears to be a function of the length of the DNA molecule; when transformants inheriting two linked markers are selected, the uptake time is increased in direct proportion to the map distance between them. Moreover, when a particular marker is selected and the inheritance of a linked, unselected marker scored as a function of time, the frequency of the latter increases sharply once the mean uptake time for the selected marker is exceeded. This is also found when DNA of high molecular weight is employed and the joint inheritance of markers which are not normally linked, but are now present on the same DNA fragments as judged by DNA saturation curves (p. 609), is scored. These findings not only indicate that transforming DNA is taken up longitudinally, but also suggest that penetration by a certain minimal length of DNA, estimated as about five genes long, may be necessary for successful transformation—possibly for recombination to occur (Strauss, 1965, 1966).

THE KINETICS OF TRANSFORMATION

The subsequent fate of DNA which has been taken up irreversibly can be traced in the following way. Imagine two transformable strains which are reciprocally resistant to two drugs, A and B, such that one strain is resistant to A but sensitive to B (*Ab*) while the other is sensitive to A but resistant to B (*aB*). Competent bacteria of one strain (say, *aB*) can be transformed to double resistance (*AB*) by DNA extracted from the other (*Ab*). In this case we will refer to *B* as the *recipient* or *resident marker* and to *A* as the *transforming marker*. Similarly, if wild type, doubly sensitive bacteria (*ab*) are exposed to DNA from strain *Ab*, *Ab* transformants will emerge.

Competent *aB* bacteria are exposed *for a very short time* to *Ab* DNA labelled with ^{32}P, and then treated with DNase to stop further fixation. The bacteria are thoroughly washed in the cold and returned to nutrient broth at 37°C. At intervals thereafter samples are removed and, after plating aliquots to assay the number of *AB* transformants, the bacteria are lysed and their DNA extracted. This DNA comprises that taken up from the original transforming preparation, carrying the marker *A* as well as the ^{32}P label whereby the efficiency of re-extraction can be judged, and also the resident DNA which carries the *B* marker. By using this DNA to transform *a new population of ab bacteria*, the biological activity of the DNA taken up in the first transformation can be followed in terms of that of the resident DNA, as a function of time of incubation after uptake, by observing the ratio of *Ab*:*aB* transformants produced by each sample.

In the case of *Pneumococcus* it is found that, immediately after treatment with DNase and before incubation, more than 50 per cent. of the ^{32}P label taken up by the recipient bacteria can be recovered in the re-extracted DNA, *but this* DNA *shows virtually no transforming ability for the donor A marker*, despite the fact that transformation with respect to the resident *B* marker is normal. Unlysed bacteria from the same sample yield the expected number of *AB* transformants on plating. On incubation of the recipient bacteria, however, the capacity of the re-extracted DNA to produce *Ab* transformants on strain *ab* increases rapidly until, after about six minutes, the ratio of *Ab* to *aB* transformants reaches a constant value, as Fig. 120 shows (Fox, 1960). Once this constant ratio is achieved it remains the same even on prolonged incubation during which the original recipient bacteria being to multiply exponentially, while the same ratio is also found in other experiments where the recipient bacteria are exposed for much longer times to the original transforming DNA. Since the transforming ability of DNA is an index of its concentration (Fig. 117) this shows that, very soon after uptake, the transforming DNA and the resident DNA increase in amount at the same rate. Moreover, this constant ratio between the transforming ability of the re-extracted transforming and resident DNA is equal to the absolute frequency of *AB* transformants issuing from the first transformation, thus showing once again that a molecule of transforming DNA has a probability of about 1·0 of producing a transformant once it has penetrated a competent cell (Fox and Hotchkiss, 1960).

The short time after uptake during which transforming DNA loses its biological activity in transformation is known as the *eclipse period* (Fig. 120). A similar eclipse has been demonstrated in *B. subtilis* transformation (Venema, Pritchard and Venema-Schröder, 1965a) but, strangely enough, is not found in the case of *Haemophilus* (Voll and Goodgal, 1961, 1965) although the kinetics as well as the general nature of incorporation of donor markers into the recipient chromosome appear to be the same in all three species. From the fact that

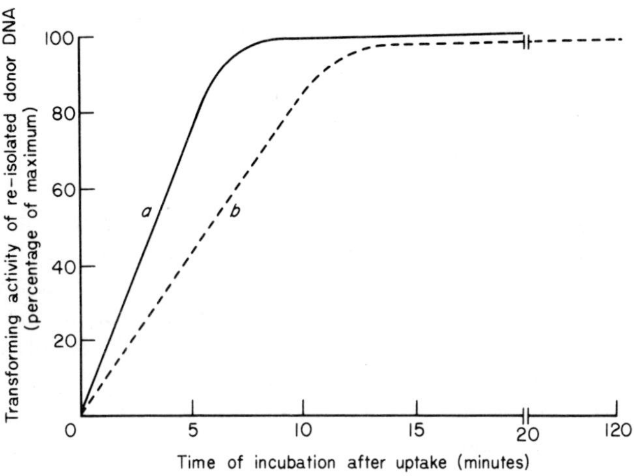

FIG. 120. The kinetics of recovery of the transforming activity of donor DNA, re-extracted from recipient bacteria after uptake.

The recipient cells are exposed for a very short time to donor DNA, treated with DNase, washed and returned to growth medium at 37°C. At intervals thereafter (abcissa), DNA is extracted from samples of the bacteria and assayed for ability to transform a new recipient strain with respect to the original donor marker as well as to a resident marker of the original recipient. The activity of the re-isolated donor DNA (ordinate) is given by the ratio
(transformants for donor marker)/(transformants for resident marker), expressed as a percentage of the highest ratio attained.

Curve *a* (continuous line) follows the recovery of the activity of the donor marker alone.

Curve *b* (interrupted line) shows the rate at which *double transformants* appear when the donor and resident markers are very closely linked, and indicates the kinetics of incorporation of the donor marker into the chromosome of the first recipient.

See text. (Data from Fox, 1960.)

radioactive phosphorus in transforming DNA is irreversibly incorporated into the DNA of recipient bacteria in proportion to the number of transformants produced (Goodgal and Herriot, 1957a; Lerman and

Tolmach, 1957), it follows that the eclipse period must be due to some change in the state of the DNA itself and not, for example, to the transfer of its genetic information to some non-transforming intermediary. What is the nature of this change of state? One hypothesis, based on the finding that re-extracted donor DNA is less dense than normal, is that the donor DNA complexes with some other cellular component such as protein, which renders it resistant to uptake by competent cells (Ephrussi-Taylor, 1960b). Alternatively the DNA could become denatured or broken down into low molecular weight fragments which are refractory to uptake, recovery from the eclipse being associated with incorporation into the resident DNA.

There is now convincing direct and indirect evidence that this is indeed what happens. The direct evidence was the discovery by Lacks (1962) that when extracts of competent pneumococci are analysed at intervals after exposure to ^{32}P-labelled DNA, one half of the bound radioactivity is present, immediately after entry, in highly polymerised single-stranded DNA, and the other half in degraded, acid-soluble fragments. The single-stranded DNA then begins to disappear *pari passu* with the incorporation of the donor radioactivity into the resident DNA, and the progress of this process parallels the appearance of transforming DNA carrying the donor marker. Since single-stranded DNA has negligible transforming ability, this finding offers an adequate and sufficient explanation of the eclipse period, which has recently been confirmed by the demonstration that the transforming activity of donor DNA, extracted during the eclipse period, can be restored by *in vitro* annealing with denatured homologous, but non-transforming, DNA (Ghei and Lacks, 1967).

Lacks (1962) proposed a model in which the penetrating end of the DNA duplex encounters a nuclease which breaks down one strand sequentially while pulling the other, intact strand into the cell. Such an enzyme would have properties similar to exonuclease III which specifically attacks double-stranded DNA from the 3′-hydroxyl ends (p. 256). Assuming that either end of a transforming DNA molecule may enter the cell first, it is clear that the two strands would have an equal probability of being left intact. The finding, mentioned above, that re-extracted donor DNA is less dense than double-stranded (Ephrussi-Taylor, 1960b), whereas single-stranded DNA is more dense, may be explained by supposing that the nuclease remains associated with the intact strand after extraction (Schaeffer, 1964).

THE NATURE OF INCORPORATION

The final step in transformation is the incorporation of the donor marker into the recipient chromosome. The fact that the transforming abilities of re-extracted donor and resident markers attain a constant, maximum ratio about six to eight minutes after DNA fixation (Fig. 120, curve *a*) and that this ratio is maintained when the bacteria divide, suggests that incorporation occurs before the first division and probably considerably earlier. The kinetics of incorporation can be approached more directly by employing two genetic markers which are so closely linked that, when present together in donor bacteria (*AB*), they are frequently transferred together on the same molecule of transforming DNA with the result that double (*AB*) transformants comprise a considerable proportion of the total (see p. 607). If, now, one of these markers is carried by the donor DNA (*Ab*) and the other by the recipient bacteria (*aB*), DNA extracted at intervals after fixation should not be able to induce double transformants (*AB*) in a new *ab* recipient until such time as the two markers become physically linked on a single DNA molecule in the original recipient bacteria.

Kinetic experiments of this sort have been performed in *Pneumococcus* using a closely linked pair of loci determining different degrees of resistance to sulphonamide. It is found that when the donor DNA carries one marker and the recipient bacteria the other, at about twelve to fifteen minutes after DNA uptake the re-isolated DNA can transmit double resistance to a new, doubly sensitive recipient with maximal efficiency (Fig. 120, curve *b*; Fox, 1960). In *Haemophilus*, in which the determinants of resistance to streptomycin and cathomycin ('Novobiocin') are similarly closely linked, about half the maximal amount of linked transformation is achieved by the re-extracted DNA in fifteen minutes and the maximal at about thirty minutes (Voll and Goodgal, 1961). These latter experiments offer a good example of the effect of segregation lag on the kinetics of transformation. When the transforming activity of DNA re-isolated from recipient cells, at intervals after uptake, is analysed under conditions of growth, the donor and recipient markers begin to increase at the same rate after about one generation time (thirty minutes). But if the kinetics of transformant production by these same recipient bacteria is followed, two or three generations are found to elapse before the number of transformants increases (Voll and Goodgal, 1961). As Fig. 53 (p. 202) shows, although the number of transformed *nuclei* may double at the first generation, proliferation of the transformed *bacteria* cannot be expected until nuclear segregation yields cells in which all the nuclei harbour the donor gene.

These kinetic experiments undoubtedly reveal the time at which the transforming marker and the resident marker become so intimately associated that they are transmitted together as linked on the same DNA fragment. However, since curves *a* and *b* of Fig. 120 do not coincide, it is evident that recovery from the eclipse can only be due in part to this association; a similar difference in the kinetics of recovery from eclipse and the establishment of linkage between donor and recipient markers has also been demonstrated in *B. subtilis* (Venema *et al.*, 1965a). The question arises whether this establishment of linkage merely represents pairing of the immigrant molecules with homologous regions of resident DNA, or the actual substitution of these regions by them. From the genetic standpoint only the latter event can be considered as true incorporation, in the sense of representing the completion of the act of recombination. In the case of *B. subtilis*, transforming DNA re-extracted shortly after uptake was found to be more resistant to thermal denaturation and to re-annealing than resident DNA, and these abnormal properties, taken to indicate incomplete incorporation, were retained for some time, even after the appearance of co-transforming DNA for linked donor and recipient markers (Venema, Pritchard and Venema-Schröder, 1965b). It was suggested that the early complex might comprise a triple-stranded DNA structure in which the single strand of transforming DNA is paired with the homologous region of the resident duplex, in a manner similar to Zubay's model for messenger RNA transcription (p. 373; Zubay, 1962).

A number of ingenious experiments have been devised in an attempt to define more precisely the nature of the incorporated fragment and the mechanism of incorporation. One approach, first devised by Herriot (1963) as a test of single-stranded incorporation in *Haemophilus*, is based on the transforming activity of hybrid DNA made by cross-annealing DNA strands of different genetic constitution. If DNA which is wild type for two closely linked markers (++) is denatured and allowed to anneal with an excess of denatured DNA which is mutant with respect to one of the markers ($a+$), the great majority of ++ strands will meet and pair with complementary $a+$ strands rather than with one of themselves, so that the two strands of each parental duplex, $\frac{++}{++}$, will now be distributed between two hybrid duplexes,

$\frac{++}{a\,+}$; that is, the number of molecules carrying the wild type markers will be doubled. It follows that if both strands of the donor molecules are incorporated, during transformation of a doubly mutant (*a b*) recipient, twice the number of transformants should arise from the hybrid as from the control, wild type preparation. In fact, in *B. subtilis*, the number of transformants remains the same. This is what would be expected if one of the two strands of the hybrid DNA was randomly chosen for incorporation—double the number of hybrid as of wild type molecules would be taken up, but from half of the hybrid molecules the mutant strand would be selected and no transformants arise (Bresler *et al.*, 1964; Venema *et al.*, 1965a). Bresler *et al.* (1964) extended their experiment by analysing the purity of the transformant colonies. If the wild type allele of marker *b* is selected and both strands of the hybrid duplex, $\frac{++}{a\,+}$, are incorporated it is clear that the first replication will segregate two types of recombinant chromosome, $++$ and $a+$. Most of the molecules in the annealed transforming mixture will have the majority constitution, homozygous $\frac{a\,+}{a\,+}$, but most of the molecules carrying both wild type markers will be hybrid. It follows that the transformant colonies arising from hybrid molecules will be those which contain wild type, $++$, recombinants. When these colonies were identified, by replica plating to minimal medium, and then analysed for the recombinant types they contained, 94 per cent. were pure cultures of $++$ transformants. It is thus clear that only one strand is normally incorporated (see also Vestri, Felicetti and Lostia, 1966).

Another type of experiment, in *Pneumococcus*, employed heavy isotopes (^{15}N and ^{2}H) and ^{32}P to label the donor DNA whose fate after uptake (8 minutes exposure before adding DNase) was then followed by density gradient centrifugation of the DNA extracted from the recipient cells (Fox and Allen, 1964). At one minute after addition of DNase, when recovery from eclipse was 21 per cent complete, 21 per cent. of the ^{32}P activity was found in a band corresponding to heavy, single-stranded DNA, 43 per cent. in acid-soluble fragments and 36 per cent. in light, native DNA. Three minutes later only five per cent of the heavy single-stranded DNA remained, while 15 minutes after adding the DNase, when recovery from eclipse was virtually complete, it had disappeared entirely and over 80 per cent. of the ^{32}P label had been transferred to native, light DNA. This disappearance of the heavy DNA moiety is more apparent than real, and is due to its physical association with much larger molecules of native, light DNA whose density is thereby not significantly affected. This can be shown by sonicating the extracted DNA before analysis so that the associated heavy fragments now become significant fractions of the smaller molecules. When this is done, the ^{32}P label shifts towards a new band of hybrid density, midway between heavy and light native DNA, and most of the donor DNA biological activity is found in this hybrid fraction (cp. p. 545). It is clear that the single strand of the heavy donor DNA has become associated with a light complementary strand, and this is confirmed, in turn, by denaturing the sonicated DNA; a fraction of the radioactivity now shifts to a density band corresponding to that of denatured, heavy DNA—part of the initial, heavy single strand has been re-isolated (Fox and Allen, 1964). The question is whether the heavy strand is actually inserted into the recipient DNA and pairs with the homologous region of the complementary strand to which, the experiment suggests, it is covalently bonded; or whether, for example, it simply acts as a template for the synthesis along it of a new complementary strand, to form a homozygous, double-stranded molecule which is then incorporated as such. The fact that the hybrid density DNA is masked by the unsonicated, light native DNA is against this, as is also the fact that recovery from eclipse as well as incorporation can occur in the absence of significant DNA synthesis (Fox, 1960; Voll and Goodgal, 1961; Bodmer, 1965).

Further evidence that only one of the two donor strands is incorporated directly into the resident DNA comes from a study of the rate at which ^{32}P decay in transforming DNA inactivates its biological activity (Fox, 1962). It turns out that the transforming ability of donor DNA, extracted from recipient bacteria at a time when incorporation is virtually complete (Fig. 120, curve *b*), declines at the same rate as does that of the original DNA. This is what is expected, provided we assume that ^{32}P disintegrations occurring in one strand of double-stranded DNA have negligible effect on the other strand. Let us suppose that, at time t after ^{32}P incorporation, there is a high probability that an inactivating disintegration will have occurred in only one of the two strands of each transforming DNA molecule. After uptake by competent bacteria, one of the two strands of each molecule is randomly destroyed. Of the remaining single strands, half will have been inactivated by ^{32}P decay while the other half will produce transformants. Now let us consider what happens if these same DNA molecules are taken up at time zero and that ^{32}P decay is taking place at the same rate in the residual single strands of each molecule, which have now been incorporated into the recipient chromosome. At time *t*, the transforming activity of half these strands will have been inactivated. Thus the re-extracted DNA will be expected to have half the transforming activity of similarly re-isolated DNA which has suffered no ^{32}P decay (Fox, 1962). On the other hand, if the ^{32}P-containing, single-stranded donor molecules, after uptake, acted as templates for the synthesis of complementary strands, these strands would carry the genetic information of the transforming strand but would not be subject to ^{32}P decay, so that the donor transforming ability of the re-extracted DNA should stabilise at half the initial rate.

An elegant investigation by Bodmer (1965) has brought to light some interesting points concerning incorporation in *B. subtilis*. Thymine-requiring competent bacteria were mixed with transforming DNA and provided with 5-bromouracil (BU) instead of thymine (T); after 90 minutes, DNase was added, and the recipient DNA extracted and banded in a caesium chloride density gradient (see p. 333). It was found that transforming activity with respect to both donor and recombinant markers was present mainly in unreplicated, light DNA which had not incorporated the dense BU, even when the competent cells were pre-incubated with the BU. It follows that the mechanism of transformation is not associated with DNA synthesis. If BU is added to a competent culture *after* transformation, at a time when the donor DNA is beginning to be replicated, donor transforming activity in the re-extracted DNA is found in a density region which lies between the light (T–T) and hybrid (T–BU) DNA. This suggested a model in which the donor DNA molecules are incorporated at the growing points of the chromosomes of competent cells which are in a state of arrested DNA synthesis; DNA synthesis, incorporating BU, would be resumed from these points, so that molecules encompassing such regions would contain both light and hybrid DNA (cp. p. 569; Bodmer, 1965).

We will conclude this section with an interesting but puzzling corollary of the idea that only one of the two strands of each transforming DNA molecule can generate transformants. We have seen that molecules of DNA which are 'hybrid' with respect to the two strands are formed if heat-denatured, single-stranded preparations of transforming and non-transforming DNA are slowly cooled together, and that these 'hybrid' molecules possess transforming activity (p. 593). Similar experiments have been carried out using two preparations of

Haemophilus DNA, each of which carries only one of a pair of closely linked markers (resistance to streptomycin and cathomycin). When these DNAs are heated and slowly cooled together, molecules arise which are able to transform to double resistance (Herriot, 1961). The inference is that these molecules are 'hybrid' for the two markers, carrying the information for streptomycin resistance on one strand and for cathomycin resistance on the other. But if one of the two strands is always destroyed on uptake, structures of this sort could not yield double transformants. In any case, the formation of stable, doubly-resistant transformant clones requires that the two markers come to lie on the *same* strand, for otherwise they would segregate at the first replication of the transformant DNA, and it is by no means obvious how this could be accomplished (see Ravin, 1961). A possible escape from the dilemma is to suppose that a proportion of the single strands break at high temperature so that, on cooling, two single-stranded 'Watson' *fragments*, each carrying a different marker, could pair with the complementary regions of an intact 'Crick' strand derived from either DNA (Doty *et al.*, 1960; Herriot, 1961); evidence has recently been presented that this explanation is the correct one (Herriot, 1965).

TRANSFORMATION AS A GENETIC PHENOMENON

Up till now we have dealt with transformation mainly in physiological and physico-chemical terms, and have not considered it from a purely genetical point of view. To what extent can the concepts of genetical analysis be applied to the phenomenon? There are three basic features of recombination as it is found in other parasexual systems in bacteria. In the first place the fragments of donor chromosome are assumed to pair with the homologous regions of the recipient chromosomes. Secondly, following pairing, the genetic determinants carried by the donor fragments replace their alleles on the recipient chromosomes to yield haploid recombinants. Thirdly, when an introduced donor fragment carries three or more markers, these segregate among recombinants in a manner consistent with their arrangement in an unambiguous sequence and at fixed positions on the chromosome, irrespective of which allele of a pair characterises either parent.

PAIRING

Indirect evidence for pairing of the molecule of transforming DNA with the homologous region of recipient chromosome, as a necessary prelude to incorporation, comes from the study of transformation to streptomycin resistance between different but related species of *Haemophilus* (Schaeffer, 1958). Let us call two such species A and B. Competent bacteria of either species are transformed at about the same frequency by DNA extracted from the same species. Similarly, competent A bacteria take up B-DNA as efficiently as the homologous A-DNA, and *vice versa*, whether the criterion is competitive inhibition between the two DNAs or the irreversible fixation of radioactive phosphorus with which the DNAs are labelled. On the contrary, the frequency of transformation to streptomycin resistance by heterologous DNA (as, for example, when DNA from streptomycin-resistant A bacteria is used to transform competent B cells) may be as much as 10^5 times lower than when homologous DNA is used. Thus although the heterologous DNA molecules penetrate the bacteria with normal efficiency, they have a very low probability of incorporation.

One explanation is that the restricted regions of the transforming molecules which determine streptomycin resistance are different in the two species, so that the donor marker itself is foreign to the recipient bacteria and, therefore, difficult to incorporate. If this is so, then those few B bacteria which *are* transformed by A-DNA should carry the foreign A marker, with the result that DNA extracted from them should prove as ineffective as the original A-DNA in transforming new B bacteria to resistance. In practice, however, transformants arise at the frequency found when homologous transforming preparations of B-DNA are used. We must infer that the initial relative failure of incorporation is not due to lack of homology within the streptomycin locus itself.

The alternative is that failure of incorporation arises from species incompatibilities involving parts of the DNA molecule other than that specifying streptomycin resistance. Such incompatibilities are most simply interpreted as differences in base sequence which prevent the proper synapsis of the DNA molecules with the corresponding region of recipient chromosome (Schaeffer, 1958). If this is so, these experiments also provide evidence that only a part of the incoming molecule of transforming DNA may be incorporated. The possibility that the results of inter-species transformation are due to host-controlled restriction has not been excluded (p. 514).

GENETIC EXCHANGE

When genetic material carrying a particular marker is transferred from one bacterium to another which lacks the marker, two kinds of recombinant inheriting the marker may be found. The usual event, which we regard as true recombination, is a genetic exchange whereby the immigrant marker replaces its allele on the recipient chromosome so that the recombinant is haploid. On the other hand the transferred genetic material may be *added* to the recipient genome; a more or less stably diploid cell, inheriting both alleles, is formed, as in the case of heterogenotes (pp. 99–104). To which category does transformation belong?

The main evidence on this point comes from the study of reciprocal crosses and of linked transformations, and shows quite clearly that the transforming marker replaces its allele in the recipient cell. Thus not only can *a* bacteria, which have lost the character A as a result of mutation, reacquire it by transformation with DNA from the wild type (*A*) strain, but either wild type or *A* transformants can in turn be transformed to *a* with about the same efficiency by DNA extracted from *a* bacteria. This is the result to be expected if the transforming marker is substituted for its recipient allele, but is inexplicable if the donor marker and its recipient allele are jointly inherited by the transformants; for in the latter case the transformants would carry both alleles (*A*/*a*) irrespective of the direction of transfer and should therefore display the same phenotype. Reciprocal transformations of this type were first demonstrated with respect to the ability of *Pneumococcus* to grow either as long chains of cells or in the normal diplococcal form (Taylor, 1949); non-capsulated pneumococci which have been transformed to the capsulated state can likewise be reverted to the non-capsulated condition by DNA from a non-capsulated strain (Ravin, 1959).

Unless the frequency of transformation is high, the demonstration of reciprocal transformation is difficult because only those transformants inheriting the resistance allele of a pair can be selected and identified. This type of evidence for genetic exchange was therefore greatly extended by the discovery that the determinants of certain characters are so closely linked that transforming DNA carrying them yields double transformants with high probability. Let us consider a theoretical case where recipient bacteria carrying the closely linked sensitivity loci, *a* and *b*, are exposed to transforming DNA from a doubly resistant, *AB* donor strain. Not only *Ab* and *aB*, but also *AB* transformants are found to arise with the same order of frequency. On the other hand if DNA from *Ab* cells is used to transform *aB* recipients, and inheritance of the donor marker *A* is selected, *Ab* transformants appear among the progeny with about the same relative frequency as *AB* transformants in the previous experiments. The event leading to incorporation of the selected *A* marker has also resulted in disappearance of the resident *B* marker and its replacement by *b* which was introduced with *A* on the same molecule of DNA.

CRITERIA OF LINKAGE

In higher organisms, linkage means that the alleles of two genes appear among recombinants in their parental combinations with a greater than random probability (50 per cent). When a pair of genes are located on the same chromosome, the factor determining whether they will appear to be linked or not is the frequency with which recombination occurs between the two parental homologues at meiosis. In all systems mediating recombination in bacteria, however, the situation is radically different, because only a fraction of the genome of the donor bacteria is transferred to the recipients to form merozygotes (p. 54). Thus the assessment of linkage depends not only on recombination but also on *pre-zygotic exclusion*. This is expecially evident in the case of transformation where the donor genetic material is fragmented into DNA molecules of relatively low molecular weight. However, the traditional measurements of genetic analysis are based exclusively on recombination, so that merozygotic systems pose the unique problem of deciding to what extent the genotype of recombinants has been determined by pre-zygotic exclusion on the one hand, or by recombination on the other. In general, we can only apply the concepts of classical genetical analysis if we know that certain groups of donor genes have actually entered the zygotes on the same fragment of donor chromosome.

In transformation, two genes are said to be linked if they are transferred together on the same molecule of transforming DNA. How may this be ascertained? We have seen that at low DNA concentrations each transformant arises from the interaction of a bacterium with a single molecule which carries the selected donor marker (p. 579; Fig. 117), but that each competent bacterium can take up more than one DNA molecule. It would seem logical to suppose that if two markers are carried by *different* molecules (that is, are *unlinked*), the probability that one bacterium will take up both molecules, and so be transformed with respect to both markers, is the product of the probabilities for the independent events. Thus if genes *A* and *B* are unlinked, and *ab* recipients exposed to DNA from *AB* donors are found to yield *Ab* and *aB* transformants with a frequency of 1 per cent, the expected frequency of double, *AB* transformants is 0·01 per cent, or one per 10,000 bacteria.

We might therefore think that a significant excess of *AB* transformants, such as one per 1000 recipients, would indicate linkage. Unfortunately the matter is more complicated than this and depends not only on the concentration of DNA but also on the proportion of competent bacteria in the recipient population. We have seen that when the concentration of DNA becomes saturating (at the plateau of Fig. 117), each competent bacterium may adsorb at least ten molecules of DNA, so that the probability that any bacterium will take up two molecules carrying independent markers increases as the DNA concentration rises to this level. Moreover, although only competent cells can take up DNA, the frequency of transformants must be expressed in terms of the recipient population as a whole. This means that the proportion of competent cells becomes an important factor in our calculations, irrespective of the DNA concentration. An example should make this clear. A limiting concentration of DNA from *AB* donor bacteria is used to transform a population of *ab* recipients. It is found that one *Ab* and one *aB* transformant arises for every 100 recipient cells; if genes *A* and *B* are unlinked, the random frequency of *AB* transformants is calculated as one per 10,000 recipients. But suppose that competent bacteria, which are the only ones involved, comprise only 10 per cent of the population. The *Ab* and *aB* transformants which are found per 100 of the population as a whole actually arose from only ten competent cells, so that the true frequency of single transformations is 10 per cent and the true expected number of doubles is 1 per cent of the *competent* population; that is, one per 1000 of the whole population or ten times greater than the number calculated without regard to the competence of the culture. Thus spurious linkage may be (and has been) inferred when the proportion of competent cells is low (see Ravin, 1959; Anagnostopoulos and Crawford, 1961).

This source of error in computing the expected number of double events from the frequencies of single ones is precisely the same as that invoked to explain the phenomenon of negative interference (p. 379). An interesting corollary of it, which is evident from our example, is that the estimated and observed frequencies of double transformants become the same when 100 per cent of the recipient bacteria are competent; conversely, the proportion of competent cells may be calculated from the disparity between the two results. This method has been used to show that the proportion of competent cells in cultures of *Pneumococcus* and of *H. influenzae* may be very high at the peak of competence (Fox and Hotchkiss, 1957; Goodgal and Herriot, 1961), while that of *B. subtilis* is low (Nester and Stocker, 1963).

What criteria, then, should we adopt to establish linkage? The answer is clear if we think primarily in terms of individual DNA molecules, which we can label as *A*, *B* and *AB*, rather than of genetic determinants. Competent bacteria of type *ab* are exposed to decreasing concentrations of DNA below the saturation level; as we have seen, the number of single transformants falls off as a direct, linear function of the DNA concentration, indicating the decreasing probability of interactions between individual bacteria and single DNA molecules. Thus, irrespective of whether we look at interactions involving *A*, *B* or *AB* molecules, we will obtain curves *which fall at the same rate*; the curves will all have the same slope (Fig. 121, curves *a* and *b*). On the other hand if there are no *AB* molecules but only *A* and *B* ones, the production of *AB* transformants depends on the interaction of two independent molecules with a single cell; for a ten-fold reduction in DNA concentration the probability of the double interaction falls, not ten times, but 100 times so that a curve of only half the slope is obtained (Fig. 121, curve *c*) (Goodgal, 1961).

The results shown in Fig. 121 can readily be demonstrated by the use of a pair of markers, *A* and *B*, which are known to be closely linked. When DNA extracted from *AB* donors is used to transform *ab* recipients, curves of the same slope are obtained whether *A*, *B* or *AB* transformants are selected (Fig. 121, curves *a*, *b*). On the other hand if a *mixture* of DNAs isolated from *Ab* and *aB* donors, in which the markers *A* and *B* are necessarily on separate molecules, is used in the same experiment, the slopes of the single transformants, *Ab* and *aB*, are the same as before, but that for the *AB* transformants falls at twice the rate (Fig. 121, curve *c*). Thus by scoring the number of single and double transformants with respect to two markers, as a function of diminishing DNA concentrations, the presence or absence of linkage can be unambiguously established, irrespective of the competence of the recipient culture or the transforming efficiency of the DNA preparation (Goodgal, 1961). It is, of course, always desirable to test the alleles of the markers in different parental combinations in order to preclude bias arising from physiological causes or from the selective methods used.

Many examples of true linkage in several transformation systems are now on record. In *Pneumococcus*, for example, the loci for streptomycin resistance and mannitol fermentation, as well as three loci determining different degrees of resistance to sulphonamide, are linked (Hotchkiss and Marmur, 1954; Hotchkiss and Evans, 1958), in

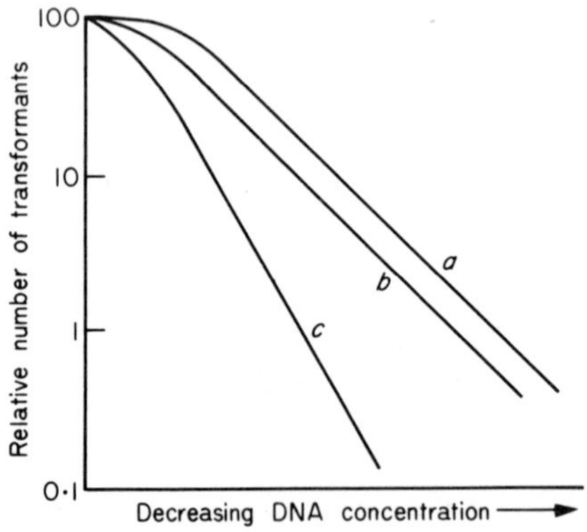

FIG. 121. Idealised curves illustrating the effect of decreasing concentration of transforming DNA on the relative number of single and double transformants for the two markers, *A* and *B*.

Curve a shows the numbers of single transformants of either *A* or *B* type.

Curve b shows the number of double transformants of type *AB* when *A* and *B* are closely linked on the same DNA molecules.

Curve c shows the number of double transformants of type *AB* when *A* and *B* are carried by *different* DNA molecules, i.e. are unlinked. (After Goodgal, 1961.)

addition to a number of loci concerned in the synthesis of capsular polysaccharides (reviews: Ravin, 1961; Jackson, 1962). In *Haemophilus*, the determinants of streptomycin- and cathomycin-resistance are closely linked (Goodgal and Herriot, 1957b). In *B. subtilis*, linkage was first discovered between the loci for indole synthesis and sucrose fermentation, as well as between those for sucrose fermentation and β-galactosidase synthesis (Spizizen, 1959). Since then, extensive linkage relationships have been found between series of loci concerned with the synthesis of various amino acids such as tryptophan, or isoleucine-valine, as well as between loci for different amino acids, purines and resistance to various antibiotics (see Fig. 122, p. 615; Nester and Lederberg, 1961; Anagnostopoulos and Crawford, 1961; Ephrati-Elizur, Srinivasan and Zamenhof, 1961; Nester, Schaffer and Lederberg, 1963; Anagnostopoulos, Barat and Schneider, 1964; Barat, Anagnostopoulos and Schneider, 1965; Dubnau *et al.*, 1967).

We have defined linkage in transformation in terms of the location of different loci on the same molecule of transforming DNA, so that the length of these molecules is a crucial factor in establishing linkage. When *B. subtilis* donor DNA is extracted under mild conditions, involving minimal shear, it turns out that linkage can be demonstrated between markers which appear unlinked when the usual transforming preparations are used. However, this linkage is sensitive to dilution of the DNA, as well as to shear and to the action of DNase, but can be stabilised if the dilution is made in a solution of non-transforming 'carrier' DNA so that the total DNA concentration remains the same (cp. p. 551). In this way, by establishing linkage between groups of markers previously unrelated, the map of *B. subtilis* can be considerably extended (see below; Kelly and Pritchard, 1965).

It has recently been found that efficient transformation occurs in *B. subtilis* when intact donor cells are mixed with competent recipient bacteria; extraction of the donor DNA is not required. It seems that the DNA is not released by lysis but is extruded from the surface of viable cells whence it is taken up by recipients. The process remains sensitive to DNase action. As might be expected, more extensive linkage is found in this system than in transformation mediated by DNA extracted in the normal way (Ephrati-Elizur, 1968).

GENETIC AND FUNCTIONAL ANALYSIS BY TRANSFORMATION

Polysaccharide synthesis

Early experiments on transformation were concerned with the ability of DNA, extracted from capsulated (*S*) pneumococci, to restore to non-capsulated (*R*) organisms of the same, original serological type the capacity to synthesise specific capsular polysaccharide; that is, the inheritance of only single donor markers was looked for. The first indication that recombination might be the process whereby this is accomplished was the discovery by Ephrussi-Taylor (1951a, b), that transformation between independently isolated mutants of *Pneumococcus* Type III, which were partially deficient in their capacity to produce polysaccharide, could yield fully capsulated, wild type transformants—the so-called *allogenic transformations*—in addition to those of donor type. Moreover, the frequency with which such transformants arose depended on the particular pairs of mutants employed as donor and recipient, so that the possibility of constructing linkage maps of the mutant sites became apparent. In fact, these studies constitute one of the first recorded attempts to analyse genetic fine structure in microorganisms. It was later demonstrated that these transformations to full encapsulation are due to single events and not to two successive transformations (Ravin, 1954), and that all these mutations involving the synthesis of Type III polysaccharide are closely linked (Ravin, 1960).

More recent studies of capsular transformation have attempted, directly or indirectly, to correlate the results of genetic analysis with the metabolic steps concerned in specific polysaccharide synthesis. For example, transformation between certain *R* mutants derived from *different* serological types, such as II and VIII, yield *both* types of encapsulated pneumococci, but none containing both polysaccharides in the same capsule. The recipient (Type II*R*, for instance) must therefore be blocked in a synthetic step common to both types, since its defect is made good by DNA from the other type (Type VIII*R*), while preserving intact those genes which determine the specificity of its polysaccharide. On the other hand, the fact that polysaccharides of either specificity, but not both, may be found in transformants means that the determinants of type specificity are strict alternatives (that is, are allelic) and are carried on the same DNA molecules as bear the common determinant (Jackson, MacLeod and Krauss, 1959). The situation can be represented thus:

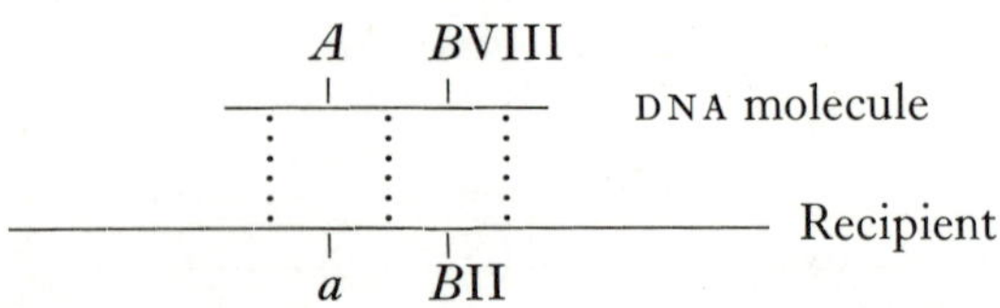

A represents the gene specifying a step common to the synthesis of both Type VIII and Type II polysaccharide, *a* being its defective allele in the recipient strain; *B*VIII and *B*II are alternative alleles determining whether the product possesses Type VIII to Type II specificity respectively. The formation of a capsulated recombinant requires two crossovers, one of which must lie to the left of *A* whose inheritance is necessary if any polysaccharide is to be synthesised at all. If the second crossover occurs in the *A—B* interval the recombinant will have the genotype *AB*II and will produce a Type II capsule; but if the crossover occurs to the right of *B*, the genotype will be *AB*VIII, yielding a type VIII capsule.

The biochemical basis of capsule formation in *Pneumococcus* has been investigated more directly by correlating transformation studies with enzymatic analysis. For instance, it has been shown that a particular pair of non-capsulated mutants of Type I, which can be inter-transformed to give capsulated Type I bacteria, lack different enzymes in the pathway of Type I polysaccharide synthesis, while the transformants yield both enzymes (Austrian *et al.*, 1959; reviews: Jackson, 1962; Mills and Smith, 1962).

These and other studies of polysaccharide synthesis, as well as the analysis of other linked markers in *Pneumococcus*, offer considerable evidence that genes determining the steps of a biochemical sequence tend to be clustered on a single molecule of transforming DNA, just as in the case of transductional analysis of amino acid metabolism (p. 163; review: Ravin, 1961). However, the most striking example of gene clustering yet to be revealed by transformation comes from analysis of *B. subtilis* which we will now consider briefly.

Amino acid synthesis: mapping of the B. Subtilis *chromosome*

The genetic markers available in *Pneumococcus* and *Haemophilus* are mainly concerned with polysaccharide synthesis, sugar fermentation and resistance to various chemotherapeutic agents and are not ideally adapted to combined genetical and biochemical studies. The situation is quite different in *B. subtilis*, of which the wild type can synthesise all its amino acids and vitamins and therefore grows on unsupplemented minimal medium. Auxotrophic mutants requiring different amino acids for growth are easily obtained, while prototrophic recombinants resulting from transformation may be selected by treating a culture of one mutant with DNA from another and then plating on minimal agar. The range and flexibility of this system is therefore equivalent to that of transduction in *Salmonella* or *E. coli* and offers the interesting possibility of comparing the genetical and biochemical relationships of two phylogenetically diverse species.

We have mentioned that the mean molecular weight of transforming DNA fragments in *B. subtilis* is about 25×10^6 (p. 592), which is adequate to carry some 30 to 50 genes, so that the existence of such phenomena as gene clustering in this organism should be easy to establish. It was early found that at least four loci concerned with tryptophan synthesis are clustered and, with one exception, arranged in the same sequence as the biochemical steps they mediate (Anagnostopoulos and Crawford, 1961). A number of loci concerned with the synthesis and regulation of the aromatic group of amino acids were also discovered to be clustered, and linked to the tryptophan loci as well as to a histidine locus (Nester and Lederberg, 1961; Nester *et al.*, 1963; Ephrati-Elizur *et al.*, 1961). Linkage of a number of loci controlling isoleucine, valine and leucine, as well as a methionine locus, were similarly identified on this same molecular linkage group (Anagnostopoulos *et al.*, 1964; Barat *et al.*, 1965). On the other hand, other isoleucine, valine, histidine and methionine loci, as well as several loci concerned with the synthesis of arginine and other amino acids, are found to be carried on other molecules.

We have explained how linkage relationships may be extended by avoiding shear in the preparation of transforming DNA (Kelly and Pritchard, 1965). In addition, three other methods of mapping are available. We have already discussed the first two of these in Chapter 19 in connection with the initiation and polarity of chromosome replication. One is the establishment of a gradient of marker frequency in unsynchronised cultures (p. 561); the other is the analysis, by transformation, of the sequence with which markers appear in newly synthesised DNA of hybrid density, when stationary phase bacteria or spores, which have incorporated heavy isotopes into their DNA, initiate a new cycle of synchronous replication in a light medium (p. 564) (Yoshikawa and Sueoka, 1963; Oishi *et al.*, 1964). We will not recapitulate these elegant methods, but merely point out that they are well adapted to establishing linkage and sequence relationships between groups of markers, within which the order of more closely linked loci can be better assessed by two- or three-point transformational crosses.

The third method is transduction. A number of phages, some of which have different strain specificities, are available for this purpose (Thorne, 1961; Takahashi, 1963). One of these, PBS1, is particularly useful since it infects the donor strain most widely used in transformation studies, and is also an unusually large phage whose transducing particles carry DNA of about $1 \cdot 7 \times 10^8$ molecular weight; as a consequence, linkage values measured by transduction with the phage are higher than those obtained by transformation so that linkages can be established between markers too distant to be co-transformed (Takahashi, 1961; Barat *et al.*, 1965; Dubnau *et al.*, 1967).

By combining all these methods, a rather detailed linkage map of *B. subtilis* can be constructed, as is shown in Fig. 122 (Dubnau *et al.*, 1967). It will be seen, from the vertical arrows which indicate linkage

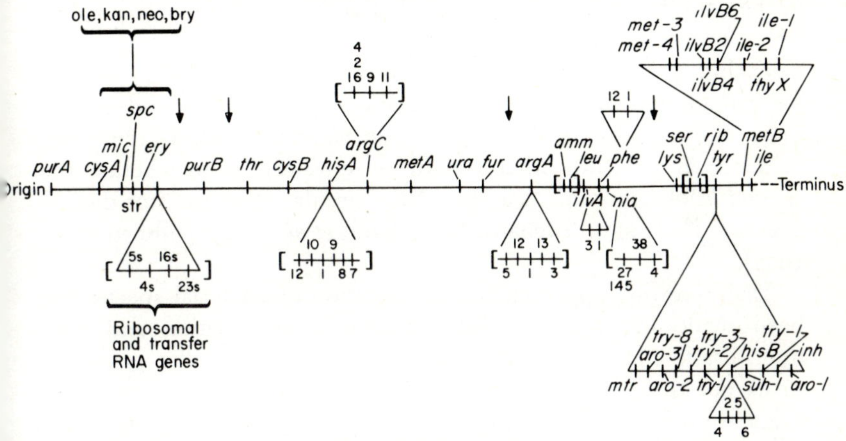

FIG. 122. Linkage map of the *B. subtilis* genome.

The map is constructed from data obtained using transformation and phage PBS1-mediated transduction in strain 168, as well as from previous reports in the literature. The orientation of several groups of markers is inferred from density transfer data obtained in strain W23. Linkage gaps appear in the intervals marked with vertical arrows. The markers enclosed within brackets have not been ordered relative to one another. The figures refer to mutant isolations. 4*s*, 5*s*, 16*s*, 23*s* indicate the sedimentation constants of the different RNA species.

amm = inability to assimilate NH_4^+
arg = arginine requirement.
aro = aromatic amino acid requirement.
bry = bryamycin resistance.
cys = cysteine requirement.
ery = erythromycin resistance.
fur = fluorouracil resistance.
his = histidine requirement.
ile = isoleucine requirement.
ilv = isoleucine & valine requirement.
kan = kanamycin resistance.
leu = leucine requirement.
lys = lysine requirement.
met = methionine requirement.
mic = micrococcin resistance.
neo = neomycin resistance.
nia = niacin requirement.
ole = oleandomycin resistance.
phe = phenylalanine requirement.
pur = adenine or guanine requirement.
r- and *t*-RNA = ribosomal and transfer RNA respectively.
rib = riboflavin.
ser = serine.
spc = spectinomycin resistance.
str = streptomycin resistance.
thr = threonine requirement.
thy = thymine requirement.
try = tryptophan requirement.
tyr = tyrosine requirement.
ura = uracil requirement.

This map is reproduced from Dubnau *et al.*, 1967, by kind permission.

gaps as yet unjoined by transduction or transformation, that this map consists of four linkage groups whose sequence is deduced only from the results of marker ratio or density transfer experiments, and it must be admitted that, despite general agreement, there are some discrepancies between the results of these two types af analysis. Although, by analogy with other bacteria and from certain consistencies in the data themselves, it seems likely that *B. subtilis* possesses only a single chromosome, on the existing evidence the possibility cannot be excluded that two or more chromosomes are present. For example, we cannot discriminate between a single chromosome and two or more chromosomes which replicate in sequence; nor is there any evidence at present, either genetic or physical, as to whether the *B. subtilis* chromosome is a linear or a circular structure (for discussion, see Kelly and Pritchard, 1965; Barat *et al.*, 1965; Dubnau *et al.*, 1967).

An interesting approach to the evolution of bacterial species has been made by means of inter-specific transformation and DNA hybridisation studies of eight members of the *Bacillus* genus. It turns out that loci concerned with the synthesis of various amino acids are not inter-transformable, while those determining resistance to certain antibiotics (streptomycin, erythromycin, micrococcin and bryamycin), as well as the regions specifying the structure of ribosomal and transfer RNA, not only are inter-transformable, but hybrid molecules formed between them are also active in transformation. These latter determinants, which possess close inter-specific homology, are found to map in two discrete regions of the chromosome and constitute a 'conserved core' of stable genetic material which is relatively resistant to genetic change (Dubnau *et al.*, 1965; Dubnau, Smith and Marmur, 1965). In this context it is perhaps not inapt to mention that DNA–DNA hybridisation techniques have also been used to establish phylogenetic relationships among higher animals (McLaren and Walker, 1965).

Anomalies of fine structure mapping

A priori, transformation would seem an ideal tool for fine structure genetic analysis because of the relative efficiency of DNA transfer, as compared with transduction for example, and its normal limitation to very small chromosomal regions. However, two recent analyses in *Pneumococcus* have shown that the interpretation of recombination frequencies in fine structure mapping may be far from straightforward. In the first study, Sicard and Ephrussi-Taylor (1965) show that mutant sites in the *recipient* strain fall into two classes with respect to the efficiency with which they participate in recombination. Thus in mutant × mutant crosses, the frequency of wild type recombinants may depend more on the efficiency of the recipient site than on the distance between the donor and recipient mutant sites. As a consequence there may be a ten-fold or greater difference in the recombination frequency shown by simple reversed crosses, so that classical mapping procedures are useless. To overcome this diffculty in the construction of a fine structure map, it is suggested that a large number of mutant donor DNAs should be crossed with a single recipient, and the order of donor mutations established by comparing the frequencies of wild type recombinants (Sicard and Ephrussi-Taylor, 1965).

Again, in an analysis of 76 mutations in the amylomaltase locus, whose order was established by a series of overlapping deletion mutants, Lacks (1966) has demonstrated that single-site mutations fall into four classes which may differ 20-fold in their efficiency of recombination, and that even different mutations *at the same site* may differ in efficiency, so that the phenomenon must be interpreted at the level of single DNA bases. Lacks (1966) suggests a model in which the presence, on the donor DNA strand, of bases which are not identical with one of the two bases at the equivalent site on the recipient duplex, leads to a decreased stability of stereochemical fit during pairing, thus reducing the probability of recombination. It is not known to what extent such vagaries of recombination may be general, or be restricted to transformation and determined by the participation of only a single donor strand in recombination.

Genetic expression following transformation.

Early studies by Lacks and Hotchkiss (1960) on the time of expression of the gene determining production of amylomaltose, following DNA uptake in *Pneumococcus*, showed that the enzyme begins to be formed at about the same time as incorporation of the gene, some six minutes after addition of the transforming DNA. A very contrasting picture has been found in the case of *B. subtilis*, in which a delay of up to four hours intervenes between DNA uptake and the expression of genes determining flagellation and synthesis of tryptophan synthetase, although incorporation requires only 30 minutes as shown by the establishment of linkage, in the re-extracted DNA, between the donor *try*$^{+}$ gene and a neighbouring *his*$^{+}$ recipient gene.

This delay is associated with an equivalent lag in the multiplication of transformants (Stocker, 1963; Nester and Stocker, 1963). Similar delays in expression have been found for *B. subtilis* genes determining synthesis of alkaline phosphatase, an inducible sucrase and a constitutive threonine deaminase (Anagnostopoulos and Spizizen, quoted by Spizizen *et al.*, 1966).

A possible explanation of this striking difference in expression time during transformation in *Pneumococcus* and *B. subtilis* may be found in the different competence patterns of these two organisms. We have seen that competent *B. subtilis* bacteria do not synthesise DNA, are resistant to thymineless death and to the action of penicillin, and do not multiply, and so may be presumed to be in a state of synthetic latency (pp. 556, 211); moreover, the duration of resistance to penicillin coincides with that of the delay in expression. It is thus reasonable to ascribe this delay to persistence of the competent state in the transformed cells (Nester and Stocker, 1963). In contrast, competence in *Pneumococcus* is sharp in onset and of very short duration in individual bacteria (p. 584).

TRANSFECTION

The knowledge derived from transformation, that large molecules of nucleic acid, of molecular weight in excess of ten million, can readily penetrate the walls and semipermeable membranes of competent bacteria suggested that nucleic acids other than bacterial or, indeed, other than DNA, might similarly gain access to cells. This was first demonstrated for purified ribonucleic acid from tobacco mosaic virus which was shown to be infective by itself, though with very low efficiency as compared with the intact virus equivalent, and to promote synthesis by the plant of new viral RNA and protein and the release of complete, infectious virus particles (see p. 229; Gierer and Schramm, 1956; Fraenkel-Conrat *et al.*, 1957). Since then there have been many examples of the infectivity of nucleic acids, from both plant and animal as well as DNA and RNA viruses.

We have already described how *E. coli* K12 may be infected by DNA extracted from phage λ, with the aid of superinfecting 'helper' phage—the first example of such an infection of bacteria (p. 544). The discovery of transformation in *B. subtilis* suggested the possibility of productively infecting competent cells of this organism with DNA from some of its phages, and this was first achieved by Földes and Trautner (1964) who gave the name 'transfection' to the phenomenon.

Only competent cultures proved infectable and the infection was sensitive to the action of DNase. Since then, DNAs from many *B. subtilis* phages have been found to be similarly infectious. Moreover, fragmented genomes can give rise to productive infection by recombination, when several molecules are taken up by the same cell (review, Spizizen *et al.*, 1966).

A remarkable development of this phenomenon, stimulated partly by the experimental evidence for the universality of the genetic code (p. 364), has been the apparently successful attempts to grow animal viruses in *B. subtilis* by exposing competent bacteria to preparations of the viral nucleic acids. In this way, complete particles of vaccinia virus are claimed to have emerged from *B. subtilis* infected with vaccinia virus DNA, although replication of the viral DNA in the bacteria has not been proven (Abel and Trautner, 1964). Polyoma virus has been similarly grown by transfection of competent *B. subtilis* bacteria (Bayreuther and Romig, 1964). Again, by employing phage λ as 'helper', it is claimed that *E. coli* has been infected with RNA from encephalomyocarditis virus, with a resulting formation of complete virus particles (Ben-Gurion and Ginsburg-Tietz, 1965). In this case also, there is as yet no evidence of replication of the viral RNA, though it is clear that the RNA can behave as a messenger in *E. coli*, determining the synthesis of specific virus protein. In all cases of phage transfection, the efficiency of the process is markedly lower than that of infection by the intact phage particles.

There has recently been an interesting report of the acquisition by pneumococci of a high level of sulphonamide resistance by treating them with RNA fractions extracted from resistant donor cells. The transforming principle proved sensitive to both DNase and RNase, suggesting that its activity is due to a DNA-RNA complex, while the transformants behaved like semi-stable heterozygotes in which the character of resistance is ultimately either lost completely, or stably inherited by individual clones (Evans, 1964; review, Spizizen, Reilly and Evans, 1966).

CHAPTER 21

Transduction

In 1951 Lederberg, Lederberg, Zinder and Lively explored the possibility that genetic transfer and recombination, similar to that previously found during conjugation in *E. coli* K-12 (Lederberg and Tatum, 1946a,b), might also occur between auxotrophic mutant strains of *Salmonella typhimurium*. Their studies revealed that small numbers of prototrophic recombinants did arise, not only when *cultures* of the two mutants were mixed but also when a culture of one mutant was treated with a *cell-free extract* of the other. They called this process *transduction*. Zinder and Lederberg (1952) then turned their attention to analysing the nature of the filterable agent which mediated transduction, and made the novel and unexpected discovery that this agent appeared to be inseparable from particles of a temperate bacteriophage, PLT-22 (now commonly called P22), which lysogenised one of the parental strains and was able to infect and lyse the other.

When this phage is grown on wild type *Salmonella* strains and the progeny particles then plated with various mutant strains, which differ from wild type with respect to growth requirements, fermentation capacity, motility (pp. 97–99) or drug resistance, wild type colonies can be selected with a frequency of one for approximately every 10^6 to 10^5 phage particles present. Conversely, mutant bacteria plated with phage propagated on the *same* (recipient) strain, yield no more wild type colonies than arise from back-mutation when the bacteria are plated alone (but see p. 222). When strains carrying several mutational markers are used as recipients, wild type transductants for each marker can be obtained with about the same frequency, but those selected with respect to any particular marker remain mutant for all the others; that is, each donor marker appears to be independently transferred (Zinder and Lederberg, 1952; Zinder, 1953; Stocker, Zinder and Lederberg, 1953). Moreover, although the faculty of transduction is evidently a function of the infective capacity of the phage preparation, it can be clearly dissociated from the ability of the phage particles to replicate or to establish lysogeny in the recipient bacteria (see below). It thus turned out that the transducing agent itself, while relying on the phage for its transfer, is not identical with the phage.

The discovery of transduction fell in the same year as the demonstration by Hershey and Chase (1952) that the essence of phage infection is the injection of the phage DNA into the bacterium (p. 423). Thus the simplest concept of the role of phage in transduction was to suppose that, during the lytic cycle, fragments of the genetic material of the host are occasionally incorporated into the progeny particles which thus become *vectors*, transporting them to other bacteria which they subsequently infect. It is surely one of the more bizarre manifestations of evolutionary adaptation, that a potentially lethal virus should acquire the redeeming function of a gamete, rescuing some of its victims' genes for posterity!

Although the word 'transduction', which literally means 'leading across', was applied to this phenomenon before its basic mechanism became known, its use is now generally and appropriately reserved for those cases of genetic transfer which are mediated by a phage vector. The occurrence of transduction has since been reported in *E. coli* (phages λ, P1 and 363: Morse, 1954; Lennox, 1955; Jacob, 1955), *Shigella* (phage P1: Lennox, 1955; Adams and Luria, 1958), *Pseudomonas* (Loutit, 1959; Holloway and Monk, 1960), *Staphylococcus* (Morse, 1959), *Proteus* (Coetzee and Sacks, 1960) and *B. subtilis* (Thorne, 1961, Takahashi, 1961, 1963).

We have already given a rather detailed account of the use and value of transduction as a tool for the analysis of genetic fine structure (Chapters 7 and 8, pp. 127, 158) and function (p. 99). We will not repeat these studies here but will confine our attention to the transduction process itself, concentrating especially on the nature of the transduced fragments of donor chromosome and the way in which they are conveyed by particles of transducing phage. At least from a superficial point of view, two distinct mechanisms of transduction may operate, depending on the particular temperate phage employed and its relation with the bacterial chromosome. Before considering these diversities, however, let us review some general features of transduction which appear to be common to both systems.

EVIDENCE FOR A PHAGE VECTOR

The *Salmonella* transducing phage, P22, is a normal temperate phage belonging to the A1 class of *Salm. typhimurium* phages in Boyd's

(1950) classification. When it became apparent, from the transducing activity of lysates produced by this phage, that the phage particles themselves might be directly implicated in the process, rigorous attempts were made to separate the transducing agent from the phage (Zinder and Lederberg, 1952; Zinder, 1953). For example we have seen that the proportion of particles of a temperate phage which lyse their host bacteria can be varied widely by altering the experimental conditions (p. 468). It was found that, irrespective of the amount of lysis which follows infection, the ratio of transducing titre to plaque titre of the lysate remains constant. Moreover this ratio is maintained during purification and concentration of the phage by centrifugation and exposure to chloroform and amyl alcohol, and is unaffected by treatment with deoxyribonuclease or ribonuclease.

Attempts at differential inactivation were likewise unsuccessful. The plaque titre and the transducing titre of the phage suspension fall off at the same rate following exposure to heat and to phage antiserum, both of which interfere with adsorption of the phage—the latter in a highly specific way. Similarly, both phage and transducing agent are proportionately removed by filtration through a gradocol membrane of critical pore size, showing that they have a common particle size. To conclude this chronicle of similarities, we will mention two further pertinent comparisons which are concerned with the attachment of phage and its specificity. Firstly, the phage receptor sites of *Salm. typhimurium* strain LT-22 (which is lysogenic for phage P22) become saturated by ten P22 particles per bacterium; when this strain is used as recipient in transduction experiments involving increasing multiplicities of phage, the number of transductants rises linearly with multiplicity to reach a plateau at ten particles per bacterium. Some sensitive strains which can adsorb many more than ten P22 particles yield a proportionately inflated maximum number of transductants at high multiplicities of infection (Zinder, 1953).

Secondly, phage P22 attaches specifically to a particular somatic antigen designated '12', and is adsorbed by a number of species of *Salmonella*, other than *Salm. typhimurium*, which possess this antigen. In spite of the fact that many of these species are neither lysed nor lysogenised by the phage, nevertheless they are transducible. On the contrary, *Salmonella* species which lack antigen 12, and so cannot adsorb the phage, are not transducible (Zinder and Lederberg, 1952; Zinder, 1953).

There is thus a perfect correlation between the properties of phage and transducing agent in every respect examined save one—transducing activity is independent of the ability of the phage either to lyse or to lysogenise the recipient bacteria. Lysogenic, immune recipients are transducible not only by temperate phage, but also by *virulent, clear-plaque mutants* of the phage which are unable to lysogenise (p. 466; Fig. 103C, p. 465). In this latter case the transductants continue to carry their original prophage (Zinder, 1953). This discrimination between the ability of a phage preparation to form plaques and to generate transductants is emphasised by other kinds of experimental approach. For example, if a suspension of transducing phage is irradiated with ultraviolet light, which does not interfere with phage adsorption, its plaque-forming ability is inactivated exponentially (p. 525), but its transforming capacity may actually rise to reach a maximum at about the one per cent survival level, and then fall with a slope about three to four times less steep than that of the survival curve (Zinder, 1953; Garen and Zinder, 1955). Similar effects follow X-irradiation and the decay of incorporated ^{32}P atoms; the plaque-forming titre may be reduced more than 1000-fold without affecting the efficiency of transduction although, in these cases, no enhancement of the frequency of transduction is found (Garen and Zinder, 1955). Again, when sensitive bacteria are employed as recipients in transduction experiments, not all transductant clones are lysogenic, showing that inheritance of the transduced donor character may be divorced from that of the phage genome (Stocker, Zinder and Lederberg, 1953; Zinder, 1955). We will discuss this aspect of transduction in more detail when we come to consider its mechanism.

GENERAL FEATURES OF TRANSDUCTION

Linkage

We have seen the evidence that a small proportion (10^{-5} to 10^{-7}) of progeny particles of phage, emerging from the lysis of donor bacteria, can transport donor markers into recipient bacteria which they subsequently infect. The transducing fragments of donor chromosome are 'passengers' carried by the virus vector (Lederberg, 1956c). In general it is found that the determinants of unrelated donor characters are located on separate fragments and so are independently transduced. But we have already encountered many examples of the joint, or linked, transduction of genes, especially of those concerned with different steps of a single biochemical pathway (pp. 158–165) and have deduced that these are so closely linked on the chromosome that they are usually transported *en bloc* on the same transducing fragment. In the case of transformation, however, we have found that spurious linkage relationships may be inferred from the uptake and incorporation of more than one transforming DNA molecule by a single, competent bacterium, so that rather stringent criteria of linkage must be observed (p. 607). We may therefore ask two questions about linkage in transduction. What criteria should we adopt to establish true linkage? What length of chromosome do transducing fragments comprise, over which we may expect to find linkage?

The answer to the first question is easy. In the case of transformation every molecule of DNA must be assumed to be potentially transforming. In transduction, however, only 10^{-5} to 10^{-7} of the phage particles are able to yield a transductant. If we suppose that one in 10^{3} particles carries a particular donor marker and that each recipient bacterium can adsorb and be infected by ten particles at high multiplicities of infection, the probability that a double transductant for two unlinked markers will arise is about 10^{-8} which, is of the same order as the probability of back-mutation with respect to most single markers. Since the frequency of genuinely linked transductions is at least 1000 times greater than this, the possibility of spurious linkage may be ignored.

The size of donor chromosome fragments

A first approximation to the size of transducing fragments can be inferred from the fact that the amount of DNA normally contained by the average phage particle is about 1/100 that of the *E. coli* or *Salmonella* chromosome. We would not expect a fragment of greater size than this to be transduced. If a transducible strain of bacterium was available in which the distances between genes on a very well marked region of the chromosome was already known in terms of the entire chromosome length, as a result of independent measurement, the maximum length of a transduced fragment could then be directly assessed, by finding the most distant pair of genes amenable to joint transduction. *E. coli* K-12 presents us with such a system. Not only have large regions of the chromosome of this organism been accurately mapped in absolute terms (Fig. 127, p. 666), but two strains of phage (P1 and 363) are available for this kind of transductional analysis. The chromosome map of *E. coli* is shown in Fig. 127 (p. 666). It turns out that a number of pairs of functionally unrelated genes are jointly transduced, and these are just the pairs that are known to be very closely linked (*thr* and *leu*, *leu* and *azi*, *str* and *mal*, *xyl* and *mtl*, *gal* and prophages 82, λ or *434*: Lennox, 1955; Jacob, 1955). Markers which are not closely linked are not transduced together.

A particularly revealing chromosome region is that carrying the markers *thr—leu—azi* (top of Fig. 127, p. 666). When phage P1-mediated transductants inheriting leu^+ from the donor strain are selected, about 50 per cent also inherit the donor marker *azi-r*, so that *leu* and *azi* are very closely linked; when thr^+ is selected, about three per cent of the transductants are also leu^+ but none of these is *azi-r*. Of 450 transductants selected for joint inheritance of both thr^+ and leu^+ from the donor, none inherited the donor marker *azi-r* (Lennox, 1955). Thus the region *thr—leu* represents the extreme length of donor fragment which can be transduced by phage P1, and it is known from conjugation experiments that the distance separating these two markers is about 1/50 the length of the whole chromosome, that is, about the proportionate length of the chromosome of the phage DNA itself. As we shall see later, this genetical estimate is borne out by physical measurements since it has been shown rather directly that transducing particles incorporate only bacterial DNA in an amount equivalent to the DNA of a normal phage genome, which has a molecular weight of 6×10^7 as compared to about 2.8×10^9 for the entire *E. coli* chromosome (Ikeda and Tomizawa, 1965a).

An interesting corollary of these experiments is raised by transductional analysis of the galactose (*gal*) region of *E. coli* K-12 by phage 363 which is very closely related to phage P1 and yields similar transduction frequencies for the *thr—leu* region. Mapping by conjugation shows that the locus gal_b is very closely linked to the sites of attachment of prophages 82, λ and 434 (Fig. 127, p. 666). When the donor strain is gal^+ and non-lysogenic, and the recipient is the doubly lysogenic gal_b^- (*82*) (*434*), among transductants selected for gal^+ about 11 per cent are non-lysogenic for phage 82 and about one per cent non-lysogenic for phage 434. Comparison of the frequency of joint transduction of gal^+ and non-lysogeny with that obtained for *thr* and *leu* implies that the distance between the gal_b and *434* loci is at the cxtreme limit of transducibility. Yet when a doubly lysogenic gal^+ (*82*)(*434*) strain is used as donor, and a non-lysogenic gal_b^- strain as recipient, not only do gal^+ lysogenic transductants arise with about the same frequency as do non-lysogenic transductants in the previous cross, but *doubly lysogenic transductants* are also found (Jacob, 1955). This means that a single particle of phage 363 can act as a vector of two complete prophage genomes in addition to a length of bacterial chromosome equivalent to its own. Even assuming that these prophage genomes, like that of phage λ (molecular weight 3×10^7), contain only half the normal phage 363 DNA content, it is difficult to escape the conclusion that some transducing particles must be able to carry more than the normal complement of DNA. In parenthesis, we should note than when inducible prophages are transduced to sensitive recipients at 37°C., zygotic induction occurs just as in conjugal crosses (p. 463), but this can be prevented by permitting the initial stages of transduction to take place at a lower temperature (20°C.) (Jacob, 1955).

Environmental influences

It might be thought that phages which can transduce a wide range of markers would transduce each separately with about the same frequency but, for reasons which are not yet understood, this is often not the case. For example, if aliquots of a recipient strain carrying several nutritional and fermentative defects are infected, at the same multiplicity, with transducing phage grown on wild type bacteria, the frequencies with which transductants for each marker arise may vary by a factor 20 or more, and the variations are usually reproducible (see Lennox, 1955; Jacob, 1955; Ozeki, 1959; see Table 23, p. 642). It is not sure, however, to what extent such discrepancies may be imposed by differences in the selective environment. Thus when selecting for fermentative markers, the inability of the recipient bacteria to utilise the carbon source provided may restrict their energy production, and interfere more with the process of recombination than deprivation of a required amino acid.

Some reports of the low or absent transducibility of certain markers may be ascribed to such unusual defects in the recipient as inversions, large deletions or double (unlinked) mutations; others to the use of selective conditions prejudicial to the expression of recessive donor markers. A good instance of the latter concerns the transduction of streptomycin-resistance which, in *E. coli* and *Salmonella*, happens to be recessive to sensitivity. Resistant transductants will only be found if application of the drug is withheld until segregation and the expression of resistance is complete. Apart from these extraneous sources of variability, however, there is some evidence that the frequency of transduction of a marker may vary with the time during the latent period at which the transducing phage is harvested, and with the location of the marker on the bacterial chromosome (Zinder, 1955; Plough, 1958).

TYPES OF TRANSDUCTION

In the phage P22-*Salmonella* system of transduction described by Zinder and Lederberg (1952), and subsequently deployed on a grand scale by Demerec and his colleagues for the analysis of genetic fine structure (Chapters 7 and 8, pp. 127, 158), any small region of donor chromosome can be transduced and the gene clusters located on it studied. This type of transduction is therefore called *unrestricted* or *generalised transduction*. In 1954, Morse made the interesting discovery that phage λ, which lysogenises *E. coli* K-12, can also mediate transduction, but this is strictly limited to a cluster of galactose loci which are contiguous to the prophage location on the bacterial chromosome (pp. 453–455). The frequency of transduction is about one transductant for every 10^6 λ particles. This is termed *restricted* or *localised transduction*.

A singular point of distinction between the two types of transduction is that only preparations of phage λ obtained by induction of lysogenic cultures, following irradiation with ultraviolet light (p. 451), possess transducing properties; unlike the phage P22-*Salmonella* system, phage emerging from the lytic infection of sensitive bacteria is quite inert. It therefore seemed that the phage must be in the prophage state for it to be able to pick up a segment of bacterial chromosome, while the segment must be located close to the site of prophage attachment. These observations inspired a spate of brilliant studies which revealed that restricted transduction is due to some kind of recombination between the bacterial and prophage chromosomes, with the result that a fragment of the bacterial chromosome becomes incorporated into the prophage structure. Since this concept carries implications of a more general kind (see Chapter 24, pp. 746–748), we must examine its origins in some detail.

THE MECHANISM OF RESTRICTED TRANSDUCTION

A culture of a galactose-negative (*gal*$^-$) mutant of *E. coli* K-12 ,which may be either sensitive or lysogenic, is infected at high multiplicity with particles of phage λ, obtained by induction of a lysogenic *gal*$^+$

strain. The infected bacteria are plated on eosin-methylene blue (EMB) agar containing galactose, so as to yield confluent growth. After incubation a small number of darkly pigmented papillae, indicating clones of *gal*$^+$ bacteria, are observed against the background of unpigmented growth (Plate 7, p. 97). Pure cultures of *gal*$^+$ transductants are then isolated by picking these papillae and re-streaking them on EMB-galactose agar so as to obtain isolated colonies.

The great majority of such *gal*$^+$ transductants differ from the original donor bacteria in two striking ways. Firstly, they turn out to be unstable for the *gal* character, segregating *gal*$^-$ progeny with a probability of about 10^{-3} per cell per generation. This implies persistence of the recipient *gal*$^-$ gene; the transductant must therefore be *diploid* for the *gal* region, carrying both the resident *gal*$^-$ and the transduced *gal*$^+$ alleles. Thus in contrast to what normally happens in unrestricted transduction, the transduced marker does not replace its recipient allele but is added to the recipient genome. Partial diploids of this sort are called *heterogenotes*. Names have also been coined to denote the components of a heterogenote, the immigrant fragment of donor chromosome being termed an *exogenote*, and the recipient chromosome an *endogenote* (Morse, Lederberg and Lederberg, 1956a, b).

Secondly, when cultures of heterogenotes (which are normally lysogenic when obtained by the method described above) are themselves induced by ultraviolet light, a lower than usual yield of phage is obtained but, unlike that emerging from the original donor strain, *about half the total number of particles is now able to transduce the gal*$^+$ *character to gal*$^-$ *recipients*. Such phage lysates are termed 'HFT' since they mediate 'high frequency transduction' (Morse *et al.*, 1956a, b).

These two features suggest that in about one cell in a million of the original, lysogenic donor population, the prophage in the process of release, following induction, picks up the neighbouring *gal* region of bacterial chromosome and incorporates it into its own structure, so that progeny particles having 'λ—*gal*$^+$' instead of normal λ chromosomes are liberated. When a *gal*$^-$ recipient bacterium is infected by one of these particles, it becomes lysogenised by the λ—*gal*$^+$ structure to form a *gal*$^-$/λ—*gal*$^+$ heterogenote. If a population of such heterogenotes, homogeneously lysogenised by λ—*gal*$^+$ prophage, is induced, every bacterium will liberate λ—*gal*$^+$ progeny particles instead of only one per million, so that the capacity of the lysate to transduce with very high frequency is explained.

In these transduction studies of Morse *et al.* (1956a, b), the recipient bacteria were usually non-lysogenic and were infected with a high multiplicity of transducing phage. Most of the resulting heterogenotes appeared to be normally lysogenic, but a few failed to release phage on induction although they were immune to infection by free λ phage, suggesting that they carried a defective prophage. This clue was followed up independently by Arber, Kellenberger and Weigle (1957) and by Campbell (1957a) who disclosed the remarkable fact that when transduction is mediated by HFT preparations of phage *at very low multiplicities of infection*, virtually all the heterogenotes are defectively lysogenic. They are immune to infection and lyse after the normal latent period following induction, but no phage particles are released and the lysate has no transducing activity. On the other hand, if the recipient bacteria are mixedly infected with transducing phage at a multiplicity of 10^{-2} together with normal particles at a multiplicity of 3, normally lysogenic heterogenotes arise which yield HFT lysates on induction. Similarly, if a culture of a defectively lysogenic heterogenote is induced and then superinfected with normal phage, and HFT lysate is obtained (Arber *et al.*, 1957).

The conclusion is that incorporation of the bacterial galactose region into the phage chromosome results in a defective phage which is able to lysogenise recipient bacteria but not to yield progeny particles on induction (p. 474). The initial transducing preparation, however, contains an excess of normal phage so that every bacterium which receives a defective transducing particle will also be infected by at least one normal particle. The result is that the transduced cells become doubly lysogenised by a defective and a normal prophage. When the normal prophage is induced it supplements the deficiency of the defective phage so that both phages multiply and are liberated in about equal numbers. In fact, only about half the total number of phage particles in an HFT lysate, counted under the electron microscope, is able to produce plaques (Arber *et al.*, 1957).

We have seen that incorporation of the bacterial *gal* region into the λ genome, to form a single unit of replication and transfer, results in the loss of one or more phage functions. The obvious explanation is that part of the phage chromosome has been replaced by the *gal* region. This was confirmed experimentally by 'marker rescue' experiments (p. 530) in which heterogenotes carrying, say, defective wild type phage are induced and superinfected with genetically marked mutant phages; the absence from the progeny phages of recombinant classes inheriting particular wild type markers indicates that these markers are missing from the exogenote. The results of such experiments show that defective transducing phages, termed λ*dg* (which stands for λ'*d*efective, *g*al'), lack about ¼ to ⅓ of their chromosome (Arber *et al.*, 1957). The missing segment is confined to the *A*—*H* region of the chromosome which determines the structural features of the phage (Fig. 101, p. 457). It invariably begins in the *b2* region (see Fig. 124c) and appears always to involve *h*. The other extremity of the deletion penetrates to a different extent into the chromosome of different isolates and may involve any series of markers between *h* and *A* (Campbell, 1959, 1964). We may mention here that the bacterial gene for biotin synthesis (*bio*: Fig. 127, p. 666), located near the other extremity of the inserted λ prophage, is independently transducible by the same mechanism (Wollman, 1963).

Is this replacement of a variable length of phage chromosome by a segment of bacterial chromosome due to a reciprocal recombination event of the kind we are familiar with, or to some other mechanism? This question was answered by measuring the densities of a number of independent isolates of transducing λ phage by centrifugation in a caesium chloride density gradient (pp. 243, 394) (Weigle, Meselson and Paigen, 1959). Since the DNAs of *E. coli* K-12 and of phage λ have identical bases and base ratios, the replacement of a particular length of phage chromosome by an equivalent length of bacterial chromosome should not alter the density of the phage. On the contrary it was found that each of the independent isolates of transducing phage had a characteristic density which differed from that of normal λ, some being heavier (+8 per cent DNA) and others lighter (−14 per cent DNA), as well as from those of the other isolates. The specific density of each strain of λ *dg* was preserved through multiplication and transfer, and was unaffected by recombination events involving the *gal* region of the heterogenote (see next Section). It follows that the deleted region of phage chromosome is not replaced by the same length of bacterial chromosome, but by a longer or shorter region (Weigle *et al.*, 1959).

We have already mentioned the temperate phages 82 and 434 which are closely related to phage λ (p. 625). The chromosomal locations of these phages relative to the *gal* region are *gal*—82—λ—434. Both phages are able to transduce the *gal* region in the way we have described but at a much lower frequency than phage λ, despite the closer proximity in the case of prophage 82. By crossing phages λ and 434, a 'hybrid' strain of phage has been obtained which possesses a large fraction of the λ chromosome but the attachment region, and therefore the chromosomal location, of phage 434 (pp. 457, 458). This 'hybrid' phage, although situated much further from the *gal* region, is able to transduce it with the same efficiency as λ (Campbell and Balbinder, 1959). All these findings suggest that the ability of the phage genome to pair with and incorporate a fragment of bacterial chromosome depends on very limited and intermittent regions of homology between the two.

Recombination in heterogenotes

The existence of heterogenotes was deduced from the segregation of stable *gal*$^-$ bacteria by transduced *gal*$^+$ populations (Morse *et al.*, 1956a, b). When these *gal*$^-$ segregants are examined for the presence of transducing phage, most of them turn out to have lost it; they are haploid *gal*$^-$, identical with the original recipient. A few *gal*$^-$ segregants, however, yield an HFT phage preparation on induction, but the phage now carries the same *gal*$^-$ allele as the endogenote—they are *homogenotes* of the type *gal*$^-$/λ—*gal*$^-$ which have resulted from a recombination event between the *gal* regions of endogenote and exogenote.

If such an HFT preparation of phage, bearing a particular *gal*$^-$ mutation (λ—gal_1^- gal_2^+), is used to infect a recipient strain having a mutation in a different gene (gal_1^+ gal_2^-) in the pathway of galactose metabolism, heterogenotes of the type gal_1^+ gal_2^-/λ—gal_1^- gal_2^+ are formed and can be recognised by their ability to ferment galactose as a result of complementation. As we have seen, *trans*-heterogenotes of this sort form the basis of a useful complementation test for ascertaining whether *gal*$^-$ mutations belong to the same or different functional loci (p. 101). As before, these heterogenotes throw off non-galactose-fermenting segregants whose genotype can be investigated by seeing whether they continue to carry transducing phage (that is, are still diploid for the *gal* region), and by crossing them with other *gal*$^-$ strains. All the recognisable recombinant types of segregant are found; some are homogenotes of the type shown in Fig. 123D, while the haploid segregants show the three possible non-functional arrangements of the parental *gal* alleles (E. M. Lederberg, 1960; Fig. 123E, F, G).

If the *gal*$^-$ alleles of recipient and transducing phage happen to belong to the *same* functional locus the heterogenotes, of course, do not ferment galactose. Nevertheless, *galactose-fermenting segregants* may be isolated from them, some of which turn out to be heterogenotes in which the *gal* alleles have recombined to yield the *cis* arrangement (Fig. 123B, C; see pp. 128, 635), while others are haploid but carry both wild type alleles (Fig. 123H). Thus heterogenotes have been

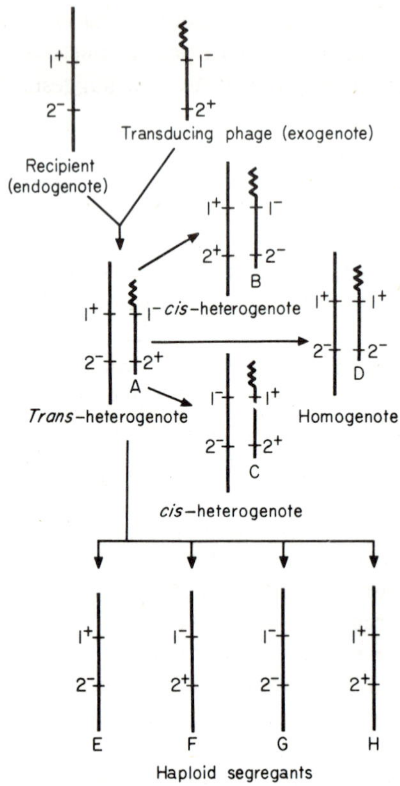

FIG. 123. Illustrating the genotypes of segregants obtainable from heterogenotes which carry two pairs of alleles within the galactose region.

'1' and '2' are different mutational sites, which may lie either in different functional loci or in the same locus (see text). '+' and '−' indicate wild type and mutant (non-functional) alleles respectively.

The shorter line of each pair represents the *gal* region of the exogenote which is incorporated into the phage genome (wavy line).

Trans-heterogenotes ferment galactose only when the mutations are in different functional loci.

All *cis*-heterogenotes ferment galactose whether or not the mutations are in the same locus.

Homogenotes do not ferment galactose.

shown to yield every possible type of recombinant for the *gal* region (Lederberg, 1960; for discussion of how homogenotes may arise, see pp. 635, 636).

The frequency with which haploid segregants appear shows that λ*dg* has a less stable association with its host than normal prophage. Moreover, the high frequency with which double lysogenisation with defective and normal prophage can occur suggests that λ*dg* relies on the homology of the *gal* regions for its insertion, so that the λ-specific attachment site is left available for insertion of the wild type chromosome. We have already seen in the case of the related phage φ80 that insertion at *its* attachment site, and presumably also its normal release, requires the operation of a phage function which is repressed in the prophage state and which is not needed for insertion of the defective transducing phage, φ80*dlac* at the *lac* region (p. 472).

Models of transducing phage formation

The evidence is indisputable that the initial event which generates a transducing λ*dg* genome is the substitution of a rather variable segment of the prophage genome by a fragment of bacterial chromosome which may be markedly longer or shorter than the segment it replaces. Campbell's (1962) model of how this substitution occurs is now generally accepted as the only one which appears to explain all the facts. It is based on the concept of a circular phage chromosome which is integrated into the bacterial chromosome by a single recombination event in the attachment region (Fig. 124A, B; see also p. 458 and Fig. 102). Fig. 124C shows how a single crossover, following illegitimate pairing between one of several regions to the left of *gal* and a prophage region to the left of *h*, could yield a circular prophage chromosome which incorporates the *gal* regions and excludes a sizable part of the phage genome. The *gal* region of such a phage now offers a major region of homology whereby it could insert itself into, or be released from, the chromosome of a new host (Fig. 124D, E). This model predicts that every chromosomal marker that lies between the prophage and the *gal* region must also be carried by the transducing phage, and this has been demonstrated experimentally in several cases (see Signer, 1966). The evidence for the more general features of the Campbell (1962) model has already been discussed in detail (pp. 458–463). Since analysis of the genomes of λ*dg* particles reveal not only a wide range of deletions of the right (*h*—*A*) half of the phage chromosome, but also an extensive series of deletions encroaching into the

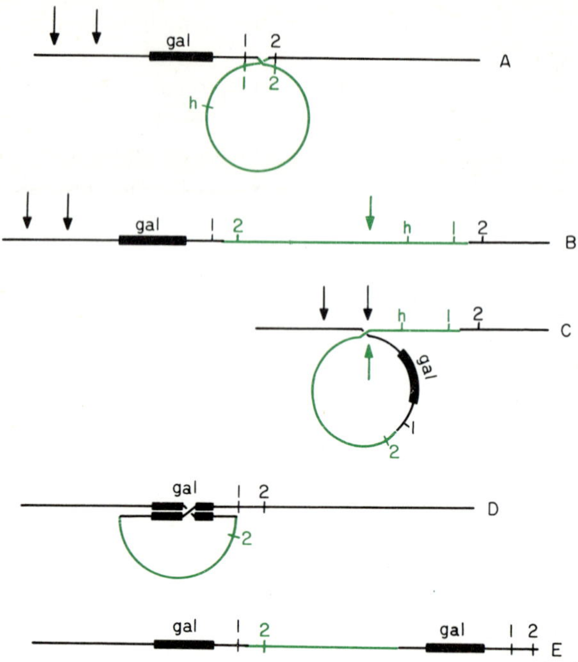

FIG. 124. A model showing how a circular prophage might incorporate regions of bacterial chromosome as a result of rare, illegitimate pairing.

The bacterial chromosome is shown black, and the prophage green. The green figures 1, 2 represent the normal attachment region (*C*) of the prophage and indicate homology with the corresponding bacterial region 1, 2 (black); the green '*h*' is the host-range locus which is missing from transducing phage.

The thick black segment represents the *gal* region. The black and green arrows point to regions of homology between the bacterial and phage chromosomes at which illegitimate pairing may occur.

A & B illustrate the insertion of prophage by recombination following proper pairing (compare Fig. 102, p. 459).

C shows how prophage incorporating the *gal* region might be formed by illegitimate pairing.

D & E demonstrate that the insertion of transducing phage into the chromosome of a new host, and its subsequent release, may be by pairing and recombination in the *gal* region which is now the main region of homology.

See text for further explanation.

(After Campbell, 1962)

galactose operon from the left (Shapiro, 1967), the amount of homology between the phage λ chromosome and the bacterial chromosome must be considerable — as, indeed, DNA hybridisation experiments have confirmed (Cowie and McCarthy, 1963).

A feature of interest is how exchange of *gal* alleles can occur between chromosome and transducing phage, and how homogenotes ($gal^-/\lambda{-}gal^-$ or $gal^+/\lambda{-}gal^+$) are formed. The Campbell model provides a satisfying explanation. As Fig. 125 shows, all that is required for an exchange of alleles is that the recombination event leading to release of the exogenote from the chromosome should

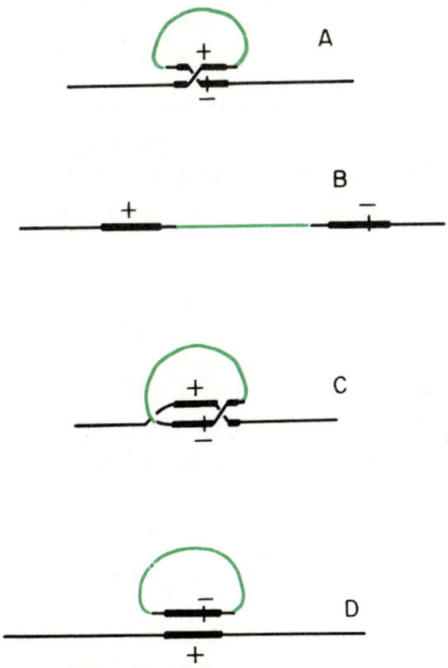

FIG. 125. A postulated mechanism for the exchange of alleles between exogenote and endogenote.

The bacterial chromosome is shown black and the phage chromosome green. The thick black segments represent the *gal* region. The position of the mutant site on the gal^- allelic region is indicated by the '−' sign.

A. The exogenote is inserted into the chromosome by a recombination event to the *left* of the mutational site on the chromosomal allele.
B. The integrated diploid structure is formed.
C. Pairing of the allelic regions, and recombination to the *right* of the mutational site, leads to emergence of the exogenote.
D. The released exogenote now carries the mutant allele, and the chromosome the wild type allele.

occur on the opposite side of the mutational site, in the recipient allele, from that which led to its insertion. If this event occurs after replication of the *gal* region, the released, defective λ—gal^- genome could become inserted into the other daughter chromosome which is still gal^-. Assuming that release and re-insertion of λ*dg* is a frequent occurrence, if this daughter chromosome had previously lost, or subsequently loses, its indigenous λ*dg*, a gal^-/λ—gal^- homogenote will be formed. It is clear that any recombinant type can arise by such a mechanism.

The question may be asked whether all prophages which have a chromosomal location are potentially able to mediate transduction of a neighbouring chromosomal fragment. Such transduction would, of course, only be manifest if the fragment happened to carry an observable marker. We have already seen that the *gal* region of *E. coli* can be transduced by three inducible prophages, 82, λ and 434; but the fact that phage 82 transduces at a much lower frequency than λ, despite its closer linkage to *gal* (p. 630), suggests that the degree of homology between prophage and bacterial chromosome is an important factor in transduction. Another inducible prophage of *E. coli* K-12, called φ80, is closely linked to the tryptophan region and can transduce it specifically. As in the case of phage λ*dg*, the transducing particles of phage φ80 are defective and yield heterogenotes from which HFT transducing preparations can be obtained (Matsushiro *et al.*, 1962; see p. 472).

There is one non-inducible prophage (18) which is known to lie between two closely linked methionine loci with which it can be jointly transduced by phage 363 (Fig. 127, p. 666; Jacob and Wollman, 1961a, p. 289), so that, presumably, it could transduce one or the other of these loci were it capable of transduction; this phage, however, appears to be non-transducing.

THE MECHANISM OF GENERALISED TRANSDUCTION

Elucidation of the hybrid nature of phages mediating restricted transduction provoked the thought that the mechanism of generalised transduction might be similar in nature. The fact that occasional transductants, mediated by phage P22 in *Salmonella*, may be non-lysogenic (Stocker *et al.*, 1953; Garen and Zinder, 1955), provided a starting point for further investigations concerning the nature of the transducing particles. Since, in generalised transduction, the frequency of transfer of a donor marker is about 10^{-5} per phage particle, a vast excess of normal, non-transducing particles must be plated with the recipient bacteria. Many of these particles produce lytic infection, so that the concentration of non-transducing phage on the plate becomes so high that the likelihood of a transductant escaping secondary infection and lysogenisation is remote. Thus the lysogenic state of a sensitive bacterium singly infected with a transducing particle is difficult to determine. Nevertheless it has been shown that the proportion of *Salmonella* transductants which are non-lysogenic increases as the multiplicity of infection by phage P22 falls, and is much higher than the proportion of non-lysogenics among the recipient population as a whole (Clowes, 1958c).

The phage P22-*Salmonella* system is rather unsuitable for this type of analysis because several generations may be required for the phage to establish lysogeny; infected bacteria segregate both sensitive and lysogenic progeny to yield impure clones (Luria *et al.*, 1958). Phage P1, on the other hand, makes its decision whether or not to lysogenise before the infected bacterium divides, so that pure lysogenic clones are formed (Adams and Luria, 1958). Moreover, this phage needs calcium for adsorption so that secondary infection on a transduction plate can be prevented by excluding this element. Thus when prototrophic transductants of *E. coli*, mediated by phage P1, are selected on *minimal* agar, the great majority may be non-lysogenic; on the contrary, most transductants for fermentative and other markers which are recognised by their growth on *nutrient* media, such as EMB-agar, turn out to be lysogenic (Lennox, 1955; Adams and Luria, 1958). All this suggests that the transducing phage particle either lacks a phage genome or possesses one so defective that it is normally unable to establish lysogeny.

How do the phage particles pick up their fragments of bacterial chromosome? The distinguishing features of generalised transduction are, 1. that any bacterial gene can be transferred; 2. that gene pick-up can occur during lytic infection as well as after the induction of lysogenic cells so that vegetative phage, and not prophage, appears to be involved; 3. that the transferred piece of donor chromosome usually becomes integrated into the recipient chromosome and is not maintained as the exogenotic fragment of a heterogenote; and 4. that (in the case of phage P1 at least) most transductants may show no trace of residual phage. Two general models may be proposed to explain these facts.

The first equates the mechanisms of generalised and localised transduction to the extent that, in both, fragments of bacterial chromosome are incorporated into the phage genome to yield defective phages. In generalised transduction, however, the phage genome may be supposed to have affinities for many regions along the host chromosome with which it can mate during vegetative multiplication. This model seemed a plausible one following the discovery that phage P1 transduction can yield heterogenotes carrying defective phage, from which HFT preparations of phage can be obtained (p. 628). In transduction by phage P1 between strains of *E. coli*, stable P1-sensitive transductants are the rule when precautions against secondary lysogenisation are taken. On the other hand, if phage P1 is used to transduce the ability to ferment lactose (lac^+) from *E. coli* to *Shigella* (which is completely devoid of the potential to ferment this sugar), or to another *E. coli* strain which has an extensive deletion in the *lac* region, unstable lac^+ clones, which segregate lac^- bacteria, arise at a very low frequency. The majority of these lac^+ transductants are immune to phage P1 infection but fail to liberate infective phage on induction. If, after induction, the transductants are superinfected with normal, non-transducing phage, lysates result in which about one per 1000 of the particles can transduce the lac^+ character (Luria *et al.*, 1958; Luria, Adams and Ting, 1960).

Analysis of the properties of a large number of lac^+ transductants of *Shigella* has revealed a much wider range of variability than exists in the case of the λ-*gal* system. For example, the transducing phage may be so little defective that, on induction, it can multiply without the help of normal, superinfecting phage to yield particles which can lyse as well as transduce; at the other extreme, not only immunity but every trace of phage activity, except the ability to transduce, may be lacking (Luria *et al.*, 1960). This phenomenon is not restricted to the *lac* region; comparable results have been obtained in transduction of the *thr—leu* region of *E. coli* by phage P1. You may remember that these two loci are jointly transduced with only about three per cent of the frequency of either locus alone (p. 625). From some joint transductants, however, HFT preparations can be made which transfer the *thr* and *leu* loci together at the same high frequency (Wilson, 1960).

These findings suggest that whether a transduced fragment is integrated to form a stable recombinant, or fails to be integrated so that a heterogenote results, depends on the balance of two factors—the extent of homology between the fragment and the recipient chromosome, and the degree of residual phage function which the transducing particle carries; the greater the former and the less the latter, the higher will be the probability of integration, and *vice versa* (Luria *et al.*, 1960). In the case of restricted transduction by phages λ and φ80, the prevalence of heterogenote formation, despite extensive genetic homology, may result from the fixed prophage location which limits the extent and region of phage chromosome that can be substituted. Either the phage moiety of the hybrid genome is big enough to interfere with integration, or else it retains some independent function which represses it. We may add that studies of such transducing particles of phage P1 in a caesium chloride density gradient have shown that they are slightly more dense than normal particles, and that HFT preparations of separate P1-*lac* isolates differ in density (compare p. 630; Ting, 1962).

There is thus no doubt that generalised transducing phage can give rise to rare transductants of heterogenotic type. More recently, however, experiments have been reported which favour a different mechanism as the more usual cause of generalised transduction. This is that, during lytic infection, fragments of bacterial chromosome are incorporated into phage heads in lieu of phage chromosomes by a process analogous to phenotypic mixing (p. 511). According to this model, at no time is there any physical association between phage and bacterial chromosomes (Clowes, 1958c). The main evidence comes from physical analysis of transducing particles of phage P1. The chromosomes of thymine-requiring *E. coli* bacteria were density-labelled with bromouracil and then infected with a clear-plaque mutant of phage P1 in the presence of thymine and radioactive phosphorus (^{32}P). When the progeny particles were analysed by density gradient centrifugation, it was found that only transducing particles separated in the denser, bromouracil band, identical with that of phage grown in bromouracil medium. In addition, these transducing particles were devoid of radioactive label. This clearly indicates that the transducing particles incorporated only fragments of bacterial DNA present at the time of infection and contained no phage genome, since the phage DNA incorporated thymine and ^{32}P (Ikeda and Tomizawa, 1965a). It was therefore possible to label transducing and infective particles differentially, by growing the phage on bacteria labelled with tritiated thymidine in a medium containing ^{32}P, and then to estimate the proportion of transducing particles from the ratio of ^{3}H to ^{32}P among the progeny. The transducing particles were found to comprise 0.3 per cent of the total. This turns out to be compatible with the calculated proportion, based on the relative molecular weights of bacterial and phage DNA molecules and on the assumptions that all regions of the bacterial chromosome have an equal probability of incorporation, and a probability of between 10^{-5} and 10^{-4} for the transduction of any given marker (Ikeda and Tomizawa, 1965a). A peculiar finding to emerge from these studies was that the DNA in transducing particles is lighter than bacterial DNA but increases to the density of bacterial DNA on treatment with proteolytic enzymes. It therefore seems that the bacterial DNA is associated with protein, but the nature or function of this protein is not known (Ikeda and Tomizawa, 1965b).

In normal populations of phage P1, about one quarter of the particles are of small size, having heads 650Å instead of 900Å in diameter and containing about 67 per cent of the protein and 40 per cent of the DNA of the majority particles. As might be expected, these small particles are defective and unproductive in single infections but their genomes are, presumably, circularly permuted because, in multiple infections, they can complement one another to yield normally infectious particles. Moreover, a proportion of these small particles are transducing, but carry a diminished length of bacterial DNA which is reflected in a reduced probability of co-transduction of bacterial genes. This reduction in both bacterial and phage DNA content clearly implicates a defect in the head structure as the primary lesion (Ikeda and Tomizawa, 1965c), and supports the hypothesis, previously discussed in the case of phage T4 (p. 434), that fixed lengths of a concatenate of phage genomes, or of bacterial chromosome, may be chopped off and incorporated into the phage head by the maturation process itself.

In the case of *B. subtilis* transducing phages there is also evidence for the incorporation of bacterial DNA alone (see Okubo *et al.*, 1963; Mahler, Cahoon and Marmur, 1964). Moreover, the *B. subtilis* transducing phage SP10, as in the case of phage P1, appears to be carried cytoplasmically by the lysogenic bacteria, while its DNA (or messenger RNA), unlike that of phage λ, shows no significant homology for the bacterial chromosome in hybridisation experiments (Bott and Strauss, 1965). However, the absence of a chromosomal site for phage insertion is not a necessary condition for generalised transduction, because the prophage of phage P22 has been found to map specifically in the proline region of the *Salm. typhimurium* chromosome (Smith and Stocker, 1966).

ABORTIVE TRANSDUCTION AND THE NATURE OF TRANSDUCED FRAGMENTS

We have already described the experimental criteria, as well as the theoretical interpretation, of *abortive transduction* in some detail (p. 96). The phenomenon remains the only reliable indicator of complementation between functionally non-allelic loci which can be applied to a wide range of markers in *Salmonella* crosses (Stocker *et al.*, 1953; Ozeki, 1956) and has also been reported in transduction of *E. coli* by phage P1 (Gross and Englesberg, 1959). The essence of the phenomenon is simply stated. A proportion of the transduced fragments of donor chromosome become integrated into the recipient chromosomes shortly after transfer, and then replicate with the chromosomes to yield clones of *complete transductants* as we have described. Those fragments which fail to be integrated early, however, are unable to replicate and, for reasons not yet understood, are not integrated subsequently although the genes they carry can express their functional activity. The result is that *these abortively transduced fragments are inherited unilinearly* by the descendants of the original recipient bacteria; at each generation a fragment is transmitted by the cell containing it to only one of the two daughters so that, at any time, only a single cell of the clone possesses the fragment. Since the fragment is functionally active, however, its products (enzymes or 'enzyme-forming systems') are synthesised in every cell through which it passes and these may be transferred cytoplasmically, and exert their effects, through several generations of daughter cells before they are diluted out (Fig. 26, p. 98). In the case of an auxotrophic recipient and a donor fragment which can perform the missing function, abortive transduction will allow sufficient multiplication on minimal medium to produce *minute colonies* which can clearly be distinguished from the large colonies of complete transductants as well as from any residual background growth of non-transduced recipients (Ozeki, 1956; Plate 5, pp. 96–97).

Apart from its importance in complementation tests, abortive transduction is valuable in providing new information about the nature of the transduction process, since the frequency of its occurrence indicates that of *transfer* of the marker, in contrast with the frequency of complete transduction which betokens integration only. A glance at Table 23 will show that, for any marker, the number of abortive transductants always exceeds that of complete transductants by a factor of between about five and twenty; that is, only a minority of those donor genes which are transferred to recipients are integrated. A closer perusal of the first column of Table 23 brings further relationships to light. First of all, the total number of abortive transductants

TABLE 23

A comparison of the frequencies of abortive and complete transduction for various markers in *Salmonella typhimurium*.

Recipient strain	Number of transductants:* Abortive	Complete	Ratio abortive/complete transductants
{ *ade*C-7	1,120	81	13·7
{ *adth*C-5	1,240	105	11·8
{ *gua*A-1	1,030	190	5·4
{ *try*D-10	3,810	346	11·0
{ *cys*B-12	3,430	503	6·8
{ *try*D-10. *cys*B-12	3,070	136	22·5
{ *ser*-1	1,440	160	9·0
{ *ser*-5	1,260	152	8·3
*his*D-39	18,880	1,988	9·5
*adth*A-4	7,160	644	10·8
*pro*A-46	1,760	376	4·7
*cys*A-1	1,700	188	9·0
*met*C-50	900	64	14·0
*adth*D-12	480	25	19·2

* In each cross the number of transductants is per 4×10^8 phage P22 particles of a preparation grown on the same wild type donor strain.

Recipient strains carrying linked markers are bracketed together.

The symbols represent the following nutritional requirements: *ade* = adenine, *adth* = adenine + thiamine (single mutational step), *cys* = cysteine, *gua* = guanine, *his* = histidine, *met* = methionine, *pro* = proline, *ser* = serine, *try* = tryptophan; the capital suffixes, A, B, C, D, indicate different functional loci, and the figures different mutant isolates. (Data from Ozeki, 1959.)

which arise per 4×10^8 phage particles may vary widely for different, unlinked markers; for example, 18,880 minute colonies were found when the mutant *his*D-39 was used as recipient, as compared with only 480 for the *adth*D-12 mutant, indicating a great disparity in the frequency of transfer and therefore, presumably, in the frequency with which the two types of fragment are incorporated into transducing particles.

On the other hand, the numbers of abortive transductants for markers which are known to be closely linked are virtually the same, even though the functions of the loci may be very different. This can be most strikingly demonstrated by crossing a doubly auxotrophic recipient strain with a wild type donor; although the number of complete transductants may vary markedly depending on whether the minimal medium is supplemented with one or the other, or neither, of the required growth factors, similar numbers of abortive transductants are found (Table 23; bracketed groups of markers). This implies that linked markers have a high probability of being transferred together to recipients on the same fragment of chromosome, however much they may be separated in subsequent recombination events (Ozeki, 1959). If chromosome fragmentation is a random process, this result can only be explained if the fragments are large and the linked loci are compressed into a small segment. In such a case one would expect that recombination in the long regions on either side of the loci, leading to their joint inheritance by complete transductants, would be much more common than recombination between them to yield transductants for single loci. Thus in the cross

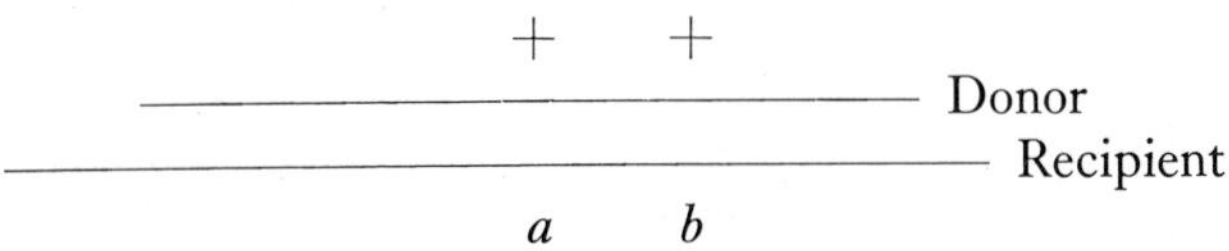

the number of ++ transductants should greatly exceed those of +*b* or *a*+ type. In fact just the reverse is found, the frequency of the double recombinant class always being smaller than either single class (Demerec and Hartman, 1956). The logical conclusion is that the fragments are comparatively short and are not randomly determined (Ozeki, 1959).

Supporting evidence for the idea that transduced fragments consist of small, pre-determined segments of donor chromosome comes from the study of reciprocal crosses with respect to pairs of linked markers (Ozeki, 1959). Let us take the reciprocal crosses shown in Fig. 126 If the donor fragments are randomly determined, the distance between

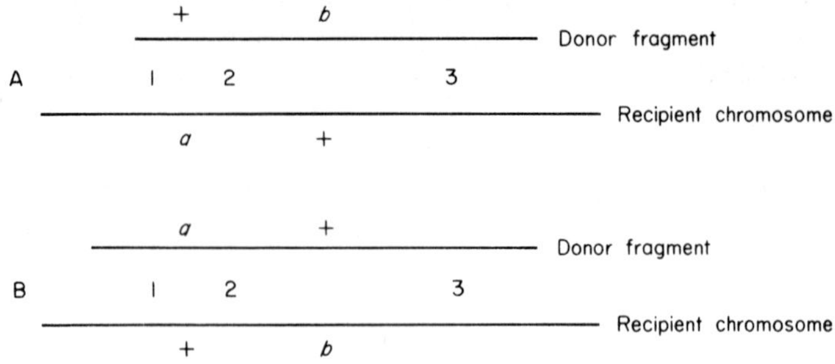

FIG. 126. Illustrating the effect of non-random fragmentation of the donor chromosome on the results of reciprocal crosses with respect to linked markers in transduction.

The markers *a* and *b*, and their wild type alleles (+), are represented at fixed positions in relation to the extremities of a pre-determined fragment, such that interval 1 is $\frac{1}{3}$ the length of interval 3.

In each cross, the ratio wild type/donor type transductants is estimated.

In both crosses the same number of donor type transductants is expected, from crossovers in intervals 1 and 3.

Wild type transductants arise from crossovers in intervals 1 and 2 in *cross A*, and in intervals 2 and 3 in *cross B*; since interval 3 is three times longer than interval 1, cross B will yield three times as many transductants.

Wild type/donor type transductants: Cross A/Cross B = $\frac{1}{3}$.

the two markers *a* and *b* (interval 2) should bear a constant relationship to the mean length of the fragments; *on the average*, interval 2 should lie no nearer one extremity than the other, so that the ratio of wild type to donor-type transductants should be the same irrespective of which parent acts as donor. On the other hand, if the two ends of the fragments are so defined that interval 1 is always one third the length of interval 3 (Fig. 126), then the reciprocal crosses will give different results. Although the numbers of donor-type transductants will be the same because, in both cases, these arise from crossovers in regions 1 and 3, the numbers of wild type transductants will not. In cross A, wild type transductants arise from crossovers in intervals 1 and 2, and in cross B from crossovers in intervals 2 and 3. Since interval 3 is three times as long as interval 1, cross B will yield three times as many wild type transductants as cross A. This is the kind of result obtained by Ozeki (1959) for reciprocal crosses involving the linked markers *adth*C-5 and *gua*A-1 (Table 24), in which the efficiency of the two crosses was strictly controlled by the number of transductants for an unlinked marker (*pro*A) (see *Ratio Test*, p. 138); similar results were found for the *try-cys*B region.

If the extremities of transduced fragments are strictly defined, at least some of the variability in the frequency with which different markers are transduced may be explained on this basis, since a gene which is always located near the end of a fragment will have a lower probability of integration by recombination than one situated at the middle. On the other hand, some experiments on the abortive transduction of flagellation by phage P22 in *Salmonella* appear to show that transducing fragments from at least one part of the chromosome may be heterogeneous in length. The locus fla^+, determining formation of flagella, is linked in transduction to the *H1* locus specifying their antigenic structure (see p. 739). When a fla^- recipient is transduced to fla^+ from a fla^+ donor, abortive transductants for motility are recognised by the occurrence of 'trails' of colonies (p. 97). If the antigenic structure of the donor flagella differs from the latent structure of the recipient flagella, specific donor antiserum inhibits trail formation only when the transducing fragment carries the donor fla^+—*H1* region so that flagella containing the donor antigen are formed. When a number of strains, yielding different frequencies of motile transductants inheriting flagella of donor type, were analysed it was found that the proportion of trails inhibited by donor antiserum decreased with the distance between the *fla* and *H1* loci. Thus fla^+ fragments of variable length are transferred to the recipients (Pearse and Stocker, 1965).

Again, a study of three markers in another region (*ileu-val*) of the *Salmonella* chromosome show that while one marker can be co-transduced with either, but not both, of the other two, and thus presumably lies between them, the two outside markers are not co-transducible. Clearly all three markers cannot lie on a single, fixed transducing fragment. Moreover, transducing particles for the middle marker turn out to have slightly different buoyant densities, those transducing the less closely linked outside marker being the denser (Roth and Hartman, 1965).

In the case of transduction of *E. coli* by phage P1, the evidence favours the transfer of random fragments. For example in the region *thr—leu—azi* the loci *thr* and *leu* on the one hand, and *leu* and *azi* on the other are jointly transducible, but transductants for all three markers are scarcely ever found (Lennox, 1955). Thus the *leu* locus can be carried by at least two distinct fragments, while the probability of joint transduction appears to be a function of the distance between the markers rather than of their containment between fixed points of fragmentation (see p. 528).

The subject of transduction has been reviewed by Hartman (1957), Hartman and Goodgall (1959), Clowes (1960), Hartman (1963) and Campbell (1964).

PHAGE CONVERSION

So far we have considered the phage-mediated transfer of genes that are obviously bacterial in origin. The first clear indication that bacteriophage itself might play a part in determining the phenotype of bacteria, however, preceded the discovery of transduction. In 1951, Freeman found that if certain strains of *Corynebacterium diphtheriae*, which fail to produce diphtheria toxin, are treated with a preparation of phage derived from virulent, toxigenic bacilli of the same species, a proportion of the survivors acquire the hereditary ability to synthesise toxin, as well as immunity to lytic infection by the phage (Freeman, 1951; Freeman and Morse, 1952). Subsequent studies of the kinetics of the phenomenon showed that conversion of the non-toxigenic strain to toxigenicity is not due to selection by the phage of a minority of toxigenic survivors, but to the establishment of lysogeny (Groman, 1953; Barksdale and Pappenheimer, 1954). Not only is there a complete correlation between lysogenisation and toxigenicity, but the ability to produce toxin is lost *pari passu* with loss of phage. Moreover, genetic crosses between the converting phage and other related strains of phage which lack converting ability may yield recombinant particles which no longer confer toxigenicity (Groman and Eaton, 1955), so that the genetic determinant of toxin synthesis resides in the phage. Nevertheless the biochemical cooperation of the host is also needed, since phage strains capable of converting avirulent strains of *C. diphtheriae* to toxigenicity may be isolated from non-toxigenic hosts (Groman, 1956).

Because of the correlation between lysogenisation and toxin production this phenomenon became known as *lysogenic conversion*, until it was discovered, in this as well as other systems which we will shortly mention, that virulent mutants of converting phages can also induce the same syntheses, very soon after infection, in cells destined to be lysed (Uetake, Luria and Burrous, 1958; Barksdale, 1959). The name 'lysogenic conversion' has therefore been replaced by the more general term, *phage conversion* (review: Barksdale, 1959).

One of the most thoroughly investigated examples of phage conversion concerns the production of certain somatic antigens by strains of *Salmonella* belonging to Group E. This group comprises three serologically different sub-groups which are distinguished by their somatic antigens; sub-group E_1 is characterised by antigens 3 and 10, sub-group E_2 by antigens 3 and 15, and sub-group E_3 by antigen 34. Organisms of sub-groups E_2 and E_3 are lysogenised by unrelated temperate phages, called ε^{15} and ε^{34} respectively. Sub-group E_1 bacteria are sensitive to phage ε^{15} and when lysogenised by it cease to produce antigen 10 and synthesise antigen 15 instead (Iseki and Sakai, 1953). If such converted $E_1(\varepsilon^{15})$ bacteria are cultivated in the presence of antiserum against antigen 15, reversions to the original E_1 phenotype, producing antigen 10, may be isolated and these no longer carry phage ε^{15} (Uetake, Nakagawa and Akiba, 1955). Sensitive E_1 cultures infected with phage ε^{15} begin to produce antigen 15 within a few minutes of infection, even when a virulent mutant of the phage is used, so that synthesis can be provoked by vegetative phage as well as by prophage (Uetake *et al.*, 1958).

Phage ε^{34}, derived from E_3 strains (antigen 34), is unable to infect E_1 bacteria since it cannot adsorb to antigens 3 or 10. It can, however, attach to antigen 15 and so is infective for naturally occurring E_2 cells as well as for E_1 cells lysogenised by phage ε^{15} which are indistinguishable from them. Following infection, doubly lysogenic bacteria $(\varepsilon^{15})(\varepsilon^{34})$, can be isolated and these now yield antigen 34 in addition to a modified form of antigens 3 and 15 (Uetake *et al.*, 1958). Thus we have a series of conversions mediated by two phages, which can be summarised in the form $\overset{E_1}{3,10} \xrightarrow{\varepsilon^{15}} \overset{E_2}{3,15} \xrightarrow{\varepsilon^{34}} \overset{E_3}{(3),(15),34}$. What would happen if phage ε^{34} could be persuaded to infect and lysogenise E_1 bacteria? Would these be induced to synthesise antigen 34? To answer this question, ingenious use was made of the fact that if E_1 bacteria are infected with phage ε^{15} at a multiplicity of about 5, a state of pseudo-lysogeny is set up among the survivors which, although infected, segregate a high proportion of sensitive (non-carrier) cells while only a minority continue to carry the phage (see p. 449).

These sensitive segregants, however, being the progeny of cells which harboured the phage and therefore synthesised antigen 15, are temporarily endowed with ε^{34} receptors and so can be infected and lysogenised with this phage (Uetake *et al.*, 1958). It turns out that such $E_1(\varepsilon^{34})$ bacteria preserve the normal E_1 phenotype (3,10) and do not produce antigen 34. If, however, they are secondarily infected or lysogenised with phage ε^{15}, antigen 34, together with the modified form of antigen 15, is synthesised while the ability to yield antigen 10 disappears. This means that the expression of the phage ε^{34} gene is dependent on the prior expression of that of the unrelated phage ε^{15}, and strongly suggests that the two genes mediate sequential steps in the biosynthesis of antigen 34 (Uetake and Hagiwara, 1960; 1961).

At the present time we have no inkling whether phage conversion is an extreme manifestation of transduction, by the descendants of phages which once incorporated bacterial genes without becoming in any way defective (see Luria *et al.*, 1958), or, alternatively, is a quite distinct phenomenon. In the latter event the gene determining production of toxin or somatic antigen would be, *de novo*, a phage gene, but one which plays no essential role in the synthesis of the phage itself; for example, $E_1(\varepsilon^{34})$ bacteria, although yielding no antigen 34, can liberate normal phage ε^{34} which can induce antigen 34 production on subsequent infection of E_2 cells (Uetake and Hagiwara, 1961).

The dependence of diphtheria bacilli on specific lysogenisation for their ability to produce toxin and, therefore, for their virulence for man, is a matter of considerable epidemiological interest and importance. More recently the ability to synthesise two other exotoxins associated with bacterial pathogenicity has been found to be controlled by a carried phage. One is the production of erythrogenic toxin, responsible for the specific skin rash of scarlet fever, by Group A haemolytic streptococci (Zabriskie, 1965); the other concerns the production of fibrinolysin by staphylococci, which is conferred by lysogenisation by a specific phage which, strangely enough, at the same time suppresses the synthesis of β-toxin (Winkler *et al.*, 1965). One may speculate to what extent we unjustly incriminate bacteria in general for the sins of their viruses! Another bacterial species which owes its pathogenicity for man to the production of a lethal exotoxin is the tetanus bacillus, *Clostridium tetani*. As long ago as 1893, before genetics had become an established science, San Felice reported that 'pseudotetanus' bacilli, which fail to make toxin, can acquire the hereditary ability to do so by growth in a medium to which metabolic products of virulent bacilli have been added (quoted by Lederberg, 1956c). It is strange that, in the light of recent discoveries, more attention has not been payed to the possible role of bacteriophages in the conferment of bacterial virulence for man and animals.

CHAPTER 22

Conjugation

One of the most important landmarks in the history of genetics was the discovery, during the early part of this century, that the theoretical model of inheritance inferred from the results of genetic analysis harmonised perfectly with the mechanical model revealed by microscopic observation. The empirical concept of linkage groups, comprising linear arrays of genes which reassorted themselves at each generation, found a material basis in the chromosomes and their behaviour at meiosis (p. 35). This cytological correlation became a great unifying factor in biology, for it was found to be a basic feature of life, shared not only by higher animals and plants but by many fungi as well.

How do bacteria fit into this picture? Although the existence of linkage in bacteria is evident from genetic analysis by transformation and transduction, the fragmentary nature of genetic transfer in these systems precludes them from providing much information about the *overall* organisation of the bacterial genetic material. The view they give of the bacterial genome has been compared with 'watching a football match through a telescope; the tactics and behaviour of the individual players (that is, the genes) can be better seen, but the strategy of the game is no longer observable' (Hayes, 1960). On the contrary, in *conjugation* in *E. coli*, with which this chapter is mainly concerned, there is actual copulation between sexually differentiated bacteria, and this may be followed by the transfer of a considerable part, and occasionally the whole, of the chromosome of one cell to the other. Thus zygotes are formed which are more nearly complete than in other systems of transfer in bacteria, with the result that genes widely scattered on the chromosome can be mapped in relation to one another.

The importance of conjugation in *E. coli* goes far beyond its capacity to extend the limits of orthodox genetic analysis in bacteria. In the first place, this system has presented new ways of translating genetics distances directly into absolute, physico-chemical terms. Secondly, the ability of *E. coli* to conjugate and transfer genetic material is promoted by the presence, in one of the two parental strains, of a determinant called the *sex factor* which turns out to be the prototype of a new kind of genetic element which has been discovered, under a variety of different disguises, in other bacterial species and genera (see Chapter 24, p. 748). Thirdly, the flexibility of conjugation in genetic and functional analysis has given birth to exciting new ideas concerning the nature of lysogeny (pp. 452–474), as well as the genetic regulation of metabolism which we will recount in the next chapter (pp. 712–735). Our present aim is to give a rather didactic account of the nature of conjugation as a genetic system, and to leave discussion of its molecular mechanism to Chapter 24.

THE DISCOVERY OF CONJUGATION

The literature of bacterial morphology contains many descriptions of microscopic observations of cell pairs which were interpreted as indicating mating and sexuality among bacteria (see Bisset, 1950; Hutchinson and Stempen, 1954). However, in the absence of any confirmatory genetic evidence these claims remained unsubstantiated and, in general, convinced only their authors. Some early but well constructed genetic experiments to demonstrate sexuality in bacteria, by scoring the characters of cells emerging from mixtures of genetically marked strains, were also unsuccessful (Sherman and Wing, 1937; Gowen and Lincoln, 1942); even had these crosses yielded recombinants it now seems likely that they would have been too few in number to have been revealed by the non-selective methods used.

The discovery of recombination in *E. coli*, strain K-12, by J. Lederberg and E. L. Tatum (1946a, b)—a discovery which Luria (1947a) thought might prove 'to be among the most fundamental advances in the whole history of bacteriological science'—was the outcome of a carefully planned study, based on the ability to *select* rare prototrophic recombinants from mixtures of auxotrophic mutant strains on unsupplemented minimal agar. This investigation illustrates well the stepwise progress of scientific advance, since the development of the necessary mutants (Gray and Tatum, 1944) was the direct outcome of an attempt to extend to bacteria the studies in biochemical genetics initiated in *Neurospora* by Beadle and Tatum (1941) (see Chapter 6, p. 112).

In their initial experiments, Lederberg and Tatum (1946a, b), plated washed cultures of mutants having triple and complementary nutritional requirements (*abcDEF* × *ABCdef*) on minimal agar, and found that colonies of prototrophic (*ABCDEF*) bacteria, which could be propagated indefinitely on minimal agar, arose with a frequency of about one per 10^6 to 10^7 parental bacteria plated. The multiple requirements of the parental strains precluded mutational reversion to wild type, while various other possibilities which might account for the colonies, such as cross-feeding (syntrophism), were also ruled out by experiment. The development of the prototrophic colonies required the co-operation of intact bacteria of both types since cultures of one type treated with cell-free filtrates or extracts of the other remained unproductive, so that actual cellular contact appeared to be necessary (Tatum and Lederberg, 1947; Davis, 1950). Moreover, when the parental strains were further marked by character differences which were not involved in prototroph selection, these unselected markers segregated among the prototrophs as if they and the selected markers were arranged on a single, linear linkage group (Lederberg and Tatum, 1946a, b; Lederberg, 1947; see also Rothfels, 1952).

Lederberg (1947) examined a number of prototrophic colonies obtained from these early crosses to obtain evidence with regard to the existence of reciprocal recombinant classes, or the involvement of more than two strands in recombination, which would imply the operation of a classical meiotic process. With very few exceptions, however, the colonies contained only one class of recombinant, suggesting not only that recombination might be of an unorthodox kind but also that the zygotes did not multiply as such, since two or more zygotes might be expected to yield more than one type of recombinant. In later studies, however, Lederberg (1949) found that crosses involving a particular mutant strain yielded an appreciable proportion of prototrophs which behaved as relatively stable heterozygotes; when grown on complete medium they segregated one or the other parental type as well as recombinants. That these prototrophic strains were indeed diploid heterozygotes was verified by pedigree analysis of single cells isolated with a micromanipulator (Zelle and Lederberg, 1951), and they provided the means for the first analysis of dominance in bacteria and the first direct demonstration of the normally haploid nature of these organisms (Lederberg, 1949, 1951a).

With the extension of genetic analysis to include a much wider range of unselected markers, a number of complicated anomalies of segregation began to appear which showed that this sexual system was, at least, an unconventional one (Lederberg *et al.*, 1951). We will briefly mention three of the most striking of these anomalies because, in retrospect, they underline the main features of conjugation as we now see it. Firstly, analysis of heterozygotes showed that a particular segment of the chromosome of one of the parents, identified by a pair of linked markers (*str–mal*, Fig. 127, p. 666), was virtually always missing from them and failed to appear among their haploid progeny; the diploids were incomplete (*hemizygous*), but it was thought at the time that the missing segment was eliminated from them after their formation (*post-zygotic exclusion*). Secondly, genetic analysis of recombinants from normal crosses revealed one main, linear linkage group, together with several subsidiary groups containing pairs of linked genes. All these groups showed evidence of linkage to one another, but the data were incompatible with a single linear structure. Thirdly when recombinants having the genotype of one of the parents, with respect to unselected markers, were back-crossed to the other parent, the distribution of these markers among the progeny deviated markedly from that found in the original parental cross.

Thus this brilliant and remarkable series of studies by Lederberg and his colleagues which proved, for the first time, that bacteria possessed a developed form of sexuality which made them amenable to formal genetic analysis, and revealed in them the existence of chromosomal organisation, culminated in the idea that the genetic mechanism involved might really be quite different from the orthodox one they had at first supposed.

SEXUAL DIFFERENTIATION AND ITS GENETIC SIGNIFICANCE

Lederberg's fruitful studies of recombination in *E. coli* K-12 were based on the working hypothesis that the two parental types of bacteria were equal partners whose fusion led to the formation of fully diploid zygotes. Furthermore, genetic analysis, especially of the persistent heterozygotes, was virtually the sole tool used to investigate the phenomenon. Vital clues to the true nature of conjugation came from a quite different approach—an examination of its physiological mechanism as opposed to its genetic outcome.

ONE-WAY GENETIC TRANSFER

The first clue was the discovery that the fertility of the standard cross depended on the continued viability of only one of the two parents which we will call A and B. When a mixture of streptomycin-sensitive (*str-s*) A and streptomycin-resistant (*str-r*) B bacteria was plated on minimal medium containing streptomycin, prototrophic recombinants arose at approximately normal frequency; but if the streptomycin markers of the two parents were reversed (A. *str-r* × B. *str-s*), the cross was completely sterile. Comparable results were obtained by pre-treating cultures of one or the other of a pair of streptomycin-sensitive strains with streptomycin, so as virtually to abolish its capacity to form colonies, and then removing the drug by washing; the treated and untreated cultures were plated together on minimal agar without streptomycin. The cross was sterile only when parent B had been treated in this way.

The conclusion was that viable B cells are essential to the fertility of the cross because they alone form the zygotes in which the whole process of recombination occurs; they are exclusively *recipients* of genetic material. The role of A cells, on the other hand, is to act as *genetic donors*; once they have performed this function they are dispensible (Hayes, 1952a). Conjugation is therefore a *heterothallic system in which recombination is mediated by the one-way transfer of genetic material from donor to recipient bacteria*. This important concept has since been fully confirmed by analysis of ex-conjugant bacteria isolated by micromanipulation (Lederberg, 1957; Anderson, 1958), as well as by study of the transfer of isotopically labelled DNA between mating cells (Skaar and Garen, 1955; Silver, 1963) and of zygotic induction (Wollman and Jacob, 1957; see p. 463). A further feature of this sexual differentiation is that the frequency of recombinant formation is enhanced as much as 50-fold if the donor bacteria are irradiated with small doses of ultraviolet light, but reduced if the recipients are irradiated (Hayes, 1952b; 1960).

At about this time, cultures of strains A and B were encountered which no longer produced recombinants when plated together. Independent analyses of these sterile pairs by Lederberg, Cavalli and Lederberg (1952) and by Hayes (1953a) provided the next clue. When each strain was crossed with the corresponding partner of known fertility, strain A turned out to be the defective one, and was designated F^-. However, the sterility of A.F^- proved to be only relative since it yielded recombinants in crosses with wild type bacteria or with recombinant strains issuing from normal A × B matings. Those bacteria which were fertile with the defective strain A.F^- were also fertile with the normal B strain and were called F^+. It then transpired that F^+ strains corresponded to donors, and F^- strains to recipients (Hayes, 1953a, b). *E. coli* K-12 therefore comprises two 'mating types' such that $F^+ \times F^-$ crosses are fertile and $F^- \times F^-$ crosses sterile. $F^+ \times F^+$ are also fertile but usually much less so than the $F^+ \times F^-$ arrangement. Put another way, for a cross to be fertile one of the parental strains must be of donor type.

THE SEX FACTOR

What determines whether a cell is a donor or a recipient? One might guess that the donor state is conferred by a gene, the loss of this state ($F^+ \rightarrow F^-$) being due to mutation. On the other hand, on account of the enhancement of fertility of donor strains by ultraviolet light as well as by analogy with transduction (which had just been discovered), Hayes (1953a) conceived the possibility that genetic transfer might be mediated by some kind of infectious vector which resides in donor cells. Once again, independent investigations concurred in revealing the remarkable fact that the donor (F^+) character is transmitted from F^+ to F^- cells with very high efficiency when young broth cultures of the two types are mixed.* Although the mixture yields only about one in a million recombinants for the usual chromosomal genes, the majority of recipient bacteria may be converted into donors after about an hour's incubation (Lederberg *et al.*, 1952; Hayes, 1953a, b; Cavalli-Sforza, Lederberg and Lederberg, 1953). Thus *the donor state is conferred by an agent called F, or the sex factor, which is readily transmitted to recipient cells independently of the bacterial chromosome*.

What is the nature of the sex factor? Its general properties can be summarised as follows.

1. All attempts to infect recipient bacteria with the sex factor by means of cell-free filtrates or extracts of donor bacteria were unsuccessful. As in the case of the transfer of chromosomal genes, cell to cell contact between donor and recipient cells is required so that the role of the sex factor is quite different from that of a phage vector in

* The frequency of transfer of the donor character depends on the strains used, and may vary between about five and 95 per cent in one hour. The system first discovered, and described here, happens to be a highly efficient one.

transduction (Lederberg *et al.*, 1952; Cavalli-Sforza *et al.*, 1953; Hayes, 1953b). It can only be transferred to recipients by conjugation.

2. When a small number of donor cells is mixed with an excess of recipients, the recipients are converted into donors at a faster rate than the bacteria multiply. This means that the sex factor can proliferate autonomously and more rapidly than the cell to which it has been transferred, so that it spreads through the recipient population in an epidemic fashion (Lederberg *et al.*, 1952; Cavalli-Sforza *et al.*, 1953; de Haan and Stouthamer, 1963). A different approach is to carry out a pedigree analysis of single, exconjugant, recipient bacteria and their progeny with the aid of a micromanipulator. In distinction to what is found when chromosomal genes are transferred, *all* the segregants of the multinucleate recipients turn out to be donors; either many sex factors are transferred to each recipient, or else the sex factor multiplies in the recipient prior to division (Lederberg, 1958).

3. Treatment of F^+ donor populations with concentrations of acridine orange too low to prevent bacterial growth nevertheless leads to loss of the sex factor; the bacteria are efficiently converted into recipients but otherwise remain unchanged (Hirota, 1960). Moreover although, as we have seen, the donor state is occasionally lost spontaneously or following irradiation, it can usually only be restored to recipients by transfer of the sex factor from donor cells, either by conjugation or by transduction (Arber, 1960). The sex factor thus behaves as a genetic determinant which, in F^+ cells, is structurally independent of the bacterial chromosome and is capable of autonomous replication. Loss of donor function normally results from loss of the sex factor as a whole, but mutations may also arise in the sex factor itself which interfere with its functions. Mutations of this sort may usually be distinguished from loss of the sex factor because the mutant sex factor, like defective prophage, retains immunity which prevents superinfection with a normal factor (p. 775; Lederberg, J. and Lederberg, 1956; Cuzin, 1962). Selective methods exist for the isolation of sex factor mutants (p. 782).

Recent experiments have thrown some light on the chemical composition and size of the sex factor. For example, non-radioactive recipient bacteria which have newly acquired the sex factor from F^+ donors into which radioactive phosphorus (^{32}P) has been incorporated, lose it again as a function of ^{32}P decay (Lavallé and Jacob, 1961). However, this loss does not occur if the ^{32}P is taken up by the original donor bacteria under conditions where DNA synthesis is inhibited by mitomycin C (Driskell and Adelberg, 1961); this means that ^{32}P is not incorporated in the absence of DNA synthesis, so that the sex factor must be presumed to consist of DNA. The amount of DNA which the sex factor contains has been assessed, from the rate at which it is inactivated by the ^{32}P decay (p. 523), as equivalent to $2{\cdot}5 \times 10^5$ base-pairs, or about the same as the DNA content of a phage genome (Driskell and Adelberg, 1961). These results are supported by autoradiographic estimations of the amount of radioactive [^{3}H]-thymidine transferred from F^+ donors to recipient bacteria (Herman and Forro, 1962; see also p. 770).

THE GENETIC CONSEQUENCES OF SEXUAL POLARITY: INCOMPLETE TRANSFER

We can now rewrite the original fertile cross in the form A.F^+ $\times$ B.F^-, of which strain A harbours the sex factor and acts as a genetic donor while strain B, lacking this factor, behaves as a recipient in which the zygotes are formed. The discovery of a variant of strain A which had lost its sex factor (A.F^-), and the fact that strain B can be converted to the donor state by conjugation (B.F^+), permitted the reciprocal crosses A.F^+ $\times$ B.F^- and A.F^- $\times$ B.F^+ to be set up, so that the effect of *reversed F polarity* on the genetic outcome of the cross could be studied. It transpired that the recombinant classes obtained from the two types of cross differ profoundly (Hayes, 1953a, b; Cavalli-Sforza *et al.*, 1953).

Two salient features were observed to follow reversal of the donor-recipient relationship. First of all, only some of the donor markers appear to be inherited with significant frequency by the recombinants, whose phenotype predominantly resembles that of the recipient parent. Secondly, those unselected donor markers which do appear among recombinants are quite different in the two crosses (Hayes, 1953a, b). We will not pursue the controversy that ensued about what these findings meant (review: Jacob and Wollman, 1961a; Hayes, 1966a) but will briefly recount that interpretation which turned out to be the correct one. It is that *donor bacteria transfer only part of their genome to recipients, so that incomplete zygotes are formed.*

This concept may best be explained by a hypothetical example. Imagine two strains of bacterium having the genotypes

Ⓐ*B* — — *C* — — *D* — — *Ef*

and

ab — — *c* — — *d* — — *e*Ⓕ

where *A* and *F* represent wild type genes, not closely linked, which are selected in crosses, while *B*, *C*, *D*, *E* and their alleles *b*, *c*, *d*, *e* serve as unselected markers, *B* being closely linked to *A*, and *e* to *F*. When strain *abcdeF* is the recipient it must acquire gene *A* from the donor to yield prototrophic recombinants, so that inheritance of a fragment of donor chromosome carrying gene *A* and the closely linked marker *B* is selected by plating on minimal agar; the genotype *ABcdeF* will predominate among recombinants. Conversely when strain *ABCDEf* is the recipient, a fragment carrying the donor nutritional marker *F* and the linked marker *e* is selected, so that recombinants will tend to have the genotype *ABCDeF*. This is just the kind of result obtained in crosses reversed with respect to *F* polarity. On the other hand, if the zygotes contained the complete genome of both parents, the types of recombinants issuing from both crosses should be the same (Hayes, 1953a, b). This concept of one-way, partial genetic transfer also accounted for the hemizygous nature of Lederberg's persistent diploid strains (p. 653) since it transpired that the missing segment was that of the donor parent. This same segment was also missing from the recombinants of a normal A. $F^+ \times$ B.F^- cross, but when the sexual polarity of the cross was reversed a different segment of donor chromosome was found to be absent (Hayes, 1953b; but see Nelson and Lederberg, 1954).

Crosses of the type A.$F^+ \times$ B.F^+ are also fertile, though usually much less so than the $F^+ \times F^-$ variety. The recombinant pattern of such crosses is found to resemble that of a mixture of the crosses A.$F^+ \times$ B.F^- and A.$F^- \times$ B.F^+; the two F^+ strains behave as if each contains a minority of cells which can act as recipients. Two kinds of evidence favour this explanation. We have mentioned that streptomycin treatment destroys the ability of sensitive cells to act as recipients, but not as donors if they harbour the sex factor (p. 654). When the A parent of an A.$F^+ \times$ B.F^+ cross is treated with streptomycin, the pattern of recombinants changes to that characteristic of a pure A.$F^+ \times$ B.F^- cross, and *vice versa*; the recipient capacity of the treated strain has been abolished (Hayes, 1953b). Again, populations of donor bacteria which have been grown to saturation density in aerated broth, or cultured overnight on a nutrient agar surface, may temporarily lose the ability to express their donor genotype and behave almost exclusively as recipients; they are known as F^- *phenocopies* since they retain their sex factor, as shown by the rapid restoration of donor function on subculture to a fresh medium (Lederberg *et al.*,

1952). The fertility of $F^+ \times F^+$ crosses suggests that such phenocopies may comprise a minority of the population of young, growing cultures.

The hypothesis of one-way transfer of only a fraction of the genetic material of donor bacteria, to recipients which contribute their entire genome, supposes that the anomalies of recombination (p. 653) reflect the mechanism of genetic transfer and not the occurrence of post-zygotic events. In fact these anomalies are very well accounted for. The effect of partial transfer in determining hemizygous diploid strains is obvious. Secondly, since recombinants from $F^+ \times F^-$ crosses nearly always inherit the sex factor and are donors, back-crosses between the donor parent and recombinants having the unselected markers of the recipient will be of $F^+ \times F^+$ type and may therefore be expected to yield a recombinant pattern different from that of the original, parental cross. Finally, the apparent linkage, but absence of linearity, between different groups of markers could be explained as follows. More extensive analyses of $F^+ \times F^-$ crosses showed that all the markers then available could be arranged in a consistent order on a single linkage group (Clowes and Rowley, 1954; Cavalli-Sforza and Jinks, 1956). Clowes and Rowley therefore proposed that the donor fragments transferred at conjugation might result from the random breakage of this linkage group; thus fragments bearing the selected marker would be heterogenerous with respect to other markers so that additive recombination frequencies could hardly be expected.

These early studies and hypotheses formed the starting point for further outstanding research which cultminated in the development of conjugation in *E. coli* as a unique genetic tool of great refinement. We owe most of this later development to the conspicuously brilliant and sustained work of E. Wollman and F. Jacob which we must now describe. So, having set the stage, let us put on the play in which the the male lead is a new type of donor strain which yields a very high frequency of recombinants in crosses with recipients, and so has been termed *Hfr*.

THE NATURE OF CHROMOSOME TRANSFER

THE PROPERTIES OF *Hfr* DONORS

All *Hfr* strains originate from F^+ populations. The first two strains to be reported were isolated by chance from the same F^+ donor (Cavalli-Sforza, 1950; Hayes, 1953b). When mated with F^- recipient bacteria they generate about one thousand times more recombinants than does the equivalent F^+ strain under the same conditions; indeed, they behave as 'super-males' so far as chromosome transfer is concerned (Hayes, 1953b). On the other hand, unlike F^+ donors, they do not convert recipients to the donor state in mixed culture, while only a minute proportion of recombinants become donors (see below). *Hfr* strains nevertheless tend to revert readily to the F^+ state so that they must retain the sex factor of the F^+ strain from which they stemmed, but in some hidden, non-transmissible form (Cavalli-Sforza *et al.*, 1953).

A comparison of the recombinants produced by equivalent *Hfr* $\times F^-$ and $F^+ \times F^-$ crosses revealed two further important characteristics of *Hfr* donors. Firstly, $F^+ \times F^-$ crosses yield roughly the same low frequency of recombinants irrespective of the donor marker whose inheritance is selected, showing that the donor bacteria can transfer any segment of their chromosome impartially to recipients. On the other hand, the *recombinants produced at high frequency by Hfr* $\times F^-$ *crosses inherit a particular segment of donor chromosome exclusively*; when markers located on other regions of the donor chromosome are selected, the frequency of recombinants falls dramatically and may be no higher than that given by $F^+ \times F^-$ crosses. Thus *Hfr* behaviour appears to be associated with transfer of a restricted region of donor chromosome—a fact which emphasises the fragmentary nature of the transfer. Secondly, *whereas the recombinants formed at high frequency are almost invariably recipients, a proportion of those formed at low frequency inherit the Hfr donor state.* When these *Hfr* recombinants are crossed with an F^- strain they behave in precisely the same way as the original *Hfr* strain (Hayes, 1953b).

These properties of *Hfr* donors suggest that they arise from F^+ cells as the consequence of an alteration in the state of the sex factor, which now determines that a particular segment of chromosome is transferred to recipients one thousand times more frequently than formerly. At the same time the altered sex factor becomes non-infective, but is inherited in association with chromosome fragments which are transferred only at low frequency. Before we can begin to understand the relationship between F^+ and *Hfr* donors as a function of the behaviour of the sex factor, we must examine more fully the nature of chromosome transfer by *Hfr* strains.

CHROMOSOME TRANSFER AS AN ORIENTED PROCESS

The way in which *Hfr* donors transfer their chromosomes to recipients was elucidated by Wollman and Jacob (1955; 1958) by means of their famous *interrupted mating experiment*. Genetic analysis of crosses involving strain *HfrH* (isolated by Hayes, 1953b) show that this donor transmits a particular group of linked genes to recombinants at high frequency. These genes, which we will call *A*, *B*, *C*, *D* and *E*, are found to be arranged in this order on the chromosome since, when gene *A* is selected in crosses with a recipient strain carrying the alleles *a*, *b*, *c*, *d* and *e*, the donor genes *B*, *C*, *D* and *E* appear among recombinants at frequencies of about 90, 75, 40 and 25 per cent respectively.

Young broth cultures of *HfrH* and the recipient strain are mixed and, at invertals thereafter, samples of the mixture are removed and subjected to violent agitation in a high-speed mixer. This treatment tears the mating cells apart with great efficiency, without affecting the viability of zygotes or parental cells. The treated samples are then diluted and plated for recombinants. It was found that samples treated at any time prior to about eight minutes after mixing yield no recombinants. At eight minutes recombinants begin to appear but these inherit only the selected donor gene, *A*. Shortly afterwards, in samples agitated at about nine and ten minutes after mixing, genes *B* and *C* respectively begin to be found in the *A* recombinants, to be followed at 17 minutes by *D* and at 25 minutes by *E*. From the time of its first appearance, the frequency of each donor gene among recombinants rises progressively until its frequency in normal crosses is attained (Fig. 15, p. 66; Fig. 131, p. 684).

The only reasonable interpretation is that the effect of agitation is to rupture the chromosome during its transit from donor to recipient bacteria; only those genes which have already penetrated the recipient bacteria at the time of treatment can subsequently appear in recombinants. In an interrupted mating experiment, therefore, the time after mixing at which recombinants containing a particular donor gene begin to be found indicates the time taken for that gene to be transferred on the donor chromosome. Thus the donor genes *A*, *B*, *C*, *D*, and *E* enter recipients at precise and characteristic time intervals and in that order which, as we have seen, is the order of their arrangement on the chromosome. Since this result is the outcome of a large number of mating events it follows that *strain HfrH comprises a homogeneous population of donor bacteria, each of which transfers its chromosome in the same specifically oriented and linear way, such that a particular extremity, designated O (for 'origine'), is always the first to penetrate the recipient bacteria* (Wollman and Jacob, 1955; 1958).

SPONTANEOUS CHROMOSOMAL BREAKAGE DURING TRANSFER; ZYGOTIC INDUCTION

The high frequency of recombinants for certain markers generated by *Hfr* donors shows that the donors must transfer these markers to recipients with great efficiency. Indeed as many as 45 recombinants per 100 *Hfr* bacteria has been reported (Jacob and Wollman, 1961a, p. 152). However, a second striking feature of *Hfr* bacteria is that most of their genes are *not* inherited by recombinants at high frequency (>one per cent). Why is this? There are two possibilities. Either the genes are not transferred to zygotes, or else they are transferred but are not integrated into recombinant chromosomes. Fortunately there is a simple way of distinguishing between these alternatives.

You may remember that *E. coli* K-12 can be lysogenised by various inducible prophages whose locations on the bacterial chromosome are known (p. 454). These prophages are susceptible to *zygotic induction* (Jacob and Wollman, 1956a; Wollman and Jacob, 1957). As soon as one of them, at its location on the chromosome of an *Hfr* bacterium, is transferred to the cytoplasm of a sensitive recipient it enters the lytic cycle, so that the zygote is destroyed with liberation of free phage particles (p. 463). Such zygotes constitute *infectious centres* whose number can be scored since each one yields a plaque when plated on sensitive, 'indicator' bacteria (see p. 413). When, in an interrupted mating experiment involving an $Hfr(ly^+) \times F^-(ly^-)$ cross, the number of infectious centres is scored instead of the number of recombinants, the location of the inducible prophage can be mapped by observing the time after mixing at which the infectious centres first begin to appear. The results correspond precisely with those inferred from straight $Hfr(ly^-) \times F^-(ly^+)$ crosses which yield no zygotic induction but in which lysogeny and non-lysogeny segregate like normal genetic markers (p. 464). Moreover when a prophage location is known to be closely linked to a bacterial gene, as in the case of prophage λ and the galactose (*gal*) region (see p. 454), both are found to enter the zygotes at the same time in interrupted mating experiments. *Inducible prophages therefore serve as genetic markers with the unique property of expressing themselves immediately on transfer, without the intervention of recombination.* They thus provide direct information regarding the transfer of those chromosomal regions that carry them.

When strains of *HfrH*, each lysogenised by a different inducible prophage, are crossed with a non-lysogenic recipient it turns out that, even when the utmost precautions are taken to *prevent* interruption of the mating pairs, the frequency of zygotic induction (number of infectious centres per 100 *Hfr* bacteria) falls exponentially with the distance of the prophage location from the extremity of the *Hfr* chromosome, *O*, which is transferred first at conjugation (Jacob and Wollman, 1958a). Markers on the *Hfr* chromosome therefore show a gradient of decreasing transfer as their distance from *O* increases. This gradient can best be explained by an increasing probability of fracture of the chromosome as transfer progresses. The fact that the probability of transfer falls exponentially with distance from *O* suggests that there is a random probability of fracture per unit length of chromosome. We thus arrive at the important conclusion that *the incomplete nature of the zygotes formed in conjugation involving Hfr donors is a consequence of random breakage of the chromosome during transfer*, and not from transfer of a predetermined segment of constant size. We will discuss the implications of this for genetic analysis in more detail later (p. 685).

THE ORIGIN OF *Hfr* STRAINS AND THE CONCEPT OF A CIRCULAR CHROMOSOME

We have seen that although F^+ donors conjugate efficiently with recipient bacteria, since they transfer the sex factor to them at high frequency, the rate at which they generate recombinants for chromosomal markers is so low as to be negligible by comparison. That this dearth of recombinants is due to failure of chromosome transfer, and not to post-zygotic effects, is shown by the inability to demonstrate zygotic induction in $F^+ \times F^-$ crosses. On the other hand, *Hfr* donors are characterised by their high rate of chromosome transfer and by the fact that they originate from F^+ cultures. It therefore seemed possible that F^+ bacteria, as such, are incapable of chromosome transfer and that the marginal fertility displayed by large populations of them is due to small numbers of *Hfr* 'mutants' which arise in these populations. The alternative is that *every* F^+ cell has a small, random probability of transferring part of its chromosome. It is possible to discriminate between these two possibilities, for in the former case there should be a clonal distribution of fertile cells in F^+ populations, whereas in the latter the distribution will be random.

Jacob and Wollman (1956b) decided the issue by means of the fluctuation test of Luria and Delbrück (1943; see p. 181). Each of a series of small, independent cultures of an F^+ strain, seeded from a few bacteria, are crossed with a recipient strain and the number of recombinants arising from each is counted. It was found that the variance between the number of recombinants issuing from these independent cultures markedly exceed that given by different samples of a single, bulk culture which received the same inoculum. Moreover, the recombinants generated by each independent culture tend to be homogeneous with respect to the unselected donor markers which they inherit, while those from different samples of the bulk culture display greater uniformity and comprise a wider range of genotypes. By using the replica plating (indirect selection) method of mutant isolation (Lederberg, J. and Lederberg, 1952; see p.   ), pure cultures of *Hfr* strains may be isolated from the majority of sites on a 'master plate', inoculated with F^+ bacteria, which yield a recombinant colony when transferred, by a velveteen pad, to a selective medium seeded with a recipient culture. Thus at least the greater part of the fertility of F^+ cultures can be ascribed to the presence of *Hfr* 'mutants' (Jacob and Wollman, 1956b).

A number of *Hfr* strains of independent origin, obtained in this way, were analysed with respect to the markers they transferred to recipients at high frequency, and to the orientation of the transfer. The outcome of this investigation was not only remarkable but quite unforeseen. It turned out that although all the bacteria of any one *Hfr* strain behave as a stable, homogeneous population in transferring a given set of markers at high frequency and in a particular sequence, different strains transfer different chromosomal segments although parts of these segments may overlap. Moreover, some strains transfer their genes in a reverse direction to others. Despite this, the relationship of the various genes to one another is the same irrespective of the *Hfr* strain used to map them, giving an unequivocal general linkage map for *E. coli*, consisting of a single linkage group. The different transfer patterns of the independent *Hfr* isolates cannot, therefore, be ascribed to chromosomal rearrangements.

Two further, important features of these *Hfr* strains then became apparent. First of all, it transpired that while each strain transfers its chromosome as a linear, oriented structure having a specific head and tail (for example, $\overset{O\ \ A\ B\ C\qquad\qquad\quad Y\ Z}{\longleftarrow\!\!\!-\!\!\!-\!\!\!-\!\!\!-\ -\ -\ -\ -\!\!\!-\!\!\!-\!\!\!-}$), another strain can always be found which transfers genes *A* and *Z* as closely linked markers (for example, $\xleftarrow{O\,Z\,A\,B\,C\qquad\qquad Y}$ or $\xleftarrow{O\,B\,A\,Z\,Y\qquad\qquad\quad}$). Thus although each *Hfr* strain transfers its chromosome as a specific, linear structure, it is impossible to define any extremities on the chromosome of the F^+ strain from which these *Hfr* strains originated. It was concluded that the chromosome of F^+ donors is continuous, or circular, and that *Hfr* strains arise by the opening up of the circle, at a point characteristic for each *Hfr* type, to yield a linear, transferable structure Jacob and Wollman, 1958b).

We have shown that the reason why *Hfr* strains transfer only part (the *proximal* part) of their chromosome at high frequency is because the chromosome tends to break randomly during transfer, so that the probability that genes located on the *distal* part of the chromosome will enter the zygote diminishes rapidly with their distance from the leading extremity, *O*. Nevertheless, rare recombinants inheriting these distal genes can be selected, and the time required for transfer of the genes assessed in interrupted mating experiments. It is found that recombinants inheriting the most distal genes, located at the tail of the *Hfr* chromosome, only begin to appear at about 110 to 120 minutes after the commencement of mating, and that their maximum frequency is about 0·1 per 100 *Hfr* bacteria under standard conditions (Jacob and Wollman, 1958a; 1961a, p. 147; Hayes, Jacob and Wollman, 1963). The essential point is that most of these recombinants for markers located near the tail of the *Hfr* chromosome also inherit the *Hfr* donor character, *irrespective of the position of this extremity on the circular F^+ linkage group* (Jacob and Wollman, 1958b). An *Hfr* strain which transfers its genes in the sequence $\xleftarrow{O\,A\,B\,C\qquad\qquad Y\,Z}$, for example, will yield the highest proportion of *Hfr* recombinants when *Z* is selected, while virtually no *A*, *B* or *C* recombinants will be *Hfr*. In the case of another *Hfr* strain of type $\xleftarrow{O\,B\,A\,Z\,Y\qquad\qquad C}$, however, the *Hfr* state is inherited only by recombinants which acquire gene *C*; recombinants inheriting gene *Z* from this strain are not donors unless gene *C* is also inherited. Thus in *Hfr* strains, the determinant of the *Hfr* donor state, which we can assume to be the sex factor, appears to be located on the chromosome at that extremity which is transferred last during conjugation.

The circular linkage map of *E. coli* K-12 is presented in Fig. 127 which shows, in addition to nutritional and prophage markers, some of the positions at which the linkage group may be opened to form the different linear structures transferred by a variety of *Hfr* strains. The reason for the linkage anomalies which were found in $F^+ \times F^-$ crosses now becomes apparent (p. 653) for, in these crosses, we are not dealing

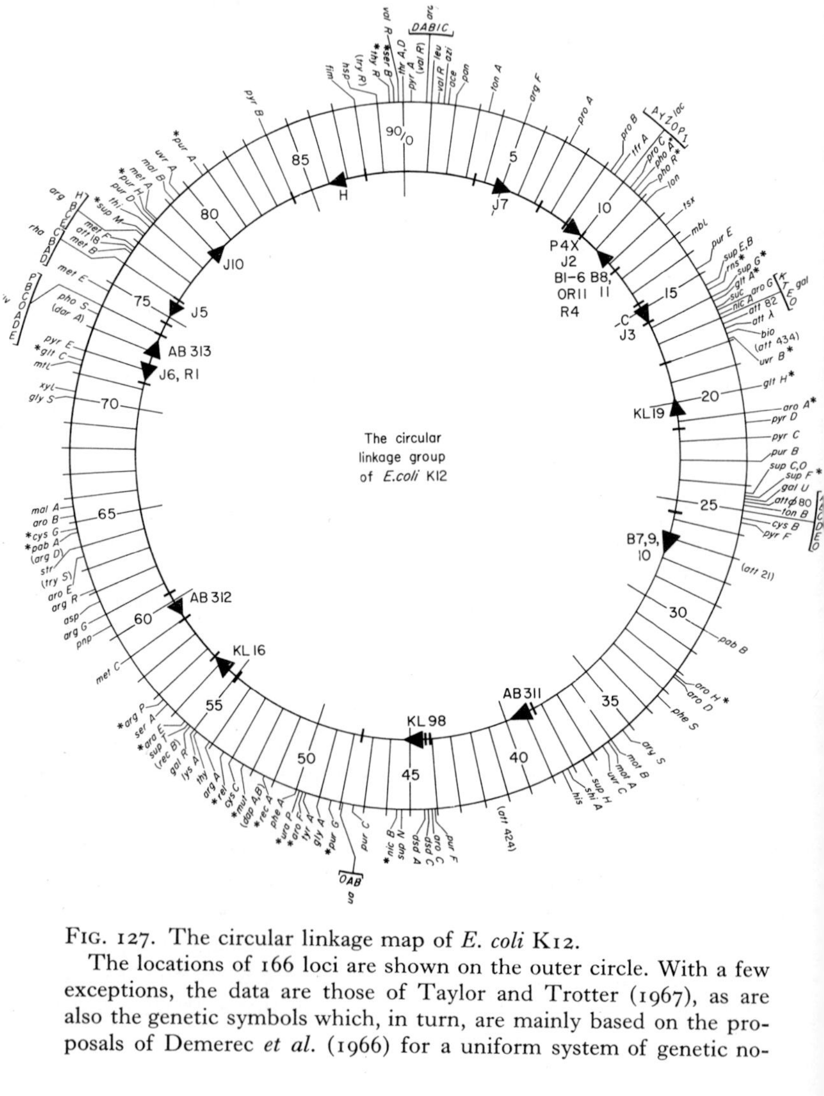

FIG. 127. The circular linkage map of *E. coli* K12.

The locations of 166 loci are shown on the outer circle. With a few exceptions, the data are those of Taylor and Trotter (1967), as are also the genetic symbols which, in turn, are mainly based on the proposals of Demerec *et al.* (1966) for a uniform system of genetic nomenclature. The map is drawn to scale in time units, the total length of 90 minutes being taken as the time required for transfer of the whole chromosome, assuming a uniform speed of transfer at 37 °C (Taylor and Thoman, 1964). The position of markers in parentheses is only approximate; the position of markers denoted by an asterisk is more accurately known, but the orders of adjacent asterisked markers is uncertain.

The arrow heads on the inner circle indicate the origin and direction of transfer of various *Hfr* strains. The short, thick radial lines on either side of each arrow head point to the first and last markers known to be transferred by particular *Hfr* strains. The data are mainly those of Taylor and Trotter (1967). It should be pointed out that the time taken by an *Hfr* strain to begin to transfer its most proximal marker may exceed by 3 to 5 minutes, or even more, the time interval between its leading extremity and the marker, as shown on the map. This 'dead time' is that required for the process of conjugation itself, which must proceed the initiation of chromosome transfer. Thus the origin of strain *HfrC* lies less than 6 minutes from the *lac* locus, but more than 10 minutes elapses between the time of mating and the commencement of *lac* transfer. The designations of the *Hfr* strains are as follows:

AB311, AB312, AB313 (Taylor and Adelberg, 1960).
B1-9 (Broda, 1967).
C (Cavalli-Sforza, 1950).
H (Hayes, 1953b).
J2-6 (Jacob and Wollman, 1958b; 1961a, p. 161).
KL16, KL19, KL98 (Brooks Low, quoted by Taylor and Trotter, 1967).
R1, R4 (P. Reeves, Ph.D. Thesis, London Univ., 1959).

Catalogue of Markers

Symbol	Map position (minutes)	Enzyme or function	Symbol	Map position (minutes)	Enzyme or function
ace	1	Acetate utilisation	*aroA*	21	Enzymes mediating common aromatic pathway through shikimate to chorismate synthesis
araA	0	Arabinose utilisation: arabinose isomerase	*aroB*	65	
araB	0	Ribulokinase	*aroC*	44	
araC	0	Arabinose regulatory gene	*aroD*	32	
araD	0	Ribulose 5-phosphate 4-epimerase	*aroE*	63	
araE	55	Arabinose permease	*aroF*	49	Isoenzymes mediating first step of aromatic pathway, specifically repressible by tyrosine (*F*), phenylalanine (*G*) and tryptophan (*H*) respectively
araI	0	Initiator	*aroG*	16	
argA	54	Arginine synthesis	*aroH*	32	
argB	77		*asp*	61	Aspartate synthesis
argC	77		*att*λ	17	Attachment site for phage λ
argD	64		*att18*	76	Attachment site for phage 18
argE	77		*att21*	27	Attachment site for phage 21
argF	5				
argG	61				
argH	77				
argP	56	Arginine permease			
argR	62	Arginine regulatory gene			
argS	35	Arginyl *t*-RNA synthetase			

Symbol	Map position (minutes)	Enzyme or function
attφ80	24	Attachment site for phage φ80
att82	17	Attachment site for phage 82
att424	41	Attachment site for phage 424
att434	17	Attachment site for phage 434
azi	1	Sodium azide resistance/sensitivity
bio	17	Biotin synthesis
cysB	25	Cysteine synthesis: reduction of thiosulphate or sulphite to sulphide
cysC	53	
cysG	64	
dapA,B	51	Diaminopimelate synthesis
darA	73	Dark repair of UV damage to DNA
dsdA	44	D-serine deaminase: D-serine resistance
dsdC	44	Regulatory gene
fim (*pil*)	88	Fimbria (pilus) synthesis
galE	17	Galactose utilisation: epimerase
galK	17	Galactokinase
galO	17	Galactose operon; operator locus
galT	17	Transferase.
galR	55	Galactose operon: regulatory gene. Old R_{gal}
galU	24	UDPG pyrophosphorylase
gltA	16	Glutamate synthesis
gltC	71	Glutamate permease
gltH	20	Glutamate synthesis
glyA	49	Glycine synthesis
glyS	70	Glycyl *t*-RNA synthetase
guaA	48	Guanine synthesis
guaB	48	
guaO	48	Guanine operon: operator locus
his	38	Histidine synthesis
hsp	89	Host specificity: DNA restriction and modification
ilvA	74	Isoleucine-valine synthesis
ilvB	74	
ilvC	74	
ilvD	74	
ilvE	74	
ilvO	74	Isoleucine-valine synthesis; operator for genes *A, D, E*
ilvP	74	Isoleucine-valine synthesis: operator for gene *B*
lacA	10	Lactose utilisation: transacetylase
lacI	10	Lactose utilisation: regulatory gene
lacO	10	Lactose operon: operator locus
lacP	10	Lactose operon: promotor locus
lacY	10	Lactose utilisation: galactoside permease
lacZ	10	Lactose utilisation: β-galactosidase
leu	1	Leucine synthesis
lon	11	Mutation leads to filament and slime formation, and to radiation-sensitivity
lysA	55	Lysine synthesis

Symbol	Map position (minutes)	Enzyme or function
malA	65	Maltose utilisation
malB	79	Maltose utilisation: possibly permease. Old *mal-5*
mbl	13	Mutation leads to sensitivity to methylene blue and to acridines, and to immunity to colicin E1
metA	78	Methionine synthesis *B* = old *met-1* *E* = old *met-B12* *F* = old *met-2*
metB	76	
metC	59	
metE	74	
metF	76	
motA	36	Motility: mutation leads to flagellar paralysis
motB	36	
mtl	71	Mannitol utilisation
mut	52	Mutability, generalised
nicA	16	Nicotinic acid synthesis
nicB	45	
pabA	64	p-Aminobenzoic acid synthesis
pabB	30	
pan	2	Pantothenic acid synthesis
pheA	50	Phenylalanine synthesis: prephenic acid dehydrase
pheS	33	Phenylalanyl *t*-RNA synthetase
phoA	10	Alkaline phosphatase
phoR	10	Alkaline phosphatase regulatory gene. Old *R1*
phoS	73	Alkaline phosphatase regulatory gene. Old *R2*
pnp	60	Polynucleotide phosphorylase
proA	7	Proline synthesis
proB	9	
proC	10	
purA	80	Purine synthesis
purB	23	
purC	47	
purD	78	
purE	14	
purF	44	
purG	48	
purH	78	
pyrA	0	Pyrimidine synthesis
pyrB	84	
pyrC	22	
pyrD	21	
pyrE	72	
pyrF	25	
recA	51	Genetic recombination and radiation-resistance
recB	55	
rel	53	Regulation of RNA synthesis; mutation leads to continuation of RNA synthesis in absence of protein synthesis
rhaA	76	Utilisation of rhamnose: isomerase
rhaB	76	Utilisation of rhamnose: rhamnulokinase
rhaC	76	Utilisation of rhamnose: regulatory gene
rhaD	76	Utilisation of rhamnose: rhamnuose-1-phosphate aldolase
rns	15	Ribonuclease I

Symbol	Map position (minutes)	Enzyme or function
serA	56	Serine synthesis
serB	89	
shiA	38	Shikimate permease
str	64	Streptomycin sensitivity/resistance/dependence
suc	16	α-ketoglutarate dehydrogenase: mutation leads to aerobic requirement for succinate, or for lysine + methionine
supB	15	Suppressors of *ochre* mutations
supC	24	
supE	15	Suppressors of *amber* mutations
supF	24	
supG	16	Suppressors of *ochre* mutations
supH	37	Suppressors
supM	77	Supressors of *ochre* mutations
supN	45	
supO	24	
supT	55	Suppressors
tfrA	9	Tfour: resistance/sensitivity to phages T4, T3, T7 and λ
thi	77	Thiamine (vitamin B1) synthesis
thrA	0	Threonine synthesis
thrD	0	
thyA	54	Thymidylate synthetase
thyR	89	Mutation determines low thymine requirement in *thyA* mutants
tonA	3	Tone: Phage T1 (and T5) resistance sensitivity
tonB	24	Phage T1 resistance/sensitivity
trpA	25	Tryptophan synthesis: tryptophan synthetase, A protein
trpB	25	Tryptophan synthesis: tryptophan synthetase, B protein
trpC	25	Tryptophan synthesis: indole-3-glycerol phosphate synthetase
trpD	25	Tryptophan synthesis: phosphoribosyl anthranilate transferase
trpE	25	Tryptophan synthesis: anthranilate synthetase
trpO	25	Tryptophan operon: operator locus
trpR	89	Tryptophan synthesis: regulatory gene. Old R_{try}
trpS	63	Tryptophan synthesis: regulatory gene
tsx	12	Tsix: Phage T6 resistance/sensitivity
tyrA	49	Tyrosine synthesis: prephenic acid dehydrogenase
uraP	49	Uracil permease
uvrA	79	Repair by excision of UV damage to DNA
uvrB	17	
uvrC	36	
valR	0	Mutation leads to resistance to inhibition of growth by valine
valR	1	
valR	89	
xyl	70	Utilisation of xylose

Further or more detailed information about the location and function of *E. coli* K12 genes may be found in Taylor and Trotter (1967). A detailed linkage map of *Salmonella typhimurium*, for comparison, is by Sanderson (1967).

with a genetically homogeneous population of donor bacteria, but with a mixture of distinctive *Hfr* clones which transfer variable lengths of different segments of the chromosome to recipients.

HOW THE SEX FACTOR DETERMINES THE DONOR STATE

We have seen that the mating system of *E. coli* comprises two types of donor, F^+ and *Hfr*, which express themselves in quite different ways. Both donor types are determined by possession of the sex factor since loss of this factor leads to an irreversible transition to the recipient type. What is the function of the sex factor, and what is the basic difference between F^+ and *Hfr* bacteria? We can obtain a partial answer to the first question by looking at the properties which both donor types have in common and which are lacking from recipient bacteria.

FUNCTIONS OF THE SEX FACTOR

The most obvious property of donor bacteria is the ability to form unions with recipients. A number of lines of evidence suggest that this is due to synthesis of a new surface component which probably acts by lowering the negative electrical charge at the surface of donor bacteria so that they can make intimate contact with recipient bacteria during random collisions. For example, not only do donor cells possess a surface antigen which recipients lack (Le Minor and LeMinor, quoted by Wollman, Jacob and Hayes, 1956; Ørskov and Ørskov, 1960), but they are also distinguished by greater precipitability at low pHs, and a different electrophoretic mobility and affinity for certain dyes (Maccacaro, 1955; Maccacaro and Comolli, 1956; Turri and Maccacaro, 1960). In addition, both F^+ and *Hfr* bacteria are susceptible to a group to RNA-containing, spherical phages (p. 440; Loeb, 1960; Loeb and Zinder, 1961; Dettori *et al.*, 1961), as well as to certain filamentous DNA phages (p. 444), to which not only F^- bacteria but also F^- phenocopies of donor strains are resistant. It turns out that these phages adsorb specifically to certain filamentous appendages called sex fimbriae (or pili) which are produced by F^+ and *Hfr* bacteria. Since the ability to produce sex fimbriae and susceptibility to donor-specific phage are lost together when F^+ bacteria are 'cured' of their sex factor by treatment with acridine orange, and are both regained when the sex factor is restored by conjugation, it is clear that synthesis of these fimbriae is a sex factor function (Crawford and Gesteland, 1964; Brinton, Gemski and Carnahan, 1964; Brinton, 1965). Moreover, it is easy to select mutants of either F^+ or *Hfr* bacteria which are resistant to the donor-specific phage, and these are found not only to have lost their sex fimbriae but also the ability to conjugate (Cuzin, 1962; Cuzin and Jacob, 1967a); often however, especially in the case of *Hfr* bacteria, the sex factor is retained as is shown by the persistence of other functions which it mediates. It follows that the presence of sex fimbriae on donor cells is a necessary condition for conjugation, but their mode of action is not yet known (see below).

Since cells which harbour mutant sex factors and cannot synthesise sex fimbriae are not efficient recipients, as are donor bacteria which have *lost* their sex factor, it seems likely that the sex factor mediates other surface functions required for conjugation. It has been known for some time that the ability of donor bacteria to mate with recipients is temporarily abolished by treating them with sodium periodate (Sneath and Lederberg, 1961). Since this treatment also inhibits adsorption of donor-specific phages (Dettori *et al.*, 1961), the periodate may be presumed to act on the sex fimbriae.

The sex (or *F*) fimbriae determined by the sex factor are far from unique structures. In fact other types of fimbriae, which are in no way associated with sexual activity and may be referred to as *common fimbriae*, are a normal feature of the *Enterobacteriaceae*. They are composed of protein, some types, at least, possess an axial hole, and they are characteristically present in rather large numbers at the surface of fimbriated bacteria (Brinton, 1965; Duguid, Anderson and Campbell, 1966). In contrast, sex fimbriae are few in number, and it has been suggested that there may be only one for every sex factor in the donor bacteria. They tend to appear longer and more flexible than common fimbriae from which they can be clearly distinguished by their coating of adsorbed virus particles in the presence of RNA phage, as Plates 19 (p. 432) and 34 (p. 773) show.

In addition to its purely conjugal functions, the sex factor has also been shown to exercise restriction on infection by phage T3 (Schell *et al.*, 1963) and some other phages which are thereby specific for recipient bacteria (Dettori *et al.*, 1961); it also possesses genes concerned with the regulation of its own replication (Cuzin and Jacob, 1967b) and with the repression of certain of its functions, including fimbria formation. As we shall see in Chapter 24, when we discuss sex factor behaviour in more detail, the sex factor *F* is normally 'derepressed', due to the presumptive mutation of a regulatory gene, so that, unlike many other factors of this kind, it expresses it conjugal functions in all of the cells which carry it.

Very little is known of the nature of the union between donor and recipient bacteria. Early electron microphotographs of mating mixtures revealed the existence of cellular bridges uniting donor and recipient cells, one of which is well shown in Plate 32. Similar bridges are seen irrespective of whether the donors are F^+ or *Hfr* bacteria (Anderson, Wollman and Jacob, 1957). Plate 33 is a more recent photograph of a section through mating *Hfr* and F^- bacteria of the same strains as in Plate 32. It seemed reasonable to assume that it is through such cellular connections that the genetic material passes at conjugation. However, the discovery of sex fimbriae and their essential role in conjugation suggested that *they* might be the real sex organ of donor bacteria, the DNA being transferred through the axial hole like phage DNA through the core of the phage tail. Moreover, it has been reported that the appearances shown in Plate 32 may also be found in mixtures of recipient bacteria and, therefore, may not be associated with genetic transfer (Brinton *et al.*, 1964; Brinton, 1965). As yet there is no expermental evidence to support this latter hypothesis. Whatever the true nature of the conjugal connection, it seems that energy production is not required for pair formation (Curtis and Stallions, 1967) although the subsequent genetic transfer is strictly dependent upon it (Fisher, 1957a, b).

STATES OF THE SEX FACTOR

Just as the similarities of F^+ and *Hfr* donors point to some of the functions of the sex factor, so we may expect the differences between them to reveal the nature of the alterations of sex factor behaviour which they display. Let us summarise these differences and see what they suggest. In F^+ cells the sex factor behaves *as an autonomous, cytoplasmic element* which can be selectively eliminated by acridine orange. On conjugation it promotes its own transfer with great efficiency but is probably incapable of initiating chromosome transfer (pp. 655–656).

In *Hfr* bacteria, on the other hand, the behaviour of the sex factor is just the opposite in every respect. On conjugation, it determines the high frequency, polarised transfer of the chromosome, but is not itself transferred except, very rarely, *as a chromosomal marker* linked to those genes which enter the recipient last at the distal extremity of the chromosome. Moreover, the sex factor is not removed from *Hfr* bacteria by treatment with acridine orange (Hirota, 1960). When we look at a number of *Hfr* strains, however, we find that, although the location of the sex factor with respect to other markers is fixed for any one strain, it may vary widely from strain to strain although it is always at the tail of the chromosome (p. 665).

The most plausible interpretation of these facts is that transition from the F^+ to the *Hfr* state is the consequence of a single event—the 'insertion' of the cytoplasmic sex factor at one of many possible sites on the circular F^+ chromosome, with concomitant opening of the chromosome at that point. That extremity of the resulting linear structure to which the sex factor remains attached becomes the tail, while the other becomes the leading extremity (*O*) in transfer. The reverse *Hfr*$\rightarrow F^+$

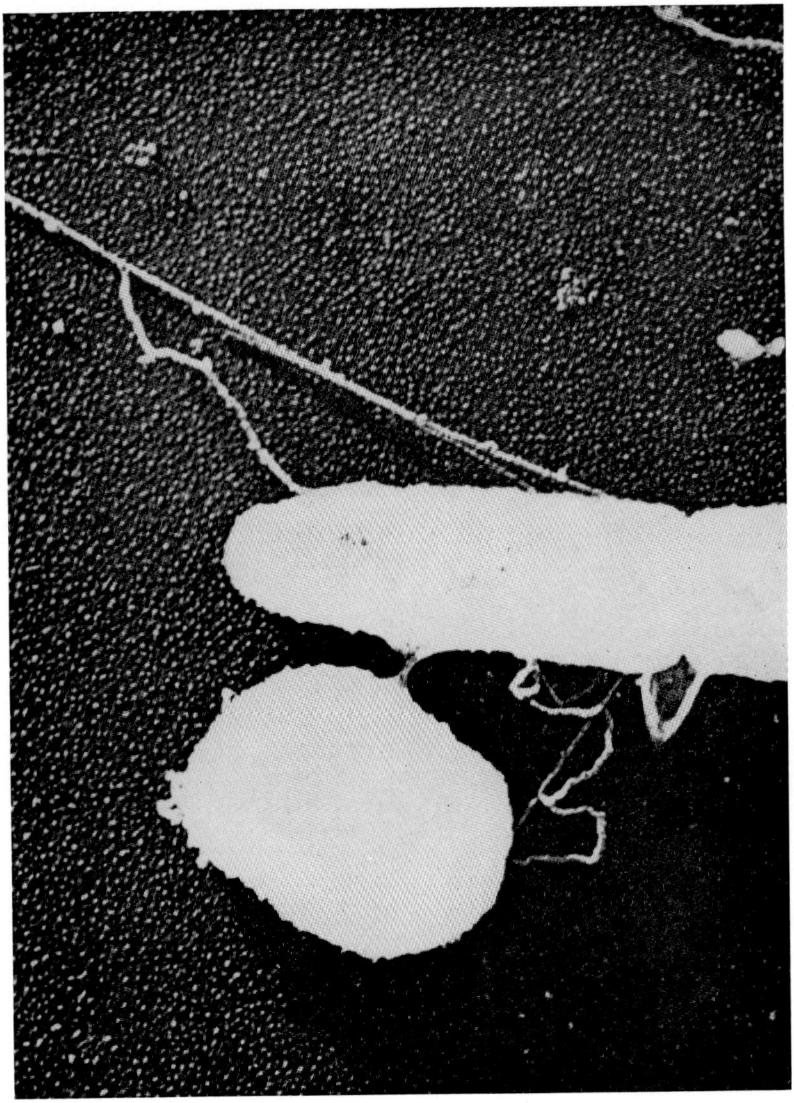

PLATE 32. Electron micrograph (shadowed) of conjugating bacteria. The upper, elongated bacterium, which is undergoing division, belongs to the *E. coli* K-12 donor strain, *HfrH*; the lower, plump bacterium is from a recipient (F^-) strain of *E. coli* C.

(Reproduced from Anderson, Wollman & Jacob, 1957, by kind permission of the authors)

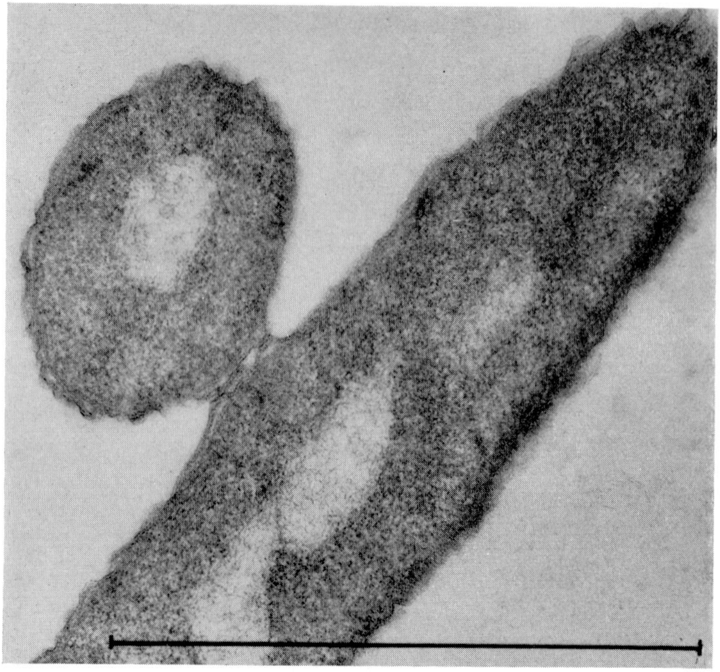

PLATE 33. Electron microphotograph of a thin section of conjugating *E. coli* bacteria. The strains are the same as those shown in Plate 32. In this cross the donor bacteria (strain K12 *Hfr*: right) and the recipients (strain C: left) have a clearly distinguishable morphology, as can be seen; however, this is a strain difference and has nothing to do with the sexual type. The less dense regions indicate the chromosomal DNA.

The photograph was taken by Maria Schnöss, and kindly provided by Dr. L. Caro.

transition is thus due to release of the sex factor from its association with the chromosome so that it returns to its free cytoplasmic state.

This elegant hypothesis, formulated by Jacob and Wollman, seems to explain the facts, but raises some new problems. One is whether insertion of the sex factor is limited to a few specific sites or occurs randomly around the chromosome. The considerable number of distinct *Hfr* types which have already been defined, points to many sites of insertion but these, in fact, are probably limited since many strains which appear to have their origins within the same small regions of chromosome have now been isolated (see Broda, 1967). Moreover, insertion of the sex factor has not been shown to produce auxotrophic or other defined mutations in *E. coli* K12 such as occur, for example, from insertion of mutator phage DNA into bacterial genes (p. 222), although some *Hfr* strains have a slower growth rate than the parental F^+ strain and so are unstable, especially in fluid culture, due to selection of F^+ revertants. However, a proportion of *Hfr* isolates from *E. coli* strain C, to which the sex factor is readily transferable from strain K12, has been shown to be markedly defective (Sasaki and Bertani, 1965).

Another question raised by sex factor insertion is whether the act of insertion itself leads to linearity of the bacterial chromosome. The answer has been decisively given by autoradiography; the chromosomes of *Hfr* bacteria, as of F^+ and F^- bacteria, are circular structures (p. 550; Cairns, 1963b). This finding is supported by genetic evidence. It is possible to cross two *Hfr* strains by employing an F^- phenocopy preparation of one of them so that it acts as recipient (p. 658). When *Hfr* strains *of the same type*, but having allelic markers at their leading (A and a) and terminal (Z and z) extremities, are crossed in this way, the same close linkage is found between A and Z as occurs in $Hfr \times F^-$ crosses. This implies that, in F^- phenocopies of *Hfr* bacteria, those markers which are at the extremities of the transferred chromosome are linked, and that the linkage group is circular (Taylor and Adelberg, 1961). If, then, we assume that the sex factor is inserted by a recombination event without disrupting the circularity of the bacterial chromosome, we must also postulate a circular sex factor. This strongly suggests that the sex factor is inserted by the same basic mechanism as that proposed for the phage λ chromosome by Campbell (1962) and that the *Hfr* chromosome is normally circular and becomes linear only for purposes of transfer. Since the sex factor is located between the first and last markers to be transferred, it could be postulated that polarised linear transfer might be initiated by opening up and replication of the sex factor itself. We shall see the impressive evidence which has now accumulated in support of these hypotheses when we explore this topic in more detail in Chapter 24 p. 789).

It is interesting to consider what might be the effect of inserting two sex factors into the chromosome at different locations. This can be done by crossing one *Hfr* type with an F^- phenocopy preparation of a *different Hfr* type, as is shown in Fig. 128a. In the figure, the inner circle represents the chromosome of the recipient *Hfr* strain in which the sex factor is located between genes *A* and *Z*, and the outer circle that of the donor *Hfr* strain, the sex factor here being located between genes *C* and *P*. If, in this cross, recombinants inheriting gene C^+ from the donor and P^+ from the recipient are selected, a proportion will also inherit the donor sex factor and will therefore carry both sex factors in their chromosomes. When such 'double males' are mated with recipients, they behave as though they possess two chromosomes which they transfer independently. In any one mating event the recipient receives either segment *A-B-C* or segment *P-Y-Z*, but not both (Clark, 1961). It is not yet known whether the double *Hfr* genome replicates as a single structure, as in Fig. 128b, or as two separate chromosomes of which at least one is under the replicative control of the sex factor (Fig. 128c). In the latter event we may find a possible speculative model for the evolution of multi-chromosomal organisms from an ancestral type containing a single chromosome, through the intervention of sex factor analogues.

INTERACTIONS BETWEEN SEX FACTOR AND BACTERIAL CHROMOSOME: SEXDUCTION

Having established the relationship between the two main types of donor bacteria, we must now introduce a third type which has properties intermediate between the other two and is therefore known as an *intermediate donor* (I). The existence of intermediate donors was discovered by Adelberg and Burns (1959; 1960) who observed that the behaviour of a particular *Hfr* strain in their stocks had altered in three major respects. Firstly, although it continued to transfer its chromosome with the same oriented sequence as before, it now did so at only about one-tenth its former efficiency. Secondly, it transferred its sex factor to recipients with the same high efficiency as F^+ strains, but the

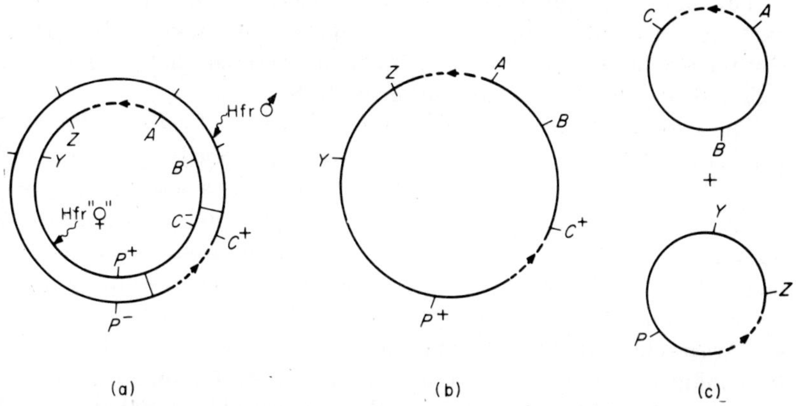

FIG. 128. The formation and behaviour of recombinant *Hfr* bacteria which carry two sex factors at different locations in their chromosomes. The continuous and broken lines represent bacterial chromosomes and sex factors respectively. The arrow heads in the broken lines indicate the points and directions from which chromosome transfer begins. The letters designate the locations of genetic loci on the bacterial chromosome. (a) The genetic structure of zygotes leading to segregation of 'double *Hfr*' recombinants. The inner circle represents the chromosome of the *Hfr* strain acting as recipient, and the outer circle that of the donor *Hfr* strain. The radial lines between the circles indicate cross-over positions. (b) Haploid recombinant having a single chromosome with two inserted sex factors. (c) Alternative possible type of haploid recombinant having two independent chromosomes (replicons), each possessing one sex factor. See text.
From Hayes, 1966b.

recipients were converted to intermediate donors, and not to the F^+ state; the sex factor had been modified so as to retain a 'memory' of its *Hfr* location on the chromosome. Thirdly, treatment of the original intermediate donor with acridine orange led to loss of the sex factor so that it became a recipient. However on re-infection with a *normal* sex factor from an F^+ strain, it reacquired its intermediate properties; the original *Hfr* location on the bacterial chromosome had evidently preserved an affinity for the sex factor (sex factor affinity, or *sfa*, locus) to which the normal sex factor was attracted (Adelberg and Burns, 1960). A similar acquisition of sex factor affinity by an erstwhile *Hfr* strain had previously been reported, but in this case the sex factor itself had disappeared, so that its modified behaviour was not observed (Richter, 1957; 1961).

The logical interpretation is that this state of affairs arises from an exchange of segments between sex factor and chromosome. The resulting sex factor, carrying a segment of bacterial chromosome, is called an *F-prime* (F') or *substituted* sex factor.

The same mechanism as leads to the production of *gal*-transducing particles of phage λ(λ*dg*) (Fig. 124, p. 634; Campbell, 1962) can be invoked for the formation of *F*-prime factors. As a result of a rare recombination event between regions of 'illegitimate' pairing on the inserted sex factor and a neighbouring region of chromosome, a sex factor is liberated which carries a chromosomal segment, while part of the sex factor is retained in the chromosome. Provided that the released sex factor retains those genes which are required for its replication and for conjugation, the joint structure can be transferred to, and propagated in recipient bacteria. Because the segment of bacterial chromosome carried by the sex factor has a virtually perfect homology for the allelic region of recipient chromosome, pairing and recombination between them will lead to insertion and release of the sex factor with a frequency vastly greater than that of the wild sex factor, so that the cell will alternate rapidly between the *Hfr* and F^+ states. The result is a population of bacteria which can transfer both autonomous sex factor and chromosome at high frequency. Up till now, *F*-prime factors defective in sex factor function, and equivalent to λ*dg* genomes, have not been identified, probably only because our knowledge of sex factor functions is so rudimentary.

As in the case of phage λ, which can independently transduce either the bacterial *gal* region or the biotin (*bio*) which are located on either side of the prophage, there is no theoretical reason why *F*-prime factors carrying either proximal or distal markers should not arise. However it happens that no factor has been found which carries a proximal marker alone, although factors jointly carrying both proximal and distal markers have been described (p. 797). On the other hand, a simple technique is available for selecting factors which carry a distal wild type gene (Z^+) closely linked to the sex factor on the *Hfr* chromosome. A streptomycin-sensitive, Z^+ *Hfr* strain is mated in broth with a streptomycin-resistant recipient carrying a recessive allele of the gene (Z^-). After about an hour, when only the proximal half of the chromosome has been transferred, the mating is interrupted by agitation in a mixer, and the *Hfr* bacteria are killed by treatment with streptomycin. Since the mating was arrested long before the Z^+ gene could have been transferred as a chromosomal marker, the only Z^+

survivors should be those recipients which were infected early by very rare F'—Z^+ factors generated in the *Hfr* population. The isolation of *F*-prime factors in this way is greatly facilitated by incubation of the interrupted mating mixture in fresh broth overnight, to allow infective spread of the *F*-prime factors through the recipient population prior to plating on selective or differential media for the recognition of Z^+ clones.

The rare Z^+ colonies which emerge from such interrupted matings are found to consist of bacteria which differ from true recombinants in two major respects. Firstly, they are unstable for the Z^+ character, segregating Z^- progeny with a probability of about 10^{-3} per cell per generation; secondly, they are intermediate donors which transfer not only proximal chromosomal markers, but also the sex factor and the Z^+ gene concomitantly and with great efficiency to Z^- recipients; sex factor and gene behave as a single unit in replication and transfer. They are, in fact, *heterogenotes* of type Z^-/F—Z^+, in every respect similar to those produced by HFT lysates in restricted transduction, except that none of the known functions of the sex factor appear to be defective. This process of genetic transfer by means of *F*-prime factors has been given the dysphonic name of *F-duction* or *sexduction*. By selecting for early transfer of distal markers from a range of different *Hfr* types, in the way we have described, *F*-prime factors carrying many different regions of chromosome have now been isolated. As our knowledge of localised transduction would suggest, diverse lengths of the same region of chromosome may be incorporated into independently isolated *F*-prime factors derived from the same *Hfr* strain; an *F*-prime factor incorporating as much as ten per cent of the bacterial chromosome and carrying at least four auxotrophic markers has been reported (Jacob and Adelberg, 1959; Jacob, Schaeffer and Wollman, 1960; E. M. Lederberg, 1960; Hirota and Sneath, 1961; Pittard, Loutit and Adelberg, 1963; Pittard and Adelberg, 1964; Berg and Curtis, 1967; a comprehensive list of *F*-prime factors reported to date is given by Scaife, 1967). It is interesting that the time taken for transfer of bacterial markers by *F*-prime factors, as well as the sequence in which they are transferred, are the same as in transfer of the same region by an *Hfr* strain (Hirota and Sneath, 1961; Pittard and Adelberg, 1964).

Heterogenotes of type Z_1^-/F—Z^+, arising from sexduction, segregate a proportion of *homogenotes* (Z_1^-/F—Z_1^-) as a result of recombination between exogenote and endogenote; these are recognised by their Z^- phenotype and their retention of a transmissible sex factor, just as in

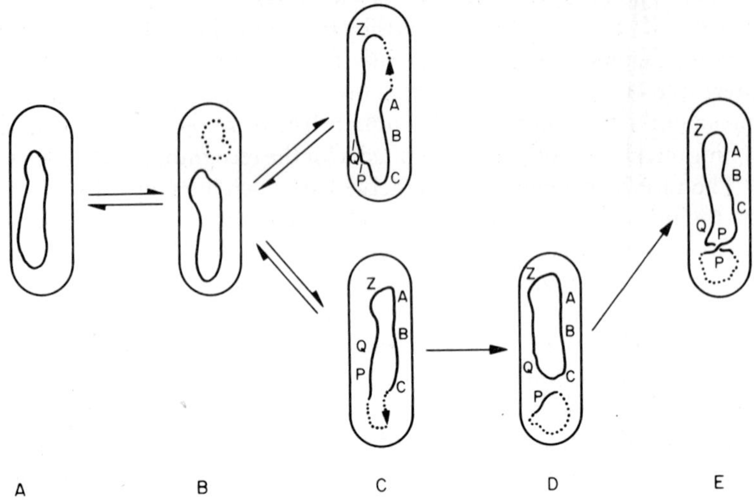

FIG. 129. The sexual types of *E. coli*.

The chromosome is indicated by the continuous line and the sex factor by the dotted line. The letters show the locations of the various genes on the chromosome.

A. No sex factor: recipient (female) bacterium.

B. Autonomous, cytoplasmic sex factor: F^+ donor (male) bacterium. The factor may be lost spontaneously or following treatment with acridine orange.

C. Sex factor inserted into the chromosome: *Hfr* donor (male) bacteria, transferring the chromosome from different points and with different polarity, depending on position and orientation of the insertion. May revert to the F^+ state by release of the sex factor from the chromosome.

D. Formation of an *F*-prime factor incorporating a segment of bacterial chromosome carrying gene *P*.

E. Transfer of *F*-prime factor to recipient bacterium. The *F*-prime factor alternates rapidly between insertion into and release from the chromosome as a result of efficient pairing and recombination between the allelic *P* regions: intermediate donor bacterium.

the case of transduction heterogenotes (p. 627). Transfer of such sex factors,carrying a particular recessive allele (F—Z_1^-), to recipient cells of the same mutant phenotype yield relatively stable heterogenotes of the general type Z_2^-/F—Z_1^-; these partial diploids will display the mutant phenotype if Z_1 and Z_2 are in the same functional unit and the wild phenotype if they are in different functional units. Sexduction therefore provides a reliable means of performing complementation tests, and of establishing relationships of dominance and recessiveness, hitherto lacking in *E. coli*. In the next chapter (pp. 713, 729) we will consider some of the important new concepts concerning the genetic regulation of metabolic processes, which have come from the application of these methods to the genetic and functional analysis of lactose fermentation. In Chapter 24 (p. 789 *et seq.*) we will examine the functions of the sex factor and its interactions with the bacterial chromosomes in greater depth and detail. Reviews embodying information about *F*-prime factors are by Adelberg and Pittard (1965), Driskell-Zamenhof (1965) and Scaife (1967).

The relationships between sex factor and chromosome which determine the various sexual types of *E. coli* are summarised in diagrammatic form in Fig. 129.

THE STAGES OF CONJUGATION

The whole process of conjugation in $Hfr \times F^-$ crosses may be divided arbitrarily into two main stages which can be defined and anlaysed by experiment. The *stage of zygote formation* begins with the initial contact between a donor and a recipient bacterium, and ends when the donor has completed the transfer of its genetic material. Then follows the *stage of recombinant formation* which comprises the processes of recombination, segregation and the expression of donor genes, and culminates in the appearance of an independent recombinant bacterium. We have already given a brief account of our scanty knowledge of the mechanism of recombination in bacteria (Chapter 15, p. 397), and will deal with genetic expression in some detail in Chapter 23 (p. 700). We therefore propose to concentrate now on the stage of zygote formation, for it is upon our understanding of this that our use of conjugation as an instrument of genetic analysis primarily rests.

THE KINETICS OF ZYGOTE FORMATION

Zygote formation may be subdivided into three successive steps—collision, union formation and chromosome transfer. *Collision* between donor and recipient bacteria is an obvious first step and appears to be random and a simple function of the population density of the parental cells (Nelson, 1951; Hayes, 1957a). The maximum number of collisions is probably achieved within a few minutes in the usual mixture of young broth cultures. The *formation of unions*, on the other hand, implies a specific interaction between donor and recipient bacteria. The specificity resides in the wall of the donor bacterium, is sensitive to pH change, is reversibly destroyed by periodate, and can be ascribed to the prior formation of an antigen under the control of the sex factor. We have seen that some of these attributes can be ascribed to the formation of sex fimbriae (p. 670), but it is not yet certain whether other features determined by the sex factor also play an essential role. The fact that energy production seems not to be needed for union formation (Curtis and Stallions, 1967) suggests that this is effected by pre-formed structures and does not involve new synthesis. On the other hand, donor cells from which the fimbriae have been sheared by agitation in a mixer (blendor) are not only unable to mate but require protein synthesis for recovery of mating ability; while this experiment confirms the essential role of sex fimbriae, it fails to clarify whether these act as tentacles which bind the bacteria together and so facilitate the formation of conjugation tubes like that shown in Plate 32 (p. 672), or whether they are themselves the organs of genetic transfer (Fisher, 1966).

The kinetics of union formation can be studied by mixing broth cultures of *Hfr* donor and recipient bacteria, removing samples at intervals thereafter, diluting them *very gently* about a thousand times into nutrient medium at 37°, and finally after allowing time for the selective marker to be transferred, plating them for recombinants. The only ostensible effect of the dilution is to separate those cells which have not yet united firmly, and to prevent the occurrence of further collisions. Those pairs which have not been separated continue to mate, so that the curve obtained by plotting the number of recombinants as a function of time should indicate the rate at which unions are formed; that is, for a given population density, the proportion of collisions that are converted into unions in a given time. We should point out, however, that the results vary according to the degree of gentleness of the dilution technique, as well as to other technical procedures, so that the method indicates only a rather arbitrary degree of adherence between mating pairs and provides no precise information about the state of bridge formation, although we may assume that this has at least been initiated.

The kind of results obtained from such an experiment are illustrated by the continuous black curves in Fig. 130. These represent the

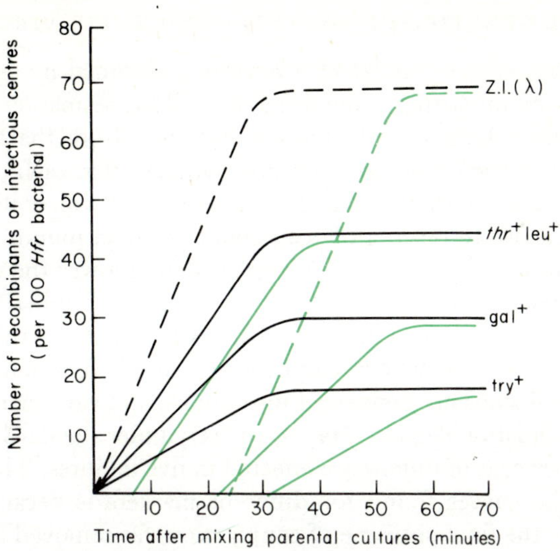

FIG. 130. A series of idealised curves illustrating the kinetic aspects of zygote formation.

The donor is wild type *HfrH*(*str-s*) which transfers the following proxima markers at the times (minutes after mixing) indicated below:

O	thr^+leu^+	$gal^+\lambda$	try^+
←	8	25 26	33

The streptomycin-sensitivity marker (*str-s*) is distal and does not appear among recombinants at high frequency.

The recipient requires threonine, leucine (*thr. leu*) and tryptophan (*try*), does not ferment galactose (*gal*) and is streptomycin-resistant (*str-r*).

When scoring for recombinants, non-lysogenic (ly^-) derivatives of both strains are used; thr^+leu^+ recombinants are selected on minimal agar containing tryptophan and streptomycin, try^+ recombinants on minimal agar containing threonine, leucine and streptomycin, and gal^+ recombinants on EMB-galactose agar containing streptomycin. The streptomycin prevents the donor bacteria from forming colonies.

The titre of zygotic induction of phage λ (infectious centres), in the $Hfr(ly^+) \times F^-(ly^-)$ cross, is scored by plating on streptomycin-nutrient agar, together with *str-r*, non-lysogenic indicator bacteria.

Young broth cultures of the parental strains are mixed, in the ratio one $Hfr{:}20F^-$ bacteria, in broth at 37°C. Samples are removed at intervals after mixing and highly diluted to prevent the formation of further unions.

Black curves: Every precaution is taken to prevent separation of mating pairs during dilution and plating, and to obtain the maximum yield of recombinants or infectious centres (see Jacob and Wollman, 1961a, p. 153).

Green curves: The samples are agitated in a high-speed mixer before diluting and plating.

The dotted curves refer to production of infectious centres resulting from zygotic induction.

The continuous curves indicate recombinant production.

For explanation, see text. (Based on data from Jacob and Wollman, 1961a.)

idealised outcome of an *HfrH* $\times$ F^- cross, obtained by selecting for recombinants inheriting three different, wild type markers, thr^+leu^+, gal^+ and try^+, located at increasing distances from the leading extremity, *O*, of the donor chromosome. Observe that each curve starts from the origin, so that unions must follow rapidly on collisions. The curves then rise to reach a plateau about 30 to 40 minutes later, but this plateau is different for each marker, being lower the further the donor marker is located from *O*. If recombinants inheriting only a *single* marker (say, gal^+) are scored in experiments of this sort, the slope and plateau of the curve appear as a simple function of the rate at which collisions are converted into unions and vary markedly with pH and population density (Hayes, 1957a); under standard conditions about 15 per cent of unions are effected in five minutes. The variation between the curves found for different markers is because what is observed is the final outcome of conjugation, far removed from union formation; we are looking at the unions through the double prism of transfer and recombination. The curves are therefore more complicated than at first appeared and give information about every stage of conjugation, although in each case the same information about union formation may be derived from them. The important point to note here is that the plateau level of recombinants achieved for each selective marker is different, but constant for a given *Hfr* type. Thus the number of recombinants inheriting the various donor markers can be arranged in a *gradient of transmission* of these markers to recombinants. This gradient coincides with the order of arrangement of the markers on the chromosome and can therefore be used in mapping. We will return to this point in the next section.

We now come to *chromosome transfer* which heralds the actual formation of zygotes and, as we shall see, determines their ultimate structure. The earliest approach to the kinetics of zygote formation was to remove samples of a mating mixture at intervals as before but, instead of diluting and plating them directly, they were first treated with a high multiplicity of a virulent phage (T6) to which the donor strain was sensitive and the recipient resistant. When recombinants inheriting the markers thr^+leu^+ from strain *HfrH* were selected, instead of arising from the time of mixing, their appearance was delayed for about eight minutes (Hayes, 1955; 1957a).

However, the real 'break-through' in our understanding of the nature of chromosomal transfer came from the kinetic studies of Wollman and Jacob (1955, 1958; Wollman *et al.*, 1956) who employed a high-speed mixer (Waring 'blendor') to separate the mating pairs and followed, as a function of time, the inheritance among recombinants of *each of a series of donor markers* whose locations on the chromosome were known. Each marker begins to enter the zygotes at a different and specific time after the commencement of mating, which is highly reproducible under standard conditions; each curve then rises to reach the same plateau as that achieved in uninterrupted matings (Fig. 130, continuous green curves). The various markers can therefore be arranged in a sequence according to their times of entry into the recipients, and this sequence is found to correspond to their order of arrangement on the chromosome. From this it follows that the chromosome of an *Hfr* donor must be transferred linearly from a particular extremity, so that *the distance between markers can be translated into a time scale which is a function of the transfer process alone and is independent of subsequent recombination events.*

Similar findings are obtained in interrupted mating experiments when recombinants for the most proximal donor marker are selected, and these are then scored for their percentage inheritance of various unselected markers as a function of time. Fig. 131 gives the result of one such experiment (Wollman and Jacob, 1955). It will be seen that the unselected donor markers, *azi*, *T*1, *lac* and *gal*, begin to appear among thr^+leu^+ recombinants in their map order on the chromosome and, at their plateaux, reach the proportions found among thr^+leu^+ recombinants derived from normal crosses; the times of first entry of these markers are given by the points at which extension of each curve intersect the time axis.*

When interrupted and uninterrupted mating experiments are performed on $Hfr(ly^+) \times F^-(ly^-)$ crosses, in which the frequency of zygotic induction (infectious centres) is scored instead of the frequency of recombinants, exactly similar results are obtained. In the case of phage λ, the uninterrupted mating yields a curve which starts from the origin and reaches a plateau at the same time as the recombinant curves for other markers (Fig. 130; black, dotted curve). Similarly the

* The astute reader may notice that the final proportion of $thr^+leu^+gal^+$ to total thr^+leu^+ recombinants in this early experiment (25 per cent) is much lower than that shown in Fig. 130 (about 66 per cent). This decrement, which also applies to the frequency of thr^+leu^+ recombinants, is probably due to the fact, not realised at the time, that immediate plating of zygotes on minimal medium interferes with the recombination process and may lead to differential loss of recombinants (Jacob and Wollman, 1961a, p. 153; J. D. Gross, personal communication).

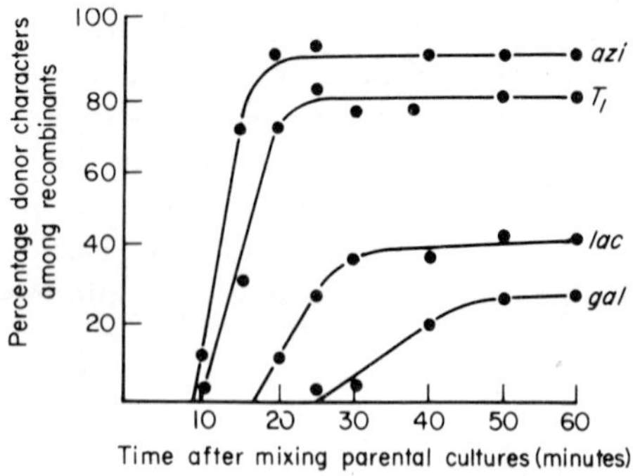

FIG. 131. The kinetics of inheritance of various unselected donor (*HfrH*) markers among recombinants selected for *thr*$^+$*leu*$^+$ inheritance, as a function of time of interruption of mating pairs after the commencement of mating.

The order of markers on the *HfrH* chromosome is *O—thr*$^+$*leu*$^+$*— azi—T*1*—lac—gal.*

thr$^+$*leu*$^+$*:* ability to synthesise threonine and leucine.
azi: sensitivity or resistance to sodium azide.
*T*1*:* senstivity or resistance to phage *T*1.
lac, gal: ability or inability to ferment lactose, galactose.

See text. Data from Jacob and Wollman, 1961a.

interrupted mating curve starts at 26 minutes, just after the curve for *gal*$^+$ recombinants (Fig. 130, green, dotted curve). This emphasises the fact that the effect of treatment in a high-speed mixer is indeed to interrupt chromosome *transfer* and not to interfere with post-zygotic events such as pairing, since *gal* and prophage λ, although closely linked on the chromosome, are markers which express themselves in quite different ways—the prophage by entering the lytic cycle and the *gal* marker through recombination.

Before considering the impact of these results on our ideas about the genetic constitution of zygotes, which is the key to interpreting the results of genetic analysis, there remains one feature of the kinetics of zygote formation to be discussed. What is the meaning of the *slopes* of the curves shown in Fig. 130? Observe that *all* the curves take about the same time to reach their plateaux (30 to 40 minutes) irrespective of the marker or of whether the matings are interrupted or not. If we suppose that the earliest points on the curves are contributed by the first mating pairs to form effective unions and to begin to transfer their chromosomes, while the plateaux indicate that the maximum possible number of unions has been achieved, then the slope of the curve with respect to any given marker should represent the rate at which unions are formed throughout the population.

However, the matter turns out to be more complicated than this. If a mating mixture is diluted a thousand-fold into broth after five minutes contact, the only unions which can subsequently yield recombinants are formed within five minutes of each other. It follows that if chromosome transfer always begins immediately, or at some constant time, after union formation, any particular marker should be transferred by all the conjugating *Hfr* donors with a scatter of not more than five minutes; an interrupted mating experiment on the diluted mixture should yield curves which reach their plateaux within five minutes. In fact, experiment shows that the curves rise to their plateaux over about a 20-minute period. There is therefore a considerable degree of asynchrony among *Hfr* bacteria that have united with recipients, with respect to the *initiation* of chromosome transfer (*transfer delay*; de Haan and Gross, 1962). It is not known whether this variable lag involves the synthesis of the connecting tube between the mating cells, or the actual mobilisation of the chromosome, but it is certainly an important component of the slope of curves expressing the kinetics of union formation and chromosome transfer. It is interesting that chromosome transfer, once initiated, is unaffected by the presence of either chloramphenicol or streptomycin to which the donor bacteria are sensitive.

THE GENETIC CONSTITUTION OF ZYGOTES

We have seen that, whatever the experimental approach, the various markers of an *Hfr* parent show a *gradient of transmission* to recombinants; the further the marker from the leading extremity of the donor chromosome, the lower is the frequency of recombinants inheriting it. The study of zygotic induction has also shown that there is a *gradient of transfer* of markers due to spontaneous, random breakage of the chromosome during its transit from donor to recipient cells. The gradient of transmission could therefore be a true reflection of the gradient of transfer or, alternatively, could be due, at least in part, to a gradient of integration whereby those markers entering the zygotes late have a lower probability of participating in recombination than have early markers.

Once again, the phenomenon of zygotic induction provides a way of deciding between these two possibilities. The frequency of zygotic induction for a given prophage indicates the frequency with which the prophage location on the donor chromosome is transferred to zygotes and, therefore, the frequency of transfer of other markers which are closely linked to it (p. 662). Thus Fig. 130 shows that the frequency of zygotic induction of prophage λ is about 70 per cent, so that the proximal and closely linked *gal* marker may be assumed also to be transferred to zygotes at this frequency. But in the equivalent $HfrH(ly^-) \times F^-(ly^-)$ cross the donor *gal* marker is transmitted to recombinants with a maximum frequency of only about 30 per cent. It follows that a donor *gal* marker which has been transferred to a zygote has a mean probability approaching 0·5 of being integrated into a recombinant chromosome. This *coefficient of integration* can also be measured for other markers which lie distally and at some distance from *gal* on the chromosome of strain *HfrH*. For example, the locations of the inducible prophages 21 and 424 are closely linked to the tryptophan (*try*) and histidine (*his*) regions respectively (Fig. 127, p. 666). In both cases it turns out that the ratio of the frequency of zygotic induction to that of recombinants inheriting the adjacent marker is about 0·5. This means that any donor marker has approximately the same probability of being incorporated into a recombinant once it has entered the zygote (Jacob and Wollman, 1961a, p. 152).

The progressive exclusion of donor markers from recombinants as their distance from the leading chromosomal extremity increases is therefore entirely prezygotic—they are excluded from the zygotes by random breakage of the chromosome during transfer. *The merozygotes formed at conjugation thus comprise a population of recipient cells, complete with their cytoplasm, which are heterogeneous in having received varying and randomly distributed lengths of donor chromosome measured from its leading extremity, O.*

GENETIC ANALYSIS BY CONJUGATION

We now come to the problem of how this unique system of conjugation may be used to map the relative order of genes on the *E. coli* chromosome and to estimate the relative distances between them. Since the techniques and methodology of genetic analysis by conjugation have recently been fully reviewed elsewhere (Jacob and Wollman, 1961a; Hayes, Jacob and Wollman, 1963; Clowes and Hayes, 1968), we will describe here only the main concepts upon which these methods are based. The essential requirements for all methods of genetic analysis are an *Hfr* strain carrying a dominant gene which can be selected, as well as the series of unselected markers to be mapped, on the proximal half of its chromosome, and a recipient strain which is allelic at all these loci.

In order to map loci situated elsewhere on the chromosome, other *Hfr* strains, which transfer them at high frequency, are used. Thus by employing a series of *Hfr* strains, each transferring different but overlapping segments of chromosome at high frequency, all known loci can be related to one another and the locations of new mutations readily defined. In fact, more than 140 loci have now been mapped in this way and have been found to be arranged on a single linkage group, so that it is certain that *E. coli* possesses only a single chromosome. The relationships of some of these loci are given in Fig. 127 (p. 666).

MAPPING BY THE GRADIENT OF TRANSMISSION

From the point of view of genetic analysis this chapter reached its climax in the conclusion of the last section, that the number of zygotes, and therefore of recombinants, which inherit a given donor marker depends on the distance of that marker from *O*; the further the marker the smaller the number of recombinants. Donor and recipient cultures are mixed in broth for about an hour; a sample is then diluted and plated to select recombinants inheriting a proximal marker of the donor parent. The frequencies with which the various unselected markers appear among the resulting recombinants are inversely related to their distances from the selected marker (see plateaux, Figs. 130, 131, pp. 681, 684), *provided* they lie distal to it. If an unselected marker lies *between* the leading extremity of the donor chromosome and the selected marker, a quite different situation arises since all the zygotes which yield recombinants will have been homogeneous in having received *both* markers from the donor (see p. 685).

Since the genetic constitution of the zygote population depends exclusively on random chromosomal breakage during transfer, the proportions of the different recombinant classes obtained by this method have little or nothing to do with the frequency of *recombination* between markers but are a simple function of the probability of prezygotic exclusion. In view of the evidence that chromosomal breakage is random, however, the method is a valid one for estimating the distances between markers as well as their relative order, although this may not hold in the case of distal markers. However, despite the definitive role of chromosome breakage in determining the gradient of inheritance of donor markers, it should be remembered that this inheritance still depends on recombination, and there is one circumstance in which recombination plays a decisive role. It turns out that markers which lie less than about 4 to 5 time units (minutes) from the origin are inherited at a lower than expected frequency; the nearer their location to the origin the greater is the effect, so that the frequency of recombinants inheriting very early markers may be less than ten per cent of that expected from the transfer gradient alone (Low, 1965). For example, the *Hfr* strain B8 transfers sensitivity to phage T6(*T6-s*) as the first marker and ade^+ next (Fig. 127, p. 666); yet when ade^+ recombinants are selected in crosses with a *T6-r.*ade^- recipient, only about ten per cent also inherit the donor T6 allele (Broda and Hayes, unpublished data)*. The reason, of course, is that inheritance of the first marker requires a recombination event in the region between it and the sex factor, as well as a second event on the other side, so that when the first region becomes very short the probability of integration falls.

The gradient obtained for a particular set of loci is very dependent on the environmental conditions of the cross. For instance, the gradient becomes steeper if the samples of a mating mixture are immediately plated on minimal medium (footnote, p. 683), or if the mixture is aerated too violently so that some of the mating pairs are prematurely pulled apart. Conversely, if the cross is made by plating the parental strains together on the surface of a solid medium, the probability of chromosome breakage is reduced and distal markers may appear among recombinants with significant frequency. The outcome of a cross also depends on the inherent physiological properties of the donor strain. For example, *Hfr* strains have been isolated which appear to transfer their terminal markers at ten to 20 times the usual frequency, perhaps due to the abnormal stability or toughness of the unions they form (Taylor and Adelberg, 1960; Clark and Adelberg, 1962). Alternatively, the low frequency with which *normal Hfr* strains transfer their terminal markers has been ascribed to the usual employment of streptomycin to prevent their multiplication (see 'contra-selective markers', below); when growth of such *Hfr* strains is suppressed by a nutritional deficiency, similarly elevated frequencies of terminal markers is observed (Alfoldi, Stent and Clowes, 1962). Again, there is evidence that *Hfr* strains may vary in the rate at which their unions with recipients suffer spontaneous metabolic breakdown while this, in

* This is the reason why the relative positions of these two markers were incorrectly transposed on the map of the *E. coli* chromosome in the first edition of this book.

turn, depends on the richness of the mating medium; when early dissolution of unions occurs, the separating *Hfr* bacteria may withdraw the transferred parts of their chromosomes if these have not already been stranded in the recipient by spontaneous rupture (de Haan and Gross, 1962; Symonds, 1962b).

A further important point about crosses in general should be stressed. The essence of genetic analysis in bacteria is to select for the inheritance of donor markers among recombinants. In conjugation, where large numbers of viable donor bacteria are plated, the selective medium must obviously be one which prevents these from growing. It follows that the donor parent must carry an auxotrophic or sensitivity marker which is *contra-selected.* If such a contra-selective marker is situated near one or more of the unselected markers to be mapped, a proportion of recombinants will inherit it and be lost, so that the apparent gradient of transmission will be exaggerated or distorted. Contra-selective markers should therefore be located as distally as possible on the *Hfr* chromosome. Finally, it should not be forgotten that the occurrence of zygotic induction may virtually eliminate recombinants for donor markers which are closely linked to the site of attachment of the prophage (Wollman and Jacob, 1954; 1957).

MAPPING IN TIME UNITS

We have seen that the sequence of genes on the proximal part of the donor chromosome can be mapped in relation to the leading extremity, by noting the specific time at which each begins to appear in recombinants in interrupted mating experiments. Since the time interval between the initial appearance of each of a pair of markers represents the time taken for the transfer of the corresponding chromosomal interval, we can translate the distance between markers into a time scale which is independent of any distortion which might be introduced by the vagaries of chromosome breakage or recombination. Such a time unit map would be equivalent to a cytological map in *Drosophila* (p. 64), but one which can be quickly and accurately constructed. The crucial question is, how is the time scale related to true chromosomal distance? Measurements in time units will only be valid if chromosome transfer proceeds at a uniform speed. Suppose that, when a particular *Hfr* strain is used, two donor markers, *A* and *B*, are found to be transferred at ten and 20 minutes respectively; that is, they are located ten minutes (time units) apart. If we now take another *Hfr* strain on whose chromosome these same two markers are situated much further from the leading extremity, what do we find? It turns out that if *A* is transferred by this strain at 25 minutes, then *B* will be transferred at 35 minutes, so that the markers are still estimated as ten time units apart. In this way it can be shown that at least the proximal half of the chromosome is transferred at a constant speed, although this tends to slow somewhat when the markers lie more distally. Thus the distances in time units between proximal markers indicates their true, relative distances apart. If the rate of progression were constant throughout, transfer of the whole chromosome would take about 90 minutes (Taylor and Thomas, 1964). We have seen, from direct autoradiographic measurement, that the *E. coli* chromosome consists of a DNA duplex of about 1,100μ long (Cairns, 1963a) so that, assuming that base-pairs are separated by 3·4Å, the chromosome contains some $3{\cdot}2 \times 10^6$ base-pairs. We can therefore express time unit measurements in physico-chemical terms; about $3{\cdot}5 \times 10^4$ base-pairs of DNA, corresponding to 30–50 genes, are transferred in one minute at 37°C.

Since chromosome transfer is a vital process performed by the donor, it is extremely sensitive to environmental conditions which influence energy production, of which temperature is, perhaps, the most important (Fisher, 1957b). Thus lowering the temperature from 37° to 32°C halves the speed of transfer so that the apparent distance in time units between two markers is doubled (Hayes, 1957a). The conditions under which interrupted mating experiments are done must therefore be rigidly controlled. The method is best suited for measuring the distance between markers which are more than a minute or so apart; at lesser distances its accuracy is restricted by the difficulty of precise manipulation, although resolution may be increased by lowering the temperature.

MAPPING BY RECOMBINATION

The unorthodoxy of genetic analysis by conjugation is evident from the fact that the only methods of measuring fairly long genetic distances are quite independent of the frequency of recombination events. To study true recombination frequencies a population of zygotes which are homogeneously diploid for a given length of chromosome must be obtained and their recombinant progeny analysed. This can only be achieved by selecting the most distal of a series of donor genes, so that all the zygotes which yield recombinants have received a length of donor chromosome extending from the leading locus to this gene.

The recombinants are then examined for inheritance of the more proximal, unselected markers which are located on this segment. It is found that proximal markers which are separated by more than about three time units behave as unlinked; they are inherited randomly with a frequency of approximately 50 per cent. This conforms with the results of zygotic induction experiments which show that such markers, once transferred, have an equal probability of appearing in recombinants (p. 686).

For markers separated by less than about two time units, mapping by the gradient of transmission or in time units becomes imprecise. On the other hand, it is just at this level that recombination analysis becomes a really effective tool which operates down to the intra-cistronic level. Various types of situation which are amenable to recombination analysis are presented in Fig. 132. To estimate the distance between two closely linked genes, *B* and *C* (Fig. 132a), the more distal *Hfr* wild type allele, *C*, is selected and the *C* recombinants are then scored, by replica plating, for inheritance of the recipient *b* allele, which requires a single crossover between them. The recombination frequency between *B* and *C*, indicating their distance apart, is given by the number of *bC* recombinants as a percentage of the total *C* recombinants ($bC/C \times 100$).

A different situation arises when three markers, *A*, *B* and *C*, are involved, since the recombination frequencies between them depend on the order of *A* and *B* relative to the selective marker *C*. For example, in Fig. 132b, if the order is *A—B—C* only two crossovers are needed to give an *ABC* recombinant; but if the order is *B—A—C* then four crossovers are required (Fig. 132c). The order must first be established by reciprocal crosses with respect to the *A* and *B* alleles (Fig. 38, p. 140). Once this has been done, the distance between *A* and *B* is given by the percentage ratio of *ABC* to total *C* recombinants (Fig. 132b). Of course the sensitivity, and therefore the resolution, of the system is greatly increased if *A* and *B*, as well as *C*, can be used as selective markers. If this is so then, in Fig. 132b, the total number of *C* recombinants is given by plating the zygotes on minimal medium supplemented with growth factors A and B, while very small numbers of *ABC* recombinants can be scored directly by plating on unsupplemented minimal medium. In this way the resolving power of the analysis becomes high enough to be applied to the measurement of intra-cistronic distances as shown in Fig. 132d. Assessing the order of closely linked markers by 3-point crosses, using conjugation, may prove

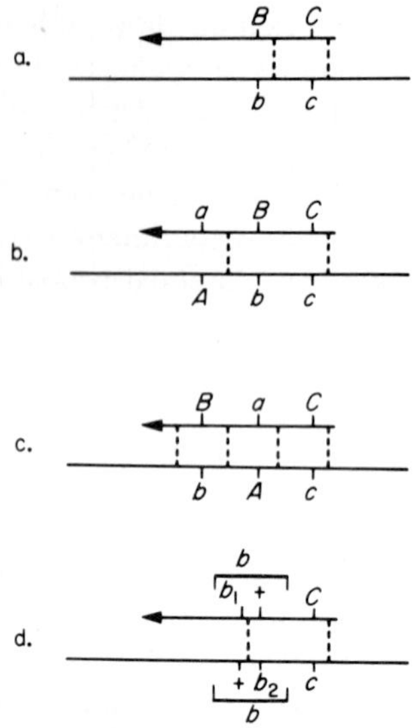

FIG. 132. Illustrating various genetic arrangements which are amenable to mapping by recombination.

The short upper line represents the donor chromosome, the arrow indicating its leading extremity.

The longer, lower line represents the recipient chromosome.

The vertical dotted lines show the regions where crossing-over must occur to yield the classes of recombinants referred to in the text.

See text for explanation.

unduly complicated since reversing the cross involves the synthesis of donor and recipient strains in which the *A* and *B* alleles of the two parents are switched. Within the relatively short regions where mapping by recombination is effective, markers are often co-transducible, so that 3-point crosses mediated by transduction may prove more feasible (Jacob and Monod, 1965); as we have seen, reversing a transductional cross is effected merely by growing the transducing phage on the previous recipient strain, which thus becomes the donor (see p. 141).

A good practical example of mapping and measurement in recombination units is an analysis by Jacob and Wollman (1961a, p. 228) of the region of the *E. coli* chromosome controlling lactose fermentation, illustrated in Fig. 133. This region comprises three loci, *y*, *z* and *i*. The *y* and *z* loci determine synthesis of the enzymes galactoside permease and β-galactosidase respectively, while *i* controls the inducibility of these enzymes by the production of a specific repressor.

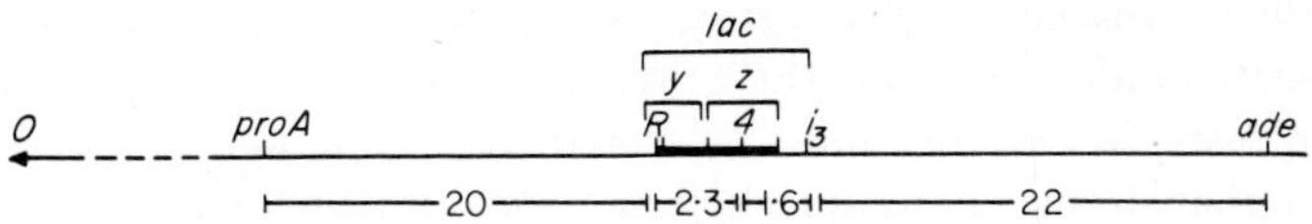

FIG. 133. Map of the *lac* region and contiguous segments of the *E. coli* chromosome, inferred from recombination analysis.

The *lac* region, concerned with lactose fermentation, is located centrally on the figure; it includes the adjacent *y* (galactoside permease) and *z* (β-galactosidase) loci, as well as the *i* locus which controls the inducibility of the enzymes. y_R, z_4 and i_3 are three mutational sites within these respective loci.

The loci *pro* and *ade* determine proline and adenine synthesis respectively.

The arrow indicates the direction of chromosome transfer by the donor strain (*HfrH*; see Fig. 127, p. 666).

The numerals show the recombination frequences obtained between the various sites.

The map is not drawn to scale.

For explanation, see text. (Data from Jacob and Wollman, 1961a).

Interrupted mating experiments show that strain *HfrH* transfers its markers in the sequence *proA—lac—ade*. The adenine (ade^+) marker of the donor was therefore used to select recombinants, and the order *y—z—i* established for the *lac* loci by means of reciprocal crosses, as described above. The distance between mutations in these three loci were: y_R—z_4, 2·3 recombination units; z_4—i_3, 1·6 units. On the other hand, the distance *pro—y* turned out to be about 20 units, and *i—ade* 22 units. The whole *pro—ade* region is therefore about 45 recombination units long. In crosses between different *z* mutants, where lac^+ade^+ recombinants can be directly selected, the smallest recombination frequency measured is 5×10^{-6}.

A very sensitive method of determining the order of mutations can be used when, as in the case of *z* mutations, wild type recombinants may be selected and scored for the segregation of an outside marker. Because of negative interference (p. 379), however, this outside marker must be very closely linked to yield clearcut results. For example, in the cross

$$\frac{z_1 \quad + \quad +}{+ \quad z_2 \quad i}$$

where *i* is the outside marker, the great majority of the z^+ recombinants are also i^+, showing that the order must be z_1—z_2—*i*. If *ade* is used as outside marker in this cross, however, its alleles are found to segregate randomly among the z^+ recombinants. Finally, complementation tests, in heterogenotes mediated by sexduction (p. 103), have shown that all the *y* mutations belong to one functional unit and all the *z* mutations to another. We will consider functional analysis of the *lac* region in detail in the next chapter (p. 712).

A factor of some practical importance in conjugal crosses is that more than one recombinant type may emerge from the growth of a single zygote. This was first revealed by pedigree analysis of the progeny of zygotes isolated by micromanipulation over many generations, which showed not only that recombinants might segregate at any generation from the third to the ninth, but that different recombinant types might issue from the different generations of a single line of descent (Anderson, 1958). In a comparable pedigree analysis using different strains, however, Lederberg (1957) found that the great majority of zygotes produced only a single recombinant at an early generation. It now turns out that this difference in behaviour depends on the particular *Hfr* strain used. Thus an analysis of the distribution of recombinant classes in individual primary clones shows that in the case of strain *HfrH*, used by Anderson, only half the clones contain a single recombinant class, as compared with over 90 per cent from crosses employing strain *HfrC* (used by Lederberg) as donor (Wood, 1967).

The fact that a high proportion of zygotes may, under certain conditions, yield several different recombinant types involving the same segment of donor chromosome, suggests that this segment replicates and segregates with the recipient chromosome to produce more than one heterogenote. Since the majority of chromosome fragments are probably incapable of self-replication, this suggests that the donor fragment becomes inserted into the recipient chromosome by a recombination event, as in the case of λ*dg* and *F*-prime factors, so that the heterogenote takes the form of a tandem duplication. Pairing and recombination between different parts of the duplicated region could thus account for the formation of different haploid recombinants, along the lines suggested in Fig. 125 (p. 635; see also p. 698).

GENETICAL AND PHYSICAL CORRELATIONS

We are now in a position to correlate the various measurements of distance which analysis by conjugation offers (Jacob and Wollman, 1958b; 1961a, p. 234). We have seen that the distance between *proA* and *ade* on the *E. coli* chromosome is equivalent to about 45 per cent recombination (Fig. 133), so that these markers appear as only weakly linked in recombination analysis. In interrupted mating experiments, however, transfer of this same interval takes about two minutes, so that one time unit is equivalent to 20 recombination units. Since the total length of the chromosome is about 90 time units, the total *map* length works out at about 1800. Again, one time unit corresponds to $3{\cdot}5 \times 10^4$ nucleotide pairs of DNA, so that there are approximately 1750 nucleotide pairs per map unit—a figure which is intermediate between similar estimates for bacteriophage T4 on the one hand and *Aspergillus* on the other (Pontecorvo, 1958; see Table 16, p. 382). These correlations are summarised in Table 24.

We have restricted this account of conjugation to that found in *E. coli* K-12, because this is the only system which has been investigated in sufficient detail to permit a rather precise model of its physiological and genetic mechanism to be built. However, conjugation is now known to occur not only in other strains of *E. coli* but in many other species,

TABLE 24

Correlation of physical and genetic measurements of the chromosome of *E. coli*.

1 time unit	= 20 recombination units
	= $3{\cdot}5 \times 10^4$ nucleotide pairs of DNA duplex
	= c. 12μ length of DNA duplex.
The whole chromosome	= 90 time units.
Map length	= 1800.
One recombination unit	= 1750 nucleotide pairs.

The correlations are modified from Jacob and Wollman (1961a) and are, of course, only approximate.

and even between the bacteria of different species. Although our knowledge of these other systems is at present limited, and despite the differences in detail that exist between them, they appear to have one feature in common; in all, the ability to conjugate is determined by the presence in one of the parental strains of an autonomous or semi-autonomous entity analogous to the sex factor of *E. coli* K-12. We will examine some of these conjugal systems, and speculate about the properties and functions of sex factors in general, in Chapter 24 (p. 748).

For readers who wish to study conjugation in *E. coli* in more detail, the lucid and complete account by Jacob and Wollman (1961a) remains the classical exposition of the subject. The methodology of conjugation is reviewed by Hayes *et al.* (1963). Reviews of a broader and more critical nature, embracing conjugal systems and sex factors other than that of *E. coli* K-12, are those by Clark and Adelberg (1961) and by Gross (1964). The story of the growth of our current ideas about sexual differentiation and conjugation in *E. coli* is told by Hayes (1966a) and Wollman (1966).

GENETIC RECOMBINATION IN STREPTOMYCES

The genus *Streptomyces* is important for two reasons. One is that many valuable antibiotics, such as streptomycin, chloramphenicol and tetracycline, are produced by *Streptomyces* species, so that systems of genetic analysis in the genus have been looked for on account of their practical value (see, for example, Bradley and Lederberg, 1956; Alikhanian and Mindlin, 1957; Alikhanian, Iljina and Lomovskaya, 1960; Alikhanian and Borisova, 1961). Secondly, *Streptomyces*, although classified as a bacterial genus on account of cellular size and cytology, resembles the fungi in growing as multinucleate and branching hyphae which form a mycelial mass. *Streptomyces* therefore shows a complexity of morphological organisation which is unique among the bacteria, so that a comparison of its genetic structure and behaviour with those of a phylogenetically remote bacterial species such as *E. coli* is of considerable theoretical interest (see Hopwood, 1967).

The only representative of the genus whose genetics has been extensively studied is a particular strain of *Streptomyces coelicolor*, so called because it produces a pigment which is blue at alkaline pH. This organism grows in a minimal synthetic medium with glucose as carbon source and an inorganic source of nitrogen, so that auxotrophic mutants for use in genetic analysis are readily obtainable. On solid medium colonies arise by the outward and downward growth of the branching mycelium to reach a diameter of 2—3 mm. after a few days at 28—30°C. During this growth, certain of the hyphal branches grow upwards to form aerial hyphae which are finally converted into chains of spores by the development of septa—a process associated with a change of the colonies from a slightly shiny to a powdery appearance. The spores are about 1–1·5 μ in diameter, contain a single chromatin body and, on plating, germinate by extruding one or more tubes which grow longitudinally and branch, to become the vegetative mycelium.

The mycelial hyphae are divided by septa into compartments containing several chromatin bodies. The complete cycle, comprising spore germination, mycelial growth, development of aerial hyphae and sporulation, takes 2–3 days at 30°C.

From a cytological point of view, *S. coelicolor* is identical with bacteria in many respects. The maximum diameter of the vegetative hyphae is about 1·0μ. The chromatin bodies stain in the same way with basic dyes throughout the division cycle and there is no evidence of mitosis. In electron-microphotographs of ultra-thin sections the chromatin bodies are less electron-dense than the cytoplasm, show a finely fibrillar appearance and lack a nuclear membrane (p. 552), while the hyphal walls appear identical to those of Gram positive bacteria and, in fact, have a similar chemical composition (Hopwood and Sermonti, 1962; Sermonti and Hopwood, 1964).

Crosses are made by seeding a complete agar medium with a mixed inoculum of spores or hyphal fragments of two genetically marked parental strains. After several days growth the spores are harvested in water and spread on plates of selective medium on which neither parent can grow. Usually, of course, auxotrophic mutants are used and the spores plated on minimal agar. Recombinant colonies arise at a frequency which may vary between 10^{-2} and 10^{-7} of the total spores plated. Although inter-strain fertility may vary markedly, it seems that no completely sterile combinations have been found in pairwise crosses among hundreds of strains derived from the original wild type, unlike the case of *E. coli* K12 strains (Hopwood, 1967; but see below). It is not known at what stage of growth genetic transfer occurs, nor is there direct evidence for the occurrence of hyphal anastamoses although this may be assumed from the fact that heterokaryons do arise in *S. coelicolor*. Since, as we shall see, recombination involves the transfer of a considerable proportion of the genome, it is almost certainly mediated by some form of conjugation.

When recombinant colonies are analysed they are found to be of two kinds which behave quite differently. One is a *haploid colony* which has grown from a haploid, recombinant spore and is genetically homogeneous. The other is termed a *heteroclone*. Heteroclones arise from incomplete heterozygotes, which may be hyphal segments or spores, and behave like bacterial heterogenotes (pp. 99, 631), segregating not only a variety of haploid recombinants but also further heteroclones. Heteroclones are therefore genetically heterogeneous. Although they may be maintained indefinitely by selection and subculture, segregation is frequent so that only a small proportion of the spores of heteroclone cultures yield heteroclones. Heteroclone colonies can be distinguished from haploid colonies by their small size and by the fact that their spores characteristically fail to develop on replica plating to the selective medium; growth of the heteroclones on this medium depends on complementation between the selected wild type loci which normally segregate in the haploid spores.

Not only the genetic constitution of heteroclones, but also analysis of the haploid recombinant colonies clearly indicate that, as in the case of conjugation in *E. coli*, only part of the genome of one of the parents is transferred to the zygote. However the situation resembles the case of crosses involving F^+ rather than *Hfr* males, since the transferred fragments, which on the average amount to about one sixth of the genome, appear to be random segments. Moreover, unlike the *E. coli* case, there is no evidence that the fragments emanate from a particular one of the two parents.

Early genetic analysis of *S. coelicolor* seemed to indicate the existence of two linkage groups (Sermonti and Hopwood, 1964), but a more extended analysis has recently shown that, in fact, all the markers map on a single, circular linkage group (Hopwood, 1965b). The random genetic constitution of the transferred chromosome segments present in heteroclones precludes the possibility of fixed chromosomal extremities, but it is not yet known whether the chromosome is a physically circular structure, or a linear one with ends which are circularly permuted among the population, as in the case of phage T4 for example (pp. 433, 537; Hopwood, 1966a).

Hopwood (1967) proposes a model for the development of haploids and heteroclones which is equally applicable to recombination in *E. coli* (p. 694). On this model, the occurrence of two recombination events between the whole chromosome and the transferred segment will substitute part of the segment for its homologue and yield a haploid, recombinant chromosome. However, if only *one* recombination event occurs between the segment and a complete circular chromosome, the result will be a linear chromosome which is terminally redundant. Assuming that this structure can replicate either in linear form or after closure to form a circle, a heteroclone will develop in which secondary recombination events will generate haploid recombinants. From the point of view of practical mapping, analysis of haploid colonies from a cross yields most information about the linkage map as a whole, due to the heterogeneity of the segments involved. On the other hand, analysis of individual heteroclones not only provides complementation data but, because all the recombinant progeny arise from a single merozygote, permits mapping in standard recombination units (Hopwood, 1967). In fact, analysis of the progeny of *S. coelicolor* heteroclones is equivalent to that of an *E. coli* cross in which a distal marker is selected and the frequency of proximal unselected markers scored (p. 690).

A peculiar feature of the circular linkage map is that two 'silent' quadrants, containing no known genes, separate what were formerly thought to be the two separate linkage groups. The existence of these quadrants was confirmed by the isolation of a considerable number of temperature-sensitive mutants and the finding that their mutational sites, which would be expected to be randomly distributed, in fact mapp on the two original linkage groups (Hopwood, 1966b). In addition to this, although the linkage map shows examples of the partial clustering of functionally related genes, there is a striking tendency for many related genes to be located diametrically opposite to one another on the circular map. A possible model to explain this circular symmetry postulates the formation of a tandem duplication as a result of recombination between two circular, homologous chromosomes, followed by loss, through mutation, of one of each pair of redundant genes from one or the other chromosome (Hopwood, 1967).

We have seen that the genetic behaviour of *S. coelicolor* resembles that of *E. coli* in several important respects such as possession of a single, circular (or circularly permuted) chromosome, the clustering of functionally related genes on the chromosome, and the formation of partially diploid zygotes. Some years ago this similarity was extended by a report that two groups of *S. coelicolor* strains could be recognised —those which were sterile, or of very low fertility, when crossed together (R^-), and those (R^+) which were fertile with each other as well as with R^- strains. In addition, there was some evidence that the R^+ character was transmitted in crosses to R^- strains at a much higher frequency than chromosomal markers, so that the situation appeared very similar to that of $F^+ \times F^-$ *E. coli* crosses (Sermonti and Casciano, 1953; Sermonti and Hopwood, 1964). However, these early reports have not yet been confirmed so far as the complete sterility of inter-strain crosses and the transmissibility of fertility are concerned (Hopwood, 1967). Extensive reviews of both the theory and method of genetic analysis of *S. coelicolor* are by Hopwood and Sermonti (1962), Sermonti and Hopwood (1964) and Hopwood (1967).

CHAPTER 23

Genetic Expression and its Control

The conjugation system of *E. coli* possesses several features which make it a tool of unique refinement for analysing both qualitative and quantitative aspects of gene function. By a judicious choice of *Hfr* strains it is possible to introduce any particular wild type gene into the majority of a population of mutant bacteria under precisely controlled conditions, and to follow the ensuing phenotypic alteration from the very moment of penetration. Thus samples of a population of merozygotes (or their progeny) can be removed at intervals after the commencement of transfer, and the time (or number of generations) required by those which have received the gene to express the donor character studied. Alternatively, if the donor gene specifies an enzyme which lends itself to easy quantitative estimation, the kinetics of enzyme production can be accurately related to the time at which the gene enters the recipient cytoplasm.

The one-way nature of genetic transfer in conjugation, and the fact that *Hfr* donors appear to transfer only their chromosomes to the zygotes (see Fisher, 1962; Silver, 1963) while the recipients contribute their cytoplasm as well, permits analysis of nucleo-cytoplasmic interactions which can be investigated in higher organisms only by means of nuclear transplantations. Finally, the ability to construct relatively stable diploid strains (heterogenotes) by means of sexduction (p. 103) offers a reliable method of performing complementation tests and of assessing the dominance or recessiveness of alleles. In this chapter we will illustrate these various attributes of conjugation by reviewing the novel and practical contributions they have made to our knowledge of the functional behaviour of genes.

THE KINETICS OF EXPRESSION OF GENE ACTIVITY

The first attempt to analyse the kinetics of phenotypic expression after conjugation involved donor genes determining such complex characters as resistance to valine, sodium azide and phage T1, all of which happen to be closely linked to a proximal nutritional marker (*thr*$^+$*leu*$^+$) on the chromosome of the donor strain *HfrH* (Hayes, 1957a, b; see Fig. 127, p. 666). The resistant donor strain is mated in broth with a sensitive, auxotrophic (*thr.leu*) recipient. When it is judged that an appreciable number of zygotes have received both the nutritional selective marker and the resistance marker under study, the donor population is killed by treatment with a virulent phage (T6) to which it alone is sensitive, and the zygotes are diluted into fresh broth. Samples are then removed at intervals and plated on minimal medium on which all the prototrophic (*thr*$^+$*leu*$^+$) recombinants will yield colonies, as well as on minimal medium containing the antibacterial agent (valine, sodium azide or phage T1) on which only those zygotes or recombinants which have inherited the donor determinant of resistance, *and in which this determinant has expressed itself*, can grow.

The results obtained in this kind of experiment are shown in Fig. 134, in which curve 1 represents the total number of prototrophic (*thr*$^+$*leu*$^+$) recombinants obtained as a function of time after zygote formation; the curve remains constant for about 100 minutes and then begins to rise exponentially, thus indicating the time at which the recombinant bacteria have segregated and begin to multiply. The constancy of the curve prior to segregation reveals that the selected nutritional markers must begin to express themselves very shortly after entering the zygotes and must therefore be dominant to their inactive, recipient alleles, for otherwise the zygotes would be unable to support the process of recombination and yield recombinants. When the abilities of the donor genes determining resistance to valine, sodium azide and phage T1 to express themselves as a function of time are compared, three quite distinct types of curve are found. That for valine resistance (curve 2) is virtually identical with curve 1, so that the gene *val-r* is dominant and rapidly expressed. In the case of resistance to sodium azide (curve 3), however, the expression of the gene is much slower; there are scarcely any resistant zygotes at the time of dilution, but thereafter their number rises exponentially, presumably as a result of some synthetic activity, so that resistance is expressed in all those that have inherited the gene just prior to the first division of recombinants. The kinetics show that resistance is expressed in the zygotes and must therefore be dominant to sensitivity, but the reason for the delay in its expression is not known. In contrast, all the zygotes which receive the determinant of

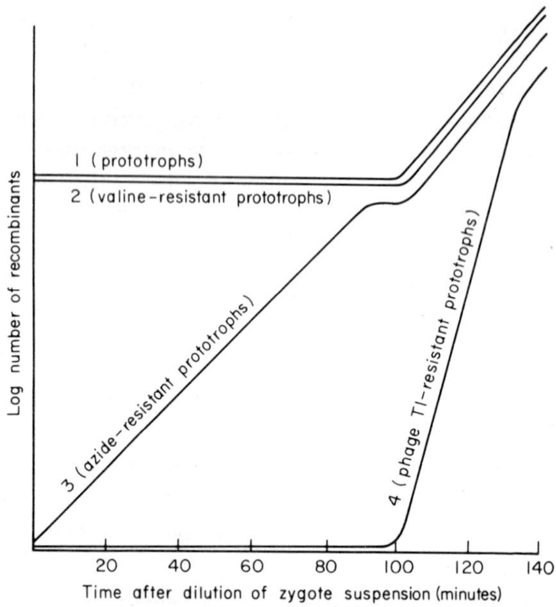

FIG. 134. Curves illustrating the kinetics of expression of three genes determining resistance to various antibacterial agents, following their transfer to sensitive recipient bacteria (zygotes).

Donor strain: HfrH carrying the proximal, wild type nutritional markers *thr*$^+$*leu*$^+$, and the closely linked determinants of resistance to valine (*val-r*), sodium azide (*azi-r*) and phage T1(*T1-r*) (see Fig. 127, p. 666); sensitive to phage T6 and to streptomycin.

Recipient strain: auxotrophic (*thr.leu*) and sensitive to valine, sodium azide and phage T1; resistant to phage T6 and streptomycin.

Broth cultures of donor and recipient are mixed at 37°C. for a short time to allow zygotes to form. The donor bacteria are then killed by addition of phage T6 at high multiplicity, and the resulting zygote suspension diluted into fresh broth at 37°C. Samples are removed at intervals and plated on streptomycin-minimal agar to select for prototrophic (*thr*$^+$*leu*$^+$) recombinants (curve 1); on streptomycin-minimal agar + valine (curve 2); on streptomycin-minimal agar + sodium azide (curve 3); and on streptomycin-minimal agar spread with a high concentration of washed phage T1 (curve 4).

On the plates containing an antibacterial agent, zygotes or recombinants can grow to yield colonies only if the donor resistance gene has been inherited *and expressed* at the time of plating. (See text.)

After Hayes, 1957a, b.

resistance to phage T1 remain sensitive to the phage; the expression of resistance commences only when the segregated recombinants start to multiply, and is not achieved in all of them until several generations later. Resistance to phage T1 is therefore recessive to sensitivity (see also Lederberg, 1949), while the delayed expression in the haploid segregants almost certainly reflects the time required for the synthesis of new cell walls in which all the attachment sites for the phage have been substituted by new molecular groupings (Hayes, 1957a; cp. p. 203).

What is observed in these experiments is the ability or inability of the bacteria to manifest certain characters whose mechanism of expression at the biochemical level is probably complex and, in any case, is not yet understood, so that the experiments contribute little to our knowledge of gene action. From the point of view of molecular genetics it would be more pertinent to examine directly the rate at which the genetic information carried by the gene is translated into protein structure, since this is very relevant to the mechanism of the translation process itself in which we are primarily interested. Instead of looking at the phenotype of recombinant clones which ultimately emerge from the zygotes, the rate of enzyme synthesis by the zygotes themselves is assayed as a function of the kinetics of transfer of the gene determining the synthesis. An ideal enzyme for this purpose is β-galactosidase whose structure is specified by the *z* gene of the *lac* region of *E. coli* (Fig. 133, p. 693) and for which a simple and sensitive assay procedure exists.

As we shall see in the next section, wild type, z^+ strains of *E. coli* are *inducible* and do not synthesise significant amounts of β-galactosidase unless the substrate (lactose) or some specific galactoside *inducer* is added to the culture; similarly, strains which have suffered a mutation in the *z* gene (z^-) fail to produce active enzyme whether inducer is present or not. Cultures of an inducible, streptomycin-sensitive, z^+ *Hfr* strain, which transfers the z^+ gene early, and of a streptomycin-resistant z^- recipient are mated, and samples of the mixture withdrawn at intervals. A moiety of each sample is agitated in a mixer and plated for lactose-fermenting recombinants in order to follow the kinetics of transfer of the z^+ gene to the zygotes (interrupted mating experiment, p. 683); the other moiety is treated with streptomycin which destroys the inducibility of the z^+ donor, and is then assayed for β-galactosidase after the addition of inducer. Since neither donor nor recipient can synthesise the enzyme under these conditions, all the β-galactosidase measured comes from those streptomycin-resistant recipients which have received the z^+ gene from the donor. The results of an experiment of this type are shown in Fig. 135. It turns out that appreciable amounts of β-galactosidase begin to be formed within a minute or so of the first entry of the z^+ gene into the zygotes. As conjugation proceeds, the enzyme accumulates at a rate

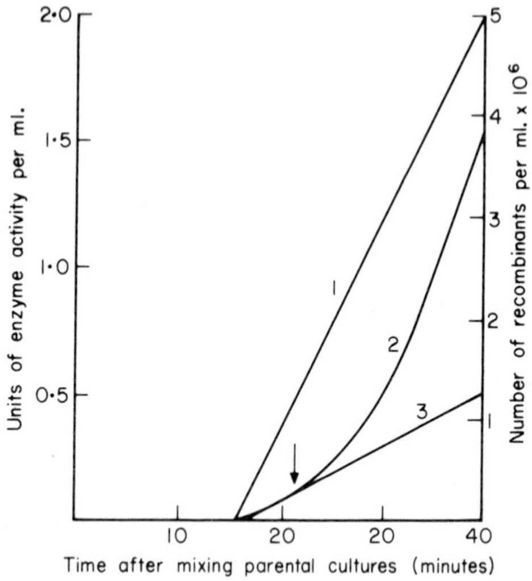

FIG. 135. Comparative kinetics of gene transfer and enzyme synthesis in *E. coli* merozygotes.

Cultures of an inducible *Hfr* strain, which begins to transfer the z^+ gene (determining β-galactosidase synthesis) at 15 minutes, and of a z^- recipient are mixed at time zero. Samples of the mixture are removed at intervals and examined as follows:

Curve 1 shows the kinetics of transfer of the z^+ gene to the zygotes, as determined by interrupting the mating and plating for z^+ recombinants.

Curve 2 follows the production of β-galactosidase in the mixture, under conditions where the enzyme can be synthesised only in merozygotes which have received the z^+ gene (see text, and original source, for further details).

Curve 3 follows the synthesis of β-galactosidase in sub-samples of a sample in which mating has been interrupted, and further mating prevented, at 5 minutes after the z^+ gene has begun to be transferred. The time of interruption is indicated by the arrow. (See text.)

After Jacob & Wollman, 1961a, p. 258; Riley, Pardee, Jacob & Monod, 1960.

which increases with time up to about 60 minutes, (Pardee Jacob & Monod, 1959). However, this overall rate of synthesis by the population as a whole reflects not only the rates of synthesis by the individual zygotes, but also the rate at which new zygotes inheriting the z^+ gene arise in the mixture (p. 685). This latter complication can be eliminated, and the true rate of synthesis in individual zygotes revealed, by following the production of enzyme in a sample in which mating has been interrupted very shortly after the initiation of z^+ transfer. Under these conditions the rate of β-galactosidase synthesis is found to remain constant and equal to the rate attained at the time of interruption (Fig. 135, curve 3). When this rate is compared with that of normal z^+ bacteria under the same conditions, and correction is made for the fact that normal, multinucleate bacteria possess several copies of the gene as against only one in each merozygote, it transpires that the rate of synthesis by the merozygotes is maximal from the start (Riley, Pardee, Jacob & Monod, 1960). This immediacy of gene action, coupled with the complementary demonstration that the rate of enzyme synthesis declines *pari passu* with destruction of the integrity of the gene by ^{32}P decay (Riley *et al.*, 1960), were the basic ingredients of the idea that an unstable 'messenger' intervenes between gene and ribosome in protein synthesis (see p. 274).

THE REGULATION OF GENE ACTIVITY

So far we have looked at how the genetic material and its ancillary apparatus specifies the synthesis of proteins much as we might investigate the operation of a motorcar factory, by examining the blueprints of the various components in relation to the machinery which makes them and assembles them into the final product. Granted the blueprints, raw materials and machines, however, the most important aspect of a factory, especially from an economic point of view in a highly competitive society, is how it is organised and controlled. The pace at which the various components are made must be coordinated, so that cars are produced at the maximum rate with a minimum wastage of material and effort.

It is not long since bacteria were thought of as mere 'bags of enzymes', to be plundered at will by biochemists! Nothing could be further from the truth. We have already seen that the high rate of multiplication of bacteria and the enormous populations they achieve renders them particularly susceptible to small evolutionary pressures, so that even minute advantages in biochemical efficiency will tend to be rapidly selected. We may therefore expect to find, and do, that bacteria have developed intricate and refined devices to streamline their biochemical economy. An obvious economic measure is to cut down the production of end-products which either are temporarily not required or are already in full supply, and here we find experimental support for the view that this kind of regulation is not only an evolutionary advantage but is probably a necessity for immediate survival. For example, it has been found in several cases that a particular enzyme, when being synthesised by bacteria at the maximal rate, may account for five to eight per cent. of the total cellular protein, and that the synthesis may be accompanied by a decreased growth rate. It is clear that the production of only a few such enzymes at maximal efficiency, out of the thousands which the cell is potentially able to synthesise, would be incompatible with the proper operation of the cell (see Davis, 1961).

Finally, since only inheritable modifications are of evolutionary value, we may confidently predict that mechanisms of biochemical regulation will be rooted in the bacterial genome and, therefore, subject to mutational alteration and amenable to genetic analysis. We will devote most of the remainder of this chapter to a description of the masterly and imaginative studies by F. Jacob and J. Monod of various types of regulatory mechanism which have been defined at the biochemical level. But first let is distinguish these mechanisms in rather general terms.

BIOCHEMICAL ASPECTS OF THE CONTROL OF SYNTHETIC PATHWAYS

Inhibition of enzyme activity

In the case of many (and perhaps all) biosynthetic pathways which lead to the formation of essential metabolites, the end-product of the pathway inhibits the activity of the first enzyme of the sequence or, where branched pathways are concerned, of the 'branch-point' enzyme, with the result that the pathway ceases to function. Metabolic pathways thus carry a built-in 'feedback' mechanism which controls their activity according to the rate at which the end-product is utilised. This form of control is called *feedback or end-product inhibition.* Its existence was initially inferred from three kinds of observation. Firstly, it was found that accumulation of a trytophan precursor (indol-3-glycerol phosphate) behind a mutational block in the tryptophan pathway of *E. coli* was abruptly arrested by the addition of tryptophan to the medium, although it could be assumed that the enzymes necessary for synthesis of the precursor were still present in the bacteria (Novick & Szilard, 1954). Secondly, in experiments involving incorporation of radioisotopes into protein, it was shown that any unlabelled amino acid that was added to the system became the exclusive source of that amino acid in the protein; endogenous synthesis of that specific amino acid was arrested (Roberts *et al.*, 1955). Finally, *in vitro* inhibition of the extracted enzymes by the end-product was demonstrated for the isoleucine-valine and pyrimidine pathways (Umbarger, 1956; Yates & Pardee, 1956), and has since been confirmed for many other pathways (review: Umbarger, 1961; see also Cold Spring Harbor Symposium on 'Synthesis and Structure of Macromolecules', Vol 28, 1963).

As a control mechanism, feedback inhibition has three outstanding characteristics—it is specific in that only the end-product (or an analogue of it) is effective, it acts on a single early enzyme of the pathway, and its action is immediate and independent of the presence of preformed enzymes. The specificity of the control is stressed by the fact that, in the case of branched pathways, the same cell may synthesise two enzymes which have identical catalytic actions but are differentially inhibited by the endproducts of the divergent pathways (see Davis, 1961; Stadtman *et al.*, 1961). However, the most remarkable and important feature of feedback inhibition is that the inhibitory end-product has no steric relationship to the normal substrate of the inhibited enzyme. The interaction of substrate and inhibitor with the enzyme is *not* a competitive one such as is found between analogues, but is of bimolecular type. Moreover, treatments which partially denature enzymes in general may result in loss of sensitivity to the inhibitor and alteration in the sedimentation velocity of the enzyme without affecting its activity for its normal substrate (Changeux, 1961, 1963; Gerhart & Pardee, 1962; Martin, 1963).

It therefore seems very probable that the evolutionary pressures towards a streamlined economy have selected, for those enzymes which are crucial in feedback control, a protein structure which carries *two non-overlapping specificities*, for which the term *allosteric* has been coined (Monod & Jacob, 1961). Mutational alteration of the sites of inhibitor attachment to such an allosteric enzyme should result in release of inhibition of the pathway and excretion of the end-product into the medium. Mutants of this type can, in fact, be selected. For instance, 5-methyltryptophan is an analogue of tryptophan which acts as inhibitor of an early enzyme in the tryptophan pathway, but cannot substitute for tryptophan in protein, so that it completely prevents bacterial growth by tryptophan starvation. Rare mutants arise, however, which are resistant to the analogue, and some of these are found to excrete tryptophan and to possess an early enzyme which is insensitive to feedback inhibition by either tryptophan or 5-methyltryptophan (Moyed, 1960; see also, Umbarger, 1961). Thus the excess yield of tryptophan is not the cause of the resistance, as the hypothesis of competitive inhibition would predict, but its result.

The primary importance of the concept of allosteric inhibition is that it offers a very flexible model which can be applied to a wide range of different situations. Since the inhibitor is not sterically related to the substrate of the enzyme, there is no reason why this type of control should not be used to coordinate the activities of *different* pathways. Moreover, the same general mechanism could operate in reverse; that is, certain enzymes, instead of being inhibited, might be *activated* by *allosteric effectors*. In fact, two cases of enzymes which are inhibited by one metabolite and activated by another have already been reported (Changeux, 1962; Gerhardt & Pardee, 1962). Thus an effector produced by one metabolic pathway could trigger off the synthetic activity of an unrelated pathway, so that the model can serve to explain the kind of sequential order of function which has already been observed to operate in the synthesis of phage particles (p. 492), as well as the coordination of cell division and DNA synthesis (Jacob & Monod, 1963a; see below, p. 618). It is clear that a most elaborate system of inter-connected 'circuits' can be built up on the basis of 'cross-feedback' inhibitors and effectors, so that the model may be applied not only to the metabolism of single cells, but also to such phenomena as hormone action, differentiation and carcinogenesis in higher organisms (Monod & Jacob, 1961).

The idea that combination of a small molecule with a large one may alter the properties of the latter by producing conformational changes in it is not a new one, and was first proposed nearly 20 years ago to account for the behaviour of haemoglobin (Wyman, 1948). In the case of haemoglobin there is evidence that, in the absence of oxygen, the four subunits interact strongly with one another and weakly with oxygen (see p. 152). However, the binding of oxygen to one of the subunits weakens the bonds between them so that oxygen uptake by another subunit is facilitated, thus leading to further relaxation of the tetramer and an increased velocity of oxygen uptake. The result is that oxygen is bound and released efficiently over a narrow range of partial pressure, yielding the typical sigmoid oxygen dissociation curve of haemoglobin. Since the oxygen-carrying haem groups are too widely separated at the surfaces of the haemoglobin subunits to interact directly, the oxygen molecules presumably exert their effect by altering the conformation of the subunit protein (Gibson, 1959; Muirhead and Perutz, 1963; Wyman, 1963).

The enzyme aspartate transcarbamylase (ATCase) is the first enzyme of the pathway of pyrimidine biosynthesis, and its activity is strongly inhibited by cytidine triphosphate (CTP), one of the end-products of the pathway, although this inhibition can be reversed by the substrate, aspartate, which attaches at a different site on the enzyme surface. It happens not only that the kinetics of this reversal, as a function of aspartate concentration, show the same type of sigmoidal curve as the oxygen dissociation curve of haemoglobin, but ATCase is also composed of four subunits which, as in the case of the monomeric myoglobin, retain their enzymatic activity after dissociation but are no longer inhibited by CTP. The inhibitory effect of the binding of CTP is interpreted as due to a strengthening of the cohesion of the subunits which thereby reduces the affinity of the tetramer for aspartate by distorting the aspartate attachment sites (Gerhart and Pardee, 1963).

Further evidence for this general concept that allosteric inhibitors act by increasing the interaction between enzyme subunits comes from the study of threonine deaminase whose activity is inhibited by L-isoleucine (see Fig. 35, p. 121). In addition, this inhibitory effect is reversed by L-valine, but the reversal is not due to competition since not only do threonine, isoleucine and valine appear to bind to different sites on the enzyme, but the presence of valine alone actually increases enzyme activity; that is, valine acts as an allosteric effector. It happens that, in the presence of low concentrations of urea, threonine deaminase establishes a reversible equilibrium between the active polymeric and an inactive monomeric state, which is very sensitive to environmental changes. Under these conditions the addition of isoleucine does not inhibit but *increases* enzyme activity, while the addition of valine reduces it. This reversal of roles is explicable if, as postulated, the binding of valine to the unstabilised protein increases the interaction between the subunits and so shifts the equilibrium towards the formation of active polymer, while valine has the opposite effect (Changeux, 1963; see also Freundlich and Umbarger, 1963).

The regulation of enzyme synthesis

Repression

Feedback inhibition is a highly efficient method for controlling the synthesis of *small molecules* such as amino acids or nucleosides but, as we have seen, it has no immediate or direct effect on the synthesis of enzymes or, presumably, of other proteins. However, as knowledge of biosynthetic pathways increased and it was possible to assay the various enzymes that mediated them, it became apparent that the *synthesis* of these enzymes could also be specifically repressed by the terminal metabolite. Cohn & Monod (1953) first showed that the production of tryptophan synthetase and methionine synthetase was specifically repressed by addition to the bacteria of exogenous tryptophan and methionine respectively. It then became apparent, in the case of arginine synthesis, that repression by arginine *affected the production of the majority, if not all, of the enzymes of the pathway simultaneously*, even when the enzymes were separated by a mutational block involving one of them (Vogel, 1957; Gorini & Maas, 1958). This shutting off of all the enzymes of a sequence by its end-product is termed *repression*, or *coordinated enzyme repression*, to distinguish its mechanism from that of feedback inhibition by the same metabolite (Vogel, 1957). The enzymes of a considerable number of metabolic pathways, including those of other amino acids such as histidine (Ames & Garry, 1959), of pyrimidines (Yates & Pardee, 1957) and of purines (Magasanik 1957), have since been found to be subject to coordinate repression. Finally, repression of a particular pathway can be released as the result of a single mutation, so that production of the enzymes is no longer restricted by the presence of the end-product.

We are thus confronted with two distinct mechanisms, each apparently operated by the same switch (the end-product of a pathway) and geared to the same end (regulation of the pathway). How are these two mechanisms coordinated or, to put it another way, what is the gear ratio? This problem was studied by Gorini (1958) by supplying a culture of a prototrophic strain of *E. coli*, growing continuously in minimal medium in a chemostat (p. 193), with increasing concentrations of arginine; at each concentration, when a steady state had been attained, the utilisation of exogenous arginine, as well as the level of a repressible enzyme of the arginine pathway, were measured. At concentrations between one and ten μg. per ml., virtually all the exogenous arginine was used up, while the level of enzyme remained constant. Repression was therefore in abeyance, and the bacteria were precisely compensating for the increasing arginine concentrations by restricting endogenous synthesis by means of feedback inhibition. On the other hand, when more than ten μg. per ml. arginine were provided, its utilisation ceased to be complete and repression of the enzyme became marked. We can thus regard feedback inhibition as a fine control for regulating the synthesis of small molecules, while repression behaves as a coarse adjustment, mainly concerned with economy of protein synthesis (Davis, 1961).

Induction

Up till now we have only considered systems of negative control in which the functioning of a synthetic pathway is switched off by excess of its end-product. In fact, the reverse behaviour whereby certain enzymes are synthesised in significant amounts only in the presence of their substrate, which may stimulate a 10,000-fold increase in production, has been recognised for a much longer time. Originally known as *enzymatic adaptation*, the phenomenon is now generally designed by the more dynamic term, *induction of enzyme synthesis* (Monod & Cohn, 1952). In bacteria, inducible enzymes appear to be mainly those which attack exogenous substrates (see Stanier, 1951), of which β-galactosidase (Monod & Cohn, 1952) penicillinase (Kogut, Pollock & Tridgell, 1956; Citri & Pollock, 1966) and the enzymes mediating galactose utilisation in *E. coli* (Kalckar, Kurahashi & Jordan, 1959) are good examples. The outstanding characteristics of induction are similar to those of repression. Firstly, the inducer is completely specific although analogues of the substrate which are not attacked by the enzyme, such as thiogalactosides in the case of β-galactosidase or 6-deoxygalactose(D-fucose) in the case of the galactose enzymes (Buttin, 1961), may be highly effective. Secondly, where a number of enzymes are involved in utilisation of the substrate, all of them are *coordinately induced* by the inducer (Pardee *et al.*, 1959; Kalckar *et al.*, 1959). Thirdly, *constitutive mutants* of inducible strains can be isolated in which the enzymes of a particular pathway are synthesised continuously in the absence of the inducer. The enzymes produced constitutively by such mutants are identical with those resulting from induction of the wild type strain (Cohn & Monod, 1953; Kogut *et al.*, 1956).

Most of our knowledge of how these systems of induction and repression operate at the genetic level comes from the brilliant analyses, by F. Jacob and J. Monod and their colleagues, of the control of lactose fermentation in *E. coli* K-12 which we must now consider in some detail.

GENETIC ANALYSIS OF THE CONTROL OF ENZYME SYNTHESIS

Three enzymes appear to be involved in the fermentation of lactose by *E. coli*—β-galactosidase and β-galactoside transacetylase, both of which have been isolated in a highly purified state, and galactoside permease which is required for the uptake of lactose and is identified by *in vivo* tests, although a protein which is probably the permease itself has recently been isolated (Fox and Kennedy, 1965). Mutations affecting the activity of β-galactosidase map in the *z* locus, and of galactoside permease in the contiguous *y* locus, of the *lac* region of the chromosome (Fig. 136). While all z^- mutants fail to produce active β-galactosidase, a proportion of them continue to yield a protein which is serologically similar to the enzyme (cross-reacting material or CRM). All *y* mutants synthesise β-galactosidase but are unable to synthesise permease, and so are phenotypically lac^- ('cryptic fermenters') because lactose cannot enter the cells. Complementation tests in heterogenotes formed by *F—lac* sexduction (p. 103) show that z^- and y^- mutations belong to different functional units. However, although no y^- mutations have been found to complement one another, certain pairs of z^- mutations are partially complementary; since extracts of the mutants show a low level of β-galactosidase activity when mixed *in vitro*, and since it is known that the enzyme comprises a number of polypeptide subunits (see Jacob & Monod, 1961b), this may validly be ascribed to interallelic complementation (p. 151).

The role of the transacetylase is unknown, but in view of the location of its determinant gene adjacent to the *y* gene, and its coordinate induction with β-galactosidase and permease, it might be thought to participate in lactose utilisation. Nevertheless, deletion mutants involving the *a* gene, which produce no transacetylase but retain β-galactosidase and permease activity, have recently been isolated and shown to grow normally with lactose as sole carbon source (Fox *et al.*, 1966). For simplicity we shall, in general, refer only to β-galactosidase and permease in what is to follow.

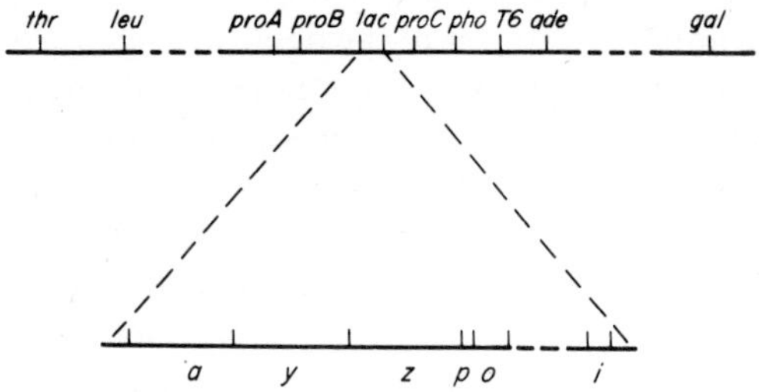

FIG. 136. A magnified genetic map of the region of the *E. coli* chromosome controlling lactose fermentation (the *lac* region).

The upper lines indicate the position of the *lac* region in relation to some other loci on the chromosome: *thr* = threonine; *leu* = leucine; *pro* = proline; *pho* = alkaline phosphatase; *T6* = phage T6 resistance/sensitivity; *ade* = adenine; *gal* = galactose utilisation. Compare Fig. 127, p. 666.

The lower line is a magnification of the *lac* region and reveals its genetic components as follows: *a*, *y* and *z* are contiguous genes specifying the structure of galactoside transacetylase, permease and β-galactosidase respectively; *p* is the promoter, and *o* the operator region; *i* is a small region, separated from *z*, which determines the inducibility of lactose utilisation. See text.

Structural genes and regulator genes

From inducible (i^+) strains of *E. coli*, constitutive (i^-) mutants can be isolated which produce both β-galactosidase and permease in the absence of a galactoside inducer. In crosses between i^+ and i^- strains, which are also distinguished by mutations in the *z* and *y* loci, recombination analysis shows that the i^- mutations are clustered within a small region clearly separated from, although closely linked to, the *z* gene (Fig. 136).

We have seen that by mating a z^+Hfr strain with a z^- recipient, the kinetics of inducible synthesis of β-galactosidase by the merozygote population can be followed (p. 703 and Fig. 135). The same method can provide important information about the functional activity of the *i* alleles. Parental strains of genotypes z^+i^+ (wild type, inducible) and z^-i^- (constitutive but unable to synthesise active enzyme), neither of which are capable of constitutive synthesis, are mated in reciprocal crosses and the ability of the z^+i^+/z^-i^- zygotes to yield β-galactosidase *in the absence of inducer* is followed. The two crosses give quite different results. When the z^-i^- genes are transferred to z^+i^+ recipients, no trace of constitutively produced enzyme is detected, showing that i^+ is dominant to i^-. On the contrary, when the recipients are z^-i^-, transfer of the donor z^+i^+ genes is immediately followed by constitutive enzyme synthesis, as if the z^+ gene alone had been transferred; about 60 minutes later, however, enzyme production falls to zero, but if inducer is now added it rises again to its maximal rate (Pardee, Jacob & Monod, 1959). The different expressions of the two crosses, although the genetic constitution of the zygotes is identical in both (z^+i^+/z^-i^-), is strictly comparable to the situation found with respect to zygotic induction of phage (p. 463; Table 17) and leads to the same conclusion—that the difference depends on the recipient cytoplasm. The failure of the zygotes to produce β-galactosidase constitutively when the recipients are z^+i^+ indicates not only that i^+ is dominant to i^-, but that *the* i^+ *gene determines the production of a cytoplasmic repressor which prevents the* z^+ *gene from functioning*. In the reverse cross where the z^+i^+ genes are transferred together to z^-i^- recipients, the initial, constitutive production of enzyme, which is later repressed, is due to a lag in the synthesis of repressor by the incoming i^+ gene (Pardee *et al.*, 1959). This delayed repression of the constitutive phenotype appears to be unaffected by the presence of inhibitors of protein synthesis, such as chloramphenicol or 5-methyltryptophan, which suggested, misleadingly as it turned out (p. 732), that the product of the i^+ gene might not be a protein (Pardee & Prestidge, 1959). We have already seen that the repressor of vegetative phage activity behaves in the same manner, since inhibition of protein synthesis greatly enhances lysogenisation (p. 468).

Similar results are obtained in heterogenotes. Irrespective of the combination of z, y and i alleles, every diploid which carries a single i^+ gene on either endogenote or exogenote is inducible with respect to both β-galactosidase and permease production. For example, in a heterogenote of type $z^+i^-/F—z^-i^+$, the i^+ gene carried by the sex factor can repress the activity of the z^+ gene on the bacterial chromosome—it exerts its effect through the cytoplasm; only homogenotes of the type $i^-/F—i^-$ are constitutive (Jacob & Wollman, 1961a, p. 263; Jacob & Monod, 1961b). Moreover, the repressibility of a gene is independent of its ability to specify an *active* enzyme. In the case of z^-_{CRM} mutants, which produce a cross-reacting protein but no active enzyme, the synthesis of CRM responds to induction in precisely the same way as does the synthesis of enzyme by wild type strains; when heterogenotes of the type $z^-_{CRM}i^+/F—z^+i^+$ are induced, the same quantitative yields of enzyme and CRM are obtained (Perrin, Jacob & Monod, 1960).

The repressor disclosed by these experiments could, theoretically, exert its effect in one or the other of two ways. Firstly, it might act directly in shutting off the expression of the genes it controls; in this case the inducer would operate by antagonising the repressor so that the genes would be released from inhibition. Alternatively, the repressor might be an enzyme which destroys an endogenous inducer. The argument was decisively settled in favour of the first hypothesis by the study of an unusual mutant of inducible, wild type *E. coli* which simultaneously lost the capacity to synthesise both β-galactosidase and permease in the presence of inducer. The loss of both enzymes is not due to a deletion covering parts of the z and y genes since the mutant yields recombinants with all z^- and y^- mutants. Instead, the mutation maps at a point in the i locus. Moreover, this mutation (designated i^s), unlike i^- mutations, turns out to be *dominant over* i^+ *in diploids*; heterogenotes of type $i^sz^+y^+/i^+z^+y^+$ do not ferment lactose and yield neither β-galactosidase nor permease on addition of an inducer.

The explanation is that the i^s mutation results in the synthesis of an *altered* repressor (*super-repressor*) which is not antagonised by galactoside inducers, so that the repression becomes permanent. The existence of such a repressor is compatible only with the first hypothesis, that the action of normal repressor is a direct one (Jacob & Monod, 1961a, b; Willson *et al.*, 1964). We might anticipate that lac^+ revertants obtained from this mutant would more often be due to mutations at other sites in the i locus than at the original i^s site itself. Such double mutants (i^-i^s) should not produce active repressor and should be *constitutive* and this, in fact, is what is found (Jacob, personal communication). In every respect, therefore, the features of i^s mutations with respect to control of lactose fermentation are identical with those of ind^- mutations of phage (p. 467)—both yield altered repressors which are not destroyed by inducers which, in the case of phage, is ultraviolet light.

These studies of the control of genetic expression in the lactose-fermentation system of *E. coli*, and in infection and lysogeny by phage, led to the recognition of two categories of gene—*structural genes* and *regulatory genes*. A structural gene carries the genetic information determining the structure, and therefore the specificity of function, of a protein. A regulatory gene (designated R), on the other hand, determines the production of a cytoplasmic substance which controls the activity of a specific structural gene, or coordinated group of structural genes (Jacob & Monod, 1961a). We shall shortly see that regulatory genes are concerned not only with inducible enzyme systems such as we have just described, but also with co-ordinated enzyme repression. However, before proceeding to generalisations about regulatory mechanisms we must extend our analysis of the lactose system a bit further. In particular, we must examine the way in which the repressor exercises its repressive action.

Operators and operons

Repressors must act by combining with some element, called an *operator* (o), which is necessary for the transcription into protein of the genetic information contained in the group of genes whose functions are repressed. The specificity of repressor action further implies an equally specific operator, whose structure is therefore genetically determined. It follows that mutational alterations of operator specificity should be found which lead to decreased susceptibility to repressor and thus to constitutive enzyme production (derepression). An ingenious way of looking for derepressed (o^c) mutants of this sort is to select for constitutive lactose fermenters among inducible heterogenotes which are homozygous for the i^+ gene ($z^+i^+/F—z^+i^+$); only the improbable occurrence of i^- mutations in *both* i^+ genes could lead to recessive depression due to inactive repressor (Jacob, Perrin, Sanchez & Monod, 1960). Genetic analysis of a number of o^c mutations reveals that they display just the characteristics we might expect from the postulated mechanism of derepression. They are dominant to o^+ in o^+/o^c heterogenotes, they promote the same level of constitutive production of both β-galactosidase and permease, and they are insensitive to the action of super-repressor (p. 715) as the constitutive behaviour of $i^so^+/F—i^+o^c$ heterogenotes shows.

In addition, o^c mutations possess two further features which point to the nature of the operator and the level of its intervention in the transcription process. Firstly, they map together in a minute region between the z and i loci, but adjacent to z (Fig. 136, p. 713). Secondly, when tested in heterogenotes, *only those genes located on the same chromosome as the o^c mutation are constitutively expressed.* In genetic language, o^c mutations function only in the *cis* position (p. 128). For example, heterogenotes of constitution $o^+z^-/F—o^cz^+$ produce β-galactosidase (and permease) constitutively, while $o^cz^-/F—o^+z^+$

heterogenotes do not. Again, we have mentioned that heterogenotes of type z^+/F—$z\bar{}_{CRM}$, which are also homozygous o^+i^+, synthesise both active enzyme and inactive protein (CRM) on induction. When an o^c allele is introduced into either endogenote or exogenote, it transpires that o^cz^+/F—$o^+z\bar{}_{CRM}$ heterogenotes constitutively produce only active enzyme, and o^+z^+/F—$o^cz\bar{}_{CRM}$ only inactive protein.

Various models of the mechanism of repression and derepression can therefore be discarded. For instance, each gene of a coordinately controlled pathway cannot have a separate operator, for in this case the production of only one enzyme would be derepressed by a single $o^+ \rightarrow o^c$ mutation. Furthermore, the operator does not determine the synthesis of some essential prosthetic group shared by all the enzymes of the pathway; if this were the case, the constituent made by a derepressed operator should be available in the cytoplasm to activate the incomplete enzymes specified by the opposite chromosome of a diploid. We are left with the concept that *a single operator switches on a sequence of genes, and that it and the structural genes it controls comprise an integrated unit at both the physical and functional levels.* Such a unit is called an *operon* (Jacob *et al.*, 1960).

What, then, is the nature of the operon? We have seen that a striking feature of the linkage maps of *E. coli* and *Salmonella* is the clustering of genes determining the enzymes serving common biosynthetic pathways (p. 163). There are now many examples of such clusters where the completeness of genetic analysis makes it virtually certain that the clustered genes are contiguous as, for example, in the case of the lactose (*lac*), galactose (*gal*), tryptophan (*try*) and, particularly, the histidine (*his*) region where the genes determining the ten enzymes of the pathway are located (p. 160). It happens that in all these cases the enzymes of the pathway are coordinately induced or repressed. We have also seen that protein synthesis is mediated by molecules of messenger RNA which transcribe the information carried in the DNA of the genes, and are then translated on the ribosomes into the amino acid sequences of proteins (Chapter 12, pp. 284–288).

The most economical and satisfying hypothesis to link these facts is that all the genes of an operon are transcribed together into a single, polygenic or *polycistronic* m-RNA molecule, and that the operator is the locus which controls initiation of m-RNA synthesis. According to this hypothesis, *the operon is a genetic unit of coordinated transcription*, the action of repressor being to block transcription by combining specifically with the operator, so that no message is transmitted to the ribosomes and synthesis of the proteins specified by the genes of the operon is arrested (Jacob and Monod, 1961a, b). In support of this it has been shown experimentally that induction of the *lac* operon in *E. coli* is accompanied by a striking increase of a rapidly-labelled RNA fraction which hybridises specifically with the DNA of an *F—lac* factor extracted from *Serratia marescens*, a species which not only cannot ferment lactose but whose DNA has a quite different GC content from that of *E. coli* (Attardi *et al.*, 1963; see also Guttman and Novick, 1963). The m-RNA molecules isolated in these experiments were found to be too large to have been transcribed from individual genes. Again, the sedimentation constant of m-RNA specifically synthesised and isolated following derepression of the histidine pathway in *Salmonella* turned out to be similar to that calculated for m-RNA corresponding to the entire histidine operon (38*s*, equivalent to about 11,000 base pairs and a molecular weight of 4×10^6) (Martin, 1963). In the case of the tryptophan region of *E. coli*, studies of the hybridisation of m-RNA with the transducing phage $\phi 80$, carrying variously deleted parts of the *trp* operon, have shown that the m-RNA molecules (33*s*) can hybridise with more than one gene of the operon (Imamoto, Morikawa and Sato, 1965). We may therefore subscribe to the 'one operon-one messenger' doctrine, that the structural information for all the proteins coded for by an operon are transcribed by a single molecule of m-RNA, and that the coordinate nature of induction and repression is a consequence of this. It follows that the synthesis of separate polypeptides corresponding to each gene of the operon occurs during translation of the message on the ribosomes, and that each structural gene contains the appropriate information for initiating and terminating chain formation. We have already seen that this is so (pp. 284–288, 361).

Polarity of the operon

The whole concept of the operon, from which polycistronic messenger RNA is transcribed from a fixed starting point, the operator, implies a polarised process, and this is borne out by analyses of many kinds. We have noted that the *lac* operon consists of three genes arranged in the sequence *a—y—z* and that the operator, as indicated by o^c mutations, appears to map at the right-hand extremity of *z*. We have also seen, from a purely biochemical point of view, that m-RNA is both transcribed by RNA polymerase, and translated on the ribosomes, from the 5′ towards the 3′ end, the 5′ end corresponding to the amino end of the derivative polypeptide (pp. 294, 285). The problem is whether this polarity can be related to the genetic map of the operon. At the level of protein synthesis, this has been achieved for the *lac* operon by studying the kinetics of production of β-galactosidase and of transacetylase following induction and de-induction. β-galactosidase begins to appear shortly after addition of inducer and reaches a plateau level in two to three minutes, while the transacetylase appears a minute or two later; on removing the inducer, the transacetylase continues to be produced for about two minutes after the β-galactosidase, suggesting that the m-RNA is also degraded from the operator end (Alpers and Tomkins, 1965). This sequential synthesis of β-galactosidase and transacetylase has been confirmed by similar kinetic studies, and has also been extended to the m-RNA level by exposing the cell to a brief pulse of 5-fluorouracil, or of X-rays, at the time of induction, to disrupt the information carried by that part of the messenger molecule to be made first. The outcome was that synthesis of β-galactosidase was inhibited while that of acetylase was not (Kepes, 1967). In the case of the tryptophan operon in *E. coli*, the operator lies at that end of the operon specifiying the first enzyme of the pathway, anthranilic synthetase (Matsushiro *et al.*, 1965); m-RNA isolated soon after derepression of tryptophan synthesis was found to hybridise with DNA from a transducing preparation of phage ϕ80 carrying the whole *trp* region, but not with DNA from transducing phage which lacked the anthranilic synthetase determinant (Imamoto *et al.*, 1965). Polarity of the kinetics of production of the enzymes of the histidine pathway in *Salmonella* has also been reported. Moreover, the time interval separating the production of two enzymes of this pathway can be abolished by the occurrence of a deletion involving the interval between their determinants (Goldberger and Berberich, 1966; see Ames *et al.*, 1967).

One of the most interesting indications of the polarity of operons is the occurrence and properties of what are called polar mutations (Franklin and Luria, 1961; Jacob and Monod, 1961b). These were first discovered in the *lac* operon as the result of a search for a type of mutation in the operator which, unlike o^c mutations which render the operator insensitive to repressor, would prevent the initiation of transcription. Such *operator negative* (o^o) mutations seemed, in fact, to have been found; they blocked the synthesis of all enzymes of the lactose pathway, mapped in the presumptive *o* region, were revertible, showed no complementation with any z^- or y^- mutations, and were recessive to o^+ in diploids. However, other mutations with similar properties were then found which did not map in the *o* region but more or less randomly along the *z* gene. These latter mutants produce no β-galactosidase but, as a rule, their ability to synthesise permease and transacetylase is not abolished but impaired markedly and to the same degree. They do not generally interfere with induction or repression of the operon.

Some mutations in the *y* gene were also found to behave in the same way in that they impaired or even abolished transacetylase production, but none of these mutations affected β-galactosidase synthesis. In other words, the characteristic of all these mutations is that only those genes of the operon located on the opposite side to the operator show impaired activity; genes between the mutation and the operator retain normal function (Jacob and Monod, 1961b). It was then found that the mutations originally thought to be o^0 were revertable by mutations located elsewhere on the chromosome, and these turned out to be suppressors of 'ochre' mutations (Beckwith, 1963; Brenner and Beckwith, 1965).

It is now known, from extensive data on the *lac*, *his* and *trp* operons, that both *amber* and *ochre* chain-terminating mutations are polar (see pp. 361–364; Yanofsky and Ito, 1966). In addition all frame-shift mutations (pp. 318, 346) are also polar (Whitfield, Martin and Ames, 1966; Martin *et al.*, 1966) but this is almost certainly due to the generation of a distal, chain-terminating 'nonsense' triplet as a result of the frame-shift, since prototrophic revertants of a frame-shift mutation, in a permissive strain carrying a 'nonsense' suppressor, may become auxotrophic and display polarity again if the suppressor is removed; on the other hand, $+-$ or $-+$ revertants in a strain which lacks a 'nonsense' suppressor are never polar (Martin, 1967; Whitfield *et al.*, 1966).

Thus the postulated o^0 mutations of Jacob and Monod (1961b) turned out to be extreme polar mutations of chain-terminating or 'nonsense' type. Since these mutations are expressed at the level of protein synthesis and not of m-RNA transcription, and terminate only the polypeptide products of the affected genes, the question of the mechanism of polarity arises. An important clue was the finding that within any gene of an operon, the closer the 'nonsense' mutation is to the operator end of the gene, the greater is the polar effect, which explains why the polar mutations which appeared to be located in the *lac* operator region mimicked the postulated o^0 type (Newton *et al.*,

1965; Martin *et al.*, 1966; Yanofsky and Ito, 1966). Furthermore, this position effect is related to the distance of the mutation from the beginning of the gene distal to it and not to its distance from the operator, because a deletion which decreases the interval between the mutation and the distal gene reduces the strength of the polarity, while a deletion between the mutation and the operator has no effect (Zipser and Newton, 1967).

On the basis of these facts, two models have been put forward to explain the effects of polar mutations. One model supposes that when a ribosome traverses a 'nonsense' condon in the m-RNA it is, as it were, derailed and is likely to fall off unless it encounters a chain-initiating sequence at the beginning of the next gene which puts it back on the rails again. Thus the closer the 'nonsense' mutation is to the beginning of the operator-distal gene, the greater is the probability that that gene will be translated. The other hypothesis is that the m-RNA tends to be broken down distal to a nonsense codon, perhaps due to an altered steric relationship between the RNA and the ribosome which is known to protect it (for discussion see Brenner and Beckwith, 1965; Martin *et al.*, 1966; Yanofsky and Ito, 1966).*

Although we have discussed the mode of action of polar mutations as if they expressed their polarity exclusively at the level of m-RNA translation, this is not the whole story since there is considerable evidence that smaller molecules of m-RNA are transcribed from operons carrying these mutations, due to a deficit distal to the mutational site (Attardi *et al.*, 1963; Imamoto, Ito and Yanofsky, 1966). In accordance with this, smaller polysomes have been reported to result from polar mutations (Kiho and Rich, 1965). All this points to some sort of inter-dependence between messenger RNA transcription and protein synthesis. We have already discussed the evidence for the hypothesis put forward by Stent (1964, 1966) that the ribosomes attach to the m-RNA as soon as it begins to be synthesised, and that this provides the energy and the mechanism for stripping it from the DNA-polymerase complex (p. 293).

* Direct evidence that polarity is expressed purely at the level of translation comes from recent studies of *in vitro* protein synthesis using, as template, RNA from an amber mutant of male-specific phage (p. 440). The mutation involves an early amino acid in the coat protein of the phage, and blocks synthesis of both coat protein and RNA polymerase following infection. In the *in vitro* system, unlike when wild type RNA is used as template, neither coat protein nor any other protein is made (Engelhardt, Webster and Zinder, 1967; Capecchi, 1967).

A unified hypothesis of the operon

The operon hypothesis emerged from studies of the regulation of the pathway of lactose breakdown, and it was in this inducible system that mutations involving the postulated operator, as well as the repressor locus determining production of cytoplasmic repressor, were first defined. The genetic regulation of the pathway of galactose utilisation, which is another inducible system, has also been studied in considerable detail. In this case too, the synthesis of three enzymes (kinase, transferase and epimerase; Fig. 29, p. 110), determined by contiguous genes, is coordinately controlled by an operator located at one extremity of the sequence (*k-t-e-o*) which, in the absence of inducer, is normally switched off by a cytoplasmic repressor produced by an unlinked gene (Buttin, 1962, 1963a, b; Shapiro, 1967). In addition, the pathways of arabinose and rhamnose utilisation have been found to be mediated by three adjacent genes which are coordinately induced; but, in these cases, the regulatory mechanism appears to be of a different kind which we will discuss later (p. 730).

Most of the gene clusters which have been reported, however, are concerned with amino acid syntheses which are not inducible but are coordinately repressed by the end-product of the pathway (p. 710). How can these be fitted into the repressor-operator model? A comparison of Figs. 137A and B shows that the model of an inducible system serves equally well for a repressible one if a rather different assumption is made about the nature of the repressor. In induction, the product of the *R* gene is thought to combine directly with the operator and to be inactivated by the inducer (Fig. 137A). In repression, on the other hand, the product of the *R* gene is regarded as a kind of *co-repressor* which cannot itself combine with the operator, but which can be *activated* by the end-product of the pathway to yield repressor. The small molecules which either inactivate or activate the product of the *R* gene (R) are given the general name of *allosteric effectors*, so that we can express the interactions of the two in the general form $R + F \rightleftharpoons RF$. In induction, R is the repressor, while in repression the repressor is RF (Jacob and Monod, 1961b, 1963a).

There is now much evidence to support this general model of the mechanism of coordinated repression. Thus of the considerable number of synthetic pathways whose genes are known to be clustered on the *E. coli* and *Salmonella* linkage maps, the enzymes of many have now been shown to respond simultaneously *and to the same degree* in repression and de-repression, and so may be classified as operons.

Among these are the pathways concerned with the synthesis of histidine, in which coordinate repression was first described (Ames, Garry and Herzenberg, 1960; Ames *et al.*, 1967), of tryptophan (Lester and Yanofsky, 1961; but see below, and Margolin & Bauerle,

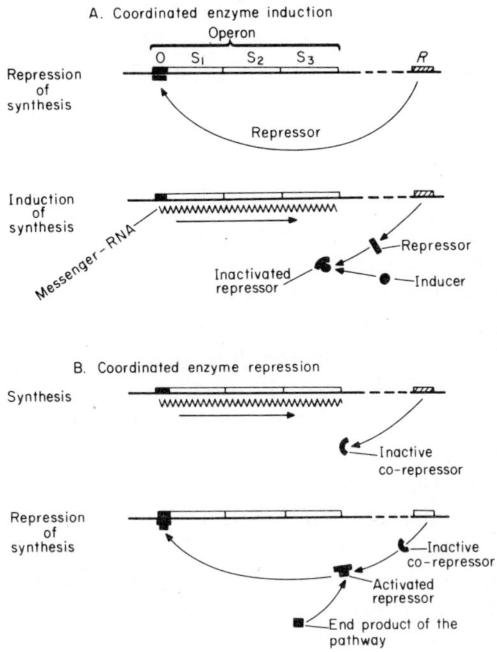

FIG. 137. Illustrating the genetic mechanism for regulating coordinated enzyme induction and repression.

S_1, S_2 and S_3 are three contiguous structural genes determining the synthesis of three enzymes serving a single metabolic pathway.

O is the operator which controls transcription of the genetic information of S_1, S_2 and S_3 into messenger RNA.

R is the regulator locus.

In *coordinated enzyme induction* (A), the product of *R* is a specific repressor which, by combining with *O*, prevents transcription of the structural genes and, therefore, synthesis of the enzymes they specify. The action of inducer is to inactivate the repressor, thus releasing the operator from repression.

In *coordinated enzyme repression* (B), the product of *R* is an inactive co-repressor which is specifically activated by the end-product of the pathway (the amino acid synthesised by the pathway).

For further explanation, see text.

Compare the similar model for the mechanism of immunity to temperate bacteriophage in Fig. 103 (p. 465).

1966), of leucine, threonine and isoleucine-valine (p. 121; Wagner & Berquist, 1960) and of pyrimidine (Beckwith *et al.*, 1962), as well as the lactose (Jacob & Monod, 1961b) and galactose (Wilson, 1966) pathways. In addition, mutants showing derepression of the galactose (Buttin, 1963b; Adhya & Echols, 1966), histidine (Roth *et al.*, 1966) and tryptophan (Cohen & Jacob, 1959; Matsushiro *et al.*, 1962) pathways have been isolated and have been found to be of two types. One type corresponds to mutants having an altered regulatory (*R*) gene which are unable to make repressor, since the mutation maps outside the operon and is recessive in diploids, which may be either transient merozygotes (p. 96) or F-prime heterogenotes (pp. 103, 674); in the other type the mutation is located at one end of the operon and is *cis*-dominant, so that this type is equivalent to the operator-constitutive (o^c) mutants of the lactose system (Review: Ames & Martin, 1964).

A prediction of the operon hypothesis is that, while deletions involving the operator and the first gene of the sequence should inactivate all the genes of the operon, a proximal extension of this deletion might physically connect these genes to another operon so that they would be activated again, but their activity would now be responsive to a different biochemical control. This phenomenon was first demonstrated in the case of the *Salmonella* histidine pathway. A mutant having a small deletion encroaching only into gene *G* at one end of the operon (Fig. 46, p. 163) had lost the ability to synthesise not only the enzyme specified by this gene (PR-ATP pyrophosphorylase) which happens to be the first enzyme, but all the other enzymes of the pathway as well (Ames *et al.*, 1960; Hartman *et al.*, 1960b). Revertants selected for growth on L-histidinol, requiring restoration of synthesis of histidinol dehydrogenase, which is the tenth enzyme of the pathway but is specified by the second gene of the operon, were found to have recovered the ability to produce all the enzymes of the pathway except the first, and these were no longer subject to histidine repression. One of the revertants had suffered an extension of the original deletion (Ames, Hartman and Jacob, 1963; see also Ames *et al.*, 1967).

Several notable cases have since been described where the fusion of two known operons has been deliberately selected for and the resulting mutants analysed in detail. In some of these the lactose operon has been connected to the neighbouring *ade* region by a deletion extending into the *lac z* gene on the left and into the second of two *ade* genes on the right (Fig. 136, p. 713). These adenine genes must constitute an operon with an operator at the right extremity since, in these deletion mutants, the synthesis of β-galactosidase permease and transacetylase occurs constitutively during growth in the presence of limiting amounts of adenine, is repressed by the addition of excess adenine, and is not induced by β-galactosides. The actual union of the lactose and adenine regions was shown by the fact that the residual genes of the two pathways were jointly transducible by phage P1 in these deletion mutants. Since deletions of this length might be expected to be lethal, they were isolated by special selection techniques in an F-prime heterogenote which was diploid for the entire *lac—ade* region (Jacob, Ullman and Monod, 1965).

By means of another ingenious but complicated technique, the widely separated lactose and tryptophan operons have been connected (see Fig. 127, p. 666). If a temperature-sensitive *F—lac*$^+$ factor is introduced into a strain of *E. coli* with an extensive deletion in the *lac* region, the only *lac*$^+$ bacteria which survive growth at high temperature are *Hfr* strains in which the *F—lac*$^+$ factor has become inserted into the chromosome and so come under chromosomal control of its replication. Since the bacterial chromosome lacks an homologous *lac* region, insertion occurs, at low frequency, at other regions which possess some non-allelic homology, as in the case with wild type sex factor. It is possible to select for insertion of the *F*-prime factor into loci determining the receptors for virulent phages, by looking for *Hfr* strains among phage-resistant mutants arising at high temperature. In this way, strains have been isolated in which the *lac* region has been transposed to the phage T1 receptor locus. This location happens to be adjacent to the attachment site for the transducing phage ϕ80, so that preparations of transducing phage carrying various parts of the *lac* region (ϕ80-*lac*) can be obtained. As a result, the *lac* region can be transferred and inserted into the ϕ80 attachment region of other strains carrying a normal *T1-rec* (*ton B* in Fig. 127) locus, at the opposite side of which lies the tryptophan operon. Now, it is known that many mutations to phage T1 resistance are due to deletions of the *T1-rec* locus (Franklin, Dove and Yanofsky, 1965) so that, by selecting for T1-resistance, we may hope to isolate some strains in which a deletion connects the *lac* and *trp* operons. Such strains have in fact been obtained in which β-galactoside permease is produced only in the absence of exogenous tryptophan, or when a mutation (R^-_{trp}) in the tryptophan regulator gene, leading to derepression of the tryptophan operon, is present as well (Beckwith, Signer and Epstein, 1966).

We have already seen that phage $\phi 80$ can transduce the neighbouring tryptophan region. Of two isolates of plaque-forming transducing phage, one was found to carry the operator and the adjacent anthranilic synthetase gene, but not the distal tryptophan synthetase genes (p. 666). When this type of phage was used to infect an *E. coli* mutant which was unable to synthesise anthranilic synthetase, synthesis of the enzyme commenced in the absence of tryptophan, but was rapidly repressed when tryptophan was added; in other words, in this transducing phage the residual tryptophan genes remained under the control of their own operator. The other transducing phage isolate lacked the operator and anthranilic synthetase region, but carried the tryptophan synthetase genes. When a non-lysogenic recipient, mutant with respect to tryptophan synthetase, was infected with this type, enzyme production started without delay regardless of the presence of tryptophan. On the other hand, a remarkably different outcome followed infection of a lysogenic recipient, for no tryptophan synthetase at all was produced; however, prophage induction in the recipient population after infection, resulted in enzyme synthesis about 30 minutes later. The most likely explanation of these findings is that, in the second type of transducing phage, the tryptophan synthetase genes were inserted into a region of the phage chromosome under the control of the phage repressor or immunity substance (Sato and Matsushiro, 1965).

It might be thought that all the operons of a chromosome would have the same polarity, and that messenger RNA transcription would be uniformly either clockwise or counterclockwise. This is normally so in the case of *E. coli* (see Fig. 127, p. 666). However, in many of the *lac* transpositions isolated by Beckwith *et al.* (1966) the *lac* region was inserted into the chromosome the wrong way round, but this did not affect expression of the operon. Similarly, some operons of *Salm. typhimurium* have different polarities (see Sanderson, 1967). In view of the evidence, previously reviewed, that only one strand of DNA is transcribed into effective messenger RNA (p.368), this strongly suggests that each operon is transcribed independently.

The promoter

The operator, as it was originally conceived, was thought to possess the dual function of being both the site of repressor action as well as the initiation point for messenger RNA synthesis. This concept seemed to be supported by the isolation of apparently o^o mutations which mapped, with o^c mutations, at one extremity of the *lac* operon and were assumed to prevent the initiation of transcription of the operon. It then turned out, as we have seen (p. 718), that these mutations are really extreme polar mutations arising within the *z* gene itself, which determines β-galactosidase structure. The operator can therefore be defined experimentally only in terms of o^c mutations which prevent repressor attachment. Three lines of evidence combine to show clearly that this operator is distinct from the *z* gene. Firstly, accurate mapping by three- and four-factor crosses define *o* as a distinct marker lying unambiguously between *z* and the repressor gene *i* (Jacob and Monod, 1965). Secondly, deletions within the operator-proximal region of the *z* gene, which involve the extreme polar pseudo-o^o mutations, restore permease function but do not interfere with the regulation of the pathway (Beckwith, 1964). Lastly, and most convincingly, a large number of o^c mutants have been isolated, all of which result from deletions which do not impair the function of the *z* or *y* genes (Jacob, Ullman and Monod, 1964).

Deletion mutants involving the *lac o* region can be selected by plating heterogenotes of the type $y^+z^+o^+i^s/F—y^+z^+o^+i^s$ on mellibiose medium. The dominant i^s mutation, determining production of 'super-repressor', not only switches off both y^+ and z^+ genes, but its diploidy makes it highly improbable that derepression will arise from further mutations (i^si^-) in both *i* genes (see p. 715); at the same time, β-galactosidase permease also acts as a permease for mellibiose which is utilised as a source of energy if it can enter the cell, so that use of this sugar as sole carbon source selects only for cells which have *y* gene activity—gene *z* function is not required for growth. Analysis of a large number of derepressed mutants obtained in this way show that they are all due to deletions involving *o*. Some of these deletions affect *o* alone, yielding ordinary o^c mutants of type $y^+x^+o^ci^s$. Others include the *i* region ($y^+z^+o^ci^-$) or both the *z* and *i* regions ($y^+z^-o^ci^-$) but, among 80 tested, none was found to involve both *z* and *o* but leave the *i* region intact ($y^+z^-o^ci^s$). This implies that deletions of this latter kind are incompatible with a functional *y* gene and, therefore, that there is a region between *z* and *o*, but distinct from *o*, which is required for expression of the operon. This region has been called the *promoter* (*p*) (Jacob *et al.*, 1964). It is assumed that the restoration of *y* gene function by deletions involving all three of the *z*, *o* and *i* regions is due to fusion with another operon. In fact, in such cases, the rate of permease and transacetylase synthesis is only a fraction of the maximal normally obtained in o^c mutants or after induction.

What, then, is the function of the promoter? We have seen that the operator is the site of action of repressor. In addition to the operator deletions mentioned above, another class has been isolated which removes the *i* region but only *part* of the operator, so that these mutants are partially repressed in the presence of a wild type i^+ gene. On induction, however, these mutants all yield β-galactosidase and permease at the maximal rate (Jacob *et al.*, 1964). It follows that, whatever ultimately controls the rate of operon activity, partial defects in the operator do not interfere with the rate of m-RNA transcription on which maximal enzyme production ultimately depends. Thus the most likely role for the promoter (*p*) is initiation of m-RNA synthesis. If this is so, since *p* lies distal to *o*, the operator cannot be transcribed into m-RNA so that, in the regulation of the operon, repression, at least, occurs at the level of the chromosome (see Beckwith, 1967).

A new class of *lac*-mutants of *E. coli* have recently been isolated which are characterised by a marked reduction in the maximal level of synthesis of all the emzymes of the pathway after induction. These mutations map between *o* and the most proximal z^- mutations, that is, in the postulated *p* region; they do not appear to be polar mutations, are dominant only in the *cis* position, and their effect is independent of the operator-repressor control system. In other words, they display all the properties expected of promoter mutations, but there is no direct evidence whether they intervene at the level of transcription or translation (Scaife and Beckwith, 1966)*. Mutants having similar phenotypes with respect to inducible penicillinase production, have been isolated in *Staphylococcus aureus* (see p. 783) and show a greatly reduced level of enzyme production, either constitutively or after induction. The mutations, which are revertable, do not map in the equivalent of the i^+ locus but between this locus and the penicillinase gene, and it is suggested that they are probably located in the peni-

* More exacting genetic analysis has now shown that these mutations in fact lie between the *i* and *o* loci, and not between *o* and *z* as was previously thought. Since these mutations reduce the level of expression of the operon, they must either reduce the amount of m-RNA synthesised, or themselves be transcribed into m–RNA and act at the level of protein synthesis (translation). In either case it may be concluded that the initiation site for m-RNA synthesis is located between *i* and *o* (Ippen *et al.*, 1968).

cillinase operator region in this organism (Richmond, 1966). Mutants having a reduced constitutive production of penicillinase have also been isolated in *Bacillus licheniformis* (Dubnau and Pollock, 1965).

Positive control and other regulatory systems

The main feature of the systems of regulation which we have so far considered is that it is the natural state of the operon to be switched on, control being exercised by a cytoplasmic repressor which switches off its activity; that is, *the control is negative.* In coordinated enzyme induction the product of the regulatory gene (R) is the repressor, which is inactivated by the inducer; in coordinated repression, the R gene product is activated to become the repressor by the end-product of the pathway. These are the two simplest variations on this theme, but more complicated ones could clearly be devised.

One such complication has been described in the case of coordinate repression of the tryptophan pathway in *Salm. typhimurium.* This pathway is now known to be mediated by five enzymes (anthranilic, phosphoribosyl, indole glycerol phosphate, and tryptophan B and A synthetases) whose genes are arranged in this same sequence on the chromosome, the operator being located at the anthranilic synthetase end. A genetic and functional analysis of a series of mutants having deletions extending to various degrees into the operon from the operator end, showed that any deletion involving the first gene (and, therefore, *o*) abolishes the activity of the second gene, as we might expect on the promoter hypothesis. Surprisingly, however, the last three genes of the operon continue to function at half their normal derepressed rate, but their activity is no longer subject to tryptophan repression. The activity of the entire operon is prevented only by deletions which extend across the junction of the second and third genes. There is clearly something in the region of this junction which is able to activate the genes distal to it when these are cut off from the operator and the presumptive promoter at the beginning of the operon; it has been suggested that this is a second promoter (Margolin and Bauerle, 1966).

A quite different kind of variation is a *positive control* mechanism in which the operon is normally turned off and requires a cytoplasmic effector to turn it on. This effector is activated or inactivated only in the presence of the appropriate small molecule so that either induction or repression can occur, just as in the case of negative control. The most characteristic genetic feature of positive control systems is the existence of *recessive mutations which prevent the activity of all the genes of the operon.* In systems of negative control, only mutations in the *R* gene leading to production of 'super-repressor', which are dominant in *trans*, or strong polar mutations which are *cis* dominant, can yield this phenotype.

Positively controlled operons have been reported in inducible systems mediating the utilisation of arabinose (Englesberg *et al.*, 1965) and of rhamnose (Power, 1967), while a positive cytoplasmic element plays a role in the expression of inducible penicillinase production in *Staphylococcus* (Richmond, 1967a) and of repressible alkaline phosphatase synthesis in *E. coli* (Garen and Echols, 1962), although both these latter cases are also subject to repressor control. As an example of this kind of system we will outline the facts about regulation of the L-arabinose pathway.

This pathway is mediated by three enzymes, an isomerase (gene *A*), a kinase (gene *B*) and an epimerase (gene *D*). In addition, there is a regular gene, *C*. All four genes lie between the *thr* and *leu* regions of the *E. coli* chromosome (Fig. 127, p. 666), and are arranged contiguously in the sequence *thr—D-A-B-C—leu*. There is also an L-arabinose permease gene (*E*) which is located elsewhere and is irrelevant to this discussion. In wild type (C^+) strains the isomerase, kinase and epimerase are coordinately inducible. Two different types of regulatory mutant have been described, all of which map within the *C* gene. In the C^c type all three enzymes are produced constitutively; in the C^- type, genes *D*, *A* and *B* are switched off and the enzymes are not inducible. By means of F-prime factors carrying the *ara* region, the dominance relationships of these three *C* alleles have been investigated. It turns out that they are quite different from those between mutations yielding the same phenotypes in the lactose pathway. Thus i^s mutations, which lead to non-inducible switching off of the *lac* operon, are dominant in *trans* to i^+ inducible and i^- constitutive mutations since they yield a cytoplasmic super-repressor which is not inactivated by inducer; on the contrary, the phenotypically equivalent *ara*C^- mutations are *recessive* in *trans* to both C^+ in the presence of arabinose, and C^c in its absence. This means that the inert behaviour of the arabinose operon in *ara*C^- mutants results from lack of a cytoplasmic substance which is made by C^+ mutants when inducer is present, and by C^c mutants in its absence. In other words, the function of the C^+ gene is to produce a cytoplasmic substance which, when activated by the inducer, switches on the operon; in the case of C^c

mutants, the structure of the *C* gene product has been so altered that it no longer requires inducer to activate it. Moreover, unlike the *araC*^{c} mutations, *o*^{c} mutations in the *lac* operon are dominant only in *cis* because they do not act through a cytoplasmic product but merely render the operator insensitive to repressor.

From all this it is reasonable to suppose that the product of the *araC* gene is an allosteric protein with two specific sites, one for the activating arabinose molecule and the other for some kind of promoter region on the operon. However, a peculiar complication in this otherwise straightforward story of positive control is that C^+, *in the absence of inducer*, turns out to be dominant over C^c in *trans*. This shows that the C^+ gene produces a cytoplasmic substance in the absence of inducer, but it is far from clear how this substance prevents activation of the operon by the C^c product. One possibility is that the C^+ product has a greater affinity for the promoter than the C^c product, and that it is activated *in situ* by the inducer; another is that an interaction occurs between the molecules of C^+ and C^c protein, or their subunits, which changes the active configuration of the latter. Deletion mutants which involve only the *C* gene abolish the function of all the genes of the *ara* operon but, as we have seen, inducible enzyme synthesis is restored by the introduction of a C^+ gene in *trans*. On the other hand, deletions which extend through gene *C* into gene *B* are *cis* dominant and render the operon insensitive to activation by C^+, so that the site of action of the activator is probably located between genes *C* and *B*, and is designated *I* in Fig 127. (Englesberg *et al.*, 1965; Sheppard and Englesberg, 1966, 1967).

The beauty of the operon concept is its great flexibility for, as we have seen, it can be adapted to explain not only induction and repression, but also both negative and positive systems of control. Are there any exceptions to the hypothesis? An apparent exception is found in the biosynthesis of arginine. The pathway of arginine synthesis is mediated by eight enzymes. In both *E. coli* and *Salm. typhimurium* the genes determining four of these (enzymes 2, 3, 5 and 8) are clustered close to the methionine *B* and *F* loci, while the remaining four genes are widely scattered on the chromosome (Fig. 127, p. 666). The synthetic activity of all these genes is repressed by arginine. Mutant strains which are resistant to the arginine analogue, canavanine, can be selected and are found to synthesise all the enzymes of the pathway in the presence of arginine. The mutational sites of these derepressed mutants all map in a small region (R_{arg} or *arg R*) close to the streptomycin locus (see Gorini, Gundersen and Burger, 1961). The conclusion is that the wild type R^{+}_{arg} gene determines the synthesis of a single co-repressor which, when activated by arginine, can switch off a series of unlinked genes, each of which must therefore be assumed to have an operator of its own. In such a system one would anticipate that the separate genes would be switched on and off at the same time, but not necessarily to the same degree as in true coordinate induction or repression, because in this case the genes are not transcribed into a single, polycistronic molecule of m-RNA. This is what is found experimentally, even among the clustered *arg* genes. An *E. coli* mutant was isolated in which two genes of the *arg* cluster, which are adjacent in the wild type, had become widely separated by a chromosomal rearrangement so that they now lay on opposite sides of the *metB* gene. However this transposition had no effect on the repression of either gene which showed the same relative differences in rates of enzyme synthesis in both wild type and mutant (Baumberg, Bacon and Vogel, 1965).

If the arginine repressor acts by combining specifically with an operator, we must assume that most, if not all, of the arginine genes are equipped with individual operators and are not part of an operon. However this exception, which may be only one of many in bacteria, does nothing to invalidate the general hypothesis of the control of gene function by the action of repressors or activators on operators but, in fact, serves to widen its potential applicability to fungi and other more highly developed forms of life which show little or no evidence of the clustering of genes with related functions. The selective disadvantage of dispersion of the genes of an operon obviously disappears if this dispersion is accompanied by reduplication of the operator so that each gene acquires its own. The fact that synchronous but not coordinate regulation may be found in the absence of gene clustering is yet another example of a prediction of the operon hypothesis which has proved correct.

The nature of repressors and operators

The fact that inhibitors of protein synthesis do not appear to interfere with the production of repressor in the regulation either of phage multiplication (p. 468) or of lactose utilisation (p. 714) at first suggested that repressors might not be protein in nature. An obvious alternative was a polynucleotide molecule which can easily recognise short base sequences in the DNA and could inhibit m-RNA synthesis by specific bonding to the operator region. Soon, however, good theoretical reasons began to be put forward for thinking that repressors should be protein or, at least, partly protein in nature. For instances, only protein molecules possess sufficient complexity to recognise directly the small metabolites which activate or inactivate repressor function; if the repressor were RNA, an enzyme similar to amino acyl-*t*-RNA transferases (p. 284), determined by another gene, would be required to effect the recognition, so that R^- mutations should normally be found to arise at two loci. Secondly, the capacity of the products of *R* genes to be activated or inactivated by small molecules, presumably by the induction of conformational changes, and their susceptibility to mutational changes in these respects, are just the properties we have found to characterise allosteric proteins (Jacob & Monod, 1963a).

The first experimental evidence for a protein repressor was the finding that *E. coli* bacteria carrying suppressors of chain-terminating mutations, which are known to operate at the level of protein synthesis (pp. 362, 719), can be lysogenised by a small proportion of clear-plaque (c_1) mutants of phage λ, whose virulence is due to de-repression (p. 466; Jacob, Sussman and Monod, 1962). Similarly suppressible amber mutations of the *i* gene (i^-) of the lactose system have also been reported (Bourgeois, Cohn & Orgel, 1965). Finally, the repressor of the *lac* operon has recently been isolated and shown directly to be at least partly protein, by purifying that fraction of *E. coli* protein which specifically binds to a radioactive, gratuitous inducer, isopropyl-thiogalactoside (IPTG). That this protein is really the product of the i^+ gene was confirmed by its absence from i^-, i^s and *i*-deletion mutants, and its altered affinity for inducer in a strain with a mutation in the *i* gene leading to increased susceptibility to the inducer. The protein has a molecular weight of the order 200,000 and about ten copies per i^+ gene appear to be present (Gilbert & Müller-Hill, 1966). Phage λ repressor molecules have also been isolated and shown to bind to λ DNA, but not to DNA from a phage λ-434 recombinant strain which carries the phage 434 immunity region (Ptashne, 1967b). The isolation and properties of both types of repressor are reviewed by Bretscher (1968).

Hitherto, the repression of amino acid pathways has been assumed to follow direct activation of the co-repressor by the end-product of the pathway, that is, by the amino acid. The studies by B. Ames and his colleagues of the genetics and biochemistry of the histidine pathway in *Salm. typhimurium* suggest that this may not be so. Mutants de-repressed with respect to the histidine biosynthetic enzymes may be obtained by selecting for resistance to the histidine analogue, triazolealanine (p. 708). Genetic analysis shows that the mutations fall into four classes. The *hisO* class are operator mutations and map at the *G* gene end of the operon (Fig. 46, p. 163). A second class, *hisS*, constitute mutations unlinked to the *his* operon and there is good evidence that they involve the structural gene for histidyl-*t*-RNA synthetase, since de-repression is associated with a decreased growth rate and a reduced amount of charged histidyl-*t*-RNA which can be reversed by increasing the internal pool of free histidine. This suggests that the true activator of the co-repressor may be histidyl-*t*-RNA rather than histidine. This hypothesis is supported by the finding that the third mutational class, *hisR* mutations, which are located far from the histidine operon, result in a lowering of the histidine *t*-RNA content to 55 per cent. of normal. Although the *t*-RNA produced by these mutants appears to be qualitatively normal, the fact that the mutants are strongly de-repressed for histidine points to an active role of *t*-RNA in repression. The fourth de-repressed class of mutation, *hisT*, maps independently of the other classes and nothing is known of its function. A possible model of repression, based on these facts, is that the histidyl-*t*-RNA might operate at the level of protein synthesis by directly blocking the site on the *m*-RNA where translation of the operon commences. Another possibility opened up by these studies is that the protein co-repressor is the amino acyl-*t*-RNA synthetase itself (Roth *et al.*, 1966; Roth and Ames, 1966; Ames *et al.*, 1967).

Finally, what do we know about the nature of the operator? Two features distinguish it from other chromosomal regions. One is that it is specifically recognised by, and binds, a protein—the repressor. The other, which certainly holds for the *lac* operon, is that all mutations affecting its function appear to be deletions. These features imply that rather a large number of base pairs is involved and that the specificity is of a kind which is affected only by relatively drastic alterations of base sequence, and suggest that the DNA may be arranged in a two- or three-dimensional conformation. A model for the operator, based on the known structure of *t*-RNA molecules, has therefore been proposed by Gierer (1966) in which the two strands of the DNA possess unpaired regions in which base homologies along individual strands produce branches and loops which have specific interactions with repressor protein (Fig. 72, p. 283). Such a folded structure might be stabilised in a particular conformation by the attachment of repressor so that the promoter site is blocked; alternatively, in positive systems, the confirmation impressed by the activator would be such as to expose the promoter site.

Reviews and assessments of various aspects of the genetic regulation of biochemical pathways, in addition to those already mentioned, are by Riley & Pardee (1962), Ames & Martin (1964), Brenner (1965), Cohen (1965), Beckwith (1967) and Vogel & Vogel (1967). The Cold Spring Harbor Symposium, 1961 (Vol. 26) is devoted exclusively to this subject, while the Symposia for 1963 (Vol. 28) and 1966 (Vol. 31) contain many important papers relevant to it.

THE REGULATION OF DNA SYNTHESIS: THE REPLICON

In Chapter 19 (p. 553 *et seq.*) we dealt in considerable detail with the replication and segregation of the bacterial chromosome and we do not intend here to do more than introduce an extension of the concept of regulators and operators to the control of DNA synthesis, as has been proposed by Jacob & Monod (1963a, b; Jacob & Brenner, 1963; Jacob, Brenner & Cuzin, 1963).

The genetic study of bacteria and their viruses has brought to light some interesting facts about the replicative behaviour of genetic material. First of all, when a fragment of donor chromosome is transferred to a recipient bacterium, although it can be integrated into the recipient chromosome by recombination it appears, in general, to be unable to replicate. This is particularly evident in the case of abortive transduction (p. 96) but is also shown by the fact that stable, partial diploids hardly ever arise, even in the case of conjugation where the greater part of the donor chromosome may enter the zygote. On the other hand, the genomes of temperate bacteriophages or of the *E. coli* sex factor, despite their small size, behave quite differently. In their autonomous state they replicate at their own pace in the bacterial cytoplasm, but when integrated with the bacterial chromosome their replication becomes coordinated with that of the host. Moreover, when a fragment of bacterial chromosome, which is unable to replicate when introduced by itself into a recipient cell, becomes incorporated into the genome of either temperate phage or sex factor, it acquires the property of autonomous replication. This leads to the idea that a complete chromosome, whether of bacterium, sex factor or phage, carries a determinant or determinants which are necessary to initiate replication; once replication has started it runs to completion, copying whatever base sequences of DNA may be included in the continuity of the chromosome, irrespective of their origin. A genetic element which is capable of independent replication may thus be regarded as a *unit of replication* or *replicon.*

The vegetative reproduction of prophage, or of superinfecting phage in lysogenic bacteria, is specifically prevented by synthesis of a repressor determined by a phage gene (p. 463). Release of the prophage from its repressor, however, leads to replication at its autonomous rate and cannot be invoked to explain its normal, coordinated division with the bacterial chromosome. It therefore seems more likely that DNA synthesis is initiated by a *positive* stimulus which is specific for each kind of replicon.

Accordingly, the following model, which is formally analogous to an activator-operator system, has been proposed (Jacob & Monod, 1963a; Jacob & Brenner, 1963). Each replicon carries two loci necessary for its replication. One determines the production of a specific, cytoplasmic element, called *initiator*, which acts on an operator region, termed the *replicator*, in such a way that the DNA duplex is opened up at this point and so can serve as a primer for new DNA synthesis. The synthesis then progresses automatically until the whole replicon has been copied, but cannot be repeated in the absence of the stimulus provided by new initiator molecules.

This model accounts for many of the features of DNA replication in bacteria and their viruses. For instance, in lysogenic bacteria the phage replicon (prophage) is prevented from multiplying at its own rate because the production of its specific initiator is inhibited by repressor; instead, it behaves as a functional part of the bacterial replicon with which it is integrated, so that its replication is controlled by the bacterial initiator and replicator which are unaffected by phage repressor. In infection by vegetative phage, on the other hand, the phage initiator is produced along with the early enzymes (p. 428) so that the phage replicon begins to be copied independently of the bacterial chromosome and at its own rate. A similar type of regulation is assumed to apply to the sex factor in its integrated and autonomous states except that, in this case, autonomous replication is self-limiting (see p. 775). Similarly, a fragment of bacterial chromosome which is incorporated into, say, a phage genome becomes part of the phage replicon. In heterogenotes its replication is controlled by the bacterial replicator, but when the prophage enters the vegetative state following induction it obeys the phage replicator. Finally, the inability of fragments of bacterial chromosome to multiply by themselves when transferred to recipient bacteria is explained by their lack of a replicator. Indeed, it has been suggested that Lederberg's (1949) persistent diploid strains of *E. coli* might have arisen from the inheritance of donor fragments which happened to carry the replicator, since such fragments would be expected to reproduce themselves in response to initiator produced by the recipient chromosome (Jacob & Monod, 1963a).

There now exists a considerable body of evidence in support of these ideas. For instance, if it were valid to generalise the information derived from different bacterial and viral systems, which we have reviewed in Chapter 19, we find that chromosomes are continuous structures, that replication begins at a specific locus, and that it requires protein synthesis for its initiation. The extent to which attachment of chromosomes to the cell membrane is important in regulation as well as in segregation is not yet known. At the moment, in several laboratories, genetic and functional analyses are being made of temperature-sensitive bacterial mutants which are defective in some aspect of chromosome replication, and there is little doubt that these studies will cast new light on how this process is controlled.

PHASE VARIATION IN SALMONELLA

Bacteria of the genus *Salmonella* characteristically possess flagella and are motile. They are classified into a large number of distinct 'serotypes' according to the specific antigens which they carry in their cell walls (O antigens*) and in their flagella (H antigens*). A minority of *Salmonella* serotypes are unambiguously defined by their specific reaction with only one O and one H antiserum; they are antigenically stable or *monophasic* types. In the case of the other serotypes, however, the flagella alternate between one and the other of two distinct antigenic structures in the following manner. The flagella of a single bacterium are exclusively of one antigenic type

* These seemingly inapt designations for somatic and flagellar antigens were derived from the original study of the distinctive types of agglutination reaction displayed by motile (flagellated) and non-motile (non-flagellated) strains of *Proteus*. Motile strains of this organism grow out from a point of inoculation on the surface of a solid medium to form a spreading *film* (German=*hauch*=H) of growth; non-motile strains grow as discrete colonies (=no film=*ohne hauch*=O) (Weil & Felix, 1917).

which we will call phase 1. When this bacterium multiplies, progeny cells arise with a certain probability which possess flagella of the alternative antigenic type—phase 2. When one of these phase 2 bacteria multiplies, it, in turn, throws off progeny whose flagella have reverted to phase 1, and so on. Thus, depending on the rates of variation in each direction, cultures of such diphasic organisms will accumulate bacteria of two distinct serological types with respect to their flagella, each agglutinable by its specific H antiserum, whose proportions will ultimately reach equilibrium after prolonged cultivation (p. 199).

The rate at which variation from one phase to the other occurs differs greatly from strain to strain. In strains studied by Stocker (1949) the probability of variation was random and ranged from 10^{-5} to 10^{-3} per bacterial division, so as to mimic the effects of forward and reverse mutation (see also, Makela, 1964). On the other hand the variation may be so rapid that the picking and serological testing of individual colonies suffices to detect it. Furthermore, diphasic strains may become stablilised in one or the other phase so that the alternative, suppressed phase can be demonstrated only by rigorous selective methods.

Flagella appear to be built up from homogeneous subunits of a globular protein, flagellin (pp. 259, 262), whose antigenic specificity depends on amino acid composition and arrangement so that the antigenic structure of flagella would be expected to be determined by a single gene (see Kerridge, 1961; McDonough, 1965). The normal restriction of phase variation to only two alternative structures, as well as the rapidity with which it may occur, suggests, *a priori*, that the phenomenon is unlikely to be due to mutational changes within a single locus but, rather, that the flagellins characteristic of the two phases are determined by separate structural genes which are alternately expressed. Genetic analysis, by means of transductional crosses between diphasic *Salmonella* strains of different serotype, has shown that this is indeed the case, and that the expression of the structural genes is regulated at other loci which can be genetically defined but whose behaviour at the biochemical level is still a mystery.

We will illustrate the main features of this system by means of two serotypes which have been widely employed in its analysis; *Salm. typhimurium* with the alternative flagellar structures *i* and *1.2* ($i \rightleftharpoons 1.2$), and *Salm. abony* ($b \rightleftharpoons e.n.x$). In transduction studies mediated by phage P22, in which either strain can be used as donor and the other as recipient, transductants inheriting one or the other donor flagellar structure can be selected in the following way. A culture of, say, *Salm. abony* ($b \rightleftharpoons e.n.x$) is infected with phage grown on *Salm. typhimurium* ($i \rightleftharpoons 1.2$) and spread on soft, nutrient gelatin-agar through which motile bacteria can migrate from the site of the inculum to yield 'flares' of growth (p. 97 and Plate 3). If, however, specific antisera against both phases of the flagella of the *recipient* bacteria are incorporated in the medium, the flagella are immobilised; only transductants inheriting flagella of donor type, for which no antibodies are present, can swarm to produce flares. Bacteria from the individual flares which appear are picked and their flagellar structure determined serologically. In this way it is found that crosses between an $i \rightleftharpoons 1.2$ donor and a $b \rightleftharpoons e.n.x$ recipient give rise to $i \rightleftharpoons e.n.x$ and $b \rightleftharpoons 1.2$, but never to $i \rightleftharpoons 1.2$ (donor type) recombinants. This means that, of the determinants of the four antigens involved, *i* and *b* (phase 1) on the one hand, and *1.2* and *e.n.x* (phase 2) on the other, are allelic, while the determinants of the two phases are unlinked in transduction. The alleles of phase 1 are said to belong to the H_1 locus, and those of phase 2 to the H_2 locus (Lederberg & Edwards, 1953).

Non-motile mutants of *Salmonella* strains arise spontaneously and can also be obtained by treatment with mutagens. These mutants usually possess no visible flagella and are designated *fla*. Transductional crosses between independent *fla* mutants, whether isolated from the same or different serotypes, give rise to motile recombinants which produce flares on soft gelatin-agar devoid of antiserum, so that the ability to synthesise flagella is determined by a small *fla* region which probably constitutes a single locus. If, now, a non-motile (*fla*) mutant derived from an $i \rightleftharpoons 1.2$ serotype (*Salm. typhimurium*) is crossed, as recipient, with a motile (fla^+), $b \rightleftharpoons e.n.x$ donor (*Salm. abony*), motile recombinants of both $b \rightleftharpoons 1.2$ and $i \rightleftharpoons 1.2$ types are found. This implies that the $H_1{}^b$ and fla^+ loci of the donor are linked in transduction, while the *fla* mutation of the recipient has not involved its H_1^i locus determining the structure of its phase 1 antigen (*i*). We can represent the situation formally as

H_1^b $+$

———————— Donor

1 2 3 ,

—————————— Recipient

H_1^i *fla*

where crossovers in intervals 1 and 3 yield motile recombinants with type b flagella, and crossovers in intervals 2 and 3 motile type *i* recombinants. On the other hand, the phase 2 antigenic structure of the donor (*e.n.x*) is never inherited with the *fla*$^+$ character among recombinants. The H_1 and *fla* loci are therefore closely linked to one another, and are unlinked to the H_2 locus (Fig. 138B, p. 743) (see Stocker *et al.*, 1953). In fact it is now known from analysis of *Salmonella* by conjugation (see next chapter) that the H_1 and H_2 loci are widely separated on the chromosome (Smith & Stocker, 1962; Makela, 1964; Sanderson & Demerec, 1965).

Further information about the genetic and functional relationships between the two flagellar phases of diphasic serotypes comes from the elegant studies of J. Lederberg and T. Iino. Young cultures of single colonies of diphasic strains usually consist almost entirely of bacteria in which only one of the two flagellar phases is expressed. It is therefore possible, by transductional crosses, to investigate whether the *expression* of one or the other phase in the recipient affects that of incoming phase determinants from the donor. The results of a typical experiment, in which inheritance of the H_1 (phase 1) locus from the donor is alone selected, are shown in Table 26. The first thing to note is that all the recombinants are diphasic; whatever phase they display on isolation, they have a normal potentiality to express the alternate phase. The most striking feature of the experiment is that the phase which the recombinants exhibit depends on the phase which is expressed in the recipient. For example, the selected donor phase 1 is expressed in all the recombinants when the recipient is also in phase 1 (Table 25, A), but is expressed in none of those which inherit it when the recipient is in phase 2 (Table 25, B). Conversely, the expression of phase 2 in the donor (Table 25, C, D) has no effect on the transduction of its phase 1 locus, but this locus is only expressed in the recombinants when the recipient is also in phase 1 (Lederberg & Iino, 1956).

The expression of phase 2 thus appears to dominate that of phase 1. Is this dominance due to some cytoplasmic factor in the recipient and, if so, is this determined at the H_2 locus or elsewhere? By the use of appropriate immobilising antisera it is possible to select for all four combinations of donor and recipient phases among recombinants. It turns out that recombinants which express the donor phase 2 are found only when the donor is itself in phase 2, and then arise at about the same frequency irrespective of the phase of the recipient.

TABLE 25

Flagellar phase variation in *Salmonella*: transductional analysis demonstrating the influence of the phases expressed in the parental strains on the flagellar phenotype of recombinants.

	Donor $b \rightleftharpoons e.n.x$ (1) (2)	Recipient $i \rightleftharpoons 1.2$ (1) (2)	No. of $b \rightleftharpoons 1.2$ recombinants in phase	
	Expressed phase	Expressed phase	1	2
A	1	1	14	0
B	1	2	0	27
C	2	1	18	0
D	2	2	0	20

In these crosses, mediated by phage P22, the donor strain was flagellated (+) and the recipient non-flagellated (*fla*). Recombinants were selected for inheritance of the donor fla^+ gene which is closely linked to the phase 1 determinant $H1^b$. Under these conditions a proportion of the fla^+ recombinants have the recipient flagellar phenotype ($i \rightleftharpoons 1.2$) but, for simplicity, only those inheriting the donor phase 1 genotype ($b \rightleftharpoons 1.2$) in addition to motility are recorded here.

The prevalent phase of the non-motile recipient strains was determined by the serological testing of aliquots from cultures of single colonies after restoration of motility by transduction (see Lederberg & Iino, 1956).

For explanation, see text. Data selected from those of Lederberg & Iino, 1956.

This means that if a recipient in phase 1 inherits the H_2 locus from a donor *in which phase 2 is expressed*, the expression of the recipient. phase 1 is inhibited in the recombinant. Assuming that only genetic material is transported from donor to recipient in transduction, it follows that the determination of phase is a function of the *state* of the H_2 locus itself, or, possibly, of some other locus very closely linked to it. Thus when the H_2 locus is active, the function of the H_1 locus is suppressed; when the H_2 locus becomes inactive, the H_1 locus automatically expresses itself again by the synthesis of phase 1 flagella (Lederberg & Iino, 1956).

What is the mechanism which switches the H_2 locus on and off? We have mentioned that strains of diphasic *Salmonella* serotypes are found which have become stabilised in one or the other phase.

While these strains behave as if they were monophasic, they still carry the determinant of the suppressed phase since this can readily be transduced to other serotypes. Transductional crosses between a strain stabilised in phase 2, and a strain of different serotype which is subject to normal variation, have revealed that stability of variation is determined at a locus (Vh_2) which is very closely linked to, but separable from, the H_2 locus (see Fig. 138). When the allele determining normal variation (Vh_2^+) is inherited by stable recipients, the recombinants undergo variation at the normal rate. In the reverse cross, when the allele Vh_2^-, which stabilises phase 2 in the stable donor, is inherited by diphasic recipients in phase 1, the recombinants turn out to be stabilised in phase 1 and not in phase 2. Since the expression of phase 1 depends on whether the H_2 gene is active or inactive, the fact that the Vh_2^- mutation can stabilise either phase indicates that it merely prevents the H_2 gene from changing its state. By implication, the wild type allele (Vh_2^+) is responsible for controlling the rate of variation (Iino, 1961a). Since the effect of the Vh_2 locus is specific, its action is quite different from that of mutator loci (p. 219).

Two other types of mutation have been described which prevent synthesis of flagella but, unlike *fla* mutations, are phase-specific. One type (Ah_1) maps at a locus very closely linked to the H_1 and *fla* loci and results in inability to form phase 1 flagella, although phase 2 flagella are normally produced. The other type (Ah_2) involves a locus closely linked to the H_2 locus and specifically blocks the synthesis of phase 2 flagella. Diphasic variation in a strain carrying a mutation of one or the other of these types is manifested as an alteration between motility and non-motility (Iino, 1961b). Finally, there is a locus in the H_1 region (*nml*) which determines the N-methylation of lysine to form the unique amino acid, ε-N-methyl lysine (p. 259), which is incorporated into the flagellins of both phases in many diphasic *Salmonella* serotypes (Stocker, McDonough & Ambler, 1961).

To what extent is this interesting type of variation explicable in terms of the regulatory systems we have discussed earlier in this chapter? We saw that a definition of these systems depends largely on a knowledge of the dominance relationships between various mutations affecting the regulation. In the case of mutations involving flagella formation, dominance can be investigated in abortive transduction, where the donor genetic fragment is transmitted unilinearly

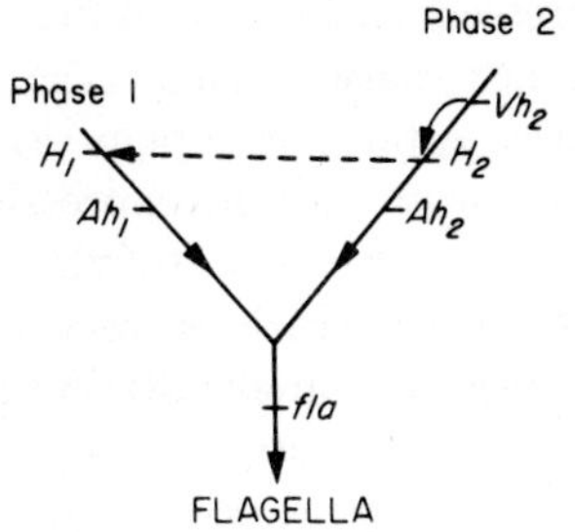

A. Functional interactions of loci determining flagellar structure and behaviour

B. Genetic map of the loci

FIG. 138. Schematic representation of the functional (A) and genetic (B) relationships between the loci determining flagellar structure and behaviour in *Salmonella*.

The various loci are as follows:

H_1 and H_2 determine the antigenic specificities of phase 1 and phase 2 flagellins respectively.

Ah_1 and Ah_2 determine the specific synthesis of flagella of phase 1 and phase 2 respectively.

Vh_2 regulates the rate at which the H_2 locus alternates between expression and non-expression.

fla determines the synthesis of flagella of both phases.

nml determines the N-methylation of lysine and the incorporation of ε-N–methyl lysine into the flagella of both phases.

In A, the arrows indicate, in a formal way, the presumed pathways of synthesis and the regulatory interactions between different loci.

In B, the continuous lines indicate linkage, and the interrupted line no linkage, between the loci in transduction. The brackets enclosing groups of loci show uncertainty as to their order.

in *trans* (Fig. 26, p. 98), by observing the inhibition of 'trail' formation in semi-solid media by specific flagellar antisera (p. 97). This sort of analysis has shown that when a recipient in phase 1 has a fla^- mutation closely linked to the H_1 locus, so that the abortively transduced donor fragment which restores motility also carries the H_1 gene determining donor phase 1 flagella, then the phase 1 flagella of both parents are expressed. In contrast, if the recipient is in phase 2, the expression of the donor phase 1 flagella is suppressed. This means that the recipient in phase 2, but not in phase 1, produces a cytoplasmic substance which can switch off the expression of the H_1 gene of the *trans* exogenote (Pearce & Stocker, 1967).

In other experiments, a recipient was used which carries an Ah_1 mutation, and so is non-motile in phase 1 but normally motile in phase 2. When this strain, in phase 1, is abortively transduced to motility by a phase 2 donor, only the donor phase 2 allele is expressed. This implies that the mechanism which switches on the H_2 region (including, presumably, the gene which produces repressor for the H_1 region) acts only in *cis*, since the recipient H_2 gene is not activated. The question arises whether the repressor of H_1 activity might not be the H_2 flagellin itself. It happens that there is an exceptional *Salmonella* strain which, in phase 1, produces flagellin of antigenic structure 1,2, which is apparently identical with that normally produced by other strains in phase 2. When this $H_1^{1,2}$ allele was abortively transduced to a recipient strain expressing the phase 1 allele H_1^b, flagella of both types were produced; it is therefore unlikely that either phase 2 flagellin or its messenger RNA acts as phase 1 repressor (Pearce & Stocker, 1967).

A model to explain phase variation has been proposed, according to which the H_2 region comprises an operon which is normally repressed by the product of a closely linked R_2 gene. The H_1 region is also subject to repressor control but, in this case, the R_1 determinant is located in the H_2 operon and is switched off, so that H_1 flagellin is normally synthesised. However, a mutation in the R_2 gene will lead to derepression of the H_2 operon and synthesis of H_1 repressor, so that phase 1—phase 2 variation results (Klein, 1964). According to this hypothesis, reversion to phase 1 would be due to mutational reversion at the original R_2^- site, but this should be a very rare event as compared with the first mutation; in fact, any hypothesis involving forward- and back-mutation is inadequate to explain the regularity of the variation and its high frequency in some systems. Moreover, the hypothesis is at variance with the facts since it predicts that an expressed phase 2 in the donor, being due to absence of H2 repressor, should not be transducible to a recipient in phase 1 which produces phase 2 repressor; as we have seen, the reverse is the case. (For discussion, see Iino & Lederberg, 1964; Pearce & Stocker, 1967).

Elucidation of the genetic mechanism underlying this type of variation is clearly of great interest as a possible model for differentiation. The various genetic relationships and functional interactions between the loci we have described are summarised schematicallly in Fig. 138. There is evidence that phase variation is not restricted to bacterial flagella. A similar type of alternation between the presence or absence of certain somatic (O) antigens of *Salmonella* has been described. The variation here may normally be very rapid, while strains which have become stabilised in either phase are known to occur (Kauffman, 1941; Edwards & Bruner, 1946; Hayes, 1947). These systems have not yet been subjected to genetic analysis. In addition, phase variation between the production of two, non-allelic types of α-amylase has been described in *B. subtilis* (Green & Colarusso, 1964).

CHAPTER 24

Sex Factors and other Plasmids

One of the most interesting, and certainly the strangest, of the discoveries to issue from the genetic study of bacteria is the existence of a new kind of genetic element, the sex factor, which determines the ability of *E. coli* K-12 bacteria to conjugate and to transfer genetic material to recipient cells. We have seen, in Chapter 22 (p. 669), that the sex factor is a small, supernumerary chromosome, serving a number of functions pertaining to the sexual process, which has the peculiar property of being able to exist in alternative states in its host cell. In one state it is inserted into the bacterial chromosome at one or another of many possible sites, behaves in inheritance as if it were a chromosomal determinant and, at conjugation, promotes the linear, oriented transfer of the chromosome in such a way that it, itself, is the last marker to be transferred at the chromosomal tail. Moreover, in this state the sex factor can participate in rare genetic exchanges with neighbouring regions of the host chromosome, incorporating bacterial genes into its own structure. In its alternative state the sex factor exists in the cytoplasm, where it replicates autonomously and appears unable to promote chromosomal transfer at conjugation.

Were the sex factor of *E. coli* K-12 unique, we might regard it merely as an intriguing novelty evolved by Nature to fill a very specialised need. In fact it turns out that the sex factor is only a special case of what appears to be a very general phenomenon—at least among the bacteria. The first indication of this was presented by the remarkable similarities which were found between the sex factor and certain temperate bacterial viruses, of which phage λ is the prototype (Chapter 17, p. 452). When in the prophage state, the genetic material of the phage is located at a specific site on the bacterial chromosome, replicates coordinately with it, and can incorporate neighbouring fragments of host chromosome into its structure. Alternatively, in a proportion of newly infected cells, or following the induction of lysogenic bacteria, the phage genome enters the vegetative state in which it replicates autonomously and expresses its viral functions. These properties epitomise those of the sex factor, save only that autonomous growth of the phage is usually unrestricted (but see p. 449) so that the host is killed and the phage genetic material transferred to a new host by way of the environment, instead of through a conjugation tube. Moreover, the genomes of both phage and sex factor are of approximately the same size, consisting of DNA equivalent to about one-hundredth that of the host chromosome, while both, when lost from host cells, can be reacquired only by infection. Finally, we may note that both prophage and sex factor may become functionally defective as a result of mutation. In the case of prophage, mutation to virulence also occurs; there is no reason to suppose that lethal sex factors may not also arise from mutations involving the mechanism which regulates their autonomous replication, though this has not yet been observed.

Thus although, at a superficial level, phage λ is clearly what has long been called a virus, while the sex factor behaves more like an additional chromosome, these two entities can now be seen to resemble each other closely from a genetic point of view. Both are independent genetic structures which are additional to the normal genome of the cell they inhabit, are transmissible by infection, and may be propagated in alternative states—either autonomously in the cytoplasm or as an integral part of the host chromosome. For this new category of genetic element Jacob and Wollman (1958c) coined the word *episome*. However, as we shall shortly see, many varieties of conjugation factor, more or less resembling the *E. coli* sex factor, have since been found but in none of these has a stably integrated chromosomal state yet been demonstrated. Furthermore the *E. coli* sex factor, although freely transmissible to and propagated by *Shigella* species, in which it promotes conjugation and its own transfer, has no chromosomally inserted state—presumably because it lacks sufficient genetic homology with the chromosome (see below). In addition, a number of autonomous cytoplasmic factors have been described which are unable to promote conjugation. Nevertheless, all these factors clearly belong to the same general class, for which Lederberg (1952) proposed the name *plasmid* to connote all extranuclear genetic structures which can reproduce autonomously, that is, which are independent replicons. We think that the word 'episome', although an excellent substitute for 'plasmid', has become a source of confusion because the existence of alternative chromosomal and cytoplasmic states was the central feature of its original usage. Like the word 'cistron' (p. 174), it has served an admirable purpose as a centre of gravity around which our ideas have revolved and interacted before settling into more stable orbits. It now seems to us that the most meaningful biological distinction is between those plasmids which promote conjugation, which we will class together as *sex factors*, and those which do not. Let us now examine some of the data on which this tentative classification is based.

OTHER SYSTEMS OF CONJUGATION

BARRIERS TO FERTILITY

A search for the existence of sexual compatibility between different strains or species of bacteria rests upon the isolation and identification of recombinants which are hybrid with respect to one or more characters of a pair of strains, whether or not the determinants of these characters are carried by the chromosome or by plasmids. While positive results are obviously meaningful, the interpretation of negative results is beset with difficulties. For example, our study of sexuality in *E. coli* K-12 has shown us that fertility in this strain depends on the presence of a sex factor; had the original strain chanced to have lost this factor, no inter-fertility between auxotrophic mutants of it would have been found. Again, conjugation and chromosome transfer between different species may occur, but no recombinants arise because of a lack of genetic homology. In addition, even though strains are otherwise potentially fertile other incompatibilities not directly related to the sexual process may intervene. Among the more important of these are the rejection of immigrant 'foreign' DNA which is part of the phenomenon of host-controlled restriction and modification (p. 514) and the production by one of the strains of a temperate bacteriophage, or a bacteriocin such as colicin (p. 755), to which the other strain is susceptible. Even when none of these barriers to fertility is present, a low frequency of recombinant formation must be anticipated and a selective system employed; but, as a rule, wild type strains are not naturally endowed with suitable selective markers so that large surveys are precluded.

CONJUGATION BETWEEN DIFFERENT E. COLI STRAINS

Many independent strains of *E. coli* have been found which yield recombinants for chromosomal genes in crosses with *E. coli* K-12.

Some were reported before the donor-recipient relationship was discovered. The first of these was one of seven naturally occurring auxotrophic strains tested (Cavalli-Sforza and Heslot, 1949). Then followed a survey of some 2000 prototrophic, streptomycin-sensitive strains, about 50 of which gave rise to streptomycin-resistant prototrophs when crossed with a streptomycin-resistant K-12 auxotrophic strain, which subsequently turned out to be a recipient (Lederberg, 1951b; Lederberg and Tatum, 1953). Thus a proportion of *E. coli* strains were shown to act as donors of presumptively chromosomal genes. In other surveys, a number of such donor strains were shown to harbour sex factors which vary in their transmissibility to K-12 recipients. *E. coli* K-12 recipients successfully infected with these heterospecific sex factors become donors which may differ from normal F^+ strains in crosses, with respect to the yield of recombinants and their genetic constitution, as well as to the efficiency and stability of sex factor transfer (Furness and Rowley, 1957; Bernstein, 1958). These independently isolated sex factors do not appear to have been investigated further, and no *Hfr* strains derived from them have been reported. Recently, however, the presence of a sex factor of *E. coli* K-12, or *F*, type in as many as six out of 26 freshly and independently isolated strains of *E. coli*, has been inferred from the sensitivity of these strains to bacteriophage MS2, which specifically infects *E. coli* 'male' bacteria (Meynell and Datta, 1966a; see p. 771 below).

A rather larger proportion of *E. coli* strains appear to be fertile with K-12 donors. In an examination of 200 independently isolated strains, 18 per cent gave recombinants with K-12 F^+ and 30 per cent with *Hfr* strains, but it is not sure to what extent these rather high figures are due to homogeneity among the isolates (Ørskov and Ørskov, 1961). Strain C yields as many recombinants as K-12 recipients in crosses with K-12 donors and can be efficiently infected with the sex factor, *F*, thereafter performing as a normal donor in crosses with strain C recipients (Lieb, Weigle and Kellenberger, 1955). Moreover, *Hfr* derivatives can readily be isolated from CF^+ strains (p. 673; Sasaki and Bertani, 1965). The pairs of mating bacteria shown in Plates 32 and 33 (pp. 672, 673) are from crosses between K-12 *Hfr* and C recipient strains.

In contrast, although *E. coli* strain B yields recombinants in crosses with K-12 F^+ and *Hfr* donors, but with a lower efficiency than strain C, it is highly refractory to infection by the sex factor. However, those rare clones which do acquire it behave as normal donors in crosses with B recipients and can transfer the sex factor back to K-12 *F* strains (deHaan, 1954a, b; 1955). This difference between strains B and C is almost certainly because strain B exercises restriction, breaking down immigrant K-12 DNA, while strain C is known to be non-restricting—at least in so far as K-12 and B DNA is concerned. For instance, phage λ grown in K-12 plates on C with an efficiency of 1·0, and can also lysogenise C normally (p. 514). The ability of crosses between K-12 donors and B recipients to yield recombinants for chromosomal genes is probably due to the relatively large amount of DNA transferred, which soon saturates the restricting nucleases.

CONJUGATION BETWEEN E. COLI AND SHIGELLA

The 'generic' name *Shigella* embraces a number of so-called species or types of Gram negative, intestinal bacteria, such as *Sh. shigae*, *Sh. flexneri*, *Sh. boydii* and *Sh. sonnei*, which may differ in fermentative capacity and antigenic structure but share a common ability to produce acute bacillary dysentery in man. Despite the absence of recombination among *Shigella* themselves, it turns out that all the *Shigella* types which have been tested are fertile and yield recombinants when mixed with *E. coli* K-12 donor strains, but not with *E. coli* recipients (Luria and Burrous, 1957). This inter-species fertility reveals some features of considerable interest.

In *E. coli Hfr* × *Shigella* crosses recombinants arise at a frequency one hundred to one thousand times lower than when *E. coli* recipients are employed. *Shigella* strains cannot normally be infected with phage λ because they lack the cell wall receptors necessary for adsorption of this phage. However, when an *E. coli Hfr* strain, lysogenic for phage λ, is mated with *Shigella*, *zygotic induction* occurs at the same frequency as in *E. coli* crosses (pp. 463, 685). It follows that *E. coli Hfr* bacteria conjugate with *Shigella* cells and transfer their chromosomes to them with great efficiency, so that the dearth of recombinants must be ascribed to poor genetic homology between the *E. coli* and *Shigella* chromosomes. Confirmation of this comes from analysis of those recombinants which do arise; some proximal genes on the chromosome of the *E. coli* donor, which can be assumed to have been transferred and which are inherited by recombinants from *E. coli* crosses, are absent from them altogether (Luria and Burrous, 1957).

From mixtures of *E. coli* F^+ and *Shigella* strains, *Shigella* recipients can be isolated which have acquired the *E. coli* sex factor and can propagate it, since they are able to convert *E. coli* recipients to the F^+

state. The sex factor therefore determines the ability of *Shigella* to conjugate. Nevertheless, no recombinants arise in crosses between these *Shigella F*+ strains and either *Shigella* or *E. coli* recipients, presumably because they are unable to mobilise the chromosome for transfer; a plausible reason for this defect is that the *Shigella* chromosome lacks regions of homology at which the sex factor can insert itself (p. 672) (Luria and Burrous, 1957).

CONJUGATION BETWEEN E. COLI AND SALMONELLA: SALMONELLA-SALMONELLA CROSSES

The *Salmonella* 'genus' comprises a rather homogeneous group of pathogenic, Gram negative bacilli which, although divisible into a large number of distinct serotypes (p. 737), show considerable uniformity in their fermentative capacity and antigenic pattern, are usually prototrophic, and characteristically produce either enteric fever or gastro-enteritis in man and animals. In the absence of transducing phage (p. 620), mixtures of genetically marked *Salmonella* strains show no fertility.

When *E. coli Hfr* strains are crossed with auxotrophic salmonellae which carry a mutator gene (*mut*; p. 219), recombinants for a range of wild type donor markers can be selected but they arise only at very low frequency. However, if these recombinants are isolated and back-crossed to the *E. coli Hfr* strain, recombinants are now formed at a relatively high rate (Baron, Carey and Spilman, 1959a; Miyake and Demerec, 1959; Zinder, 1960a). It appears that the original *Salmonella* population contained only a few mutant bacteria which were capable of acting as effective recipients, probably because they no longer restrict; the recombinants issuing from them inherit the mutation and thus present no barrier to fertility in further matings. The alternative possibility, that the initial barrier is due to a lack of genetic homology which is partly restored in the rare recombinants, either by chromosomal rearrangements or by the inheritance of regions of donor chromosome, is unlikely for two reasons. In the first place, the increased fertility of the recombinants involves *Salmonella* markers which were not implicated in the first cross; secondly, fertile *Salmonella* mutants can be isolated by indirect selection, using the replica plating technique, without having participated in recombination (p. 187) (Miyake, 1959).

Analysis of *E. coli-Salmonella* hybrids reveals that the chromosomes of the two species in fact show a considerable degree of homology as judged by the range of *E. coli* genes which can be inherited. However, the hybrids also lend themselves to a novel demonstration that this gross homology does not extend to the level of the genes themselves. Wild type *E. coli* and *Salmonella* have a number of *characters* in common, such as the ability to utilise certain carbohydrates and to synthesise a range of amino acids. By conjugal crosses between appropriate mutant strains, hybrids can be made in which the source of the genes determining these various characters is known. When these hybrids are employed as donors in *transductional* crosses mediated by phage P22, it turns out that wild type transductants do not arise at significant frequency when the source of the 'alleles' is different in the two strains; mutant *Salmonella* genes cannot be transduced to wild type by transfer of the corresponding wild type *E. coli* gene and *vice versa*, although the genes determine the same phenotype (Zinder, 1960a). Thus there appears to be an appreciable lack of homology over small chromosomal regions—the corresponding genes are not truly allelic—although enough chromosomal homology exists to allow genetic exchange when larger segments are transferred by conjugation.

When *E. coli* F^+ and *Salmonella* bacteria are plated together it is possible, by a special technique, to recognise sites on the plate where conjugation has occurred between F^+ donors and recipient *Salmonella* cells. From these sites *Salmonella* clones can be isolated which carry the sex factor, but no other *E. coli* determinant. These *Salmonella* F^+ strains not only transmit the sex factor back to *E. coli* recipients but also, unrestrictedly, to *Salmonella* cells. Mixtures of F^+ and F^- *Salmonella* strains generate recombinants at about the same frequency as do equivalent *E. coli* crosses (Zinder, 1960b). Furthermore, *Hfr* mutants can be isolated from irradiated *Salmonella* F^+ strains by replica plating (p. 187), and these transfer large segments of their chromosome at high frequency to *Salmonella* recipients, as well as to *E. coli* and *Shigella* (Baron, Carey and Spilman, 1959b; Zinder, 1960b). However it seems that, in *Salmonella*, the mating bacteria form rather labile unions which are unstable in broth, so that crosses are best performed on the surface of a solid medium.

Genetic analysis of *Salmonella* can also be performed by conjugation mediated by the colicinogenic factor *colI*, and it was by this means that the circularity of the linkage group of *Salm. typhimurium* was first demonstrated (see p. 759; Smith and Stocker, 1962). More than 130 auxotrophic and other genetic markers have now been mapped in *Salm. typhimurium* by *Hfr* $\times$ F^- as well as by transductional crosses. In interrupted mating experiments, transfer of the whole chromosome takes 138 minutes at 37°, as compared with about 90 minutes in *E. coli*, but since the matings were carried out on the surface of membranes instead of in liquid media, it is difficult to assess the validity of the comparison (Sanderson and Demerec, 1965). It is interesting that the sequence and relative positions of 59 loci, whose equivalents have also been mapped in *E. coli*, turn out to be the same in both organisms. The only disparities are the *lac* region which is missing altogether in *Salmonella* and the opposite polarities of some operons (cp. Taylor and Trotter, 1967 and Sanderson, 1967). There is thus good genetic evidence that *Escherichia*, *Salmonella* and *Shigella* are phylogenetically related (Sanderson and Demerec, 1965; see also Zinder, 1960a; Miyake, 1962; Falkow, Rownd and Baron, 1962; Taylor and Demerec, 1963).

CONJUGATION IN PSEUDOMONAS AERUGINOSA (PYOCYANEA)

This ubiquitous species of Gram negative bacillus, which derives its alternative name from the characteristic production of two striking pigments, pyocyanin (blue) and fluorescin (fluorescent yellow), is classified in a family phylogenetically distinct from the *Enterobacteriaceae* to which *Escherichia*, *Salmonella* and *Shigella* belong. Holloway (1955) discovered that *Ps. aeruginosa* exhibits a system of conjugation and genetic recombination which is mediated by a sex factor of its own (*FP*) and is, in essence, similar to that of *E. coli*.

Four independent strains of this organism were studied. Genetically marked mutants of three of the strains, called FP^+, are fertile with one another as well as with those of the fourth strain. Crosses between mutants of the fourth strain, however, are invariably sterile (FP^-). The highest yield of recombinants (about one per 10^7 parental bacteria) is given by $FP^+ \times FP^-$ crosses, while fertility can be shown to depend on cellular contact (Holloway, 1955, 1956). There is substantial evidence that conjugation is followed by the one-way transfer of genetic material from FP^+ to FP^- bacteria, for when the FP^+ parent alone is specifically killed by addition of virulent phage 30 minutes after the commencement of mating, the yield of recombinants is little affected; but if the FP^- parent is sensitive to the phage the cross is sterile (see p. 654). Moreover, analysis of the recombinants reveals that most of the unselected markers are derived from the FP^- parent, so that genetic transfer from the FP^+ parent may be assumed to be partial (see p. 657). The inheritance of seven markers has been studied in $FP^+ \times FP^-$ crosses and all have been found to be linked in a manner consistent with a linear arrangement on a single chromosome (Holloway and Fargie, 1960).

The sex factors of two of the FP^+ donor strains which have been investigated appear to possess different properties. In fact, in one case there is no direct evidence for a sex factor at all since its transmission has not been demonstrated. The sex factor of the other FP^+ strain is transmissible but unstable in recipients. However, a variable proportion of recombinants inherit the donor state and some of these have been found to transmit their sex factor with high efficiency, in the manner of *E. coli* F^+ strains (Holloway and Jennings, 1958), but the sex factor is not eliminated by treatment with acridine orange (p. 656). All attempts to obtain *Hfr* strains from FP^+ cultures, by replica plating after irradiation, have been unsuccessful; nor has it proved possible to infect FP^- strains of *Pseudomonas* with the *E. coli* K-12 sex factor although, to our knowledge, the use of *F—lac*$^+$ factors which permit very rare clones of infected *lac*$^-$ bacteria to be selected has not yet been tried (see next section) (Holloway and Fargie, 1960).

A more recent study of conjugation in this organism reveals that too high population densities of parental cells suppress the development of recombinants, of which the number may increase markedly when diluted cultures are mixed. Under optimal conditions as many as 1·7 recombinants for a particular marker per 100 FP^+ bacteria were obtained—a frequency considerably higher than is found in *E. coli* $F^+ \times F^-$ crosses (Loutit and Pearce, 1965).

CONJUGATION IN SERRATIA MARESCENS

Serratia marescens is a small, prototrophic Gram negative bacillus, originally called *Chromobacterium prodigiosum* because of the red pigment which it characteristically produces. It is phylogenetically unrelated to any of the organisms we have discussed above. When auxotrophic mutants of a single strain are crossed, prototrophic recombinants arise at a very low frequency (one per 10^8 parental bacteria) and their formation depends on cellular contact (Belser and Bunting, 1956). There is some evidence of sexual differentiation in this strain, but the existence of a sex factor has not been demonstrated. For example, some of the auxotrophic mutants behave like donors since they are fertile when crossed together while others, analogous to recipients, yield no recombinants. As in the case of *E. coli* F^+ strains, irradiation of the presumptive donor strains with ultraviolet light enhances their fertility in a striking way (see p. 654), but some increase in the frequency of recombinants may also be found when 'recipient' strains are irradiated.

A similar ambiguity is apparent when recombinants are scored for their inheritance of unselected markers; although these are usually derived from the presumptive recipient, in a proportion of recombinants (which may be high) the 'donor' genotype predominates, suggesting that genetic material may be transferred in both directions. In comparison with the conjugal systems we have discussed so far, the amount of transfer is very small; only in the case of two out of many markers could any evidence of linkage be established (Belser and Bunting, 1956).

CONJUGATION IN VIBRIO

A particular strain of the cholera vibrio (*V. cholerae*) was found to give rise to a small number of recombinants (about one per 10^6 parental cells) when mixed with various other strains (Bhaskaran, 1958). This fertile strain possesses the additional property of producing an agent which lyses other *V. cholerae* isolates. The lytic agent, at first thought to be a bacteriophage, was then found to be non-reproducible and to resemble a bacteriocin (see next section). The ability to produce bacteriocin is transmissible to other strains by contact and invariably confers fertility upon them, so that the determinant of bacteriocin synthesis (P) seems also to play the role of a sex factor. As in the *E. coli* and *Pseudomonas* systems, crosses between P^+ and P^- vibrios give the highest recombinant yield, $P^+ \times P^+$ crosses are less fertile, and $P^- \times P^-$ crosses are completely sterile. P^+ cells appear to act as donors and P^- cells as recipients, but linkage between donor markers is rarely found among recombinants so that the chromosomal fragments transferred by the donors are probably very small (Bhaskaran, 1960; Bhaskaran and Iyer, 1961).

COLICIN DETERMINANTS AS SEX FACTORS

Colicins comprise a range of bactericidal agents, probably protein in nature since they are destroyed by proteolytic enzymes, which are produced by some strains of *Enterobacteriaceae* and kill, but do not lyse, other strains within this family (*Escherichia*, *Shigella*, *Salmonella*). Since the synthesis of such lethal antibiotics is a not uncommon feature of bacteria in general, the name *bacteriocin* has been proposed to denote them all (reviews: Ivanovics, 1962; Reeves, 1965; Nomura, 1967). Different colicins are designated by capital letters, as B, E1, E2, I, K, V, and may be distinguished experimentally by their diffusibility and host specificity. Like the protein of phage tails, they adsorb to specific receptors in the cell walls of sensitive bacteria; *resistant* bacterial mutants can readily be selected which have lost the ability to adsorb a particular colicin and are resistant to its action. Colicin-producing bacteria are termed *colicinogenic* and may carry the determinants (col^+) of more than one colicin; like lysogenic bacteria, they are specifically *immune* to the lethal effect of their own colicins.

Indeed, there are many similarities between colicinogeny and lysogeny except, of course, that colicin molecules, unlike phage particles, have no capacity for self-reproduction and simply kill susceptible bacteria to which they adsorb. For example, some col^+ determinants are inducible, the yield of colicin being greatly increased by ultraviolet irradiation of cultures carrying them. Moreover, there is good evidence that production of some colicins is due to a lethal synthesis; colicinogenic (col^+) bacteria carry the *potentiality* to synthesise colicin, but this is normally expressed in a random way by only a small proportion of cells in the population which die in the process (Ozeki, Stocker and de Margerie, 1959; see also Jacob, Siminovitch and Wollman, 1953, and below). Reviews of the mode of action of colicins are by Nomura (1963, 1967).

We are mainly interested here with the nature and behaviour of colicin determinants. Colicinogeny is a very stable character but, when lost, it cannot be recovered by mutation. However, the determinants of certain colicins can be transmitted from col^+ to col^- bacteria by cellular contact, not only among the strains of the same species but also between *Salmonella*, *Shigella* and *E. coli* strains which we have already seen to be compatible in conjugation mediated by the *E. coli* sex factor (Frédéricq, 1957). The behaviour of the determinants of two colicins, I and *E*1, in *E. coli* K-12 crosses reveals some essential facts about their nature. When two F^- strains, one of which carries the determinant *colI* while the other is col^- but resistant to colicin I, are grown together, a proportion of the col^- bacteria are converted to *colI* (Clowes, 1961). Thus the *colI* determinant appears to promote conjugation and its own transfer since, in its absence, $F^- \times F^-$ crosses do not conjugate. No transfer of $colE_1$ occurs under similar conditions. However, in $F^+colE_1 \times F^-col^-$ crosses, $colE_1$ transfer is effected rapidly and with great efficiency, and is unassociated with that of any other donor marker save the sex factor itself (Frédéricq, 1957, 1958). Furthermore, when the progeny of the zygotes are isolated by micromaniupulation, all are found to be *col*+; since bacteria are multinucleate, either several *col*+ elements are transferred to each zygote, or else a single transferred element multiplies therein (Alfoldi, Jacob and Wollman, 1957). Thus these two *col*+ determinants behave as autonomous cytoplasmic elements, while the *colI* factor appears to have some of the properties of a sex factor as well. An analysis of $Hfr \times F^-$ crosses involving the $colE_1$ factor at one time suggested that this determinant has an alternative location on the bacterial chromosome and therefore qualifies as an episome (Alfoldi *et al.*, 1958), but recent work has undermined the assumptions on which these experiments were based (Clowes, 1963). There is, as yet, no valid, direct evidence that any colicin determinant occupies a chromosomal locus. It has recently been found that only in the case of nontransmissible *col* factors is colicin production a lethal synthesis (Herschman and Helinski, 1967).

Further details about individual colicins and their genetic determinants are comprehensively reviewed by P. Frédéricq (1957;1958) to whom we owe the greater part of our knowledge of them (see also, Reeves, 1965; Nomura, 1967). In the remainder of this section we will concentrate on some studies, primarily by B. A. D. Stocker and his colleagues, which show that certain *col*+ factors are able to promote chromosomal transfer by the cells which harbour them (review, Smith and Stocker, 1962).

Salmonella strains, whether or not they produce colicins, are naturally resistant to them and are therefore well adapted to investigation of the conjugal transfer of *col*+ determinants. *Salm. typhimurium* strain LT-2, which has been extensively used in genetic analysis by transduction, is naturally non-colicinogenic but can easily be rendered colicinogenic for one or more *col*+ determinants by mixed growth with *E. coli*, *Shigella* or other *Salmonella* strains which harbour them (Frédéricq, 1957). When singly colicinogenic derivatives are grown overnight in mixed culture with the *col*− strain, the various *col*+ determinants are found to be transferred to the *col*− bacteria at very different rates. For example, the efficiency of transmission of *colI* and *colB* factors is high (30 to 70 per cent), that of the $colE_1$ factor is very low (see page 769), while the $colE_2$ and *colK* factors appear not to be transmitted at all (Ozeki, Stocker and Smith, 1962).

Although a high proportion of *col*− cells may inherit the *colI* factor after *overnight* incubation with *colI* donors, the rate of acquisition during the first five hours or so is very slow (<0·5 per cent), due to the fact that only a very small fraction of the donor cells (10^{-3} to 10^{-4}) can transfer the factor at any one time. The subsequent rapid increase in the spread of colicinogeny among the *col*⁻ population suggested that efficient transfer might be restricted to bacteria *newly infected* with the *colI* factor. This, in fact, is the case, for if the overnight mixture is diluted and incubated in fresh broth, not only are virtually all the remaining (30 to 70 per cent) *col*⁻ bacteria converted to colicinogeny within two hours, but the mixture is now able to transmit the *colI* factor to 50 per cent of freshly added *col*⁻ bacteria *in one hour*. This property of *high frequency transfer of colicinogeny* (HFCT) is maintained by newly converted cells for only three to seven generations, so that HFCT cultures rapidly revert to a state in which only rare cells are capable of transfer although all the cells remain colicinogenic (Stocker, Smith and Ozeki, 1963). *E. coli* K-12 *F*⁻ strains, infected with the *colI* factor, behave in a similar fashion (Monk and Clowes, 1964a).

A striking and easily observed feature of HFCT preparations of *Salm. typhimurium* is that, unlike broth cultures of established *colI* and *col*⁻ strains, the bacteria have a high tendency to adhere together in pairs or clumps. It therefore seems likely that newly transferred *colI* factors mediate their own transfer by determining the synthesis of a surface antigen which allows conjugation to occur and which, as we shall see, turns out to be a sex fimbria or pilus (p. 774). Stocker *et al.* (1963) suggested that HFCT behaviour might be due to the fact that conjugal functions are expressed only by autonomous factors which are rapidly switched off by chromosomal 'integration'. However, since HFCT preparations also show a marked increase in colicin production, while induction of colicin synthesis by exposure to ultraviolet light is accompanied by a simultaneous increase in the efficiency of *colI* transfer, it seemed more likely that the reversion to low frequency transfer after a few generations is due to production of a repressor which shuts off a number of *col* factor functions, in the same way that expression of the i^+ gene of the lactose region imposes inducibility on β-galactocidase production some time after transfer of the z^+ and i^+ genes to z^-i^- recipients (p. 714; Monk and Clowes, 1964b).

We have said that *Salm. typhimurium* strains carrying the $colE_1$ or $colE_2$ determinants transfer them to *col*⁻ recipients at a very low frequency or not at all. It turns out that if such *colE*$_1$ or *colE*$_2$ bacteria are newly infected with the *colI* factor, they now transfer these resident determinants, as well as the *colI* factor, with high efficiency to *col*$^-$ cells. Interrupted mating experiments show that transfer is initiated within about two minutes of mixing the donor and recipient cultures, but the different factors (*colI*, *colE*$_1$, *colE*$_2$) are randomly inherited; that is, they are not linked in transfer. The inability of the *colE* factors to promote their own transfer is therefore due to their failure to initiate conjugation; if effective unions are created by other agents, such as *F* or *colI*, the barrier to their transfer is removed (Smith, Ozeki and Stocker, 1963).

The behaviour of the *colI* determinant as a sex factor which initiates conjugation and promotes the transfer of other genetic determinants, suggested to Stocker and his colleagues that it might also provoke chromosomal transfer and this, indeed, was found to be so. When two genetically marked *Salm. typhimurium* strains are crossed by mixing a *colI*-HFCT preparation of one with a *col*$^-$ culture of the other, recombinants arise with a frequency of about 10^{-8}. In contrast, no recombinants are found when established *colI* strains, or strains carrying *colE*$_1$ or *colE*$_2$, are used as presumptive donors. Strangely enough, if the HFCT donor harbours the *colE*$_1$ factor in addition to its newly acquired *colI*, the frequency of recombination is enhanced 100-fold, so that *colE*$_1$ must play some accessory role in promoting transfer of the chromosome (Stocker, 1960; Ozeki and Howarth, 1961).

When recombinants inheriting a range of different donor markers are selected in this system, and the inheritance of unselected markers scored among them, it is found that segments of donor chromosome of considerable size and representing every part of it are transmitted to the *col*$^-$ recipients. The order of a considerable number of genes has thus been mapped on the *Salm. typhimurium* chromosome; all have been shown to be arranged on a single linkage group, in the same sequence as revealed by $Hfr \times F^-$ crosses in this organism (p. 751). Moreover, the data indicate that the linkage group is a continuous (circular) one since, by appropriate selection, close linkage may be demonstrated between any pair of donor markers which might be postulated, from other crosses, to lie at the extremities of an open chromosome (Ozeki and Howarth, 1961; Smith and Stocker, 1962).

Similar results are obtained when the *colI* factor is introduced into one of a pair of F^- strains of *E. coli* K-12. Although such *col*-mediated crosses yield recombinants with a frequency (10^{-8}) about 100-fold lower than equivalent $F^+ \times F^-$ crosses, the same relative frequencies of the various recombinant classes are obtained from both under the same selective conditions, so that both agents may be assumed to promote the transfer of equivalent chromosomal segments. Unlike *Salm. typhimurium*, however, strains of *E. coli* retain much of their fertility after colicinogeny has become established in them, while the concomitant presence of the $colE_1$ factor in *colI* donors does not increase the frequency of recombinants (Clowes, 1961).

RESISTANCE TRANSFER FACTORS

From 1957 onwards a large and increasing number of strains of *Shigella*, isolated from cases of bacillary dysentery in Japan, were found to be resistant to one or more of the commonly used chemotherapeutic agents, streptomycin, chloramphenicol, tetracycline and sulphonamide. The first such multiply-resistant strain to be isolated was from a patient who had just returned from Hong Kong in 1955. The majority of the resistant strains proved to be concomitantly resistant to two or more of these drugs, about ten per cent of all strains isolated in 1959, for example, being resistant to all four (see Watanabe and Fukasawa, 1961a). Moreover, strains of *E. coli* obtained from these dysentery cases often carried a similar pattern of resistance to that of the infecting *Shigella* strain (see Mitsuhashi, Harada and Hashimoto, 1960). It was then discovered, and reported (in Japanese) by Ochiai and by Akiba, that this multiple resistance is transferable *en bloc* to sensitive bacteria, not only among *Shigella* strains but among the *Enterobacteriaceae* as a whole, and that transfer depends on contact between living cells (see Akiba *et al.*, 1960). Subsequently, transfer to such phylogenetically remote species as *Vibrio cholerae*, *Past. pestis*, *Serratia*, *Proteus* and *Pseudomonas* has been achieved (see Watanabe, 1963; Meynell, Meynell and Datta, 1968). There is evidence that inter-species transfer of this sort is limited by restriction (p. 748). The intensive studies of several groups of Japanese workers then brought to light the remarkable fact that the transfer of multiple drug resistance is mediated by genetic elements, called *resistance transfer factors* (*RTF*), to which the resistance determinants are linked. As we shall see, these elements closely resemble the *colI* factor and the *E. coli* sex factor.

The dramatic epidemiological effects of the evolution of resistance transfer factors is evident from the fact that, in three major Japanese cities, virtually no drug-resistant strains of *Shigella* were isolated in 1956, while about 50 per cent of isolates in 1964 were resistant to all the four drugs mentioned above (Watanabe, 1966b). Outside Japan, Datta (1962) was the first to identify a resistance transfer factor, among strains of *Salm. typhimurium* isolated from an outbreak of gastroenteritis at a London hospital, and, between then and the winter of 1964–1965, the incidence of resistant strains of this species rose from about three per cent to 61 per cent (Anderson and Lewis, 1965; Anderson, 1967, 1968). Since then, resistance transfer factors have been found wherever they have been sought, and have a world-wide distribution (Watanabe, 1966). The increase in incidence of multiply resistant bacteria has been matched by an increase in the number of resistance determinants carried by the resistance transfer factors. Thus, in addition to resistance to streptomycin, chloramphenicol, tetracycline and sulphonamide, strains resistant to kanamycin and neomycin (Lebek, 1963) and, subsequently, to penicillin, due to constitutive production of penicillinase (Anderson and Datta, 1965), were isolated. Resistance determinants to as many as seven or eight antibiotics have been found to be carried by a single factor (Anderson and Datta, 1965; Anderson, 1967; Kontomichalou, 1967).

Resistance factors are not only of theoretical interest but also of great medical importance, since the speed with which they acquire the determinants of resistance to antibiotics newly introduced into the environment, as well as the efficiency of their spread among the flora of the intestinal tract, prevents the effective therapy not only of *Salmonella* and *Shigella* infections, but also of urinary tract and other non-epidemic infections caused by *E. coli* and other 'coliform' species. The mass of information which has accumulated during the past few years is far too extensive for review here, and compels us to select only a few features which bear strikingly on the topic of sex factors and their behaviour. In this section we will look at some genetic features which appear to be peculiar to resistance transfer factors, and will then later consider the place of these factors among sex factors as a whole (p. 769).

The components of resistance transfer factors

It happens that the association between the determinants of resistance to various antibiotics on a given factor is not necessarily a stable one, dissociation occurring either spontaneously or during transfer by conjugation or transduction. Multiple resistance can be jointly transduced from quadruply resistant donors to sensitive strains, by phage P1 in the case of *E coli* K-12 (Nakaya, Nakamura and Murata, 1960) and by phage P22 in *Salm. typhimurium* (Watanabe and Fukasawa, 1960, 1961c). However, these two systems yield quite different results. Virtually all *E. coli* transductants, mediated by phage P1, inherit quadruple resistance which they can thereafter transfer by conjugation in the usual way; the RTF is transferred intact. On the contrary, the P22-mediated *Salm. typhimurium* transductants generally show segregation of resistance into streptomycin-chloramphenicol-sulphonamide-resistant (*c.* 87 per cent) and tetracycline-resistant (*c.* 13 per cent) types. Moreover, in the great majority of transductants, the break-up of the resistance pattern is associated with loss of the ability to transfer resistance by conjugation. This loss is due to a defect in the transfer factor and not in the bacteria, which are capable of normal transfer when the same RTFs are introduced into them by conjugation (Watanabe and Fukasawa, 1960, 1961b). Moreover, the resistance genes are not immobilised because they become integrated into the bacterial chromosome, since infection of the transductants with the sex factor, *F*, restores the ability of the bacteria to transfer resistance, independently of chromosome transfer (Harada *et al.*, quoted by Watanabe, 1963). It follows that these resistance determinants, which have become separated from the determinant of transfer in the process of transduction, continue to exist as autonomous, cytoplasmic factors—they are replicons.

When resistance transfer factors carrying quadruple resistance are introduced into *Salmonella* by conjugation, they also tend to be unstable and to segregate among the bacterial progeny into much the same restricted patterns of resistance as result from phage P22 transduction. In this case, however, the majority of segregants retain the capacity to transfer their residual resistance determinants by conjugation. Sensitive bacteria can be rendered quadruply resistant by mixedly infecting them with two complementary resistance factors from such segregants. These tend to be unstable unless maintained on media containing the four drugs, but from them emerge stable strains which transfer all four resistance determinants on one structure; the original, quadruply resistant factor has been reconstituted from its segregated components, presumably as a result of recombination (Watanabe and Lyang, 1962). Similar results have been reported in sensitive strains of *E. coli* following mixed infection with naturally occurring RTFs carrying complementary patterns of resistance; in this case the new factors bearing four resistance determinants were found to be transduced as a single unit by phage P1 (Mitsuhashi *et al.*, 1962).

Studies of resistance transfer factors isolated from strains of *Salm. typhimurium* in Britain have revealed a superficially different pattern of behaviour (Anderson and Lewis, 1965; Anderson, 1967). In the case of a strain resistant to ampicillin (*A*), streptomycin (*S*) and tetracycline (*T*), the determinants of the three resistances were found to be transferred independently and at different rates during conjugation with a sensitive strain of *E. coli*; thus while the transfer frequencies of *A* and *S* alone were about 10^{-2}, and that of their joint transfer (*AS*) about 10^{-3}, that of *T* was as low as 10^{-6}. However, as a result of *T* transfer new associations became established since although, in subsequent matings, the frequencies of transfer of *A*, *S* and *AS* remained as before, that of *T* was now dramatically increased to about 5×10^{-1}; all the *A*, *S* and *AS* recombinants also received *T*. This new pattern turned out to be stable and to persist through transfers from *E. coli* to *Salmonella* and back again, and resembles that initially described by the Japanese workers. By interrupting these matings it proved possible again to separate *A* and *S* from *T*, but these resistances were no longer transferable. The existence of a *transfer factor* may therefore be postulated, which interacts with independent *resistance factors* and mediates their transfer. In the case of the resistant determinant *T* this interaction is rare but, when it does occur, the two factors become part of a single transferable structure. Moreover the fact that, in all these experiments, the frequency of joint transfer of *A* and *S* (10^{-3}) is ten-fold higher than would be expected from chance, suggests that these two independent factors frequently become associated on the same structure.

The existence of transfer factors in *sensitive* bacterial strains, which have the potential to promote the transfer of resistance factors, can be demonstrated by means of a 'ménage à trois' experiment. Cultures of the drug-sensitive strain to be tested (A) and of a strain carrying an immobile resistance factor (B) are mixed and incubated; a drug-sensitive 'indicator' strain (C), carrying a specific selective marker, is then added and the mixture incubated further. If strain A carries a transfer factor, this will be transferred by conjugation to strain B where it will 'pick up' the resistance factor and transfer it, in turn, to strain C—neither A nor B alone can transfer resistance to C. This sequence of events can be summarised as follows:

$$1\ A(t)+B(R)\rightarrow B(tR);$$

$$2.\ B(tR)+C\rightarrow C(tR),$$

where *t* indicates the transfer factor and *R* the immobile resistance determinant. Thus when the mixture is plated on the selective medium for strain C, containing the drug to which only strain B is resistant, the appearance of colonies indicates the presence of a transfer factor in strain A. A preliminary survey of 90 sensitive strains of *Salm. typhimurium* revealed that 63 per cent carried transfer factors (Anderson, 1965a, b). Since one of the two major groups of factor associated with resistance transfer has been found to produce sex fimbriae which permit infection by the *E. coli* 'male-specific' RNA phage (see below, p. 772), sensitivity to this phage also serves to indicate the carriage of a transfer or sex factor. A survey of independently isolated strains of drug-sensitive *E. coli*, using this method, has shown six out of 26 to harbour a transfer factor (Meynell and Datta, 1966a).

We mentioned above that resistance determinants which have lost their mobility are probably not integrated into the chromosome, since they can often be independently mobilised by the addition of a transfer factor. However, in cases where the segregation of resistance determinants follows transduction by phage P22, the situation may frequently be more complicated than this, since integration of a transduced, defective *R* factor at the prophage P22 attachment region on the *Salm. typhimurium* chromosome has been demonstrated (Dubnau and Stocker, 1964). Chromosomal integration of resistance determinants has also been reported following transduction by phage ε_{15} (p. 647) (Harada *et al.*, 1967). In these cases, however, the resistance is not subsequently transmissible by superinfecting sex factors.

The origin and biochemical nature of transferable resistance

It is reasonable to imagine that the first step in the development of transferable drug resistance would be the selection of resistant bacterial mutants in the intestine, as a result of the presence of a particular drug in the environment. Transfer factors acquired by such cells would then 'pick up' the resistance determinant with a low probability, and transfer it to sensitive cells. The efficiency of this mode of spread of resistance would ensure that bacterial populations carrying RTFs would rapidly come to dominate the intestinal flora. These factors would thus be well placed to incorporate the determinants of resistance to other antibiotics as these arose by mutation, following substitution of the first drug by another, or the concurrent employment of several drugs in the therapeutic environment, so that factors carrying multiple resistance would arise. Alternatively, factors carrying multiple drug resistance might arise by recombination between independent factors, carrying different patterns of resistance, infecting the same cell. In any case it would be expected that the types of resistance involved would be genetically and biochemically similar to the types arising from mutation, among those coliform organisms which predominate in the intestine and to which transfer factors are most readily transmissible.

There is some evidence that this sequence of events may occur. For example, Ginoza and Painter (1964) reported the isolation of an *E. coli* mutant which was resistant to 25μg/ml tetracycline. This resistance was not transferable, but became so when the strain was infected with an RTF carrying resistance determinants for streptomycin and chloramphenicol, but not for tetracycline. Subsequent recipients of these markers could themselves transmit them all at a low frequency, but the tetracycline determinant proved unstable. On the other hand, it seems unlikely that the majority of resistance determinants carried by RTFs arises hy mutation in *E. coli*, *Salmonella* or *Shigella*, from which they are usually isolated, since their integration into the chromosomes of these species, as a result of genetic homology, has never been demonstrated—except for those cases mentioned above where the determinants were associated with the genome of transducing phage and were integrated at the phage attachment region.

From a biochemical point of view the type of transferable resistance often appears different from that normally acquired by mutation in *E. coli*. A striking example is the case of resistance to streptomycin which, in *E. coli*, usually arises by a single mutation to a high level, and is recessive to sensitivity in diploids. The effect of the mutation is to alter the structure of the ribosomes so that, unlike those of sensitive bacteria, they are resistant to the presence of streptomycin in both *in vivo* and *in vitro* protein synthesis (Brock, 1966). In contrast, the streptomycin resistance carried by RTFs is low level, does not result in altered ribosomes and, as is evident from its expression in infected cells, is dominant to sensitivity (Rosenkranz, 1964). Nevertheless mutations leading to low levels of streptomycin resistance do arise naturally in *E. coli* and *Salmonella*, but virtually nothing is known about their properties. If resistance to this type is similar to that determined by R factors, then its ability to express itself would lead to its selection by transfer factors in preference to the high level, recessive type.

Again, a variety of penicillinases are produced by different resistance factors, and these can be distinguished from one another, as well as from that normally produced by *E. coli*, by their action spectra against a range of penicillins. Unlike the staphylococcal and certain other penicillinases (see Citri and Pollock, 1966; Pollock, 1967), they are not exo-enzymes and are non-inducible; one type resembles a penicillinase produced by *Klebsiella* (Datta and Richmond, 1966; Egawa, Sawai and Mitsuhashi, 1967). In the case of tetracycline there is considerable evidence that both transmissible resistance, as well as that acquired by mutation in the laboratory, are due to specific alterations in permeability (Izaki, Kiuchi and Arima, 1966; Franklin, 1967), while transmissible resistance to kanamycin, dihydrostreptomycin and chloramphenicol appears to be associated with enzymes which can inactivate these drugs by acetylation (Okamoto and Suzuki, 1965; Shaw, 1967).

It is impossible, on such scanty information, to form an opinion concerning the origin of the resistance determinants of *R* factors. Indeed, there may be a number of independent sources. The evidence suggests that, in most cases at least, the determinants do not originate in the chromosome of *E. coli* or other related coliform bacilli. What, then, is their source? There are two promising lines of enquiry. One depends on the likelihood that *E. coli* and similar species, despite their apparent predominance in the usual faecal cultures, in fact are only a minority among the multitudinous flora of the gut, and that many other genera, including anaerobes, greatly exceed them in number. If such genera are susceptible to infection by transfer factors, genes determining drug resistance picked up from them are unlikely to bear any genetic resemblance to those studied in *E. coli*, although they may behave functionally in the same general way.

Secondly, it has been shown that the genetic determinants of resistance to penicillin and erythromycin, as well as to a number of metallic ions such as mercury, are located on non-transmissible plasmids in

Staphylococcus aureus, and not on the chromosome (see p. 783). The fact that a determinant for mercury resistance has recently been located on a number of resistance transfer factors of coliform bacilli points to an ecological similarity between these factors and the staphylococcal plasmids (Novick, 1967). It is therefore possible that certain cellular functions which are correlated with sensitivity to various agents, are normally mediated by plasmids which can become incorporated into the structure of transfer factors by recombination. In such a case the whole process of the development of resistance and its transmissibility would be extra-chromosomal.

The ecology of transferable drug resistance

We have already noted the dramatic rise, during recent years, in the proportion of *Escherichia*, *Salmonella* and *Shigella* strains carrying transferable drug resistance factors, as well as in the range of drugs to which these strains are resistant. There can now be no doubt that the basic reason for this phenomenon is the selective pressure imposed by the increasing and widespread use of these drugs. An analysis of strains of *Salm. typhimurium* isolated in Britain in recent years not only confirms this but sheds new and important light on the ecology of transferable resistance (Anderson and Lewis, 1965; Anderson, 1967, 1968). The dramatic rise in incidence of resistant *Salm. typhimurium* strains during 1964–65, which we have previously mentioned (p. 761), was associated with an increase in the proportion of a particular phage type, type 29, which rose from about ten per cent or less of the total *Salm. typhimurium* strains isolated, to more than 60 per cent. Not only this, but whereas all of the small proportion of type 29 strains isolated in 1962 and early 1963 were sensitive, virtually all of those isolated in 1965 carried transmissible resistance to five or more drugs. When the sources of these type 29 resistant strains were explored, the great majority were found to have been isolated from calves and comprised 73 per cent of all bovine strains; in contrast, only about three per cent of strains isolated from other animals belonged to type 29, and an appreciable proportion of these were sensitive. It also turned out that many of the resistant strains of human origin could be directly related to bovine disease, since they were also resistant to the drug furazolidone which is widely used in attempts to control the disease of scours in calves, but not in the treatment of human infections.

It is thus evident that the main reservoir of *Salm. typhimurium* strains carrying transferable drug resistance is cattle and, particularly, calves. It further transpired from epidemiological studies that the infected calves came, almost exclusively, from farms specialising in 'intensive breeding', into which the infection was initially introduced by animals newly bought from particular dealers who serve as centres of dissemination. These 'intensive farms' are characterised by crowding of the animals under very poor hygienic conditions, which greatly facilitates the spread of infection, together with the extensive use of antibiotics, both to improve growth as well as in attempts to prevent and control infectious disease. Such conditions are clearly ideal for the evolution and spread of resistance transfer factors.

What is the reason for the involvement of a particular phage type of *Salmonella*, initially a minority one, in the evolution of resistance? You may remember that the phage types of *Salm. typhi* and *Salm. typhimurium* are in part due to prophages they carry which, though unrelated to the typing phages, determine the specific pattern of susceptibility to them, probably by 'restricting' their infecting DNA (p. 521). Resistance transfer factors frequently carry genes which specifically restrict, and sometimes modify, the DNA of various phages (see Watanabe *et al.*, 1966; Meynell, Meynell and Datta, 1968), and it has been shown that infection by certain transfer factors either determines or alters the phage type of some *Salmonella* strains (Anderson and Lewis, 1965; Guinée, Scholtens and Willems, 1967). *Salm. typhimurium* type 29 is such a case in which the type is determined by a transfer factor, which has been identified by the 'ménage-à-trois' method (see above) in an antibiotic-sensitive strain of this type isolated as long ago as 1947. Laboratory experiments show that if cultures of this sensitive type 29, and of *E. coli* carrying an immobile *R* factor, are mixed, a resistant type 29 can subsequently be isolated from the mixture; the transfer factor can migrate into the resistant cells, pick up the *R* factor, and then re-enter its parental host, conferring resistance upon it (Anderson, 1967). It is clear how *Salm. typhimurium* type 29 became selected in the antibiotic environment, from association of the transfer factor which defines the type with the resistance determinant which is selected. In a quite different context, this is an example of the same kind of associative evolution which we encountered early in this book in the relation between sickle cell anaemia and malaria (pp. 27, 28).

As Anderson (1967) points out, the spread to man of *E. coli* strains of animal origin, carrying resistance transfer factors, is not only very much commoner and more widespread than that of *Salm. typhimurium* but constitutes a greater threat, since these non-pathogenic bacteria may be harboured indefinitely in the body, to pass on their resistance determinants to subsequently infecting and dangerous pathogens such as *Salm. typhi*, the cause of typhoid fever, so that otherwise effective treatment is rendered worthless. A comprehensive review of the ecology of transferable drug resistance is by Anderson (1968).

SEX FACTORS AS A CLASS

So far we have considered separately a number of systems which have in common the property of promoting conjugation and genetic transfer between bacteria. We have seen that, so far as the *Enterobacteriaceae* are concerned, the central feature of these systems is a group of cytoplasmic elements which are self-replicating, are responsible for the formation of conjugal unions, and determine the donor state. In general, we have tended to classify all these elements as sex factors but have distinguished between them on the basis of the type of determinant whose transfer they mediate and by which their existence was first recognised. Thus we have colicin factors, resistance transfer factors, and the sex factor *F* which mediates the transfer of chromosomal genes. On the other hand, we have found that resistance transfer factors, for example, are separable into transfer factors and resistance factors which can exist independently. The sex factor, *F*, is able to promote the efficient transfer of the otherwise non-transmissible colicin factors *E1* and *E2*, while the apparent transmissibility of a particular *colE1* factor turns out to be due to its natural association with an independent transfer factor which appears to express no function other than conjugation (Meynell *et al.*, 1968). Again, we have already seen that the *colI* and *colB* factors can promote chromosome transfer (Ozeki and Howarth, 1961; Clowes, 1961), while two colicin V factors do so with an efficiency approaching to that of the sex factor *F* (Kahn and Helinski, 1964; Macfarren and Clowes, 1967). In this catalogue of factors which show overlapping behaviour, we may also include *F*-prime factors which carry and transfer particular bacterial determinants as part of their structure.

In addition to these well recognised, if overlapping, entities, transferable factors conferring other characters have been described. Among these are two factors resembling, but distinct from, F-prime factors, which carry genes determining lactose fermentation. They are the F_0—*lac* factor isolated from *Salm. typhi* (Baron, Carey and Spilman, 1959), and *P—lac* from a strain of *Proteus mirabilis* (Falkow *et al.*, 1964), both initially recognised because neither species from which they were obtained normally ferments lactose. More recently, three new transferable factors have been reported; one carries the determinant of a surface antigen (K88) of *E. coli* strains which cause enteritis in pigs (Stirm *et al.*, 1967), another (*Hly*) is associated with a soluble haemolysin, and the third (*Ent*) with enterotoxin, the two latter also being characteristic of strains of *E. coli* pathogenic for pigs (Smith and Halls, 1967; Smith, 1968).

All these facts imply the existence of a versatile range of sex factors which can mobilise a variety of bacterial genetic determinants, located either on the chromosome or on plasmids, and transfer them to recipient bacteria. Under the influence of environmental pressures, the association of appropriate character determinants with the sex factors is selected. Since the presence of colicins is a natural factor in the environment of *Enterobacteriaceae*, an association of colicin resistance with resistance transfer factors might be expected and has, in fact, been demonstrated in *E. coli* (Siccardi, 1966). In the remainder of this section we propose to examine and compare purely as sex factors, the properties of the various transfer factors we have described, ignoring the particular determinants which happen to be linked to them.

Composition of sex factors

We have already seen that the *E. coli* sex factor, *F*, is composed of DNA of the order 10^5 base-pairs long (p. 657). In the case of colicin factors, quantitative correlations between the transfer to recipients of various combinations of *colI*, *colE1* and *colE2*, and of [^{14}C]-thymidine with which the donor bacteria were labelled, indicate that each of these factors likewise consists of DNA of a length equivalent to 10^4 to 10^5 nucleotide pairs, *colE2* being somewhat smaller than the other two (Silver and Ozeki, 1962). Resistance transfer factors have also been shown to be composed of DNA and estimates of their size, by studying the kinetics of their inactivation by the decay of incorporated ^{32}P (p. 523) and other biophysical methods, work out within the same range (*c.* 5×10^4 base pairs, corresponding to a molecular weight of *c.* 2×10^7) (Watanabe, 1963; Falkow *et al.*, 1966; Rownd, Nakaya and Nakamura, 1966).

Interest has centred especially on the base composition of RTFs in the hope that, by its correspondence to that of the DNA of some particular bacterial species, it might shed light on the origin of transfer factors or resistance determinants. The usual method used is to compare the density profiles of DNA extracted from bacterial cultures, with and without the factor, after centrifuging to equilibrium in a caesium chloride density gradient. Providing that the DNA of the factor has a different base composition and, therefore, density from that of the host bacterium, it can be seen as a small peak distinct from the main peak of bacterial DNA, and from its density the G+C content can be estimated. By using a variety of host bacteria having different DNA compositions, a range of minority DNA components may be identified. For example, DNA containing 58 per cent G+C can be identified in *E. coli*, *Salmonella* or *Shigella* (50 to 52 per cent G+C) but not in *Serratia*, *Klebsiella* or *Aerobacter* (56 to 58 per cent G+C), while DNA of 50 per cent G+C content is recognisable in *Serratia* but not in *E. coli*; both types would be seen against a background of *Proteus* DNA (38 per cent G+C).

When the sex factor *F* is examined in this way, nine-tenths of its DNA is found to have a G+C content of 50 per cent, like *E. coli* DNA, and one-tenth 44 per cent G+C. Moreover, isolated *F* factor DNA is able to form DNA–DNA hybrids with *E. coli* DNA, while that of the *P—lac* factor cannot although it also consists predominantly of 50 per cent G+C molecules. Resistance transfer factors also turn out to possess two DNA moieties within the one structure, one containing 52 per cent and the other 58 per cent G+C; the proportion of these two components may vary markedly between different factors, but this is not associated with different resistance determinants which the factors may carry. On the other hand, loss of resistance determinants by some factors has been correlated with density changes which show that, in these cases, resistance to tetracycline and chloramphenicol were associated, respectively, with DNA fractions of 50 and 56 per cent G+C composition (Falkow *et al.*, 1966; Rownd *et al.*, 1966).

Cytoplasmic interactions and cross-repression: a basis for classification

When resistance transfer factors were first described they seemed to have little in common with the sex factor *F*, beyond the ability to promote conjugation and their own transfer. Indeed, they seemed to behave more like the *colI* factor. For example, they can coexist in the same cell with *F*, unlike an *F*-prime factor (p. 775), and can mediate their own transfer between *E. coli* strains without regard to the *F* status of these strains (Mitsuhashi *et al.*, 1960). Transfer occurs very rapidly, within one minute of mixing the parental cultures (Watanabe and Fukasawa, 1961a). On the other hand, the frequency of transfer is initially very low compared to that of *F*, but is greatly enhanced from newly infected cells, which thus constitute 'high frequency transfer' (HFT) systems (p. 758). Moreover while resistance transfer factors can, like *F*, be eliminated from cells which harbour them by treatment with acridines, the efficiency of curing is low and acriflavine appears to be more effective than acridine orange (p. 656; Watanabe and Fukasawa, 1961b; Mitsuhashi, Harada and Kameda, 1961). Finally, cultures of cells carrying RTFs are not lysed by 'male-specific' phage.

In view of these differences, a surprising discovery was that the majority of resistance transfer factors profoundly influence the *expression* of *F* in cells in which the two factors coexist. The introduction of such factors, termed fi^+ (for 'fertility inhibition'), into F^+ or intermediate (*F*-prime) donor bacteria enormously reduces their capacity to transfer the *F* factor, while the frequency of chromosome transfer by *Hfr* strains is reduced to one per cent or less of its normal value. In all cases this functional inhibition of donor properties is accompanied by inability of the bacteria to adsorb 'male-specific' RNA phages. These defects are not due to elimination of the *F* factor since rare segregants are found which have lost their RTF, and these manifest full recovery of *F* activity. In contradistinction to this suppression of *F* function, fi^+ resistance transfer factors have no effect on transfer of the *colI* factor when both are present in the same cell (Watanabe and Fukasawa, 1962; Watanabe, Fukasawa and Takano, 1962). One of several hypotheses put forward to explain this phenomenon, was that the fi^+ transfer factor produces a cytoplasmic repressor which acts on *F*, switching off its conjugal functions but not its ability to replicate normally (Egawa and Hirota, 1962).

Meynell and Datta (1965) conceived the idea that this repression indicates a close relationship between *F* and fi^+ resistance transfer factors, and that the observed differences between the two could be accounted for if the repressor determined by the fi^+ factor acts not only on *F* but also on itself. Under these circumstances the fi^+ factor might actually determine the synthesis of sex fimbriae similar to *F*-type fimbriae, as well as sensitivity to 'male-specific' phage, but repression would mask these effects since the factor would normally

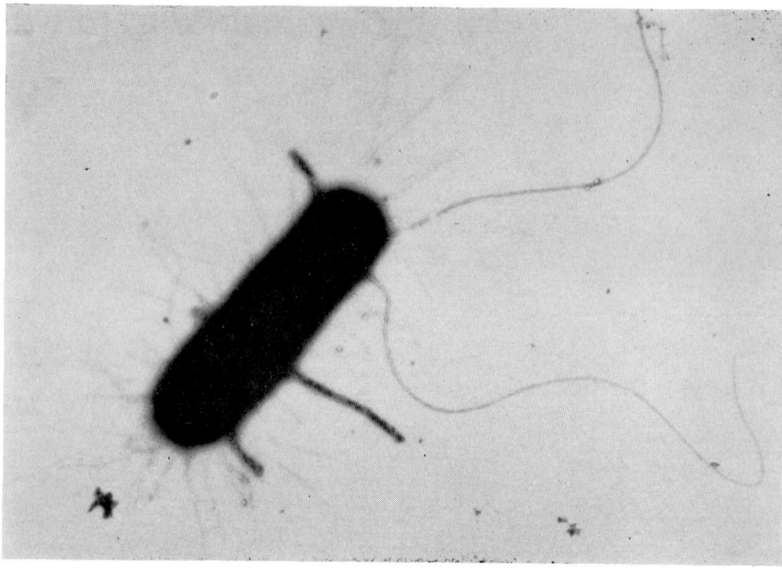

A

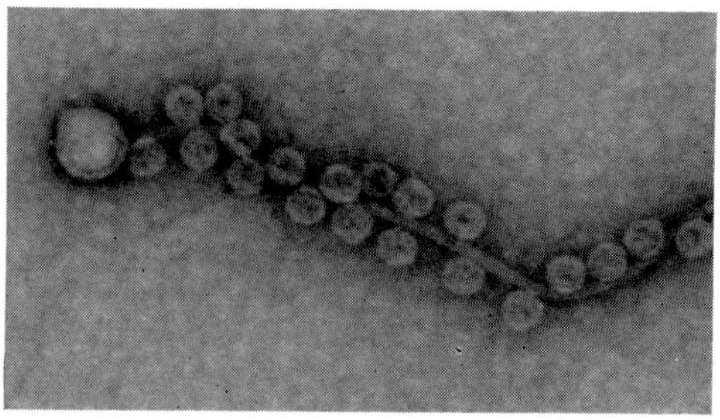

B

Plate 34. A. Electron microphotograph of an *E. coli* bacterium carrying a cytoplasmic resistance transfer factor (*fi*$^{+}$). The two very long appendages of intermediate thickness are flagella. The numerous, much finer and shorter appendages are common (Type I) fimbriae which play no sexual role. The thick, darkly stained projections are fimbriae determined by the sex factor, which are densely coated with adsorbed particles of the spherical, male-specific RNA-containing virus, MS2. Magnification about ×42,000.

B. This electron microphotograph shows a sex fimbria similar to those in A. above, with adsorbed virus particles, at a magnification of about ×340,000. The terminal 'knob', frequently found on fimbriae determined by resistance transfer factors, but not by the sex factor *F*, is well seen.

Both preparations were made by negative contrast staining, using uranyl acetate.

The original photographs were kindly provided by Dr. Alan Lawn. (From Datta, Lawn and Meynell, 1966, by kind permission.)

express itself only in a minute proportion of the population. This indeed turned out to be so, since it was shown not only that F^- populations infected with fi^+ resistance transfer factors are able to support some increase of F-specific RNA phage, but that newly infected populations, which display HFT behaviour, are readily infectable by the phage (Meynell and Datta, 1965, 1966b). Finally a correlation was found, in such F^-R^+ HFT preparations, between sensitivity to RNA phage, the frequency of transfer of the factor and the proportion of bacteria possessing F-like fimbriae on electron-microscopy—expression of these three functions is coordinately derepressed (Datta, Lawn and Meynell, 1966).

These findings implied that fi^+ resistance transfer factors and the sex factor F are closely similar, at least in the sex fimbriae they produce, the main difference between them being that the F factor is *normally* derepressed so that its functions are expressed by every bacterium which harbours it. Derepressed, mutant resistance transfer factors were accordingly sought by screening for clones, either of *Hfr* bacteria carrying an fi^+ factor which yield a high frequency of recombinants for chromosomal genes, or which mediate high frequency resistance transfer. A number of derepressed mutants have now been isolated. F^- bacteria which harbour them show a several hundred-fold increase in conjugation frequency and visible lysis by the F-specific RNA phage, while 50 to 90 per cent of the bacteria are fimbriated. When these mutant factors are transferred to F^+ or *Hfr* bacteria they no longer repress F function. The mutants remain sensitive to repression by wild type repressor so that, from a genetic point of view, they are recessive in *trans* and, therefore, equivalent to i^- mutations of the lactose regulatory system, and not to O^c mutations (Meynell and Datta, 1967; Nishimura *et al.*, 1967).

There thus exists a group of resistance transfer factors which share with the factor F the ability to synthesise sex fimbriae which specifically adsorb 'male-specific' RNA phages such as f2 or MS2 (p. 440), and we have already seen that many freshly and independently isolated drug-sensitive strains of *E. coli* are also infectable by these phages and, by inference, possess fimbriae of similar structure (Meynell and Datta, 1966a). Sex fimbriae produced by a derepressed RTF, and coated with RNA phage, are well shown in Plate 34. Many resistance transfer factors, however, neither repress F function nor mediate infection by these 'male' phages. Bacteria newly infected with these so-called fi^- factors generally show HFT behaviour and, in the case of normally non-fimbriate strains, can be seen to produce fimbriae which are morphologically distinct from *F* fimbriae. Derepressed mutants of these *fi*$^{-}$ factors have been isolated which transmit their drug resistance with high efficiency, but the host bacteria remain insensitive to *F*-specific phage. It then turned out that the fimbriae characteristic of *fi*$^{-}$ factors closely resemble those produced by HFT cultures of bacteria carrying a *colI* factor. A filamentous phage was subsequently isolated which lyses bacteria carrying a derepressed *fi*$^{-}$ mutant factor, and this was found also to infect HFT preparations containing *colI* and other *fi*$^{-}$ resistance transfer factors. In contrast, cultures carrying *F* or derepressed *fi*$^{+}$ factors were not infected (G. G. Meynell and Lawn, 1967; Lawn *et al.*, 1967).

It is thus clear that two categories of sex factor exist, each of which determines the structure of a distinct type of fimbria—the *F*-like and the *I*-like. Apart from their differential adsorption of the *F*- and *I*-type sex phages, the two kinds of fimbria are morphologically and serologically distinguishable. Whereas *F*-type fimbriae tend to be much longer (up to 20μ) than common fimbriae (1·5μ), those of *I*-type are short (*c.* 2μ), thinner and more delicate in appearance; unlike common fimbriae, sex fimbriae of both types often terminate in knobs whose origin and function, if any, remains unknown (see Plate 34b). The antigenic specificity of the two types is best revealed by electron-microscopic observation of adsorbed antibody of one or the other type on the fimbrial surface (Lawn *et al.*, 1967).

Of 60 transmissible plasmids which have been tested for their association with susceptibility to *F*- or *I*-type phages, at least 47 fall clearly into one or the other category; five bacterial strains harbouring an *fi*$^{+}$ factor were sensitive to both types of phage, probably due to the frequency with which mixed infection with *fi*$^{+}$ and *fi*$^{-}$ factors occurs (Romero and Meynell, 1968). Eight of the strains (two *fi*$^{+}$ and six *fi*$^{-}$) failed to be lysed by either phage, possibly because they did not yield good HFT preparations which are needed for the test. Alternatively, of course, there may turn out to be more than two types of sex factor. Apart from resistance transfer factors, two *colV*, two *colB*, and the factors F_o—*lac* and *Hly* (p. 770) belong to the *F* type, while four *colI* factors and the factor mediating transfer of *colE1* (p. 769) were of *I*-type (Lawn *et al.*, 1967). We shall see in the following sections that factors belonging to the same fimbrial type are not necessarily identical, but often differ in many important properties. Nevertheless it is interesting that the great majority of sex factors so far examined do belong to one or the other of two archetypes, the members of each of which are almost certainly related phylogenetically.

Superinfection immunity and the control of plasmid replication

It is a general feature of the regulatory mechanism which controls the replication of plasmids, that bacteria which harbour a plasmid cannot normally be stably superinfected with a mutant of the same or a closely related element. This phenomenon is known as *superinfection immunity*, and was demonstrated for the sex factor *F* by comparing what happens when an *F—lac*$^+$ factor is transferred by conjugation to nonlactose-fermenting (*lac*$^-$) *F*$^-$ and *F*$^+$ bacteria; it is assumed that the *lac*$^+$ gene which serves as a marker for the *F*-prime factor in no way alters the way in which replication of the factor is regulated. As expected, infected *F*$^-$ bacteria yield unsectored colonies of lactose-fermenting cells, since unregulated replication of the sex factor on transfer leads to its migration to all the nuclear compartments, and its subsequent inheritance by all the segregants, of a newly infected bacterium (p. 656). On the contrary, transfer of the *F—lac*$^+$ factor to *F*$^+$ recipients is predominantly manifested by *sectored colonies* in which only a proportion of the bacteria inherit the sex factor and ferment lactose; the sex factor no longer spreads to all the segregants of those bacteria which receive it, implying that its replication is restricted by the presence of the resident, autonomous factor. Moreover, in none of those bacteria which acquire the *F—lac*$^+$ factor can the original *F* factor be demonstrated by transfer experiments. The *F* and *F—lac*$^+$ factors mutually exclude one another (Scaife and Gross, 1962).

When *lac*$^-$ *Hfr* bacteria were used as the recipients of an *F—lac*$^+$ factor, the proportion of *lac*$^+$ clones obtained is the same as when *F*$^+$ bacteria are infected, but the character of the clones is quite different. In the case of the *F*$^+$ recipients, all the *lac*$^+$ cells are heterogenotes; if the *F*-prime factor is eliminated by acridine orange treatment, they become *lac*$^-$. On the contrary, all the *lac*$^+$*Hfr* recipients turn out to be true haploid recombinants, which carry no *F*-prime factor and in which the *lac*$^+$ gene is incorporated into the chromosome. Here again, there is competition between the two factors as to which will be rejected, but the resident one, being entrenched in the chromosome, always wins (Scaife and Gross, 1962). This same mutual exclusion has been demonstrated in the case of superinfection of *lac⁻gal⁻* bacteria carrying an *F—lac*⁺ factor with an *F—gal*⁺ factor; virtually all the segregants of the superinfected cells carry either the *F—lac*⁺ or the *F—gal*⁺ element, but not both (deHaan and Stouthamer, 1963).

This mutual exclusion between two plasmids of the same type suggests that there is only one plasmid per chromosome, and there is independent evidence to support this. Firstly, the amount of β-galactosidase produced by *lac⁻/F—lac*⁺ heterogenotes is about twice that yielded by haploid *lac*⁺ bacteria (Jacob, Schaeffer and Wollman, 1960); this is what would be expected if a single *F—lac*⁺ factor replicated once, rapidly, at the beginning of every division cycle so that two copies of its *lac*⁺ gene are present for most of the time, as compared with a single copy of the chromosomal gene. Secondly, the DNA of a resistance transfer factor, separated from the DNA of its *E. coli* host by density gradient centrifugation (p. 771), turns out to have a molecular weight about 0·5 per cent that of a single chromosome, while the total amount of factor DNA is about the same proportion of the total chromosomal DNA (Rownd, Nakaya and Nakamura, 1966). Thirdly, a single plasmid per chromosome is suggested by the kinetics of elimination of a temperature-sensitive *F—lac*⁺ factor from an *E. coli* population at the inhibitory temperature (Cuzin and Jacob, 1967b).

The existence of superinfection immunity probably indicates a very close relationship between the two factors. On the other hand, there is no barrier to the co-existence of many different plasmids, such as *F*, *fi*⁺ and *fi*⁻ RTFs, and various *col* factors, all of which can exist stably together in the same cell. It is, perhaps, slightly paradoxical that plasmids which can co-exist in this way, because they are different, are referred to as *compatible*, while similar or identical plasmids which cannot co-exist are said to be *incompatible*. There is probably a wide spectrum of intermediate degrees of interaction between these two extremes, in which pairs of factors can co-exist in bacterial populations with different degrees of stability. An example of this is the exclusion exhibited between pairwise combinations of the sex factor, *F*, and two *col* factors, *colV2* and *colV3*, which are closely related to it (p. 769). Cultures of a *lac*⁻ strain of *E. coli*, mixedly infected with either *col* factor and an *F—lac*⁺ factor, showed segregation of the *col*⁺ and *lac*⁺ characters. However, in the case of *colV2* it was the *F*-prime factor which was excluded, whereas the *F*-prime factor, in turn, excluded *colV3* irrespective, in both cases, of which was the resident factor at the time of infection (Macfarren and Clowes, 1967). Although many compatible combinations may be found within either the *F*-like or the *I*-like groups of factors, as is shown by the fact that all fi^+ RTFs are compatible with *F*, among these fi^+ factors many are known to be incompatible with each other, and it may be that all show some degree of incompatibility (Dr Elinor Meynell, personal communication). In the patterns of superinfection immunity, therefore, we have another indication of phylogenetic relationships, more refined than cross-repression of fimbria formation.

The question of the nature of superinfection immunity between sex factors then arises. In the case of some temperate phages, such as λ, we have seen that immunity to superinfection, as well as the establishment and maintenance of lysogeny, is due to production by a phage gene of a cytoplasmic repressor which prevents replication of the phage genome. Is this type of mechanism also responsible for the incompatibilities between sex factors?

We must admit that, at the moment, we do not know and can only speculate from indirect evidence. We have, of course, seen that negative regulation by repressor action does indeed operate to control the synthesis of fimbriae by many sex factors. In the case of phage λ the action of the repressor is presumed to be continuous, the phage genome being replicated by the bacterial replication system as an integral part of the bacterial chromosome. This may also happen in *Hfr* bacteria where the sex factor *F* is similarly integrated into the chromosome. However, a basic distinction between *F* and λ is that, in F^+ bacteria, *F* can clearly replicate in the cytoplasmic, autonomous state and yet exclude a superinfecting factor in the same way as in *Hfr* strains. If this system is negatively controlled by a repressor, repression must be presumed to be lifted once every generation time to permit the factor to replicate; it is difficult to understand why a superinfecting factor should not also replicate under these conditions, the two factors thereafter continuing to do so in synchrony. At the same time, the unrestrained replication of sex factors in general in newly infected cells, prior to the establishment of superinfection immunity, strongly suggests repression. We should make it clear, in contrasting the behaviour of *F* and phage λ, that we are not generalising with respect to sex factors and temperate phages, for you may remember that phage P1 appears to have no chromosomal location and, in F^+ cells, can be transferred to recipients during conjugation as an extrachromosomal factor (p. 478).

In Chapter 19 we saw the evidence that the bacterial chromosome is attached to the cell membrane, which mediates the segregation of daughter chromosomes by its outward growth from the equatorial region of chromosomal attachment (p. 568). We also saw that the autonomous sex factor, *F*, does not segregate randomly but remains associated with a particular parental DNA strand of the chromosome (p. 569). All this suggests that *F*, and probably all plasmids, behave more like small supernumerary chromosomes than like viruses, and are subject to the same mechanisms of control and of segregation as is the chromosome itself. It seems likely that chromosome replication is regulated by a positive system which is also based on a specific site of membrane attachment (p. 735). An alternative hypothesis therefore postulates that superinfection immunity is due to competition between plasmids of the same type for attachment to a single, specific site, on the bacterial membrane. We will see later that the most likely current hypothesis of chromosome transfer assumes that the inserted sex factor of *Hfr* bacteria is also attached to the *F*-specific membrane site so that it can continue to exercise exclusion. The fact that bacterial mutants have been isolated which are unable to support replication of normal *F*-prime factors, lends weight to the participation of a bacterial structure in plasmid replication (Jacob, Brenner and Cuzin, 1963). While we can be fairly confident that superinfection immunity is associated with the mechanism of the regulation of replication and of segration in plasmids, it will probably turn out that both cytoplasmic and membranous factors play a role in the phenomenon. (For review and discussion see Scaife, 1967.)

Genetic interactions

The genetic interactions of sex factors may involve recombination with the bacterial chromosome or with other plasmids, or complementation effects between defective factors. We will briefly discuss each of these aspects.

Chromosome transfer

There is now a considerable weight of evidence that chromosome transfer by F^+ *E. coli* populations is due to insertion of the sex factor into the chromosome so that an *Hfr* cell, which may be stable or abortive, is formed. Conjugation of such donor cells is then followed by opening up of the chromosome at the site of insertion, and its polarised transfer from a particular one of the extremities so produced.

However, there is also some reason to think that, as in the case of transduction, a second mechanism may also operate, but at a very low frequency. We will deal with these matters in detail later on in this Chapter (p. 789 *et seq.*) and, for the moment, will merely compare the behaviour of various factors with respect to chromosome transfer.

Many transmissible factors, including both fi^+ and fi^- resistance transfer factors as well as some *colI* and *colV* factors, have now been shown to mediate transfer of chromosomal genes in *E. coli*. However, in most of these cases the frequency of tranfer is about 1000-fold lower than that of comparable F^+ strains (Sugino and Hirota, 1962). This was at first thought to signify a different relationship of these factors to the chromosome until it was realised that it could be accounted for by the low frequency with which these factors promote conjugation as a result of repression. In those cases where chromosome transfer by resistance transfer factors failed to be demonstrated, the frequency of transfer of the autonomous factor itself turns out to be extremely low. Conversely, derepressed factors transfer not only themselves but also chromosomal genes at about the same frequency as *F*. Thus two *F*-like and three *I*-like derepressed resistance transfer factors behave just like *F* and transfer different chromosomal segments with about the same frequency (Meynell and Cooke, 1968).

We have seen that stable *Hfr* strains result from the integration of *F* at a limited number of specific sites on the *E. coli* chromosome. No *Hfr* strains arising from the insertion of other factors into the chromosome have yet been reported. However, a derepressed mutant of a particular resistance transfer factor has been described which is unusual in mediating the transfer of a specific group of linked genes in the tryptophan region of the *E. coli* chromosome, at a frequency about 30 times higher than the genes of other regions. In addition, the genes transferred at high frequency show a gradient of transmission to recombinants (*pyrF*. *trp*. *purB*) virtually identical with that given by an *Hfr* strain (B10) which transfers the *pyrF* locus as the first marker (see Fig. 127, p. 666). This behaviour is quantitatively and qualitatively similar to that of an *F*-prime factor which has incorporated a region of the *E. coli* chromosome close to the site where *F* is inserted in the *Hfr* strain B10. This region does not carry a genetic marker whereby its incorporation into a sex factor could be recognised, but a high degree of homology between the *RTF* and the chromosome may be inferred from its behaviour, and this homology is associated with the transfer element and not with the resistance genes (Pearce and Meynell, 1968).

Two *colV* factors, producing the same colicin but isolated from different strains and distinguishable by their genetic behaviour, have been described which appear to be unusually close relatives of the *E. coli F* factor. For example, they are naturally derepressed and efficiently transferred, promote transfer of the bacterial chromosome and are readily eliminated by treatment with acridine orange, while host bacteria adsorb and are infected by RNA *F*-specific phages (Frédéricq, 1963; Kahn and Helinski, 1964; Macfarren and Clowes, 1967). On the other hand, the frequency of chromosome transfer by these two *colV* factors is only about one to five per cent of that mediated by *F*, when the behaviour of all three factors is compared in the same *E. coli* K12 strain. Although each factor appears to transmit individual markers distributed around the chromosome with about the same relative efficiency, one of the *col* factors (*colV3*) differs from the others in showing asymmetrical linkage effects when the inheritance of certain unselected markers was examined, suggesting different sites of insertion and polarities of transfer (Macfarren and Clowes, 1967).

In the case of the *colI* factor, we have already seen that physiologically derepressed (HFT) cultures transfer chromosomal genes to give much the same recombinant patterns as F^+ cultures, but the frequency of transfer is only about 10^{-8} to 10^{-9}, that is, about 10,000 times lower. We will see later on that transfer by this and some other factors may not depend on direct interactions between factor and chromosome (p. 805).

Recombination between plasmids

The occurrence of recombination between sex factors has been most clearly shown in the case of incompatible resistance transfer factors which carry resistance determinants for different groups of drugs. Such factors are unstable in mixedly infected cells, one or the other, together with its resistance genes, being eliminated. If, however, doubly infected cultures are grown in the presence of drugs of both groups, clones which maintain both sets of resistance determinants in a stable condition are selected. These turn out to be associated with a single transferable structure, as is shown by their joint transduction by phage P1 (Mitsuhashi *et al.*, 1962; Watanabe *et al.*, 1964). Similarly, simultaneous infection of *E. coli* cultures with the two *F*-prime factors, *F—lac*$^+$ and *F—gal*$^+$, yields rare, stable clones which ferment both lactose and galactose and in which the lac$^+$ and *gal*$^+$ determinants are located on a single replicon. Of course, every cell from cultures infected with a pair of *compatible* plasmids carrying different character determinants would stably inherit both plasmids as independent elements, and would express both sets of characters. Although there would be no bar to the formation of a joint structure, provided the plasmids possessed some genetic homology, this might be difficult to recognise and isolate in the absence of any selective pressure favouring it.

Recombination has indeed been demonstrated to occur between various different plasmids and other elements, with the formation of hybrid structures. For example, a partially defective resistance transfer factor, resulting from phage P22 transduction (p. 762), when introduced into an F^+ cell, has been shown to recombine with the sex factor *F* to produce a stable and fully functional factor, which is hybrid with respect to certain other character differences between the sex factor and the original RTF (Watanabe and Ogata, 1966).

We have previously noted that phage P22 transduction in *Salmonella* may separate the resistance determinants of RTFs from a functional transfer factor, and that these determinants may become integrated into the bacterial chromosome at the phage P22 attachment region; the only reasonable explanation is that part of the resistance transfer factor was incorporated into the phage genome by recombination (Dubnau and Stocker, 1964; see also Harada *et al.*, 1967). In the same way, the integration of resistance genes into the genomes of phages P1 and ε15, to yield fully functional transducing phages, has been reported. Every bacterium infected and lysogenised with these phages is converted to resistance (Kondo and Mitsuhashi, 1964; Kameda *et al.*, 1965).

Again, if bacteria transduced to immobile resistance by phages ε15 and ε34, from a *Salmonella* donor carrying a resistance transfer factor, are subsequently infected with an *F*-prime factor (*F*′13), the resistance determinants link up with the *F*-prime factor to form a new resistance transfer factor. Not only is this factor transducible as a single structure by phage P1 but, unlike the original RTF but like the *F*-prime factor, it can be eliminated by treatment with acridine orange (p. 656; Harada *et al.*, 1964).

We have seen that immobile and non-integrated resistance determinants, as well as non-transmissible *col* factors, can usually be readily 'picked up' and transferred by the sex factor *F* and other transfer factors (pp. 762, 756). It is possible that such transfer is due to independent migration of the otherwise immobile plasmids through conjugation tubes determined by the sex factor. However, it seems more likely, in view of the facts mentioned above, that transfer results from integration of the immobile element into the transfer factor by recombination. Since segregated resistance determinants frequently seem to behave as replicons, they probably retain a considerable part of the transfer factor to which they were initially linked, which provides regions of genetic homology with related sex factors. These potentialities for recombination between the sex factor *F*, transfer factors associated with resistance, colicinogenic and other determinants, and the genomes of transducing phages, offers susceptible bacteria an enormous flexibility in their genetic adaptation to environmental changes, and plasmids a wide scope for variation.

Complementation between plasmids

The main barrier to performing complementation tests in plasmids is the problem of establishing stable diploids, due to the incompatibility of similar elements. One solution is to use compatible plasmids which appear to have some functions in common. For example, the sex factor *F* and fi^+ resistance transfer factors can co-exist stably in the same cell, while both determine the production of fimbriae of very similar, if not identical, structure. Defective mutants of *F* or of derepressed fi^+ factors, which no longer produce sex fimbriae, can readily be obtained by selecting for resistance to 'male-specific' RNA phage, or for inability to transfer resistance. In bacteria infected with two such compatible, defective factors it is usually found that the capacity to produce sex fimbriae is restored, but the fimbriae are of *F* type as judged by a superior affinity for RNA phage. On the other hand, defective *F* and fi^- resistance transfer factors show no complementation (Hirota, Fujii and Nishimura, 1966; Nishimura *et al.*, 1967). These results imply that the formation of sex fimbriae depends on a number of functions, and not just on the production of a single type of protein subunit.

An attempt has been made to define the functions required for plasmid replication by means of a complementation test involving a defective, lac^-Hfr strain, unable to mate, infected with a temperature-sensitive mutant of an F—lac^+ factor, unable to replicate at 42°C. The test depends on the isolation of rare clones in which, for unknown reasons, the *F*-prime factor has escaped exclusion and co-exists with the resident, integrated factor. These diploids are then grown at 42°, when retention of the ability to ferment lactose is taken to

300 $500

1/4

$500 (N.Z) to cover NZ & Australia

$450 (USA) for return ticket

HyTONE®

tele-notes

REFILL FOR NO. 9940

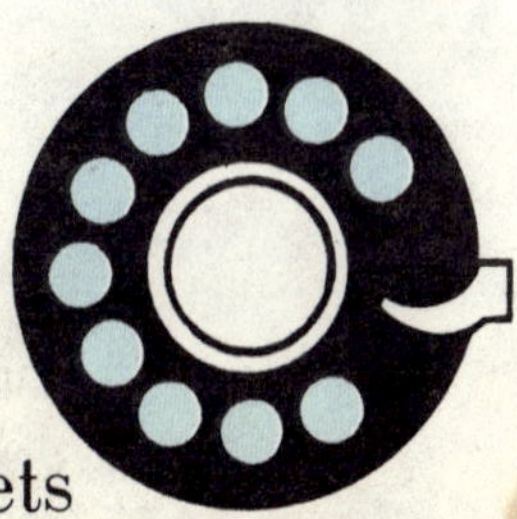

29¢

No. 8946 160 Sheets

A PRODUCT OF Westab DAYTON 2, OHIO

indicate complementation. A number of temperature-sensitive *F—lac*$^+$ mutants tested in this way were found to fall into three complementation groups, so that at least three functions appear to be necessary for autonomous replication of the *F* factor (Cuzin and Jacob, 1965). An interesting by-product of this investigation is the evidence it provides that an integrated sex factor seems to continue to make the positive requirements for its cytoplasmic reproduction.

In other experiments, restoration of function has been claimed between two defective mutants of the same derepressed plasmid, which are used in mixed infections where they are incompatible. In yet another type of experiment, cells carrying a residual fragment of sex factor in their chromosomes, as the result of the formation of an *F*-prime factor in an ancestral *Hfr* cell (sex factor affinity or *sfa* locus; p. 797), are infected with either a mutant or a wild type *F* factor, and a revival of the original *Hfr* activity looked for. In both these types of case there is a high degree of genetic homology between the two elements, on which there may be superimposed a selection against the restored phenotype due to exclusion, so that it is difficult to distinguish between repair due to complementation and that following recombination or, in the case where the bacterial chromosome possesses an *sfa* locus, chromosomal integration of the immigrant sex factor (Meynell *et al.*, 1968).

NON-TRANSMISSIBLE PLASMIDS IN STAPHYLOCOCCUS AUREUS

Staphylococcus aureus (*pyogenes*) is a Gram positive coccus which owes its specific name to the typical golden colour of its colonies. It causes boils, and is among the commonest causes of wound infection and pus production in man. Since the introduction of antibiotic therapy this organism has revealed a striking ability to acquire resistance to the drugs currently in use, so that the multiply-resistant 'hospital staphylococcus' has become a serious clinical problem. The first antibiotic to which resistance was acquired was penicillin, and by 1950 the great majority of strains of *Staph. aureus* isolated in hospitals all over the world were penicillin-resistant, and still remain so. By contrast, the majority of strains isolated from the population at large are sensitive, although the proportion of resistants among these now appears to be increasing. The effect of antibiotic selection on the development and spread of resistance is evident from the fact that multiply resistant strains tend to belong to a few notorious phage types, whereas sensitive strains are of many types (see Garrod and O'Grady, 1968).

Although penicillin-tolerant mutants of *Staph. pyogenes* can be isolated in the laboratory by 'multiple-step' selection through increasing concentrations of the antibiotic, the type of resistance selected naturally, and the only type of clinical importance, is due to production of an exo-enzyme, penicillinase, which destroys penicillin by opening up its β-lactam ring. It is interesting that penicillinase production by bacteria long preceded the development of penicillin as a therapeutic agent, since strains of *B. licheniformis*, isolated from spores preserved with the soil of plant specimens sealed in 1689, have been shown to produce the same penicillinase as one of the two types now synthesised by this species. There is evidence that sufficient penicillin is produced by fungi in the soil to account for the evolution of penicillinase determinants in this case at least (see Pollock, 1967).

The first clue that staphylococcal penicillinase might not be determined by a chromosomal gene was the observation that penicillinase-producing (p^+) staphylococci tend to become penicillinase-negative (p^-) at a relatively high rate (c. one per 1000 generations), and that these penicillinase-negative strains, like naturally occurring ones, are incapable of reverting to penicillinase production (Barber, 1949). Genetic analysis by means of transduction then showed that a number of other markers, notably resistance to mercury ions (Hg^r) and to erythromycin (em^r), are very closely linked to the penicillinase determinant, being transduced together with a frequency of over 99 per cent, and that all three markers are spontaneously lost *en bloc*, and are regained together by transduction. Moreover, ultraviolet irradiation of the transducing phage preparation inactivates all the markers together at the same rate and with the same single-hit kinetics (p. 525) as any one marker alone, and at about the same rate as the plaque-forming ability of the phage. In addition, segregation of markers among the resulting transductants is rare, suggesting not only that the whole transduced region behaves like a replicon in being inactivated by a single hit, but that no homologous region exists on the recipient chromosome whereby undamaged markers can be rescued by recombination. It was therefore proposed that the penicillinase determinant and the loci linked to it are located on a plasmid, and all the evidence since has confirmed this hypothesis (Novick, 1963; Novick and Richmond, 1965).

However, acquisition of the plasmid does not confer conjugal fertility on host cells, the plasmid being transmissible by transduction alone. Although its rate of loss may be increased by high temperature and by treatment with a number of drugs, the effect of acridines appears doubtful (Novick and Richmond, 1965). Density gradient analysis has failed to reveal, in strains carrying penicillinase plasmids, a DNA fraction distinguishable from that of staphylococcal DNA (Novick, 1967). In addition to those plasmid loci already mentioned, others have recently been described which are concerned with resistance to arsenate, arsenite, lead, cadmium and zinc ions (Novick, 1967). Some other loci will be described below.

Staphylococcal penicillinase is an inducible enzyme whose rate of synthesis increases 30-fold or more from its basal level in the presence of a penicillin inducer. By the use of mutagens, many different types of mutant plasmid have been isolated in which either the amount or the inducibility of penicillinase production is affected. A few of these make altered enzymes of low specific activity which cross-react with antiserum against the wild type enzyme, so that they probably result from a mutation (p^-) in the structural gene for penicillinase; others lead to a very low level of inducible production of normal penicillin, and behave either like polar mutations (p. 718) in the *p* gene or as 'promoter' mutations (p. 726). Another class of mutation, which we will shortly discuss, results in constitutive production of normal penicillin (Novick, 1963; Richmond, 1965c, 1968).

Penicillinase plasmids, isolated from different staphylococcal strains, fall into a number of types which can be distinguished both by the natural markers they carry as well as by producing penicillins of different chemical structure and immunological specificity (Richmond, 1965a). The characters of three of these types, designated α, β, γ, δ, ε and so on, are compared in Table 26. By means of transduction, plasmids of one type can be transferred to cells already carrying another type. For example, plasmid type γ can be transduced into strains carrying either α or β plasmids, and its inheritance recognised by selecting for erythromycin resistance. It turns out that plasmids α and β, and γ and β, are compatible and form relatively stable diploids which, in the case of α and β for example, produce both types A and C penicillin on induction. Recombination is found rarely to occur between genetically marked, compatible plasmids.

On the other hand, plasmids α and γ are incompatible and segregate at the first division with a probability of 95 per cent. However, with such incompatible pairs of plasmids recombination is frequent. Thus

TABLE 26

The characters and compatibility patterns distinguishing staphylococcal plasmids

Characters	Plasmid type α	β	γ
Penicillinase production	+	+	+
Erythromycin resistance	−	−	+
Mercury (Hg) resistance	+	+	+
Penicillinase type	A	C	A

Compatible pairs	Incompatible pairs
α × β	α × α
γ × β	β × β
	γ × γ
	α × γ

this system lends itself admirably to genetic studies since dominance relations can be studied between compatible plasmids, while suitably marked, incompatible plasmids permit the construction of linkage maps (Novick and Richmond, 1965). When the constitutive penicillinase mutants mentioned above were tested in diploids with a compatible wild type plasmid, inducibility was found to be dominant to constitutivity—the diploids all have the wild, inducible phenotype. This means that the production of penicillin is regulated by a repressor, the constitutive mutants corresponding to i^-, rather than to o^c, mutants in the case of lactose utilisation in *E. coli* (pp. 715, 716) (Richmond, 1965b).

This analogy between the penicillinase and lactose systems was extended by recombinational mapping. For instance, using $\gamma.i^-p^+em^r$ as the immigrant (donor) plasmid and $\alpha.i^+p^-em^s$ as the resident one, erythromycin-resistant (em^r) transductants can be selected and scored for micro-constitutive (i^-p^-) and wild type (i^+p^+) penicillinase production. In this way, some 0·2 per cent of the transductants were found to result from exchanges between *i* and *p*, as compared to about 16 per cent from exchanges between *p* and *em*. Thus the *i* locus is closely linked to, but distinct from, the *p* locus which its product controls (Novick and Richmond, 1965; Richmond, 1968).

The existence of mutations in the *p* and *i* genes permitted the compatibility of isogenic plasmids ($\alpha\times\alpha$, $\beta\times\beta$) to be tested, since stable i^+p^-/i^-p^+ diploids, for example, would display the inducible, wild phenotype. In fact, isogenic plasmids proved to be incompatible, rapidly segregating the parental plasmids as well as a proportion of recombinants (Table 26). As might be expected from such isogenic crosses, as many as 8 per cent of transductants may be recombinant for the *i* and *p* markers—a much higher proportion than is reported for heterogenic crosses (see above; Novick and Richmond, 1965).

An extended recombination analysis by Richmond, employing mutant strains of α and γ plasmids, has given a linkage map with the following sequence of markers,

—*i*—*p*——*Hg*————*em*—,

in which *Hg* is closer to *p* than to *em* (see Novick, 1967). This map has been confirmed by the isolation of a series of deletion mutants of the γ plasmid, which have also revealed that all the loci determining resistance to metallic ions lie between *p* and *Hg*. A study of these deletion mutants has yielded some interesting information about the nature of plasmid incompatibility. A large class of deletions entering the plasmid from the left (*p* end) and deleting the *Hg* locus, leaving only the em^r determinant, were found to confer two new properties on the plasmid. Firstly, it now produces a large proportion of abortive transductants ('minute' colonies; p. 97) and only a minority of complete transductants with respect to erythromycin resistance, as compared with normal plasmids which yield no abortives on transduction; the inference is that the deletion has removed a region of chromosome which is needed for autonomous replication. However, UV-irradiation increases the number of complete transductants at the expense of the abortives, suggesting that the former arise from integration of the em^r gene into the chromosome.

The second characteristic of this deletion class is that, whatever its relation with the chromosome may be, it no longer confers incompatibility on previously incompatible plasmids. This is probably due to absence of a locus directly determining maintenance compatibility (*mc*) and not from some other effect such as absence of a repressor, since all attempts to 'rescue' the original *mc* gene, by superinfection and genetic exchange with a previously incompatible plasmid, have failed. Finally, a number of deletions were isolated which enter the plasmid from the right extremity, deleting the *em* but not the *Hg* locus, and all these behave like normal plasmids with respect to transduction and compatibility. Thus the postulated maintenance compatibility (*mc*) locus lies between *em* and *Hg*, and probably close to *Hg*.

Occasional mutant strains of *Staph. pyogenes* have been isolated in which one of the two plasmid types is highly unstable. We have seen that analogous *E. coli* mutants have been reported which cannot support replication of the sex factor *F* (p. 571), and that these have been postulated to lack a specific membrane site for *F* attachment which is needed for replicon activity (Jacob *et al.*, 1963). According to this hypothesis, plasmids of each type must carry a type-specific region which can recognise its attachment site on the membrane; in the mutant bacteria this site would be so altered as to be unrecognisable. The question is, Are the plasmid recognition region and the maintenance compatibility (*mc*) locus the same? In an attempt to answer this question, the stability of a number of recombinants between a pair of well marked compatible plasmids, growing in normal bacteria, was tested in the mutant bacteria; at the same time, the compatibility type of the recombinants was checked. The outcome was that in no recombinant were the parental determinants of stability or instability in the mutant host segregated from the compatibility determinants—recombination failed to separate the two loci (Novick, 1967). The number of recombinants tested in this experiment was small, but if we accept its tentative conclusion that the membrane attachment region and the compatibility locus are the same, this strongly suggests that the incompatibility of related plasmids is due to competition for a single site of membrane attachment.

Richmond (1967b) has studied another mutant strain of *Staph. aureus* in which the β plasmid is stable but the α plasmid highly unstable. When this strain is made diploid for both plasmids, it is found that the β plasmid is also lost at a high rate. Since the two plasmids are co-transducible from these diploids at a relative high frequency, it is likely that the instability conferred on the β plasmid in the presence of the unstable α is due to a union of the two structures by recombination. As we have seen in the case of prophage insertion (p. 458), and will reiterate in the next section, the efficient formation of a single genetic entity from two independent ones by recombination, is good evidence that both are circular structures. The fact that recombinants between the α and β plasmids arise frequently in the mutant host, but very rarely in a normal host in which they are stable, suggests that the production of recombinants may depend on the formation of a tandem diploid structure, as has been postulated in the case of *E. coli* (p. 694) and *Streptomyces* (p. 698).

Although penicillinase production seems to be a plasmid-determined activity in most strains of *Staph. aureus*, there is evidence that this may not be so in some strains. Two features of the transduction of plasmid genes, as compared with that of chromosomal genes, are that the frequency of transduction is relatively high and is not increased by UV-irradiation of the phage. In contrast, a number of staphylococcal strains have been reported in which the penicillinase determinant behaves like a chromosomal gene in transduction; the frequency is low, but markedly stimulated by UV-irradiation (Asheshov, 1966; Poston, 1966). We have already seen that, in some plasmid deletions which appear to involve its replicative function, the plasmid determinant for erythromycin resistance may become inserted into the chromosome.

Brief reviews on staphylococcal plasmids are by Richmond (1965c) and by Novick (1967). A comprehensive review, which includes much new work not reported here, is by Richmond (1968).

THE NATURE OF SEX FACTOR ACTIVITY

In Chapter 22 we examined the attributes and behaviour of the *E. coli* sex factor, *F*, only in so far as was necessary to understand how the conjugation system which this factor mediates may be applied to the problems of genetic and functional analyses. We have now surveyed the wide field of plasmids in general, and sex factors in particular, and have seen that, in the *Enterobacteriaceae* at least, it is hard to place them in clearcut or meaningful categories on the basis of what we know of their properties and behaviour.

For instance, a single mutational step may change a transmissible plasmid (sex factor) into a non-transmissible one; a sex factor which has a chromosomal location, or transfers the chromosome, in one bacterial host may fail to do so in another from lack of genetic homology; sex factors with vastly different efficiencies of transfer may differ by only a single mutation which has led to derepression of the conjugal functions of one of them; a transmissible colicin factor or resistance factor, in which a sex factor has incorporated a recognisable genetic determinant of unknown origin, is in no way basically different from an *F*-prime factor. In fact, however important or interesting may be the ability of a plasmid to transfer itself to other cells, together with any heterologous DNA that may have become integrated into its structure, the basic feature of all plasmids is that they are independent replicons with a wide range of potentialities.

When we first encountered resistance transfer factors they seemed very different from the sex factor, *F*, both in what they did and in what they didn't do. But we now see that derepressed *fi*$^+$ factors behave in a manner highly similar to *F*, even with respect to the efficiency of chromosomal transfer. Although, therefore, no *Hfr* strain due to the chromosomal insertion of a resistance transfer factor has yet been reported, we can confidently predict that it will be as soon as someone seriously attempts to isolate it. From a genetic point of view, one of the most interesting aspects of sex factors is their ability to interact with the chromosome, and the various phenomena which stem from these interactions. In the following sections, therefore, we will study their mechanism in more detail, as they have been revealed in the case of the factor *F*, the prototype of sex factors.

INSERTION INTO THE CHROMOSOME

Let us begin by briefly re-stating the evidence that the sex factor of *Hfr* bacteria is really inserted into the chromosome. Firstly, genetic analysis shows that the sex factor of *Hfr* bacteria, instead of being independently transmissible as in *F*$^+$ strains, is linked to those chromosomal genes which are transferred last during conjugation, so that the majority of recombinants are precluded, by chromosome breakage, from inheriting the *Hfr* donor state. Secondly, the sex factor is not eliminated from *Hfr* bacteria by acridine orange treatment, as it is from *F*$^+$ cells, suggesting that its replication may no longer be under its own control (Hirota, 1960). In confirmation of this, temperature-sensitive mutants of the sex factor have been isolated which are unable to replicate at 42°C; although they are eliminated from *F*$^+$ bacteria at this temperature, they remain stable in the *Hfr* state, presumably because they have become an integral part of the bacterial replicon. Assuming this is so, how is the insertion achieved?

We have reviewed the overwhelming evidence now available in support of Campbell's (1962) hypothesis to explain the establishment of lysogeny—that the circular phage chromosome is inserted into the continuity of the circular bacterial chromosome by a single, reciprocal recombination event (p. 458 *et seq.*). The Campbell model is, clearly, equally applicable to insertion of the sex factor, and although the evidence here is less complete, it was a study of chromosome transfer by *F*-prime factors that provided the first indication that recombination was indeed involved in insertion. *F*-prime factors are formed when a sex factor, during release from the chromosome of an *Hfr* strain, acquires a neighbouring chromosomal fragment so that the factor, on transfer to another bacterium, now carries a region of perfect homology for part of its chromosome. This confers on the factor the ability to transfer the chromosome with, apparently, the same origin and polarity as the original *Hfr* strain. Thus if the original *Hfr* strain transferred gene *A* first, gene *B* next and gene *Z* last, infection of recipient bacteria with a derivative *F*-prime factor which has incorporated gene *Z*, converts them to intermediate donors which transfer their chromosomal genes in the sequence *A—B*. It was generally assumed that transfer was the same as that by the original *Hfr* strain, the chromosomal *Z* allele being transferred last; however, this is difficult to check because the dominant, wild type allele, Z^+, is usually carried by the sex factor (p. 676) and so spreads independently to virtually all the recipients during the prolonged mating necessary for transfer of the tail markers.

To circumvent this difficulty, Scaife and Gross (1963) employed the reciprocal arrangement in which the dominant alleles, A^+ and Z^+, were on the chromosome, while the sex factor carried the recessive Z^- allele, so that the transfer of both chromosomal markers can be studied in crosses with A^-Z^- recipients. When such matings were interrupted early, it was found that the donor chromosomal gene Z^+ is actually transferred as a *proximal* marker, and may be found in recombinants to which the sex factor has not been transferred. Since the Z^+ gene is never transferred as a proximal chromosomal marker by the original *Hfr* strain, the promotion of its transfer by the F—Z^- sex factor can best be ascribed to pairing and recombination in the new region of homology conferred on the sex factor by its *Z* allele (Scaife and Gross, 1963). This should lead to insertion of the sex factor into the chromosome, and it is assumed that *Hfr* strains have a similar mechanism of formation in F^+ populations, but with a very much lower probability because of the poor homology between the wild sex factor and its chromosomal locations.

A single, reciprocal recombination event between a linear sex factor and a circular chromosome produces a single linear chromosome as

Fig. 139A demonstrates; in fact, this model was first proposed in 1961 by C. Stern to explain not only insertion of the sex factor but also the opening up of the chromosome for linear transfer (see Discussion of Hayes *et al.*, 1963). However, we now know from the autoradiographic studies of Cairns (1963a, b) that the chromosome of *Hfr* bacteria, like those of F^+ and F^- cells, is also a circular loop of DNA. It follows that the sex factor must be circular too, if the circularity of the chromosome is to be preserved after insertion (Fig. 139B). Although the sex factor *F* has not yet been visualised, direct electron-microscopic evidence of the circularity of the DNA of a colicin

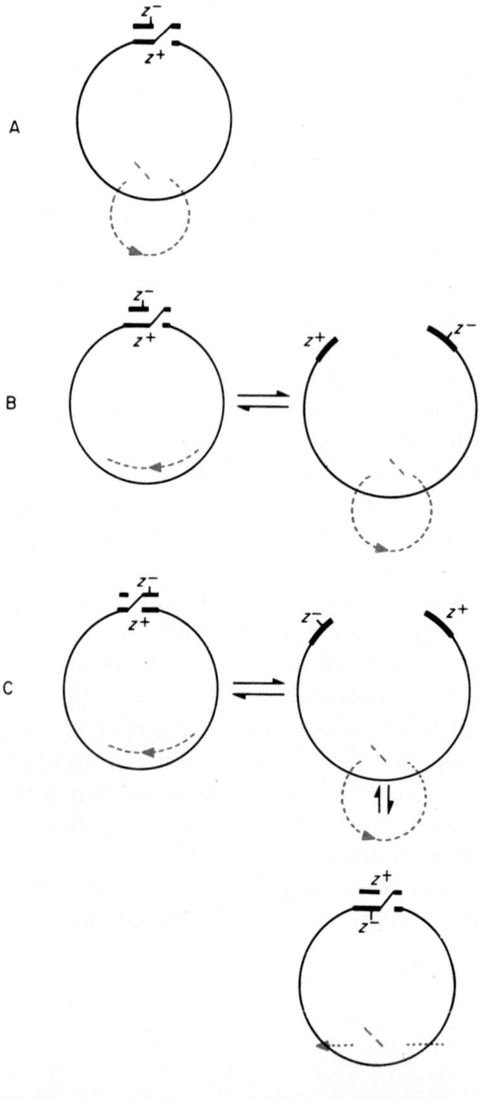

factor, isolated from the DNA of its host, *Proteus mirabilis*, in a density gradient (p. 771), has recently been reported (Roth and Helinski, 1967).

As Fig. 139B shows, insertion yields a structure in which the sex factor is flanked on either side by the two alleles of the *Z* gene. Three predictions can be made about the behaviour of such a structure. Firstly, since it carries two allelic regions of bacterial chromosome, Z^+ and Z^-, frequent pairing and recombination between these may be expected, leading to re-emergence or 'recombination out' of the *F*-prime factor, so that bacteria of this type should alternate rapidly between the *Hfr* and F^+ states. As we have seen, this is just the behaviour which populations of intermediate males do display, transferring both chromosome and autonomous sex factor with high efficiency (p. 676). Secondly, as is shown in Fig. 139B where the arrow in the sex factor indicates the polarity of chromosome transfer, the Z^+ allele will be transferred first and Z^- last if the recombination event inserting the *F*-prime factor occurs to the right of the Z^- mutation. Alternatively, Fig. 139C demonstrates that if the recombination event occurs to the left of the mutation, then Z^- will be transferred first and Z^+ last. Thus the frequency of Z^+ recombinants

---

FIG. 139. Genetic interactions between an *F*-prime factor and the chromosome.

The continuous lines represent the bacterial chromosome, and the interrupted lines the sex factor. The arrow heads indicate the origin and direction of chromosome transfer. The thickened lines are allelic regions on the chromosome and the sex factor which carry the gene *Z*, the location of the Z^- mutation in the gene being marked by a small vertical line. In each case the illustrated crossover may be construed as either inserting or releasing the sex factor.

A. The *F*-prime factor is a linear structure. Recombination yields a linear chromosome.

B. The *F*-prime factor is circular. Insertion by a single reciprocal recombination event to the right of the Z^- mutation, yields a circular chromosome in which the sex factor is located between the two *Z* alleles, Z^+ being transferred first and Z^- last (tandem duplication).

C. Insertion of the *F*-prime factor by a recombination event to the left of the Z^- mutation leads to an invertion of the transfer sequence shown at B.; the first allele transferred is now Z^- and the last Z^+. Release of the *F*-prime factor by recombination on the opposite side of the Z^- mutation from that which led to insertion, results in an exchange of alleles between sex factor and chromosome, as the figure on the right shows.

will depend on the position of the Z^- mutation in the *F*-prime factor. That either allele may be transferred first has been confirmed experimentally (Cuzin and Jacob, 1963; Jacob, Brenner and Cuzin, 1963). Thirdly, as the sequence of Fig. 139C reveals, if the recombination event leading to release of the *F*-prime factor occurs on the opposite side of the mutation from that leading to insertion, there will be an exchange of alleles between *F*-prime factor and chromosome—the original $F—Z^-$ factor becomes an $F—Z^+$ one. This exchange of alleles is a well known fact (Jacob, Schaeffer and Wollman, 1960) which is also a feature of transductional heterogenotes (pp. 632, 635).

This hypothesis, that the *F*-prime factor and, by inference, the normal sex factor, insert themselves into the chromosome to yield *Hfr* bacteria by recombination, has received strong support from the discovery of mutant, recipient bacteria in which loss of the capacity to repair irradiation damage is correlated with inability to yield recombinants in crosses (p. 334; Clark and Margulies, 1965; Howard-Flanders and Boyce, 1966; Howard-Flanders and Thériot, 1966). When such 'recombination-less' (rec^-) bacteria are infected with an *F*-prime factor or a normal sex factor, the factors propagate and transfer themselves in the usual way, but their capacity for chromosome transfer is either enormously reduced or abolished. This provides good evidence that chromosome transfer by F^+ populations is mediated primarily, if not exclusively, by insertion of the sex factor into the chromosome, even though some of the *Hfr* bacteria so formed may be unstable (but see p. 805).

Since pairing is an essential prelude to recombination, chromosome transfer should not occur in the absence of homology between sex factor and chromosome. There are many variations on this theme, one of the most striking being that if an *F*-prime factor is introduced into a bacterium whose chromosome has a deletion for the region carried by the factor, then the bacterium behaves as a normal F^+ donor and not as an intermediate one (P. Driskell, quoted by Clark and Adelberg, 1962).

THE FORMATION OF F-PRIME FACTORS

Hfr bacteria tend to be unstable and to revert to the F^+ state, the frequency of reversion differing from one strain to another, even when they appear to have similar origins (Broda, 1967). Reversion is presumably due to a reversal of the events that led to insertion. In the case of phage λ, you may remember that insertion is a very efficient process, due to good homology between phage and chromosomal attachment regions, but depends on the activity of a specific phage product needed for recombination. The fact that spontaneous prophage release is a rare event is ascribed to prophage repression so that the enzyme required for 'recombination out' is missing (p. 472). In the case of *F* we have no evidence on these aspects of insertion. The low frequency of its occurrence implies poor genetic homology. On the other hand, the fact that insertion does occur in *E. coli* K12, but not to a significant extent in *Shigella* whose DNA has a similar guanine+cytosine content, suggests that the regions of sex factor homology on the chromosome of *E. coli* are not due to the chance equivalence of DNA base sequences. This is supported by the absence of *Hfr* strains in which *F* has become inserted into a known gene.

Our present ideas about the mechanism of *F*-prime factor formation can best be illustrated by examining some pairing interactions between the chromosome and the inserted sex factor of an *Hfr* bacterium which will lead to recombination. These are shown in Fig. 140, in which the circles on the left represent the chromosome of an *Hfr* strain, the inserted sex factor being indicated by the dotted line. The pairs of radial arrows point to postulated regions of homology which will tend to pair and between which recombination will occur. The most common event is shown in the top row of diagrams (Fig. 140a) where the paired regions which originally led to insertion are re-established. A normal sex factor is released and a normal chromosome left behind.

In the second row (Fig. 140b) pairing is assumed to occur between two different regions of homology, one within the sex factor itself and the other on the chromosome between genes *Y* and *Z*. In this case recombination results in release of an *F*-prime factor carrying the *Z* region of chromosome, while the bacterial chromosome acquires a fragment of the sex factor. Since this is a rare event compared with normal sex factor release, it is possible that the homology promoting it is due to a random correspondence of base sequences. Provided the recombination does not retain in the chromosome some sex factor gene necessary for replication or transfer, the released factor will be functionally normal. If the *F*-prime factor is now eliminated by treatment with acridine orange, the cell which carried it should behave as an F^- recipient. However, if it is re-infected with a normal, wild type sex factor, it should not behave as an F^+ donor but as an intermediate donor which transfers its chromosome at high frequency and

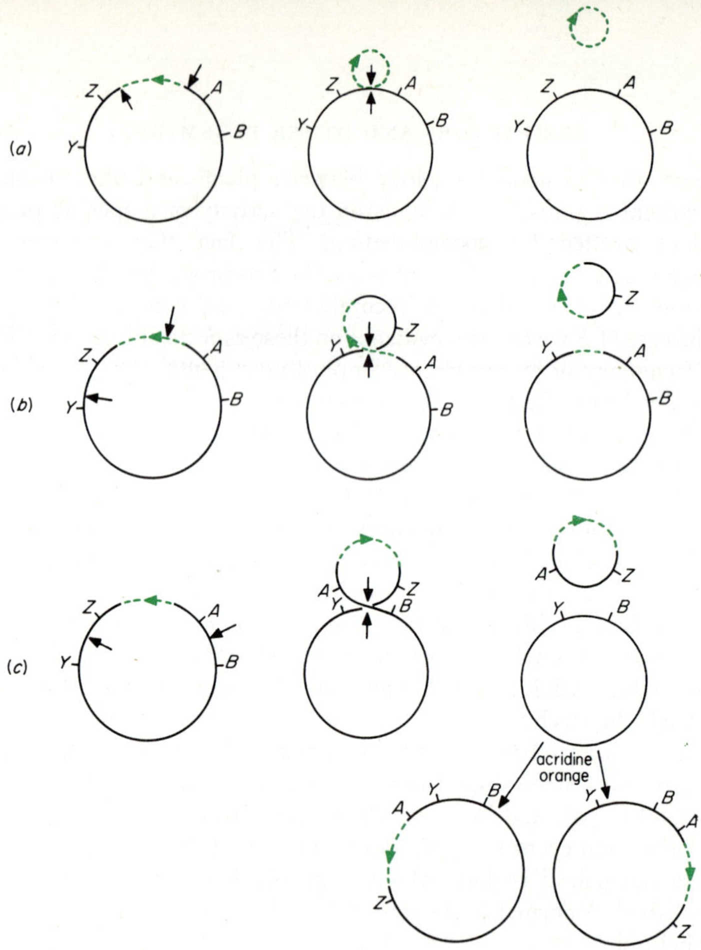

FIG. 140. The origin of *F*-prime factors. Models of possible pairing interactions between various regions of an *Hfr* chromosome, and the events which may follow recombination in each case. The continuous and interrupted lines represent bacterial chromosome and sex factor respectively. The arrow head in the interrupted line indicates not only the point and direction from which chromosome transfer is thought to begin, but also the location of the sex factor replicator. The letters designate genetic loci on the bacterial chromosome. The radial arrows point to presumptive regions of homology on the chromosome between which pairing and recombination may take place.
Row (a). Recombination between the regions which initially led to insertion of the sex factor.

Row (b). Recombination between regions of the bacterial and sex factor chromosomes.

Row (c). Recombination between regions of the bacterial chromosome located on either side of the sex factor.

For explanation, see text.

From Hayes, 1966b.

with an *A—B* gene sequence, because the sex factor fragment retained in the chromosome offers a region of good homology for integration of the immigrant factor (*sex factor affinity* or *sfa* locus). Two cases of this kind have now been reported (Adelberg and Burns, 1960; Jacob *et al.*, 1960; Richter, 1961).

A third possibility, illustrated in Fig. 140c, is that two regions of the *bacterial* chromosome on either side of the sex factor, one between *Y* and *Z* and the other between *A* and *B*, should pair. In this case recombination yields an *F*-prime factor which carries both gene *A* and gene *Z*. Two apparently identical *F*-prime factors of this sort have been obtained from independently isolated strains of the same *Hfr* type (Hirota and Sneath, 1961; Broda, Beckwith and Scaife, 1964), and the genetic behaviour of one of these has been studied in considerable detail. It happened that the parental, initially *Hfr* strain in which the *F*-prime factor arose was isolated and this displayed three remarkable features. Firstly, although the strain transferred the *F*-prime factor efficiently to recipient bacteria, so far as chromosome transfer was concerned it behaved like an F^+ donor, transferring genes *B* and *Y* at low frequency only. This, indeed, is what would be expected from the fact that a segment of the chromosome has, in effect, been deleted and inserted into the sex factor so that the cell remains haploid and possesses no region of good homology between *F*-prime factor and chromosome to mediate re-insertion. Secondly, when the *F*-prime factor is eliminated by acridine orange treatment, no surviving recipient (F^-) bacteria are found since a vital part of the bacterial chromosome, transferred to the sex factor, is now lacking. It was predicted that the only surviving cells would be those in which the *F*-prime factor acquired resistance to the acridine orange by insertion into the chromosome and, experimentally, the treatment resulted in the isolation of a small number of stable *Hfr* clones (Broda *et al.*, 1964; Scaife and Pekhov, 1964; Scaife, 1966; Berg and Curtis, 1967; see also Cuzin and Jacob, 1964).

Genetic analysis of these *Hfr* clones has revealed some interesting new facts which add weight to the general hypothesis. First, it turns out that, in different *Hfr* isolates, the *F*-prime factor has become inserted at different regions of the chromosome and sometimes with a different orientation, as the pair of right-hand diagrams of Fig. 140c show. Secondly, when these *Hfr* strains are used to measure the distance between genes *Y* and *B* in interrupted mating experiments, this distance is found to be much smaller than normal, as is to be expected from the deletion and transposition of the entire *Z—A* region which formerly intervened between them. Thirdly, a number of the *Hfr* isolates examined transfer gene *A* first and gene *Z* last, as in the original *Hfr* strain. This means that the acts of pairing and recombination which led to integration must have been between the *bacterial* segment of the *F*-prime factor and different regions of the bacterial chromosome. In other words, we have evidence here that there are many regions of the bacterial chromosome which are capable of undergoing 'non-allelic' pairing and recombination with one another, though with a very low probability. In the absence of a sex factor, a proportion of such recombination events could be responsible for deletions, as was first suggested in a general way by M. Demerec.

When an *F*-prime factor of this type is transferred to recipient bacteria which carry allelic *Z* and *A* regions, it behaves like a normal *F*-prime factor, not only transferring itself to other cells but also promoting chromosome transfer in the sequence *AB—Z* just like the parental *Hfr* strain. However, these intermediate donors show one predictable departure from normal behaviour—they segregate *stable Hfr* bacteria of parental type with moderate frequency. This is because both the *A* and *Z* regions, on either side of the sex factor, can pair with their homologues on the bacterial chromosome. Recombination in *both* these regions, which may be simultaneous or sequential, will delete one *A* and one *Z* allele and insert the sex factor between the others. If the deleted *A—Z* region has no replication mechanism of its own it will be discarded, leaving only a haploid *Hfr* chromosome (Broda, personal communication).

THE MECHANISM OF CONJUGATION AND GENETIC TRANSFER

The conjugal transfer of genetic material confronts us with three distinct questions, all of which are controversial and unresolved. What is the nature of the conjugal union? How does the genetic material so readily find the union between the mating bacteria, since virtually every conjugation is followed by genetic transfer? What is the mechanism that propels the genetic material from the donor into the recipient bacterium?

We have already discussed the problem of the conjugal unions, and there is little to add to what we have said (p. 671). Before the discovery of sex fimbriae it was generally accepted that the genetic material passed through the kind of connection shown in Plates 32 and 33 (p. 672). However, it has since been claimed that similar appearances may be found in populations of recipient bacteria, while the absolute dependence of genetic transfer on the presence of sex fimbriae, together with the existence, at least in the *F*-type, of an axial hole which could accommodate a DNA double helix, has raised the possibility that these fimbriae may represent the male sex organ of bacteria (Brinton, Gemski and Carnahan, 1964; Brinton, 1965). Alternatively, the fimbriae could conceivably act as grappling arms without which the close contact needed for localised cellular fusion could not be achieved. Only further experiment can clarify this problem.

The question of the efficiency of transfer has a clear topographical solution, even though there is no direct evidence to support it. It was first suggested by J. Lederberg on theoretical grounds, and later presented in more concrete form by Jacob, Brenner and Cuzin (1963) in the light of the increasing evidence that bacterial chromosomes and plasmids are connected to the cell membrane (pp. 568, 569). According to this hypothesis, the sex factor is attached to a specific site on the membrane and determines the local production there of an antigen, whether fimbrial or of some other kind. It is at this point that intimate contact with a recipient bacterium is made and a connection between them established. Thus the sex factor is waiting at the door when the door is opened. If the sex factor, or *F*-prime factor, is autonomous, it alone is transferred.

At present, the only really plausible hypothesis of how transfer is effected is that developed by Jacob and Brenner from the hypothesis of the replicon and its regulation (Jacob and Brenner, 1963; Jacob, Brenner and Cuzin, 1963). This is illustrated in Fig. 141, where the upper row of diagrams show contact being established between various donor and recipient bacteria at the site of sex factor attachment. Some stimulus provoked by the conjugation is then presumed to act as a specific initiator of replication of the sex factor, one of the two daughter replicas being directed into the recipient bacterium as shown in the lower row of diagrams (Fig. 141A). If the sex factor happens to be inserted into the chromosome (Fig. 141B), the replication initiated in the sex factor proceeds around the complete, integrated structure so that the chromosome is transferred as well. Since the direction of sex factor replication is polarised, the polarity of chromosome transfer will depend on which way round the sex factor is inserted. Finally, if the sex factor carries a chromosomal fragment, so that it is repeatedly being inserted into and released from the allelic region of host chromosome by recombination, then whether the

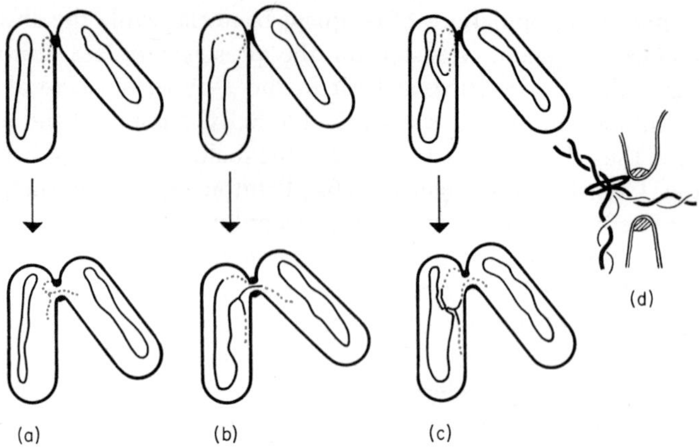

FIG. 141. Diagram of the mechanism of conjugation and genetic transfer in *E. coli*, according to the hypothesis of Jacob and Brenner (1963). In each pair of bacteria the donor is shown on the left and the recipient on the right. The bacterial chromosome is represented by the continuous line, and the sex factor by the interrupted line. The black blob at the point of bacterial contact represents the conjugal antigen synthesised under sex factor control.

A. The sex factor is autonomous and extra-chromosomal.

B. The sex factor is inserted into the chromosome.

C. The sex factor carries a chromosomal fragment (*F*-prime factor).

D. A magnified representation of the transfer process. The ring indicates the sex factor DNA polymerase attached to the membrane, along (or through) which the DNA passes as it is replicated, one replica entering the female cell. The old, parental strands are shown as heavy lines, and the newly synthesised strands as light ones.

For further explanation, see text.

From Hayes, 1966c.

F-prime factor alone, or the chromosome as well, is transferred depends on the state of the sex factor at the time of conjugation. In Fig. 141C a genetic exchange inserting the sex factor is shown to have occurred, so that when replication reaches this point it will proceed along, and transfer, the chromosome. As has been pointed out by Gross and Caro (1965), an *F*-prime factor and an *Hfr* chromosome may be viewed as resulting from the insertion of a part of the chromosome, and of the whole chromosome, respectively, into the sex factor, so that the transfer of each simply becomes a special case of transfer of the sex factor alone.

In this scheme, the energy required for propelling one daughter

DNA duplex into the recipient is assumed to be that needed for the replication process itself. As is shown in Fig. 141D, the polymerase is visualised as attached to the membrane so that the DNA moves through it, rather than that the polymerase runs along the DNA, during replication. The paradox that transfer of the chromosome takes about 90 minutes in broth as well as in synthetic medium, whereas normal chromosome replication is accomplished in about 20 minutes in broth but may require three or four times as long in synthetic medium, is ascribed to the different properties of the bacterial and sex factor DNA polymerases (Jacob *et al.*, 1963).

The elegance and simplicity of this model are almost too good not to be true. However, it turns out that while the majority of experiments confirm predictions of the model, there are some which appear to refute them. If the hypothesis is correct, two basic predicates must follow. The first is that chromosome transfer is absolutely dependent on simultaneous DNA synthesis. Secondly, all the transferred DNA is newly synthesised and double-stranded, consisting of one old and one new strand.

Let us begin by reviewing some of the experiments which favour the hypothesis. The most convincing of these, devised by Gross and Caro (1965, 1966), utilises quantitative autoradiography to reveal the strand constitution of the transferred DNA. The DNA of thymine-requiring *Hfr* bacteria was labelled with radioactive ^{3}H-thymine either before or during mating with unlabelled recipients, depending on the information required. These recipients were morphologically distinguishable from the donors (as in Plates 32 and 33, p. 672), and required adenine and some amino acids for growth, so that they were unable to take up and incorporate ^{3}H-thymine into their DNA in the mating medium. After mating the bacterial pairs were separated by agitation, and the average amount of radioactive DNA transferred to recipient bacteria was measured by grain counts after autoradiography.

Three kinds of experiment were performed. In one, donor bacteria, whose DNA was already labelled in both strands, were mated with recipients in the presence of either labelled or unlabelled thymine. As Fig. 142 (experiment 1A) shows, in the presence of label both strands of the transferred DNA are bound to be labelled by any hypothesis; however, in the absence of label, the model predicts the transfer of newly synthesised DNA of which one strand is old and labelled and the other new and unlabelled (Fig. 142, experiment 1B).

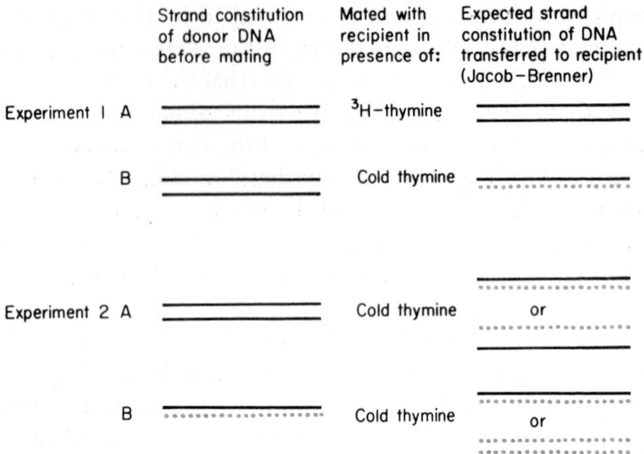

FIG. 142. Diagram illustrating the DNA strand constitution in *Hfr* donor bacteria, before mating in the presence or absence of tritium label, and the expected strand constitution of the DNA transferred to recipient bacteria according to the Jacob-Brenner hypothesis (Gross and Caro, 1966). Tritium labelled (hot) strands are shown as continuous lines, and unlabelled (cold) strands as interrupted lines. For a description of the experiments, see text.

Experimentally, the amount of radioactive label transferred to the recipients mated in the presence of unlabelled thymine turned out to be almost precisely half that found when the label was present throughout the mating period. The experiment therefore accords with the hypothesis that the DNA transferred during mating is not that existing before mating but a newly synthesised replica of it.

This was substantiated by kinetic experiments in which unlabelled donors and recipients were mated in medium devoid of label, the label being added 20 minutes later when chromosome transfer could be assumed to have begun in all the mating pairs. Analysis of samples removed at intervals thereafter revealed that the label began to appear in the recipients immediately after addition to the medium, and increased in amount with time. However, if the label was withdrawn after ten minutes, labelling of the recipients stopped at once. In a complementary experiment, where fully labelled donors were first mated for 20 minutes in the presence of radioactive label, the rate of transfer of the label fell to half its former value immediately after dilution into medium containing cold thymine. These experiments exclude an alternative hypothesis that completion of the replication cycle is a necessary prelude to transfer of the chromosome from its recently completed extremity (see Adelberg and Pittard, 1965).

Finally, an experiment was performed which shows that only one of the two strands of the DNA pre-existing in the *Hfr* donor prior to mating is transferred to the recipient. The amount of label transferred by fully labelled donors in the presence of 'cold' thymine (Fig. 142, experiment 2A) was compared with that transferred by an aliquot of the same culture which had been grown in unlabelled medium for just over a generation before mating, so that not more than one strand of all the DNA molecules would remain labelled (Fig. 142, experiment 2B). The model now predicts that, in this second aliquot, either half-labelled or unlabelled molecules will be transferred with equal probability. In fact, in this second cross, only half as many recipients received label as in the first cross, but the amount of label transferred to individual recipient bacteria in the two crosses was the same (Gross and Caro, 1966).

We will briefly describe three different kinds of experiments which support these findings. We have previously reported how Ptashne (1965a) showed, by density gradient centrifugation, that a proportion of particles of phage λ, released by induction of lysogenic bacteria fully labelled with heavy isotopes, might carry both DNA strands of the original prophage (p. 472). He repeated this type of experiment, using intermediate donor cells infected with an *F*-prime factor carrying prophage λ, the DNA of which was fully labelled with heavy isotopes. When recipient cells to which the *F*-prime factor had been transferred by conjugation in 'light' medium were induced, and the density of the progeny phage particles assessed as in the previous experiment, many particles containing DNA with one labelled strand were found, but none with two (Ptashne, 1965b).

Again, it has recently been shown that naladixic acid, which has been reported to be a rapid and specific inhibitor of DNA synthesis, completely arrests chromosome transfer by *Hfr* strains, even if added to the mating mixture after transfer has begun (Hollom and Pritchard, 1965). Furthermore, assuming the specificity of action of the drug, the use of naladixic acid-resistant mutants as donor and recipient strains has confirmed that DNA synthesis in the donor, but not in the recipient, is required for *F*-prime factor transfer; on the other hand, once an *F—lac*$^+$ factor has been transferred, naladixic acid does not prevent β-galactosidase synthesis in the recipients, even though DNA synthesis is completely suppressed (Barbour, 1967). Thirdly, we have seen that treatment with acridine orange leads to the elimination of autonomous *F* factors, probably as a result of specific inhibition of *F* replication (p. 656). If this is so, acridine orange should prevent the initiation of chromosome transfer by *Hfr* strains, since the hypothesis attributes this to the conjugal triggering of sex factor replication—a function normally performed by the chromosome replicator when the sex factor is integrated. Again, this prediction has been confirmed by experiment (Jacob *et al.*, 1963; Cuzin and Jacob, 1966).

Despite the appealing logic of the hypothesis, and the convincing nature of the experiments we have recounted in its favour, we must admit that there are some experiments, involving the prevention of DNA synthesis, which raise doubts in our mind. For example, Pritchard (1963) has studied chromosome transfer by thymine-requiring *Hfr* bacteria in the absence of thymine. In the case of one strain, the frequency and kinetics of transfer remained virtually unchanged although the rate of DNA synthesis was reduced to six per cent of normal. In another strain, which synthesised DNA at 12 per cent of the usual rate in the absence of thymine, 30 per cent or more of the population transferred their chromosomes normally, while failure of the rest to do so was shown to be due to their failure to initiate transfer.

Even more paradoxical results have been obtained in experiments employing a particular pair of temperature-sensitive mutants of *Hfr* and recipient strains, which behave normally at 37° but are unable to synthesise DNA at 42°, although RNA and protein synthesis remain unaffected. It turns out that the normal number of recombinants arise from matings at 42° when the *Hfr* strain is temperature-sensitive, but none when the recipient is the temperature-sensitive strain (Bonhoeffer, 1966). Not only this, but whereas in control crosses at 42° the *Hfr* lac^+ region is transferred early to lac^- recipients, which thus increasingly acquire the ability to synthesise β-galactosidase, no β-galactosidase is produced by the temperature-sensitive recipient population under the same conditions. This suggests that the absence of recombinants in the cross at the high temperature is not due to a failure of recombination but, rather, of transfer (Bonhoeffer, Hösselbarth and Lehmann, 1967). In support of these experiments is an independent report that transfer from both *Hfr* and intermediate donors to an adenine-requiring recipient strain is greatly reduced in the absence of adenine (Freifelder, 1967).

A resolution of these conflicting findings must await a more precise understanding of the way in which these various methods of interfering with DNA synthesis operate, and of what other cellular activities may be affected in the process. It may well be that the recipient plays a more active role in conjugation than was formerly thought. Two points are worth making here. The first is that none of the experiments we have described exclude the possibility that only a single DNA strand is transferred from the donor by some as yet unknown process, so that synthesis of a second strand in the recipient might be necessary for either expression of function or recombination. Secondly, the possibility that increased DNA breakdown may be associated with some derangement of synthesis should not be forgotten.

A genetic analysis of recombinants derived from $Hfr \times F^-$ crosses under conditions favouring transfer of the complete chromosome has produced evidence that when transfer of the whole chromosome is accomplished, a second round of transfer is initiated as a continuation of the first, without interruption or a break in the continuity of the transferred DNA. That is, the transferred structure can be represented as *A—B——Z—A—B——* (Fulton, 1965). This has been interpreted on the basis of the Jacob-Brenner hypothesis by assuming that only one strand of the circular chromosomal DNA of the donor opens during conjugation, and that this is the strand which is replicated into the recipient. Preservation of the circularity of the other strand presents a continuous template for the second round of replication (Fulton, 1965).

Whatever the usual mechanism for chromosome transfer may be, Clowes and Moody (1966) have put forward the idea that a quite different mechanism, not dependent on insertion of a sex factor, may mediate very low frequency chromosome transfer. We have already described the occurrence of mutant strains of *E. coli* (rec^-) which are unable to yield recombinants when used as recipients in crosses (p. 334). However, these strains can support the normal growth and conjugal transfer of autonomous sex factors. It was found that the transfer of chromosomal genes by high frequency transfer (HFT) preparations of the *colI* factor (p. 758) occurs at the same frequency from both rec^+ and rec^- strains, suggesting that chromosome transfer by this plasmid does not depend on genetic interactions with the chromosome. It was then discovered that rec^- bacteria infected with other factors such as *F*, *F*-prime and *colV*, which normally promote chromosome transfer at variable but much higher frequencies, continue to transfer chromosomal genes, but at the uniform low frequency characteristic of the *colI* factor (10^{-9} to 10^{-8}).

Two facts suggest that this residual low level of transfer is not due to the occurrence of occasional recombination as a result of 'leaky' *rec*$^-$ mutations. The first is that *F* and *F*-prime factors, which have strikingly different pairing affinities for the chromosome and promote very different rates of chromosome transfer in *rec*$^+$ strains, are reduced to the same low level of efficiency in *rec*$^-$ bacteria. Secondly, if the residual degree of chromosome transfer displayed by *F*-prime factors in *rec*$^-$ bacteria were due to rare insertions into the chromosome, the genes transferred would be expected to show the same polarity of transfer as from *rec*$^+$ donors carrying the factor. In fact, no polarity is found, genes lying proximal and distal to the normal site of *F*-prime factor insertion being transferred at the same frequencies. These experiments suggest that rare and spontaneously arising chromosome fragments may find their way through conjugation tubes in the absence of any direct interaction with resident sex factors (Clowes and Moody, 1966).

PLASMIDS AND EVOLUTION

In this Chapter we have been preoccupied with a class of plasmid to which the name 'sex factor' has been given because it manifests itself primarily as an agent responsible for conjugation and chromosome transfer in bacteria. Because sexuality is the principal means of variation and evolution in higher organisms we are, perhaps, apt to regard sex factors from a rather teleological point of view, as having been developed by bacteria to enhance their own evolutionary flexibility. In fact, it seems rather unlikely that the acquisition by bacteria of mechanisms of genetic transfer and recombination, especially of the rather rudimentary type which appears to prevail among them, offers much advantage in their struggle for survival. Were sexuality a real asset we might anticipate that it would have become a more common property of bacteria in general, and be naturally manifested in more efficient ways than experiment has so far revealed. Besides, one might expect that sexuality would begin to acquire over-riding importance as an evolutionary factor only in rather slowly growing organisms which have to contend with a changing environment, or in multicellular, diploid creatures in which most potentially favourable mutations are either not expressed or cannot be inherited. Bacteria, on the other hand, are unicellular, haploid organisms which usually multiply with great rapidity and achieve vast populations so that, in them, mutation is certainly a very efficient and flexible means of variation beside which any advantages which might accrue from an incompetent sexual system would be expected to be insignificant. In fact, apart from the special case of resistance transfer and some other factors, there is no evidence that species variation which could have been due to the operation of a parasexual process, such as the multiplicity of serotypes in *Salmonella*, is not as likely to have evolved by mutation alone.

On the whole, therefore, it seems more likely that transmissible plasmids may have been selected on their own merits and without regard to their effect on bacterial behaviour. The sexuality which sex factors confer on their host cells could be incidental to the evolution of the sex factors themselves. All its features can be explained as benefiting the sex factor at least as much as the bacterium. Thus the advantages to the sex factor of autonomous replication and the ability to spread from host to host are obvious. The homology which exists between sex factors and the chromosomes of their hosts, which we have postulated as instrumental in chromosome transfer, could be a vestigal reflection of a chromosome of which the sex factor was once a part or, alternatively, could have been evolved as a mechanism to ensure stability of the factor in host populations. Nevertheless it remains possible that the appearance of sex factor analogues at an early stage of evolution may have provided unicellular organisms with the potentiality to evolve towards a multicellular existence and the true sexuality which this entails. In other words, symbiosis between a primitive sex factor and its host could be the prototype from which all sexual organisms have developed, while in the bacteria we are observing a sterile evolutionary backwater of this relationship.

Whatever the origin of sex factors there is little to distinguish them, in a general way, from temperate bacteriophages. Both have affinities for the host chromosome, can replicate autonomously in the cytoplasm and have evolved mechanisms to ensure their infective transfer. Moreover, transducing phages also behave as sex factors in the sense that they mediate genetic transfer and recombination between bacteria. In addition sex factors, like phages, fulfil all the criteria of Lwoff's (1959) famous definition of viruses. They are infectious, they depend on the host cell for their metabolism, and they possess only one type of nucleic acid. We think it logical to regard them as viruses which, instead of killing their host cells and elaborating a complex protein to protect their nucleic acid from the environment and to ensure its infectivity, have developed the conjugation mechanism as an elegant way of promoting their efficient dissemination among a wide range of bacterial species and genera.

In conclusion we must mention the striking evolutionary flexibility of plasmid systems (see Jacob *et al.*, 1960; Jacob and Wollman, 1961a). In the case of temperate phages, mutations in the c_1 locus, which block the synthesis of repressor, transform the virus into a virulent one; it cannot establish lysogeny and is compelled to adopt a purely cytoplasmic existence. At the other extreme, mutations involving prophage genes, whose function is required for autonomous reproduction, result in loss of the viral properties of the phage; it can only survive in permanent association with its locus on the bacterial chromosome where it mimics a bacterial gene conferring immunity against specific phage infection. Similarly, the sex factor in stable *Hfr* strains of *E. coli* presents itself as a region of chromosome determining various conjugal functions. It seems likely (though this has not been observed) that a single mutation affecting the regulation of its autonomous replication could turn this non-pathogenic agent into a virulent one, transmissible by conjugation.

Another feature of many plasmids is their tendency to incorporate host genes into their structure. In this way, genes of chromosomal origin can acquire not only autonomy, so that several copies are present in otherwise haploid cells, but also the infectivity characteristic of viral genes. Finally, the phenomenon of phage conversion (p. 646) has shown us that some typical bacterial characters of taxonomic importance turn out to be determined by lysogenising phages. Thus in bacteria it is no longer possible to draw a firm line of demarcation between chromosomal and cytoplasmic genetic determinants, between viral and non-viral elements, or even between viral and bacterial genes. All can merge into one another as a result of mutational and recombinational events.

References

The italicised figures, included in brackets after each reference, indicate the pages of this book where the work described in the paper is mentioned. It is hoped that this will supplement the index, and the cross-references in the text, in assisting the reader to abstract the maximum information on any particular topic.

ABEL P. & TRAUTNER T. A. (1964). Formation of an animal virus within a bacterium. *Z. Vererbungsl.*, **95**, 66 (*619*).

ABELSON J. & THOMAS C. A. Jnr. (1966). The anatomy of the T5 bacteriophage DNA molecule. *J. Mol. Biol.*, **18**, 262 (*342*).

ABRAM D. & KOFFLER H. (1964). The *in vitro* formation of flagella-like filaments and other structures from flagellin. *J. Mol. Biol.*, **9**, 168 (*262*).

ADAMS J. N. & LURIA S. E. (1958). Transduction by bacteriophage P1: abnormal phage function of the transducing particles. *Proc. Natl. Acad. Sci., Wash.*, **44**, 590 (*621, 637*).

ADAMS M. H. (1959). *Bacteriophages*. Interscience Publishers, Inc., New York (*410, 487, 523*).

ADELBERG E. A. (1955). The biosynthesis of isoleucine, valine and leucine. In *A Symposium on Amino Acid Metabolism* (eds. McElroy W. D. & Glass H. B.), p. 419. John Hopkins Press, Baltimore (*122*).

ADELBERG E. A. & BURNS S. N. (1959). A variant sex factor in *Escherichia coli*. *Genetics*, **44**, 497 (*104, 674*).

ADELBERG E. A. & BURNS S. N. (1960). Genetics variation in the sex factor of *Escherichia coli*. *J. Bacteriol.*, **79**, 321 (*104, 674, 675, 797*).

ADELBERG E. A. & PITTARD J. (1965). Chromosome transfer in bacterial conjugation. *Bacterial Rev.*, **29**, 161 (*104, 679, 803*).

ADELBERG E. A. & UMBARGER H. E. (1953). Isoleucine and valine metabolism in *Escherichia coli*. V.α-Ketoisovaleric acid accumulation. *J. Biol. Chem.*, **205**, 475 (*123*).

ADELBERG E. A., MANDEL M. & CHEIN CHING CHEN G. (1965). Optimal conditions for mutagenesis by N-methyl-N'-nitro-N-nitrosoguanidine in *Escherichia coli* K12. *Biochem. biophys. Res. Comm.*, **18**, 788 (*213*).

ADHYA S. & ECHOLS H. (1966). Glucose effect and the galactose enzymes of *Escherichia coli*: correlation between glucose inhibition of induction and inducer transport. *J. Bacteriol.*, **92**, 601 (*724*).

ADLER H. I. (1966). The genetic control of radiation sensitivity in micro-organisms. In *Advances in Radiation Biology* (eds. Augenstein L. G., Mason R. D. & Zelle M. R.), Vol. 2, p. 167. Academic Press, Inc., New York and London (*338*).

AKIBA T., KOYAME K., ISHIKI Y. & KUMIRA S. (1960). On the mechanism of the development of multiple drug-resistant clones of *Shigella*. *Japan J. Microbiol.*, **4**, 219 (*760*).

ALEXANDER H. E. & LEIDY G. (1951). Determination of inherited traits of *H. influenzae* by desoxyribonucleic acid fractions isolated from type-specific cells. *J. Exptl. Med.*, **93**, 345 (*577*).

ALEXANDER H. E., LEIDY G. & HAHN E. (1954). Studies on the nature of *Haemophilus influenzae* cells susceptible to heritable changes by desoxy-ribonucleic acids. *J. Exptl. Med.*, **99**, 505 (*581*).

ALFERT M. (1957). Some cytogenetical contributions to genetic chemistry. In *The Chemical Basis of Heredity* (eds. McElroy W. D. & Glass B.), p. 186. The John Hopkins Press, Baltimore (*534*).

ALFOLDI L., JACOB F. & WOLLMAN E. L. (1957). Zygose létale dans des croisements entre souches colicinogènes et non-colicinogènes d'*Escherichia coli*. *C. R. Acad. Sci., Paris*, **244**, 2974 (*757*).

ALFOLDI L., JACOB F., WOLLMAN E. L. & MAZÉ R. (1958). Sur le détermin-isme génétique de la colicinogénie. *C. R. Acad. Sci., Paris*, **246**, 3531 (*757*).

ALFOLDI L., STENT G. S. & CLOWES R. C. (1962). The chromosomal site of the RNA control (*RC*) locus in *Escherichia coli*. *J. Mol. Biol.*, **5**, 348 (*688*).

ALIKHANIAN S. I. & BORISOVA L. N. (1961). Recombination in *Actinomyces aureofaciens*. *J. gen. Microbiol.*, **26**, 19 (*696*).

ALIKHANIAN S. L. & ILJINA T. S. (1960). Mutagenic action of actinophage. *J. Genet.*, **57**, 11 (*222*).

ALIKHANIAN S. I. & MINDLIN S. Z. (1957). Recombinations in *Streptomyces rimosus*. *Nature, Lond.*, **180** (*696*).

ALIKHANIAN S. I., ILJINA T. S. & LOMOVSKAYA N. D. (1960). Transduction in Actinomycetes. *Nature*, **188**, 245 (*696*).

ALIKHANIAN S. I., ILJINA T. S., KALTAEVA E. S., KAMENEVA S. V. & SUKHODOLEC V. V. (1965). Mutants of *Escherichia coli* K12 lacking thymine. *Nature, Lond.*, **206**, 848 (*573*).

ALIKHANIAN S. I., GRINBERG C. N., KRYLOV V. N., MAISOURIAN A. N. & OGANESIAN M. G. (1966a). The temperature sensitive (ts) mutations of the bacteriophage T4B. *J. Genetics*, **59**, 283 (*485*).

ALKHANIAN S. I., ILJINA T. S., KALIAEVA E. S., KAMENEVA S. V. & SUKHODOLEC V. V. (1966b). A genetical study of thymineless mutants of *E. coli* K12. *Genet. Res., Camb.*, **8**, 83 (*573*).

ALLISON A. C. (1954). Protection afforded by sickle-cell trait against sub-tertian malarial infection. *Brit. med. J.*, **1**, 290 (*28*).

ALLOWAY J. L. (1933). Further observations on the use of pneumococcus extracts in effecting transformation of type *in vitro*. *J. Exptl. Med.*, **57**, 265 (*575*).

ALPERS D. H. & TOMKINS G. M. (1965). The order of induction and de-induction of the enzymes of the lactose operon in *E. coli*. *Proc. Natl. Acad. Sci., Wash.*, **53**, 797 (*719*).

AMES B. N. & GARRY B. (1959). Coordinate repression of the synthesis of four histidine biosynthetic enzymes by histidine. *Proc. Natl. Acad. Sci., Wash.*, **45**, 1453 (*710*).

AMES B. N. & MARTIN R. G. (1964). Biochemical aspects of genetics: the operon. *Ann. Rev. Biochem.*, **33**, 235 (*724, 735*).

AMES B. N. & WHITFIELD A. J., Jnr. (1966). Frameshift mutagenesis in *Salmonella. Cold Spr. Harb. Symp. quant. Biol.*, **31**, 221 (*355*).

AMES B. N., DUBIN D. T. & ROSENTHAL S. M. (1958). Presence of polyamines in certain bacterial viruses. *Science*, **127**, 814 (*420*).

AMES B. N., GARRY B. & HERZENBERG L. A. (1960). The genetic control of the enzymes of histidine biosynthesis in *Salmonella typhimurium. J. gen. Microbiol.*, **22**, 369 (*150, 160, 722, 724*).

AMES B. N., HARTMAN P. E. & JACOB B. F. (1963). Chromosomal alterations affecting the regulation of histidine biosynthetic enzymes in *Salmonella. J. Mol. Biol.*, **7**, 23 (*724*).

AMES B., GOLDBERGER R., HARTMAN P., MARTIN R. & ROTH J. (1967). The histidine operon. In *Regulation of nucleic acid and protein biosynthesis* (Koningsberger V. V. & Bosch L., eds.), p. 271. Elsevier Publishing Co., Amsterdam (*719, 722, 724, 734*).

AMMANN J., DELIUS H. & HOFSCHNEIDER P. H. (1964). Isolation and properties of an intact phage specific form of RNA phage M12. *J. Mol. Biol.*, **10**, 557 (*443*).

ANAGNOSTOPOULOS C. & CRAWFORD I. P. (1961). Transformation studies on the linkage of markers in the tryptophan pathway in *Bacillus subtilis. Proc. Natl. Acad. Sci., Wash.*, **47**, 378 (*608, 611, 614*).

ANAGNOSTPOULOS C., BARAT M. & SCHNEIDER R. (1964). Étude, par transformation, de deux groupes de gènes régissant la biosynthèse de l'isoleucine, de la valine et de la leucine chez *Bacillus subtilis. C. R. Acad. Sci., Paris*, **258**, 749 (*164, 611, 614*).

ANDERSON E. S. (1957). Visual observation of deoxyribonucleic acid changes in bacteria during growth of bacteriophage. *Nature, Lond.*, **180**, 1336 (*548*).

ANDERSON E. S. (1962). The genetic basis of bacteriophage typing. *Brit. Med. Bull.*, **18**, 64 (*522*).

ANDERSON E. S. (1965a). A rapid screening test for transfer factors in drug-sensitive *Enterobacteriaceae. Nature, Lond.*, **208**, 1016 (*441, 764*).

ANDERSON E. S. (1965b). Origin of transferable drug-resistant factors in the *Enterobacteriaceae. Brit. med. J.*, **2**, 1289 (*764*).

ANDERSON E. S. (1967). Facteurs de transfert et résistance aux antibiotiques chez les enterobactéries. *Ann. Inst. Pasteur*, **112**, 547 (*761, 763, 767, 768*).

ANDERSON E. S. (1968). The ecology of transferable drug resistance in the Enterobacteria. *Ann. Rev. Microbiol.*, **22** (*761, 767, 769*).

ANDERSON E. S. & DATTA N. (1965). Resistance to penicillin and its transfer in *Enterobacteriaceae. Lancet*, **1**, 407 (*761*).

ANDERSON E. S. & FELIX A. (1953). The Vi type-determining phages carried by *Salmonella typhi. J. gen. Microbiol.*, **9**, 65 (*522*).

ANDERSON E. S. & FRASER A. (1956). The statistical distribution of phenotypically modifiable particles and host-range mutants in populations of Vi-phage II. *J. gen. Microbiol.*, **15**, 225 (*521*).

ANDERSON E. S. & LEWIS M. J. (1965). Characterisation of a transfer factor associated with drug resistance in *Salmonella typhimurium*. *Nature, Lond.*, **208**, 843 (*761, 763, 767, 768*).

ANDERSON E. S., ARMSTRONG J. A. & NIVEN J. S. F. (1959). Fluorescence microscopy: observation of virus growth with aminoacridines. *Symp. Soc. gen. Microbiol.*, **9**, 224 (*548*).

ANDERSON T. F. (1948). The activation of the bacterial virus T4 by L-tryptophan. *J. Bacteriol.*, **55**, 637 (*422*).

ANDERSON T. F. (1953). The morphology and osmotic properties of bacteriophage systems. *Cold Spr. Harb. Symp. quant. Biol.*, **18**, 197 (*417*).

ANDERSON T. F. (1958). Recombination and segregation in *E. coli*. *Cold Spr. Harb. Symp. quant. Biol.*, **23**, 47 (*654, 694*).

ANDERSON T. F. (1960). On the fine structures of the temperate bacteriophages P1, P2 and P22. *Proc. European Regional Conf. on Electron Microscopy, Delft*, Vol. 2, p. 1008 (*417, 418*).

ANDERSON T. F. & DOERMANN A. H. (1952). The intracellular growth of bacteriophages. III. The growth of T3 studied by sonic disintegration and by T6 cyanide lysis of infected cells. *J. Gen. Physiol.*, **35**, 657 (*424*).

ANDERSON T. F., RAPPAPORT C. & MUSCATINE N. A. (1953). On the structure and osmotic properties of phage particles. *Ann. Inst. Pasteur*, **84**, 5 (*420*).

ANDERSON T. F., WOLLMAN E. L. & JACOB F. (1957). Sur les processus de conjugaison et de recombinaison chez *E. coli*. III. Aspects morphologiques en microscopie électronique. *Ann. Inst. Pasteur*, **93**, 450 (*671, 672*).

ANFINSEN C. B. (1959). *The Molecular Basis of Evolution*. John Wiley & Sons, Inc., New York (*263, 264, 268*).

APHOSIAN H. V. & KORNBERG A. (1962). Enzymatic synthesis of deoxyribonucleic acid. IX. The polymerase formed after T2 bacteriophage infection of *Escherichia coli*: a new enzyme. *J. Biol. Chem.*, **237**, 519 (*247, 428*).

APPLEYARD R. K. (1954). Segregation of lambda lysogenicity during bacterial recombination in *Escherichia coli* K-12. *Genetics*, **39**, 429 (*454*).

ARBER W. (1960). Transduction of chromosomal genes and episomes in *E. coli*. *Virology*, **11**, 273 (*656*).

ARBER W. (1962). Specificites biologiques de l'acide desoxyribonucleique. *Path. Microbiol., Lausanne*, **25**, 668 (*522*).

ARBER W. (1964). Host specificity of DNA produced by *Escherichia coli*. III. Effects on transduction mediated by *λdg*. *Virology*, **23**, 173 (*522*).

ARBER W. (1965a). Host-controlled modification of bacteriophage. *Ann. Rev. Microbiol.*, **19**, 365 (*515, 521, 522*).

ARBER W. (1965b). Host specificity of DNA produced by *Escherichia coli*. V. The role of methionine in the production of host specificity. *J. Mol. Biol.*, **11**, 247 (*519*).

ARBER W. & DUSSOIX D. (1962). Host specificity of DNA produced by *Escherichia coli*. I. Host controlled modification of bacteriophage λ. *J. Mol. Biol.*, **5**, 18 (*516, 517, 518*).

ARBER W., KELLENBERGER G. & WEIGLE J. (1957). La défectuosité du phage lambda transducteur. *Schweiz. z. für Path. Bakt.*, **20**, 659 (*629, 630*).

ARLINGHAUS R., SCHAEFFER J. & SCHWEET R. (1964). Mechanism of peptide bond formation in polypeptide synthesis. *Proc. Natl. Acad. Sci., Wash.*, **51**, 1291 (*286*).

ARNSTEIN H. R. V. (1965). Mechanism of protein biosynthesis. *Brit. med. Bull.*, **21**, 217 (*284, 287, 296*).

ASHESHOV E. H. (1966). Chromosomal location of the genetic elements controlling penicillinase production in a strain of *Staphylococcus aureus*. *Nature, Lond.*, **210**, 804 (*789*).

ATTARDI G., NAONO S., ROUVIERE J., JACOB F. & GROS F. (1963). Production of messenger RNA and regulation of protein synthesis. *Cold Spr. Harb. Symp. quant. Biol.*, **28**, 363 (*718, 721*).

ATTARDI G. (1967). The mechanism of protein synthesis. *Ann. Rev. Microbiol.*, **21**, 383 (*296*).

ATWOOD K. C., SCHNEIDER L. K. & RYAN F. J. (1951). Selective mechanisms in bacteria. *Cold Spr. Harb. Symp. quant. Biol.*, **16**, 345 (*200*).

AUERBACH C. & ROBSON J. M. (1946). Chemical production of mutations. *Nature, Lond.*, **157**, 302 (*204*).

AUERBACH C., ROBSON J. M. & CARR J. R. (1948). The chemical production of mutations. *Science*, **105**, 243 (*204*).

AUGUST J. T., COOPER S., SHAPIRO L. & ZINDER N. D. (1963). RNA phage induced RNA polymerase. *Cold Spr. Harb. Symp. quant. Biol.*, **28**, 95 (*291, 442*).

AUSTRIAN R. (1952). Bacterial transformation reactions. *Bacteriol. Rev.*, **16**, 31 (*578*).

AUSTRIAN R., BERNHEIMER H. P., SMITH E. E. B. & MILLS G. T. (1959). Simultaneous production of two capsular polysaccharides by pneumococcus. II. The genetic and biochemical bases of binary capsulation. *J. Exptl. Med.*, **110**, 585 (*613*).

AVERY O. T., MACLEOD C. M. & MCCARTY M. (1944). Studies on the chemical nature of the substance inducing transformation of pneumococcal types. I. Induction of transformation by a desoxyribonucleic acid fraction isolated from pneumococcus type III. *J. Exptl. Med.*, **79**, 137 (*228, 575*).

BAGLIONI C. & INGRAM V. M. (1961). Four adult haemoglobin types in one person. *Nature, Lond.*, **189**, 465 (*153*).

BALDWIN R. L., BARRAND P., FRITSCH A., GOLDTHWAIT D. A. & JACOB F. (1966). Cohesive sites on the deoxyribonucleic acids from several temperate coliphages. *J. Mol. Biol.*, **17**, 343 (*462, 543*).

BALTIMORE D. & FRANKLIN R. M. (1963). Properties of the mengovirus and poliovirus RNA polymerases. *Cold Spr. Harb. Symp. quant. Biol.*, **28**, 105 (*291*).

BARAT M., ANAGNOSTOPOULOS C. & SCHNEIDER A. M. (1965). Linkage relationships of genes controlling isoleucine, valine and leucine biosynthesis in *Bacillus subtilis*. *J. Bacteriol.*, **90**, 357 (*611, 614, 616*).

BARBER M. (1949). The incidence of penicillin-sensitive variant colonies in penicillinase-producing strains of *Staphylococcus pyogenes*. *J. gen. Microbiol.*, **3**, 274 (*784*).

BARBOUR S. D. (1967). Effect of naladixic acid on conjugal transfer and expression of episomal *lac* genes in *Escherichia coli* K12. *J. Mol. Biol.* **28**, 373 (*803*).

Barksdale L. (1959). Lysogenic conversions in bacteria. *Bacteriol. Rev.*, **23**, 202 (*647*).

Barksdale W. L. & Pappenheimer A. M. Jnr. (1954). Phage-host relationships in non-toxigenic and toxigenic diphtheria bacilli. *J. Bacteriol.*, **67**, 220 (*646*).

Barner H. D. & Cohen S. S. (1957). The isolation and properties of amino acid requiring mutants of a thymineless bacterium. *J. Bacteriol.*, **74**, 350 (*556*).

Barnett L., Brenner S., Crick F. H. C., Shulman R. G. & Watts-Tobin R. J. (1967). Phase-shift mutants and other mutants in the first part of the rIIB cistron of bacteriophage T4. *Phil. Trans. Roy. Soc. B.*, **252**, 487 (*354*).

Baron L. S., Carey W. F. & Spilman W. M. (1959a). Genetic recombination between *Escherichia coli* and *Salmonella typhimurium. Proc. Natl. Acad. Sci., Wash.*, **45**, 976 (*751*).

Baron L. S., Carey W. F. & Spilman W. M. (1959b). Characteristics of a high frequency of recombination (Hfr) strain of *Salmonella typhosa* compatible with *Salmonella*, *Shigella* and *Escherichia* species. *Proc. Natl. Acad. Sci., Wash.*, **45**, 1752 (*752*, *770*).

Barratt R. W., Newmeyer D., Perkins D. D. & Garnjobst L. (1954). Map construction in *Neurospora crassa. Advances in Genet.*, **6**, 1 (*165*).

Barricelli N. A. (1956). A 'chromosomic' recombination theory for multiplicity reactivation in phages. *Acta. Biotheoretica* (*Leiden*), **11**, 107 (*529*).

Barricelli N. A. & Doermann A. H. (1960). An analytical approach to the problems of phage recombination and reproduction. II. High negative interference. *Virology*, **11**, 136 (*382*, *497*, *504*).

Barrington L. F. & Kozloff L. M. (1954). Action of T2r^+ bacteriophage on host-cell membranes. *Science*, **120**, 110 (*423*).

Baumberg S., Bacon D. F. & Vogel H. J. (1965). Individually repressible enzymes specified by clustered genes of arginine synthesis. *Proc. Natl. Acad. Sci., Wash.*, **53**, 1029 (*732*).

Bautz-Freese E. (1961). Transitions and transversions induced by depurinating agents. *Proc. Natl. Acad. Sci., Wash.*, **47**, 540 (*315*, *317*, *319*).

Bautz-Freese E. & Freese E. (1961). Induction of reverse mutations and cross-reactivation of nitrous acid-treated phage T4. *Virology*, **13**, 19 (*305*, *314*, *315*).

Bautz E. & Freese E. (1960). On the mutagenic effect of alkylating agents. *Proc Natl. Acad. Sci., Wash.*, **46**, 1585 (*308*, *315*).

Bautz E. K. F. & Hall B. D. (1962). The isolation of T4-specific RNA on a DNA-cellulose column. *Proc. Natl. Acad. Sci., Wash.*, **48**, 400 (*371*).

Baylor M. B., Hurst D. D., Allen S. L. & Bertani E. T. (1957). The frequency and distribution of loci affecting host range in the coliphage T2H. *Genetics*, **42**, 104 (*482*).

Bayreuther R. K. E. & Romig W. R. (1964). Polyoma virus: production in *Bacillus subtilis. Science*, **146**, 778 (*619*).

BEADLE G. W. (1945a). Biochemical genetics. *Chem. Rev.*, **37**, 15 (*108, 112*).

BEADLE G. W. (1945b). Genetics and metabolism in *Neurospora*. *Physiol. Rev.*, **25**, 643 (*108, 112*).

BEADLE G. W. (1957). The role of the nucleus in heredity. In *The Chemical Basis of Heredity* (eds. McElroy W. D. & Glass B)., p. 3. Johns Hopkins Press, Baltimore (*176*).

BEADLE G. W. & COONRADT V. L. (1944). Heterocaryosis in *Neurospora crassa*. *Genetics*, **29**, 291 (95).

BEADLE G. W. & EPHRUSSI B. (1936). The differentiation of eye pigments in *Drosophila* as studied by transplantation. *Genetics*, **21**, 225 (*111*).

BEADLE G. W. & TATUM E. L. (1941). Genetic control of biochemical reactions in *Neurospora*. *Proc. Natl. Acad. Sci., Wash.*, **27**, 499 (*108, 112, 652*).

BEALE G. H. (1948). A method for the measurement of mutation rate from phage sensitivity to phage resistance in *Escherichia coli*. *J. gen. Microbiol.*, **2**, 131 (*197*).

BECKWITH J. R. (1963). Restoration of operon activity by suppressors. *Biochim. Biophys. Acta.*, **76**, 162 (*720*).

BECKWITH J. R. (1964). A deletion analysis of the *lac* operator region in *Escherichia coli*. *J. Mol. Biol.*, **8**, 427.

BECKWITH J. R. (1967). Regulation of the *lac* operon. *Science*, **156**, 597 (*728, 735*).

BECKWITH J. R., SIGNER E. R. & EPSTEIN W. (1966). Transposition of the *lac* region of *E. coli*. *Cold Spr. Harb. Symp. quant. Biol.*, **31**, 393 (*726*).

BECKWITH J. R., PARDEE A. B., AUSTRIAN R. & JACOB F. (1962). Coordination of the synthesis of the enzymes in the pyrimidine pathway of *E. coli*. *J. Mol. Biol.*, **5**, 168 (*164, 724*).

DE BEER G. (1964). Mendel, Darwin and Fisher (1865-1965). *Notes and Records, Roy. Soc., Lond.*, **19**, 192 (*25, 42*).

DE BEER G. (1966). Genetics: the centre of science. *Proc. Roy. Soc., B.*, **164**, 154 (*25*).

BELLING J. (1931). Chromomeres in liliaceous plants. *Univ. Calif. (Berkeley) Publs. Botany*, **16**, 153 (*377*).

BELSER W. L. & BUNTING M. I. (1956). Studies on mechanism providing for genetic transfer in *Serratia marescens*. *J. Bacteriol.*, **72**, 852 (*754, 755*).

BENDET I. J., GOLDSTEIN D. A. & LAUFFER M. A. (1960). Evidence for internal organisation of nucleic acid in T2 bacteriophage. *Nature, Lond.*, **187**, 781 (*537*).

BEN-GURION R. (1967). On the induction of a recombination deficient mutant of *Escherichia coli* K-12. *Genet. Res., Camb.*, **9**, 377 (*474*).

BEN-GURION R. & GINSBURG-TIETZ Y. (1965). Infection of *Escherichia coli* K12 with RNA of encepholomyocarditis virus. *Biochem. biophys. Res. Comm.*, **18**, 226 (*619*).

BENZER S. (1952). Resistance to ultraviolet light as an index to the reproduction of bacteriophage. *J. Bacteriol.*, **63**, 59 (*531*).

BENZER S. (1953). Induced synthesis of enzymes in bacteria analysed at the cellular level. *Biochim. et Biophys. Acta.*, **11**, 383 (*427*).

BENZER S. (1955). Fine structure of a genetic region in bacteriophage. *Proc. Natl. Acad. Sci., Wash.*, **41**, 344 (*105, 132, 169*).

BENZER S. (1957). The elementary units of heredity. In *The Chemical Basis of Heredity*, (eds. McElroy W. D. & Glass B.), p. 70. John Hopkins Press, Baltimore (*105, 132, 134, 136, 145, 147, 167, 168, 169, 173, 299, 313*).

BENZER S. (1959). On the topology of the genetic fine structure. *Proc. Natl. Acad. Sci., Wash.*, **45**, 1607 (313).

BENZER S. (1961). On the topography of the genetic fine structure. *Proc. Natl. Acad. Sci., Wash.*, **47**, 403 (*147, 167, 168, 171, 173, 175, 313, 314, 319, 325, 326, 350*).

BENZER S. & CHAMPE S. P. (1961). Ambivalent *rII* mutants of phage T4. *Proc. Natl. Acad. Sci., Wash.*, **47**, 1025 (*216, 369*).

BENZER S. & CHAMPE S. P. (1962). A change from nonsense to sense in the genetic code. *Proc. Natl. Acad. Sci., Wash.*, **48**, 1114 (*362*).

BENZER S. & JACOB F. (1953). Étude du développement du bactériophage au moyen d'irradiations par la lumière ultra-violette. *Ann. Inst. Pasteur*, **84**, 186 (*412, 487*).

BERG C. M. (1966). Inverted and non-inverted transposition in the genome of *Escherichia coli* K12. *Bacteriol. Proc.*, p. 26 (*219, 374*).

BERG C. M. & CARO L. G. (1967). Chromosome replication in *Escherichia coli*. 1. Lack of influence of the integrated F factor. *J. Mol. Biol.*, **29**, 419 (*566, 567*).

BERG C. M. & CURTIS R., III. (1967). Transposition derivatives of an Hfr strain of *Escherichia coli* K-12. *Genetics*, **56**, 503 (*219, 374, 677, 797*).

BERG P. & OFENGAND E. J. (1958). An enzymatic mechanism for linking amino acids to RNA. *Proc. Natl. Acad. Sci., Wash.*, **44**, 78 (*280*).

BERNSTEIN A. (1957). Multiplicity reactivation of ultra-violet irradiated Vi-phage II of *Salmonella typhi*. *Virology*, **3**, 286 (*529*).

BERNSTEIN H. L. (1958). Fertility factors in *Escherichia*. *Symp. Soc. exptl. Biol.*, **12**, 93 (*749*).

BERTANI G. (1951). Studies on lysogenesis. I. The mode of phage liberation by lysogenic *E. coli*. *J. Bacteriol.*, **62**, 293 (*448*).

BERTANI G. (1953). Infections bactériophagiques secondaires des bactéries lysogènes. *Ann. Inst. Pasteur*, **84**, 273 (*409, 447, 468*).

BERTANI G. (1955). The role of phage in bacterial genetics. *Brookhaven Symposia in Biol.*, **8**, 50 (*455*).

BERTANI G. (1958). Lysogeny. *Advances in Virus Res.*, **5**, 151 (*477*).

BERTANI G. & WEIGLE J. J. (1953). Host controlled variation in bacterial viruses. *J. Bacteriol.*, **65**, 113 (*516*).

BERTANI L. E. (1957). The effect of the inhibition of protein synthesis on the establishment of lysogeny. *Virology*, **4**, 53 (*468*).

BESSMAN M. U. (1963). In *Molecular Genetics* (ed. Taylor J. H.), Pt. 1, p. 1. Academic Press Inc., New York (*252*).

BEUKERS R. & BERENDS W. (1960). Isolation and identification of the irradiation product of thymine. *Biochim. et Biophys. Acta*, **41**, 550 (*329*).

BHASKARAN K. (1958). Genetic recombination in *Vibrio cholerae*. *J. gen. Microbiol.*, **19**, 71 (*755*).

BHASKARAN K. (1960). Recombination of characters between mutant stocks of *Vibriocholerae*, strain 162. *J. gen. Microbiol.*, **23**, 47 (*755*).

BHASKARAN K. & IYER S. S. (1961). Genetic recombination in *Vibriocholerae. Nature, Lond.*, **189**, 1030 (*755*).

BISHOP D. H. L. & BRADLEY D. E. (1965). Determination of base ratios of six ribonucleic acid bacteriophages specific to *Escherichia coli. Biochem. J.*, **95**, 82 (*441*).

BISHOP J., LEAHY J. & SCHWEET R. (1960). Formation of the peptide chain of hemoglobin. *Proc. Natl. Acad. Sci., Wash.*, **46**, 1030 (*286, 343*).

BISSET K. A. (1950). *The Cytology and Life History of Bacteria*. Livingstone, Edinburgh (*651*).

BLAKE C. C. F. (1966). How does lysozyme work? *New Scientist*, **29**, 333 (*265*).

BLAKE C. C. F., KOENIG D. F., MAIR G. A., NORTH A. C. T., PHILLIPS D. C. & SARMA V. R. Structure of hen egg-white lysozyme. *Nature, Lond.*, **206**, 757 (*265*).

BLAKE C. C. F., MAIR G. A., NORTH A. C. T., PHILLIPS D. C. & SARMA V. R. (1967). On the conformation of the hen egg-white lysozyme molecule. *Proc. Roy. Soc. B.*, **167**, 365 (*265*).

BLAKE C. C. F., JOHNSON L. N., MAIR G. A., NORTH A. C. T., PHILLIPS D. C. & SARMA V. R. (1967b). Crystallographic studies of the activity of hen egg-white lysozymes. *Proc. Roy. Soc. B.*, **167**, 378 (*265*).

BODE V. C. & KAISER A. D. (1965). Repression of the C II and C III cistrons of phage lambda in a lysogenic bacterium. *Virology*, **25**, 111 (*467*).

BODMER W. F. (1965). Recombination and integration in *Bacillus subtilis* transformation: involvement of DNA synthesis. *J. Mol. Biol.*, **14**, 534 (*586, 602, 603*).

BOICE L. & LURIA S. (1961). Transfer of transducing prophage P1*dl* upon bacterial mating. *Bacteriol. Proc.*, (**V80a**), 197 (*478*).

BOLLUM F. J. (1963). In *Progress in Nucleic Acid Research* (eds. Davidson J. N. & Cohn W. E.), vol. 1, p. 1. Academic Press, Inc., New York (*247*).

BOLTON E. T. & MCCARTHY B. J. (1962). A general method for the isolation of RNA complementary to DNA. *Proc. Nat. Acad. Sci., Wash.*, **48**, 1390 (*538*).

BONHOEFFER F. (1966). DNA transfer and DNA synthesis during bacterial conjugation. *Z. Vererbungsl.*, **98**, 141 (*804*).

BONHOEFFER F., HÖSSELBARTH R. & LEHMANN K. (1967). Dependence of the conjugal DNA transfer on DNA synthesis. *J. Mol. Biol.*, **29**, 539 (*804*).

BONNER D. (1946a). Production of biochemical mutations in Penicillium. *Amer. J. Bot.*, **33**, 788 (*115*).

BONNER D. (1946b). Biochemical mutations in *Neurospora. Cold Spr. Harb. Symp. quant. Biol.*, **11**, 14 (*115, 122*).

BONNER D. M. (1951). Gene enzyme relationships in *Neurospora. Cold Spr. Harb. Symp. quant. Biol.*, **16**, 143 (*129*).

BONNER D. M., SUYAMA T. & DEMOSS J. A. (1960). Genetic fine structure and enzyme formation. *Fed. Proc.*, **19**, 926 (*118, 155, 166*).

BOTT K. & STRAUSS B. (1965). The carrier state of *Bacillus subtilis* infected with the transducing bacteriophage SP10. *Virology*, **25**, 212 (*450, 479, 640*).

BOURGEOIS S., COHN M. & ORGEL L. E. (1965). Suppression of and complementation among mutants of the regulatory gene of the lactose operon of *Escherichia coli*. *J. Mol. Biol.*, **14**, 300 (*733*).

BOYCE R. P. & HOWARD-FLANDERS P. (1964a). Release of ultraviolet light-induced thymine dimers from DNA in *Escherichia coli K12*. *Proc. Natl. Acad. Sci., Wash.*, **51**, 293 (*255, 333*).

BOYCE R. P. & HOWARD-FLANDERS P. (1964b). Genetic control of DNA breakdown and repair in *Escherichia coli K12* treated with mitomycin C or ultraviolet light. *Z. Vererbungslehre*, **95**, 345 (*334*).

BOYD J. S. K. (1950). The symbiotic bacteriophages of *Salmonella typhimurium*. *J. Path. Bact.*, **62**, 501 (*621*).

BOYER H. (1964). Genetic control of restriction and modification in *Escherichia coli*. *J. Bacteriol.*, **88**, 1652 (*518*).

BRACHET J. (1957). *Biochemical Cytology*. Academic Press Inc., New York (*269*).

BRADLEY D. E. (1964). The structure of some bacteriophages associated with male strains of *Escherichia coli*. *J. gen. Microbiol.*, **35**, 471 (*444, 445*).

BRADLEY D. E. (1965a). The structure of the head, collar and base-plate of the 'T-even' type bacteriophages. *J. gen. Microbiol.*, **38**, 395 (*418*).

BRADLEY D. E. (1965b). The morphology and physiology of bacteriophages as revealed by the electron microscope. *J. Roy. microsc. Soc.*, **84**, 257 (*418, 419*).

BRADLEY D. E. (1966). The structure and infective process of a *Pseudomonas aeruginosa* bacteriophage containing ribonucleic acid. *J. gen. Microbiol.*, **45**, 83 (*441*).

BRADLEY S. G. & LEDERBERG J. (1956). Heterokaryosis in *Streptomyces*. *J. Bacteriol.*, **72**, 219 (696).

BREMER H. & CONRAD M. W. (1964). A complex of enzymatically synthesised RNA and template DNA. *Proc. Natl. Acad. Sci., Wash.*, **51**, 801 (*293*).

BREMER H., KONRAD M. W., GAINES K. & STENT G. S. (1965). Direction of chain growth in enzymic RNA synthesis. *J. Mol. Biol.*, **13**, 540 (*294*).

BRENNER S. (1955). Tryptophan biosynthesis in *Salmonella typhimurium*. *Proc. Natl. Acad. Sci., Wash.*, **41**, 862 (*158*).

BRENNER S. (1957a). On the impossibility of all overlapping triplet codes in information transfer from nucleic acids to proteins. *Proc. Natl. Acad. Sci., Wash.*, **43**, 687 (*340*).

BRENNER S. (1957b). Genetic control and phenotypic mixing of the adsorption cofactor requirement in bacteriophages T2 and T4. *Virology*, **3**, 560 (*105, 513*).

BRENNER S. (1959). Physiological aspects of bacteriophage genetics. *Advance in Virus Res.*, **6**, 137 (*410, 422, 513*).

BRENNER S. (1965). Theories of gene regulation. *Brit. Med. Bull.*, **21**, 244 (*135*).

BRENNER S. (1966). Colinearity and the genetic code. *Proc. Roy. Soc. B.*, **164**, 170 (*362, 363, 368*).

Brenner S. & Barnett L. (1959). Genetic and chemical studies on the head protein of bacteriophages T2 and T4. *Brookhaven Symposia in Biology*, **12**, 86 (*486, 514*).

Brenner S. & Beckwith J. R. (1965). Ochre mutants, a new class of suppressible nonsense mutant. *J. Mol. Biol.*, **13**, 629 (*216, 362, 720, 721*).

Brenner S. & Horne R. W. (1959). A negative staining method for high resolution electron microscopy of viruses. *Biochim. biophys. Acta*, **34**, 103 (*417*).

Brenner S. & Stent G. S. (1955). Bacteriophage growth in protoplasts of *Bacillus megaterium*. *Biochim. biophys. Acta*, **17**, 473 (*424*).

Brenner S., Benzer S. & Barnett L. (1958). Distribution of proflavin-induced mutations in the genetic fine structure. *Nature, Lond.*, **182**, 983 (*147, 319*).

Brenner S., Jacob F. & Meselson M. (1961). An unstable intermediate carrying information from genes to ribosomes for protein synthesis. *Nature, Lond.*, **190**, 576 (*278*).

Brenner S., Stretton A. O. W. & Kaplan S. (1965). Genetic code: the 'nonsense' triplets for chain termination and their suppression. *Nature, Lond.*, **206**, 994 (*363*).

Brenner S., Barnett L., Crick F. H. C. & Orgel A. (1961). The theory of mutagenesis. *J. Mol. Biol.*, **3**, 121 (*319, 320*).

Brenner S., Barnett L., Katz E. R. & Crick F. H. C. (1967). UGA: a third nonsense triplet in the genetic code. *Nature, Lond.*, **213**, 449 (*363*).

Brenner S., Champe S. P., Streisinger G. & Barnett L (1962). On the interaction of adsorption cofactors with bacteriophages T2 and T4. *Virology*, **17**, 30 (*422*).

Brenner S., Streisinger G., Horne R. W., Champe S. P., Barnett L., Benzer S. & Rees M. W. (1959). Structural components of bacteriophage. *J. Mol. Biol.*, **1**, 281 (*416, 417, 418, 420*).

Bresch C. (1953). Genetical studies on bacteriophage T1. *Ann. Inst. Pasteur*, **84**, 157 (*482*).

Bresch C. (1955). Zum Paarungsmechanismus von Bakterophagen. *Z. Naturforsch.*, **10b**, 545 (*495*).

Bresch C. & Trautner T. (1955). Zur Kinetik der Rekombinantenbildung bei T1 Bakteriophagen. *Z. Naturforsch.*, **10b**, 436 (*509*).

Bresler S. E., Kreneva R. A., Kushev V. V. & Mosevitskii M. I. (1964). Molecular mechanism of genetic recombination in bacterial transformation. *Z. Vererbungsl.*, **95**, 288 (*601*).

Bresler S. E. & Lanzov V. A. (1967). Mechanism of genetic recombination during conjugation of *Escherichia coli* K-12. II. Incorporation of the donor DNA fragment into the recombinant chromosome. *Genetics*, **56**, 117 (*397*).

Bretscher M. S. (1968). How repressor molecules function. *Nature, Lond.*, **217**, 509 (*733*).

Bretscher M. S. & Marcker K. A. (1966). Polypeptidyl-s ribonucleic acid and amino-acyl-ribonucleic acid binding sites on ribosomes. *Nature, Lond.*, **211**, 380 (*288*).

Brinton C. C., Jnr. (1965). The structure, formation, synthesis and genetic control of bacterial pili and a molecular model for DNA and RNA transport in gram negative bacteria. *Trans. N.Y. Acad. Sci.* (Ser. II), **27**, 1003 (*434, 440, 445, 670, 671, 672, 799*).

BRINTON C. C. JNR., GEMSKI P. JNR. & CARNAHAN J. (1964). A new type of bacterial pilus genetically controlled by the fertility factor of *E. coli* K12 and its role in chromosome transfer. *Proc. Natl. Acad. Sci., Wash.*, **52**, 776 (*441, 670, 672, 799*).

BROCK T. D. (1966). Streptomycin. *Symp. Soc. gen. Microbiol.*, **16**, 131 (*765*).

BRODA P. (1967). The formation of Hfr strains in *Escherichia coli* K12. *Genet. Res. Camb.*, **9**, 35 (*667, 673, 794*).

BRODA P., BECKWITH J. R. & SCAIFE J. (1964). The characterisation of a new type of F-prime factor in *Escherichia coli* K12. *Genet. Res., Camb.*, **5**, 489 (*797*).

BROOKS K. (1965). Studies in the physiological genetics of some suppressor-sensitive mutants of bacteriophage λ. *Virology*, **26**, 489 (*475*).

BROOKS K. & CLARK A. J. (1967). Behaviour of λ bacteriophage in a recombination-deficient strain of *Escherichia coli. J. Virol.*, **1**, 283 (*474*).

BROOKES P. & LAWLEY P. D. (1960). The reaction of mustard gas with nucleic acids *in vitro* and *in vivo*. *Biochem. J.*, **77**, 478 (*307*).

BROWN D. M. & TODD A. R. (1955). Evidence on the nature of the chemical bonds in nucleic acids. In *The Nucleic Acids, I* (eds. Chargaff E. & Davidson J. N.), p. 409. Academic Press, New York (*231*).

BROWN G. L. (1960). DNA and specific protein synthesis. *Symp. Soc. gen. Microbiol.*, **10**, 208 (*284*).

BROWN G. L. & LEES S. (1965). Amino-acid transfer ribonucleic acid: structure and function. *Brit. Med. Bull.*, **21**, 236 (*275, 282, 285, 365*).

BURNET F. M. (1929). A method for the study of bacteriophage multiplication in broth. *Brit. J. Exp. Path.*, **10**, 109 (*415*).

BURNET F. M. & MCKIE M. (1929). Observations on a permanently lysogenic strain of *B. enteritidis gaertner*. *Austral. J. Exp. Biol. Med. Sci.*, **6**, 277 (*450*).

BURTON K. (1955). The relation between the synthesis of deoxyribonucleic acid and the synthesis of protein in the multiplication of bacteriophage T2. *Biochem. J.*, **61**, 473 (*428*).

BURTON K. & PETERSEN G. B. (1960). The frequencies of certain sequences of nucleotides in deoxyribonucleic acid. *Biochem. J.*, **75**, 17 (*252*).

BUSSARD A., NAONO S., GROS F. & MONOD J. (1960). Effets d'un analogue de l'uracile sur les propriétés d'une protéine synthétisée en sa présence. *C. R. Acad. Sci., Paris*, **250**, 4049 (*371*).

BUTTIN G. (1961). Some aspects of regulation in the synthesis of the enzymes governing galactose metabolism. *Cold Spr. Harb. Symp. quant. Biol.*, **26**, 213 (*711*).

BUTTIN G. (1962). Sur la structure de l'opéron galactose chez *E. coli* K-12. *C. R. Acad. Sci., Paris*, **255**, 1233 (*722*).

BUTTIN G. (1963a). Méchanismes régulateurs dans la biosynthèse des enzymes du métabolisme du galactose chez *Escherichia coli* K-12. I. La biosynthèse induit de la galactosidase et l'induction simultanée de la séquence enzymatique. *J. Mol. Biol.* **7**, 164 (*722*).

BUTTIN G. (1963b). Méchanismes régulateurs dans la biosynthèse des enzymes du métabolisme du galactose chez *Escherichia coli* K-12. II. La déterminisme génétique de la régulation. *J. Mol. Biol.*, **7**, 183 (*722*, *724*).

CAIRNS J. (1961). An estimate of the length of the DNA molecule of T2 bacteriophage by autoradiography. *J. Mol. Biol.*, **3**, 756 (*537*, *544*).

CAIRNS J. (1962). A minimum estimate for the length of the DNA of *Escherichia coli* obtained by autoradiography. *J. Mol. Biol.*, **4**, 407 (*551*).

CAIRNS J. (1963a). The bacterial chromosome and its manner of replicating as seen by autoradiography. *J. Mol. Biol.*, **6**, 208 (*254*, *551*, *566*, *690*, *792*).

CAIRNS J. (1963b). The chromosome of *Escherichia coli*. *Cold Spr. Harb. Symp. quant. Biol.*, **28**, 43 (*254*, *544*, *551*, *555*, *556*, *673*, *792*).

CALEF, E. (1957). Effect on linkage maps of selection of crossovers between closely linked markers. *Heredity*, **11**, 265 (*130*).

CALEF E. & LICCIARDELLO G. (1960). Recombination experiments on prophage-host relationships. *Virology*, **12**, 81 (*458*, *460*).

CALEF E., MARCHELLI C. & GUERRINI F. (1965). The formation of super-infection-double lysogens of phage λ in *Escherichia coli* K12. *Virology*, **27**, 1 (*462*).

CALLAN H. G. (1963). The nature of lampbrush chromosomes. *Int. Rev. Cytol.*, **15**, 1 (*534*).

CAMPBELL A. (1957a). Transduction and segregation in *Escherichia coli* K-12. *Virology*, **4**, 366 (*629*).

CAMPBELL A. (1957b). Synchronisation of cell division. *Bacteriol. Rev.*, **21**, 263 (*554*).

CAMPBELL A. (1959). Ordering of genetic sites in bacteriophage λ by the use of galactose-transducing defective phages. *Virology*, **9**, 293 (*630*).

CAMPBELL A. (1961). Sensitive mutants of bacteriophage λ. *Virology*, **14**, 22 (*362*, *456*, *457*, *474*, *484*).

CAMPBELL A. (1962). Episomes. *Advanc. in Genetics*, **11**, 101 (*459*, *460*, *462*, *633*, *634*, *673*, *676*, *790*).

CAMPBELL A. (1963). Segregants from lysogenic heterogenotes carrying recombinant lambda prophages. *Virology*, **20**, 344 (*461*).

CAMPBELL A. (1964). Transduction. In *The Bacteria* (eds. Gunsalus I. C. & Stanier R. Y.), vol. 5, p. 49. Academic Press, Inc., New York and London (*630*, *646*).

CAMPBELL A. (1965). The steric effect in lysogenisation by bacteriophage lambda. II. Chromosomal attachment of the *b2* mutant. *Virology*, **27**, 340 (*471*).

CAMPBELL A. & BALBINDER E. (1959). Transduction of the galactose region of *Escherichia coli* K-12 by the phages λ and λ-434 hybrid. *Genetics*, **44**, 309 (*631*).

CANNON M., KRUG R. & GILBERT W. (1963). The binding of SRNA by *Escherichia coli* ribosomes. *J. Mol. Biol.*, **7**, 360 (*285*).

CAPECCHI M. R. (1967). Polarity *in vitro*. *J. Mol. Biol.*, **30**, 213 (*721*).

CARO L. (1965). The molecular weight of lambda DNA. *Virology*, **25**, 226 (*544*).

CARO L. & BERG C. (1968). In preparation. *Cold Spr. Harb. Symp. quant. Biol.*, **33** (personal communication). (*567*)

CARO L. G. & SCHNÖS M. (1966). The attachment of the male-specific bacteriophage F1 to sensitive strains of *Escherichia coli*. *Proc. Natl. Acad. Sci., Wash.*, **56**, 126 (445).

CARO L. G., VAN TUBERGEN R. P. & FORRO F. (1958). The localisation of deoxyribonucleic acid in *Escherichia coli*. *J. Biophys. Biochem. Cytol.*, **4**, 491 (*548*).

CARUSI E. A. & SINSHEIMER R. L. (1961). The protein subunit of bacteriophage φX 174. *Fed. Proc.*, **20**, 438 (*437*).

CATCHESIDE D. G. (1951). *The genetics of microorganisms*. Sir Isaac Pitman & Sons Ltd., London (*108*, *199*).

CATCHESIDE D. G. (1960). Relation of genotype to enzyme content. *Symp. Soc. gen. Microbiol.*, **10**, 181 (*157*).

CATCHESIDE D. G. (1964). Interallelic complementation. *Brookhaven Symp. Biol.*, **17**, 1 (*157*).

CATCHESIDE D. G. & OVERTON A. (1958). Complementation between alleles in heterokaryons. *Cold Spr. Harb. Symp. quant. Biol.*, **23**, 137 (*153*).

CATCHESIDE D. G., JESSOP A. P. & SMITH D. R. (1964). Recombination. Genetic controls of allelic recombination in *Neurospora*. *Nature, Lond.*, **202**, 1242 (*402*).

CAVALLI-SFORZA L. L. (1950). La sessulita nei batteri. *Boll. Ist. Sieroter. Milano*, **29**, 281 (*659*, *667*).

CAVALLI-SFORZA L. L. & HESLOT H. (1949). Recombination in bacteria outcrossing *E. coli* K-12. *Nature, Lond.*, **164**, 1057 (*749*).

CAVALLI-SFORZA L. L. & JINKS J. L. (1956). Studies on the genetic system of *E. coli* K-12. *J. Genet.*, **54**, 87 (*659*).

CAVALLI-SFORZA L. L. & LEDERBERG J. (1956). Isolation of preadaptive mutants in bacteria by sib selection. *Genetics*, **41**, 367 (*190*, *191*, *192*).

CAVALLI-SFORZA L. L., LEDERBERG J. & LEDERBERG E. M. (1953). An infective factor controlling sex compatibility in *Bacterium coli*. *J. gen. Microbiol.*, **8**, 89 (*384*, *655*, *656*, *657*, *660*).

CHAMBERLIN M. & BERG P. (1963). Studies on DNA-directed polymerase: formation of DNA-RNA complexes with single-stranded φX174 DNA as template. *Cold Spr. Harb. Symp. quant. Biol.*, **28**, 67 (*290*).

CHAMBERLIN M. & BERG P. (1964). Mechanism of RNA polymerase action: formation of DNA-RNA hybrids with single-stranded templates. *J. Mol. Biol.*, **8**, 297 (*290*).

CHAMPE S. P. (1963). Bacteriophage reproduction. *Ann. Rev. Microbiol.*, **17**, 87 (*428*).

CHAMPE S. P. & BENZER S. (1962a). An active cistron fragment. *J. Mol. Biol.*, **4**, 288 (*351*).

CHAMPE S. P. & BENZER S. (1962b). Reversal of mutant phenotypes by 5-fluorouracil: an approach to nucleotide sequences in messenger RNA. *Proc. Natl. Acad. Sci., Wash.*, **48**, 532 (*318*, *369*, *371*).

CHANGEUX J. P. (1961). The feedback control mechanism of biosynthetic L-threonine deaminase by L-isoleucine. *Cold Spr. Harb. Symp. quant. Biol.*, **26**, 313 (*707*).

CHANGEUX J. P. (1962). Effet des analogues de la L-thréonine et de la L-isoleucine sur la L-thréonine désaminase. *J. Mol. Biol.*, **4**, 220 (*708*).

CHANGEUX J. P. (1963). Allosteric interactions on biosynthetic L-threonine deaminase from *E. coli* K12. *Cold Spr. Harb. Symp. quant. Biol.*, **28**, 497 (*707, 710*).

CHANTHRENNE H. (1953). Problems of protein synthesis. *Symp. Soc. gen. Microbiol.*, **3**, 1 (*269*).

CHAPEVILLE F., LIPMANN F., VAN EHRENSTEIN G., WEISBLUM B., RAY W. J. JNR. & BENZER S. On the role of soluble ribonucleic acid in coding for amino acids. *Proc. Natl. Acad. Sci., Wash.*, **48**, 1086 (*285*).

CHARGAFF E. (1955). Isolation and composition of the deoxypentose nucleic acids and of the corresponding nucleoproteins. In *The Nucleic Acids I* (eds. Chargoff E. & Davidson J. N.), p. 307. Academic Press, New York (*231*).

CHARPAK M. & DEBONDER R. (1965). Production d'un 'facteur de compétence' soluble par *Bacillus subtilis* Marburg *ind*-168. *C. R. Acad. Sci., Paris*, **260**, 5638 (*589*).

CHASE M. & DOERMANN A. H. (1958). High negative interference over short segments of the genetic structure of bacteriophage T4. *Genetics*, **43**, 332 (*145, 382, 497, 504*).

CITRI N. & POLLOCK M. R. (1966). The biochemistry and function of β-lactamase (penicillinase). *Advanc. Enzymol.*, **28**, 238 (*711, 766*).

CLARK A. J. (1961). Preparation of a strain of *Escherichia coli* K-12 containing two linkage groups. *Bacteriol. Proc.*, **98** (*674, 696*).

CLARK A. J. (1964). Recurrent nucleotide sequences as explanation of the selfing phenomenon in *Salmonellà typhimurium*. *Z. Vererbl.*, **4**, 368 (*223*).

CLARK A. J. & ADELBERG E. A. (1962). Bacterial conjugation. *Ann. Rev. Microbiol.*, **16**, 289 (*688, 794*).

CLARK A. J. & MARGULIES A. D. (1965). Isolation and characterisation of recombination deficient mutants of *E. coli* K12. *Proc. Natl. Acad. Sci., Wash.*, **53**, 451 (*334, 794*).

CLARK B. F. C. & MARCKER K. A. (1966a). The role of N-formylmethionyl-s RNA in protein biosynthesis. *J. Mol. Biol.*, **17**, 394 (*288*).

CLARK B. F. C. & MARCKER K. A. (1966b). N-formylmethionyl-s ribonucleic acid and chain initiation in protein biosynthesis. Polypeptide synthesis directed by a bacteriophage nucleic acid in a cell-free system. *Nature, Lond.*, **211**, 378 (*288*).

CLAUDE A. (1943). Distribution of nucleic acids in the cell and morphological constitution of cytoplasm. *Cold Spr. Harb. Symp. quant. Biol.*, **10**, 111 (*270*).

CLOWES R. C. (1958a). Nutritional studies on cysteineless mutants of *Salmonella typhimurium*. *J. gen. Microbiol.*, **18**, 154 (166).

CLOWES R. C. (1958b). Investigation of the genetics of cysteineless mutants of *Salmonella typhimurium*. *J. gen. Microbiol.*, **18**, 154 (*166*).

CLOWES R. C. (1958c). The nature of the vector involved in bacterial transduction. *Abstracts VIIIth Int. Cong. Microbiol.*, p. 53. Stockholm: Almqvist & Wiksell (*637, 639*).

Clowes R. C. (1960). Fine genetic structure as revealed by transduction. *Symp. Soc. gen. Microbiol.*, **10**, 92 (*165*, *646*).

Clowes R. C. (1961). Colicine factors as fertility factors in bacteria: *Escherichia coli. Nature, Lond.*, **190**, 986 (*756*, *760*, *769*).

Clowes R. C. (1963). Colicin factors and episomes. *Genet. Res. Camb.*, **4**, 163 (*757*).

Clowes R. C. & Hayes W. (1968). *Experiments in microbial genetics*. Blackwell Scientific Publications Ltd., Oxford (*686*).

Clowes R. C. & Moody E. E. M. (1966). Chromosomal transfer from 'recombination deficient' strains of *Escherichia coli* K-12. *Genetics*, **53**, 717 (*805*, *806*).

Clowes R. C. & Rowley D. (1954). Some observations on linkage effects in genetic recombination in *E. coli* K-12. *J. gen. Microbiol.*, **11**, 250 (*659*).

Coddington A. & Fincham J. R. S. (1965). Proof of hybrid enzyme formation in a case of inter-allelic complementation in *Neurospora crassa. J. Mol. Biol.*, **12**, 152 (*157*).

Coetzee J. N. & Sacks T. G. (1960). Intrastrain transduction in *Proteus mirabilis. Nature, Lond.*, **185**, 869 (621).

Cohen G. N. (1965). Regulation of enzyme activity in microorganisms. *Ann. Rev. Microbiol.*, **19**, 105 (*735*).

Cohen G. N. & Hirsch M. L. (1954). Threonine synthase, a system synthesising L-threonine from L-homoserine. *J. Bacteriol.*, **67**, 182 (*121*).

Cohen G. N. & Jacob F. (1959). Sur la répression de la synthèse des enzymes intervenant dans la formation de tryptophane chez *E. coli. C. R. Acad. Sci., Paris*, **248**, 3490 (*724*).

Cohen L. W. & Levine M. (1966). Detection of proteins synthesised during the establishment of lysogeny with phage P22. *Virology*, **28**, 208 (*493*).

Cohen S. N., Maitra U. & Hurwitz J. (1967). Role of DNA in RNA synthesis XI. Selective transcription of λ DNA segments *in vitro* by RNA polymerase of *Escherichia coli. J. Mol. Biol.*, **26**, 19 (*475*).

Cohen S. S. (1948). Synthesis of bacterial viruses; origin of phosphorus found in desoxyribonucleic acids of T2 and T4 bacteriophages. *J. Biol. Chem.*, **174**, 295 (*427*).

Cohn M. & Monod J. (1953). Specific inhibition and induction of enzyme biosynthesis. *Symp. Soc. gen. Microbiol.*, **3**, 132 (*710*, *712*).

Collins J. F. (1965). Antibiotics, proteins and nucleic acids. *Brit. Med. Bull.*, **21**, 223 (*288*).

Colson C., Glover S. W., Symonds N. & Stacey K. A. (1965). The location of the genes for host-controlled modification and restriction in *Escherichia coli* K-12. *Genetics*, **52**, 1043 (*518*, *519*).

Cook A. & Lederberg J. (1962). Recombination studies of lactose non-fermenting mutants of *Escherichia coli* K-12. *Genetics*, **47**, 1335 (*217*).

Cooper S. & Zinder N. D. (1963a). The growth of an RNA bacteriophage: the role of DNA synthesis. *Virology*, **18**, 405 (*442*).

Cooper S. & Zinder N. D. (1963a). The growth of an RNA bacteriophage: the role of protein synthesis. *Virology*, **20**, 605 (*442*, *444*).

Cowie D. B. & McCarthy B. J. (1963). Homology between bacteriophage λ DNA and *E. coli* DNA. *Proc. Natl. Acad. Sci., Wash.*, **50**, 537 (*635*).

CRAWFORD E. M. & GESTELAND R. F. (1964). The adsorption of bacteriophage R-17. *Virology*, **22**, 165 (*441*, *570*).

CRAWFORD I. P. & YANOFSKY C. (1958). On the separation of the tryptophan synthetase of *Escherichia coli* into two protein components. *Proc. Natl. Acad. Sci., Wash.*, **44**, 1161 (*118*, *119*, *155*).

CRICK F. H. C. (1954). The structure of the hereditary material. *Scientific American*, **191**, No. 4 (*233*).

CRICK F. H. C. (1958). On protein synthesis. *Symp. Soc. exp. Biol.*, **12**, 138 (*264*, *270*, *280*, *339*).

CRICK F. H. C. (1966a). The genetic code: III. *Scientific American*, **215**, 55 (*284*, *358*, *365*).

CRICK F. H. C. (1966b). Codon-anti-codon pairing: the wobble hypothesis. *J. Mol. Biol.*, **19**, 548 (*283*, *284*, *365*).

CRICK F. H. C. (1967a). An error in model building. *Nature, Lond.*, **213**, 798 (*368*)

CRICK F. H. C. (1967b). The Croonian Lecture, 1966: The genetic code. *Proc. Roy. Soc. B.*, **167**, 331 (*368*).

CRICK F. H. C. & KENDREW J. C. (1957). X-ray analysis and protein structure. *Advances in Protein Chem.*, **12**, 133 (*232*).

CRICK F. H. C. & WATSON J. D. (1953). A structure for deoxyribose nucleic acid. *Nature, Lond.*, **171**, 737 (*232*).

CRICK F. H. C., GRIFFITH J. S. & ORGEL L. E. (1957). Codes without commas. *Proc. Natl. Acad. Sci., Wash.*, **43**, 416 (*341*).

CRICK F. H. C., BARNETT L., BRENNER S. & WATTS-TOBIN R. J. (1961). General nature of the genetic code for proteins. *Nature, Lond.*, **192**, 1227 (*214*, *216*, *320*, *324*, *346*, *347*, *349*, *350*, *352*, *387*).

CURTIS R. III & STALLIONS D. R. (1967). Energy requirements for specific pair formation during conjugation in *Escherichia coli* K-12. *J. Bacteriol.*, **94**, 490 (*680*).

CUZIN F. (1962). Mutants défectifs de l'épisome sexuel chez *Escherichia coli* K-12. *C. R. Acad. Sci., Paris*, **255**, 1149 (*656*, *670*).

CUZIN F. & JACOB F. (1963). Intégration réversible de l'épisome sexuel F′ chez *Escherichia coli* K12. *C. R. Acad. Sci., Paris*, **257**, 795 (*794*).

CUZIN F. & JACOB F. (1964). Délétions chromosomiques et intégration d'un épisome sexuel F-lac$^+$ chez *Escherichia coli* K-12. *C. R. Acad. Sci., Paris*, **258**, 1350 (*219*, *374*, *797*).

CUZIN F. & JACOB F. (1965). Analyse génétique fonctionnelle de l'épisome sexuel d'*Escherichia coli* K12. *C. R. Acad. Sci., Paris*, **260**, 2087 (*782*).

CUZIN F. & JACOB F. (1966). Inhibition par les acridines du transfert génétique par les souches donatrices d'*Escherichia coli* K12. *Ann. Inst. Pasteur*, **111**, 427 (*804*).

CUZIN F. & JACOB F. (1967a). Mutations de l'épisome F d'*Escherichia coli* K-12. 1. Mutations défectives. *Ann. Inst. Pasteur*, **112**, 1 (*670*).

CUZIN F. & JACOB F. (1967b). Mutations de l'épisome F d'*Escherichia coli* K12. 2. Mutants à réplication thermosensible. *Ann. Inst. Pasteur*, **112**, 397 (*671*, *776*).

CUZIN F. & JACOB F. (1967c). Existence chez *Escherichia coli* K12 d'une unité génétique formée de différents réplicons. *Ann. Inst. Pasteur*, **112**, 529 (*571*).

DARLINGTON C. D. (1935). The time, place and action of crossing-over. *J. Genet.*, **31**, 185 (*376*).

DATTA N. (1962). Transmissible drug resistance in an epidemic strain of *Salmonella typhimurium. J. Hyg., Camb.*, **60**, 301 (*761*).

DATTA N. & RICHMOND M. H. (1966). The purification and properties of a penicillinase whose synthesis is mediated by an R factor in *Escherichia coli. Biochem. J.*, **98**, 204 (*766*).

DATTA N., LAWN A. M. & MEYNELL E. (1966). The relationship of F type piliation and F phage sensitivity to drug resistance transfer in R^+F^- *Escherichia coli* K12. *J. gen. Microbiol.*, **45**, 365 (*773*).

DAVERN C. I. & MESELSON M. (1960). The molecular conservation of ribonucleic acid during bacterial growth. *J. Mol. Biol.*, **2**, 153 (*275*).

DAVIDSON J. N. (1947). Some factors influencing the nucleic acid content of cells and tissues. *Cold Spr. Harb. Symp. quant. Biol.*, **12**, 50 (*269*).

DAVIDSON J. N. & COHN W. E. (EDS.) (1963a). *Progress in nucleic acid research.* Vol. 1. Academic Press, Inc., New York and London (*246, 296*).

DAVIDSON J. N. & COHN W. E. (EDS.) (1963b). *Progress in nucleic acid research.* Vol. 2. Academic Press, Inc., New York and London (*246, 296*).

DAVIDSON J. N. & COHN W. E. (EDS.) (1964). *Progress in nucleic acid research and molecular biology.* Vol. 3. Academic Press, Inc., New York and London (*246, 296*).

DAVIDSON J. N. & COHN W. E. (EDS.) (1965). *Progress in nucleic acid research and molecular biology.* Vol. 4. Academic Press, Inc., New York and London (*246, 296*).

DAVIDSON J. N. & COHN W. E. (EDS.) (1966). *Progress in nucleic acid research and molecular biology.* Vol. 5. Academic Press, Inc., New York and London (*246, 296*).

DAVIS B. D. (1948). Isolation of biochemically deficient mutants of bacteria by penicillin. *J. Amer. Chem. Soc.*, **70**, 4267 (*211*).

DAVIS B. D. (1950). Non-filterability of the agents of genetic recombination in *E. coli. J. Bacteriol.*, **60**, 507 (*652*).

DAVIS B. D. (1955). Biosynthesis of the aromatic amino acids. In *A Symposium on Amino Acid Metabolism* (eds. McElroy W. D. & Glass H. B.), p. 799. Johns Hopkins Press, Baltimore (*119, 120*).

DAVIS B. D. (1961). The teleonomic significance of biosynthetic control mechanisms. *Cold Spr. Harb. Symp. quant. Biol.*, **26**, 1 (*706, 707, 711*).

DAVIS J. E., STRAUSS J. H. JNR. & SINSHEIMER R. L. (1961). Bacteriophage MS2: another RNA phage. *Science*, **134**, 1427 (*441*).

DAVISON P. F. (1959). The effect of hydrodynamic shear on the deoxyribonucleic acid from T2 and T4 bacteriophages. *Proc. Natl. Acad. Sci., Wash.*, **45**, 1560 (*536*).

DAVISON P. F., FREIFELDER D., HEDE R. & LEVINTHAL C. (1961). The structural unity of the DNA of T2 bacteriophage. *Proc. Natl. Acad. Sci., Wash.*, **47**, 1123 (*536*).

DAWSON G. W. P. & SMITH-KEARY P. F. (1963). Episomic control of mutation in *Salmonella typhimurium. Heredity*, **18**, 1 (*220, 221*).

DAWSON M. H. & SIA R. H. P. (1931). A technique for inducing transformation of pneumococcal types *in vitro. J. Exptl. Med.*, **54**, 681 (*575*).

De, D. N. (1964). A new chromosome model. *Nature, Lond.*, **203**, 343 (*534*).

Dean A. C. R. (1960). Chloramphenicol resistance of *Bact. lactis aerogenes* (*Aerobacter aerogenes*). II. Production of highly resistant strains in non-selective conditions. *Proc. Roy. Soc. B.*, **153**, 329 (*193*).

Dean A. C. R. & Broadbridge P. H. (1963). The accelerated training of *Bact. lactis aerogenes* to 5-aminoacridine. *Proc. Roy. Soc. B.*, **158**, 279 (*193*).

Dean A. C. R. & Hinshelwood C. (1957). Aspects of the problem of drug resistance in bacteria. In Ciba Foundation Symposium, *Drug Resistance in Microorganisms* (eds. Wolstenholme G. E. W. & O'Connor C. M.), p. 4. J. A. Churchill, London (*192*).

Dean A. C. R. & Hinshelwood C. (1963). Integration of cell reaction. *Nature, Lond.*, **199**, 7 (*193*).

Dean A. C. R. & Hinshelwood C. (1964a). Some basic aspects of cell regulation. *Nature, Lond.*, **201**, 232 (*193*).

Dean A. C. R. & Hinshelwood C. (1964b). What is heredity? *Nature, Lond.*, **202**, 1046 (*193*).

Dean A. C. R. & Hinshelwood C. (1965). Cell division. *Nature, Lond.*, **206**, 546 (*193*).

Delbrück M. (1940). The growth of bacteriophage and lysis of the host. *J. gen. Physiol.*, **23**, 643 (*414, 435*).

Delbrück M. (1945). The burst size distribution in the growth of bacterial viruses (bacteriophages). *J. Bacteriol.*, **50**, 131 (*416*).

Delbrück M. (1948). Biochemical mutants of bacterial viruses. *J. Bacteriol.*, **56**, 1 (*422*).

Delbrück M. & Bailey W. T. Jnr. (1946). Induced mutations in bacterial viruses. *Cold Spr. Harb. Symp. quant. Biol.*, **11**, 33 (*486, 513*).

Delbrück M. & Stent G. S. (1957). On the mechanism of DNA replication. In *The Chemical Basis of Heredity*, p. 699 (eds. McElroy W. D. & Glass B.). Johns Hopkins Press, Baltimore (*240, 241*).

DeMars R. I. (1953). Chemical mutagenesis in bacteriophage T2. *Nature, Lond.*, **172**, 964 (*318*).

DeMars R. I. (1955). The production of phage-related materials when bacteriophage development is interrupted by proflavine. *Virology*, **1**, 83 (*430*).

Demerec M. (1946). Induced mutations and possible mechanisms of the transmission of heredity in *Escherichia coli*. *Proc. Natl. Acad. Sci., Wash.*, **32**, 36 (*506*).

Demerec M. (1953). Reaction of genes of *E. coli* to certain mutagens. *Symp. Soc. exptl. Biol.*, **7**, 43 (*205*).

Demerec M. (1960). Frequency of deletions among spontaneous and induced mutations in *Salmonella*. *Proc. Natl. Acad. Sci., Wash.*, **48**, 1075 (*218*).

Demerec M. (1963). Selfer mutants of *Salmonella typhimurium*. *Genetics*, **48**, 1519 (*222, 223*).

Demerec M. (1964). Clustering of functionally related genes in *Salmonella typhimurium*. *Proc. Natl. Acad. Sci., Wash.*, **51**, 1057 (*164*).

DEMEREC M. & DEMEREC Z. (1956). Analysis of linkage relationships in *Salmonella* by transduction techniques. *Brookhaven Symposia in Biol.*, **8**, 75 (*165*).

DEMEREC M. & HARTMAN Z. (1956). Tryptophan mutants in *Salmonella typhimurium*. *Carnegie Inst., Wash. Publ.*, **612**. Genetics studies with bacteria, p. 5 (*131, 158, 643*).

DEMEREC M. & HARTMAN P. E. (1959). Complex loci in microorganisms. *Ann. Rev. Microbiol.*, **13**, 377 (*128, 132, 147, 166, 217*).

DEMEREC M. & LATARJET R. (1946). Mutations in bacteria induced by radiations. *Cold Spr. Harb. Symp. quant. Biol.*, **11**, 38 (*204*).

DEMEREC M., GILLESPIE D. H. & MIZOBUCHI K. (1963). Genetic structure of the CYS-C region of the *Salmonella* genome. *Genetics*, **48**, 997 (*217*).

DEMEREC M., ADELBERG E. A., CLARK A. J. & HARTMAN P. E. (1966). A proposal for a uniform nomenclature in bacterial genetics. *Genetics*, **54**, 61 (*666*).

DEMEREC M., MOSER H., CLOWES R. C., LAHR E. L., OZEKI H. & VIELMETTER W. (1956). Genetic studies with bacteria. *Carnegie Inst. Wash. Year Book*, **55**, 301 (*166*).

DEMEREC M., LAHR E. L., MIYAKE T., GOLDMAN I., BALBINDER E., BANIC S., HASHIMOTO K., GLANVILLE E. V. & GROSS J. D. (1958). Bacterial Genetics. *Carnegie Inst. Wash. Year Book*, **57**, 390 (*164*).

DENHARDT D. T. & SILVER R. B. (1966). An analysis of the clone size distribution of φX174 mutants and recombinants. *Virology*, **30**, 10 (*439*).

DENHARDT D. T. & SINSHEIMER R. L. (1965a). The process of infection with bacteriophage φX174. IV. Replication of the viral DNA in a synchronised infection. *J. Mol. Biol.*, **12**, 647 (*437, 439*).

DENHARDT D. T. & SINSHEIMER R. L. (1965b). The process of infection with bacteriophage φX174. V. Inactivation of the phage bacterium complex by decay of ^{32}p incorporated in the infecting particle. *J. Mol. Biol.*, **12**, 663 (*439*).

DETTORI R. & NERI M. G. (1965). Batteriofago filamentoso specifice per cellule Hfr e F^+ di *E. coli* K12. *Giornale Microbiol.*, **12**, 111 (*444*).

DETTORI R., MACCACARO G. A. & PICCININ G. L. (1961). Sex-specific bacteriophages of *Escherichia coli* K-12. *Giornale di Microbiol.*, **9**, 141 (*441, 449, 670, 671*).

DINTZIS H. M. (1961). Assembly of the peptide chains of hemoglobin. *Proc. Natl. Acad. Sci., Wash.*, **47**, 247 (*286, 344*).

DOBRZANSKI W. T. & OSOWIECKI H. (1967). Isolation and some properties of the competence factor from group H streptococcus strain CHALLIS. *J. gen. Microbiol.*, **48**, 299 (*589*).

DOERMANN A. H. (1948). Lysis and lysis inhibition with *E. coli* bacteriophage. *J. Bacteriol.*, **55**, 257 (*415, 481*).

DOERMANN A. H. (1952). The intracellular growth of bacteriophages. I. Liberation of intracellular bacteriophage T4 by premature lysis with another phage or with cyanide. *J. gen. Physiol.*, **35**, 645 (*424, 425*).

DOERMANN A. H. (1953). The vegetative state in the life cycle of bacteriophage: evidence for its occurrence, and its genetic characterisation. *Cold Spr. Harb. Symp. quant. Biol.*, **18**, 3 (*506*).

DOERMANN A. H. & BOEHNER L. (1964). An experimental analysis of bacteriophage T4 heterozygotes. II. Distribution in a density gradient. *J. Mol. Biol.*, **10**, 212 (*499*).

DOERMANN A. H. & HILL M. B. (1953). Genetic structure of bacteriophage T4 as described by recombination studies of factors influencing plaque morphology. *Genetics*, **38**, 79 (*481*, *505*).

DOERMANN A. H., CHASE M. & STAHL F. W. (1955). Genetic recombination and replication in bacteriophage. *J. Cell. Comp. Physiol.*, **45**, Suppl. 2, 51 (*530*).

DOI R. H. & SPIEGELMAN S. (1962). Homology test between the nuclei of an RNA virus and the DNA in the host cell. *Science*, **138**, 1270 (*442*).

DOTY P., MARMUR J., EIGNER J. & SCHILDKRAUT C. (1960). Strand separation and specific recombination in deoxyribonucleic acids: physical chemical studies. *Proc. Natl. Acad. Sci., Wash.*, **46**, 461 (*276*, *604*).

DOVE W. F. (1966). Action of the lambda chromosome. I. Control of functions late in bacteriophage development. *J. Mol. Biol.*, **19**, 187 (*457*).

DOVE W. F. (1967). The synthesis of the lambda chromosome: the role of the prophage termini. In *Molecular Biology of Viruses* (ed. Colter J.). Academic Press, New York (*471*, *475*).

DOVE W. F. & WEIGLE J. J. (1965). Intracellular state of the chromosome of bacteriophage lambda. I. The eclipse of infectivity of the bacteriophage DNA. *J. Mol. Biol.*, **12**, 620 (*463*).

DRAKE J. W. (1963). Properties of ultraviolet-induced r_{II} mutants of bacteriophage T4. *J. Mol. Biol.*, **6**, 268 (*337*).

DRAKE J. W. (1966). Heteroduplex heterozygotes in bacteriophage T4 involving mutations of various dimensions. *Proc. Natl. Acad. Sci., Wash.*, **55**, 506 (*499*).

DRISKELL P. J. & ADELBERG E. A. (1961). Inactivation of the sex factor of *Escherichia coli* K-12 by the decay of incorporated radiophosphorus. *Bacteriol. Proc.*, **186** (*657*).

DRISKELL-ZAMENHOF P. (1964). Bacterial episomes. In *The Bacteria*, vol. 5 (eds. Gunsalus I. C. & Stanier R. Y.), p. 155. Academic Press, Inc., New York and London (*679*).

DUBININ N. P. (1932). Step-allelomorphism and the theory of centres of the gene, achaete-scute. *J. Genet.*, **26**, 37 (*128*).

DUBININ N. P. (1933). Step-allelomorphism in *Drosophila melanogaster*. *J. Genet.*, **27**, 443 (*128*).

DUBNAU D. A. & POLLOCK M. R. (1965). The genetics of *Bacillus licheniformis* penicillinase: a preliminary analysis from studies on mutation and on inter-strain transformation. *J. gen. Microbiol.*, **41**, 7 (*729*).

DUBNAU D., SMITH I. & MARMUR J. (1965). Gene conservation in *Bacillus* species. II. The location of genes concerned with the synthesis of ribosomal components and soluble RNA. *Proc. Natl. Acad. Sci., Wash.*, **54**, 724 (*616*).

DUBNAU D., SMITH I., MORELL P. & MARMUR J. (1965). Gene conservation in *Bacillus* species. I. Genetic and nucleic acid base sequence homologies. *Proc. Natl. Acad., Sci., Wash.*, **54**, 491 (*616*).

DUBNAU D., GOLDTHWAITE, SMITH I. & MARMUR J. (1967). Genetic mapping in *Bacillus subtilis*. *J. Mol. Biol.*, **27**, 163 (*611, 615, 616*).

DUBNAU E. & STOCKER B. A. D. (1964). Genetics of plasmids in *Salmonella typhimurium*. *Nature, Lond.*, **204**, 1112 (*764, 781*).

DUBOS R. J. (1945). In *The Bacterial Cell*, Chapter 2. Harvard Univ. Press, Cambridge, Mass., U.S.A. (*546*).

DUGUID J. P. & ANDERSON E. S. (1967). Terminology of bacterial fimbriae, or pili, and their types. *Nature, Lond.*, **215**, 89 (*434, 440*).

DUGUID J. P., ANDERSON E. S. & CAMPBELL I. (1966). Fimbriae and their adhesive properties in salmonellae. *J. Path. Bact.*, **92**, 107 (*434, 440, 671*).

DUNN D. B., SMITH J. D. & SPAHR P. F. (1960). Nucleotide composition of soluble ribonucleic acid from *Escherichia coli*. *J. Mol. Biol.*, **2**, 113 (*281*).

DUPRAW E. J. (1965). The organisation of nuclei and chromosomes in honeybee embryonic cells. *Proc. Natl. Acad. Sci., Wash.*, **53**, 161 (*534*).

DUPRAW E. J. (1966). Evidence for a 'folded fibre' organisation in human chromosomes. *Nature, Lond.*, **209**, 577 (*534*).

DUPRAW E. J. & RAE P. M. M. (1966). Polytene chromosome structure in relation to the 'folded fibre' concept. *Nature, Lond.*, **212**, 598 (*534*).

DUSSOIX D. & ARBER W. (1962). Host specificity of DNA produced by *Escherichia coli*. II. Control over acceptance of DNA from infecting phage λ. *J. Mol. Biol.*, **5**, 37 (*516, 517, 522*).

EBERLE H. & LARK K. G. (1966). Chromosome segregation in *Bacillus subtilis*. *J. Mol. Biol.*, **22**, 183 (*573*).

EDGAR R. S. (1966). Conditional lethals. In *Phage and the origins of Molecular Biology* (eds. Cairns J., Stent G. S. & Watson J. D.), p. 166. Cold Spring Harbor Laboratory of Quantitative Biology, Cold Spring Harbor, New York (*484*).

EDGAR R. S. & LIELAUSIS I. (1964). Temperature-sensitive mutants of bacteriophage T4D: their isolation and genetic characterisation. *Genetics*, **49**, 649 (*484*).

EDGAR R. S. & WOOD W. B. (1966). Morphologies of bacteriophage T4 in extracts of mutant infected cells. *Proc. Natl. Acad. Sci., Wash.*, **55**, 498 (*433*).

EDGAR R. S., DENHARDT G. H. & EPSTEIN R. H. (1964). A comparative genetic study of conditional lethal mutations of bacteriophage T4D. *Genetics*, **49**, 635 (*485*).

EDGELL M. H. & GINOZA W. (1965). The fate during infection of the coat protein of the spherical bacteriophage R17. *Virology*, **27**, 23 (*442*).

EDWARDS P. R. & BRUNER D. W. (1946). Form variation in *Salmonella pullorum* and its relation to X strains. *Cornell Veterinarian*, **36**, 318 (*745*).

EGAWA R. & HIROTA Y. (1962). Inhibition of fertility by multiple drug resistance factor in *Escherichia coli* K12. *Japan. J. Genet.*, **37**, 66 (*772*).

EGAWA R., SAWAI T. & MITSUHASHI S. (1967). Drug resistance of enteric bacteria. XII. Unique substrate specificity of penicillinase produced by R factor. *Jap. J. Microbiol.*, **11**, 173 (*766*).

EISEN H. A., FUERST C. R., SIMINOVITCH L., THOMAS R., LAMBERT L., PEREIRA DA SILVA L. & JACOB F. (1966). Genetics and physiology of defective lysogeny in K12 (λ): studies of early mutants. *Virology*, **30**, 224 (*475*).

Ellis E. L. & Delbrück M. (1939). The growth of bacteriophage. *J. Gen. Physiol.*, **22**, 365 (*412, 415*).

Engelhardt D. L. & Zinder N. D. (1964). Host-dependent mutants of the bacteriophage f2. III. Infective RNA. *Virology*, **23**, 582 (*510*).

Engelhardt D. L., Webster R. E. & Zinder N. D. (1967). Amber mutants and polarity *in vitro*. *J. Mol. Biol.*, **29**, 45 (*721*).

Englesberg E., Irr J., Power J. & Lee N. (1965). Positive control of enzyme synthesis by gene C in the L-arabinose system. *J. Bacteriol.*, **90**, 946 (*730, 731*).

Ephrati-Elizur E. (1968). Spontaneous transformation in *Bacillus subtilis*. *Genet. Res. Camb.*, **11**, 83 (*611*).

Ephrati-Elizur E., Srinivasan P. A. & Zamenhof S. (1961). Genetic analysis by means of transformation of histidine linkage groups in *Bacillus subtilis*. *Proc. Natl. Acad. Sci., Wash.*, **47**, 56 (*164, 611, 614*).

Ephrussi-Taylor H. E. (1951a). Transformations allogènes du pneumocoque. *Exptl. Cell. Res.*, **2**, 589 (*576, 611*).

Ephrussi-Taylor H. E. (1951b). Genetic aspects of transformations of pneumococci. *Cold Spr. Harb. Symp. quant. Biol.*, **16**, 445 (*576, 611*).

Ephrussi-Taylor H. E. (1955). Current status of bacterial transformations. *Advanc. Virus Res.*, **3**, 275 (*578*).

Ephrussi-Taylor H. E. (1960). Recombination analysis in microbial systems. In *Growth: Molecule, Cell, Organism*. Basic Books, New York (*578*).

Ephrussi-Taylor H. & Freed B. A. (1964). Incorporation of thymidine and amino acids into deoxyribonucleic acid and acid insoluble cell structures in pneumococcal cultures synchronised for competence to transform. *J. Bacteriol.*, **87**, 1211 (*586*).

Epstein C. J. (1966). Role of the amino-acid code and of selection for conformation in the evolution of proteins. *Nature, Lond.*, **210**, 25 (*367*).

Epstein R. H., Bolle A., Steinberg C. M., Kellenberger E., Boy de la Tour E., Chevalley R., Edgar R. S., Slisman M., Denhardt G. H. & Lielausis A. (1963). Physiological studies of conditional lethal mutants of bacteriophage T4D. *Cold Spr. Harb. Symp. quant. Biol.*, **28**, 375 (*432, 433, 485, 492*).

Erikson R. L. & Szybalski W. (1964). The Cs_2SO_4 equilibrium density gradient and its application for the study of T-even phage DNA: glucosylation and replication. *Virology*, **22**, 111 (*520*).

Erskine J. (1966). Ph.D. Thesis, London University (*334*).

Eskridge R. W., Weinfield H. & Paigen K. (1967). Susceptibility of different coliphage genomes to host-controlled variation. *J. Bacteriol.*, **93**, 835 (*521*).

Evans R. H. (1964). Introduction of specific drug resistance properties by purified RNA-containing fractions from *Pneumococcus*. *Proc. Natl. Acad. Sci., Wash.*, **52**, 1442 (*619*).

Falkow S., Rownd R. & Baron L. S. (1962). Genetic homology between *Escherichia coli* K-12 and *Salmonella*. *J. Bacteriol.*, **84**, 1303 (*753*).

Falkow S., Wohlhieter J. A., Citarella R. V. & Baron L. S. (1964). Transfer of episomic elements to *Proteus*. 2. Nature of *lac*⁺ *Proteus* strains isolated from clinical specimens. *J. Bacteriol.*, **88**, 1598 (*770*).

Falkow S., Citarella R. V., Wohlhieter J. A. & Watanabe T. (1966). The molecular nature of R factors. *J. mol. Biol.*, **17**, 102 (*770, 771*).

Fargie B. & Holloway B. W. (1965). Absence of clustering of functionally related genes in *Pseudomonas aeroginosa*. *Genet. Res., Camb.*, **6**, 284 (*165*).

Fattig W. D. & Lanni F. (1965). Mapping of temperature-sensitive mutants in bacteriophage T5. *Genetics*, **51**, 157 (*538*).

Feary T., Fisher E. Jnr. & Fisher T. (1963). A small RNA-containing *Pseudomonas aeruginosa* bacteriophage. *Biochem. biophys. Res. Comm.*, **10**, 359 (*441*).

Felkner I. C. & Wyss O. (1964). A substance produced by competent *Bacillus cereus* 569 cells that affects transformability. *Biochem. biophys. Res. Comm.*, **16**, 94 (*589*).

Felsenfeld G. & Rich A. (1957). Studies on the formation of two and three-stranded polyribonucleotides. *Biochim. biophys. Acta.*, **26**, 457 (*273*).

Feughelman M., Langridge R., Seeds W. E., Stokes A. R., Wilson H. R., Hooper C. W., Wilkins M. F. H., Barclay R. K. & Hamilton L. D. (1955). Molecular structure of deoxyribose nucleic acid and nucleoprotein. *Nature, Lond.*, **175**, 834 (*534*).

Fincham J. R. S. (1959). On the nature of the glutamic dehydrogenase produced by interallelic complementation at the *am* locus of *Neurospora crassa*. *J. gen. Microbiol.*, **21**, 600 (*153*).

Fincham J. R. S. (1960). Genetically controlled differences in enzyme activity. *Advanc. Enzymol.*, **22**, 1 (*153*).

Fincham J. R. S. (1962a). Genes and enzymes in microorganisms. *Brit. Med. Bull.*, **18**, 14 (*151, 153*).

Fincham J. R. S. (1962b). Genetically determined multiple forms of glutamic dehydrogenase in *Neurospora crassa*. *J. Mol. Biol.*, **4**, 257 (*156*).

Fincham J. R. S. (1966). *Genetic complementation*. W. A. Benjamin, Inc., New York and Amsterdam (*157*).

Fincham J. R. S. & Coddington A. (1963). Complementation at the *am* locus of *Neurospora crassa*: a reaction between different mutant forms of glutamate dehydrogenase. *J. Mol. Biol.*, **6**, 361 (*157*).

Fincham J. R. S. & Day P. R. (1963). *Fungal genetics*. Blackwell Scientific Publications, Ltd., Oxford (*157*).

Fischer-Fantuzzi L. & di Girolamo M. (1961). Triparental matings in *Escherichia coli*. *Genetics*, **46**, 1305 (*382*).

Fisher K. W. (1957a). The role of the Krebs cycle in conjugation in *Escherichia coli* K-12. *J. gen. Microbiol.*, **16**, 120 (*672*).

Fisher K. W. (1957b). The nature of the endergonic processes in conjugation in *Escherichia coli* K-12. *J. gen. Microbiol.*, **16**, 136 (*672, 690*).

Fisher K. W. (1959). Bacteriophage penetration and its relation to host cell wall structure. *Proc. Roy. Physical Soc., Edinburgh*, **28**, 91 (424).

Fisher K. W. (1962). Conjugal transfer of immunity to phage λ multiplication in *Escherichia coli* K12. *J. gen. Microbiol.*, **28**, 711 (*469, 700*).

FISHER K. W. (1966). Mechanically caused damage to Hfr cells of *Escherichia coli* K-12. *Genet. Res., Camb.*, **7**, 267 (*680*).

FISHER R. A. (1936). 'Has Mendel's work been rediscovered?' *Ann. Sci.*, **1**, 115 (*24*).

FLAKS J. G. & COHEN S. S. (1959). Virus-induced acquisition of metabolic function. I. Enzymatic formation of 5-hydroxymethyl-deoxycytidylate. *J. Biol. Chem.*, **234**, 1501 (*428*).

FLAKS J. G., LICHTENSTEIN J. & COHEN S. S. (1959). Virus-induced acquisition of metabolic functions. II. Studies on the origin of the deoxycytidylate hydroxymethylase of bacteriophage-infected *E. coli*. *J. Biol. Chem.*, **234**, 1507 (*428*).

FLUKE D., DREW R. & POLLARD E. (1952). Ionising particle evidence for the molecular weight of the pneumococcus transforming principle. *Proc. Natl. Acad. Sci., Wash.*, **38**, 180 (*578*).

FÖLDES J. & TRAUTNER T. A. (1964). Infectious DNA from a newly isolated *B. subtilis* phage. *Z. Vererbungsl.*, **95**, 57 (*618*).

FOSTER R. A. C. (1948). An analysis of the action of proflavine on bacteriophage growth. *J. Bacteriol.*, **56**, 795 (*430*).

FOX C. F. & KENNEDY E. P. (1965). Specific labelling and partial purification of the M protein, a component of the β-galactoside transport system of *E. coli*. *Proc. Natl. Acad. Sci., Wash.*, **54**, 891 (*712*).

FOX C. F., BECKWITH J. R., EPSTEIN W. & SIGNER E. (1966). Transposition of the *lac* region of *Escherichia coli*. II. On the role of the galactoside transacetylase in lactose metabolism. *J. Mol. Biol.*, **19**, 576 (*712*).

FOX M. S. (1955). Mutation rates of bacteria in steady state populations. *J. gen. Physiol.*, **39**, 267 (*193*).

FOX M. S. (1960). Fate of transforming deoxyribonucleate following fixation by transformable bacteria, II. *Nature, Lond.*, **187**, 1004 (*596, 597, 599, 602*).

FOX M. S. (1962). The fate of transforming deoxyribonucleate following fixation by transformable bacteria, III. *Proc. Natl. Acad. Sci., Wash.*, **48**, 1043 (*602, 603*).

FOX M. S. & ALLEN M. K. (1964). On the mechanism of deoxyribonucleate integration in pneumococcal transformation. *Proc. Natl. Acad. Sci., Wash.*, **52**, 412 (*397, 601, 602*).

FOX M. S. & HOTCHKISS R. D. (1957). Initiation of bacterial transformation. *Nature, Lond.*, **179**, 1322 (*587, 588, 594, 609*).

FOX M. S. & HOTCHKISS R. D. (1960). Fate of transforming deoxyribonucleate following fixation by transformable bacteria. I. *Nature, Lond.*, **187**, 1002 (*596*).

FRAENKEL-CONRAT H. & SINGER B. (1957). Virus reconstitution: combination of protein and nucleic acid from different strains. *Biochim. biophys. Acta*, **24**, 541 (*229*).

FRAENKEL-CONRAT H. & WILLIAMS R. C. (1955). Reconstitution of active tobacco mosaic virus from its active protein and nucleic acid components. *Proc. Natl. Acad. Sci., Wash.*, **41**, 695 (*229, 262*).

FRAENKEL-CONRAT H., SINGER B. & WILLIAMS R. C. (1957). Infectivity of viral nucleic acid. *Biochim. biophys. Acta*, **25**, 87 (*229, 618*).

FRANKEL F. R. (1966a). The absence of mature phage DNA molecules from the replicating pool of T-even infected *Escherichia coli*. *J. Mol. Biol.*, **18**, 109 (*434*).
FRANKEL F. R. (1966b). Studies on the nature of replicating DNA in T4 infected *Escherichia coli*. *J. Mol. Biol.*, **18**, 127 (*434*).
FRANKLIN N. C. & LURIA S. E. (1961). Transduction by bacteriophage P1 and the properties of the *lac* genetic region in *E. coli* and *S. dysenteriae*. *Virology*, **15**, 299 (*719*).
FRANKLIN N. C., DOVE W. F. & YANOFSKY C. (1965). The linear insertion of a prophage into the chromosome of *E. coli* shown by deletion mapping. *Biochem. biophys. Res. Comm.*, **18**, 910 (*457, 461, 725*).
FRANKLIN R. E. & GOSLING R. G. (1953). Molecular configuration in sodium thymonucleate. *Nature, Lond.*, **171**, 740 (*232*).
FRANKLIN T. J. (1967). Resistance of *Escherichia coli* to tetracyclines. Changes in permeability to tetracyclines in *Escherichia coli* bearing transferable resistance factors. *Biochem. J.*, **105**, 371 (*766*).
FRASER D. K. (1957). Host range mutants and semitemperate mutants of bacteriophage T3. *Virology*, **3**, 527 (*449*).
FREDERICQ P. (1957). Colicins. *Ann. Rev. Microbiol.*, **11**, 7 (*756, 757*).
FREDERICQ P. (1958). Colicins and colicinogenic factors. *Symp. Soc. Exp. Biol.*, **12**, 104 (*757*).
FREDERICQ P. (1963). On the nature of colicinogenic factors: a review. *J. Theoret. Biol.*, **4**, 159 (*780*).
FREEMAN V. J. (1951). Studies on the virulence of bacteriophage-infected strains of *Corynebacterium diphtheriae*. *J. Bacteriol.*, **61**, 675 (*646*).
FREEMAN V. J. & MORSE I. U. (1952). Further observations on the change to virulence of bacteriophage-infected avirulent strains of *Corynebacterium diphtheriae*. *J. Bacteriol.*, **63**, 407 (*646*).
FREESE E. (1959a). The specific mutagenic effect of base analogues on phage T4. *J. Mol. Biol.*, **1**, 87 (*303, 304*).
FREESE E. (1959b). On the molecular explanation of spontaneous and induced mutations. *Brookhaven Symposia in Biology*, **12**, 63 (*304, 308, 315, 319*).
FREESE E. (1959c). The difference between spontaneous and base analogue induced mutations of phage T4. *Proc. Natl. Acad. Sci., Wash.*, **45**, 622 (*309, 315*).
FREESE E., BAUTZ E. & BAUTZ-FREESE E. (1961a). The chemical and mutagenic specificity of hydroxylamine. *Proc. Natl. Acad. Sci., Wash.*, **47**, 845 (*307, 311, 315*).
FREESE E., BAUTZ-FREESE E. & BAUTZ E. (1961b). Hydroxylamine as a mutagenic and inactivating agent. *J. Mol. Biol.*, **3**, 133 (*307, 311, 314, 315*).
FREIFELDER D. (1967). Role for the female in bacterial conjugation in *Escherichia coli*. *J. Bacteriol.*, **94**, 396 (*804*).
FRESCO R. J., ALBERTS B. M. & DOTY P. (1960). Some molecular details of the secondary structure of ribonucleic acid. *Nature, Lond.*, **188**, 98 (*273*).
FREUNDLICH M. & UMBARGER H. E. (1963). The effects of analogues of threonine and of isoleucine on the properties of threonine deaminase. *Cold Spr. Harb. Symp. quant. Biol.*, **28**, 505 (*710*).

Fuerst C. R. & Stent G. S. (1956). Inactivation of bacteria by decay of incorporated radioactive phosphorus. *J. gen. Physiol.*, **40**, 73 (*524*).

Fuhs G. W. (1962). Experimental results concerning the physical and chemical state of the nucleoids in *Bacillus subtilis*. *8th Int. Cong. Microbiol., Montreal* (*544, 546, 553*).

Fuhs G. W. (1965). Symposium on the fine structure and replication of bacteria and their parts. I. Fine structure and replication of bacterial nucleoids. *Bacteriol. Rev.*, **29**, 277 (*546, 553*).

Fukasawa T. (1964). The course of infection with abnormal bacteriophage T4 containing non-glucosylated DNA on *Escherichia coli* strains. *J. Mol. Biol.*, **9**, 525 (*521*).

Fulton C. (1965). Continuous chromosome transfer in *Escherichia coli*. *Genetics*, **52**, 55 (*805*).

Furness G. & Rowley D. (1957). The presence of the transmissible agent F in non-recombining strains of *E. coli*. *J. gen. Microbiol.*, **17**, 550 (*749*).

Gamow G. (1954). Possible relation between deoxyribonucleic acid and protein structures. *Nature, Lond.*, **173**, 318 (*340*).

Ganesan A. T. & Lederberg J. (1965). A cell-membrane bound fraction of bacterial DNA. *Biochem. biophys. Res. Comm.*, **18**, 824 (*569*).

Garen A. (1960). Genetic control of the specificity of the bacterial enzyme, alkaline phosphatase. *Symp. Soc. gen. Microbiol.*, **10**, 239 (*299*).

Garen A. & Echols H. (1962). Genetic control of induction of alkaline phosphatase synthesis in *E. coli*. *Proc. Natl. Acad. Sci., Wash.*, **48**, 1398 (*730*).

Garen A. & Kozloff L. M. (1959). The initiation of bacteriophage infection. In *The Viruses* (eds. Burnet F. M. & Stanley W. M.), Vol. 2, p. 203. Academic Press, Inc., New York (*410, 424*).

Garen A. & Siddiqi O. (1962). Suppression of mutations in the alkaline phosphatase structural cistron of *E. coli*. *Proc. Natl. Acad. Sci., Wash.*, **48**, 1121 (*362*).

Garen A. & Zinder N. D. (1955). Radiological evidence for partial genetic homology between bacteriophage and host bacteria. *Virology*, **1**, 347 (*524, 623, 636*).

Garrod A. E. (1909, 1923). *Inborn errors of metabolism*. Henry Frowde, London (*108*).

Garrod L. P. & O'Grady S. W. (1968). *Antibiotics and Chemotherapy*. E. and S. Livingstone, Edinburgh and London (*784*).

Gefter M., Hausmann R., Gold M. & Hurwitz J. (1966). The enzymatic methylation of ribonucleic acid and deoxyribonucleic acid. X. Bacteriophage T3 induced S-adenosylmethionine cleavage. *J. Biol. Chem.*, **241**, 1995 (*520*).

Geiduschek E. P., Tocchini-Valenti G. P. & Sarnat M. T. (1964). Asymmetric synthesis of RNA *in vitro*: dependence on RNA continuity and conformation. *Proc. Natl. Acad. Sci., Wash.*, **52**, 486 (*291*).

Gerhart J. C. & Pardee A. B. (1962). The enzymology of control by feedback inhibition. *J. Biol. Chem.*, **237**, 891 (*707, 708*).

Gerhart J. C. & Pardee A. B. (1963). The effect of the feedback inhibitor,

CTP, on subunit interactions in aspartate transcarbamylase. *Cold Spr. Harb. Symp. quant. Biol.*, **28**, 491 (*709*).

GHEI O. K. & LACKS S. A. (1967). Recovery of donor deoxyribonucleic acid marker activity from eclipse in pneumococcal transformation. *J. Bacteriol.*, **93**, 816 (*598*).

GIACOMONI D. & SPIEGELMAN S. (1962). Origin and biologic individuality of the genetic dictionary. *Science*, **138**, 1328 (*289*).

GIBSON Q. H. (1959). The kinetics of reactions between haemoglobins and gasses. *Progress in Biophys. and Biophys. Chem.*, **9**, 1 (*709*).

GIERER A. (1960). Ribonucleic acid as genetic material of viruses. *Symp. Soc. gen. Microbiol.*, **10**, 248 (*229, 314*).

GIERER A. (1966). Model for DNA and protein interactions and the function of the operator. *Nature, Lond.*, **212**, 1480 (*734*).

GIERER A. & SCHRAMM G. (1956). Infectivity of ribonucleic acid from tobacco mosaic virus. *Nature, Lond.*, **177**, 702 (*229, 618*).

GILBERT W. (1963a). Polypeptide synthesis in *Escherichia coli*. I. Ribosomes and the active-complex. *J. Mol. Biol.*, **6**, 374 (*278, 285*).

GILBERT W. (1963b). Polypeptide synthesis in *Escherichia coli*. II. The polypeptide chain and S-RNA. *J. Mol. Biol.*, **6**, 389 (*285, 286*).

GILBERT W. & MÜLLER-HILL B. (1966). Isolation of the lac repressor. *Proc. Natl. Acad. Sci., Wash.*, **56**, 1891 (*469, 733*).

GILES N. H. (1951). Studies on the mechanism of reversion in biochemical mutants of *Neurospora crassa*. *Cold Spr. Harb. Symp. quant. Biol.*, **16**, 283 (*129*).

GINOZA H. S. & PAINTER R. B. (1964). Genetic recombination between the resistance transfer factor and the chromosome of *Escherichia coli*. *J. Bacteriol.*, **87**, 1339 (*765*).

GLANVILLE E. V. & DEMEREC M. (1960). Threonine, isoleucine and isoleucine-valine mutants of *Salmonella typhimurium*. *Genetics*, **45**, 1359 (*164*).

GLOVER S. W. (1956). A comparative study of induced reversions in *Escherichia coli*. 'Genetic studies with bacteria', p. 121. *Carnegie Inst., Wash. Publ.*, **612** (*205*).

GLOVER S. W., SCHELL J., SYMONDS N. & STACEY K. A. (1963). The control of host-induced modification by phage P1. *Genet. Res., Camb.*, **4**, 480 (*518*).

GOLDBERGER R. F. & BERBERICH M. A. (1966). Studies on the mechanism of derepression of the histidine operon in *S. typhimurium*. *Fed. Proc.*, **25**, 337 (*719*).

GOLDTHWAIT D. & JACOB F. (1964). Sur le méchanisme de l'induction du développement du prophage chez les bactéries lysogènes. *C. R. Acad. Sci., Paris*, **259**, 661 (473).

GOULIAN M., KORNBERG A. & SINSHEIMER R. L. (1967). Enzymatic synthesis of DNA. XXIV. Synthesis of infectious phage φX174 DNA. *Proc. Natl. Acad. Sci., Wash.*, **58**, 2321 (*257*).

GOODGAL S. H. (1961). Studies on transformation of *Hemophilus influenzae*. IV. Linked and unlinked transformations. *J. gen. Physiol.*, **45**, 205 (*609, 610*).

GOODGAL S. H. & HERRIOT R. M. (1957a). Studies on transformation of *Haemophilus influenzae*. In *The Chemical Basis of Heredity* (eds. McElroy W. D. & Glass B.), p. 336. Johns Hopkins Press, Baltimore (*575, 578, 579, 580, 597*).

GOODGAL S. H. & HERRIOT R. M. (1957b). A study of linked transformations in *Haemophilus influenzae*. *Genetics*, **42**, 372 (*610*).

GOODGAL S. H. & HERRIOT R. M. (1961). Studies on transformation of *Haemophilus influenzae*. I. Competence. *J. gen. Physiol.*, **44**, 1201 (*587, 609*).

GORINI L. (1958). Régulation en retour (feedback control) de la synthèse de l'arginine chez *Escherichia coli*. *Bull. Soc. Chim. Biol.*, **40**, 1939 (*710*).

GORINI L., GUNDERSEN W. & BERGER M. (1961). Genetics of regulation of enzyme synthesis in the arginine biosynthetic pathway of *Escherichia coli*. *Cold Spr. Harb. Symp. quant. Biol.*, **26**, 173 (*732*).

GORINI L. & KAUFMAN H. (1960). Selecting bacterial mutants by the penicillin method. *Science*, **131**, 604 (*212*).

GORINI L. & MAAS W. (1958). Feedback control of the formation of biosynthetic enzymes. In *The Chemical Basis of Development* (eds. McElroy W. D. & Glass B.), p. 469. Johns Hopkins Press, Baltimore (*710*).

GOWEN J. W. & LINCOLN R. E. (1942). A test for sexual fusion in bacteria. *J. Bacteriol.*, **44**, 551 (*651*).

GRANBOULAN N. & FRANKLIN R. M. (1966). Electron microscopy of viral RNA, replicative form and replicative intermediate of the bacteriophage R17. *J. Mol. Biol.*, **22**, 173 (*444*).

GRAY C. H. & TATUM E. L. (1944). X-ray induced growth factor requirements in bacteria. *Proc. Natl. Acad. Sci., Wash.*, **30**, 404 (*651*).

GREEN D. M. & COLARUSSO L. J. (1964). The physical and genetic characterisation of a transformable enzyme: *Bacillus subtilis* α-amylase. *Biochim. biophys. Acta*, **89**, 277 (*745*).

GREEN M. H. (1966). Inactivation of the prophage λ repressor without induction. *J. Mol. Biol.*, **16**, 134 (*474*).

GRIFFITH F. (1928). Significance of pneumococcal types. *J. Hyg., Camb.*, **27**, 113 (*574, 575*).

GROMAN N. B. (1953). Evidence for the induced nature of the change from nontoxigenicity to toxigenicity in *Corynebacterium diphtheriae* as a result of exposure to specific bacteriophage. *J. Bacteriol.*, **66**, 184 (*646*).

GROMAN N. B. (1956). Conversion in *Corynebacterium diphtheriae* with phages originating from non-toxigenic strains. *Virology*, **2**, 843 (*646*).

GROMAN N. B. & EATON M. (1955). Genetic factors in *Corynebacterium diphtheriae* conversion. *J. Bacteriol.*, **70**, 637 (*646*).

GROS F., GILBERT W., HIATT H., KURLAND C. G., RISEBROUGH R. W. & WATSON J. D. (1961). Unstable ribonucleic acid revealed by pulse labelling of *Escherichia coli*. *Nature, Lond.*, **190**, 581 (*276*).

GROSS J. D. (1964). Conjugation in bacteria. In *The Bacteria* (eds. Gunsalus I. C. & Stanier R. Y.), Vol. 5, p. 1. Academic Press, Inc., New York and London (*696*).

GROSS J. D. & CARO L. (1965). Genetic transfer in bacterial mating. *Science*, **150**, 1679 (*800, 801*).

GROSS J. D. & CARO L. (1966). DNA transfer in bacterial conjugation. *J. Mol. Biol.*, **16**, 269 (*801*, *802*, *803*).
GROSS J. D. & ENGLESBERG E. (1959). Determination of the order of mutational sites governing L-arabinose utilisation in *Escherichia coli* B/r by transduction with phage P1*bt*. *Virology*, **9**, 314 (*164*, *384*, *641*).
GRUNBERG-MANAGO M., ORTIZ P. J. & OCHOA S. (1956). Enzymic synthesis of polynucleotide. I. Polynucleotide phosphorylase in *Azobacter vinelandii*. *Biochim. Biophys. Acta*, **20**, 269 (*292*).
GUINEE P. A. M., SCHOLTENS R. TH. & WILLEMS H. M. C. C. (1967). Influence of resistance factors on the phage types of *Salmonella panama*. *Antonie van Leeuwenhoek*, **33**, 30 (*768*).
GUTTMAN B. S. & NOVICK A. (1963). A messenger RNA for β-galactosidase in *Escherichia coli*. *Cold Spr. Harb. Symp. quant. Biol.*, **28**, 373 (*718*).
DE HAAN P. G. (1954a). Genetic recombination in *E. coli* B. I. The transfer of the F agent to *E. coli* B. *Genetica*, **27**, 293 (*750*).
DE HAAN P. G. (1954b). Genetic recombination in *E. coli* B. II. The cross resistance of *E. coli* B to the phages T3, T4 & T7. *Genetica*, **27**, 300 (*750*).
DE HAAN P. G. (1955). Genetic recombination in *E. coli* B. III. The influence of experimental conditions on the transfer of unselected markers. *Genetica*, **27**, 364 (*750*).
DE HAAN P. G. & GROSS J. D. (1962). Transfer delay and chromosome withdrawal during conjugation in *E. coli*. *Genet. Res. Camb.*, **3**, 188 (*685*, *689*).
DE HAAN P. G. & STOUTHAMER A. H. (1963). F-prime transfer and multiplication of sexduced cells. *Genet. Res. Camb.*, **4**, 30 (*656*, *776*).
HAGGIS G. H., MICHIE D., MUIR A. R., ROBERTS K. B. & WALKER P. B. M. (1964). *Introduction to molecular biology*. Longmans, Green and Co., Ltd., London (*232*).
HAHN F. E., WISSEMAN C. L. JNR. & HOPPS H. E. (1954). Mode of action of chloramphenicol, II. *J. Bacteriol.*, **67**, 674 (*208*).
HALL B. D. & SPIEGELMAN S. (1961). Sequence complementarity of T2-DNA and T2-specific RNA. *Proc. Natl. Acad. Sci., Wash.*, **47**, 137 (*276*, *289*).
HANAWALT P. C. & RAY D. S. (1964). Isolation of the growing point in the bacterial chromosome. *Proc. Natl. Acad. Sci., Wash.*, **52**, 125 (*569*).
HANAWALT P. C., MAALØE O., CUMMINGS D. J. & SCHAECHTER M. (1961). The normal DNA replication cycle. *J. Mol. Biol.*, **3**, 156 (*557*).
HARADA K., KAMEDA M., SUZUKI M. & MITSUHASHI S. (1964). Drug resistance of enteric bacteria. III. Acquisition of transferability of non-transmissible R (TC) factor in co-operation with F factor and formation of FR (TC). *J. Bacteriol.*, **88**, 1257 (*781*).
HARADA K., KAMEDA M., SUZUKI M., SHIGENARA S. & MITSUHASHI S. (1967). Drug resistance of enteric bacteria. VIII. Chromosomal location of non-transferable R factor in *Escherichia coli*. *J. Bacteriol.*, **93**, 1236 (*764*, *781*).
HARM W. (1959). Untersuchungen zur Wirkungsweise eines die UV-Empfindlichkeit bestimmenden Gens der Bakteriophagen T2 und T4. *Z. Vererbungsl.*, **90**, 428 (*527*).

HARRIS A. W., MOUNT D. W. A., FUERST C. R. & SIMINOVITCH L. (1967). Mutations in bacteriophage lambda affecting host cell lysis. *Virology*, **32**, 553 (*475*).

HARTMAN P. E. (1956). Linked loci in the control of consecutive steps in the primary pathway of histidine synthesis in *Salmonella typhimurium*. In *Genetic Studies with Bacteria. Publ. Carneg. Inst.*, **612**, 35 (*143*).

HARTMAN P. E. (1957). Transduction: a comparative review. In *The Chemical Basis of Heredity* (eds. McElroy W. D. & Glass B.), p. 408. The Johns Hopkins Press, Baltimore, U.S.A. (*646*).

HARTMAN P. E. (1963). Methodology in transduction. In *Methodology in Basic Genetics* (ed. Burdette W. L.), p. 103. Holden-Day, Inc., San Francisco (*646*).

HARTMAN P. E. & GOODGAL S. H. (1959). Bacterial Genetics (with particular reference to genetic transfer). *Ann. Rev. Microbiol.*, **13**, 465 (*646*).

HARTMAN P. E., HARTMAN Z. & ŠERMAN D. (1960). Complementation mapping by abortive transduction of histidine-requiring *Salmonella* mutants. *J. gen. Microbiol.*, **22**, 354 (*96, 138, 139, 149, 150, 160*).

HARTMAN P. E., LOPER J. C. & ŠERMAN D. (1960). Fine structure mapping by complete transduction between histidine-requiring *Salmonella* mutants. *J. gen. Microbiol.*, **22**, 323 (*131, 134, 149, 150, 160, 724*).

HARUNA I. & SPIEGELMAN S. (1965). Specific template requirements of RNA replicases. *Proc. Natl. Acad. Sci., Wash.*, **54**, 579 (*292, 443*).

HASHIMOTO K. (1960). Streptomycin resistance in *Escherichia coli* analysed by transduction. *Genetics*, **45**, 49 (*216*).

HASTINGS P. J. & WHITEHOUSE H. C. K. (1964). A polaron model of genetic recombination by the formation of hybrid deoxyribonucleic acid. *Nature, Lond.*, **201**, 1052 (*401*).

HATTMAN S. (1964). The functioning of T-even phages with unglucosylated DNA in restricting *Escherichia coli* host cells. *Virology*, **24**, 333 (*521*).

HATTMAN S. & FUKASAWA T. (1963). Host-induced modification of T-even phages due to defective glucosylation of their DNA. *Proc. Natl. Acad. Sci., Wash.*, **50**, 297 (*520*).

HAUSMANN R. & BRESCH C. (1960). Zum Problem der Genetischen Rekombination von Bakteriophagen. II. Versuch einer experimenteller Unterscheidung von paarweiser und kompletter Kooperation. *Z. Vererbl.*, **91**, 266 (*382, 504*).

HAYASHI M., HAYASHI M. N. & SPIEGELMAN S. (1963). Restriction of genetic *in vivo* transcription to one of the complementary strands of DNA. *Proc. Natl. Acad. Sci., Wash.*, **50**, 664 (*439*).

HAYASHI M., HAYASHI M. N. & SPIEGELMAN S. (1964). DNA circularity and the mechanism of strand selection in the generation of genetic messages. *Proc. Natl. Acad. Sci., Wash.*, **51**, 351 (*291*).

HAYES D. (1967). Mechanisms of nucleic acid synthesis. *Ann. Rev. Microbiol.*, **21**, 369 (*246, 296*).

HAYES W. (1947). The nature of somatic phase variation and its importance in the serological standardisation of *O* suspensions of *Salmonella*. *J. Hyg., Camb.*, **45**, 111 (*745*).

Hayes W. (1952a). Recombination in *Bact. coli* K-12: unidirectional transfer of genetic material. *Nature, Lond.*, **169**, 118 (*654*).

Hayes W. (1952b). Genetic recombination in *Bact. coli* K-12: analysis of the stimulating effect of ultraviolet light. *Nature, Lond.*, **169**, 1017 (*654*).

Hayes W. (1953a). Observations on a transmissible agent determining sexual differentiation in *Bact. coli. J. gen. Microbiol.*, **8**, 72 (*654, 655, 657, 658*).

Hayes W. (1953b). The mechanism of genetic recombination in *Escherichia coli. Cold Spr. Harb. Symp. quant. Biol.*, **18**, 75 (*655, 656, 657, 658, 659, 660, 661, 667*).

Hayes W. (1955). A new approach to the study of kinetics of recombination in *E. coli* K-12. Proceedings of the Society for General Microbiology, *J. gen. Microbiol.*, **13**, ii (*682*).

Hayes W. (1957a). The kinetics of the mating process in *E. coli. J. gen. Microbiol.*, **16**, 97 (*203, 679, 682, 690, 701, 702, 703*).

Hayes W. (1957b). The phenotypic expression of genes determining various types of drug resistance following their inheritance by sensitive bacteria. In *Drug resistance in microorganisms* (eds. Wolstenholme G. E. W. & O'Connor C. M.), p. 197: Ciba Foundation Symposium. J. & A. Churchill, London (*701, 702*).

Hayes W. (1960). The bacterial chromosome. *Symp. Soc. gen. Microbiol.*, **10**, 12 (*549, 650, 654*).

Hayes W. (1966a). Genetic transformation: a retrospective appreciation. (1st Griffith Memorial Lecture). *J. gen. Microbiol.*, **45**, 385 (*562, 565, 574*).

Hayes W. (1966b). Sexual differentiation in bacteria. In *Phage and the origins of molecular biology* (eds. Cairns J., Stent G. S. & Watson J. D.), p. 201. Cold Spring Harbor Laboratory of Quantitative Biology, Long Island, New York (*657, 696*).

Hayes W. (1966c). Genetic Society Mendel Lecture: sex factors and viruses. *Proc. Roy. Soc., B.*, **164**, 230 (*675, 796*).

Hayes W. (1966d). The Leeuwenhoek Lecture, 1965: some controversial aspects of bacterial sexuality. *Proc. Roy. Soc., B.*, **165**, 1 (*800*).

Hayes W., Jacob F. & Wollman E. L. (1963). Conjugation in bacteria. In *Methodology in Basic Genetics* (ed. Burdette W. L.), p. 129. Holden-Day, Inc., San Francisco (*665, 686, 696, 792*).

Haywood A. M. & Sinsheimer R. L. (1963). Inhibition of protein synthesis in *E. coli* protoplasts by Actinomycin-D. *J. Mol. Biol.*, **6**, 247 (*442*).

Herschman H. R. & Helinski D. R. (1962). Comparative study of the events associated with colicin induction. *J. Bacteriol.*, **94**, 691 (*757*).

Hecht L. I., Stephenson M. L. & Zamecnik P. C. (1959). Binding of amino acids to the end group of a soluble ribonucleic acid. *Proc. Natl. Acad. Sci., Wash.*, **45**, 505 (*281*).

Heden C.-G. (1951). Studies of the infection of *E. coli* with the bacteriophage T2. *Acta. Pathol. Microbiol. Scand. Suppl.*, **88** (*414*).

Helinski D. R. & Yanofsky C. (1962). Correspondence between genetic data and the position of amino acid alteration in a protein. *Proc. Natl. Acad. Sci., Wash.*, **48**, 173 (*300, 341*).

HERMAN R. K. & FORRO F. JNR. (1962). Autoradiographic study of the sex factor of *E. coli* K-12 after genetic transfer. *Abstr. Biophysical Soc.*, **FB-11** (*657*).

HERRIOT R. M. (1961). Formation of heterozygotes by annealing a mixture of transforming DNAS. *Proc. Natl. Acad. Sci., Wash.*, **47**, 146 (*604*).

HERRIOT R. M. (1963). The mechanism of renaturation of *Hemophilus* transforming DNA. *Biochem. Z.*, **338**, 179 (*600*).

HERRIOT R. M. (1965). Structure of the hybrid transforming unit. *Genetics*, **52**, 1235 (*604*).

HERSHEY A. D. (1946). Spontaneous mutations in bacterial viruses. *Cold Spr. Harb. Symp. quant. Biol.*, **11**, 67 (*415, 482, 486*).

HERSHEY A. D. (1953). Nucleic acid economy in bacteria infected with bacteriophage T2. II. Phage precursor nucleic acid. *J. gen. Physiol.*, **37**, 1 (*423, 426*).

HERSHEY A. D. (1955). An upper limit to the protein content of the germinal substance of bacteriophage T2. *Virology*, **1**, 108 (*420*).

HERSHEY A. D. (1957a). Some minor components of bacteriophage T2 particles. *Virology* **4**,, 237 (*420*).

HERSHEY A. D. (1957b). Bacteriophages as genetic and biochemical systems. *Advanc. in Virus Res.*, **4**, 25 (*410*).

HERSHEY A. D. (1958). Production of recombinants in phage crosses. *Cold Spr. Harb. Symp. quant. Biol.*, **23**, 19 (*382, 504*).

HERSHEY A. D. & BURGI E. (1956). Genetic significance of the transfer of nucleic acid from parental to offspring phage. *Cold Spr. Harb. Symp. quant. Biol.*, **21**, 91 (*423*).

HERSHEY A. D. & BURGI E. (1960). Molecular homogeneity of the deoxyribonucleic acid of phage T2. *J. Mol. Biol.*, **2**, 143 (*536*).

HERSHEY A. D. & BURGI E. (1965). Complementary structure of interacting sites at the ends of lambda DNA molecules. *Proc. Natl. Acad. Sci., Wash.*, **53**, 325 (*462, 543*).

HERSHEY A. D. & CHASE M. (1951). Genetic recombination and heterozygosis in bacteriophage. *Cold Spr. Harb. Symp. quant. Biol.*, **16**, 471 (*496, 497*).

HERSHEY A. D. & CHASE M. (1952). Independent functions of viral protein and nucleic acid in growth of bacteriophage. *J. gen. Physiol.*, **36**, 39 (*228, 423, 621*).

HERSHEY A. D. & MELECHEN N. E. (1957). Synthesis of phage-precursor nucleic acid in the presence of chlormaphenicol. *Virology*, **3**, 207 (*414, 415, 428*).

HERSHEY A. D. & ROTMAN R. (1949). Genetic recombination between host range and plaque-type mutants of bacteriophage in single bacterial cells. *Genetics*, **34**, 44 (*481, 486, 487, 489, 495, 502, 504*).

HERSHEY A. D., BURGI E. & INGRAHAM L. (1963). Cohesion of DNA molecules isolated from phage lambda. *Proc. Natl. Acad. Sci., Wash.*, **49**, 748 (*462, 543*).

HERSHEY A. D., DIXON J. & CHASE M. (1953). Nucleic acid economy in bacteria infected with bacteriophage T2: I. Purine and pyrimidine composition. *J. gen. Physiol.*, **36**, 777 (*275, 426, 429*).

HERSHEY A. D., BURGI E., GAREN A. & MELECHEN N. E. (1955). Growth and inheritance in bacteriophage. *Carnegie Inst. Wash. Yearbook*, **54**, 216 (*427*).

HERSHEY A. D., GAREN A., FRASER D. K. & HUDIS J. D. (1954). *Carnegie Inst. Wash. Yearbook*, **53**, 210 (*427*).

HERSHEY A. D., KAMEN M. D., KENNEDY J. W. & GEST H. (1951). The mortality of bacteriophage containing assimilated radioactive phosphorus. *J. gen. Physiol.*, **34**, 305 (*523*).

HERTMAN I. & LURIA S. E. (1967). Transduction studies on the role of a *rec*$^+$ gene in the ultraviolet induction of prophage lambda. *J. Mol. Biol.*, **23**, 117 (*474*).

HILL L. R. (1966). An index to deoxyribonucleic base compositions of bacterial species. *J. gen. Microbiol.*, **44**, 419 (*366*).

HILL R. F. (1958). A radiation-sensitive mutant of *Escherichia coli*. *Biochim. Biophys. Acta*, **30**, 636 (*332*, *527*).

HILL R. F. (1965). Ultraviolet induced lethality and reversion to prototrophy in *Escherichia coli* strains with normal and reduced dark repair ability. *Photochem. Photobiol.*, **4**, 563 (*337*).

HINSHELWOOD C. N. (1946). *The Chemical Kinetics of the Bacterial Cell*. Oxford University Press (Clarendon), London (*192*).

HIROTA Y. (1960). The effect of acridine dyes on mating type factors in *Escherichia coli*. *Proc. Natl. Acad. Sci., Wash.*, **46**, 57 (*656*, *672*, *790*).

HIROTA Y. & SNEATH P. H. A. (1961). F′ and F mediated transduction in *Escherichia coli* K-12. *Japanese J. Genetics*, **36**, 307 (*677*, *797*).

HIROTA Y., FUJII T. & NISHIMURA Y. (1966). Loss and repair of conjugal fertility and infectivity of the resistance factor and sex factors in *Escherichia coli*. *J. Bacteriol.*, **91**, 1298 (*782*).

HOAGLAND M. B., ZAMECNICK P. C. & STEPHENSON M. L. (1957). Intermediate reactions in protein biosynthesis. *Biochim. biophys. Acta*, **24**, 215 (*280*).

HOAGLAND M. B., STEPHENSON M. L., SCOTT J. F., HECHT R. J. & ZAMECNIK, P. C. (1958). A soluble ribonucleic acid intermediate in protein synthesis. *J. biol. Chem.*, **231**, 241 (*280*).

HOEKSTRA W. P. M. & DEHAAN P. G. (1965). The location of the restriction locus for λ.K in *Escherichia coli* B. *Mutation Res.*, **2**, 204 (*518*).

HOFFMAN-BERLING H. & MAZE R. (1964). Release of male-specific bacteriophages from surviving host bacteria. *Virology*, **22**, 305 (*446*).

HOFSCHNEIDER P. H. (1963). Untersuchungen über 'kleine' *E. coli* bakteriophagen. *Z. Naturforsch.*, **18b**, 203 (*419*, *444*).

HOFSCHNEIDER P. H. & PREUSS A. (1963). M13 bacteriophage liberation from intact bacteria as revealed by electron microscopy. *J. Mol. Biol.*, **7**, 450 (*446*).

HOGNESS D. S. & SIMMONS J. R. (1964). Breakgage of λ*dg* DNA: chemical and genetic characterisation of each isolated half molecule. *J. Mol. Biol.*, **9**, 411 (*544*).

HOGNESS D., DOERFLER W., EGAN J. & BLACK L. (1966). The position and orientation of genes in λ and λ*dg* DNA. *Cold Spr. Harb. Symp. quant. Biol.*, **31** (*403*, *404*, *545*).

HOLLEY R. W., APGAR J., EVERETT G. A., MADISON J. T., MARQUISER M., MERRILL S. H., PENSWICK J. R. & ZAMIR A. (1965). Structure of a ribonucleic acid. *Science*, **147**, 1462 (*282*).

HOLLIDAY R. (1956). A new method for the identification of biochemical mutants of micro-organisms. *Nature, Lond.*, **178**, 987 (*211*).

HOLLIDAY R. (1964). A mechanism for gene conversion in fungi. *Genet. Res., Camb.*, **5**, 282 (*400, 403*).

HOLLOM S. & PRITCHARD R. H. (1965). Effect of inhibition of DNA synthesis on mating in *Escherichia coli* K12. *Genet. Res., Camb.*, **6**, 479 (*803*).

HOLLOWAY B. W. (1955). Genetic recombination in *Pseudomonas aeruginosa. J. gen. Microbiol.*, **13**, 572 (*753*).

HOLLOWAY B. W. (1956). Self-fertility in *Pseudonomonas aeruginosa. J. gen. Microbiol.*, **15**, 221 (*753*).

HOLLOWAY B. W. & FARGIE B. (1960). Fertility factors and genetic linkage in *Pseudomonas aeruginosa. J. Bacteriol.*, **80**, 362 (*754*).

HOLLOWAY B. W. & JENNINGS P. A. (1958). An infectious fertility factor for *Pseudomonas aeruginosa. Nature, Lond.*, **181**, 855 (*754*).

HOLLOWAY B. W. & MONK M. (1960). Transduction in *Pseudomonas aeroginosa. Nature, Lond.*, **184**, 1426 (*621*).

HOPWOOD D. A. (1965a). New data on the linkage map of *Streptomyces coelicolor. Genet., Res., Camb.*, **6**, 248 (*164*).

HOPWOOD D. A. (1965b). A circular linkage map in the Actinomycete *Streptomyces coelicolor. J. Mol. Biol.*, **12**, 514 (*164, 698*).

HOPWOOD D. A. (1966a). Lack of constant genome ends in *Streptomyces coelicolor. Genetics*, **54**, 1177 (*698*).

HOPWOOD D. A. (1966b). Nonrandom location of temperature-sensitive mutants on the linkage map of *Streptomyces coelicolor. Genetics*, **54**, 1169 (*699*).

HOPWOOD D. A. (1967). Genetic analysis and genome structure in *Streptomyces coelicolor. Bacteriol. Rev.*, **31**, 373 (*696, 697, 698, 699*).

HOPWOOD D. A. & SERMONTI G. (1962). The genetics of *Streptomyces coelicolor. Advanc. Genetics*, **11**, 273 (*697, 699*).

HORIUCHI K., LODISH H. F. & ZINDER N. D. (1966). Mutants of the bacteriophage f2. VI. Homology of temperature-sensitive and host-dependent mutants. *Virology*, **28**, 438 (*510*).

HOTCHKISS R. D. (1949). Études chimiques sur le facteur transformant du pneumocoque. In *Unités Biologiques Douées de Continuité Génétique. Colloq. int. Cent. Nat. Rech. Sci.*, **8**, 57 (*228, 575*).

HOTCHKISS R. D. (1952). The role of deoxyribonucleates in bacterial transformations. In *Phosphorus Metabolism* (eds. McElroy W. D. & Glass B.), vol. 2, p. 426. Johns Hopkins Press, Baltimore (*575*).

HOTCHKISS R. D. (1954). Cyclical behaviour in pneumococcal growth and transformability occasioned by environmental changes. *Proc. Natl. Acad. Sci., Wash.*, **40**, 49 (*581, 585, 593, 594*).

HOTCHKISS R. D. (1955). The biological role of the deoxypentose nucleic acids. In *Nucleic Acids* (eds. Chargaff E. & Davidson J. N.), vol. II, p. 435. Academic Press, New York (*578*).

Hotchkiss R. D. (1957). Criteria for quantitative genetic transformation of bacteria. In *The Chemical Basis of Heredity* (eds. McElroy W. D. & Glass B.), p. 321. Johns Hopkins Press, Baltimore, U.S.A. (*579, 581*).

Hotchkiss R. D. (1966). Gene, transforming principle, and DNA. In *Phage and the origins of molecular biology* (eds. Cairns J., Stent G. S. & Watson J. D.), p. 180. Cold Spring Harbor Laboratory of Quantitative Biology, Long Island, New York (*574*).

Hotchkiss R. D. & Evans A. H. (1958). Analysis of the complex sulfonamide resistance locus of pneumococcus. *Cold Spr. Harb. Symp. quant. Biol.*, **23**, 85 (*610*).

Hotchkiss R. D. & Marmur J. (1954). Double marker transformations as evidence of linked factors in desoxyribonucleate transforming agents. *Proc. Nat. Acad. Sci., Wash.*, **40**, 55 (*610*).

Howard A. & Pelc S. R. (1951). Nuclear incorporation of ^{32}P as demonstrated by autoradiographs. *Exp. Cell. Res.*, **2**, 178 (*389*).

Howard B. D. & Tessman I. (1964a). Identification of the altered bases in mutated single-stranded DNA. II. *In vivo* mutagenesis by 5-bromodeoxyuridine and 2-aminopurine. *J. Mol. Biol.*, **9**, 364 (*318, 510*).

Howard B D. & Tessman I. (1964b). Identification of the altered bases in mutated single-stranded DNA. III. Mutagenesis by ultraviolet light. *J. Mol. Biol.*, **9**, 372 (*337, 570*).

Howard-Flanders P. (1964). In discussion of Setlow R. B. (1964). Physical changes and mutagenesis. *J. Cell. Comp. Physiol.*, **64** (Suppl. 1), 51 (*334*).

Howard-Flanders P. & Boyce R. P. (1966). DNA repair and genetic recombination: studies on mutants of *Escherichia coli* defective in these processes. *Radiation Research Suppl.*, **6**, 156 (*335, 400, 794*).

Howard-Flanders P. & Theriot L. (1962). A method for selecting radiation-sensitive mutants of *Escherichia coli. Genetics*, **47**, 1219 (*332*).

Howard-Flanders P. & Theriot L. (1966). Mutants of *Escherichia coli* K12 defective in DNA repair and in genetic recombination. *Genetics*, **53**, 1137 (*334, 335, 474, 794*).

Howard-Flanders P., Boyce R. P. & Theriot L. (1962). The mechanism of sensitisation to ultra-violet light of T1 bacteriophage by the incorporation of 5-bromodeoxyuridine or by pre-irradiation of the host cell. *Nature, Lond.*, **195**, 51 (*527*).

Howard-Flanders P., Boyce R. P. & Theriot L. (1966). Three loci in *Escherichia coli* K-12 that control the excision of pyrimidine dimers and certain other mutagen products from DNA. *Genetics*, **53**, 1119 (*333, 334*).

Howarth S. (1958). Suppressor mutations in some cysteine-requiring mutants of *Salmonella typhimurium. Genetics*, **43**, 404 (*216*).

Hurwitz J. & Furth J. J. (1962). Messenger RNA. *Scientific American*, **206**, 41 (*278*).

Hurwitz J., Bresler A. & Diringer R. (1960). The enzymic incorporation of ribonucleotides and the effect of DNA. *Biochem. biophys. Res. Comm.*, **3**, 15 (*290*).

Hurwitz J., Evans A., Babinet C. & Skalka A. (1963). On the copying of DNA in the RNA polymerase reaction. *Cold Spr. Harb. Symp. quant. Biol.*, **28**, 59 (*290*).

HURWITZ J., FURTH J. J., ANDERS M., ORITZ P. J. & AUGUST J. T. (1961). The enzymatic incorporation of ribonucleotides into RNA and the role of DNA. *Cold Spr. Harb. Symp. quant. Biol.*, **26**, 91 (*290*).

HUTCHISON C. III, & SINSHEIMER R. (1967). The process of infection with bacteriophage φX174. X. Mutations in a φX lysis gene. *J. Mol. Biol.* (*440*).

HUTCHINSON W. G. & STEMPEN H. (1954). Sex in bacteria: evidence from morphology. In *Sex in Micro-organisms*, p. 29. Amer. Assoc. for the Advancement of Science, Washington, D.C. (*651*).

HUXLEY H. E. & ZUBAY G. (1960). Electron microscope observations on the structure of microsomal particles from *Escherichia coli. J. Mol. Biol.*, **2**, 10 (*277*).

IINO T. (1961a). A stabiliser of antigenic phases in *Salmonella abortus-equi. Genetics*, **46**, 1465 (*742*).

IINO T. (1961b). Genetic analysis of O—H variation in *Salmonella. Japanese J. Genetics*, **36**, 268 (*742*).

IINO T. & LEDERBERG J. (1964). Genetics of *Salmonella*. In *The World problem of Salmonellosis* (ed. van Oye E.). *Monographiae biol.*, **13**, 111 (*744*).

IKEDA H. & TOMIZAWA J. (1965a). Transducing fragments in generalised transduction by phage P1. I. Molecular origin of the fragments. *J. Mol. Biol.*, **14**, 85 (*625, 639, 640*).

IKEDA H. & TOMIZAWA J. (1965b). Transducing fragments in generalised transduction by phage P1. II. Association of DNA and protein in the fragments. *J. Mol. Biol.*, **14**, 110 (*640*).

IKEDA H. & TOMIZAWA J. (1965c). Transducing fragments in generalised transduction by phage P1. III. Studies with small phage particles. *J. Mol. Biol.*, **14**, 120 (*640*).

IMAMOTO F., MORIKAWA N. & SATO K. (1965). On the transcription of the tryptophan operon in *Escherichia coli*. III. Multicistronic messenger RNA and polarity for transcription. *J. Mol. Biol.*, **13**, 169 (*718, 719*).

IMAMOTO F., ITO J. & YANOFSKY C. (1966). Polarity in the tryptophan operon of *E. coli. Cold Spr. Harb. Symp. quant. Biol.*, **31**, 235 (*119, 721*).

INGRAM V. M. (1957). Gene mutations in human haemoglobin: the chemical difference between normal and sickel cell haemoglobin. *Nature, Lond.* (in press) **180**, 326 (*174, 238, 300, 341*).

INGRAM V. M. (1965). *The Biosynthesis of Macromolecules*. W. A. Benjamin, Inc., New York and Amsterdam (*284*).

IPPEN K., MILLER J., SCAIFE J. & BECKWITH J. R. (1968). The identification of a new controlling element in the *lac* operon of *E. coli. Nature, Lond.* (*729*).

ISEKI S. & SAKAI T. (1953). Artificial transformation of *O* antigens in *Salmonella* group E. II. Antigen transforming factor in bacilli of group E_2. *Proc. Japan Acad.*, **29**, 127 (*647*).

ISHIBASHI M., SUGINO Y. & HIROTA Y. (1964). Chromosomal location of thymine and arginine genes in *Escherichia coli* and F′ incorporating them. *J. Bacteriol.*, **87**, 554 (*573*).

ITANO H. A. & ROBINSON E. A. (1960). Genetic control of the α and β-chains of hemoglobin. *Proc. Natl. Acad. Sci., Wash.*, **46**, 1492 (*153*).

Iterson W. van & Robinow C. F. (1961). Observations with the electron microscope on the fine structure of the nuclei of two spherical bacteria. *J. Biophys. Biochem. Cytol.*, **9**, 171 (*552*).

Ivanovics G. (1962). Bacteriocins and bacteriocin-like substances. *Bacteriol. Rev.*, **26**, 108 (*756*).

Iyer V. N. & Szybalski W. (1964). Mitomycin and porfiromycin: chemical mechanisms of activation and cross-linking of DNA. *Science*, **145**, 55 (*334*).

Izaki K., Kiuchi K. & Arima K. (1966). Specificity and mechanism of tetracycline resistance in a multiple drug resistant strain of *Escherichia coli*. *J. Bacteriol.*, **91**, 628 (*766*).

Jackson S. (1962). Genetic aspects of capsule formation in the pneumococcus. *Brit. Med. Bull.*, **18**, 24 (*610*, *613*).

Jackson S., MacLeod C. M. & Krauss M. R. (1959). Determination of type in capsulated transformants of pneumococcus by the genome of non-capsulated donor and recipient strains. *J. Expl. Med.*, **109**, 429 (*612*).

Jacob F. (1955). Transduction of lysogeny in *Escherichia coli*. *Virology*, **1**, 207 (*464*, *621*, *625*, *626*).

Jacob F. & Adelberg E. A. (1959). Transfert de caractères génétiques par incorporation au facteur sexuel d'*Escherichia coli*. *C. R. Acad. Sci., Paris*, **249**, 189 (*677*).

Jacob F. & Brenner S. (1963). Sur la régulation de la synthèse du DNA chez les bactéries: l'hypothèse du réplicon. *C. R. Acad. Sci., Paris*, **256**, 298 (*735*, *736*).

Jacob F. & Campbell A. (1959). Sur le système de répression assurant l'immunité chez les bactéries lysogènes. *C. R. Acad. Sci., Paris*, **248**, 3219 (*467*).

Jacob F. & Fuerst C. R. (1958). The mechanism of lysis by phage studied with defective lysogenic bacteria. *J. gen. Microbiol.*, **18**, 518 (*435*).

Jacob F. & Monod J. (1961a). Genetic regulatory mechanisms in the synthesis of proteins. *J. Mol. Biol.*, **3**, 318 (*104*, *166*, *275*, *715*, *716*, *718*).

Jacob F. & Monod J. (1961b). On the regulation of gene activity. *Cold Spr. Harb. Symp. quant. Biol.*, **26**, 193 (*104*, *156*, *166*, *712*, *714*, *715*, *718*, *719*, *720*, *722*, *724*).

Jacob F. & Monod J. (1963a). Elements of regulatory circuits in bacteria. *In* Unesco Symposium on *Biological Organisation* (ed. Harris R. J. C.). Academic Press, Inc., New York (*708*, *722*, *733*, *735*, *736*, *737*).

Jacob F. & Monod J. (1963b). Genetic repression, allosteric inhibition and cellular differentiation. In *Cytodifferentiation and Macromolecular Synthesis*, 21st Growth Symposium, p. 30. Academic Press, Inc., New York (*735*).

Jacob F. & Monod J. (1965). Genetic mapping of the elements of the lactose region in *Escherichia coli*. *Biochem. biophys. Res. Comm.*, **18**, 693 (*692*, *727*).

Jacob F. & Wollman E. L. (1953). Induction of phage development in lysogenic bacteria. *Cold Spr. Harb. Symp. quant. Biol.*, **18**, 101 (*452*, *453*).

Jacob F. & Wollman E. L. (1954). Étude génétique d'un bactériophage tempéré d'*Escherichia coli*: I. Le système génétique du bactériophage λ. *Ann. Inst. Pasteur*, **87**, 653 (*456*, *468*, *481*, *482*).

Jacob F. & Wollman E. L. (1956a). Sur les processus de conjugaison et de recombinaison génétique chez *E. coli*: I. L'induction par conjugaison ou induction zygotique. *Ann. Inst. Pasteur*, **91**, 486 (*452, 464, 662*).

Jacob F. & Wollman E. L. (1956b). Recombinaison génétique et mutants de fertilité chez *E. coli* K-12. *C. R. Acad. Sci., Paris*, **242**, 303 (*664*).

Jacob F. & Wollman E. L. (1957). Genetic aspects of lysogeny. In *The Chemical Basis of Heredity* (eds. McElroy W. D. & Glass H. B.), p. 468. Johns Hopkins Press, Baltimore (*452, 454, 456*).

Jacob F. & Wollman E. L. (1958a). Sur les processus de conjugaison et de recombinaison chez *E. coli*. IV. Prophages inductibles et mesure des segments génétiques transférés au cours de la conjugaison. *Ann. Inst. Pasteur*, **95**, 497 (*663, 665*).

Jacob F. & Wollman E. L. (1958b). Genetic and physical determinations of chromosomal segments in *E. coli*. *Symp. Soc. exp. Biol.*, **12**, 75 (*665, 667, 694*).

Jacob F. & Wollman E. L. (1958c). Les épisomes, éléments génétiques ajoutés. *C. R. Acad. Sci., Paris*, **247**, 154 (*222, 747*).

Jacob F. & Wollman E. L. (1959). Lysogeny. In *The Viruses* (eds. Burnet F. N. & Stanley W. M.), vol. 2, p. 319. Academic Press, New York (*410, 430, 451, 456, 468*).

Jacob F. & Wollman E. L. (1961a). *Sexuality and the Genetics of Bacteria.* Academic Press, New York (*145, 218, 456, 468, 470, 478, 636, 657, 662, 665, 667, 681, 683, 684, 686, 692, 693, 694, 695, 704, 714, 808*).

Jacob F. & Wollman E. L. (1961b). Viruses and genes. *Scientific American*, *204*, 92 (*432*).

Jacob F., Brenner S. & Cuzin F. (1963). On the regulation of DNA replication in bacteria. *Cold Spr. Harb. Symp. quant. Biol.*, **28**, 329 (*256, 569, 570, 572, 573, 735, 778, 788, 794, 799, 801, 804*).

Jacob F.. Fuerst C. R. & Wollman E. L. (1957). Recherches sur les bactéries lysogènes défectives. II. Les types physiologiques liés aux mutations du prophage. *Ann. Inst. Pasteur*, **93**, 724 (*474, 502*).

Jacob F., Ryter A. & Cuzin F. (1966). On the association between DNA and the membrane in bacteria. *Proc. Roy. Soc. B.*, **164**, 267 (*256, 568, 569, 571, 573*).

Jacob F., Schaeffer P. & Wollman E. L. (1960). Episomic elements in bacteria. *Symp. Soc. gen. Microbiol.*, **10**, 67 (*104, 409, 677, 776, 794, 797, 808*).

Jacob F., Siminovitch L. & Wollman E. L. (1953). Comparaison entre la biosynthèse induite de la colicine et des bactériophages et entre leur mode d'action. *Ann. Inst. Pasteur*, **84**, 313 (*756*).

Jacob F., Sussman R. & Monod J. (1962). Sur la nature du répresseur assurant l'immunité des bactéries lysogènes. *C. R. Acad. Sci., Paris*, **254**, 4214 (*468, 733*).

Jacob F., Ullman A. & Monod J. (1964). Le promoteur, élément génétique nécessaire à l'expression d'un opéron. *C. R. Acad. Sci., Paris*, **258**, 3125 (*727, 728*).

Jacob F., Ullman A. & Monod J. (1965). Délétions fusionnant l'opéron lactose et un opéron purine chez *Escherichia coli*. *J. Mol. Biol.*, **13**, 704 (*725*).

JACOB F., PERRIN D., SANCHEZ C. & MONOD J. (1960). L'opéron: groupe de gènes à expression coordonné par un opérateur. *C. R. Acad. Sci., Paris,* **250**, 1727 (*716, 717*).

JESSOP A. P. & CATCHESIDE D. G. (1965). Interallelic recombination at the *his*-1 locus in *Neurospora crassa* and its genetic control. *Heredity,* **20**, 237 (*402*).

JOHNSON L. N. & PHILLIPS D. C. (1965). Structure of some crystalline lysozymes—inhibition complexes determined by x-ray analysis at 6Å resolution. *Nature, Lond.,* **206**, 761 (*265*).

JORDAN E. (1965). The location of the b_2 deletion of bacteriophage lambda. *J. Mol. Biol.,* **10**, 341 (*471*).

JOSET F., MOUSTACCHI E. & MARCOVICH H. (1966). Approaches to the determinisation of the initial sites of action of radiations in *Escherichia coli* and yeast. In *Current topics in radiation research* (eds. Ebert M. & Howard A.), vol. II, p. 249. North Holland Publishing Co., Amsterdam (*338*).

JOSSE J., KAISER A. D. & KORNBERG J. (1961). Enzymatic synthesis of deoxyribonucleic acid. VIII. Frequencies of nearest neighbor base sequences in deoxyribonucleic acid. *J. Biol. Chem.,* **236**, 864 (*250*).

JOYNER A., ISAACS L. N., ECHOLS H. & SLY W. S. (1966). DNA replication and messenger RNA production after induction of wild-type λ bacteriophage and λ mutants. *J. Mol. Biol.,* **19**, 174 (*475*).

KAERNER H. C. & HOFFMAN-BERLING (1964). Synthesis of double-stranded RNA in RNA-phage infected *E. coli* cells. *Nature, Lond.,* **202**, 1012 (*443*).

KÄFER E. (1960). High frequency of spontaneous and induced somatic segregation in *Aspergillus nidulans. Nature, Lond.,* **186**, 619 (*82*).

KAHN P. & HELINSKI D. R. (1964). Relationship between colicinogenic factors E1 and V and an F factor in *Escherichia coli. J. Bacteriol.,* **88**, 1573 (*769, 780*).

KAISER A. D. (1955). A genetic study of the temperate coliphage λ. *Virology,* **1**, 424 (*456, 509*).

KAISER A. D. (1957). Mutations in a temperate bacteriophage affecting its ability to lysogenise *E. coli. Virology,* **3**, 42 (*457, 466*).

KAISER A. D. (1962). The production of phage chromosome fragments and their capacity for genetic transfer. *J. Mol. Biol.,* **4**, 275 (*544*).

KAISER A. D. (1966). On the physical basis of genetic structure in bacteriophage. In *Phage and the origins of Molecular Biology* (eds. Cairns J., Stent G. S. & Watson J. D.), p. 150. Cold Spring Harbor Laboratory of Quantitative Biology, Long Island, New York (*544*).

KAISER A. D. & INMAN R. B. (1965). Cohesion and the biological activity of bacteriophage lambda DNA. *J. Mol. Biol.,* **13**, 78 (*544*).

KAISER A. D. & JACOB F. (1957). Recombination between related temperate bacteriophages and the genetic control of immunity and prophage localisation. *Virology,* **4**, 509 (*457, 458*).

KAISER A. D. & HOGNESS D. S. (1960). The transformation of *Escherichia coli* with deoxyribonucleic acid isolated from bacteriophage λ*dg*. *J. Mol. Biol.,* **2**, 392 (*463, 544, 587*).

KAJI A. & KAJI H. (1963). Specific interaction of soluble RNA with polyribonucleic acid induced polysomes. *Biochem. biophys. Res. Comm.*, **13**, 186 (*286*).

KALCKAR H. M. (1957). Biochemical mutations in man and microorganisms. *Science*, **125**, 105 (*110*).

KALCKAR H. M., KURAHASHI K. & JORDAN E. (1959). Hereditary defects in galactose metabolism. I. Determination of enzyme activities. *Proc. Natl. Acad. Sci., Wash.*, **45**, 1776 (*711*).

KALLEN R. G., SIMON M. & MARMUR J. (1962). The occurrence of a new pyrimidine base replacing thymine in a bacteriophage DNA. *J. Mol. Biol.*, **5**, 248 (*248*, *272*).

KAMEDA M., HARADA K., SUZUKI M. & MITSUHASHI S. (1965). Drug resistance of enteric bacteria. V. High frequency of transduction of R factors with bacteriophage epsilon. *J. Bacteriol.*, **90**, 1174 (*781*).

KAPLAN S., STRETTON A. O. W. & BRENNER S. (1965). *Amber* suppressors: efficiency of chain propagation and suppressor-specific amino acids. *J. Mol. Biol.*, **14**, 528 (*364*).

KARAMATA D., KELLENBERGER E., KELLENBERGER G. & TERZI M. (1962). Sur une particule accompagnant le développement du coliphage λ. *Pathol. Microbiol.*, (*Suisse*), **25**, 575 (*432*).

KAUFMANN F. (1941). A typhoid variant and a new serological variation in the *Salmonella* group. *J. Bacteriol.*, **41**, 127 (*745*).

KEIR H. M., BINNIE B. & SMELLIE R. M. S. (1962). Factors affecting the primer for deoxyribonucleic acid polymerase. *Biochem. J.*, **82**, 493 (*247*, *253*)

KELLENBERGER E. (1960). The physical state of the bacterial nucleus. *Symp. Soc. gen. Microbiol.*, **10**, 39 (*546*, *547*, *552*, *553*).

KELLENBERGER E. (1961). Vegetative bacteriophage and the maturation of the virus particles. *Advanc. Virus Res.*, **8**, 1 (*410*, *418*, *419*, *420*, *428*, *431*, *435*).

KELLENBERGER E. (1966a). Control mechanisms of bacteriophage morphopoiesis. In *Principles of Biomolecular Organisation* (Ciba Symposium, eds. Wolstenholme G. E. W. & O'Connor M.), p. 192. J. & A. Churchill, London (*433*, *434*, *492*).

KELLENBERGER E. (1966b). The genetic control of the shape of a virus. *Scientific Amer.*, **215**, 32 (*416*, *433*, *434*, *492*).

KELLENBERGER E. & ARBER W. (1955). Die Struktur des Schwarzes der Phagen T2 and T4 und der Mechanismus der irreversiblen Adsorption. *Z. Naturforsch.*, **10b**, 698 (*417*).

KELLENBERGER E. & ARBER W. (1957). Electron microscopical studies of phage multiplication. I. A method for quantitative analysis of particle suspensions. *Virology*, **3**, 245 (*411*).

KELLENBERGER E. & SECHAUD J. (1957). Electron microscopical studies of phage multiplication. II. Production of phage-related structures during multiplication of phages T2 and T4. *Virology*, **3**, 256 (*422*).

KELLENBERGER E., RYTER A. & SECHAUD J. (1958). Electron microscope study of DNA-containing plasms. II. Vegetative and mature phage DNA as compared with normal bacterial nucleoids in different physiological states. *J. biophys. biochem. Cytol.*, **4**, 671 (*552*, *553*).

KELLENBERGER E., SECHAUD J. & RYTER A. (1959). Electron microscopical studies of phage multiplication. IV. The establishment of the DNA-pool of vegetative phage and the maturation of phage particles. *Virology*, **8**, 478 (*431*).

KELLENBERGER E., BOLLE A., BOY DE LA TOUR E., EPSTEIN R. H., FRANKLIN N. C., JERNE N. K., REALE-SCAFATI A., SECHAUD J., BENDET I., GOLDSTEIN D. & LAUFFER N. R. (1965). Functions and properties related to the tail fibres of bacteriophage T4. *Virology*, **26**, 419 (*422*).

KELLENBERGER G., SYMONDS N. & ARBER W. (1966). Host specificity of DNA produced by *Escherichia coli*. 8. Its acquisition by phage λ and its persistence through consecutive growth cycles. *Z. Vererbungsl.*, **98**, 247 (*519*).

KELLENBERGER G., ZICHICHI M. L. & EPSTEIN H. T. (1962). Heterozygosis and recombination of bacteriophage λ. *Virology*, **17**, 44 (*499*).

KELLENBERGER G., ZICHICHI M. L. & WEIGLE J. (1961a). Exchange of DNA in the recombination of bacteriophage lambda. *Proc. Natl. Acad. Sci., Wash.*, **47**, 869 (*394*).

KELLENBERG G., ZICHICHI M. L. & WEIGLE J. (1961b). A mutation affecting the DNA content of bacteriophage lambda and its lysogenising abilities. *J. Mol. Biol.*, **3**, 399 (*218*, *470*).

KELLY M. S. & PRITCHARD R. H. (1965). Unstable linkage between genetic markers in transformation. *J. Bacteriol.*, **89**, 1314 (*592*, *611*, *614*, *616*).

KELNER A. (1949). Photoreactivation of ultraviolet-irradiated *Escherichia coli* with special reference to the dose-reduction principle and to ultraviolet induced mutations. *J. Bacteriol.*, **58**, 511 (*205*, *331*).

KELNER A. (1953). Growth, respiration and nucleic acid synthesis in ultraviolet-irradiated and in photoreactivated *Escherichia coli*. *J. Bacteriol.*, **65**, 252 (*209*, *329*).

KENDREW J. C. (1960). The structure of globular proteins. In *Biological Structure and Function* (eds. Goodwin T. W. & Lindberg O.), vol. 1, p. 5. Academic Press, Inc., London and New York (*152*, *262*).

KEPES A. (1967). Sequential transcription and translation in the lactose operon of *Escherichia coli*. *Biochim. biophys. Acta*, **138**, 107 (*719*).

KERRIDGE D. (1961). The effect of environment on the formation of bacterial flagella. *Symp. Soc. gen. Microbiol.*, **11**, 41 (*259*, *738*).

KIHO Y. & RICH A. (1964). Induced enzyme formed on bacterial polyribosomes. *Proc. Natl. Acad. Sci., Wash.*, **51**, 111 (*294*).

KIHO Y. & RICH A. (1965). A polycistronic RNA associated with β-galactosidase induction. *Proc. Natl. Acad. Sci., Wash.*, **54**, 1751 (*721*).

KIRCHNER C. E. J. (1960). The effects of the mutator gene on molecular changes and mutation in *Salmonella typhimurium*. *J. Mol. Biol.*, **2**, 331 (*219*).

KIRCHNER C. E. J. & RUDDEN M. J. (1966). Location of a mutator gene in *Salmonella typhimurium* by transduction. *J. Bacteriol.*, **92**, 1453 (*219*).

KLEIN A. & SAUERBIER W. (1965). Role of methylation in host-controlled modification of phage T1. *Biochem. biophys. Res. Comm.*, **18**, 440 (520).

KLEIN R. (1964). A hypothesis on the genetic mechanism governing phase variation in *Salmonella*. *Z. Vererbl.*, **95**, 167 (*744*).

KLEINSCHMIDT A., LANG D. & ZAHN R. K. (1961). Über die intrazellulare Formation von Bakterien-DNS. *Z. Naturforsch.*, **16b**, 730 (*544, 551*).

KLEINSCHMIDT A. K., LANG D., JACHERTS D. & ZAHN R. K. (1962). Darstellung und Langenmessungen des gesamten Desoxyribonucleinsäureinhaltes von T2-Bakteriophagen. *Biochim. biophys. Acta*, **61**, 857 (*420, 537, 544*).

KNOX W. E. (1967). Sir Archibald E. Garrod, 1857-1936. *Genetics*, **56**, 1 (*108*).

KNOX W. E. & BEHRMAN E. J. (1959). Amino acid metabolism. *Ann. Rev. Biochem.*, **28**, 223 (*122*).

KOCH G. & HERSHEY A. D. (1959). Synthesis of phage-precursor protein in bacteria infected with T2. *J. Mol. Biol.*, **1**, 260 (*428, 431, 494*).

KOCH G. & WEIDEL W. (1956). Abspaltung chemischer Komponenten der *Coli*-Membran durch daran adsorbierte Phagen. I. Mitt.: Allgemeine Charakterisierung des Effekts und Partialanalyse einer der abgespaltenen Komponenten. *Z. Naturforsch.*, **11b**, 345 (423).

KOGUT M., POLLOCK M. & TRIDGELL E. J. (1956). Purification of penicillin-induced penicillinase of *Bacillus cereus* NRRL569. A comparison of its properties with those of a similarly purified penicillinase produced spontaneously by a constitutive mutant strain. *Biochem. J.*, **62**, 391 (*711, 712*).

KOHOUTOVÁ M. (1965). Infection of the recipient cell by transforming DNA. The stimulation and inhibition of infection. *Symp. Biol., Hung.*, **6**, 65 (*589*).

KOHOUTOVÁ M. & MÁLEK I. (1966). Stimulation of transformation frequency by sterile filtrates from Pneumococcus (I). Competence-inducing ability of sterile filtrates from Pneumococcus (II). In *The physiology of gene and mutation expression* (eds. Kohoutová, M. & Hubáček J.), p. 195. Academia, Prague (*589, 590*).

KOIBONG L. I., BARKSDALE L. & GARMISE L. (1961). Phenotypic alterations associated with the bacteriophage carrier state of *Shigella dysenteriae. J. gen. Microbiol.*, **24**, 355 (*449, 450*).

KONDO E. & MITSUHASHI S. (1964). Active transducing bacteriophage P1CM produced by the combination of R factor with bacteriophage P1. *J. Bacteriol.*, **88**, 1266 (*781*).

KONTOMICHALOU P. (1967). Studies on resistance transfer factors. II. Transmissible resistance to eight antibacterial drugs in a strain of *Escherichia coli. Path. Microbiol., Lausanne*, **30**, 185 (*761*).

KORN D. & WISSBACH A. (1962). Thymineless induction of *Escherichia coli* K12 (λ). *Biochim. biophys. Acta*, **61**, 775 (*451, 473*).

KORNBERG A. (1957). Pathways of enzymatic synthesis of nucleotides and polynucleotides. In *The Chemical Basis of Heredity* (eds. McElroy W. D. & Glass B.), p. 579. Johns Hopkins Press, Baltimore (*246*).

KORNBERG A. (1960). Biologic synthesis of deoxyribonucleic acid. *Science*, **131**, 1503 (*246, 250*).

KORNBERG A. (1961). Enzymatic synthesis of DNA. *Ciba Lecture in Microbial Chemistry*. John Wiley and Sons, New York (*246, 249*).

KORNBERG A., ZIMMERMAN S. B., KORNBERG S. R. & JOSSE J. (1959). Enzymatic synthesis of deoxyribonucleic acid. VI. Influence of bacteriophage T2 on the synthetic pathway in host cells. *Proc. Natl. Acad. Sci., Wash.*, **45**, 772 (*428*).

KOZINSKI A. W. (1961a). Fragmentary transfer of P^{32}-labelled parental DNA to progeny phage. *Virology*, **13**, 124 (*439*).

KOZINSKI A. W. (1961b). Uniform sensitivity to P^{32} decay among progeny of P^{32}-free phage φX174 grown on P^{32}-labelled bacteria. *Virology*, **13**, 377 (*545*).

KOZLOFF L. M. & HENDERSON K. (1955). Action of complexes of the zinc group metals on the tail protein of bacteriophage T2*r*$^{+}$. *Nature, Lond.*, **176**, 1169 (*417*).

KOZLOFF L. M. & LUTE M. (1957). Viral invasion. II. The role of zinc in bacteriophage invasion. *J. Biol. Chem.*, **228**, 529 (*422*).

KOZLOFF L. M., LUTE M. & HENDERSON K. (1957). Viral invasion. I. Rupture of thiol ester bonds in the bacteriophage tail. *J. Biol. Chem.*, **228**, 511 (*420*).

KRIEG D. R. (1959). Induced reversion of T4r_{II} mutants by ultraviolet irradiation of extracellular phage. *Virology*, **9**, 215 (*337*).

KURASHI K. & WAHBE A. J. (1958). Interference with growth of certain *Escherichia coli* mutants by galactose. *Biochim. biophys. Acta*, **30**, 298 (*110*).

KURLAND C. G. (1960). Molecular characterisation of ribonucleic acid from *Escherichia coli* ribosomes. I. Isolation and molecular weights. *J. Mol. Biol.*, **2**, 83 (*277*).

LACKS S. (1962). Molecular fate of DNA in genetic transformation of *Pneumococcus*. *J. Mol. Biol.*, **5**, 119 (*598*).

LACKS S. (1966). Integration efficiency and genetic recombination in pneumococcal transformation. *Genetics*, **53**, 207 (*617*).

LACKS S. & HOTCHKISS R. D. (1960). Formation of amylomaltase after genetic transformation of *Pneumococcus*. *Biochim. biophys. Acta*, **45**, 155 (*617*).

LACOUR L. F. & PELC S. R. (1958). Effect of colchicine on the utilisation of labelled thymidine during chromosomal reproduction. *Nature, Lond.*, **182**, 506 (*245*).

LACOUR L. F. & PELC S. R. (1959). Effect of colchicine on the utilisation of thymidine labelled with tritium during chromosome reproduction. *Nature, Lond.*, **183**, 1455 (*245*).

LANGRIDGE R., WILSON H. R., HOOPER C. W., WILKINS M. F. H. & HAMILTON L. D. (1960a). The molecular configuration of deoxyribonucleic acid. I. X-ray diffraction study of a crystalline form of the lithium salt. *J. Mol. Biol.*, **2**, 19 (*236*).

LANGRIDGE R., MARVIN D. A., SEEDS W. E., WILSON H. R., HOOPER C. W., WILKINS M. F. H. & HAMILTON L. D. (1960b). The molecular configuration of deoxyribonucleic acid. II. Molecular models and their Fourier transforms. *J. Mol. Biol.*, **2**, 38 (*236*).

LANNI Y. T. & MCCORQUODALE D. J. (1963). DNA metabolism in T5-infected *Escherichia coli*: biochemical function of a presumptive genetic fragment of the phage. *Virology*, **19**, 72 (*542*).

LANNI F. & LANNI Y. T. (1953). Antigenic structure of bacteriophage. *Cold Spr. Harb. Symp. quant. Biol.*, **18**, 159 (*420*).

LANNI Y. T., MCCORQUODALE J. & WILSON C. M. (1964). Molecular aspects of DNA transfer from phage T5 to host cells. II. Origin of first-step-transfer DNA fragments. *J. Mol. Biol.*, **10**, 19 (*543*).

LARK K. G. (1966). Regulation of chromosome replication and segregation in bacteria. *Bacteriol. Rev.*, **30**, 3 (*559, 567, 573*).

LARK K. G. & BIRD R. (1965a). Premature chromosome replication induced by thymine starvation: restriction of replication to one of the two partially completed replicas. *J. Mol. Biol.*, **13**, 607 (*559*).

LARK K. G. & BIRD R. (1965b). Segregation of the conserved units of DNA in *Escherichia coli. Proc. Natl. Acad. Sci., Wash.*, **54**, 1444 (*573*).

LARK K. G. & LARK C. (1965). Regulation of chromosome replication in *Escherichia coli*: alternate replication of two chromosomes at slow growth rates. *J. Mol. Biol.*, **13**, 105 (*568*).

LARK K. G., REPKO T. & HOFFMAN E. J. (1963). The effect of amino acid deprivation on subsequent DNA replication. *Biochim. biophys. Acta*, **76**, 9 (*557, 559*).

LAVALLE R. & JACOB F. (1961). Sur la sensibilité des épisomes sexuels et colicinogènes d'*E. coli* K-12 à la désintégration du radiophosphore. *C. R. Acad. Sci., Paris*, **252**, 1678 (*656*).

LAWN A. M., MEYNELL E., MEYNELL G. G. & DATTA N. (1967). Sex pili and the classification of sex factors in the *Enterobacteriaceae. Nature, Lond.*, **216**, 343 (*774*).

LEBEK G. (1963). Über die Enstehung mehrfachresistanter Salmonellen. Ein experimenteller Beitrag. *Zbl. Bakt., Abt. 1, Orig.*, **188**, 494 (*761*).

LEDERBERG E. M. (1951). Lysogenicity in *E. coli* K-12. *Genetics*, **36**, 560 (*448*).

LEDERBERG E. M. (1960). Genetic and functional aspects of galactose metabolism in *Escherichia coli* K-12. *Symp. Soc. gen. Microbiol.*, **10**, 115 (*101, 104, 110, 631, 633, 677*).

LEDERBERG E. M. & LEDERBERG J. (1953). Genetic studies of lysogenicity in interstrain crosses in *E. coli. Genetics*, **38**, 51 (*100, 454*).

LEDERBERG J. (1947). Gene recombination and linked segregations in *Escherichia coli. Genetics*, **32**, 505 (*60, 652*).

LEDERBERG J. (1949). Aberrant heterozygotes in *Escherichia coli. Proc. Natl. Acad. Sci., Wash.*, **35**, 178 (*203, 652, 703, 737*).

LEDERBERG J. (1950). Isolation and characterisation of biochemical mutants of bacteria. In *Methods in Medical Research* (ed. Comrie J. H., Jnr.), vol. 3, p. 5. Year Book Publishers, Chicago (*211*).

LEDERBERG J. (1951a). Streptomycin resistance: a genetically recessive mutation. *J. Bacteriol.*, **61**, 549 (*203, 210, 652*).

LEDERBERG J. (1951b). Prevalence of *E. coli* strains exhibiting genetic recombination. *Science*, **114**, 68 (*749*).

LEDERBERG J. (1952). Cell genetics and hereditary symbiosis. *Physiol. Rev.*, **32**, 403 (*747*).

LEDERBERG J. (1955). Recombination mechanisms in bacteria. *J. cell. comp. Physiol.*, **45** (suppl. 2), 75 (*377, 388, 394*).

LEDERBERG J. (1956a). Linear inheritance in transductional clones. *Genetics*, **41**, 845 (*97*).
LEDERBERG J. (1956b). Bacterial protoplasts induced by penicillin. *Proc. Natl. Acad. Sci., Wash.*, **42**, 574 (*212*).
LEDERBERG J. (1956c). Genetic transduction. *American Scientist*, **44**, 264 (*623*, *649*).
LEDERBERG J. (1957). Sibling recombinants in zygote pedigrees of *Escherichia coli. Proc. Natl. Acad. Sci., Wash.*, **43**, 1060 (*654*, *694*).
LEDERBERG J. (1958). Extracellular transmission of the F compatibility factor in *E. coli. Abstr. Communs. 7th Intern. Congr. Microbiol.*, p. 59. Stockholm. Almqvist & Wiksell (*656*).
LEDERBERG J. & EDWARDS P. R. (1953). Serotypic recombination in *Salmonella. J. Immunol.*, **71**, 232 (*739*).
LEDERBERG J. & IINO T. (1956). Phase variation in *Salmonella. Genetics*, **41**, 743 (*740*, *741*).
LEDERBERG J. & LEDERBERG E. M. (1952). Replica plating and indirect selection of bacterial mutants. *J. Bacteriol.*, **63**, 399 (*187*, *188*, *190*, *664*).
LEDERBERG J. & LEDERBERG E. M. (1956). Infection and heredity. In *Cellular Mechanisms in Differentiation and Growth* (ed. Rudnick R.), p. 101. Princeton Univ. Press, Princeton, N.J. (*656*).
LEDERBERG J. & TATUM E. L. (1946a). Novel genotypes in mixed cultures of biochemical mutants of bacteria. *Cold Spr. Harb. Symp. quant. Biol.*, **11**, 113 (*620*, *651*, *652*).
LEDERBERG J. & TATUM E. L. (1946b). Gene recombination in *E. coli. Nature, Lond.*, **158**, 558 (*620*, *651*, *652*).
LEDERBERG J. & TATUM E. L. (1953). Sex in bacteria; genetic studies, 1945–1952. *Science*, **118**, 169 (*749*).
LEDERBERG J., CAVALLI L. L. & LEDERBERG E. M. (1952). Sex compatability in *E. coli. Genetics*, **37**, 720 (*654*, *655*, *658*).
LEDERBERG J., LEDERBERG E. M., ZINDER N. D. & LIVELY E. R. (1951). Recombination analysis of bacterial heredity. *Cold Spr. Harb. Symp. quant. Biol.*, **16**, 413 (*620*, *653*).
LEDERBERG S. (1957). Suppression of the multiplication of heterologous bacteriophages in lysogenic bacteria. *Virology*, **3**, 496 (*515*, *516*).
LENGYEL P., SPEYER J. F. & OCHOA S. (1961). Synthetic polynucleotides and the amino acid code. I. *Proc. Natl. Acad. Sci., Wash.*, **47**, 1936 (*356*).
LENNOX E. S. (1955). Transduction of linked genetic characters of the host by bacteriophage P1. *Virology*, **1**, 190 (*384*, *621*, *625*, *626*, *637*, *646*).
LERMAN L. S. (1961). Structural considerations in the interactions of DNA and acridines. *J. Mol. Biol.*, **3**, 18 (*322*).
LERMAN L. S. (1963). The structure of the DNA-acridine complex. *Proc. Natl. Acad. Sci., Wash.*, **49**, 94 (*322*).
LERMAN L. S. & TOLMACH L. J. (1957). Genetic transformation. I. Cellular incorporation of DNA accompanying transformation in *Pneumococcus. Biochim. biophys. Acta*, **26**, 68 (*575*, *580*, *588*, *591*, *593*, *594*, *597*).
LERMAN L. S. & TOLMACH L. J. (1959). Genetic transformation. II. The significance of damage to the DNA molecule. *Biochim. biophys. Acta*, **33**, 371 (*588*, *593*).

LESTER G. & YANOFSKY C. (1961). Influence of 3-methylanthranilic and anthranilic acids on the formation of tryptophan synthetase in *Escherichia coli*. *J. Bacteriol.*, **81**, 81 (*722*).

LEUPOLD U. (1958). Studies on recombination in *Schizosaccharomyces pombe*. *Cold Spr. Harb. Symp. quant. Biol.*, **23**, 161 (*130*).

LEVINE M. (1957). Mutations in the temperate phage P22 and lysogeny in *Salmonella*. *Virology*, **3**, 22 (*457*).

LEVINTHAL C. (1954). Recombination in phage T2: its relationship to heterozygosis and growth. *Genetics*, **39**, 169 (*388, 394, 497, 498*).

LEVINTHAL C. (1959). Bacteriophage genetics. In *The Viruses* (eds. Burnet F. M. & Stanley W. M.), vol. 2, p. 281. Academic Press, New York (*410, 502, 509*).

LEVINTHAL C. & VISCONTI N. (1953). Growth and recombination in bacterial viruses. *Genetics*, **38**, 500 (*507*).

LEVINTHAL C., HOSODA J. & SHUB D. (1967). The control of protein synthesis after phage infection. In *Molecular Biology of Viruses* (ed. Colter J.). Academic Press, New York (*492, 493, 494*).

LEWIS E. B. (1951). Pseudoallelism and gene evolution. *Cold Spr. Harb. Symp. quant. Biol.*, **16**, 159 (*128*).

LEWIS I. M. (1934). Bacterial variation with special reference to behaviour of some mutable strains of colon bacteria in synthetic media. *J. Bacteriol.*, **28**, 619 (*180*).

LHOAS P. (1961). Mitotic haploidization by treatment of *Aspergillus niger* diploids with *para*-fluorophenylalanine. *Nature, Lond.*, **190**, 744 (*79, 83*).

LIEB M. (1953a). The establishment of lysogeny in *E. coli*. *J. Bacteriol.*, **65**, 642 (*409, 468*).

LIEB M. (1953b). Studies on lysogenisation in *Escherichia coli*. *Cold Spr. Harb. Symp. quant. Biol.*, **18**, 71 (*409, 468*).

LIEB M. (1966). Studies of heat-inducible λ phage. III. Mutations in cistron N affecting heat induction. *Genetics*, **54**, 835 (*474*).

LIEB M., WEIGLE J. J. & KELLENBERGER E. (1955). A study of hybrids between two strains of *E. coli*. *J. Bacteriol.*, **69**, 468 (*749*).

LISSOUBA P., MOUSSEAU J., RIZET G. & ROSSIGNOL J. L. (1962). Fine structure of genes in the ascomycete *Ascobolus immersus*. *Adv. Genetics*, **11**, 343 (*398, 400*).

LITMAN R. & EPHRUSSI-TAYLOR H. (1959). Inactivation et mutation des facteurs génétiques de l'acide desoxyribonucléique du pneumocoque par l'UV et par l'acide nitreux. *C. R. Acad. Sci., Paris*, **249**, 838 (*305, 337*).

LITT M., MARMUR J., EPHRUSSI-TAYLOR H. E. & DOTY P. (1958). The dependence of pneumococcal transformation on the molecular weight of deoxyribose nucleic acid. *Proc. Natl. Acad. Sci., Wash.*, **44**, 144 (*588, 592*).

LITTLEFIELD J. W., KELLER E. B., GROSS G. & ZAMECNIK P. (1955). Studies on cytoplasmic ribonucleoprotein particles from the liver of the rat. *J. biol. Chem.*, **217**, 111 (*270*).

LODISH H. F. & ZINDER N. D. (1966a). Replication of the RNA of bacteriophage f2. *Science*, **152**, 372 (*443*).

LODISH H. F. & ZINDER N. D. (1966b). Mutants of the bacteriophage f2. VIII. Control mechanisms for phage-specific syntheses. *J. Mol. Biol.*, **19**, 333 (*444*).

LODISH H. F., COOPER S. & ZINDER N. D. (1964). Host-dependent mutants of the bacteriophage f2. IV. On the biosynthesis of a viral RNA polymerase. *Virology*, **24**, 60. (*444*).

LOEB T. (1960). Isolation of a bacteriophage for the F^+ and Hfr mating types of *Escherichia coli* K-12. *Science*, **131**, 932 (*440, 444, 670*).

LOEB T. & ZINDER N. D. (1961). A bacteriophage containing RNA. *Proc. Natl. Acad. Sci., Wash.*, **47**, 282 (*419, 440, 670*).

LOPER J. C. (1961). Enzyme complementation in mixed extracts of mutants from the *Salmonella* histidine B locus. *Proc. Natl. Acad. Sci., Wash.*, **47**, 1440 (*155*).

LOPER J. C., GRABNER M., STAHL R. C., HARTMAN Z. & HARTMAN P. E. (1964). Genes and proteins involved in histidine biosynthesis in *Salmonella*. *Brookhaven Symp. Biol.*, **17**, 15 (*154, 160, 161, 162, 163*).

LOUTIT J. S. (1958). A transduction-like process within a single strain of *Pseudomonas aeruginosa*. *J. gen. Microbiol.*, **18**, 315 (*621*).

LOUTIT J. S. & PEARCE L. E. (1965). Mating in *Pseudomonas aeruginosa*. *Nature, Lond.*, **205**, 822 (*754*).

LOVELESS A. (1958). Increased rate of plaque-type and host-range mutation following treatment of bacteriophage *in vitro* with ethyl methane sulphonate. *Nature, Lond.*, **181**, 1212 (*307*).

LOVELESS A. & HOWARTH S. (1959). Mutation of bacteria at high levels of survival by ethyl methane sulphonate. *Nature, Lond.*, **184**, 1780 (*205, 307*).

LOW B. (1965). Low recombination frequency for markers very near the origin in conjugation in *E. coli*. *Genet. Res., Camb.*, **6**, 469 (*688*).

LURIA S. E. (1947a). Recent advances in bacterial genetics. *Bacteriol. Rev.*, **11**, 1 (*180, 651*).

LURIA S. E. (1947b). Reactivation of irradiated bacteriophage by transfer of self-reproducing units. *Proc. Natl. Acad. Sci., Wash.*, **33**, 253 (*528, 529*).

LURIA S. E. (1951). The frequency distribution of spontaneous bacteriophage mutants as evidence for the exponential rate of phage reproduction. *Cold Spr. Harb. Symp. quant. Biol.*, **16**, 463 (*502*).

LURIA S. E. (1953). Host-induced modifications of viruses. *Cold Spr. Harb. Symp. quant. Biol.*, **18**, 237 (*515, 516, 522*).

LURIA S. E. & BURROUS J. W. (1957). Hybridisation between *Escherichia coli* and *Shigella*. *J. Bacteriol.*, **74**, 461 (*750, 751*).

LURIA S. E. & DELBRÜCK M. (1943). Mutations of bacteria from virus sensitivity to virus resistance. *Genetics*, **28**, 491 (*181, 183, 184, 185, 193, 195, 196, 197, 664*).

LURIA S. E. & HUMAN M. L. (1950). Chromatin staining of bacteria during bacteriophage infection. *J. Bacteriol.*, **59**, 551 (*412, 520, 532*).

LURIA S. E. & HUMAN M. L. (1952). A non-hereditary host-induced variation of bacterial viruses. *J. Bacteriol.*, **64**, 557 (*515*).

LURIA S. E. & LATARJET R. (1947). Ultraviolet irradiation during intracellular growth. *J. Bacteriol.*, **53**, 149 (*530, 531*).

LURIA S. E., ADAMS J. N. & TING R. C. (1960). Transduction of lactose-utilising ability among strains of *E. coli* and *S. dysenteriae* and the properties of the transducing phage particles. *Virology*, **12**, 348 (*638*).

LURIA S. E., WILLIAMS R. C. & BACKUS R. C. (1951). Electron micrographic counts of bacteriophage particles. *J. Bacteriol.*, **61**, 179 (*411*).

LURIA S., FRASER D., ADAMS J. & BURROUS J. (1958). Lysogenisation, transduction, and genetic recombination in bacteria. *Cold Spr. Harb. Symp. quant. Biol.*, **23**, 71 (*449, 637, 638, 639, 648*).

LUZZATTI V., MASSON F. & LERMAN L. S. (1961). Interaction of DNA and proflavine: a small-angle x-ray scattering study. *J. Mol. Biol.*, **3**, 634 (*322*).

LWOFF A. (1959). Bacteriophage as a model of host-virus relationship. In *The Viruses* (eds. Burnet F. M. & Stanley W. M.), vol. 2, p. 187. Academic Press, New York (*408, 807*).

LWOFF A. & GUTMANN A. (1950). Recherches sur un *Bacillus mégathérium* lysogène. *Ann. Inst. Pasteur*, **78**, 711 (*450, 451, 452*).

LWOFF A., SIMINOVITCH L. & KJELGAARD N. (1950). Induction de la production de bactériophages chez une bactérie lysogène. *Ann. Inst. Pasteur*, **79**, 815 (*451*).

MAALØE O. (1960). The nucleic acids and the control of bacterial growth. *Symp. Soc. gen. Microbiol.*, **10**, 272 (*269*).

MAALØE O. (1961). The control of normal DNA replications in bacteria. *Cold Spr. Harb. Symp. quant. Biol.*, **26**, 45 (*557*).

MAALØE O. & HANAWALT P. C. (1961). Thymine deficiency and the normal DNA replication cycle. *J. Mol. Biol.*, **3**, 144 (*556, 557*).

MAALØE O. & KJELDGAARD N. O. (1966). *Control of macromolecular synthesis*, p. 158. W. A. Benjamin, Inc., New York and Amsterdam (*548, 554, 573*).

MAALØE O. & SYMONDS N. D. (1953). Radioactive sulfur tracer studies on the reproduction of T4 bacteriophage. *J. Bacteriol.*, **65**, 177 (*427*).

McCARTY M. & AVERY O. T. (1946). Studies on the chemical nature of the substance inducing transformation of pneumococcal types. II. Effect of desoxyribonuclease on the biological activity of the transforming substance. *J. exptl. Med.*, **83**, 89 (*575*).

McCARTY M., TAYLOR H. E. & AVERY O. T. (1946). Biochemical studies of environmental factors essential in transformation of pneumococcal types. *Cold Spr. Harb. Symp. quant. Biol.*, **11**, 177 (*583*).

McCLINTOCK B. (1956). Controlling elements and the gene. *Cold Spr. Harb. Symp. quant. Biol.*, **21**, 197 (*221*).

McDONOUGH M. W. (1965). Amino acid composition of antigenically distinct *Salmonella* flagellar proteins. *J. Mol. Biol.*, **12**, 342 (*738*).

McFALL E. & STENT G. S. (1959). Continuous synthesis of deoxyribonucleic acid in *Escherichia coli*. *Biochim. biophys. Acta*, **34**, 580 (*554*).

MACFARREN A. C. & CLOWES R. C. (1967). A comparative study of two F-like colicin factors, *col V2* and *col V3*, in *Escherichia coli* K12. *J. Bacteriol.*, **94**, 365 (*769, 776, 780*).

MACHATTIE L. A. & THOMAS C. A. JNR. (1964). DNA from bacteriophage lambda: molecular length and conformation. *Science*, **144**, 1142 (*543, 544*).

McHattie L. A., Ritchie D. A., Thomas C. A. Jnr. & Richardson C. C. (1967). Terminal repetition in permuted T2 bacteriophage DNA molecules. *J. Mol. Biol.*, **23**, 355 (*433, 540, 542*).

McLaren A. & Walker P. M. B. (1965). Genetic discrimination by means of DNA/DNA binding. *Genet. Res., Camb.*, **6**, 230 (*616*).

McLaughlin C. S. & Ingram V. M. (1964). Amino-acyl position in amino-acyl-sRNA. *Science*, **145**, 942 (*284*).

McQuillen K. (1961). Protein synthesis *in vivo*; the involvement of ribosomes in *Escherichia coli*. In *Protein Biosynthesis* (ed. Harris, R. J. C.), p. 263. Academic Press, New York and London (*277*).

McQuillen K. (1965). The physical organisation of nucleic acid and protein synthesis. *Symp. Soc. gen. Microbiol.*, **15**, 134 (*276, 296*).

McQuillen K., Roberts R. B. & Britten R. J. (1959). Synthesis of nascent protein by ribosomes of *Escherichia coli*. *Proc. Natl. Acad. Sci., Wash.*, **45**, 1437 (*271*).

Mackal R. P., Werninghaus B. & Evans E. A. Jnr. (1964). The formation of λ bacteriophage by λ DNA in disrupted cell preparations. *Proc. Natl. Acad., Sci., Wash.*, **51**, 1172 (*476*).

Maccacaro G. A. (1955). Cell surface and fertility in *E. coli*. *Nature, Lond.*, **176**, 125 (*670*).

Maccacaro G. A. & Comolli R. (1956). Surface properties correlated with sex compatibility in *E. coli*. *J. gen. Microbiol.*, **15**, 121 (*670*).

Maccacaro G. A. & Hayes W. (1961). Pairing interaction as a basis for negative interference. *Genet. Res. Camb.*, **2**, 406 (*145, 384, 386*).

Madison J. T., Everett G. E. & Kung H. (1966). On the nucleotide sequence of yeast tyrosine tRNA. *Cold Spr. Harb. Symp. quant. Biol.*, **31** (*282, 283*).

Magasanik B. (1957). Nutrition of bacteria and fungi. *Ann. Rev. Microbiol.*, **11**, 221 (*710*).

Magni G. E. (1964). Origin and nature of spontaneous mutations in meiotic organisms. *J. Cell. comp. Physiol.*, **64**, suppl. a, 165 (*324, 325*).

Magni G. E. & von Borstel R. C. (1962). Different rates of spontaneous mutation during mitosis and meiosis in yeast. *Genetics*, **47**, 1097 (*324*).

Magni G. E., von Borstel R. C. & Sora S. (1964). Mutagenic action during meiosis and antimutagenic action during mitosis by 5-aminoacridine in yeast. *Mutation Res.*, **1**, 227 (*319, 325*).

Mahler H. R. & Fraser D. (1961). The replication of T2 bacteriophage. *Advanc. Virus Res.*, **8**, 63 (*410, 428, 429*).

Mahler I., Cahoon M. & Marmur J. (1964). *Bacillus subtilis* deoxyribonucleic acid transfer in PBS2 transduction. *J. Bacteriol.*, **87**, 1423 (*640*).

Makela P. H. (1964). Genetic homologies between flagellar antigens of *Escherichia coli* and *Salmonella abony*. *J. gen. Microbiol.*, **35**, 503 (*738, 740*).

Margolin P. & Bauerle R. H. (1966). Determinants for regulation and initiation of expression of tryptophan genes. *Cold Spr. Harb. Symp. quant. Biol.*, **31**, 311 (*722, 729*).

Markert A. & Zillig W. (1965). Studies on the lysis of *Escherichia coli* by bacteriophage φX174. *Virology*, **25**, 88 (*510*).

MARMUR J. & LANE D. (1960). Strand separation and specific recombination in deoxyribonucleic acids: biological studies. *Proc. Natl. Acad. Sci., Wash.*, **46**, 453 (*593*).

MARMUR J., ROWND R. & SCHILDKRAUT C. L. (1963). Denaturation and renaturation of deoxyribonucleic acid. In *Progress in Nucleic Acid Research* (eds. Davidson J. N. & Cohn W. E.), vol. 1, p. 231. Academic Press, Inc., New York and London (*244*).

MARMUR J., ANDERSON W. F., MATTHEWS L., BERNS K., GAJEWSKA E., LANE D. & DOTY P. (1961). The effects of ultraviolet light on the biological and physical chemical properties of deoxyribonucleic acids. *J. Cell. comp. Physiol.*, **58** (Suppl. 1), 33 (*329, 330, 332*).

MARMUR J., GREENSPAN C. M., PALACE K. E., KAHAN F. M., LEVINE J. & MANDEL M. (1963). Specificity of the complementary RNA formed by *Bacillus subtilis* infected with bacteriophage SP8. *Cold Spr. Harb. Symp. quant. Biol.*, **28**, 191 (*291, 372*).

MARTIN R. G. (1963). The first enzyme in histidine biosynthesis: the nature of feedback inhibition by histidine. *J. biol. Chem.*, **238**, 257 (*707*).

MARTIN R. G. (1963). The one operon-one messenger theory of transcription. *Cold Spr. Harb. Symp. quant. Biol.*, **28**, 357 (*718*).

MARTIN R. G. (1967). Frameshift mutants in the histidine operon of *Salmonella typhimurium*. *J. Mol. Biol.*, **26**, 311 (*720*).

MARTIN R. G., MATTHAEI J. H., JONES O. W. & NIRENBERG M. W. (1962). Ribonucleotide composition of the genetic code. *Biochem. biophys. Res. Comm.*, **6**, 410 (*356*).

MARTIN R. G., SILBERT D. F., SMITH D. W. E. & WHITFIELD H. J. JNR. (1966). Polarity in the histidine operon. *J. Mol. Biol.*, **21**, 357 (*720, 721*).

MARVIN D. A. & HOFFMAN-BERLING H. (1963). Physical and chemical properties of two new small bacteriophages. *Nature, Lond.*, **197**, 517 (*419, 444*).

MARVIN D. A., SPENCER M., WILKINS M. F. H. & HAMILTON L. D. (1961). The molecular configuration of deoxyribonucleic acid. III. X-ray diffraction of the C form of the lithium salt. *J. Mol. Biol.*, **3**, 547 (*236*).

MASON D. J. & POWELSON D. W. (1956). Nuclear division as observed in live bacteria by a new technique. *J. Bacteriol.*, **71**, 474 (*547*).

MATHER K. (1938). Crossing-over. *Biol. Rev.*, **13**, 252 (*53*).

MATSUSHIRO A. (1963). Specialised transduction of tryptophan markers in *Escherichia coli* K12 by bacteriophage φ80. *Virology*, **19**, 475 (*461*).

MATSUSHIRO A., KIDA S., ITO J., SATO K. & IMAMOTO F. (1962). The regulatory mechanisms of enzyme synthesis in the tryptophan biosynthetic pathway of *Escherichia coli* K-12. *Biochem. biophys. Res. Comm.*, **9**, 204 (*636, 724*).

MATSUSHIRO A., SATO K., ITO J., KIDA S. & IMAMOTO F. (1965). On the transcription of the tryptophan operon in *Escherichia coli*. *J. Mol. Biol.*, **11**, 54 (*719*).

MATTERN I. E., VAN WINDEN M. P. & RÖRSCH A. (1965). The range of action of genes controlling radiation sensitivity in *Escherichia coli*. *Mutation Res.*, **2**, 111 (*334*).

MESELSON M. (1963). The duplication and recombination of genes. In *Ideas in Modern Biology* (ed. Moore J. A.). The Natural History Press, Garden City, New York, 1965 (*400*).

MESELSON M. (1964). On the mechanism of recombination between DNA molecules. *J. Mol. Biol.*, **9**, 734 (*397*).

MESELSON M. (1967). The molecular basis of genetic recombination. In *Heritage from Mendel* (ed. Brink R. A.), p. 81. University of Wisconsin Press, Madison and London (*396, 397, 398, 400, 495*).

MESELSON M. & STAHL F. W. (1958a). The replication of DNA in *Escherichia coli*. *Proc. Natl. Acad. Sci., Wash.*, **44**, 671 (*233, 243, 545, 554, 555*).

MESELSON M. & STAHL F. W. (1958b). The replication of DNA. *Cold Spr. Harb. Symp. quant. Biol.*, **23**, 9 (*243, 244, 545, 554*).

MESELSON M. & WEIGLE J. J. (1961). Chromosome breakage accompanying genetic recombination in bacteriophage. *Proc. Natl. Acad. Sci., Wash.*, **47**, 857 (*394, 395, 397, 546*).

MEYNELL E. & COOKE M. (1968). Genetic recombination mediated by R factors (in preparation: personal communication) (*773*).

MEYNELL E. & DATTA N. (1965). Functional homology of the sex-factor and resistance transfer factors. *Nature, Lond.*, **207**, 884 (*772, 773*).

MEYNELL E. & DATTA N. (1966a). The nature and incidence of conjugation factors in *Escherichia coli*. *Genet. Res., Camb.*, **7**, 141 (*441, 749, 764, 773*).

MEYNELL E. & DATTA N. (1966b). The relation of resistance transfer factors to the F factor of *Escherichia coli* K12. *Genet. Res., Camb.*, **7**, 134 (*773*).

MEYNELL E. & DATTA N. (1967). Mutant drug-resistance factors of high transmissibility. *Nature, Lond.*, **214**, 885 (*773*).

MEYNELL G. G. & LAWN A. M. (1967). Sex pili and common pili in the conjugal transfer of colicin factor Ib by *Salmonella typhimurium*. *Genet. Res., Camb.*, **9**, 359 (*774*).

MEYNELL E., MEYNELL G. G. & DATTA N. (1968). Phylogenetic relationships of drug-resistance factors and their bacterial plasmids. *Bacteriol. Rev.*, **32**, 55 (*760, 768, 769, 783*).

MILLER I. L. & LANDMAN O. E. (1966). On the mode of entry of transforming DNA into *Bacillus subtilis*. In *The physiology of gene and mutation expression* (eds. Kohoutová M. & Hubáček J.), p. 187. Academia, Prague (*591*).

MILLS D. R., PACE N. R. & SPIEGELMAN S. (1966). The *in vitro* synthesis of a non-infectious complex containing biologically active viral RNA. *Proc. Natl. Acad. Sci., Wash.*, **56**, 1778 (*443*).

MILLS G. T. & SMITH E. E. B. (1962). Biosynthetic aspects of capsule formation in the pneumococcus. *Brit. Med. Bull.*, **18**, 27 (*613*).

MITCHELL H. K. & LEIN J. (1948). A *Neurospora* mutant deficient in the enzymatic synthesis of tryptophan. *J. Biol. Chem.*, **175**, 481 (*117*).

MITCHELL M. B. (1955). Aberrant recombination of pyridoxin mutants of *Neurospora*. *Proc. Natl. Acad. Sci., Wash.*, **41**, 215 (*382*).

MITRA S. (1958). Effect of X-rays on chromosomes of *Lilium longiflorum* during meiosis. *Genetics*, **43**, 771 (*389*).

MITRA S., REICHARD P., INMAN R. B., BERTSCH L. L. & KORNBERG A. (1967). Enzymatic synthesis of deoxyribonucleic acid. XXII. Replication of a circular single-stranded DNA template by DNA polymerase of *Escherichia coli. J. Mol. Biol.*, **24**, 429 (*555*).

MITSUHASHI S., HARADA K. & HASHIMOTO H. (1960). Multiple resistance of bacteria and transmission of drug-resistance to other strains by mixed cultivation. *Japanese J. Exp. Med.*, **30**, 179 (*760, 772*).

MITSUHASHI S., HARADA K. & KAMEDA M. (1961). Elimination of transmissible drug resistance by treatment with acriflavin. *Nature, Lond.*, **189**, 947 (*772*).

MITSUHASHI S., HARADA K., HASHIMOTO H., KAMEDA M. & SUZUKI M. (1962). Combination of two types of transmissible drug-resistance factors in a host bacterium. *J. Bacteriol.*, **84**, 9 (*780*).

MIYAKE T. (1959). Fertility factor in *Salmonella typhimurium*. *Nature, Lond.*, **184**, 657 (*751*).

MIYAKE T. (1960). Mutator factor in *Salmonella typhimurium*. *Genetics*, **45**, 11 (*219*).

MIYAKE T. (1962). Exchange of genetic material between *Salmonella typhimurium* and *Escherichia coli* K-12. *Genetics*, **47**, 1043 (*753*).

MIYAKE T. & DEMEREC M. (1959). *Salmonella-Escherichia* hybrids. *Nature, Lond.*, **183**, 1586 (*751*).

MONK M. & CLOWES R. C. (1964a). Transfer of the colicin I factor in *Escherichia coli* K12 and its interaction with the *F* fertility factor. *J. gen. Microbiol.*, **36**, 365 (*758*).

MONK M. & CLOWES R. C. (1964b). The regulation of colicin synthesis and colicin factor transfer in *Escherichia coli* K12. *J. gen. Microbiol.*, **36**, 385 (*758*).

MONOD J. & COHN M. (1952). La biosynthèse induite des enzymes (adaptation enzymatique). *Advances in Enzymol.*, **13**, 67 (*711*).

MONOD J. & JACOB F. (1961). Teleonomic mechanisms in cellular metabolism, growth and differentiation. *Cold Spr. Harb. Symp. quant. Biol.*, **26**, 389 (*707, 708*).

MORGAN A. R., WELLS R. D. & KHORANA H. G. (1966). Studies on polynucleotides, LIX. Further codon assignments from amino-acid incorporations directed by ribopolynucleotides containing repeating trinucleotide sequences. *Proc. Natl. Acad. Sci., Wash.*, **56**, 1899 (*359*).

MORSE M. L. (1954). Transduction of certain loci in *Escherichia coli* K-12. *Genetics*, **39**, 984 (*621, 627*).

MORSE M. L. (1959). Transduction by staphylococcal bacteriophage. *Proc. Natl. Acad. Sci., Wash.*, **45**, 722 (*621*).

MORSE M. L., LEDERBERG E. M. & LEDERBERG J. (1956a). Transduction in *Escherichia coli* K-12. *Genetics*, **41**, 142 (*100, 628, 629, 631*).

MORSE M. L., LEDERBERG E. M. & LEDERBERG J. (1956b). Transductional heterogenotes in *Escherichia coli*. *Genetics*, **41**, 758 (*97, 100, 101, 628, 629, 631*).

MOSELEY B. E. B. & LASER H. (1965). Repair of X-ray damage in *Micrococcus radiodurans*. *Proc. Roy. Soc., B.*, **162**, 210 (*336*).

MOSIG G. (1963). Genetic recombination in bacteriophage T4 during replication of DNA fragments. *Cold Spr. Harb. Symp. quant. Biol.*, **28**, 35 (*434*).

MOYED H. S. (1960). False feedback inhibition; inhibition of tryptophan biosynthesis by 5-methyltryptophan. *J. Biol. Chem.*, **235**, 1098 (*708*).
MUIRHEAD H. & PERUTZ M. F. (1963). Structure of reduced human haemoglobin. *Cold Spr. Harb. Symp. quant. Biol.*, **28**, 451 (*709*).
MULLER H. J. (1928). The production of mutations by X-rays. *Proc. Natl. Acad. Sci., Wash.*, **14**, 714 (*204*).
MUNDRY K. W. & GRIERER A. (1958). Die Erzengung von Mutationen des Tabakmosaikvirus durch chemische Behandlung seiner Nukleinsäure *in vitro. Z. Vererbungsl.*, **89**, 614 (*305*).
MUNSON R. J. & MACLEAN F. I. (1961). The nature and radiation sensitivity of the long forms of *Escherichia coli* strain B/r. *J. gen. Microbiol.*, **25**, 29 (*548*).
MURRAY N. E. (1963). Polarised recombination and fine structure within the *me*-2 gene of *Neurospora crassa. Genetics*, **48**, 1163 (*400*).
NAGATA T. (1963a). The molecular synchrony and sequential replication of DNA in *Escherichia coli. Proc. Natl. Acad. Sci., Wash.*, **49**, 551 (*564, 566*).
NAGATA T. (1963b). The sequential replication of *E. coli* DNA. *Cold Spr. Harb. Symp. quant. Biol.*, **28**, 55 (*564, 566*).
NAKADA D. (1960). Involvement of newly formed protein in the synthesis of deoxyribonucleic acid. *Biochim. biophys. Acta*, **44**, 241 (*561*).
NAKAI S. & SAEKI T. (1964). Induction of mutation by photodynamic action in *Escherichia coli. Genet. Res., Camb.*, **5**, 158 (*325*).
NAKAYA R., NAKAMURA A. & MURATA Y. (1960). Resistance transfer agents in *Shigella. Biochem. biophys. Res. Comm.*, **3**, 654 (*762*).
NATHANS D. & LIPMANN F. (1961). Amino acid transfer from aminoacyl-ribonucleic acids to protein ribosomes of *Escherichia coli. Proc. Natl. Acad. Sci., Wash.*, **47**, 497 (*281, 285*).
NELSON T. C. (1951). Kinetics of genetic recombination in *E. coli. Genetics*, **36**, 162 (*679*).
NELSON T. C. & LEDERBERG J. (1954). Postzygotic elimination of genetic factors in *E. coli. Proc. Natl. Acad. Sci., Wash.*, **40**, 415 (*658*).
NESTER E. W. (1964). Penicillin resistance of competent cells in deoxyribonucleic acid transformation of *Bacillus subtilis. J. Bacteriol.*, **87**, 867 (586).
NESTER E. W. & LEDERBERG J. (1961). Linkage of genetic units of *Bacillus subtilis* in DNA transformation. *Proc. Natl. Acad. Sci., Wash.*, **47**, 52 (*611, 614*).
NESTER E. W. & STOCKER B. A. D. (1963). Biosynthetic latency in early stage of deoxyribonucleic acid transformation in *Bacillus subtilis. J. Bacteriol.*, **86**, 785 (*587, 609, 618*).
NESTER E. W., SCHAFER M. & LEDERBERG J. (1963). Gene linkage in DNA transfer: a cluster of genes concerned with aromatic biosynthesis in *Bacillus subtilis. Genetics*, **48**, 529 (*164, 611, 614*).
NEWCOMBE H. W. (1948). Delayed phenotypic expression of spontaneous mutations in *Escherichia coli. Genetics*, **33**, 447 (*197*).
NEWCOMBE H. B. (1949). Origin of bacterial variants. *Nature, Lond.*, **164**, 150 (*185, 186*).
NEWTON W. A., BECKWITH J. R., ZIPSER D. & BRENNER S. (1965). Nonsense mutants and polarity in the *lac* operon of *Escherichia coli. J. Mol. Biol.*, **14**, 290 (*721*).

NIRENBERG N. & LEDER P. (1964). RNA codewords and protein synthesis. The effect of trinucleotides upon the binding of sRNA to ribosomes. *Science*, **145**, 1399 (*357*).

NIRENBERG M. W. & MATTHAEI J. H. (1961). The dependence of cell-free protein synthesis in *Escherichia coli* upon naturally occurring or synthetic polyribonucleotides. *Proc. Natl. Acad. Sci., Wash.*, **47**, 1588 (*287, 355, 356*).

NIRENBERG M. W., MATTHAEI J. J. & JONES O. W. (1962). An intermediate in the biosynthesis of polyphenylalanine directed by synthetic template RNA. *Proc. Natl. Acad. Sci., Wash.*, **48**, 104 (*356*).

NIRENBERG M. W., JONES O. W., LEDER P., CLARK B. F. C., SLY W. S. & PESTKA S. (1963). On the coding of genetic information. *Cold Spr. Harb. Symp. quant. Biol.*, **28**, 549 (*356*).

NIRENBERG M., LEDER P., BERNFIELD M., BRIMACOMBE R., TRUPIN J., ROTMAN F. & O'NEAL S. (1965). RNA codewords and protein synthesis. VII. On the general nature of the RNA code. *Proc. Natl. Acad. Sci., Wash.*, **53**, 1161 (*359, 367*).

NISHIMURA S., JONES D. S. & KHORANA H. G. (1965). Studies on polynucleotides. XLVIII. The *in vitro* synthesis of a co-polypeptide containing two amino acids in alternating sequence dependent upon a DNA-like polymer containing two nucleotides in alternating sequence. *J. Mol. Biol.*, **13**, 302 (*359*).

NISHIMURA S., JONES D. S., OHTSUKA E., HAYATSU H., JACOB T. M. & KHORANA H. G. (1965). Studies on polynucleotides. XLVII. The *in vitro* synthesis of homopeptides as directed by a ribopolynucleotide containing a repeating trinucleotide sequence. New codon sequences for lysine, glutamic acid and arginine. *J. Mol. Biol.*, **13**, 283 (*359*).

NISHIMURA Y., ISHIBASHI M., MEYNELL E. & HIROTA Y. (1967). Specific pilation directed by a fertility factor and a resistance factor of *Escherichia coli*. *J. gen. Microbiol.*, **49**, 89 (*773, 782*).

NISMAN B. & PELMONT J. (1964). De novo protein synthesis *in vitro*. In *Progress in nucleic acid research and molecular biology* (eds. Davidson J. N. & Cohn W. E.), vol. 3, p. 235. Academic Press, Inc., New York and London (*355*).

NOMURA M. (1963). The mode of action of colicins. *Cold Spr. Harb. Symp. quant. Biol.*, **28**, 315 (*756*).

NOMURA M. (1967). Colicins and related bacteriocins. *Ann. Rev. Microbiol.*, **21**, 257 (*756, 757*).

NOMURA M. & BENZER S. (1961). The nature of the 'deletion' mutants in the r_{II} region of phage T4. *J. Mol. Biol.*, **3**, 684 (*218, 434, 499*).

NOMURA M., HALL B. D. & SPIEGELMAN S. (1960). Characterisation of RNA synthesised in *Escherichia coli* after bacteriophage T2 infection. *J. Mol. Biol.*, **2**, 306 (*278, 369*).

NOTANI G. W., ENGELHARDT D. L., KONIGSBERG W. & ZINDER N. D. (1965). Suppression of a coat protein mutant of the bacteriophage f2. *J. Mol. Biol.*, **12**, 439 (*510*).

NOVICK A. (1956). Mutagens and antimutagens. *Brookhaven Symposia in Biology*, **8**, 201 (*205*).

NOVICK A. & SZILARD L. (1950). Experiments with the chemostat on spontaneous mutations in bacteria. *Proc. Natl. Acad. Sci., Wash.*, **36**, 708 (*193*).

NOVICK A. & SZILARD L. (1951). Virus strains of identical phenotype but different genotype. *Science*, **113**, 34 (*513*).

NOVICK A. & SZILARD L. (1954). Experiments with the chemostat on the rates of amino acid synthesis in bacteria. In *Dynamics of Growth Processes*, p. 21. Princetown University Press (*707*).

NOVICK R. P. (1963). Analysis by transduction of mutations affecting penicillinase formation in *Staphylococcus aureus*. *J. gen. Microbiol.*, **33**, 121 (*784, 785*).

NOVICK R. P. & RICHMOND M. H. (1965). Nature and interactions of the genetic elements governing penicillinase synthesis in *Staphylococcus aureus*. *J. Bacteriol.*, **90**, 467 (*784, 785, 786, 787*).

NOVICK R. P. (1967). Penicillinase plasmids of *Staphylococcus aureus*. *Fed. Proc.*, **27**, 29 (*767, 785, 787, 788, 789*).

OCHOA S. & HEPPEL L. A. (1957). Polynucleotide synthesis. In *The Chemical Basis of Heredity* (eds. McElroy W. D. & Glass B.), p. 615. Johns Hopkins Press, Baltimore (*292, 356*).

OGAWA H. & TOMIZAWA J. (1967). Bacteriophage lambda DNA with different structures formed in infected cells. *J. Mol. Biol.*, **23**, 265 (*463, 543*).

OHTAKA Y. & SPIEGELMAN S. (1963). Translational control of protein synthesis in a cell-free system directed by a polycistronic viral RNA. *Science*, **142**, 493 (*280*).

OISHI M., YOSHIKAWA H. & SUEOKA N. (1964). Synchronous and dichotomous replications of the *Bacillus subtilis* chromosome during spore germination. *Nature, Lond.*, **204**, 1069 (*561, 563, 564, 614*).

OKADA T., YANAGISAWA K. & RYAN F. J. (1960). Elective production of thymineless mutants. *Nature, Lond.*, **188**, 944 (*573*).

OKADA T., YANAGISAWA K. & RYAN F. J. (1961). A method for securing thymineless mutants from strains of *E. coli*. *Z. Vererbungsl.*, **92**, 403 (*573*).

OKADA Y., TERZAGHI E., STREISINGER G., EMRICH J., INOUYE M. & TSUGITA A. (1966). A frame-shift mutation involving the addition of two base pairs in the lysozyme gene of phage T4. *Proc. Natl. Acad. Sci., Wash.*, **56**, 1692 (*346*).

OKAMOTO S. & SUZUKI Y. (1965). Chloramphenicol, dihydrostreptomycin and Kanamycin-inactivating enzymes from multiple drug-resistant *Escherichia coli* carrying episome 'R'. *Nature, Lond.*, **208**, 1301 (*766*).

OKAZAKI T. & KORNBERG A. (1964). Enzymatic synthesis of deoxyribonucleic acid. XV. Purification properties of a polymerase from *Bacillus subtilis*. *J. Biol. Chem.*, **239**, 259 (*247*).

OKUBO S., STODOLSKY M., BOTT K. F. & STRAUSS B. (1963). Separation of the transforming and viral deoxyribonucleic acids of a transducing bacteriophage of *Bacillus subtilis*. *Proc. Natl. Acad. Sci., Wash.*, **50**, 679 (*640*).

ONO J., WILSON R. G. & GROSSMAN L. (1965). Effects of ultraviolet light on the template properties of polycytidylic acid. *J. Mol. Biol.*, **11**, 600 (*338*).

OPPENHEIM A. B. & RILEY M. (1966). Molecular recombination following conjugation in *Escherichia coli. J. Mol. Biol.*, **20**, 331 (*397*).

ØRSKOV I. & ØRSKOV F. (1960). An antigen termed f^+ occurring in F^+ *E. coli* strains. *Acta Pathol. Microbiol. Scand.*, **48**, 37 (*670*).

ØRSKOV F. & ØRSKOV I. (1961). The fertility of *Escherichia coli* antigen test strains in crosses with K-12. *Acta Pathol. Microbiol. Scand.*, **51**, 280 (*749*).

OTAKA E., OSAWA S. & SIBATANI A. (1964). Stimulation of ^{14}C-leucine incorporation into protein *in vitro* by ribosomal RNA of *Escherichia coli. Biochem. Biophys. Res. Comm.*, **15**, 568 (*294*).

OVERBY L. R., BARLOW G. H., DOI R. H., JACOB M. & SPIEGELMAN S. (1966). Comparison of two serologically distinct ribonucleic acid bacteriophages. *J. Bacteriol.*, **91**, 442 (*292, 441*).

OZEKI H. (1956). Abortive transduction in purine-requiring mutants of *Salmonella typhimurium. Carnegie Inst. Wash. Publ.*, **612**, 97 (*97, 641*).

OZEKI H. (1959). Chromosome fragments participating in transduction in *Salmonella typhimurium. Genetics*, **44**, 457 (*626, 642, 643, 645*).

OZEKI H. & HOWARTH S. (1961). Colicine factors as fertility factors in bacteria: *Salmonella typhimurium*, strain LT2. *Nature, Lond.*, **190**, 986 (*759, 769*).

OZEKI H., STOCKER B. A. D. & DEMARGERIE H. (1959). Production of colicine by single bacteria. *Nature, Lond.*, **184**, 337 (*756*).

OZEKI H., STOCKER B. A. D. & SMITH S. M. (1962). Transmission of colicinogeny between strains of *Salmonella typhimurium* grown together. *J. gen. Microbiol.*, **28**, 671 (*757*).

PAKULA R. (1965). Kinetics of provoked competence in streptococcal cultures and its specificity. *Symp. Biol., Hung.*, **6**, 33 (*589*).

PAKULA R. & WALCZAK W. (1963). On the nature of competence of transformable streptococci. *J. gen. Microbiol.*, **31**, 125 (*589*).

PAKULA R., HULANICKA E. & WALCZAK W. (1960). Inhibition of transformation in *Streptococcus sbe* by desoxyribonucleic acids of different origin. *Bull. Acad. Polon. Sci.*, **8**, 49 (*593*).

PALADE G. E. (1955). A small particulate component of the cytoplasm. *J. Biochem. biophys. Cytol.*, **1**, 59 (*270*).

PARDEE A. B. (1954). Nucleic acid precursors and protein synthesis. *Proc. Natl. Acad. Sci., Wash.*, **40**, 263 (*269*).

PARDEE A. B. & PRESTIDGE L. S. (1956). The dependence of nucleic acid syntheses on the presence of amino acids in *Escherichia coli. J. Bacteriol.*, **71**, 677 (*269*).

PARDEE A. B. & PRESTIDGE L. S. (1959). On the nature of the repressor of β-galactosidase synthesis in *E. coli. Biochim. biophys. Acta*, **36**, 545 (*714*).

PARDEE A. B., JACOB F. & MONOD J. (1959). The genetic control and cytoplasmic expression of 'inducibility' in the synthesis of β-galactosidase by *E. coli. J. Mol. Biol.*, **1**, 165 (*705, 711, 714*).

PAULING L. & COREY R. B. (1951a). Atomic coordinates and structure factors for two helical configurations of polypeptide chains. *Proc. Natl. Acad. Sci., Wash.*, **37**, 235 (*261*).

PAULING L. & COREY R. B. (1951b). Configurations of polypeptide chains with favored orientations around single bonds: two new pleated sheets. *Proc. Natl. Acad. Sci., Wash.*, **37**, 729 (*261*).

PAULING L., COREY R. B. & BRANSON H. R. (1951). The structure of proteins: two hydrogen-bonded helical configurations of the polypeptide chain. *Proc. Natl. Acad. Sci., Wash.*, **37**, 205 (*261*).

PEARCE L. E. & MEYNELL E. (1968). Specific chromosomal affinity of a resistance factor. *J. gen. Microbiol.*, **50**, 159 (*779*).

PEARSE U. & STOCKER B. A. D. (1965). Variation in composition of chromosome fragments transduced by phage P22. *Virology*, **27**, 290 (*645*).

PEARSE U. B. & STOCKER B. A. D. (1967). Phase variation of flagellar antigens in *Salmonella*; abortive transduction studies. *J. gen. Microbiol.*, **49**, 335 (*744*).

PELC S. R. (1965). Correlation between coding triplets and amino-acids. *Nature, Lond.*, **207**, 597 (*367*).

PELC S. R. & WELTON M. G. E. (1966). Stereochemical relationship between coding triplets and amino acids. *Nature, Lond.*, **209**, 868 (*368*).

PELLING C. (1966). A replicative and synthetic chromosomal unit—the modern concept of the chromomere. *Proc. Roy. Soc., B.*, **164**, 279 (*534*).

PERKINS D. D. (1962). The frequency in *Neurospora* tetrads of multiple exchanges within short intervals. *Genet. Res., Camb.*, **3**, 315 (*383*).

PERRIN D., JACOB F. & MONOD J. (1960). Biosynthèse induite d'une protéine génétiquement modifiée, ne présentant pas d'affinité pour l'inducteur. *C. R. Acad. Sci., Paris*, **251**, 155 (*715*).

PERUTZ M. F. (1951). New x-ray evidence on the configuration of polypeptide chains. Polypeptide chains in poly-γ-benzyl-L-glutamate, keratin and haemoglobin. *Nature, Lond.*, **167**, 1053 (*261*).

PERUTZ M. F. (1962). In *Proteins and nucleic acids: structure and function.* Elsevier Publishing Co., Amsterdam, London and New York (*232*).

PERUTZ M. F., KENDREW J. C. & WATSON H. C. (1965). Structure and function of haemoglobin. II. Some relations between polypeptide chain configuration and amino acid sequence. *J. Mol. Biol.*, **13**, 669 (*265, 367*).

PERUTZ M. F., ROSSMANN M. G., CULLIS A. F., MUIRHEAD H., WILL G. & NORTH A. C. T. (1960). Structure of haemoglobin. *Nature, Lond.*, **185**, 416 (*152*).

PETTIJOHN D. & HANAWALT P. (1964). Evidence for repair-replication of ultraviolet damaged DNA in bacteria. *J. Mol. Biol.*, **9**, 395 (*333*).

PFEIFER D. (1961). Genetic recombination in bacteriophage φX174. *Nature, Lond.*, **189**, 422 (*510*).

PIEKARSKI G. (1937). Cytologische Untersuchungen an Paratyphus und Colibakterein. *Arch. Mikrobiol.*, **8**, 428 (*546, 547*).

PITTARD J. & ADELBERG E. A. (1964). Gene transfer by F′ strains of *Escherichia coli* K12. III. An analysis of the recombination events occurring in the F′ male and in the zygotes. *Genetics*, **49**, 995 (*677*).

PITTARD J., LOUTIT J. S. & ADELBERG E. A. (1963). Gene transfer by F′ strains of *Escherichia coli* K12. I. Delay in initiation of chromosome transfer. *J. Bacteriol.*, **85**, 1394 (*677*).

PLOUGH H. H. (1958). Linear order of gene loci in *Salmonella* in relation to prophage site. *Cold Spr. Harb. Symp. quant. Biol.*, **23**, 69 (*627*).

POLLOCK M. R. (1967). The origin and function of penicillinase: a problem in biochemical evolution. *Brit. Med. J.*, **4**, 71 (*766*, *784*).

PONTECORVO G. (1946). Genetic systems based on heterocaryosis. *Cold Spr. Harb. Symp. quant. Biol.*, **11**, 193 (*78*, *95*).

PONTECORVO G. (1950). New fields in the biochemical genetics of micro-organisms. *Biochem. Soc. Symp.*, **4**, 40 (*130*, *165*).

PONTECORVO G. (1952a). Genetical analysis of cell organisation. *Symp. Soc. exp. Biol.*, **6**, 218 (*124*).

PONTECORVO G. (1952b). The genetic formulation of gene structure and action. *Advanc. Enzymol.*, **13**, 121 (*130*, *165*, *173*).

PONTECORVO G. (1958). *Trends in genetic analysis*. Columbia Univ. Press, New York. (1959: Oxford Univ. Press) (*142*, *157*, *326*, *380*, *382*, *695*).

PONTECORVO G. (1962). Methods of microbial genetics in an approach to human genetics. *Brit. Med. Bull.*, **18**, 81 (*83*).

PONTECORVO G. & KÄFER E. (1958). Genetic analysis based on mitotic recombination. *Advanc. Genet.*, **9**, 71 (*78*, *83*).

PONTECORVO G. & ROPER J. A. (1952). Genetic analysis without sexual reproduction by means of polyploidy in *Aspergillus nidulans*. *J. gen. Microbiol.*, **6**, vii (*78*).

POSTON S. M. (1966). Cellular location of the genes controlling penicillinase production and resistance to streptomycin and tetracycline in a strain of *Staphylococcus aureus*. *Nature, Lond.*, **210**, 802 (*789*).

POWER J. (1967). The L-rhamnose genetic system in *Escherichia coli* K12. *Genetics*, **55**, 557 (*730*).

PRATT D., TZAGOLOFF H. & ERDAHL W. S. (1966). Conditional lethal mutants of the small filamentous coliphage M13. I. Isolation, complementation, cell killing, time of cistron action. *Virology*, **30**, 397 (*446*, *510*).

PRELL H. H. (1965). DNA transfer from prophage to phage progeny after zygotic induction. *J. Mol. Biol.*, **13**, 329 (*472*).

PRITCHARD R. H. (1955). The linear arrangement of a series of alleles in *Aspergillus nidulans*. *Heredity*, **9**, 343 (*130*, *145*, *379*, *381*, *383*).

PRITCHARD R. H. (1960a). Localised negative interference and its bearing on models of gene recombination. *Genet. Res., Camb.*, **1**, 1 (*379*, *383*, *390*, *393*).

PRITCHARD R. H. (1960b). The bearing of recombination analysis at high resolution on genetic fine structure in *Aspergillus nidulans* and the mechanism of recombination in higher organisms. *Symp. Soc. gen. Microbiol.*, **10**, 155 (*379*, *381*, *382*, *383*, *390*, *392*, *393*).

PRITCHARD R. H. (1963). In *Genetics Today*: Proc. 11th Intern. Congr. Genetics, The Hague, p. 55. Pergamon Press, London (*804*).

PRITCHARD R. H. (1966). Replication of the bacterial chromosome. *Proc. Roy. Soc. B.*, **164**, 258 (*473*, *573*).

PRITCHARD R. H. & LARK K. G. (1964). Induction of replication by thymine starvation at the chromosome origin in *Escherichia coli*. *J. Mol. Biol.*, **9**, 288 (*558*, *559*).

PROMPTOV A. N. (1932). The effect of short ultraviolet rays on the appearance of hereditary variations in *Drosophila melanogaster*. *J. Genet.*, **26**, 59 (*204*).

PTASHNE M. (1965a). The detachment and maturation of conserved lambda prophage DNA. *J. Mol. Biol.*, **11**, 90 (*472*, *803*).

PTASHNE M. (1965b). Replication and host modification of DNA transferred during bacterial mating. *J. Mol. Biol.*, **11**, 829 (*803*).
PTASHNE M. (1967a). Isolation of the λ phage repressor. *Proc. Natl. Acad. Sci., Wash.*, **57**, 306 (*468*).
PTASHNE M. (1967b). Specific binding of the λ phage repressor to λ DNA. *Nature, Lond.*, **214**, 232 (*469, 733*).
RAVIN A. W. (1954). A quantitative study of autogenic and allogenic transformations in pneumococcus. *Exptl. Cell. Res.*, **7**, 58 (*612*).
RAVIN A. W. (1959). Reciprocal capsular transformations of pneumococci. *J. Bacteriol.*, **77**, 296 (*606, 608*).
RAVIN A. W. (1960). Linked mutations borne by deoxyribonucleic acid controlling the synthesis of capsular polysaccharide in pneumococcus. *Genetics*, **45**, 1387 (*612*).
RAVIN A. W. (1961). The genetics of transformation. *Advances in Genetics*, **10**, 61 (*576, 578, 604, 610, 613*).
REEVES P. (1965). The bacteriocins. *Bacteriol. Rev.*, **29**, 25 (*756, 757*).
RICH A. & DAVIES D. R. (1956). A new two-stranded helical structure; polyadenylic acid and polyuridylic acid. *J. Amer. Chem. Soc.*, **78**, 3548 (*273, 356*).
RICH A. & GREEN D. W. (1961). X-ray studies of compounds of biological interest. *Ann. Rev. Biochem.*, **30**, 93 (*232*).
RICH A., WARNER J. R. & GOODMAN H. M. (1963). The structure and function of polyribosomes. *Cold Spr. Harb. Symp. quant. Biol.*, **28**, 269 (*278, 279*).
RICHARDSON C. C., SCHILDKRAUT C. L., APHOSIAN H. V. & KORNBERG A. (1964a). Enzymatic synthesis of deoxyribonucleic acid. XIV. Further purification and properties of deoxyribonucleic acid polymerase of *Escherichia coli. J. Biol. Chem.*, **239**, 222 (*247, 253*).
RICHARDSON C. C., INMAN R. B. & KORNBERG A. (1964b). Enzymatic synthesis of deoxyribonucleic acid. XVIII. The repair of partially single-stranded DNA templates by DNA polymerase. *J. Mol. Biol.*, **9**, 46 (*253, 256*).
RICHMOND M. H. (1965a). Wild-type variants of exo-penicillinase from *Staphylococcus aureus. Biochem. J.*, **94**, 584 (*785*).
RICHMOND M. H. (1965b). The dominance of the inducible state in strains of *Staphylococcus aureus* containing two distinct penicillinase plasmids. *J. Bacteriol.*, **90**, 370 (*786*).
RICHMOND M. H. (1965c). Penicillinase plasmids in *Staphylococcus aureus. Brit. Med. Bull.*, **21**, 260 (*785, 789*).
RICHMOND M. H. (1966). The genetic constitution of certain penicillinase micro-mutants in *Staphylococcus aureus. J. gen. Microbiol.*, **45**, 51 (*729*).
RICHMOND M. H. (1967a). A second regulatory region involved in penicillinase synthesis in *Staphylococcus aureus. J. Mol. Biol.*, **26**, 357 (*730*).
RICHMOND M. H. (1967b). Associated diploids involving penicillinase plasmids in *Staphylococcus aureus. J. gen. Microbiol.*, **46**, 85 (*788*).
RICHMOND M. H. (1968). The plasmids of *Staphylococcus aureus* and their relation to other extra-chromosomal elements in bacteria. In *Recent*

Advances in Microbial Physiology, vol. 2 (eds. Rose A. H. & Wilkinson J. F.). Academic Press, New York and London (*785, 786, 789*).

RICHTER A. (1957). Complementary determinants of an Hfr phenotype in *E. coli* K-12. *Genetics*, **42**, 391 (*675*).

RICHTER A. (1961). Attachment of wild-type F factor to a specific chromosomal region in a variant strain of *Escherichia coli* K-12: the phenomenon of episome alternation. *Genet. Res., Camb.*, **2**, 333 (*675, 797*).

RILEY M. & PARDEE A. B. (1962). Gene expression: its specificity and regulation. *Ann. Rev. Microbiol.*, **16**, 1 (*735*).

RILEY M., PARDEE A. B., JACOB F. & MONOD J. (1960). On the expression of a structural gene. *J. Mol. Biol.*, **2**, 216 (*269, 275, 704, 705*).

RIS H. (1957). Chromosome structure. In *The Chemical Basis of Heredity* (eds. McElroy W. D. & Glass B.), p. 23. The Johns Hopkins Press, Baltimore (*534*).

RIS H. (1966). Fine structure of chromosomes. *Proc. Roy. Soc., B.*, **164**, 246 (*534*).

RIS H. & CHANDLER B. L. (1963). The ultrastructure of genetic systems in prokaryotes and eukaryotes. *Cold Spr. Harb. Symp. quant. Biol.*, **28**, 1 (*462, 534, 535, 543*).

RITCHIE D. A. (1964). Mutagenesis with light and proflavin in phage T4. *Genet. Res., Camb.*, **5**, 168 (*325*).

RITCHIE D. A. (1965). Mutagenesis with light and proflavin in phage T4. II. Properties of the mutants. *Genet. Res., Camb.*, **6**, 474 (*325*).

RITCHIE D. A., THOMAS C. A. JNR., MACHATTIE L. A. & WENSINK P. C. (1967). Terminal repetition in non-permuted T3 and T7 bacteriophage DNA molecules. *J. Mol. Biol.*, **23**, 365 (*539, 540, 541, 542, 544*).

ROBERTS R. B., ABELSON P. H., COWIE D. B., BOLTON E. T. & BRITTEN R. J. (1955). Studies of biosynthesis in *Escherichia coli*. *Carnegie Inst. Wash. Publ.*, **607**, Washington, D.C. (*707*).

ROBINOW C. F. (1942). A study of the nuclear apparatus of bacteria. *Proc. Roy. Soc., B.*, **130**, 299 (*546, 547*).

ROBINOW C. F. (1944). Cytological observations on *Bact. coli*, *Proteus vulgaris* and various aerobic, spore-forming bacteria with special reference to the nuclear structures. *J. Hyg., Camb.*, **43**, 413 (*546*).

ROBINOW C. F. (1945). Addendum to *The Bacterial Cell* by R. J. Dubos. Harvard Univ. Press, Cambridge, Mass., U.S.A. (*546*).

ROBINOW C. F. (1956a). The chromatin bodies of bacteria. *Symp. Soc. gen. Microbiol.*, **6**, 181 (*546*).

ROBINOW C. F. (1956b). The chromatin bodies of bacteria. *Bacteriol. Rev.*, **20**, 207 (*546*).

ROBINOW C. F. (1960). In *The Cell: biochemistry, physiology, morphology* (eds. Brachet J. & Mirsky A. E.), vol. 4, pt. 1, p. 45. Academic Press, New York (*546, 547*).

ROBINOW C. F. (1962). Morphology of the bacterial nucleus. *Brit. Med. Bull.*, **18**, 31 (*544, 546, 553*).

ROGERS H. J. (1965). The outer layers of bacteria: the biosynthesis of structure. *Symp. Soc. gen. Microbiol.*, **15**, 186 (*421*).

ROMAN H. (1956). Studies of gene mutation in *Saccharomyces*. *Cold Spr. Harb. Symp. quant. Biol.*, **21**, 175 (*130, 132*).

ROMAN H. (1958). Sur les recombinaisons non-réciproques chez *Saccaromyces cerevisiae* et sur les problèmes posés par ces phénomènes. *Ann. Génétiques*, **1**, 11 (*393*).

ROMAN H. & JACOB F. (1958). A comparison of spontaneous and ultraviolet-induced allelic recombination with reference to the recombination of outside markers. *Cold Spr. Harb. Symp. quant. Biol.*, **23**, 155 (*500*).

ROMERO E. & MEYNELL E. (1968). Covert *fi*$^-$R factors in fi$^+$ R$^+$ bacteria (in preparation: personal communication) (*774*).

ROPER J. A. (1952). Production of heterozygous diploids in filamentous fungi. *Experientia*, **8**, 14 (*79, 124, 130*).

ROSENBERG B. H., SIROTNAK F. M. & CAVALIERI L. F. (1959). On the size of genetic determinants in pneumococcus and the nature of the variables involved in transformation. *Proc. Natl. Acad. Sci., Wash.*, **45**, 144 (*592*).

ROSENKRANZ H. S. (1964). Basis of streptomycin resistance in *Escherichia coli* with a 'multiple drug resistance' episome. *Biochim. biophys. Acta*, **80**, 342 (*766*).

ROTH J. R. & HARTMAN P. E. (1965). Heterogeneity in P22 transducing particles. *Virology*, **27**, 297 (*645*).

ROTH J. R., ANTON D. N. & HARTMAN P. E. (1966). Histidine regulatory mutants in *Salmonella typhimurium*. I. Isolation and general properties. *J. Mol. Biol.*, **22**, 305 (*724, 734*).

ROTH J. R. & AMES B. N. (1966). Histidine regulatory mutants in *Salmonella typhimurium*. II. Histidine regulatory mutants having altered histidyl-*t* RNA synthetase. *J. Mol. Biol.*, **22**, 325 (*734*).

ROTH T. F. & HAYASHI M. (1966). Allomorphic forms of bacteriophage φX174 replicative DNA. *Science*, **154**, 658 (*440*).

ROTH T. F. & HELINSKI D. R. (1967). Evidence for circular DNA forms of a bacterial plasmid. *Proc. Natl. Acad. Sci., Wash.*, **58**, 650 (*793*).

ROTHFELS K. H. (1952). Gene linearity and negative interference in crosses of *Escherichia coli*. *Genetics*, **37**, 297 (*652*).

ROTHMAN J. L. (1965). Transduction studies on the relation between prophage and host chromosome. *J. Mol. Biol.*, **12**, 892 (*458, 460, 461*).

ROUNTREE P. M. (1949). The phenomenon of lysogenicity in staphylococci. *J. gen. Microbiol.*, **3**, 153 (*447*).

ROWND R., NAKAYA R. & NAKAMURA A. (1966). Molecular nature of the drug-resistance factors of the *Enterobacteriaceae*. *J. Mol. Biol.*, **17**, 376 (*770, 771, 776*).

RUBENSTEIN I., THOMAS C. A. JNR. & HERSHEY A. D. (1961). The molecular weights of T2 bacteriophage DNA and its first and second breakage products. *Proc. Natl. Acad. Sci., Wash.*, **47**, 1113 (*536*).

RUECKERT R. R. & ZILLIG W. (1962). Biosynthesis of virus protein in *Escherichia coli* C following infection with bacteriophage φX174. *J. Mol. Biol.*, **5**, 1 (*437*).

RUPERT C. S. (1961). Repair of ultraviolet damage in cellular DNA. *J. Cell comp. Physiol.*, **58** (Suppl. 1), 57 (*331, 332*).

RUPERT C. S. (1964). Questions regarding the presumed role of thymine dimers in photoreactivable U.V. damage to DNA. *Photochem. Photobiol.*, **3**, 399 (*332*).

RUPERT C. S. & HARM W. (1966). Reactivation after photobiological damage. In *Advances in Radiation Biology* (eds. Augenstein L. G., Mason R. & Zelle M. R.), vol. 2, p. 1. Academic Press, Inc., New York and London (*338*).

RYAN F. J. & SCHNEIDER L. K. (1949). The consequences of mutation during the growth of biochemical mutants of *Escherichia coli*. IV. The mechanism of inhibition of histidine-independent bacteria by histidineless bacteria. *J. Bacteriol.*, **58**, 201 (*203*).

RYAN F. J. & WAINWRIGHT L. K. (1954). Nuclear segregation and the growth of clones of spontaneous mutants of bacteria. *J. gen. Microbiol.*, **11**, 364 (*202*).

RYTER A. & JACOB F. (1964). Étude au microscope électronique de la liaison entre noyau et mésosome chez *Bacillus subtilis*. *Ann. Inst. Pasteur*, **107**, 384 (*545, 568, 572*).

RYTER A. & JACOB F. (1966). Ségrégation des noyaux chez *Bacillus subtilis* au cours de la germination des spores. *C. R. Acad. Sci., Paris*, **263**, 1176 (*573*).

RYTER A. & KELLENBERGER E. (1958a). Étude au microscope électronique de plasma contenant de l'acide désoxyribonucléique. I. Les nucléoides des bactéries en croissance active. *Z. Naturforsch.*, **13b**, 597 (*552*).

RYTER A. & KELLENBERGER E. (1958b). L'inclusion au polyester pour l'ultramicrotomie. *J. Ultrastructure Research*, **2**, 200 (*552*).

SALAS M., SMITH M., STANLEY W. M. JNR., WAHBA A. & OCHOA S. (1965). Direction of reading of the genetic message. *J. Biol. Chem.*, **240**, 3988 (*285*).

SALIVAR W. O., TZAGOLOFF H. & PRATT D. (1964). Some physical-chemical and biological properties of the rod-shaped coliphage M13. *Virology*, **24**, 359 (*444, 445*).

SAMBROOK J. F., FAN D. P. & BRENNER S. (1967). A strong suppressor specific for UGA. *Nature, Lond.*, **214**, 452 (*363*).

SANDERSON K. E. (1967). Revised linkage map of *Salmonella typhimurium*. *Bacteriol. Rev.*, **31**, 354 (*167, 375, 669, 726, 753*).

SANDERSON K. E. & DEMEREC M. (1965). The linkage map of *Salmonella typhimurium*. *Genetics*, **51**, 897 (*740, 753*).

SANGER F. (1956). In *Currents in Biochemical Research* (ed. Green D. E.). Interscience Publishers, New York (*265*).

SARABHAI A. S., STRETTON A. O. W., BRENNER S. & BOLLE A. (1964). Colinearity of the gene with the polypeptide chain. *Nature, Lond.*, **201**, 13 (*287, 345, 346, 361, 362*).

SASAKI I. & BERTANI G. (1965). Growth abnormalities in Hfr derivatives of *Escherichia coli* strain C. *J. gen. Microbiol.*, **40**, 365 (*673, 749*).

SATO K. & MATSUSHIRO A. (1965). The tryptophan operon regulated by phage immunity. *J. Mol. Biol.*, **14**, 608 (*726*).

SCAIFE J. (1966). F-prime factor formation in *E. coli* K12. *Genet. Res., Camb.*, **8**, 189 (*797*).

SCAIFE J. (1967). Episomes. *Ann. Rev., Microbiol.*, **21**, 601 (*104, 677, 679, 778*).

SCAIFE J. & BECKWITH J. R. (1966). Mutational alteration of the maximal level of *lac* operon expression. *Cold Spr. Harb. Symp. quant. Biol.*, **31**, 403 (*728*).

SCAIFE J. & GROSS J. D. (1962). Inhibition of multiplication of an *F-lac* factor in Hfr cells of *Escherichia coli* K12. *Biochem. biophys. Res. Comm.*, **7**, 403 (*775*).

SCAIFE J. & GROSS J. D. (1963). The mechanism of chromosome mobilisation by an F-prime factor in *Escherichia coli* K-12. *Genet. Res., Camb.*, **4**, 328 (*779*).

SCAIFE J. & PEKHOV A. P. (1964). Deletion of chromosome markers in association with F-prime factor formation in *Escherichia coli* K12. *Genet. Res., Camb.*, **5**, 495 (*797*).

SCHACKMANN H. K., PARDEE A. B. & STANIER R. J. (1952). Studies on macromolecular organisations of microbial cells. *Arch. Biochem. Biophys.*, **38**, 245 (*270*).

SCHAECHTER M., BENTZON M. W. & MAALØE O. (1959). Synthesis of deoxyribonucleic acid during the division cycle of bacteria. *Nature, Lond.*, **183**, 1207 (*554*).

SCHAECHTER M., MAALØE O. & KJELGAARD N. O. (1958). Dependency on medium and temperature of cell size and chemical composition during balanced growth of *Salmonella typhimurium*. *J. gen. Microbiol.*, **19**, 592 (*548*).

SCHAEFFER P. (1956). Transformation interspécifique chez des bactéries du genre *Hemophilus*. *Ann. Inst. Pasteur*, **91**, 192 (*584*).

SCHAEFFER P. (1957). L'inhibition de la transformation comme moyen de mesure de la compétence bactérienne. *C. R. Acad. Sci., Paris*, **245**, 451 (*579, 581, 582, 587*).

SCHAEFFER P. (1958). Interspecific reactions in bacterial transformation. In *The Biological Replication of Macromolecules, Symp. Soc. Exptl. Biol.*, **12**, 60 (*580, 588, 605*).

SCHAEFFER P. (1964). Transformation. In *The Bacteria* (Gunsalus I. C. & Stanier R. Y.), vol. V, p. 87. Academic Press, Inc., New York and London (*576, 578, 593, 598*).

SCHAEFFER P. & IONESCO H. (1959). Sur la transformation de *Bacillus subtilis*. *C. R. Acad. Sci., Paris*, **249**, 481 (*584*).

SCHAEFFER P., EDGAR R. S. & ROLFE R. (1960). Sur l'inhibition de la transformation bactérienne par des désoxyribonucléates de compositions variées. *C. R. Soc. Biol.*, **154**, 1978 (*593*).

SCHELL J. & GLOVER S. W. (1966a). The effect of heat on host-controlled restriction of phage λ in *Escherichia coli* K (P1). *J. gen. Microbiol.*, **45**, 61 (*516, 521*).

SCHELL J. & GLOVER S. W. (1966b). On the localisation of a factor responsible for host-controlled restriction in *Escherichia coli* K (P1). *Genet. Res., Camb.*, **7**, 277 (*521*).

SCHELL J., GLOVER S. W., STACEY K. A., BRODA P. M. A. & SYMONDS N. (1963). The restriction of phage T3 by certain strains of *Escherichia coli*. *Genet. Res., Camb.*, **4**, 483 (*671*).

SCHILDKRAUT C. L., RICHARDSON C. C. & KORNBERG A. (1964). Enzymic synthesis of deoxyribonucleic acid. XVII. Some unusual physical properties of the product primed by native DNA templates. *J. Mol. Biol.*, **9**, 24 (*253, 254*).

SCHLESINGER M. J. (1964). *In vitro* complementation and subunit structure of *E. coli* alkaline phosphatase. *Brookhaven Symp. Biol.*, **17**, 66 (*156*).

SCHLESINGER M. J. & LEVINTHAL C. (1963). Hybrid protein formation of *E. coli* alkaline phosphatase leading to *in vitro* complementation. *J. Mol. Biol.*, **7**, 1 (*156*).

SCHLESINGER M. J. & LEVINTHAL C. (1965). Complementation at the molecular level of enzyme interaction. *Ann. Rev. Microbiol.*, **19**, 267 (*156, 157*).

SCHMIDT J. M. & STANIER R. Y. (1965). Isolation and characterisation of bacteriophage active against stalked bacteria. *J. gen. Microbiol.*, **39**, 95 (*441*).

SCHWARTZ F. M. & ZINDER N. D. (1963). Crystalline aggregates in bacterial cells infected with the RNA bacteriophage f2. *Virology*, **21**, 276 (*442*).

SCHWARZ V., GOLDBERG L., KOMROWER G. M. & HOLZEL A. (1956). Some disturbances of erythrocyte metabolism in galactosemia. *Biochem. J.*, **62**, 34 (*110*).

SCOTT D. W. (1965). Serological cross reactions among the RNA coliphages. *Virology*, **26**, 85 (*441*).

SECHAUD J. & KELLENBERGER E. (1956). Lyse précoce, provoquée par le chloroforme, chez les bactéries infectées par du bactériophage. *Ann. Inst. Pasteur*, **90**, 102 (*424*).

SECHAUD J., STREISINGER G., EMRICH J., NEWTON J., LANFORD H., REINHOLD H. & STAHL M. M. (1965). Chromosome structure in phage T4. II. Terminal redundancy and heterozygosis. *Proc. Natl. Acad. Sci., Wash.*, **54**, 1333 (*499, 500*).

SERMONTI G. & CASCIANO S. (1963). Sexual polarity in *Streptomyces coelicolor. J. gen. Microbiol.*, **33**, 293 (*699*).

SERMONTI G. & HOPWOOD D. A. (1964). Genetic recombination in Streptomyces. In *The Bacteria* (eds. Gunsalus I. C. & Stanier R. Y.), vol. 5, p. 223. Academic Press, New York and London (*697, 698, 699*).

SETLOW J. K. (1966). The molecular basis of biological effects of ultraviolet irradiation and photoreactivation. In *Current topics in radiation research* (eds. Ebert M. & Howard A.), vol. II, p. 195. North-Holland Publishing Co., Amsterdam (*329, 330, 337, 338*).

SETLOW J. K. & BOLLING M. E. (1965). The resistance of *Micrococcus radiodurans* to ultraviolet radiation. II. Action spectra for killing delay in DNA synthesis, and thymine dimerisation. *Biochim. biophys. Acta*, **108**, 259 (*336*).

SETLOW R. B. & CARRIER W. L. (1963). Identification of ultraviolet-induced thymine dimers in DNA by absorbance measurements. *Photochem. Photobiol.*, **2**, 49 (*330, 338*).

SETLOW R. B. & CARRIER W. L. (1964). The disappearance of thymine dimers from DNA: an error correcting mechanism. *Proc. Natl. Acad. Sci., Wash.*, **51**, 226 (*255, 333*).

SETLOW J. K. & DUGGAN D. E. (1964). The resistance of *Micrococcus radiodurans* to ultraviolet radiation. *Biochim. biophys. Acta*, **87**, 664 (*336*).

SETLOW R., CARRIER W. & BOLLUM F. (1965). Pyrinidine dimers in UV-irradiated dI:dC. *Proc. Natl. Acad. Sci., Wash.*, **53**, 1111 (*329, 332*).

SHALITIN C. & STAHL F. W. (1965). Additional evidence for two kinds of heterozygotes in phage T4. *Proc. Natl. Acad. Sci., Wash.*, **54**, 1340 (*500*).

SHAPIRO J. (1967). Personal communication (*461, 635, 722*).

SHAPIRO J. A. (1967). The galactose operon. Ph.D. Thesis, Univ. Cambridge, England (*110*).

SHAW W. V. (1967). The enzymatic acetylation of chloramphenicol by extracts of R factor-resistant *Escherichia coli*. *J. biol. Chem.*, **242**, 687 (*766*).

SHEDLOVSKY A. & BRENNER S. (1963). A chemical basis for the host-induced modification of T-even bacteriophages. *Proc. Natl. Acad. Sci., Wash.*, **50**, 300 (*520*).

SHEPPARD D. & ENGLESBERG E. (1966). Positive control in the L-arabinose gene-enzyme complex of *Escherichia coli* B/r as exhibited with stable merodiploids. *Cold Spr. Harb. Symp. quant. Biol.*, **31**, 345 (*731*).

SHEPPARD D. E. & ENGLESBERG E. (1967). Further evidence for positive control of the L-arabinose system by gene *ara C*. *J. Mol. Biol.*, **25**, 443(*731*).

SHERMAN J. M. & WING H. U. (1937). Attempts to reveal sex in bacteria; with some light on fermentative variability in the coli-aerogenes group. *J. Bacteriol.*, **33**, 315 (*651*).

SHUGAR D. (1960). Photochemistry of nucleic acids and their constituents. In *The Nucleic Acids* (eds. Chargoff E. & Davidson J. N.), vol. III, p. 39. Academic Press, Inc., New York (*329*).

SICARD A. M. & EPHRUSSI-TAYLOR H. (1965). Genetic recombination in DNA-induced transformation of *Pneumococcus*. II. Mapping the *ami A* region. *Genetics*, **52**, 1207 (*617*).

SICCARDI A. G. (1966). Colicin resistance associated with resistance factors in *Escherichia coli*. *Genet. Res., Camb.*, **8**, 219 (*770*).

SIDDIQI O. (1963). Incorporation of parental DNA into genetic recombinants of *E. coli*. *Proc. Natl. Acad. Sci., Wash.*, **49**, 589 (*397*).

SIDDIQI O. (1965). Interallelic complementation *in vivo* and *in vitro*. *Brit. Med. Bull.*, **21**, 249 (*157*).

SIDDIQI O. & PUTRAMENT A. (1963). Polarised negative interference in the *paba*-1 region of *Aspergillus nidulans*. *Genet. Res., Camb.*, **4**, 12 (*400*).

SIGNER E. R. (1964). Recombination between coliphages λ and φ80. *Virology*, **22**, 650 (*461*).

SIGNER E. R. (1966). Interaction of prophages at the att_{80} site with the chromosome of *E. coli*. *J. Mol. Biol.*, **15**, 243 (*461, 633*).

SIGNER E. & BECKWITH J. (1966). Transposition of the *lac* region of *E. coli*. III. The mechanism of attachment of coliphage φ80 to the bacterial chromosome. *J. Mol. Biol.*, **22**, 33 (*473*).

SIGNER E. R., BECKWITH J. R. & BRENNER S. (1965). Mapping of suppressor loci in *Escherichia coli*. *J. Mol. Biol.*, **14**, 153 (*461*).

SILVER S. D. (1963). Transfer of material during mating in *Escherichia coli*. Transfer of DNA and upper limits on the transfer of RNA and protein. *J. Mol. Biol.*, **6**, 349 (*654, 700*).

Silver S. & Ozeki H. (1962). Transfer of deoxyribonucleic acid accompanying the transmission of colicinogenic properties by cell mating. *Nature, Lond.*, **195**, 873 (*770*).

Simpkin J. L. & Work T. S. (1957). Protein synthesis in guinea pig liver. Incorporation of radioactive amino acids into proteins of the microsome fraction *in vivo*. *Biochem. J.*, **65**, 307 (*271*).

Sinsheimer R. L. (1959a). Purification and properties of bacteriophage φX174. *J. Mol. Biol.*, **1**, 37 (*419, 437*).

Sinsheimer R. L. (1959b). A single-stranded DNA from bacteriophage φX174. *J. Mol. Biol.*, **1**, 43 (*437*).

Sinsheimer R. L. (1960). The nucleic acids of the bacterial viruses. In *The Nucleic Acids* (eds. Chargoff E. & Davidson J. N.), vol. 3, p. 187. Academic Press, Inc., New York (*410*).

Sinsheimer R. L. (1962). Single-stranded DNA. *Scientific American*, **207**, 109 (*436, 437*).

Sinsheimer R. L. & Lawrence M. (1964). *In vitro* synthesis and properties of a φX DNA-RNA hybrid. *J. Mol. Biol.*, **8**, 289 (*290*).

Sinsheimer R. L., Starman B., Nagler C. & Guthrie S. (1962). The process of infection with bacteriophage φX174. I. Evidence for a 'replicative form'. *J. Mol. Biol.*, **4**, 142 (*437, 439*).

Sironi G., Gallucci E. & Maccacaro G. A. (1964). Possibility of conjugal infection by a male-specific bacteriophage in *Escherichia coli*. *Giornale di Microbiol.*, **12**, 195 (*442*).

Skalka A. (1968). Regional and temporal control of genetic transcription in phage lambda. *Proc. Natl. Acad. Sci., Wash.*, **55**, 1190 (*475*).

Six E. (1961). Inheritance of prophage P_2 in superinfection experiments. *Virology*, **14**, 220 (*477*).

Skaar D. & Garen A. (1955). Transfer of DNA accompanying genetic recombination in *Escherichia coli* K-12. *Genetics*, **40**, 596 (*654*).

Slotnick C. J., Visser D. W. & Rittenberg S. C. (1953). Growth inhibition of purine-requiring mutants of *Escherichia coli* by 5-hydroxuridine. *J. biol. Chem.*, **203**, 647 (*269*).

Sly W. S., Echols H. & Adler J. (1965). Control of viral messenger RNA after lambda phage infection and induction. *Proc. Natl. Acad. Sci., Wash.*, **53**, 378 (*466*).

Smellie R. M. S. (1963). The biosynthesis of ribonucleic acid in animal systems. In *Progress in Nucleic Acid Research* (eds. Davidson J. N. & Cohn W. E.), vol. 1, p. 27. Academic Press, New York and London (*290*).

Smellie R. M. S. (1965). Biochemistry of deoxyribonucleic acid and ribonucleic acid replication. *Brit. Med. Bull.*, **21**, 195 (*246, 252, 255, 290, 296*).

Smith B. R. (1966). Genetic controls of recombination. I. The recombination-2 gene of *Neurospora crassa*. *Heredity*, **21**, 481 (*402*).

Smith D. W. E. & Ames B. N. (1964). Intermediates in the early steps of histidine biosynthesis. *J. Biol. Chem.*, **239**, 1848 (*160*).

Smith D. W. E. & Ames B. N. (1965). Phosphoribosyladenosine monophosphate, an intermediate in histidine biosynthesis. *J. Biol. Chem.*, **240**, 3056 (*160*).

SMITH D. W. & HANAWALT P. C. (1967). Properties of the growing point region in the bacterial chromosome. *Biochim. Biophys. Acta*, **194**, 519 (*569*).

SMITH M. G. & SKALKA A. (1966). Some properties of the DNA from phage-infected bacteria. *J. gen. Physiol.*, **49**, No. 6, Part 2, 127 (*543*).

SMITH S. M. & STOCKER B. A. D. (1962). Colicinogeny and recombination. *Brit. Med. Bull.*, **18**, 46 (*740, 752, 757, 759*).

SMITH S. M. & STOCKER B. A. D. (1966). Mapping of prophage P22 in *Salmonella typhimurium*. *Virology*, **28**, 413 (*479, 640*).

SMITH S. M., OZEKI H. & STOCKER B. A. D. (1963). Transfer of *colE1* and *colE2* during high frequency transmission of *colI* in *Salmonella typhimurium*. *J. Gen. Microbiol.*, **33**, 231 (*759*).

SMITH H. WILLIAMS & HALLS S. (1967). The transmissible nature of the genetic factor in *Escherichia coli* that controls haemolysin production. *J. gen. Microbiol.*, **47**, 153 (*770*).

SMITH H. WILLIAMS & HALLS S. (1968). The transmissible nature of the genetic factor in *Escherichia coli* that controls enterotoxin production. *J. gen. Microbiol.*, **52**, 319 (*770*).

SMITH-KEARY P. F. (1960). A suppressor of leucineless in *Salmonella typhimurium*. *Heredity*, **14**, 61 (*216*).

SMITH-KEARY P. F. (1966). Restricted transduction by bacteriophage P22 in *Salmonella typhimurium*. *Genet. Res., Camb.*, **8**, 73 (*221*).

SMITH-KEARY P. F. & DAWSON G. W. P. (1964). Episomic suppression of phenotype in *Salmonella*. *Genet. Res., Camb.*, **5**, 269 (*221, 223*).

SNEATH P. H. A. & LEDERBERG J. (1961). Inhibition by periodate of mating in *Escherichia coli* K-12. *Proc. Natl. Acad. Sci., Wash.*, **47**, 86 (*671*).

SÖLL D., OHTSUKA E., JONES D. S., LOHRMANN R., HAYATSU H., NISHIMURA S. & KHORANA H. G. (1965). Studies on polynucleotides, XLIX. Stimulation of the binding of aminoacyl-sRNAs to ribosomes by ribotrinucleotides and survey of codon assignments for 20 amino acids. *Proc. Natl. Acad. Sci., Wash.*, **54**, 1378 (*359*).

SONNEBORN R. M. (1950). The cytoplasm in heredity. *Heredity*, **4**, 11 (*453*).

SONNEBORN R. M. (1965). In *Evolving genes and proteins* (eds. Bryson V. & Vogel H. J.), p. 377. Academic Press, New York and London (*367*).

SPENCER H. T. & HERIOT R. M. (1965). Development of competence of *Hemophilus influenzae*. *J. Bacteriol.*, **90**, 911 (*586*).

SPEYER J. F. (1965). Mutagenic DNA polymerase. *Biochem. biophys. Res. Comm.*, **21**, 6 (*219, 256*).

SPEYER J. F., LENGYEL P., BASILIO C. & OCHOA S. (1962a). Synthetic polynucleotides and the amino acid code. II. *Proc. Natl. Acad. Sci., Wash.*, **48**, 63 (*356, 357*).

SPEYER J. F., LENGYEL P., BASILIO C. & OCHOA S. (1962b). Synthetic polynucleotides and the amino acid code, IV. *Proc. Natl. Acad. Sci., Wash.*, **48**, 441 (*356*).

SPEYER J. F., LENGYEL P., BASILIO C., WAHBA A. J., GARDNER R. S. & OCHOA S. (1963). Synthetic polynucleotides and the amino acid code. *Cold Spr. Harb. Symp. quant. Biol.*, **28**, 559 (*356*).

SPIEGELMAN S. (1957). Nucleic acids and the synthesis of proteins. In *The*

Chemical Basis of Heredity (eds. McElroy W. D. & Glass B.), p. 232. Johns Hopkins Press, Baltimore (*270*).

SPIEGELMAN S. & DOI R. H. (1963). Replication and translation of RNA genomes. *Cold Spr. Harb. Symp. quant. Biol.*, **28**, 109 (*291*, *442*, *444*).

SPIEGELMAN S. & HAYASHI M. (1963). The present status of the transfer of genetic information and its control. *Cold Spr. Harb. Symp. quant. Biol.*, **28**, 161 (*289*).

SPIEGELMAN S., HARUNA I., HOLLAND I. B., BEAUDREAU G. & MILLS D. (1965). The synthesis of a self-propagating and infectious nucleic acid with a purified enzyme. *Proc. Natl. Acad. Sci., Wash.*, **54**, 919 (*443*).

SPIZIZEN J. (1958). Transformation of a biochemically deficient strain of *Bacillus subtilis* by deoxyribonucleate. *Proc. Natl. Acad. Sci., Wash.*, **44**, 1072 (*577*).

SPIZIZEN J. (1959). Genetic activity of deoxyribonucleic acid in the reconstitution of biosynthetic pathways. *Fed. Proc.*, **18**, 957 (*584*, *610*).

SPIZIZEN J., REILLY B. E. & EVANS A. H. (1966). Microbial transformation and transfection. *Ann. Rev. Microbiol.*, **20**, 371 (*576*, *587*, *590*, *591*, *594*, *618*).

SRB A. M. & HOROWITZ N. H. (1944). The ornithine cycle in *Neurospora* and its genetic control. *J. biol. Chem.*, **154**, 129 (*114*, *115*).

SRB A. M., FINCHAM J. R. S. & BONNER D. (1950). Evidence from gene mutations in *Neurospora* for close metabolic relationships among ornithine, proline, and α-amino-δ-hydroxyvaleric acid. *Amer. J. Bot.*, **37**, 533 (*116*).

STACEY K. A. (1965a). The biosynthesis of nucleic acids and their roles in protein synthesis. *Symp. Soc. gen. Microbiol.*, **15**, 159 (*256*, *296*).

STACEY K. A. (1965b). Intracellular modification of nucleic acids. *Brit. Med. Bull.*, **21**, 211 (*281*, *520*).

STACEY K. A. & SIMSON E. (1965). Improved method for the isolation of thymine-requiring mutants of *Escherichia coli*. *J. Bacteriol.*, **90**, 554 (*573*).

STACEY K. A., SYMONDS N., GLOVER S. W. & SCHELL J. (1964). The chemical changes of phage DNA involved in host-induced modification. In *Struktur und Funktion des genetischen Materials* (ed. Strebbe H.) (Abh. dt. Akad. Wiss. Berl. Klasse für Medizin, No. 4, p. 35). Akademie-Verkg, Berlin (*520*).

STADLER L. J. (1951). Spontaneous mutation in maize. *Cold Spr. Harb. Symp. quant. Biol.*, **16**, 49 (*129*).

STADTMAN E. R., COHEN G. N., LEBRAS G. & DEROBICHON-SZULMAJSTER H. (1961). Selective feedback inhibition and repression of two aspartokinases in the metabolism of *Escherichia coli*. *Cold Spr. Harb. Symp. quant. Biol.*, **26**, 319 (*707*).

STAHL F. W. (1959). Radiobiology of bacteriophage. In *The Viruses* (eds. Burnet F. M. & Stanley W. M.), vol. 2, p. 353. Academic Press, New York (*410*, *428*, *523*, *530*).

STAHL F. W. & STEINBERG C. M. (1964). The theory of formal phage genetics for circular maps. *Genetics*, **50**, 531 (*490*).

STAHL F. W., EDGAR R. S. & STEINBERG J. (1964). The linkage map of bacteriophage T4. *Genetics*, **50**, 539 (*145*).

Stahl F. W., Crasemann J. M., Okun L., Fox E. & Laird C. (1961). Radiation sensitivity of bacteriophage containing 5-bromodeoxyuridine. *Virology*, **13**, 98 (*527*).

Stahl F. W., Modersohn H., Terzaghi B. E. & Crasemann J. M. (1965). The genetic structure of complementation heterozygotes. *Proc. Natl. Acad. Sci., Wash.*, **54**, 1342 (*500*).

Steinberg C. M. & Edgar R. S. (1962). A critical test of a current theory of genetic recombination in bacteriophage. *Genetics*, **47**, 187 (*497, 498, 508*).

Stent G. S. (1955). Decay of incorporated radioactive phosphorus during reproduction of bacteriophage T2. *J. gen. Physiol.*, **38**, 853 (*531*).

Stent G. S. (1958). Mating in the reproduction of bacterial viruses. *Advanc. in Virus Res.*, **5**, 95 (*419, 523*).

Stent G. S. (1959). Intracellular multiplication of bacterial viruses. In *The Viruses* (eds. Burnet F. M. & Stanley W. M.), vol. 2, p. 237. Academic Press, New York (*410, 487*).

Stent G. S. (1963). *Molecular biology of bacterial viruses.* W. H. Freeman and Co., San Francisco and London (*407, 410, 440*).

Stent G. S. (1964). The operon on its third anniversary. *Science*, **144**, 816 (*721*).

Stent G. S. (1966). Genetic transcription. *Proc. Roy. Soc., B.*, **164**, 181 (*293, 294, 721*).

Stent G. S. & Fuerst C. R. (1955). Inactivation of bacteriophages by decay of incorporated radioactive phosphorus. *J. gen. Physiol.*, **38**, 441 (*437, 524, 525*).

Stent G. S. & Fuerst C. R. (1960). Genetic and physiological effects of the decay of incorporated radioactive phosphorus in bacterial viruses and bacteria. *Advances in Biol. Med. Phys.*, **7**, 1 (*523*).

Stent G. S. & Maaløe O. (1953). Radioactive phosphorus tracer studies on the reproduction of T4 bacteriophage. II. Kinetics of phosphorus assimilation. *Biochim. biophys. Acta*, **10**, 55 (*427*).

Stent G. S., Fuerst C. R. & Jacob F. (1957). Inactivation d'un prophage par la désintégration du radiophosphore. *C. R. Acad. Sci., Paris*, **244**, 1840 (*455*).

Stern C. (1936). Somatic crossing-over and segregation in *Drosophila melanogaster*. *Genetics*, **21**, 625 (*78*).

Stirm S., Ørskov F., Ørskov I. & Mansa B. (1967). Episome-carried surface antigen K88 of *Escherichia coli*. II. Isolation and chemical analysis. *J. Bacteriol.*, **93**, 731 (*770*).

Stocker B. A. D. (1949). Measurements of rate of mutation of flagellar antigenic phase in *Salmonella typhimurium*. *J. Hyg., Camb.*, **47**, 398 (*199, 200, 738*).

Stocker B. A. D. (1956). Abortive transduction of motility in *Salmonella*, a non-replicated gene transmitted through many generations to a single descendant. *J. gen. Microbiol.*, **15**, 575 (*97, 99*).

Stocker B. A. D. (1960). Microorganisms in genetics. *Symp. Soc. gen. Microbiol.*, **10**, 1 (*759*).

Stocker B. A. D. (1963). Transformation of *Bacillus subtilis* to motility and prototrophy: micro-manipulative isolation of bacteria of transformed phenotype. *J. Bacteriol.*, **86**, 797 (*618*).

STOCKER B. A. D., MCDONOUGH M. W. & AMBLER R. P. (1961). A gene determining presence or absence of N-methyl-lysine in *Salmonella* flagellar protein. *Nature, Lond.*, **189**, 556 (*742*).

STOCKER B. A. D., SMITH S. M. & OZEKI H. (1963). High infectivity of *Salmonella typhimurium* newly infected by the *colI* factor. *J. gen. Microbiol.*, **30**, 201 (*758*).

STOCKER B. A. D., ZINDER N. D. & LEDERBERG J. (1953). Transduction of flagellar characters. *J. gen. Microbiol.*, **9**, 410 (*96, 97, 620, 623, 636, 641, 740*).

STRACK H. B. & KAISER A. D. (1965). On the structure of the ends of lambda DNA. *J. Mol. Biol.*, **12**, 36 (*462, 544, 588*).

STRAUSS B. & OKUBO S. (1960). Protein synthesis and the induction of mutations in *Escherichia coli* by alkylating agents. *J. Bacteriol.*, **79**, 464 (*210*).

STRAUSS N. (1965). Configuration of transforming deoxyribonucleic acid during entry into *Bacillus subtilis*. *J. Bacteriol.*, **89**, 288 (*595*).

STRAUSS N. (1966). Further evidence concerning the configuration of transforming deoxyribonucleic acid during entry into *Bacillus subtilis*. *J. Bacteriol.*, **91**, 702 (*595*).

STREISINGER G. (1956a). The genetic control of ultraviolet sensitivity levels in bacteriophage T2 and T4. *Virology*, **2**, 1 (*527*).

STREISINGER G. (1956b). Phenotypic mixing of host range and serological specificities in bacteriophages T2 and T4. *Virology*, **2**, 388 (*105, 107, 483, 513*).

STREISINGER G. & BRUCE V. (1960). Linkage of genetic markers in phages T2 and T4. *Genetics*, **45**, 1289 (*489*).

STREISINGER G. & FRANKLIN N. C. (1956). Mutation and recombination at the host range genetic region of phage T2. *Cold Spr. Harb. Symp. quant. Biol.*, **21**, 103 (*483*).

STREISINGER G., EDGAR R. S. & DENHARDT G. H. (1964). Chromosome structure in phage T4. I. Circularity of the linkage map. *Proc. Natl. Acad. Sci., Wash.*, **51**, 775 (*433, 434, 489*).

STREISINGER G., EDGAR R. & HARRAR G. (1961). Data quoted in F. Stahl, *J. Chimie Physique*, **58**, 1072 (*489*).

STREISINGER G., EMRICH J. & STAHL M. M. (1967). Chromosome structure in phage T4, III. Terminal redundancy and length determination. *Proc. Natl. Acad. Sci., Wash.*, **57**, 292 (*434*).

STREISINGER G., MUKAI F., DREYER W. J., MILLER B. & HORIUCHI S. (1961). Mutations affecting the lysozyme of phage T4. *Cold Spr. Harb. Symp. quant. Biol.*, **26**, 25 (*299, 342, 435*).

STRETTON A. O. W. (1965). The genetic code. *Brit. Med. Bull.*, **21**, 229 (*287, 360, 363, 368*).

STRETTON A. O. W. & BRENNER S. (1965). Molecular consequences of the amber mutation and its suppression. *J. Mol. Biol.*, **12**, 456 (*345, 362*).

STRIGINI P. (1964). Sul meccanismo della reversione di mutanti r_{II} nel fago T4. *Atti Assn. gen. It., Pavia*, **9**, 57 (*324*).

STURTEVANT A. H. (1913). The linear arrangement of six sex-linked factors in *Drosophila* as shown by their mode of association. *J. exp. Zool.*, **14**, 43 (*36*).

STUY J. H. (1962). Transformability of *Haemophilus influenzae*. *J. gen. Microbiol.*, **29**, 537 (*588*).

STUY J. H. & STERN D. (1964). The kinetics of DNA uptake by *Haemophilus influenzae*. *J. gen. Microbiol.*, **35**, 391 (*595*).

SUBAK-SHARPE H., SHEPHERD W. M. & HAY J. (1966). Studies on SRNA coded by herpes virus. *Cold Spr. Harb. Symp. quant. Biol.*, **31**, 583 (*366*).

SUEOKA N. (1961a). Variation and heterogeneity of base composition of deoxyribonucleic acids: a compilation of old and new data. *J. Mol. Biol.*, **3**, 31 (*231*).

SUEOKA N. (1961b). Correlation between base composition of deoxyribonucleic acid and amino acid composition of protein. *Proc. Natl. Acad. Sci., Wash.*, **47**, 1141 (*366*).

SUEOKA N. & YOSHIKAWA H. (1963). Regulation of chromosome replication in *Bacillus subtilis*. *Cold Spr. Harb. Symp. quant. Biol.*, **28**, 47 (*561, 563, 564*).

SUGINO Y. & HIROTA Y. (1962). Conjugal fertility associated with resistance factor *R* in *Escherichia coli*. *J. Bacteriol.*, **84**, 902 (*779*).

SUSKIND S. R. & KUREK L. I. (1959). On a mechanism of suppressor gene regulation of tryptophan synthetase activity in *Neurospora crassa*. *Proc. Natl. Acad. Sci., Wash.*, **45**, 193 (*215*).

SUSMAN M., PIECHOWSKI M. M. & RITCHIE D. A. (1965). Studies on phage development. I. An acridine-sensitive clock. *Virology*, **26**, 163 (*430*).

SUSSMAN R. & JACOB F. (1962). Sur un système de répression thermosensible chez le bactériophage λ d'*Escherichia coli*. *C. R. Acad. Sci., Paris*, **254**, 1517 (*467*).

SYMONDS N. D. (1957). Effects of ultraviolet light during the second half of the latent period on bacteria infected with phage T2. *Virology*, **3**, 485 (*424*).

SYMONDS N. D. (1958). The properties of a star mutant of phage T2. *J. gen. Microbiol.*, **18**, 330 (*481*).

SYMONDS N. (1962a). The effect of pool size on recombination in phage. *Virology*, **18**, 334 (*507*).

SYMONDS N. (1962b). The kinetics of chromosome transfer in *E. coli*: a mathematical treatment. *Genet. Res., Camb.*, **3**, 273 (*689*).

SYMONDS N. D. & MCCLOY E. W. (1958). The irradiation of phage-infected bacteria: its bearing on the relationship between functional and genetic radiation damage. *Virology*, **6**, 649 (*531*).

SYMONDS N. D. & RITCHIE D. A. (1961). Multiplicity reactivation after the decay of incorporated radioactive phosphorus in phage T4. *J. Mol. Biol.*, **3**, 61 (*532*).

SYMONDS N., STACEY K. A., GLOVER S. W., SCHELL J. & SILVER S. (1963). The chemical basis for a case of host-induced modification in phage T2. *Biochem. biophys. Res. Comm.*, **12**, 220 (*520*).

SZYBALSKI W. (1958). Special microbiological systems. II. Observations on chemical mutagenesis in microorganisms. *Ann. New York Acad. Sci.*, **76**, 475 (*204*).

SZYBALSKI W. & OPARA-KUBINSKA Z. (1965). Physico-chemical and biological properties of genetic markers in transforming DNA. *Symp. Biol., Hung.*, **6**, 43 (*592*).

TAKAHASHI I. (1961). Genetic transduction in *Bacillus subtilis. Biochem. biophys. Res. Comm.*, **5**, 171 (*615, 621*).

TAKAHASHI I. (1963). Transducing phages for *Bacillus subtilis. J. gen. Microbiol.*, **31**, 211 (*614, 621*).

TAKANAMI M. & OKAMOTO T. (1963). Interaction of ribosomes and synthetic polyribonucleotides. *J. Mol. Biol.*, **7**, 323 (*285*).

TAKANO T., WATANABE T. & FUKASAWA T. (1966). Specific inactivation of infectious λ DNA by sonicates of restrictive bacteria with R factors. *Biochem. biophys. Res. Comm.*, **25**, 192 (*521*).

TAKEYA N. & AMAKO K. (1966). A rod-shaped *Pseudomonas* phage. *Virology*, **28**, 163 (*419, 445*).

TATUM E. L. & BONNER D. (1944). Indole and serine in the biosynthesis and breakdown of tryptophan. *Proc. Natl. Acad. Sci., Wash.*, **30**, 30 (*117*).

TATUM E. L. & LEDERBERG J. (1947). Gene recombination in the bacterium *Escherichia coli. J. Bacteriol.*, **53**, 673 (*652*).

TATUM L., BONNER D. & BEADLE G. W. (1944). Anthranilic acid and the biosynthesis of indole and tryptophan by *Neurospora. Arch. Biochem.*, **3**, 477 (*117*).

TATUM E. L., GROSS S. R., EHRENSVARD G. & GARNJOBST L. (1954). Synthesis of aromatic compounds by *Neurospora. Proc. Natl. Acad. Sci., Wash.*, **40**, 271 (*119*).

TAYLOR A. L. (1963). Bacteriophage-induced mutation in *Escherichia coli. Proc. Natl. Acad. Sci., Wash.*, **50**, 1043 (*222*).

TAYLOR A. L. & ADELBERG E. A. (1960). Linkage analysis with very high frequency males of *Escherichia coli. Genetics*, **45**, 1233 (*667, 688*).

TAYLOR A. L. & ADELBERG E. A. (1961). Evidence for a closed linkage group in Hfr males of *Escherichia coli* K-12. *Biochem. biophys. Res. Comm.*, **5**, 400 (*673*).

TAYLOR A. L. & DEMEREC M. (1963). Personal communication (*753*).

TAYLOR A. L. & THOMAN M. S. (1964). The chromosome map of *Escherichia coli* K12. *Genetics*, **50**, 659 (*667*).

TAYLOR A. L. & TROTTER C. D. (1967). Revised linkage map of *Escherichia coli. Bacteriol. Rev.*, **31**, 332 (*164, 666, 667, 669, 753*).

TAYLOR H. E. (1949). Transformations réciproques des formes R et ER chez le pneumocoque. *C. R. Acad. Sci., Paris*, **228**, 1258 (*606*).

TAYLOR J. H. (1957). The time and mode of duplication of chromosomes. *Amer. Nat.*, **91**, 209 (*389*).

TAYLOR J. H. (1963). In *Molecular genetics* (ed. Taylor J. H.), p. 65. Academic Press, Inc., New York and London (*534*).

TAYLOR J. H. (1965). Distribution of tritium-labeled DNA among chromosomes during meiosis. I. Spermatogenesis in the grasshopper. *J. Cell. Biol.*, **25**, 57 (*390*).

TAYLOR J. H., WOODS P. S. & HUGHES W. L. (1957). The organisation and duplication of chromosomes as revealed by autoradiograph studies using tritium-labelled thymidine. *Proc. Natl. Acad. Sci., Wash.*, **43**, 122 (*245, 390*).

TEAS H. J., HOROWITZ N. H. & FLING M. (1948). Homoserine as a precursor of threonine and methionine in *Neurospora. J. biol. Chem.*, **172**, 651 (*121*).

TERZAGHI E., OKADA Y., STREISINGER G., EMRICH J., INOUYE M. & TSUGITA A. (1966). Change of a sequence of amino acids in phage T4 lysozyme by acridine-induced mutations. *Proc. Natl. Acad. Sci., Wash.*, **56**, 500 (*342, 355, 361*).

TESSMAN E. S. (1965). Complementation groups in phage S13. *Virology*, **25**, 303 (*510*).

TESSMAN E. S. & TESSMAN I. (1959). Genetic recombination in phage S13. *Virology*, **7**, 465 (*510*).

TESSMAN I. (1959). Mutagenesis in phages φX714 and T4 and properties of the genetic material. *Virology*, **9**, 375 (*305*).

TESSMAN I., PODDAR R. K. & KUMAR S. (1964). Identification of the altered bases in mutated single-stranded DNA. I. *In vitro* mutagenesis by hydroxylamine, ethyl methane-sulphonate and nitrous acid. *J. Mol. Biol.*, **9**, 352 (*318, 510*).

THOMAS C. A. JNR. (1963). The arrangements of nucleotide sequences in T2 and T5 DNA molecules. *Cold Spr. Harb. Symp. quant. Biol.*, **28**, 395 (*539*).

THOMAS C. A. JNR. & MACHATTIE L. A. (1964). Circular T2 DNA molecules. *Proc. Natl. Acad. Sci., Wash.*, **52**, 1297 (*434, 539*).

THOMAS C. A. JNR. & MACHATTIE L. A. (1967). The anatomy of viral DNA molecules. *Ann. Rev. Biochem.*, **36** (Part 1), 485 (*542*).

THOMAS R. (1955). Recherches sur la cinétique des transformations bactériennes. *Biochim. biophys. Acta*, **18**, 467 (*583, 584, 585, 586, 590*).

THOMAS R. (1959). Effects of chloramphenicol on genetic replication in bacteriophage λ. *Virology*, **9**, 275 (*429*).

THOMAS R. (1962a). Recherches sur la structure et les fonctions des acides désoxyribonucléiques: études génétiques et chimiques. *Actualités Biochimiques*, **21**. Masson et Cie, Paris (*578, 585*).

THOMAS R. (1962b). Recherches sur la multiplication du matériel génétique du bactériophage tempéré lambda. *Actualités Biochimiques*, **21**, 86 (*429*).

THORNE C. B. (1961). Transduction in *Bacillus subtilis. Fed. Proc.*, **20**, 254 (*614, 621*).

TING R. C. (1962). The specific gravity of transducing particles of bacteriophage P1. *Virology*, **16**, 115 (*639*).

TISSIERES A., SCHLESSINGER D. & GROS F. (1960). Amino acid incorporation into proteins by *Escherichia coli* ribosomes. *Proc. Natl. Acad. Sci., Wash.*, **46**, 1450 (*277*).

TISSIERES A., WATSON J. D., SCHLESSINGER D. & HOLLINGWORTH B. R. (1959). Ribonucleoprotein particles from *Escherichia coli. J. Mol. Biol.*, **1**, 221 (*277*).

TOMASZ A. (1965). Control of the competent state in *Pneumococcus* by a hormone-like cell product: an example of a new type of regulatory mechanism in bacteria. *Nature, Lond.*, **208**, 155 (*586, 589*).

TOMASZ A. & HOTCHKISS R. D. (1964). Regulation of the transformability of pneumococcal cultures by macromolecular cell products. *Proc. Natl. Acad. Sci., Wash.*, **51**, 480 (*589, 590*).

TOMIZAWA J. & ANRAKU N. (1964). Molecular mechanisms of recombination in bacteriophage. II. Joining of parental DNA molecules in phage T4. *J. Mol. Biol.*, **8**, 516 (*398*).

TOMIZAWA J. & SUNAKAWA S. (1956). The effect of chloramphenicol on deoxyribonucleic acid synthesis and the development of resistance to ultraviolet irradiation in *E. coli* infected with bacteriophage T2. *J. gen. Physiol.*, **39**, 553 (*428*).

TRAUTNER T. A. (1958). Untersuchungen an Heterozygoten des Phagen T1. *Z. Vererbungsl.*, **89**, 266 (*497*).

TRAUTNER T. A. (1960). The influence of the multiplicity of infection on crosses with bacteriophage T1. *Z. Vererbungsl.*, **91**, 259 (*507*).

TRAUTNER T. A., SWARTZ M. N. & KORNBERG A. (1962). Enzymatic synthesis of deoxyribonucleic acid, X. Influence of bromouracil substitutions on replication. *Proc. Natl. Acad. Sci., Wash.*, **48**, 449 (*303*).

TREFFERS H. P., SPINELLI V. & BELSER N. O. (1954). A factor (or mutator gene) influencing mutation rates in *Escherichia coli. Proc. Natl. Acad., Sci., Wash.*, **40**, 1064 (*219, 220*).

TROMANS W. J. & HORNE R. W. (1961). The structure of bacteriophage φX174. *Virology*, **15**, 1 (*417, 419*).

TSUGITA A. & FRAENKEL-CONRAT H. (1960). The amino acid composition and C-terminal sequence of a chemically evoked mutant of TMV. *Proc. Natl. Acad. Sci., Wash.*, **46**, 636 (*238, 300, 341, 360*).

TSUGITA A. & FRAENKEL-CONRAT H. (1962). The composition of proteins of chemically evoked mutants of TMV RNA. *J. Mol. Biol.*, **4**, 73 (*238, 300, 341, 360*).

TULASNE R. & VENDRELY R. (1947). Demonstration of bacterial nuclei with ribonuclease. *Nature, Lond.*, **160**, 225 (*547*).

TURRI M. & MACCACARO G. A. (1960). Osservazioni microelettroforetiche su cellule di *E. coli* K-12 di diversa compatibilite sessuale. *Giorn. di Microbiol.*, **8**, 1 (*670*).

TZAGALOFF H. & PRATT D. (1964). The initial steps in infection with coliphage M13. *Virology*, **24**, 372 (*445*).

UETAKE H. & HAGIWARA S. (1960). Somatic antigen 15 as a precursor of antigen 34 in *Salmonella. Nature, Lond.*, **186**, 261 (*648*).

UETAKE H. & HAGIWARA S. (1961). Genetic co-operation between unrelated phages. *Virology*, **13**, 500 (*648*).

UETAKE H., NAKAGAWA T. & AKIBA T. (1955). The relationship of bacteriophage to antigenic changes in group E *Salmonella. J. Bacteriol.*, **69**, 571 (*647*).

UETAKE H., LURIA S. E. & BURROUS J. W. (1958). Conversion of somatic antigens in *Salmonella* by phage infection leading to lysis or lysogeny. *Virology*, **5**, 68 (*647, 648*).

UMBARGER H. E. (1956). Evidence for a negative-feedback mechanism in the synthesis of isoleucine. *Science*, **123**, 848 (*707*).

UMBARGER H. E. (1961). Feedback control by endproduct inhibition. *Cold Spr. Harb. Symp. quant. Biol.*, **26**, 301 (*707, 708*).

VALENTINE R. C., ENGELHARDT D. L. & ZINDER N. D. (1964). Host-dependent mutants of the bacteriophage f2. Rescue and complementation of mutants. *Virology*, **23**, 159 (*510*).

VAN DE PUTTE P., ZWENK H. & RÖRSCH A. (1966). Properties of four mutants of *Escherichia coli* defective in genetic recombination. *Mutation Res.*, **3**, 381 (*335*).

VAN DE PUTTE P., VAN SLUIS C. A., VAN DILLEWIJN J. & RÖRSCH A. (1965). The location of genes controlling radiation sensitivity in *Escherichia coli*. *Mutation Res.*, **2**, 97 (*334*).

VENEMA G., PRITCHARD R. H. & VENEMA-SCHRÖDER T. (1965a). Fate of transforming deoxyribonucleic acid in *Bacillus subtilis*. *J. Bacteriol.*, **89**, 1250 (*596, 600, 601*).

VENEMA G., PRITCHARD R. H. & VENEMA-SCHRÖDER T. (1965b). Properties of newly introduced transforming deoxyribonucleic acid in *Bacillus subtilis*. *J. Bacteriol.*, **90**, 343 (*600*).

VESTRI R., FELICETTI L. & LOSTIA O. (1966). Transformation by hybrid DNA in *Bacillus subtilis*. *Nature, Lond.*, **209**, 1154 (*601*).

VIDAVER G. A. & KOZLOFF L. M. (1957). The rate of synthesis of deoxyribonucleic acid in *Escherichia coli* B infected with $T2r^+$ bacteriophage. *J. biol. Chem.*, **225**, 335 (*426*).

VIELMETTER W. & SCHUSTER H. (1960). The base specificity of mutation induced by nitrous acid in phage T2. *Biochem. biophys. Res. Comm.*, **2**, 324 (*305, 310, 311*).

VINOGRAD J., LEBOWITZ J., RADLOFF R., WATSON R. & LAIPIS P. (1965). The twisted circular form of polyoma viral DNA. *Proc. Natl. Acad. Sci., Wash.*, **53**, 1104 (*440*).

VISCONTI N. (1966). Mating theory. In *Phage and the origins of molecular biology* (eds. Cairns J., Stent G. S. & Watson J. D.), p. 142. Cold Spring Harbor Laboratory of Quantitative Biology, Cold Spring Harbor, Long Island, New York (*501*).

VISCONTI N. & DELBRUCK M. (1953). The mechanism of genetic recombination in phage. *Genetics*, **38**, 5 (*62, 145, 501, 503, 505, 509*).

VOGEL H. J. (1955). On the glutamate-proline-ornithine interrelation in various organisms. In *A Symposium on Amino Acid Metabolism* (eds. McElroy W. D. & Glass H. B.), p. 335. Johns Hopkins Press, Baltimore (*116*).

VOGEL H. J. (1957). Repression and induction as control mechanisms of enzyme biogenesis: the adaptive formation of acetylornithinase. In *The Chemical Basis of Heredity* (eds. McElroy W. D. & Glass B.), p. 276. Johns Hopkins Press, Baltimore (*710*).

VOGEL H. J. & VOGEL R. H. (1967). Regulation of protein synthesis. *Ann. Rev. Biochem.*, **36**, 519 (*735*).

VOLKIN E. & ASTRACHAN L. (1957). RNA metabolism in T2-infected *Escherichia coli*. In *The Chemical Basis of Heredity* (eds. McElroy W. D. & Glass B.), p. 686. Johns Hopkins Press, Baltimore (*275, 429*).

VOLL M. J. & GOODGAL S. H. (1961). Recombination during transformation in *Hemophilus influenzae. Proc. Natl. Acad. Sci., Wash.*, **47**, 505 (*597, 599, 602*).

VOLL M. J. & GOODGAL S. H. (1965). Loss of activity of transforming deoxyribonucleic acid after uptake by *Haemophilus influenzae. J. Bacteriol.*, **90**, 873 (*597*).

DEWAARD A., PAUL A. V. & LEHMAN I. R. (1965). The structural gene for deoxyribonucleic acid polymerase in bacteriophages T4 and T5. *Proc. Natl. Acad. Sci., Wash.*, **54**, 1241 (*255*).

WACKER A., DELLWEG H. & JACHERTS D. (1962). Thymin-demerisierung und Überlebensrate bei Bakterien. *J. Mol. Biol.*, **4**, 410 (*329*).

WACKER A., DELLWEG H. & LODEMANN E. (1961). Strahlenchemische Veränderung der Nucleinsäuren. *Angew. Chem.*, **73**, 64 (*329*).

WAGNER R. P. & BERGQUIST A. (1960). The nature of the genetic blocks in the isoleucine-valine mutants of *Salmonella. Genetics*, **45**, 1375 (*123, 724*).

WAHBA A. J., GARDNER R. S., BASILIO C., MILLER R. S., SPEYER J. F. & LENGYEL P. (1963). Synthetic polynucleotides and the amino acid code. VIII. *Proc. Natl. Acad. Sci., Wash.*, **49**, 116 (*356*).

WANG S. Y. (1961). A study of photochemical reactions in frozen solutions. *Nature, Lond.*, **190**, 690 (*329*).

WARNER J. R. & RICH R. (1964). The number of soluble RNA molecules on reticulocyte polyribosomes. *Proc. Natl. Acad. Sci., Wash.*, **51**, 1134 (*285*).

WARNER J., RICH A. & HALL C. (1962). Electron microscope studies of ribosomal clusters synthesising haemoglobin. *Science*, **138**, 1399 (*278*).

WARNER J. R., KNOPF P. M. & RICH A. (1963). A multiple ribosomal structure in protein synthesis. *Proc. Natl. Acad. Sci., Wash.*, **49**, 122 (*278*).

WARNER R. C. (1956). Ultraviolet spectra of enzymatically synthesised polynucleotides. *Fed. Proc.*, **15**, 379 (*273, 356*).

WATANABE T. (1963). Infectious heredity of multiple drug resistance in bacteria. *Bacteriol. Rev.*, **27**, 87 (*760, 762, 770*).

WATANABE T. (1966a). Restriction of some phages by fi^- R factors. *Abstracts 9th Intern. Cong. Microbiol., Moscow*, p. 30 (*521*).

WATANABE T. (1966b). Infectious drug resistance in enteric bacteria. *New England J. Med.*, **275**, 888 (*761*).

WATANABE T. & FUKASAWA T. (1960). Resistance transfer factor, an episome in *Enterobacteriaceae. Biochem. biophys. Res. Comm.*, **3**, 660 (*762*).

WATANABE T. & FUKASAWA T. (1961a). Episome-mediated transfer of drug resistance in *Enterobacteraceae*. I. Transfer of resistance factors by conjugation. *J. Bacteriol.*, **81**, 669 (*760, 772*).

WATANABE T. & FUKASAWA T. (1961b). Episome-mediated transfer of drug resistance in *Enterbacteriaceae*. II. Elimination of resistance factors with acridines. *J. Bacteriol.*, **81**, 679 (*762, 772*).

WATANABE T. & FUKASAWA T. (1961c). Episome-mediated transfer of drug resistance in *Enterbacteriaceae*. III. Transduction of resistance factors. *J. Bacteriol.*, **82**, 202 (*762*).

WATANABE T. & FUKASAWA T. (1962). Episome-mediated transfer of drug resistance of *Enterobacteriaceae*. IV. Interactions between resistance transfer factor and F-factor in *Escherichia coli* K-12. *J. Bacteriol.*, **83**, 727 (*772*).

WATANABE T. & LYANG K. W. (1962). Episome-mediated transfer of drug resistance in *Enterobacteriaceae*. V. Spontaneous segregation and recombination of resistance factors in *Salmonella typhimurium*. *J. Bacteriol.*, **84**, 422 (*762*).

WATANABE T. & OGATA C. (1966). Episome-mediated transfer of drug resistance. IX. Recombination of an R factor with F. *J. Bacteriol.*, **91**, 43 (*781*).

WATANABE T., FUKASAWA T. & TAKANO T. (1962). Conversion of male bacteria of *Escherichia coli* K-12 to resistance to *f* phages by infection with the episome 'resistance transfer factor'. *Virology*, **17**, 218 (*772*).

WATANABE T., NISHIDA H., OGATA C., ARAI T. & SATO S. (1964). Episome-mediated transfer of drug resistance in *Enterobacteriaceae*. VII. Two types of naturally occurring R factors. *J. Bacteriol.*, **88**, 716 (*780*).

WATANABE T., TAKANO T., ARAI T., NISHIDA N. & SATO S. (1966). Episome-mediated transfer of drug-resistance in *Enterobacteriaceae*. X. Restriction and modification of phages by fi^- R factors. *J. Bacteriol.*, **92**, 477 (*768*).

WATSON J. D. & CRICK F. H. C. (1953a). The structure of DNA. *Cold Spr. Harb. Symp. quant. Biol.*, **18**, 123 (*232, 238, 240, 302*).

WATSON J. D. & CRICK F. H. C. (1953b). Genetic implications of the structure of deoxyribonucleic acid. *Nature, Lond.*, **171**, 964 (*240, 554*).

WEIBULL C. (1953). The isolation of protoplasts from *Bacillus megaterium* by controlled treatment with lysozyme. *J. Bacteriol.*, **66**, 688 (*421*).

WEIDEL W. & PRIMOSIGH J. (1958). Biochemical parallels between lysis by virulent phage and lysis by penicillin. *J. gen. Microbiol.*, **18**, 513 (*422, 424*).

WEIDEL W., KOCH G. & LOHSS F. (1954). Über die Zellmembran von *Escherichia coli* B. II. Der Rezeptorkomplex für die Bakteriophagen T3, T4 und T7. Vergleichende chemisch-analytische Untersuchungen. *Z. Naturforsch.*, **9b**, 398 (*423*).

WEIGERT M. G. & GAREN A. (1965). Base composition of nonsense codons in *E. coli*. Evidences from amino acid substitutions at a tryptophan site in alkaline phosphatase. *Nature, Lond.*, **206**, 992 (*363*).

WEIGLE J. J. (1960). Mutations affecting the density of bacteriophage λ. *Nature, Lond.*, **187**, 161 (*218*).

WEIGLE J. (1966). Assembly of phage lambda *in vitro*. *Proc. Natl. Acad. Sci., Wash.*, **55**, 1462 (*433*).

WEIGLE J., MESELSON M. & PAIGEN K. (1959). Density alterations associated with transducing ability in the bacteriophage lambda. *J. Mol. Biol.*, **1**, 379 (*630*).

WEIL E. & FELIX A. (1917). Weitere Untersuchungen über das Wegen der Fleckfieberagglutination. *Wien klin. Wschr.*, **30**, 1509 (*737*).

WEISBERG R. A. & GALLANT J. A. (1967). Dual function of the λ prophage repressor. *J. Mol. Biol.*, **25**, 537 (*474*).

WEISS S. B. (1960). Enzymatic incorporation of ribonucleotide triphosphates into the interpolynucleotide linkages of ribonucleic acid. *Proc. Natl. Acad. Sci., Wash.*, **46**, 1020 (*290*).

WEISS S. B. & NAKAMOTO T. (1961). On the participation of DNA in RNA biosynthesis. *Proc. Natl. Acad. Sci., Wash.*, **47**, 694 (*290*).

WEISS B. & RICHARDSON C. C. (1967). Enzymatic breakage and joining of deoxyribonucleic acid. I. Repair of single-strand breaks in DNA by an enzyme system from *Escherichia coli* infected with T4 bacteriophage. *Proc. Natl. Acad. Sci., Wash.*, **57**, 1021 (*400*).

WEISSMAN C., SIMON L., BORST P. & OCHOA S. (1963). Induction of RNA synthetase in *E. coli* after infection by the RNA phage MS2. *Cold Spr. Harb. Symp. quant. Biol.*, **28**, 99 (*291, 442*).

WEISSMAN C., BORST T., BURDON R. H., BILLETER M. A. & OCHOA S. (1964a). Replication of viral RNA. III. Double-stranded replicative form of MS2 phage RNA. *Proc. Natl. Acad. Sci., Wash.*, **51**, 682 (*443*).

WEISSMAN C., BORST T., BURDON R. H., BILLETER M. A. & OCHOA S. (1964b). Replication of viral RNA, IV. Properties of RNA synthetase and enzymatic synthesis of MS2 phage RNA. *Proc. Natl. Acad. Sci., Wash.*, **51**, 890 (*443*).

WELTON M. G. E. & PELC S. R. (1966). Specificity of the stereochemical relationship between ribonucleic acid triplets and amino-acids. *Nature, Lond.*, **209**, 870 (*368*).

WESTERGAARD M. (1957). Chemical mutagenesis in relation to the concept of the gene. *Experientia*, **13**, 224 (*205*).

WHITE F. H. (1961). Regeneration of the native secondary and tertiary structures by air oxidation of reduced ribonuclease. *J. biol. Chem.*, **236**, 1353 (*264*).

WHITEHOUSE H. L. K. (1963). A theory of crossing-over by means of hybrid deoxyribonucleic acid. *Nature*, **199**, 1034 (*400, 401*).

WHITEHOUSE H. L. K. (1966). An operator model of crossing-over. *Nature, Lond.*, **211**, 708 (*402*).

WHITEHOUSE H. L. K. & HASTINGS P. J. (1965). The analysis of genetic recombination on the polaron hybrid DNA model. *Genet. Res., Camb.*, **6**, 27 (*402*).

WHITFIELD H. J. JNR., MARTIN R. G. & AMES B. N. (1966). Classification of aminotransferase (*C* gene) mutants in the histidine operon. *J. Mol. Biol.*, **21**, 335 (*355, 720*).

WHITFIELD J. F. (1962). Lysogeny. *Brit. Med. Bull.*, **18**, 56 (*412*).

WIBERG J. S. (1966). Mutants of bacteriophage T4 unable to cause breakdown of host DNA. *Proc. Natl. Acad. Sci., Wash.*, **55**, 614 (*494*).

WIBERG J. S., DIRKSEN M., EPSTEIN R. H., LURIA S. E. & BUCHANAN J. M. (1962). Early enzyme synthesis and its control in *E. coli* infected with some amber mutants of bacteriophage T4. *Proc. Natl. Acad. Sci., Wash.*, **48**, 293 (*492*).

WILDY P. & ANDERSON T. F. (1964). Clumping of susceptible bacteria by bacteriophage tail fibres. *J. gen. Microbiol.*, **34**, 273 (*422*).

WILKINS M. F. H. (1956). Physical studies of the molecular structure of deoxyribose nucleic acid and nucleoprotein. *Cold Spr. Harb. Symp. quant. Biol.*, **18**, 75 (*232*).

WILKINS M. H. F. (1961). The molecular structure of DNA. *Chimie Physique*, **58**, 891 (*236*).

WILKINS M. H. F. & RANDALL J. T. (1953). Crystallinity in sperm heads: molecular structure of nucleoprotein *in vivo*. *Biochim. biophys. Acta*, **10**, 192 (*232*).

WILKINS M. F. H., STOKES A. R. & WILSON H. R. (1953). Molecular structure of deoxypentose nucleic acids. *Nature, Lond.*, **171**, 738 (*232*).

WILLIAMS R. C. (1953). The shapes and sizes of purified viruses as determined by electron microscopy. *Cold Spr. Harb. Symp. quant. Biol.*, **18**, 185 (*417, 420*).

WILLIAMS R. C. & FRAZER D. (1953). Morphology of the seven T-bacteriophages. *J. Bacteriol.*, **66**, 458 (*417*).

WILLIAMS R. C. & FRAZER D. (1956). Structural and functional differentiation in T2 bacteriophage. *Virology*, **2**, 289 (*417, 422*).

WILLSON C., PERRIN D., COHN M., JACOB F. & MONOD J. (1964). Non-inducible mutants of the regulator gene in the lactose system of *Escherichia coli*. *J. Mol. Biol.*, **8**, 582 (*715*).

WILSON D. (1960). The effects of ultraviolet light and ionising radiation on the transduction of *Escherichia coli* by phage P1. *Virology*, **11**, 533 (*638*).

WILSON D. B. (1966). The enzymes of the galactose operon in *Escherichia coli*. Ph.D. Thesis, Stanford University, California (*724*).

WINKLER K. C., DE WAART J., GROOTSEN C., ZEGERS B. J. M., TELLIER N. F. & VERTREGT C. D. (1965). Lysogenic conversion of staphylococci to loss of β-toxin. *J. gen. Microbiol.*, **39**, 321 (*648*).

WISSEMAN C. L. JNR., SMADEL J. E., HAHN F. E. & HOPPS H. E. (1954). Mode of action of chloramphenicol, I. *J. Bacteriol.*, **67**, 662 (*208, 269*).

WITKIN E. M. (1951). Nuclear segregation and the delayed appearance of induced mutants in *Escherichia coli*. *Cold Spr. Harb. Symp. quant. Biol.*, **16**, 357 (*548*).

WITKIN E. M. (1953). Effects of temperature on spontaneous and induced mutations in *Escherichia coli*. *Proc. Natl. Acad. Sci., Wash.*, **39**, 427 (*209*).

WITKIN E. M. (1956). Time, temperature and protein synthesis: a study of ultraviolet-induced mutation in bacteria. *Cold Spr. Harb. Symp. quant. Biol.*, **21**, 123 (*208, 209, 506, 507*).

WITKIN E. M. (1958). Post-metabolism and the timing of ultraviolet-induced mutations in bacteria. *Proc. X Intern. Cong. Genetics*, **1**, 280 (*209*).

WITKIN E. M. & THEIL E. C. (1960). The effect of post-treatment with chloramphenicol on various ultraviolet-induced mutations in *Escherichia coli*. *Proc. Natl. Acad. Sci., Wash.*, **46**, 226 (*210*).

WITTMAN H. G. (1962). Proteinuntersuchungen an Mutanten des Tabakmosaikvirus als Beitrag zum Problem des genetischen Codes (with English summary). *Z. Vererbungsl.*, **93**, 491 (*300, 341, 360*).

WITTMAN H. G. & WITTMAN-LIEBOLD B. (1963). Tobacco mosaic virus mutants and the genetic coding problem. *Cold Spr. Harb. Symp. quant. Biol.*, **28**, 589 (*360*).

WOESE C. R. (1962). Nature of the biological code. *Nature, Lond.*, **194**, 1114 (*282*).

WOESE C. R. (1965). Order in the genetic code. *Proc. Natl. Acad. Sci., Wash.*, **54**, 71 (*367*).

WOESE C. R., DUGRE D. H., SAXINGER W. C. & DUGRE S. A. (1966). The molecular basis for the genetic code. *Proc. Natl. Acad. Sci., Wash.*, **55**, 966 (*367*).

WOLLMAN E. (1928). Bactériophagie et processus similaires. Hérédité ou infection? *Bull. Inst. Pasteur*, **26**, 1 (*450*).

WOLLMAN E. L. (1953). Sur le déterminisme génétique de la lysogénie. *Ann. Inst. Pasteur*, **84**, 281 (*100, 454*).

WOLLMAN E. L. (1963). Transduction spécifique du marqueur biotine par le bactériophage λ. *C. R. Acad. Sci., Paris*, **257**, 4225 (*630*).

WOLLMAN E. L. (1966). Bacterial conjugation. In *Phage and the origins of molecular biology* (eds. Cairns J., Stent G. S. & Watson J. D.), p. 216. Cold Spring Harbor Laboratory of Quantitative Biology, Long Island, New York (*696*).

WOLLMAN E. L. & JACOB F. (1954). Lysogénie et recombinaison génétique chez *E. coli* K-12. *C. R. Acad. Sci., Paris*, **239**, 455 (*689*).

WOLLMAN E. L. & JACOB F. (1955). Sur le méchanisme du transfer de matériel génétique au cours de la recombination chez *E. coli* K-12. *C. R. Acad. Sci., Paris*, **240**, 2449 (*660, 661, 682, 683*).

WOLLMAN E. L. & JACOB F. (1957). Sur les processus de conjugaison et de recombinaison chez *E. coli*. II. La localisation chromosomique du prophage λ et les conséquences génétiques de l'induction zygotique. *Ann. Inst. Pasteur*, **93**, 323 (*464, 654, 662, 689*).

WOLLMAN E. L. & JACOB F. (1958). Sur les processus de conjugaison et de recombinaison chez *E. coli*. V. Le méchanisme du transfert de matériel génétique. *Ann. Inst. Pasteur*, **95**, 641 (*660, 661, 682*).

WOLLMAN E. L., JACOB F. & HAYES W. (1956). Conjugation and genetic recombination in *Escherichia coli*. *Cold Spr. Harb. Symp. quant. Biol.*, **21**, 141 (*54, 670, 682*).

WOLSTENHOLME D. R., VERMEULEN C. A. & VENEMA G. (1966). Evidence for the involvement of membranous bodies in the process leading to genetic transformation in *Bacillus subtilis*. *J. Bacteriol.*, **92**, 1111 (*591*).

WOOD T. H. (1967). Genetic recombination in *Escherichia coli*: clone heterogeneity and the kinetics of segregation. *Science*, **157**, 319 (*694*).

WOOD W. B. (1965). Mutations in *E. coli* affecting host-controlled modification of bacteriophage λ. *Pathol. Microbiol.* (Basel), **28**, 73 (*518, 519*).

WOOD W. B. (1966). Host specificity of DNA produced by *Escherichia coli*: bacterial mutations affecting the restriction and modification of DNA. *J. Mol. Biol.*, **16**, 118 (*518*).

WOOD W. B. & EDGAR R. S. (1967). Building a bacterial virus. *Scientific American*, **217**, 60 (*433*).

WOOF J. B. & HINSHELWOOD C. (1960). Chloramphenicol resistance of *Bact. lactis aerogenes* (*Aerobacter aerogenes*). I. Adaptive and lethal processes in liquid media and on agar plates. *Proc. Roy. Soc.*, B, **153**, 321 (*193*).

WYATT G. R. & COHEN S. S. (1953). The bases of the nucleic acids of some bacterial and animal viruses: the occurrence of 5-hydroxymethylcystosine. *Biochem. J.*, **55**, 774 (*426*).

WYMAN J. (1948). Heme proteins. In *Advances in protein chemistry* (eds. Anson N. L. & Edsall J. T.), vol. 4, p. 407 (*708*).

WYMAN J. (1963). Allosteric effects in hemoglobin. *Cold Spr. Harb. Symp. quant. Biol.*, **28**, 483 (*709*).

YANOFSKY C. (1960). The tryptophan synthetase system. *Bacteriol. Rev.*, **24**, 221 (*117, 119, 155*).

YANOFSKY C. (1963). Amino acid replacements associated with mutation and recombination in a gene and their relationship to *in vitro* coding data. *Cold Spr. Harb. Symp. quant. Biol.*, **28**, 581 (*360*).

YANOFSKY C. (1964). Gene-enzyme relationships. In *The Bacteria* (eds. Gunsalis I. C. & Stanier R. Y.), vol. 5, p. 373. Academic Press, Inc., New York and London (*155*).

YANOFSKY C. & ITO J. (1966). Nonsense codons and polarity in the tryptophan operon. *J. Mol. Biol.*, **21**, 313 (*360, 720, 721*).

YANOFSKY C. & LENNOX E. S. (1959). Linkage relationship of the genes controlling tryptophan synthesis in *Escherichia coli*. *Virology*, **8**, 425 (*164*).

YANOFSKY C., COX E. C. & HORN V. (1966). The unusual mutagenic specificity of an *E. coli* mutator gene. *Proc. Natl. Acad. Sci. Wash.*, **55**, 274 (*220*).

YANOFSKY C., ITO J. & HORN V. (1966). Amino acid replacements and the genetic code. *Cold Spr. Harb. Symp. quant. Biol.*, **31**, 151 (*360*).

YANOFSKY C., HENNING U., HELINSKI D. & CARLTON B. (1963). Mutational alteration of protein structure. *Fed. Proc.*, **22**, 75 (*301, 342*).

YANOFSKY C., CARLTON B. C., GUEST J. R., HELINSKI D. R. & HENNING U. (1964). On the colinearity of gene structure and protein structure. *Proc. Natl. Acad. Sci., Wash.*, **51**, 266 (*344*).

YANKOFSKY S. A. & SPIEGELMAN S. (1962a). The identification of the ribosomal RNA cistron by sequence complementarity. I. Specificity of complex formation. *Proc. Natl. Acad. Sci., Wash.*, **48**, 1069 (*289*).

YANKOFSKY S. A. & SPIEGELMAN S. (1962b). The identification of the ribosomal RNA cistron by sequence complementarity. II. Saturation of a competitive interaction at the RNA cistron. *Proc. Natl. Acad. Sci., Wash.*, **48**, 1466 (*289*).

YANKOFSKY S. A. & SPIEGELMAN S. (1963). Distinct cistrons for the two ribosomal RNA components. *Proc. Natl. Acad. Sci., Wash.*, **49**, 538 (*289*).

YATES R. A. & PARDEE A. B. (1956). Control of pyrimidine biosynthesis in *Escherichia coli* by a feedback mechanism. *J. biol. Chem.*, **221**, 757 (*707*).

YATES R. A. & PARDEE A. B. (1957). Control by uracil of formation of enzyme required for orotate synthesis. *J. biol. Chem.*, **227**, 677 (*710*).

YČAS M. & VINCENT W. S. (1960). A ribonucleic acid fraction of yeast related in composition to desoxyribonucleic acid. *Proc. Natl. Acad. Sci., Wash.*, **46**, 804 (*276*).

YI-YUNG HSIA D. (1959). *Inborn Errors of Metabolism*. The Year Book Publishers, Inc., Chicago (*108, 109, 110*).

YOSHIKAWA H. (1965). DNA synthesis during germination of *Bacillus subtilis* spores. *Proc. Natl. Acad. Sci., Wash.*, **53**, 1476 (*257*).

YOSHIKAWA H. & SUEOKA N. (1963). Sequential replication of *Bacillus subtilis* chromosome. I. Comparison of marker frequencies in exponential and stationary growth phases. *Proc. Natl. Acad. Sci., Wash.*, **49**, 559 (*561, 563, 614*).

YOUNG E. T. II & SINSHEIMER R. L. (1964). Novel intra-cellular forms of lambda DNA. *J. Mol. Biol.*, **10**, 562 (*463, 543*).

YOUNG F. E. & SPIZIZEN J. (1961). Physiological and genetic factors affecting transformation of *Bacillus subtilis*. *J. Bacteriol.*, **81**, 823 (*591*).

YOUNG I. E. & FITZ-JAMES P. C. (1959). Chemical and morphological studies of bacterial spore formation. I. The formation of spores in *Bacillus cereus*. *J. Biophys. biochem. Cytol.*, **6**, 467 (*548*).

YURA T. (1956). Suppressor mutations in purine requiring mutants of *Salmonella typhimurium*. In *Genetic studies with bacteria, Carnegie Inst., Wash., Publ.*, **612**, 63 (*216*).

ZABRISKIE J., quoted by FREIMER E. H. & MCCARTY M. (1965). Rheumatic fever. *Scientific Amer.*, **213**, No. 6, 66 (*648*).

ZACHAU H. G., DUTTING D., FELDMANN H., MELCHERS F. & KARAU W. (1966). Serine-specific transfer ribonucleic acids. XIV. Comparison of nucleotide sequences and secondary structure models. *Cold Spr. Harb. Symp. quant. Biol.*, **31**, 417 (*282, 283*).

ZALOKAR M. (1959a). Nuclear origin of ribonucleic acid. *Nature, Lond.*, **183**, 1330 (*271*).

ZALOKAR M. (1959b). Primary gene product: protein or RNA. *Proc. X Intern. Congr. Genetics*, vol. 2, p. 330. Univ. of Toronto Press, Toronto (*271*).

ZAMENHOF S., ALEXANDER H. E. & LEIDY G. (1953). Studies on the chemistry of the transforming activity. I. Resistance to physical and chemical agents. *J. exptl. Med.*, **98**, 373 (*575*).

ZELLE M. R. & LEDERBERG J. (1951). Single cell isolations of diploid heterozygous *E. coli*. *J. Bacteriol.*, **61**, 351 (*652*).

ZGAGA V. (1967). Formation of bacteriophage lambda infectious particles from lambda DNA in the presence of the crude extract of *Escherichia coli* K12S. *Virology*, **31**, 559 (*476*).

ZGAGA V. & MILETIC B. (1965). Superinfection with homologous phage of *Escherichia coli* K12 (λ) treated with 6-azauracil. *Virology*, **27**, 205 (*451*).

ZICHICHI M. L. & KELLENBERGER G. (1963). Two distinct functions in the lysogenisation process: the repression of phage multiplication and the incorporation of the prophages in the bacterial genome. *Virology*, **19**, 450 (*471*).

ZINDER N. D. (1953). Infective heredity in bacteria. *Cold Spr. Harb. Symp. quant. Biol.*, **18**, 261 (*620, 622, 623*).

ZINDER N. D. (1955). Bacterial transduction. *J. Cell. comp. Physiol.*, **45**, suppl. 2, 23 (*623, 627*).

ZINDER N. D. (1958). Lysogenisation and superinfection immunity in *Salmonella*. *Virology*, **5**, 291 (*449*).

ZINDER N. D. (1960a). Hybrids of *Escherichia* and *Salmonella*. *Science*, **131**, 813 (*518, 751, 752, 753*).

ZINDER N. D. (1960b). Sexuality and mating in *Salmonella*. *Science*, **131**, 924 (*752*).

ZINDER N. D. (1965). RNA phages. *Ann. Rev. Microbiol.*, **19**, 455 (*410, 441, 442*).

ZINDER N. D. & ARNDT W. F. (1956). Production of protoplasts of *Escherichia coli* by lysozyme treatment. *Proc. Natl. Acad. Sci., Wash.*, **42**, 586 (*421*).

ZINDER N. D. & COOPER S. (1964). Host-dependent mutants of the bacteriophage f2. I. Isolation and primary classification. *Virology*, **23**, 152 (*510*).

ZINDER N. D. & LEDERBERG J. (1952). Genetic exchange in *Salmonella*. *J. Bacteriol.*, **64**, 679 (*620, 622, 627*).

ZINDER N. D., VALENTINE R. C., ROGER M. & STOECKENIUS W. (1963). f1, a rod-shaped male-specific bacteriophage that contains DNA. *Virology*, **20**, 638 (*419, 444, 445*).

ZIPSER D. & NEWTON A. (1967). The influence of deletions on polarity. *J. Mol. Biol.*, **25**, 567 (*721*).

ZUBAY G. (1962). A theory on the mechanism of messenger-RNA synthesis. *Proc. Natl. Acad. Sci., Wash.*, **48**, 456 (*372, 373, 374, 600*).

ZUBAY G. & WILKINS M. F. H. (1960). X-ray diffraction studies of the structure of ribosomes from *Escherichia coli*. *J. Mol. Biol.*, **2**, 105 (*273*).

Index

"The man who publishes a book without an index ought to be damned ten miles beyond hell, where the Devil himself cannot get for stinging nettles".
John Baynes (1758–1787)

Numbers indicated in bold type show where a particularly definitive account of the item may be found.

C

D

H

I

K

L

M

S

T

U

V

W

X

Y

Z